Research on Plant Genomics and Breeding

Research on Plant Genomics and Breeding

Editors

Zhiyong Li
Jian Zhang

Basel • Beijing • Wuhan • Barcelona • Belgrade • Novi Sad • Cluj • Manchester

Editors
Zhiyong Li
State Key Lab of Rice Biology
and Breeding
China National Rice Research
Institute
Hangzhou
China

Jian Zhang
State Key Lab of Rice Biology
and Breeding
China National Rice Research
Institute
Hangzhou
China

Editorial Office
MDPI
St. Alban-Anlage 66
4052 Basel, Switzerland

This is a reprint of articles from the Special Issue published online in the open access journal *International Journal of Molecular Sciences* (ISSN 1422-0067) (available at: www.mdpi.com/journal/ijms/special_issues/Genomics_Breeding).

For citation purposes, cite each article independently as indicated on the article page online and as indicated below:

Lastname, A.A.; Lastname, B.B. Article Title. *Journal Name* **Year**, *Volume Number*, Page Range.

ISBN 978-3-0365-9379-1 (Hbk)
ISBN 978-3-0365-9378-4 (PDF)
doi.org/10.3390/books978-3-0365-9378-4

© 2023 by the authors. Articles in this book are Open Access and distributed under the Creative Commons Attribution (CC BY) license. The book as a whole is distributed by MDPI under the terms and conditions of the Creative Commons Attribution-NonCommercial-NoDerivs (CC BY-NC-ND) license.

Contents

Jie Huang, Zhiyong Li and Jian Zhang
Research on Plant Genomics and Breeding
Reprinted from: *Int. J. Mol. Sci.* **2023**, *24*, 15298, doi:10.3390/ijms242015298 1

Luomiao Yang, Peng Li, Jingguo Wang, Hualong Liu, Hongliang Zheng and Wei Xin et al.
Fine Mapping and Candidate Gene Analysis of Rice Grain Length QTL *qGL9.1*
Reprinted from: *Int. J. Mol. Sci.* **2023**, *24*, 11447, doi:10.3390/ijms241411447 6

Shamseldeen Eltaher, Mostafa Hashem, Asmaa A. M. Ahmed, P. Stephen Baenziger, Andreas Börner and Ahmed Sallam
Effectiveness of *TaDreb-B1* and *1-FEH w3* KASP Markers in Spring and Winter Wheat Populations for Marker-Assisted Selection to Improve Drought Tolerance
Reprinted from: *Int. J. Mol. Sci.* **2023**, *24*, 8986, doi:10.3390/ijms24108986 21

Karolina Szala, Marta Dmochowska-Boguta, Joanna Bocian, Waclaw Orczyk and Anna Nadolska-Orczyk
Transgenerational Paternal Inheritance of *TaCKX* GFMs Expression Patterns Indicate a Way to Select Wheat Lines with Better Parameters for Yield-Related Traits
Reprinted from: *Int. J. Mol. Sci.* **2023**, *24*, 8196, doi:10.3390/ijms24098196 35

Juan Zhao, Xin Meng, Zhaonian Zhang, Mei Wang, Fanhao Nie and Qingpo Liu
OsLPR5 Encoding Ferroxidase Positively Regulates the Tolerance to Salt Stress in Rice
Reprinted from: *Int. J. Mol. Sci.* **2023**, *24*, 8115, doi:10.3390/ijms24098115 59

Guangzhen Zhou, Qiyuan An, Zheng Liu, Yinglang Wan and Wenlong Bao
Systematic Analysis of NRAMP Family Genes in *Areca catechu* and Its Response to Zn/Fe Deficiency Stress
Reprinted from: *Int. J. Mol. Sci.* **2023**, *24*, 7383, doi:10.3390/ijms24087383 71

Zheng Liu, Xiaoai Fu, Hao Xu, Yuxin Zhang, Zhidi Shi and Guangzhen Zhou et al.
Comprehensive Analysis of bHLH Transcription Factors in *Ipomoea aquatica* and Its Response to Anthocyanin Biosynthesis
Reprinted from: *Int. J. Mol. Sci.* **2023**, *24*, 5652, doi:10.3390/ijms24065652 86

Xiaobin Wang, Runlong Zhang, Kaijing Zhang, Lingmei Shao, Tong Xu and Xiaohua Shi et al.
Development of a Multi-Criteria Decision-Making Approach for Evaluating the Comprehensive Application of Herbaceous Peony at Low Latitudes
Reprinted from: *Int. J. Mol. Sci.* **2022**, *23*, 14342, doi:10.3390/ijms232214342 104

Jianjian Chen, Jinming Cao, Yunlong Bian, Hui Zhang, Xiangnan Li and Zhenxing Wu et al.
Identification of Genetic Variations and Candidate Genes Responsible for Stalk Sugar Content and Agronomic Traits in Fresh Corn via GWAS across Multiple Environments
Reprinted from: *Int. J. Mol. Sci.* **2022**, *23*, 13490, doi:10.3390/ijms232113490 127

Zhiquan Fan, Guanwen Huang, Yourong Fan and Jiangyi Yang
Sucrose Facilitates Rhizome Development of Perennial Rice (*Oryza longistaminata*)
Reprinted from: *Int. J. Mol. Sci.* **2022**, *23*, 13396, doi:10.3390/ijms232113396 146

Guilian Xiao, Junzhi Zhou, Zhiheng Huo, Tong Wu, Yingchun Li and Yajing Li et al.
The Shift in Synonymous Codon Usage Reveals Similar Genomic Variation during Domestication of Asian and African Rice
Reprinted from: *Int. J. Mol. Sci.* **2022**, *23*, 12860, doi:10.3390/ijms232112860 161

Qiqi Ling, Jiayao Liao, Xiang Liu, Yue Zhou and Yexiong Qian
Genome-Wide Identification of Maize Protein Arginine Methyltransferase Genes and Functional Analysis of *ZmPRMT1* Reveal Essential Roles in *Arabidopsis* Flowering Regulation and Abiotic Stress Tolerance
Reprinted from: *Int. J. Mol. Sci.* **2022**, *23*, 12793, doi:10.3390/ijms232112793 178

Zhiyuan Bai, Xianlong Ding, Ruijun Zhang, Yuhua Yang, Baoguo Wei and Shouping Yang et al.
Transcriptome Analysis Reveals the Genes Related to Pollen Abortion in a Cytoplasmic Male-Sterile Soybean (*Glycine max* (L.) Merr.)
Reprinted from: *Int. J. Mol. Sci.* **2022**, *23*, 12227, doi:10.3390/ijms232012227 207

Emil Khusnutdinov, Alexander Artyukhin, Yuliya Sharifyanova and Elena V. Mikhaylova
A Mutation in the *MYBL2-1* Gene Is Associated with Purple Pigmentation in *Brassica oleracea*
Reprinted from: *Int. J. Mol. Sci.* **2022**, *23*, 11865, doi:10.3390/ijms231911865 226

Xin Yao, Meiliang Zhou, Jingjun Ruan, Yan Peng, Chao Ma and Weijiao Wu et al.
Physiological and Biochemical Regulation Mechanism of Exogenous Hydrogen Peroxide in Alleviating NaCl Stress Toxicity in Tartary Buckwheat (*Fagopyrum tataricum* (L.) Gaertn)
Reprinted from: *Int. J. Mol. Sci.* **2022**, *23*, 10698, doi:10.3390/ijms231810698 240

Xuan Zhao, Tingting Wu, Shixian Guo, Junling Hu and Yihua Zhan
Ectopic Expression of *AeNAC83*, a NAC Transcription Factor from *Abelmoschus esculentus*, Inhibits Growth and Confers Tolerance to Salt Stress in *Arabidopsis*
Reprinted from: *Int. J. Mol. Sci.* **2022**, *23*, 10182, doi:10.3390/ijms231710182 261

Jiajia Wang, Yiting Liu, Songping Hu, Jing Xu, Jinqiang Nian and Xiaoping Cao et al.
LEAF TIP RUMPLED 1 Regulates Leaf Morphology and Salt Tolerance in Rice
Reprinted from: *Int. J. Mol. Sci.* **2022**, *23*, 8818, doi:10.3390/ijms23158818 279

Qibao Liu, Zhen Feng, Chenjue Huang, Jia Wen, Libei Li and Shuxun Yu
Insights into the Genomic Regions and Candidate Genes of Senescence-Related Traits in Upland Cotton via GWAS
Reprinted from: *Int. J. Mol. Sci.* **2022**, *23*, 8584, doi:10.3390/ijms23158584 297

Huayu Xu, Shufan Li, Bello Babatunde Kazeem, Abolore Adijat Ajadi, Jinjin Luo and Man Yin et al.
Five Rice Seed-Specific *NF-YC* Genes Redundantly Regulate Grain Quality and Seed Germination via Interfering Gibberellin Pathway
Reprinted from: *Int. J. Mol. Sci.* **2022**, *23*, 8382, doi:10.3390/ijms23158382 312

Nazir Ahmad, Bin Su, Sani Ibrahim, Lieqiong Kuang, Ze Tian and Xinfa Wang et al.
Deciphering the Genetic Basis of Root and Biomass Traits in Rapeseed (*Brassica napus* L.) through the Integration of GWAS and RNA-Seq under Nitrogen Stress
Reprinted from: *Int. J. Mol. Sci.* **2022**, *23*, 7958, doi:10.3390/ijms23147958 324

Huanhuan Qi, Feng Yu, Jiao Deng and Pingfang Yang
Studies on Lotus Genomics and the Contribution to Its Breeding
Reprinted from: *Int. J. Mol. Sci.* **2022**, *23*, 7270, doi:10.3390/ijms23137270 342

Editorial

Research on Plant Genomics and Breeding

Jie Huang, Zhiyong Li * and Jian Zhang *

State Key Laboratory of Rice Biology and Breeding, China National Rice Research Institute, Hangzhou 311400, China; huangjie@caas.cn
* Correspondence: lizhiyong@caas.cn (Z.L.); zhangjian@caas.cn (J.Z.)

1. Introduction

In recent years, plant genomics has made significant progress following the development of biotechnology. Based on traditional genetics, functional genomics covers research on the genome, transcriptome, proteome, metabolome, bioinformatics, and other interdisciplinary disciplines. Researching many genes in the genome and their expression reveals the gene function, subcellular localization, and interaction, which provides a more systematic and comprehensive study means for the in-depth understanding of plants [1,2]. At the same time, crop breeding is crucial for agricultural development. Many high-yield and environmental adaptability crop varieties have been bred using plant genomics and gene diversity [3]. Here, we summarized some genomics studies of different species recently published in *IJMS* in the hopes of contributing to the advancement of crop breeding.

2. Genomics Studies in Major Food Crops

2.1. In Rice

Rice is one of the most important food crops in the world. Investigating the genes controlling rice yield, quality, and stress resistance using genomics leads to significant improvement. The grain length of rice seed is a crucial trait determining rice yield. In contrast, japonica rice in the Heilongjiang region of China is significantly shorter than the indica rice varieties in southern China. Two japonica rice varieties, Pin20 and Songjing 15, with significant differences in grain shape, were used to map a new QTL *qGL9.1* controlling the grain length, which promotes the study of the molecular mechanism controlling the grain length of japonica rice and lays the theoretical foundation for japonica rice improvement in northern China [4]. The salt tolerance of rice is important to improve the adaptation and yield of rice in saline–alkali soil. Low phosphate root 5 (OsLPR5) and ferroxidase activity levels were expressed during salt stress. Then, the physiological role of *OsLPR5* in salt stress was demonstrated via overexpression and using CRISPR/Cas9-mediated mutation techniques. This study identified a novel molecular mechanism in which *OsLPR5* positively regulates salt tolerance in rice [5]. Perennial rice plays a vital role in protecting the ecological environment and alleviating the labor shortage of farmers. The development of axillary buds is a crucial trait of perennial rice (regenerative rice). It was reported that sucrose promotes the growth of axillary buds and rhizomes, which might be related to fructose and glucose. The increased osmotic pressure of the cells results in the absorption of water, which encourages cell growth and eventually leads to the development of axillary buds and rhizomes [6]. The domestication of wild rice is caused by genomic variation, including the synonymy codon usage bias (SCUB) due to synonymy nucleotide substitution. SCUB was affected in the same way during the process of domestication of Asian and African rice. The analysis of cytosine to thymine transformation mediated via SCUB and DNA methylation showed the close relationship between genetic variation and epigenetic variation in domestication [7]. *LEAF TIP RUMPLED 1* (*LTR1*) positively regulates the rice yield and salt tolerance. This study showed that *LTR1* maintained the salt tolerance of rice by mediating the activities of aquaporins and ion transporters. The research on *LTR1* provides a theoretical basis for studying salt tolerance and high rice yields [8]. NF-YCs family transcription factors, including seed development, play a key role in plant growth and development. *NF-YC8, 9, 10, 11,* and *12* are homologous genes that are expressed at highly similarity levels in seeds. Systematic genetic experiments suggested that these five genes synergistically regulate GA and ABA responses, thus affecting the rice seed quality and germination [9].

2.2. In Wheat

Wheat is the world's largest cultivated and most widely distributed food crop. Unlike rice, wheat is a dryland crop. Therefore, it is important to improve wheat's drought tolerance. Two previously reported KASP markers, TaDreb-B1 and 1-FEH w3, were employed for the marker-assisted selection (MAS) of drought tolerance. These two molecular markers were used to detect drought tolerance in winter and spring wheat populations, and the correlation of TaDreb-B1 with drought tolerance was better than that of marker 1-FEH w3 [10]. Cytokinin is an important hormone that determines the wheat yield. Members of the *TaCKX* gene family (GFMs) encode the cytokinin oxygenase/dehydrogenase enzyme (CKX), which causes the irreversible degradation of cytokinin. Thus, these genes might regulate wheat yield. It has been speculated that it is feasible to select high-yielding wheat families using the *TaCKX* GFMs cross-generation expression pattern [11].

2.3. In Maize

Maize is the world's most productive food crop per unit area, reaching up to 30,000 kg per hectare. While abiotic stresses such as drought and salinization severely restrict maize yield, it is important to resolve the mechanism of maize flowering time and abiotic stress to improve maize yield and resistance. It has been reported that protein arginine methyltransferase (PRMTs) is mainly responsible for the histone methylation of specific plant arginine residues. Ling et al. found that the overexpression of the maize *ZmPRMT* gene in *Arabidopsis* can promote early flowering and heat tolerance. First, it was reported that *ZmPRMT1* gene regulates the flowering time and resistance to heat stress in plants, which will provide an essential theoretical basis to further reveal the functions of *ZmPRMT* gene and epigenetic regulation mechanism of maize growth and development and response to abiotic stress [12]. Corn stalks can be used as an animal feed, and sugar in the stalk is also a carbon source for corn seeds. Increasing the sugar content of corn stalks can not only promote the development of animal husbandry, but also increase the yield. Chen et al. investigated 188 waxy, sweet, and hybrid corn resources with different stalk sugar contents for GWAS analysis. The results showed that the expression levels of six candidate genes were significantly different among the other stalk sugar-containing materials, which provides a significant insight into the genomic footprints of stalk sugar content in fresh corn and facilitates the breeding of corn cultivars with a higher stalk sugar content [13].

3. Genome Research in Cash Crops

The production of cash crops (like cotton, soybean, rapeseed, *Brassica oleracea,* etc.) plays a decisive role in industry development, especially the light industry, which is also the main source of exports, foreign exchange earnings, and increasing national economic income.

3.1. In Oil Crops

Soybean sterile line SXCMS5A is an H3A cytoplasmic male sterile line (CMS) developed from JY20. Bai et al. conducted the transcriptome analysis of the sterile line SXCMS5A and the maintainer line SXCMS5B to find the differentially expressed genes (DEGs) and the metabolic pathways related to pollen sterilization. These findings might provide useful information to facilitate soybean hybrid breeding [14]. Excellent root development is an important factor for plants' high nitrogen-use efficiency (NUE). Ahmad et al. revealed 16 genes involved in rapeseed root development under low nitrogen (LN) stress through the integrated analysis of GWAS, a weighted gene co-expression network, and DEGs. Seven of these genes have been previously reported to be associated with root development and NUE. This study provides genetic/SNPs resources for studying low nitrogen tolerance in rapeseed [15].

3.2. In Horticultural Crops

Khusnutdinov et al. found that the DNA binding structure of transcription factor *MYBL2-1* (a negative regulator of anthocyanin synthase *Dihydroflavonol-4-reductase* (*DFR*)) of purple varieties of *Brassica oleracea* L. had two SNP mutations, leading to an increase in *DFRs* expression. This is the reason for the high anthocyanin content of purple cabbage [16]. *Ipomoea aquatica* is a leafy vegetable rich in essential amino acids, flavonoids, and various mineral elements (calcium, potassium, phosphorus, etc.). The role of *IabHLHs* family transcription factors in the biosynthesis of anthocyanins in *I. aquatica* was analyzed using bioinformatics combined with molecular experiments. This work provides valuable clues to further explore *IabHLH*'s function and facilitating the breeding of anthocyanin-rich varieties of *I. aquatica* [17]. Improving the cold tolerance of herbaceous peonies (*Paeonia lactiflora*) is essential to plant them at low latitudes. Wang et al. modified the previously reported multi-criteria decision-making (MCDM) model. This model was implemented to analyze 15 peony varieties at different latitudes. As a result, 'Meiju', 'Hang Baishao', 'Hongpan Tuojin', and 'Bo Baishao' were excellent varieties with strong environmental adaptability. This research can provide a reference for the breeding and cultivating of perennial herbs [18]. NAC transcription factors are crucial in plant growth, development, and stress responses. Zhao et al. found that *AeNAC83* positively regulates the salt tolerance of okra (*Abelmoschus esculentus*). Exactly as the expression of *AeNAC83* gene is up-regulated in okra under salt stress, the down-regulation of *AeNAC83* is caused by virus-induced gene silencing enhanced plant sensitivity to salt stress [19]. Lotus (*Nelumbo nucifera*), which is in the Nelumbonaceae family, is a relict plant with crucial scientific research and economic values. Such as the sequencing of the lotus genome and the assembly of several high-quality genomes, the investigation of lotus functional genome has been extensively promoted. The resequencing of natural and genetic populations and different levels of genomic studies benefit the classification of different lotus germplasm resources and the identification of genes controlling various traits [20].

3.3. In Other Crops

Liu et al. analyzed the genetic basis of senescence in cotton using 355 upland cotton accessions planted in multiple environments for GWAS. From the candidate genes that have been linked, *GhMKK9* silencing improves the drought resistance of cotton, whereas *GhMKK9* overexpression accelerates senescence in *Arabidopsis* [21]. *Areca catechu* is a commercially important medicinal plant widely cultivated in tropical regions. Zhou et al. investigated 12 *natural resistance-associated macrophage protein* (*NRAMP*) genes from the whole genome of *A. catechu* with Fe and Zn deficiency, and their sequence characteristics, gene structure, phylogeny, promoter sequence, and collinearity were analyzed. This research revealed the regulatory response of *NRAMP* family genes in *A. catechu* under Fe and Zn deficiency stress [22]. Yao et al. treated tartary buckwheat (*Fagopyrum tataricum* L.) leaves with different concentrations of H_2O_2, and then analyzed the growth, photosynthesis, antioxidant enzyme activity, and related gene expression under salt stress. This study explored the mechanism of H_2O_2 enhancing salt tolerance of Tartary buckwheat from a physiological point of view [23].

4. Conclusions and Perspectives

In summary, we reviewed 20 plant genomics articles on different plant species. These functional gene studies consider the yield, quality, fertility, abiotic stress, and evolution of rice, wheat, maize, and cash crops. These researchers have made outstanding contributions to plant genomics and provided valuable genetic resources for crop breeding.

Based on these genetic resources controlling different traits, gene-editing methods, such as CRISPR-cas9 technology, can alter gene expression to improve the crop yield, quality, and resistance. Finally, strengthening the interface between the application and industrialization of functional genomics research would help to accurately quantify traits, such as the crop yield, stress resistance, quality, and nutrient content. Meanwhile, promoting more efficient and precise crop breeding is expected to bring about new transformations in the global agricultural field.

Author Contributions: All the authors participated in the editing of this Research Topic. J.H. wrote the draft, and all the other authors provided suggestive comments on the Editorial. All authors have read and agreed to the published version of the manuscript.

Conflicts of Interest: The authors declare no conflict of interest.

References

1. You, F. Plant genomics-advancing our understanding of plants. *Int. J. Mol. Sci.* **2023**, *24*, 11528. [CrossRef]
2. Ghazal, H.; Adam, Y.; Idrissi Azami, A.; Sehli, S.; Nyarko, H.N.; Chaouni, B.; Olasehinde, G.; Isewon, I.; Adebiyi, M.; Ajani, O.; et al. Plant genomics in Africa: Present and prospects. *Plant J.* **2021**, *107*, 21–36. [CrossRef] [PubMed]
3. Wang, H.; Cimen, E.; Singh, N.; Buckler, E. Deep learning for plant genomics and crop improvement. *Curr. Opin. Plant Biol.* **2020**, *54*, 34–41. [CrossRef] [PubMed]
4. Yang, L.; Li, P.; Wang, J.; Liu, H.; Zheng, H.; Xin, W.; Zou, D. Fine mapping and candidate gene analysis of rice grain length QTL qGL9.1. *Int. J. Mol. Sci.* **2023**, *24*, 11447. [CrossRef] [PubMed]
5. Zhao, J.; Meng, X.; Zhang, Z.; Wang, M.; Nie, F.; Liu, Q. OsLPR5 encoding ferroxidase positively regulates the tolerance to salt stress in rice. *Int. J. Mol. Sci.* **2023**, *24*, 8115. [CrossRef]
6. Fan, Z.; Huang, G.; Fan, Y.; Yang, J. Sucrose facilitates rhizome development of perennial rice (*Oryza longistaminata*). *Int. J. Mol. Sci.* **2022**, *23*, 13396. [CrossRef]
7. Xiao, G.; Zhou, J.; Huo, Z.; Wu, T.; Li, Y.; Li, Y.; Wang, Y.; Wang, M. The shift in synonymous codon usage reveals similar genomic variation during domestication of Asian and African rice. *Int. J. Mol. Sci.* **2022**, *23*, 12860. [CrossRef]
8. Wang, J.; Liu, Y.; Hu, S.; Xu, J.; Nian, J.; Cao, X.; Chen, M.; Cen, J.; Liu, X.; Zhang, Z.; et al. LEAF TIP RUMPLED 1 regulates leaf morphology and salt tolerance in rice. *Int. J. Mol. Sci.* **2022**, *23*, 8818. [CrossRef]
9. Xu, H.; Li, S.; Kazeem, B.B.; Ajadi, A.A.; Luo, J.; Yin, M.; Liu, X.; Chen, L.; Ying, J.; Tong, X.; et al. Five rice seed-specific NF-YC genes redundantly regulate grain quality and seed germination via interfering gibberellin pathway. *Int. J. Mol. Sci.* **2022**, *23*, 8382. [CrossRef]
10. Eltaher, S.; Hashem, M.; Ahmed, A.A.M.; Baenziger, P.S.; Börner, A.; Sallam, A. Effectiveness of TaDreb-B1 and 1-FEH w3 KASP markers in spring and winter wheat populations for marker-assisted selection to improve drought tolerance. *Int. J. Mol. Sci.* **2023**, *24*, 8986. [CrossRef]
11. Szala, K.; Dmochowska-Boguta, M.; Bocian, J.; Orczyk, W.; Nadolska-Orczyk, A. Transgenerational paternal inheritance of TaCKX GFMs expression patterns indicate a way to select wheat lines with better parameters for yield-related traits. *Int. J. Mol. Sci.* **2023**, *24*, 8196. [CrossRef] [PubMed]
12. Ling, Q.; Liao, J.; Liu, X.; Zhou, Y.; Qian, Y. Genome-wide identification of maize protein arginine methyltransferase genes and functional analysis of ZmPRMT1 reveal essential roles in *Arabidopsis* flowering regulation and abiotic stress tolerance. *Int. J. Mol. Sci.* **2022**, *23*, 12793. [CrossRef] [PubMed]
13. Chen, J.; Cao, J.; Bian, Y.; Zhang, H.; Li, X.; Wu, Z.; Guo, G.; Lv, G. Identification of genetic variations and candidate genes responsible for stalk sugar content and agronomic traits in fresh corn via GWAS across multiple environments. *Int. J. Mol. Sci.* **2022**, *23*, 13490. [CrossRef] [PubMed]
14. Bai, Z.; Ding, X.; Zhang, R.; Yang, Y.; Wei, B.; Yang, S.; Gai, J. Transcriptome analysis reveals the genes related to pollen abortion in a cytoplasmic male-sterile soybean (*Glycine max* (L.) Merr.). *Int. J. Mol. Sci.* **2022**, *23*, 12227. [CrossRef]
15. Ahmad, N.; Su, B.; Ibrahim, S.; Kuang, L.; Tian, Z.; Wang, X.; Wang, H.; Dun, X. Deciphering the genetic basis of root and biomass traits in rapeseed (*Brassica napus* L.) through the integration of GWAS and RNA-Seq under nitrogen stress. *Int. J. Mol. Sci.* **2022**, *23*, 7958. [CrossRef]
16. Khusnutdinov, E.; Artyukhin, A.; Sharifyanova, Y.; Mikhaylova, E.V. A mutation in the MYBL2-1 gene is associated with purple pigmentation in *Brassica oleracea*. *Int. J. Mol. Sci.* **2022**, *23*, 11865. [CrossRef]
17. Liu, Z.; Fu, X.; Xu, H.; Zhang, Y.; Shi, Z.; Zhou, G.; Bao, W. Comprehensive analysis of bHLH transcription factors in *Ipomoea aquatica* and its response to anthocyanin biosynthesis. *Int. J. Mol. Sci.* **2023**, *24*, 5652. [CrossRef]
18. Wang, X.; Zhang, R.; Zhang, K.; Shao, L.; Xu, T.; Shi, X.; Li, D.; Zhang, J.; Xia, Y. Development of a multi-criteria decision-making approach for evaluating the comprehensive application of herbaceous peony at low latitudes. *Int. J. Mol. Sci.* **2022**, *23*, 14342. [CrossRef]
19. Zhao, X.; Wu, T.; Guo, S.; Hu, J.; Zhan, Y. Ectopic expression of AeNAC83, a NAC transcription factor from *Abelmoschus esculentus*, inhibits growth and confers tolerance to salt stress in *Arabidopsis*. *Int. J. Mol. Sci.* **2022**, *23*, 10182. [CrossRef]

20. Qi, H.; Yu, F.; Deng, J.; Yang, P. Studies on lotus genomics and the contribution to its breeding. *Int. J. Mol. Sci.* **2022**, *23*, 7270. [CrossRef] [PubMed]
21. Liu, Q.; Feng, Z.; Huang, C.; Wen, J.; Li, L.; Yu, S. Insights into the genomic regions and candidate genes of senescence-related traits in upland cotton via GWAS. *Int. J. Mol. Sci.* **2022**, *23*, 8584. [CrossRef] [PubMed]
22. Zhou, G.; An, Q.; Liu, Z.; Wan, Y.; Bao, W. Systematic analysis of NRAMP family genes in *Areca catechu* and its response to Zn/Fe deficiency stress. *Int. J. Mol. Sci.* **2023**, *24*, 7383. [CrossRef] [PubMed]
23. Yao, X.; Guo, H.; Zhou, M.; Ruan, J.; Peng, Y.; Ma, C.; Wu, W.; Gao, A.; Weng, W.; Cheng, J. Physiological and biochemical regulation mechanism of exogenous hydrogen peroxide in alleviating NaCl stress toxicity in Tartary buckwheat (*Fagopyrum tataricum* (L.) Gaertn). *Int. J. Mol. Sci.* **2022**, *23*, 10698. [CrossRef] [PubMed]

Disclaimer/Publisher's Note: The statements, opinions and data contained in all publications are solely those of the individual author(s) and contributor(s) and not of MDPI and/or the editor(s). MDPI and/or the editor(s) disclaim responsibility for any injury to people or property resulting from any ideas, methods, instructions or products referred to in the content.

Article

Fine Mapping and Candidate Gene Analysis of Rice Grain Length QTL *qGL9.1*

Luomiao Yang [†], Peng Li [†], Jingguo Wang, Hualong Liu, Hongliang Zheng, Wei Xin and Detang Zou *

Key Laboratory of Germplasm Enhancement, Physiology and Ecology of Food Crops in Cold Region, Ministry of Education, Northeast Agricultural University, Harbin 150030, China
* Correspondence: wrathion@neau.edu.cn
[†] These authors contributed equally to this work.

Abstract: Grain length (GL) is one of the crucial determinants of rice yield and quality. However, there is still a shortage of knowledge on the major genes controlling the inheritance of GL in *japonica* rice, which severely limits the improvement of japonica rice yields. Here, we systemically measured the GL of 667 F_2 and 1570 BC_3F_3 individuals derived from two cultivated rice cultivars, Pin20 and Songjing15, in order to identify the major genomic regions associated with GL. A novel major QTL, *qGL9.1*, was mapped on chromosome 9, which is associated with the GL, using whole-genome re-sequencing with bulked segregant analysis. Local QTL linkage analysis with F_2 and fine mapping with the recombinant plant revealed a 93-kb core region on *qGL9.1* encoding 15 protein-coding genes. Only the expression level of *LOC_Os09g26970* was significantly different between the two parents at different stages of grain development. Moreover, haplotype analysis revealed that the alleles of Pin20 contribute to the optimal GL (9.36 mm) and GL/W (3.31), suggesting that Pin20 is a cultivated species carrying the optimal GL variation of *LOC_Os09g26970*. Furthermore, a functional-type mutation (16398989-bp, G>A) located on an exon of *LOC_Os09g26970* could be used as a molecular marker to distinguish between long and short grains. Our experiments identified *LOC_Os09g26970* as a novel gene associated with GL in *japonica* rice. This result is expected to further the exploration of the genetic mechanism of rice GL and improve GL in rice *japonica* varieties by marker-assisted selection.

Keywords: *Oryza sativa* L.; grain length; re-sequencing; fine mapping; P450 protein

Citation: Yang, L.; Li, P.; Wang, J.; Liu, H.; Zheng, H.; Xin, W.; Zou, D. Fine Mapping and Candidate Gene Analysis of Rice Grain Length QTL qGL9.1. *Int. J. Mol. Sci.* 2023, 24, 11447. https://doi.org/10.3390/ijms241411447

Academic Editor: Zsófia Bánfalvi

Received: 20 June 2023
Revised: 7 July 2023
Accepted: 13 July 2023
Published: 14 July 2023

Copyright: © 2023 by the authors. Licensee MDPI, Basel, Switzerland. This article is an open access article distributed under the terms and conditions of the Creative Commons Attribution (CC BY) license (https://creativecommons.org/licenses/by/4.0/).

1. Introduction

Grain size, a complex quantitative trait involving grain length (GL), grain width (GW), grain thickness, and the grain length/width ratio (GL/W), is one of the determinants of grain weight, which not only affects the yield, but also the appearance quality of rice [1,2]. As an important factor affecting rice yield and quality, mining grain-shape-related genes is an important means to understanding their molecular mechanism and genetic basis. As of 2023, at least 201 rice grain shape genes have been identified. They are located on all chromosomes of the rice genome, and most are distributed on chromosomes 1, 2, 5, 6, and 7 (https://pubmed.ncbi.nlm.nih.gov/) (accessed on 7 May 2023). Most of these 201 genes directly regulate the rice grain shape, and the remaining genes indirectly regulate the grain size through an interaction network between genes.

Previous studies have found that the major factors affecting grain size include the ubiquitination-protease pathway, G-protein signaling, mitogen-activated protein kinase signaling, phytohormone regulation, and various transcriptional regulators [3]. GRAIN WIDTH 2 (*GW2*), the first QTL cloned in rice, encodes a RING-type E3 ubiquitin ligase with ubiquitination and autoubiquitination activity located in the cytoplasm and nucleus [4]; G-proteins that regulate grain size in rice include the a-subunit encoded by *RGA1/D1* [5]; and the β-subunit encoded by *DEP1* [6]. In addition, synthetic-hormone-related genes, such as *OsTAR1* [7] and *TGW6* [8], are also involved in the regulation of seed size. Moreover, several other transcriptional regulatory modules, such as *OsmiR396-OsGRFs* [9] and *AP2/ERF* modules [10], also play a key role in rice grain shape determination. In sum, rice grain shape is regulated by multiple factors. Even so, the current molecular network of rice grain types is still insufficient to explain all of the genetic variations. It is of great importance to explore new major genes and allelic variations for the high-yield breeding of rice.

It is well known that the grain length of *indica* rice is longer than that of *japonica* rice. Therefore, the excellent allelic variations of some important GL genes cloned at present are from *indica* rice. For example, the loss-of-function variations of *GS3* [11] and *TGW6* [8] are mostly from *indica* rice. *GS5* is a large grain allelic variation that was retained during the domestication of *indica* rice [12]. The overexpression of *LG3* [13] and *GLW7* [14] in *indica* rice can increase the grain length. Using these allelic variations to improve the grain shape of *japonica* rice, in addition to the method of breeding offspring through *indicia–japonica* hybridization, mutants can be directly obtained through gene editing. However, from the perspective of the geographical adaptability of subspecies, when using the offspring of *indica–japonica* hybridization, it is difficult to obtain materials with an excellent background of *japonica* rice, therefore, it is difficult to produce and apply unless it is included in a large number of molecular breeding works. The materials obtained by gene editing also cannot be used as cultivated varieties due to a certain degree of growth defects [15]. Considering this comprehensively, it is wise and efficient to clone new grain shape genes from *japonica* rice and apply them to *japonica* rice breeding.

Heilongjiang Province, as the main production area of early maturing *japonica* rice in China, had a planting area of about 6.43 million hectares in 2022, accounting for more than 15.5% of the national rice area. However, the grain length value of *japonica* rice in this region is low compared to *indica* rice in southern China, which hinders yield improvement. Therefore, the identification of new alleles controlling grain length from local sources is significant to increasing rice yield. Recent efforts combining QTL-seq and linkage analysis have led to the localization of several candidate genes in rice [16–19]. In this study, we used two *japonica* rice varieties, Pin20 and Songjing 15 (SJ15), with significant differences in grain shape, as parents in order to develop the $F_{2:3}$ population and the BC_3F_3 population. Through QTL-seq, linkage analysis, and fine mapping strategies, we identified the long-grain genes in the large-grain-variety of Pin20. Furthermore, the KASP molecular markers were developed to identify plants with different grain types. This will facilitate the in-depth study of grain type improvement and regulatory mechanisms in *japonica* rice varieties.

2. Results
2.1. Screening and Evaluation of Plant Height

The phenotypic detection showed that there were significant differences in the grain length (GL) and length–width ratio (GL/W) between SJ 15 and Pin 20 (Table 1, Figure 1A,B). For the $F_{2:3}$ population, the variation ranges of GL, grain width (GW), and GL/W were 0.68–1.09 cm, 0.33–0.43 cm, and 1.88–3.09 cm (Table 1, Figure 1C–E), respectively. Except for grain width, the absolute values of the skewness and kurtosis of GL and GL/W were less than 1 (Table 1), showing continuous variation and normal distribution, indicating that these two traits conform to the genetic model of quantitative traits.

Table 1. Phenotypic analysis of grain-shape-related traits of parents and populations.

Traits	SJ15	Pin20	$F_{2:3}$ Population			
			Mean ± SD	Range	Skewness	Kurtosis
GL (cm)	0.69	1.06 **	0.85 ± 0.00	0.68~1.09	0.61	0.95
GW (cm)	0.36	0.40	0.37 ± 0.00	0.33~0.43	0.27	1.46
GL/W	1.92	2.65 **	2.55 ± 0.02	1.88~3.09	0.33	−0.05

** indicates the significant difference detected at $p < 0.01$ level.

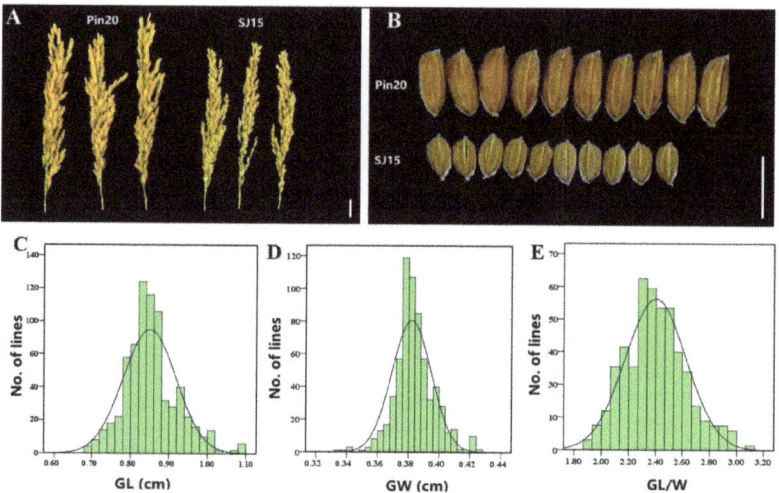

Figure 1. Phenotypic characteristics of grain shape of the parents and F$_{2:3}$ population. (**A**) Performance of the mature spikes of the two parents; (**B**) Comparison of the mature grain types of the two parents; and (**C–E**) Frequency distribution of grain length (GL), grain width (GW), and length–width ratio (GL/W) of rice grains in the F$_{2:3}$ population. The scale bars for (**A**,**B**) were 1 cm.

2.2. Phenotypic Analysis of Extreme DNA Pools of Grain Type

The differences in individual traits had an impact on the genotype frequency analysis of the DNA hybrid pool on the whole genome. To clarify the differences in the genetic background between the two DNA pools, 30 long-grain and 30 short-grain lines were analyzed for GL, GW, GL/W, spike number (PN), number of grains per spike (NGS), and spike weight per plant (SW). The results showed that there were highly significant differences in GL and GL/W between the long-grain and short-grain pools, while there were no significant differences in GW, PN, NGS, and SW (Figure 2). Therefore, the phenotypic differences between these two DNA mixing pools are distributed only in the GL and GL/W; thus, we selected 30 representative long-grain individuals and 30 short-grain individuals to prepare the GL-pool and GS-pool in order to map the candidate genomic loci using bulked segregant analysis (BSA) and re-sequencing analyses, respectively.

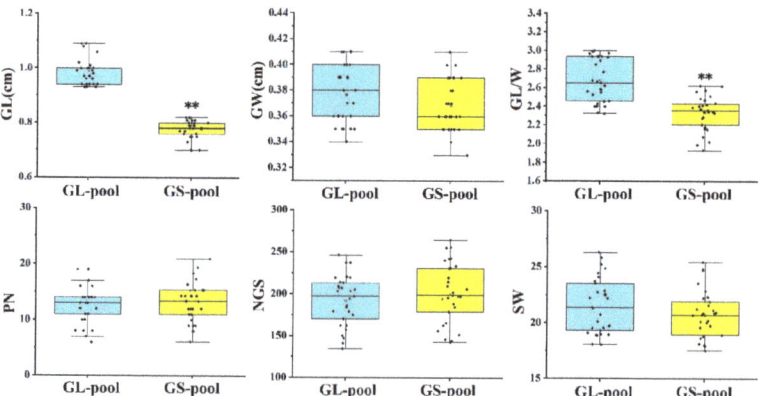

Figure 2. Box plot of panicle traits of individuals with significant differences in grain length. GL-pool, long-grain DNA pool; GS-pool, short-grain DNA pool; GL, grain length; GW, grain width; GL/W, length–width ratio; PN, spike number; NGS, number of grains per spike; and SW, spike weight per plant.

** indicates the significant difference detected at $p < 0.01$ level. Each black dot on the box plot represents the phenotypic value corresponding to a single independent plant.

2.3. Identification of a Major QTL Controlling GL in Rice Using QTL-Seq

A total of 315,277,920 clean reads and 47,291,688,000 bases were obtained by re-sequencing and data quality control of the two DNA mix pools and both parents (Supplementary Table S1). In addition, the ED (Euclidean distance) and two-tailed Fisher's exact test for each bulk were calculated by aligning the sequence with the Nipponbare reference genome. After calculating a statistical confidence interval of $p < 0.01$ between the two extreme phenotypic blocks, a 4.21 Mb (14,240,001 bp–18,445,701 bp) genomic region on chromosome 9 was identified by overlapping the results of the three algorithms (Table 2, Figure 3). We designated this QTL as *qGL9.1*.

Table 2. *qGL9.1* association results based on the G-statistic value, ED algorithm, and Fisher algorithm.

QTL	Algorithm	Start (bp)	End (bp)	Size (Mb)	Gene Number	Threshold
qGL9.1	G-statistic	14,185,459	18,445,701	4.26	623	0.01
	ED	14,240,001	18,760,000	4.52	669	0.01
	Fisher algorithm	13,560,001	19,030,000	5.47	807	0.05

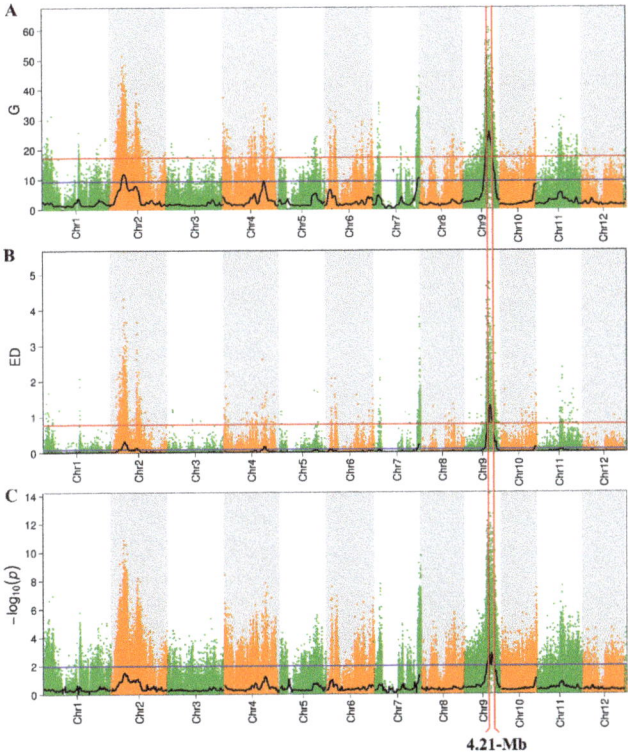

Figure 3. The results of the QTL-Seq analysis. (**A**) The G-statistic value to map *qGL9.1* based on SNP. (**B**) The Euclidean distance algorithm to map *qGL9.1*. (**C**) The two-tailed Fisher's exact test to map *qGL9.1* based on Indel. The blue lines and the red line represent the threshold line, with confidence levels of 0.95 and 0.99, respectively. The number on the horizontal coordinate represents the chromosome number. The red wireframe represents the intervals covered by *qGL9.1* in different computational modes.

2.4. Narrowing of qGL9.1 to a Fine Region

For pyramid *qGL9.1*, eight KASP (kompetitive allele-specific PCR) markers were developed for linkage analysis based on the base information provided by re-sequencing data from Pin20, SJ15, and the two pools, and a significant peak interval was detected in a 448.7-kb region between SNP5 and SNP6 on chromosome 9 when the threshold was 3.0 (Figure 4). The *qGL9.1* contributed to 20.09% of the phenotypic variation for GL (Table 3). The positive-effect allele of *qGL9.1* was derived from Pin20.

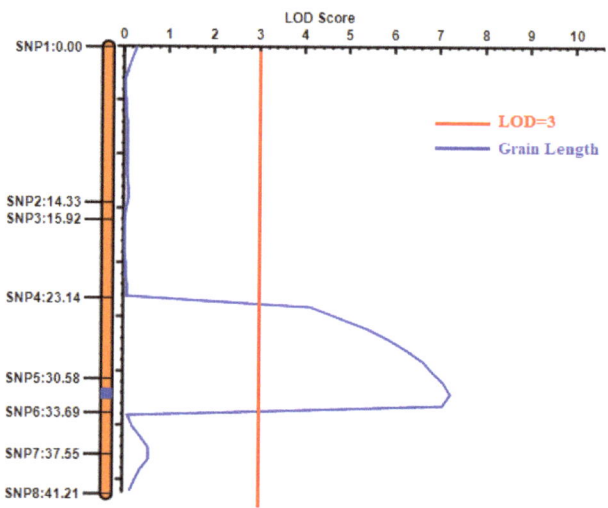

Figure 4. Linkage analysis of grain length.

Table 3. QTL association results by linkage analysis.

Trait	Chr.	Position (cM)	LOD	PVE (%)	Add	Left CI (cM)	Right CI (cM)
GL	9	32.00	7.23	20.09	0.03	27.5	33.5

Note: GL, grain length; Chr., chromosome; cM, centimorgan; LOD, logarithm of the maximum likelihood; PVE, phenotypic variation explained; Add, additive effect; Left CI, confidence interval on the left side of the linkage map; and Rigth CI, confidence interval on the right side of the linkage map.

In order to finely localize *qGL9.1*, we constructed the BC_3F_2 population and genotyped the BC_3F_2 population using the linkage markers of *qGL9.1* and six KASP markers consistent with the genetic background of SJ15 (Figure 5) and finally screened to two recombinants and obtained the BC_3F_3 population containing 1570 lines after self-crossing. For the fine mapping of *qGL9.1*, three KASP markers between SNP5 and SNP6 were developed from the re-sequencing data (Supplementary Table S2, Figure 6A). A total of 21 recombinants were identified by scanning the genotypes of 1570 BC_3F_3 individuals, and these 21 recombinants were classified into seven groups (Figure 6B). After progeny tests, the grain lengths of recombinant groups one and two were biased toward the short-grain parent SJ15, and the remaining recombinant groups were close to the long-grain parent Pin20. *qGL9.1* was delimited to the 93.0 Kb interval between the SNP10 and SNP11 markers (16.29–16.39 Mb). According to the MSU Rice Genome Annotation Project Release 7 [20], there are 15 protein-coding genes on the *qGL9.1* locus (Figure 6C), and information on SNP/InDel in the *qGL9.1* region (upstream, UTR3, downstream, and exonic) is listed in Supplementary Table S3. We found that 10 of the 15 genes had sequence differences in the promoter, exon, or downstream regions.

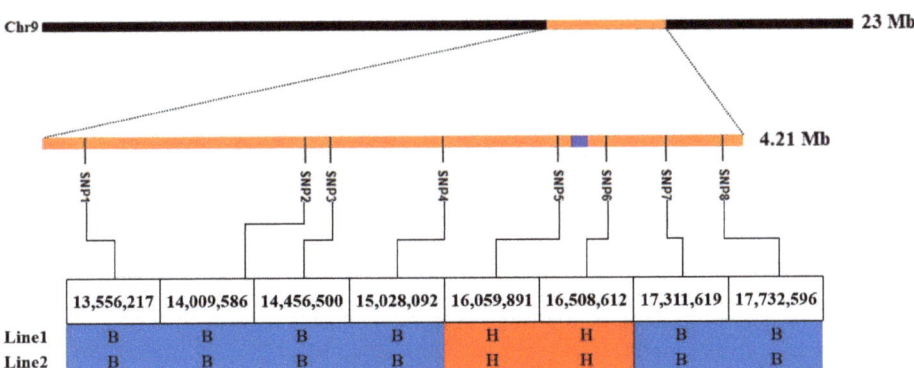

Figure 5. Screening of heterozygous lines in the BC₃F₂ population. Mb, million base pair; B, the genotype is consistent with SJ15; and H, the genotype is heterozygous.

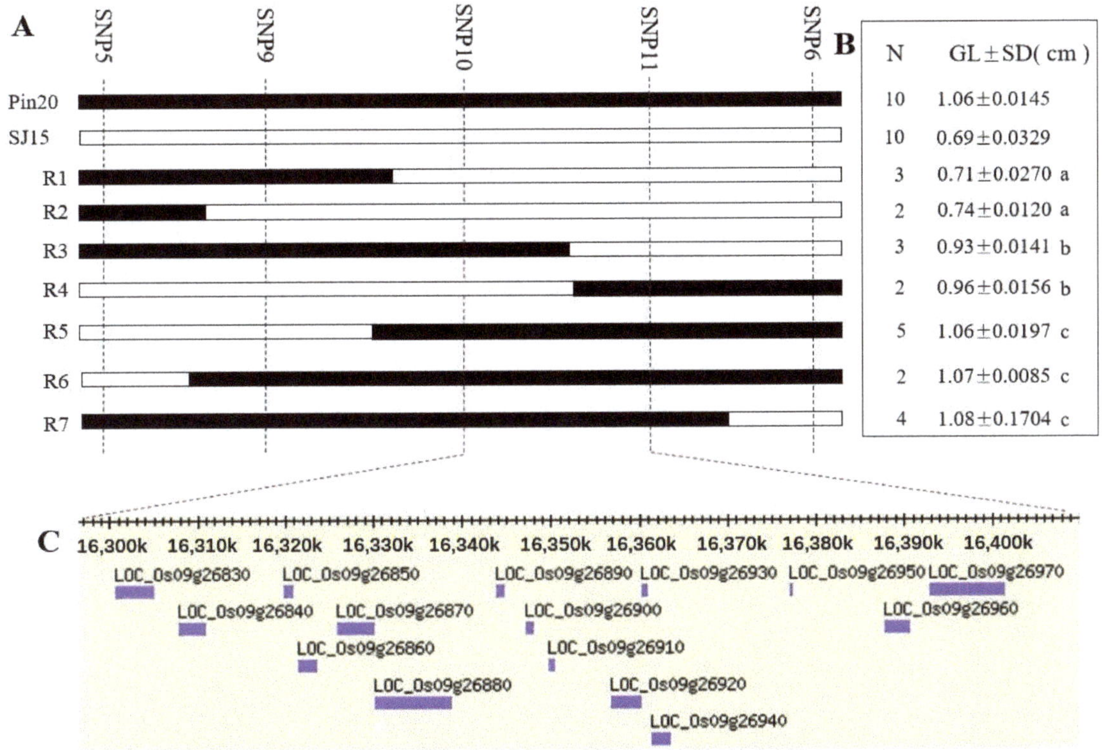

Figure 6. Fine mapping of *qGL9.1*. (**A**) The genotype of the recombinant plant. (**B**) Grain length statistics of recombinant plants. (**C**) A total of 15 genes in the *qGL9.1* region were obtained through the annotation information on the *Nipponbare* genome. Letters a, b and c indicate significant differences between groups, significant difference detected at $p < 0.05$ level.

2.5. Candidate Gene Analysis

Through the qRT-PCR analysis of 10 genes with sequence variation (Figure 7), it was found that significant differences in the relative expression of *LOC_O09g26970* between Pin20 and SJ15 occurred in samples from the 2 cm, 5 cm, and 7 cm panicles, while no significant differences were found in the relative expression of the 13 cm panicles. The results showed that the relative expression of *LOC_Os09g26970* in Pin 20 was higher than that in SJ15 at the early stage of panicle development. The expression levels of the other nine genes were not significantly different at the grain development stage. We further analyzed the structural domains of 15 genes through the Pfam database, annotated them using the Ensembl database, and found that *LOC_Os09g26970* encodes a cytochrome P450 family protein CYP92A8 (Supplementary Table S4). In addition, using the results of gene annotation based on the re-sequencing data, through pathway significant enrichment analysis, it was found that 616 genes in the 4.2 Mb interval were significantly enriched in arginine and proline metabolism (ko00330), nitrogen metabolism (ko00910), cysteine and methionine metabolism (ko00270), pentose phosphate pathway (ko00030), and glycolysis/gluconeogenesis (ko00010) (Figure 8). Among them, five genes (*LOC_Os09g26940*, *LOC_Os09g26950*, *LOC_Os09g26960*, *LOC_Os09g26970*, and *LOC_Os09g26980*) in the candidate interval were significantly enriched in brassinolide biosynthesis (ko00905) (Supplementary Table S5). The genes encoding the cytochrome P450 family proteins have been shown to play an important role in regulating rice grain shape, especially *D11* [21], *GW10* [22], and other proteins encoding the cytochrome P450 family, which plays an active role in controlling grain size through the BR pathway. Therefore, as a P450 family protein significantly enriched in the BR pathway, we believe that *LOC_Os09g26970* is a candidate gene for *qGL9.1*.

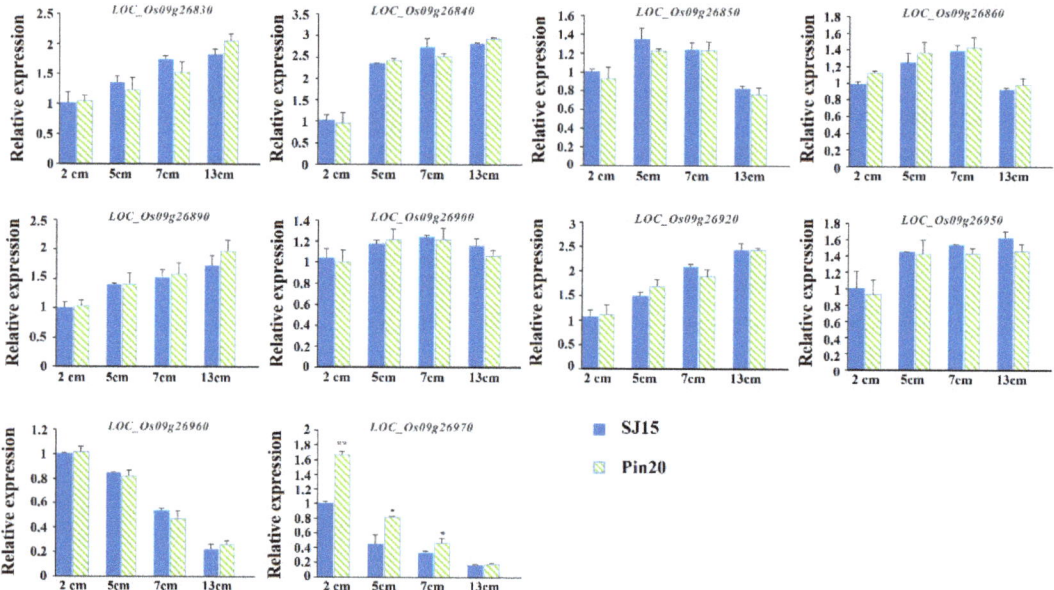

Figure 7. Expression of candidate genes during the development of the young panicles of both parents. The 2 cm, 5cm, 7cm, and 13 cm on the horizontal coordinates indicate the length of the developing young spike. The results were statistically analyzed using Student's *t*-test (*, $p < 0.05$; **, $p < 0.01$).

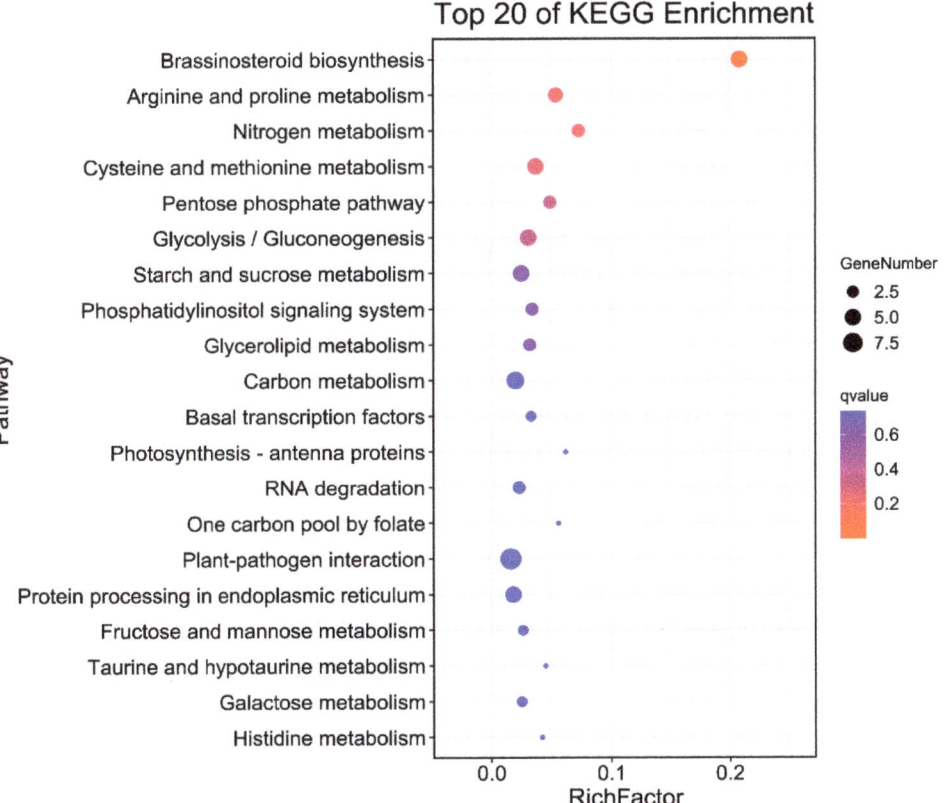

Figure 8. KO enrichment bubble plots for genes (Supplementary Table S5) in the *qGL9.1* interval. Horizontal coordinate: enrichment factor (number of differences in this pathway divided by all numbers); vertical coordinate: pathway name; bubble area size: number of genes belonging to this pathway in the target gene set; bubble color: enrichment significance. The redder the color, the smaller the P/Q value.

2.6. The Significant Association of LOC_Os09g26970 with the SNP

Sanger sequencing analysis identified 13 nSNPs on *LOC_Os09g26970* (Supplementary Table S6). The 13 SNPs were previously identified in the 3010 Rice Genome Project and the Rice Functional and Genomic Breeding (RFGB) v2.0 database [23,24]. Among them, 10 SNPs (Chr9-16397163, Chr9-16397736, Chr9-16397760, Chr9-16397792, Chr9-16398197, Chr9-16398200, Chr9-16398479, Chr9-16398989, Chr9-16399274, and Chr9-16399673) constituted nine haplotypes, and Hap1, Hap5, Hap6, Hap8, and Hap9 were mainly distributed in *indica* rice. Hap2, Hap3, and Hap4 were mainly distributed in *japonica* rice (Supplementary Table S7). The germplasm of Hap9, consistent with the Pin 20 genotype, and the Hap2, consistent with the SJ15 genotype, differed significantly between GL and GW, and the other haplotypes caused significant phenotypic differences in GL, GW, and GL/W. (Supplementary Table S8). It is worth noting that the haplotype Hap9 contributes to the optimal GL (9.36 mm) and GL/W (3.31), suggesting that Pin20 is a cultivated species carrying the optimal grain length variation of *LOC_Os09g26970*.

In order to obtain a molecular marker that could distinguish the grain length phenotype, we designed a KASP marker for an nSNP of the *LOC_Os09g26970*. SNP10 accurately divided the genotypes of 92 individuals in the 94 BC_3F_3 lines into Pin20 and SJ15 genotypes (Figure 9). These clustering results clearly distinguished the two alleles, therefore, the KASP8 marker was used to genotype the rice plants. Of the plants with the Pin20 allele, SNP10 identified 89.8% of the plants showing long-grain phenotypes. In contrast, SNP10 was able to identify 86.0% of the short-grain phenotype plants carrying the SJ15 genotype (Supplementary Table S9). This result implies that SNP10 can effectively distinguish the grain length of rice and can be used as an important molecular marker for breeding improvement.

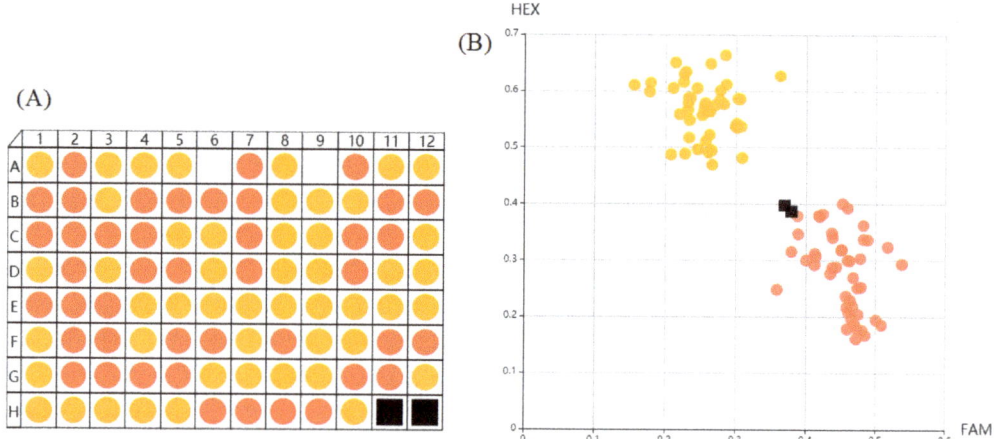

Figure 9. A total of 94 BC$_2$F$_4$ lines were genotyped with SNP10. (**A**) A demonstration of the genotyping effect of SNP10. Yellow fluorescence and red fluorescence indicate individuals with genotypes of Pin20 and SJ15, respectively. Black squares represent spotting holes without samples. The calls for the *LOC_Os09g26970-Pin20* allele were clustered near the Y-axis, while the calls for the *LOC_Os09g26970-SJ15* allele were clustered on the X-axis. (**B**) HEX, Hexachlorofluorescein; and FAM, 6-Carboxyfluorescein.

3. Discussion

3.1. QTL-Seq Analysis Combined with a Screening of Recombinant Plants Can Efficiently Fine-Map Candidate Genes

Grain length is a significant factor that limits rice yield. Improving and utilizing the large-effect genomic loci associated with GL is essential to increase rice yield. The authors of previous studies have carried out extensive QTL analyses and localized a group of genes that are associated with GL in rice. For example, *PGL1* [25] and *BG1* [26] positively regulated GL by increasing the cell size, whereas *SG1* [27], *SDF5* [28], *OsGDI1* [29], and *TGW6* [8] negatively regulated rice GL by reducing the cell size. However, the strategy of isolating genes by map-based cloning is time-consuming and labor-intensive. In recent years, with the development and application of biological high-throughput sequencing technology and bioinformatics analysis technology, the efficiency of mining QTL has significantly improved. The combination of traditional QTL mapping and QTL-seq can effectively and quickly identify the GL major QTL interval. For example, the GL locus *qTGW5.3* was mapped to a 5 Mb physical interval by QTL-seq. Furthermore, the recombinants and the progeny tests delimited the candidate region of *qTGW5.3* to 1.13 Mb [30]. Due to the lack of further mapping populations and recombinant plants, the candidate genes of *qTGW5.3* have not been identified. In this study, *qGL9.1* was isolated from Pin20 by using the QTL-seq strategy based on the ED, Fisher algorithm, and G value method, and *qGL9.1* was associated with a single strong peak in the three calculation models (Figure 3). This shows a significant difference in the allele ratio between the two mixed pools. To fine map the *qGL9.1* candidate gene, several approaches have been used to narrow down the genomic region associated with *qGL9.1*. Firstly, *qGL9.1* was fine-mapped to a 93 Kb interval containing 15 annotated genes by using the recombinant plants to optimize the target interval (Figure 3). *LOC_Os09g26970* was further anchored as the most reliable candidate for *qGL9.1* by expression analysis and functional annotation of the candidate genes. Therefore, *qGL9.1* can be considered the most significant target for GL in exploring candidate genes. Our study is a good example of using QTL-seq combined with fine mapping to mine candidate genes to obtain major QTL intervals.

3.2. LOC_Os09g26970 on qGL9.1 Links to Grain Length in Rice

LOC_O09g26970 is a cytochrome P450 structural domain (PF00067) gene, and the cytochrome P450 gene family is one of the largest supergene families in plants [31]. There are 356 P450 genes in the rice genome, and P450 plays an important role in various biochemical pathways that produce primary and secondary metabolites [32], some of which are essential for controlling plant cell proliferation and expansion. The proteins encoding cytochrome P450 families such as *D11* [21], *GW10* [22], *BSR2* [33], *GL3.2* [34], and *GE* [35] play an important role in regulating rice grain shape. In particular, the P450 family proteins encoded by *D11* and *GW10* play an active role in controlling the grain size through the biosynthetic pathway of brassinolide. In plants, BR is an essential steroid hormone that regulates many processes during plant development. It is involved in various biological reactions, such as stem elongation and vascular differentiation [36], especially in the regulation of grain size. Based on the re-sequencing data, this study found that *LOC_Os09g26970* was significantly enriched by KEGG enrichment analysis, which was related to the biosynthesis of brassinolide (ko00905). Therefore, it is speculated that the pathway of *qGL9.1* regulating grain shape is likely to be similar to that of *D11* and *GW10*. We further sequenced the CDS region of *LOC_Os09g26970* and found that there were 13 SNPs within it. Some of the haplotypes showed significant differences in grain length and grain width. We speculated that this locus was functional for grain length and grain width, but only showed differences in grain length in the genetic population of this study, which may have been caused by limited genetic variation. Next, we will construct various transgenic materials, such as knockout, overexpression, and complementation of *LOC_Os09g26970*, to verify its biological function in regulating rice grain length and analyze whether the effect of *LOC_Os09g26970* on grain length is affected by the BR pathway by applying exogenous BR.

3.3. Breeding Value and Potential of qGL9.1

In general, the cooking and eating quality of *japonica* rice is better than that of *indica* rice. While *indica* rice has longer grains and a better appearance quality than *japonica* rice, the quality of indica rice with long grains is often inferior to that of *japonica* rice [2,37]. In recent years, the molecular breeding and utilization of grain shape genes in *indica* and *japonica* rice completed several important tasks. New *indica* hybrid rice varieties, Taifengyou 55 and Taifengyou 208, with an improved grain yield and quality were developed by pyramiding semi-dominant *GS3* and *GW7^{TFA}* alleles from tropical *japonica* rice varieties [38]. The *GW8* and *GS3* alleles were polymerized into HJX74 to produce short and wide grains, resulting in the breeding of Huabiao 1 [12]. Using the deletion of *TGW6* and its alleles in the functional region, the functional marker CAPs6-1 of *TGW6* was developed and screened in order to quickly screen rice varieties carrying *TGW6* [39]. In this study, allelic variation A from *japonica* Pin20 was present in a small number of indica rice samples, but has not been identified in other *japonica* rice samples, indicating that *LOC_Os09g26970* may be a rare grain shape regulator in *japonica* rice germplasm. In addition, 10 SNPs in the coding region of *LOC_Os09g26970* had nine haplotypes in 3010 rice germplasms. The grain length of the germplasm containing the Pin20 genotype was 9.36 mm, while the average grain length of the other haplotypes was 8.60 mm. Therefore, Hap9 is the optimal haplotype of grain length, and Hap9 has the largest GL/W, which is an ideal allelic variation related to grain length. In addition, the GL/W of the germplasm corresponding to the nine haplotypes showed long-grain characteristics (the minimum GL/W was 2.52). Therefore, the nine haplotypes of *LOC_O09g26970* are helpful to determine the grain length of rice germplasm. Moreover, we selected one SNP from Hap9 as a molecular marker to analyze the individuals with significant differences in grain length in the BC$_3$F$_3$ population and found that SNP10 could be used as a target for marker grain length. Next, in *India–japonica* hybrid breeding, the selection of Hap9 can not only retain the excellent quality traits of japonica rice, but also help to improve its grain length. The KASP marker used in the study is a marker that is closely linked to GL, and it can be directly used for molecular-assisted selection. In addition, this locus can be inherited by offspring by inter-*japonica* hybridization, and long-grain varieties can be directly selected by conventional breeding methods.

4. Materials and Methods

4.1. Plant Materials

In this study, two japonica varieties, short-grain female parent SJ15 and long-grain male parent Pin20, were used as parental lines to develop 667 F_2 individuals and the corresponding $F_{2:3}$ population. The F_2 population was planted during the normal growing season (from April to October) and the mature seeds were harvested subsequently. All 667 $F_{2:3}$ lines were used for grain type identification after maturity. To fine map the target gene, one F_2 individual plant with long grains was selected to obtain BC_3F_1 seeds by backcrossing with SJ15, and a BC_3F_1 individual plant with a long-grain phenotype was self-crossed to generate the BC_3F_2 (725 individuals) and BC_3F_3 (1570 individuals) populations. The 667 $F_{2:3}$ individuals were used for QTL-seq and QTL mapping, and BC_3F_2 and BC_3F_3 were used to fine-map the *qGL9.1* candidate gene. All of the lines and their parents were planted at the Northeast Agricultural University experimental station (Heilongjiang Province, China; 47°98 N, 128°08 E; 128 m above sea level).

4.2. Evaluation of Grain Type for Rice

The grain size of the $F_{2:3}$ and BC_3F_3 populations was investigated when the rice was fully mature. We collected all of the spikes of each line in envelopes, placed them in natural light to dry, and then put them in an oven at 37 °C for one week. Three main spikes of each line, with approximately the same appearance, were randomly selected and used to measure the spike length and spike grain number. The grain length (GL) and grain width (GW) of 10 seeds of each line were measured with vernier calipers and the ratio of GL to GW was calculated. The phenotypic data for each line were measured in three replicates, and their average was used for data statistics.

4.3. Construction of Segregating Pools and Whole-Genome Re-Sequencing

Young leaves from 667 individuals of the F_2 population were collected separately for total genomic DNA extraction using a modified cetyltrimethylammonium bromide (CTAB) method [40]. Then, the genomic DNA of 30 extremely GL-type and 30 extremely GW-type individuals were selected as two bulked pools. To simplify the following description, we abbreviated the GL-type DNA pool as GL-pool, and the GW-type DNA pool as GW-pool. For GL-pool, GW-pool, and the two parents, isolated DNA was quantified using a Nanodrop 2000 spectrophotometer (Thermo Scientific, Fremont, CA, USA). All DNA from the GL-pool and GW-pool was quantified at precise concentrations with a Qubit® 2.0 Fluorometer (Life Technologies, Carlsbad, CA, USA). Equal amounts of DNA from the GL-pool and GW-pool plants were mixed. The four DNA libraries were sequenced on the Illumina MiSeq platform using the MiSeq Reagent Kit v2 (500 cycles) (Illumina Inc., San Diego, CA, USA).

4.4. QTL-Seq Analysis

The raw sequencing data were filtered using an internal Perl script, provided by Biomarker Technology Co. Ltd. (Beijing, China). These high-quality data were then mapped to the Nipponbare-Reference-IRGSP-1.0 [41] using the Burrows–Wheeler aligner [42]. Using the Picard tool (https://sourceforge.net/projects/picard/) (accessed on 8 June 2021), repeat reads were removed based on the clean reads located in the reference genome. The SNP and InDel (1–5 bp) calling was realized with GATK [43], using the default settings. A series of filters were also used to obtain highly accurate SNP and InDel sets [44]. The association analysis was performed using the ED [45], calculation of the G statistic [46,47], and two-tailed Fisher's exact test [48] based on SNP. Finally, the overlapping interval of the three methods was used as the QTL interval.

4.5. Further Mapping of the qGL9.1

To further delimit the position of *qGL9.1*, we developed KASP markers linked to the *qGL9.1* interval, and then KASP marker primers were designed with Primer 5 software (Premier Biosoft International, Corina Way, Palo Alto, CA, USA) based on the re-sequencing data of the two parents. The 5′ end of each KASP marker forward primer was ligated with FAM (5′-GAAGGTGACCAAGTTCATGCT-3′) and HEX (5′-GAAGGTCGGAGTCAACGGATT-3′) linker sequences. All polymorphic markers between the parents were selected and 667 F_2 individuals were genotyped using polymorphic markers to construct linkage maps and narrow down the candidate regions using the inclusive composite interval mapping (ICIM) module of QTL IciMapping 4.2. (http://www.isbreeding.net) (accessed on 19 June 2023) and are listed in Supplementary Table S2. The threshold of the LOD score for declaring the presence of a significant QTL was determined by a permutation test with 1000 repetitions at $p < 0.001$. Then, 725 BC_3F_2 individuals and 1570 BC_3F_3 individuals were used to screen the recombinants across the kompetitive allele-specific PCR (KASP) markers between the target regions. Each KASP marker contained two allele-specific forward primers and one common reverse primer. The reaction mixture was prepared according to the protocol described by KBiosciences (http://www.ksre.ksu.edu/igenomics) (accessed on 19 June 2023). All of the KASP primers are listed in Supplementary Table S2.

4.6. Fine Mapping and Candidate Gene Screening of qGL9.1

To fine map *qGL9.1*, the plants with interval heterozygous *qGL9.1* in the BC_3F_2 population were identified by the KASP marker, and the BC_3F_3 secondary population was obtained by selfing. The BC_3F_3 population was genotyped, and the recombinant plants were screened to achieve fine-mapped *qGL9.1*. The main methods of mining candidate genes were as follows: (1) Ensembl (http://ensemblgenomes.org/) (accessed on 19 June 2023) was used to annotate the candidate genes, and the possible domains of candidate genes were detected by the Pfam database (http://pfam.xfam.org/) (accessed on 19 June 2023). (2) Mutant genes were screened according to sequencing information. (3) The genes with sequence variation were analyzed by using qRT-PCR. When the young panicles began to differentiate, those of Pin 20 and SJ15 were sampled at the lengths of 2cm, 5cm, 7cm, and 13cm. The expression characteristics of the candidate genes between the parents were analyzed by using qRT-PCR.

The total RNA of the rice was extracted according to the steps of the GeneCopoeia-BlazeTaq™ SYBR® Green qPCR Mix 2.0 extraction kit, and RNA purification and reverse transcription were carried out according to the steps of SIMGEN of Hangzhou Xinjing Biological Reagent Co., Ltd (No.8, Xiyuan 1st Road, Xihu District, Hangzhou, Zhejiang, China). Amplification was performed with a Roche LightCycler96 fluorescence quantitative PCR instrument at Northeastern Agricultural University. According to the transcription sequence of the gene, the specific primers of the candidate gene were designed with Premier 5.0 software, and the sequence is shown in Supplementary Table S10. The original *Actin1* in the rice was used as the internal reference [49], and the specificity of the primers was based on the standard melting curve. Three replicates were set for each sample, and the relative expression of genes in tissues was calculated using the the 2-ΔΔCt method. qRT-PCR analysis was performed as previously described [50].

4.7. Haplotype Analysis of Candidate Genes

According to the RFGB database (Haplotype analysis module of https://www.rmbreeding.cn/index.php (accessed on 19 June 2023)), the differential bases in the coding region of candidate genes between parents were searched, and the haplotypes of these differential bases in 3010 rice varieties and the variation of each haplotype in the different rice germplasms were analyzed. The phenotypic data of the grain length, grain width, and aspect ratio in the RFGB database and their genomic information were used to analyze the differences between the different haplotypes of the candidate genes.

4.8. Development of KASP Markers and Validation of GL

To verify the above-identified *LOC_Os09g26970* with GL potential, two non-synonymous SNPs (nSNPs) were screened from the exons of *LOC_Os09g26970*, and the corresponding KASP markers were developed. The upstream and downstream 100-bp sequences of the target nSNPs were extracted from the Nipponbare genome sequence. Each KASP marker contained two allele-specific forward primers and a common reverse primer. The reaction mixture was prepared according to the instructions of KBiosciences (http://www.ksre.ksu.edu/igenomics (accessed on 19 June 2023)), and the KASP primers are shown in Supplementary Table S2.

5. Conclusions

In this study, we used F_2 and BC_3F_3 populations to identify a major QTL *qGL9.1* controlling rice grain length from long-grain variety Pin20 by re-sequencing and fine mapping. Furthermore, combined with functional annotation, variation detection, and qRT-PCR analysis, the gene *LOC_Os09g26970* encoding a P450 protein was identified as a candidate gene for *qGL9.1*. *LOC_Os09g26970* and was divided into nine haplotypes in 3010 rice germplasm, and Hap9, which was consistent with the genotype of Pin20, contributed the most to the grain length among all of the haplotypes. In summary, we found a new grain length gene in early-maturing *japonica* rice in the northernmost part of China, and the molecular breeding application of this gene will hopefully assist in tackling the difficult situation of improving the yield of early-maturing *japonica* rice.

Supplementary Materials: The following supporting information can be downloaded at: https://www.mdpi.com/article/10.3390/ijms241411447/s1.

Author Contributions: L.Y., P.L. and D.Z. conceived and designed the research. L.Y. and P.L. participated in data analysis. L.Y., P.L., J.W., H.L., H.Z. and W.X. performed the material development, sample preparation, and data analysis. L.Y. and P.L. wrote the manuscript. D.Z. corrected the manuscript. All authors have read and agreed to the published version of the manuscript.

Funding: This research was financially supported by the Heilongjiang Province Key R&D Program (2022ZX02B03), the Natural Science Foundation of Heilongjiang Province, China (LH2022C021), and the Postdoctoral Fund to Pursue Scientific Research of Heilongjiang Province, China (LBH-Q21097). All grants were provided by L.Y.

Data Availability Statement: The data presented in this study are available on request from the corresponding author. The data are not publicly available due to much unpublished genomic information from the resequencing data.

Conflicts of Interest: There are no conflict of interest to declare.

References

1. Sakamoto, T.; Matsuoka, M. Identifying and exploiting grain yield genes in rice. *Curr. Opin. Plant Biol.* **2008**, *11*, 209–214. [CrossRef] [PubMed]
2. Harberd, N.P. Shaping Taste: The Molecular Discovery of Rice Genes Improving Grain Size, Shape and Quality. *J. Genet. Genom. Yi Chuan Xue Bao* **2015**, *42*, 597–599. [CrossRef]
3. Ren, D.; Ding, C.; Qian, Q. Molecular bases of rice grain size and quality for optimized productivity. *Sci. Bull.* **2023**, *68*, 314–350.
4. Choi, B.S.; Kim, Y.J.; Markkandan, K.; Koo, Y.J.; Song, J.T.; Seo, H.S. GW2 Functions as an E3 Ubiquitin Ligase for Rice Expansin-Like 1. *Int. J. Mol. Sci.* **2018**, *19*, 1904. [CrossRef]
5. Utsunomiya, Y.; Samejima, C.; Takayanagi, Y.; Izawa, Y.; Yoshida, T.; Sawada, Y.; Fujisawa, Y.; Kato, H.; Iwasaki, Y. Suppression of the rice heterotrimeric G protein β-subunit gene, RGB1, causes dwarfism and browning of internodes and lamina joint regions. *Plant J.* **2011**, *67*, 907–916. [CrossRef] [PubMed]
6. Kunihiro, S.; Saito, T.; Matsuda, T.; Inoue, M.; Kuramata, M.; Taguchi-Shiobara, F.; Youssefian, S.; Berberich, T.; Kusano, T. Rice DEP1, encoding a highly cysteine-rich G protein γ subunit, confers cadmium tolerance on yeast cells and plants. *J. Exp. Bot.* **2013**, *64*, 4517–4527.

7. Abu-Zaitoon, Y.M.; Bennett, K.; Normanly, J.; Nonhebel, H.M. A large increase in IAA during development of rice grains correlates with the expression of tryptophan aminotransferase OsTAR1 and a grain-specific YUCCA. *Physiol. Plant* **2012**, *146*, 487–499. [CrossRef]
8. Ishimaru, K.; Hirotsu, N.; Madoka, Y.; Murakami, N.; Hara, N.; Onodera, H.; Kashiwagi, T.; Ujiie, K.; Shimizu, B.; Onishi, A.; et al. Loss of function of the IAA-glucose hydrolase gene *TGW6* enhances rice grain weight and increases yield. *Nat. Genet.* **2013**, *45*, 707–711. [CrossRef]
9. Duan, P.; Ni, S.; Wang, J.; Zhang, B.; Xu, R.; Wang, Y.; Chen, H.; Zhu, X.; Li, Y. Regulation of *OsGRF4* by *OsmiR396* controls grain size and yield in rice. *Nat. Plants* **2015**, *2*, 15203.
10. Schmidt, R.; Schippers, J.H.; Mieulet, D.; Watanabe, M.; Hoefgen, R.; Guiderdoni, E.; Mueller-Roeber, B. SALT-RESPONSIVE ERF1 is a negative regulator of grain filling and gibberellin-mediated seedling establishment in rice. *Mol. Plant* **2014**, *7*, 404–421. [CrossRef]
11. Takano-Kai, N.; Doi, K.; Yoshimura, A. GS3 participates in stigma exsertion as well as seed length in rice. *Breed. Sci.* **2011**, *61*, 244–250. [CrossRef]
12. Wang, S.; Wu, K.; Yuan, Q.; Liu, X.; Liu, Z.; Lin, X.; Zeng, R.; Zhu, H.; Dong, G.; Qian, Q.; et al. Control of grain size, shape and quality by OsSPL16 in rice. *Nat. Genet.* **2012**, *44*, 950–954. [CrossRef] [PubMed]
13. Yu, J.; Xiong, H.; Zhu, X.; Zhang, H.; Li, H.; Miao, J.; Wang, W.; Tang, Z.; Zhang, Z.; Yao, G.; et al. OsLG3 contributing to rice grain length and yield was mined by Ho-LAMap. *BMC Biol.* **2017**, *15*, 28. [CrossRef]
14. Si, L.; Chen, J.; Huang, X.; Gong, H.; Luo, J.; Hou, Q.; Zhou, T.; Lu, T.; Zhu, J.; Shangguan, Y.; et al. OsSPL13 controls grain size in cultivated rice. *Nat. Genet.* **2016**, *48*, 447–456.
15. Xu, Y.; Wang, R.; Wang, Y.; Zhang, L.; Yao, S. A point mutation in *LTT1* enhances cold tolerance at the booting stage in rice. *Plant Cell Environ.* **2020**, *43*, 992–1007. [CrossRef]
16. Kodama, A.; Narita, R.; Yamaguchi, M.; Hisano, H.; Adachi, S.; Takagi, H.; Ookawa, T.; Sato, K.; Hirasawa, T. QTLs maintaining grain fertility under salt stress detected by exome QTL-seq and interval mapping in barley. *Breed. Sci.* **2018**, *68*, 561–570. [CrossRef] [PubMed]
17. Tiwari, S.; Sl, K.; Kumar, V.; Singh, B.; Rao, A.R.; Mithra Sv, A.; Rai, V.; Singh, A.K.; Singh, N.K. Mapping QTLs for Salt Tolerance in Rice (*Oryza sativa* L.) by Bulked Segregant Analysis of Recombinant Inbred Lines Using 50K SNP Chip. *PLoS ONE* **2016**, *11*, e0153610. [CrossRef]
18. Wu, F.; Yang, J.; Yu, D.; Xu, P. Identification and validation a major QTL from "Sea Rice 86" seedlings conferred salt tolerance. *Agronomy* **2020**, *10*, 410. [CrossRef]
19. Shamaya, N.J.; Shavrukov, Y.; Langridge, P.; Roy, S.J.; Tester, M. Genetics of Na(+) exclusion and salinity tolerance in Afghani durum wheat landraces. *BMC Plant Biol.* **2017**, *17*, 209. [CrossRef]
20. Kawahara, Y.; de la Bastide, M.; Hamilton, J.P.; Kanamori, H.; McCombie, W.R.; Ouyang, S.; Schwartz, D.C.; Tanaka, T.; Wu, J.; Zhou, S.; et al. Improvement of the *Oryza sativa* Nipponbare reference genome using next generation sequence and optical map data. *Rice* **2013**, *6*, 4. [CrossRef]
21. Zhu, X.; Liang, W.; Cui, X.; Chen, M.; Yin, C.; Luo, Z.; Zhu, J.; Lucas, W.J.; Wang, Z.; Zhang, D. Brassinosteroids promote development of rice pollen grains and seeds by triggering expression of Carbon Starved Anther, a MYB domain protein. *Plant J.* **2015**, *82*, 570–581. [CrossRef] [PubMed]
22. Zhan, P.; Wei, X.; Xiao, Z.; Wang, X.; Ma, S.; Lin, S.; Li, F.; Bu, S.; Liu, Z.; Zhu, H.; et al. GW10, a member of P450 subfamily regulates grain size and grain number in rice. *Theor. Appl. Genet.* **2021**, *134*, 3941–3950. [CrossRef] [PubMed]
23. Wang, C.C.; Yu, H.; Huang, J.; Wang, W.S.; Faruquee, M.; Zhang, F.; Zhao, X.Q.; Fu, B.Y.; Chen, K.; Zhang, H.L.; et al. Towards a deeper haplotype mining of complex traits in rice with RFGB v2.0. *Plant Biotechnol. J.* **2020**, *18*, 14–16. [CrossRef] [PubMed]
24. Wang, W.; Mauleon, R.; Hu, Z.; Chebotarov, D.; Tai, S.; Wu, Z.; Li, M.; Zheng, T.; Fuentes, R.R.; Zhang, F.; et al. Genomic variation in 3010 diverse accessions of Asian cultivated rice. *Nature* **2018**, *557*, 43–49. [CrossRef] [PubMed]
25. Heang, D.; Sassa, H. Antagonistic actions of HLH/bHLH proteins are involved in grain length and weight in rice. *PLoS ONE* **2012**, *7*, e31325. [CrossRef]
26. Liu, L.; Tong, H.; Xiao, Y.; Che, R.; Xu, F.; Hu, B.; Liang, C.; Chu, J.; Li, J.; Chu, C. Activation of Big Grain1 significantly improves grain size by regulating auxin transport in rice. *Proc. Natl. Acad. Sci. USA* **2015**, *112*, 11102–11107. [CrossRef]
27. Nakagawa, H.; Tanaka, A.; Tanabata, T.; Ohtake, M.; Fujioka, S.; Nakamura, H.; Ichikawa, H.; Mori, M. Short grain1 decreases organ elongation and brassinosteroid response in rice. *Plant Physiol.* **2012**, *158*, 1208–1219. [CrossRef]
28. Yang, Y.; Li, J.; Li, H.; Xu, Z.; Qin, R.; Wu, W.; Wei, P.; Ding, Y.; Yang, J. SDF5 Encoding P450 Protein Is Required for Internode Elongation in Rice. *Rice Sci.* **2021**, *28*, 313–316.
29. Ali Shad, M.; Wang, Y.; Zhang, H.; Zhai, S.; Shalmani, A.; Li, Y. Genetic analysis of GEFs and GDIs in rice reveals the roles of OsGEF5, OsGDI1, and OsGEF3 in the regulation of grain size and plant height. *Crop J.* **2023**, *11*, 345–360. [CrossRef]
30. Qin, Y.; Cheng, P.; Cheng, Y.; Feng, Y.; Huang, D.; Huang, T.; Song, X.; Ying, J. QTL-Seq Identified a Major QTL for Grain Length and Weight in Rice Using Near Isogenic F_2 Population. *Rice Sci.* **2018**, *25*, 121–131.
31. Nelson, D.R. Cytochrome P450 nomenclature. *Methods Mol. Biol. (Clifton N.J.)* **1998**, *107*, 15–24.
32. Schuler, M.A.; Werck-Reichhart, D. Functional genomics of P450s. *Annu. Rev. Plant Biol.* **2003**, *54*, 629–667. [CrossRef] [PubMed]
33. Maeda, S.; Sasaki, K.; Kaku, H.; Kanda, Y.; Ohtsubo, N.; Mori, M. Overexpression of Rice BSR2 Confers Disease Resistance and Induces Enlarged Flowers in Torenia fournieri Lind. *Int. J. Mol. Sci.* **2022**, *23*, 4735. [CrossRef]

34. Xu, F.; Fang, J.; Ou, S.; Gao, S.; Zhang, F.; Du, L.; Xiao, Y.; Wang, H.; Sun, X.; Chu, J.; et al. Variations in CYP78A13 coding region influence grain size and yield in rice. *Plant Cell Environ.* **2015**, *38*, 800–811. [CrossRef]
35. Nagasawa, N.; Hibara, K.; Heppard, E.P.; Vander Velden, K.A.; Luck, S.; Beatty, M.; Nagato, Y.; Sakai, H. GIANT EMBRYO encodes CYP78A13, required for proper size balance between embryo and endosperm in rice. *Plant J.* **2013**, *75*, 592–605. [CrossRef]
36. Abe, Y.; Mieda, K.; Ando, T.; Kono, I.; Yano, M.; Kitano, H.; Iwasaki, Y. The SMALL AND ROUND SEED1 (SRS1/DEP2) gene is involved in the regulation of seed size in rice. *Genes Genet. Syst.* **2010**, *85*, 327–339. [CrossRef] [PubMed]
37. Hirose, T.; Aoki, N.; Harada, Y.; Okamura, M.; Hashida, Y.; Ohsugi, R.; Akio, M.; Hirochika, H.; Terao, T. Disruption of a rice gene for α-glucan water dikinase, *OsGWD1*, leads to hyperaccumulation of starch in leaves but exhibits limited effects on growth. *Front. Plant Sci.* **2013**, *4*, 147. [CrossRef]
38. Wang, S.; Li, S.; Liu, Q.; Wu, K.; Zhang, J.; Wang, S.; Wang, Y.; Chen, X.; Zhang, Y.; Gao, C.; et al. The *OsSPL16-GW7* regulatory module determines grain shape and simultaneously improves rice yield and grain quality. *Nat. Genet.* **2015**, *47*, 949–954. [CrossRef]
39. Wang, J.; Yang, J.; Xu, X.; Zhu, J.; Fan, F.; Li, W.; Wang, F.; Zhong, W. Development and Application of a Functional Marker for Grain Weight Gene TGW6 in Rice. *Chin. J. Rice Sci.* **2014**, *28*, 473–478.
40. Murray, M.G.; Thompson, W.F. Rapid isolation of high molecular weight plant DNA. *Nucleic Acids Res.* **1980**, *8*, 4321–4326. [CrossRef]
41. Sasaki, T. The map-based sequence of the rice genome. *Nature* **2005**, *436*, 793–800. [CrossRef]
42. Li, H.; Durbin, R. Fast and accurate short read alignment with Burrows–Wheeler transform. *Bioinformatics* **2009**, *25*, 1754–1760. [CrossRef]
43. McKenna, A.; Hanna, M.; Banks, E.; Sivachenko, A.; Cibulskis, K.; Kernytsky, A.; Garimella, K.; Altshuler, D.; Gabriel, S.; Daly, M.; et al. The Genome Analysis Toolkit: A MapReduce framework for analyzing next-generation DNA sequencing data. *Genome Res.* **2010**, *20*, 1297–1303. [CrossRef] [PubMed]
44. Reumers, J.; De Rijk, P.; Zhao, H.; Liekens, A.; Smeets, D.; Cleary, J.; Van Loo, P.; Van Den Bossche, M.; Catthoor, K.; Sabbe, B.; et al. Optimized filtering reduces the error rate in detecting genomic variants by short-read sequencing. *Nat. Biotechnol.* **2011**, *30*, 61–68. [CrossRef]
45. Hill, J.T.; Demarest, B.L.; Bisgrove, B.W.; Gorsi, B.; Su, Y.C.; Yost, H.J. MMAPPR: Mutation mapping analysis pipeline for pooled RNA-seq. *Genome Res.* **2013**, *23*, 687–697. [CrossRef] [PubMed]
46. Magwene, P.M.; Willis, J.H.; Kelly, J.K. The Statistics of Bulk Segregant Analysis Using Next Generation Sequencing. *PLoS Comp. Biol.* **2011**, *7*, e1002255. [CrossRef] [PubMed]
47. Mansfeld, B.N.; Grumet, R. QTLseqr: An R Package for Bulk Segregant Analysis with Next-Generation Sequencing. *Plant Genome* **2018**, *11*, 1–5. [CrossRef]
48. Fisher, R.A. On the interpretation of χ^2 from contingency tables, and the calculation of P. *J. R. Stat. Soc.* **1922**, *85*, 87–94. [CrossRef]
49. Siahpoosh, M.R.; Sanchez, D.H.; Schlereth, A.; Scofield, G.N.; Furbank, R.T.; van Dongen, J.T.; Kopka, J. Modification of *OsSUT1* gene expression modulates the salt response of rice Oryza sativa cv. Taipei 309. *Plant Sci.* **2012**, *182*, 101–111. [CrossRef]
50. Zhang, Z.Y.; Li, J.J.; Tang, Z.S.; Sun, X.M.; Zhang, H.L.; Yu, J.P.; Yao, G.X.; Li, G.L.; Guo, H.F.; Li, J.L.; et al. Gnp4/LAX2, a RAWUL protein, interferes with the *OsIAA3-OsARF25* interaction to regulate grain length via the auxin signaling pathway in rice. *J. Exp. Bot.* **2018**, *69*, 4723–4737. [CrossRef]

Disclaimer/Publisher's Note: The statements, opinions and data contained in all publications are solely those of the individual author(s) and contributor(s) and not of MDPI and/or the editor(s). MDPI and/or the editor(s) disclaim responsibility for any injury to people or property resulting from any ideas, methods, instructions or products referred to in the content.

Article

Effectiveness of *TaDreb-B1* and *1-FEH w3* KASP Markers in Spring and Winter Wheat Populations for Marker-Assisted Selection to Improve Drought Tolerance

Shamseldeen Eltaher [1], Mostafa Hashem [2], Asmaa A. M. Ahmed [2], P. Stephen Baenziger [3], Andreas Börner [4] and Ahmed Sallam [2,4,*]

1. Department of Plant Biotechnology, Genetic Engineering and Biotechnology Research Institute (GEBRI), University of Sadat City (USC), Sadat City 32897, Egypt; shams.eltaher@gebri.usc.edu.eg
2. Department of Genetics, Faculty of Agriculture, Assiut University, Assiut 71526, Egypt; mostafa.14224423@agr.aun.edu.eg (M.H.); asmaa.14219580@edu.aun.edu.eg (A.A.M.A.)
3. Department of Agronomy & Horticulture, University of Nebraska-Lincoln, Lincoln, NE 68583, USA; pbaenziger1@unl.edu
4. Department Genebank, Leibniz Institute of Plant Genetics and Crop Plant Research (IPK), 06466 Gatersleben, Germany
* Correspondence: sallam@ipk-gatersleben.de or amsallam@aun.edu.eg

Abstract: Due to the advances in DNA markers, kompetitive allele-specific PCR (KASP) markers could accelerate breeding programs and genetically improve drought tolerance. Two previously reported KASP markers, *TaDreb-B1* and *1-FEH w3*, were investigated in this study for the marker-assisted selection (MAS) of drought tolerance. Two highly diverse spring and winter wheat populations were genotyped using these two KASP markers. The same populations were evaluated for drought tolerance at seedling (drought stress) and reproductive (normal and drought stress) growth stages. The single-marker analysis revealed a high significant association between the target allele of *1-FEH w3* and drought susceptibility in the spring population, while the marker–trait association was not significant in the winter population. The *TaDreb-B1* marker did not have any highly significant association with seedling traits, except the sum of leaf wilting in the spring population. For field experiments, SMA revealed very few negative and significant associations between the target allele of the two markers and yield traits under both conditions. The results of this study revealed that the use of *TaDreb-B1* provided better consistency in improving drought tolerance than *1-FEH w3*.

Keywords: kompetitive allele-specific PCR; genetic validation; water deficit; MAS

1. Introduction

Wheat (*Triticum aestivum* L.) is the main source of nutrition for one-third of the world's population [1]. It is now necessary, despite limited resources, to improve wheat production to fulfill the demand for food. To meet this need, the breeding of high-yielding cultivars is essential. Years of conventional and molecular breeding in wheat have greatly improved wheat yield. Understanding genetic principles, phenotypic evaluation, and selection through traditional breeding methods has resulted in significant improvements in wheat productivity [2,3]. However, increasing productivity in the face of global warming and weather variability will be a difficult task. Due to unpredictable crop loss, abiotic stresses influence wheat yield and can lead to a food supply deficit. Drought and heat are the two most significant abiotic factors that limit wheat yield [4–6]. Many researchers have identified drought tolerance in wheat over the past few decades, but they have not been able to improve the crops' drought tolerance due to a variety of causes. First, the diverse changes in several physiological parameters of the plant caused by drought need to be quantified and better understood. Second, the selection process is impacted by the genotype x environment interaction (GE), where the environment is often variable. Third, drought is a very complicated feature that is regulated by numerous genes, the majority of which contribute only minimally genetically. To improve wheat's ability to tolerate drought, all of these three reasons should be taken into consideration. Additionally, the complexity of the wheat genome is one of the most significant elements reducing the effectiveness of the process of increasing drought tolerance in wheat.

The wheat breeding program aims to screen as many as wheat genotypes for their tolerance to drought to select promising drought-tolerant genotypes that can be crossed to produce wheat cultivars that have a high tolerance to drought stress. However, phenotypic evaluation is labor and takes a lot of time to achieve [7,8]. DNA markers linked to target traits can be used through marker-assisted selection (MAS) to accelerate breeding programs. The development of molecular markers has three main limitations: (1) a functional marker must be developed based on the gene sequence, or the markers must be tightly linked or co-segregated with the genes controlling the traits; (2) they should also be broadly applicable to various cultivars and geographical areas; and (3) when used in breeding programs, the markers should be inexpensive [9]. Unfortunately, most traditional molecular markers, including those developed over the past 30 years (such as RAPD, AFLP, SSR, and RFLP) are not on target gene sequences and frequently exist at high genetic distances from genes. In addition, cultivars showing desired marker genotypes may not necessarily have the targeted genes—a phenomenon known as "false positives" in MAS practice. Additionally, while choosing markers, many markers have low polymorphism [9,10].

Single nucleotide polymorphisms (SNPs) are base pair substitutions, deletions, or insertions that occur as point mutations in a genome. An efficient, homogenous, and fluorescence-based method called kompetitive allele-specific PCR (KASP) genotyping allows for the detection and calculation of SNPs such as insertions and deletions [11–13]. KASP is a uniplex and flexible genotyping platform that efficiently and cheaply provides high throughput [14]. Breeding program improvement could be greatly sped up by converting traditional functional markers (FMs) to KASP tests.

KASP is rarely used to investigate the effects of drought on wheat at the seedling stage, two KASP markers, namely *TaDreb-B1* and *1-FEH w3* from *Dreb* and *Fehw3* genes, were reported with their association with drought tolerance [15]. Only two studies tested these markers with their association with yield under normal conditions and drought tolerance at the germination stage, and those studies were by Rehman et al. [16] and Mohamed et al. [17], respectively. Therefore, insufficient information is provided about these two markers for drought tolerance to be used in MAS, in which big germplasm can be screened. Validating DNA markers is an important step before utilizing them in MAS to achieve fruitful improvement for target traits. Such validation of markers should be significantly tested in a different genetic background, or their effects can still be significantly detected when tested in different locations and/or years [18]. Moreover, the marker–trait association should be tested in diverse populations that have high genetic variation, which is very important to truly investigate the association of the tested markers with the target traits for MAS. The aims of this study are to test the association between drought tolerance and two KASP markers (*TaDreb-B1* and *1-FEH w3* markers) and investigate their effectiveness for marker-assisted selection to improve drought tolerance in wheat.

2. Results

2.1. KASP Genotyping

The distribution of two wheat populations based on *TaDreb-B1* and *1-FEH w3* KASP markers is shown in Figure 1. In general, the blue dots are the FAM- homozygous allele and refer to the marker's target allele. The red dots are the HEX homozygous allele (non-target allele). Green dots are heterozygous alleles, and black dots stand for no call.

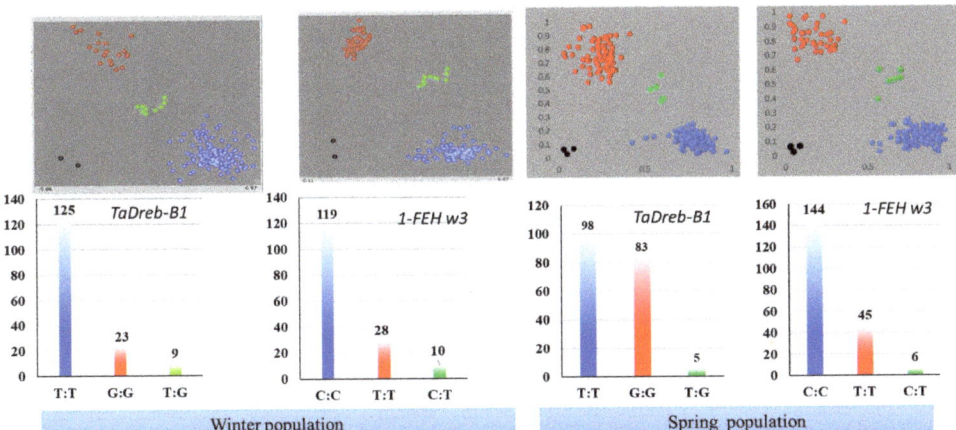

Figure 1. The result of KASP genotyping in spring and winter populations using *TaDreb-B1* and *1-FEH w3* KASP markers. T:T (blue) and G:G (red) homozygous allele and T:G (green) heterozygous allele for *TaDreb-B1*. C:C (blue) and T:T (red) homozygous allele and C:T (green) heterozygous allele for *1-FEH w3*.

According to Figure 1, the winter population (WP) had 125 genotypes containing the homozygous allele (T:T) for *TaDreb-B1* and 119 genotypes containing the homozygous allele (C:C) for *1-FEH w3*. Additionally, the red groups in the WP containing 23 genotypes had homozygous alleles (G:G) for *TaDreb-B1*, and 28 genotypes had homozygous alleles (T:T) for *1-FEH w3*. The heterozygous alleles (T:G and C:T) in the WP were shown in 9 and 10 genotypes for *TaDreb-B1* and *1-FEH w3*, respectively.

The spring population (SP) had 98 genotypes with the homozygous allele (T:T) in the *TaDreb-B1* and 144 genotypes with the homozygous allele (C:C) in the *1-FEH w3*. Additionally, 83 genotypes had the homozygous allele (G:G) for *TaDreb-B1*, and 45 genotypes had the homozygous allele (T:T) for *1-FEH w3* in the red groups. The heterozygous alleles (T:G and C:T) in the SP were present in 5 and 6 genotypes for *TaDreb-B1* and *1-1-FEH w3*, respectively.

The genotypes in each population were classified into four groups: *TaDrebB1* (genotypes had only the *TaDreb-B1* target allele), *1-FEH w3* (genotype had only the *1-FEH w3* target allele), *TaDrebB1&1-FEH w3* (genotype had the two target alleles), and Non (genotype did not have any target alleles) (Figure 2). The spring population had 61 genotypes with only TaDreB1, and this was higher than those genotypes in the winter population (5 genotypes). For genotypes having only *1-FEH w3*, a set of 27 genotypes in the winter population was found to have only *1-FEH w3*, and this number was higher than those in the spring population (23 genotypes). Interestingly, 62% (98 genotypes) of genotypes in the winter population possessed both favorable marker alleles together, while this percentage was only 37% in the spring population (75 genotypes). A total of 25 genotypes had no favorable marker allele in the spring genotypes, while only five genotypes in the winter population had no favorable marker allele.

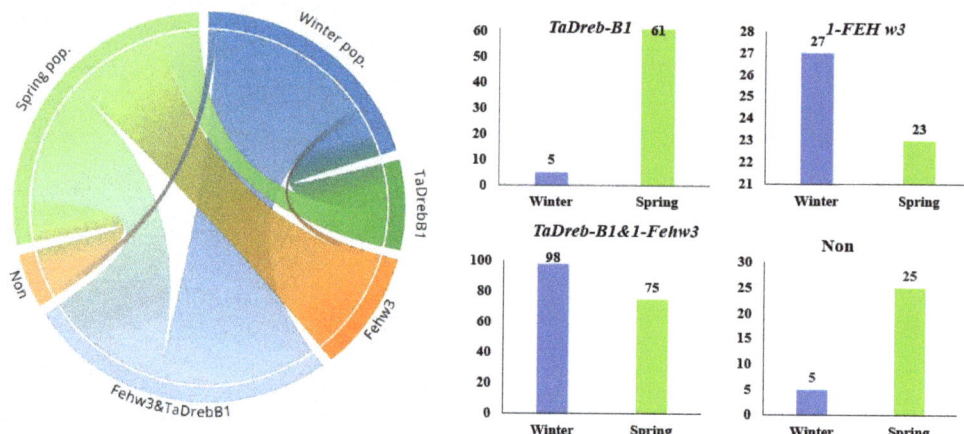

Figure 2. Number of genotypes having only gene; only *TaDreb-B1* allele, only *1-FEH w3* allele, both genes (TaDreb-B1 and 1-FEH w3) and no target allele (Non group).

2.2. The Diversity of TaDreb-B1 and 1-FEH w3 Markers

The marker allele frequency, marker allele diversity, and PIC for both marker alleles in the two wheat populations are presented in Figure 3. In the WP and SP, the *TaDreb-B1* gene frequency was detected at 0.84 and 0.54, respectively. Moreover, the gene diversity of *TaDreb-B1* in the WP and SP was 0.26 and 0.50, respectively. PIC values for *TaDreb-B1* were 0.23 and 0.37 in WP and SP, respectively.

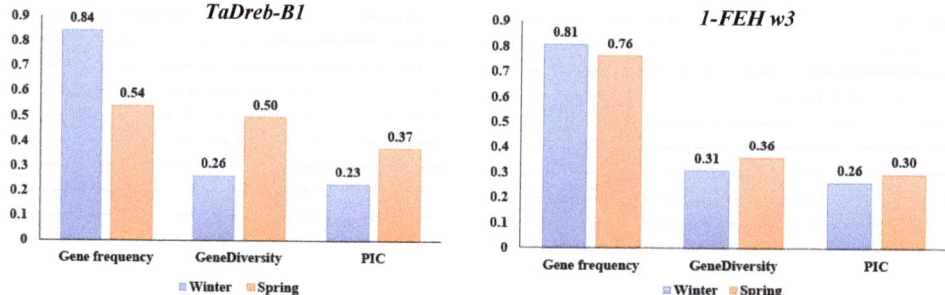

Figure 3. Diversity features. Gene frequency, gene diversity, and polymorphism information content (PIC) of *TaDreb-B1* and *1-FEH w3* markers in each population.

The gene frequency of *1-FEH w3* was 0.81 in the WP and 0.76 in the SP. Furthermore, the gene diversity of *1-FEH w3* in the WP and SP was 0.31 and 0.36, respectively, while the PIC values for *1-FEH w3* were 0.26 and 0.30 in WP and SP, respectively.

2.3. Single-Marker Analysis

2.3.1. Seedling Stage

Highly significant variation was found among genotypes in each population for all phenotypic traits scored [19,20]. In this study, a non-significant effect was found between the two markers and seedling traits of the WP. On the other hand, in the SP, the *TaDreb-B1* marker was not significantly associated with all traits except the SLW (p-value < 0.05). The *1-FEH w3* marker had a highly significant effect on all traits except DTR, with p-values of 0.00001, 0.00000022, and 0.0004 for SLW, DTW, and RB, respectively.

The allele effect of the two markers on the seedling traits in both populations is presented in Figure 4. In the SP, the allele C and T of the *1-FEH w3* and *Ta-Dreb-B1*, respectively, increased SLW and DTR and decreased DTW and RB. In the WP, the target allele of the *TaDreb-B1* marker had, on average, decreased SLW and RB and increased DTW and DTR. The target allele of the *1-FEH w3* in the WP, on the other hand, increased SLW, DTW, and DTR, while, it decreased RB.

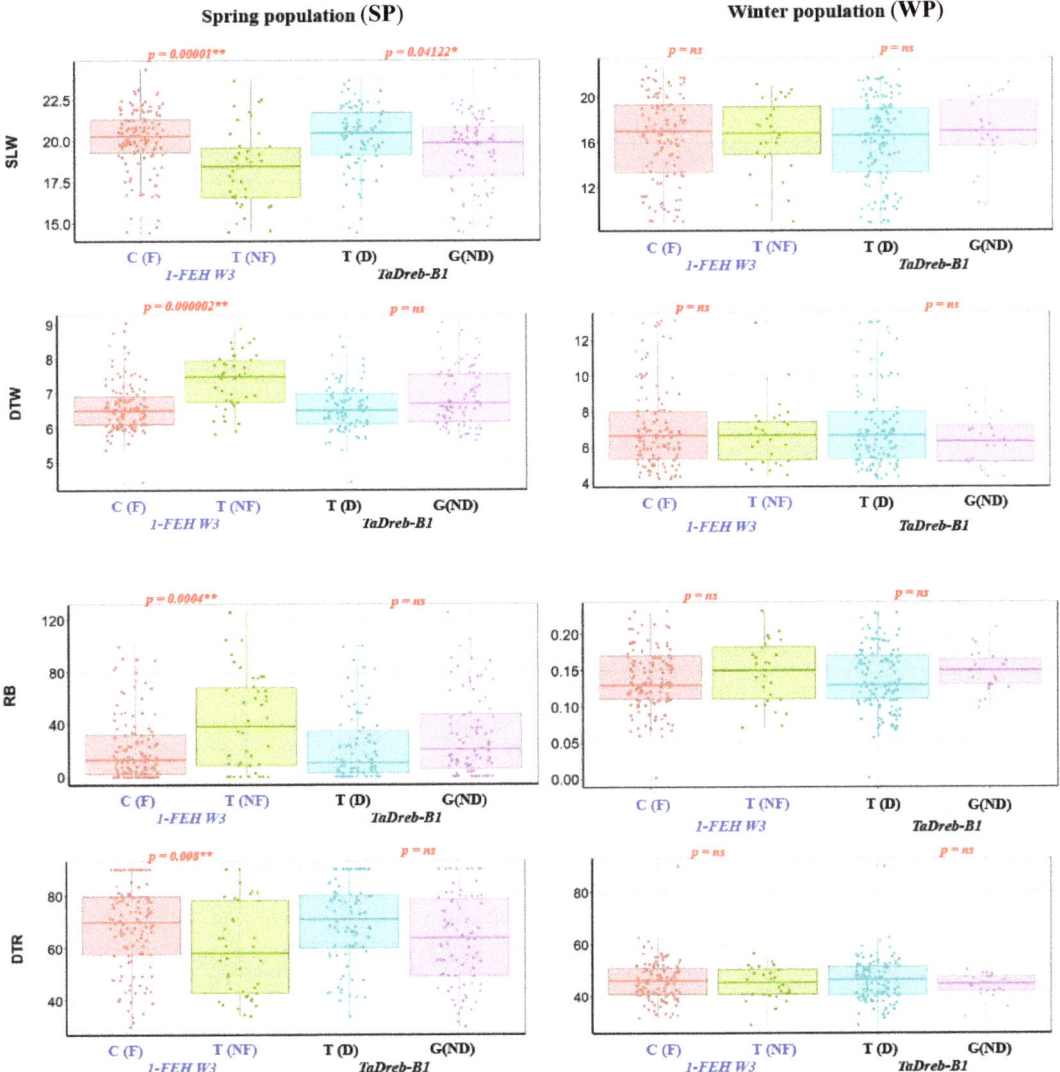

Figure 4. Box plots showing the difference between genotypes with the target allele and without the target allele for each marker in both populations scored at the seedling stage. F, NF, D, and ND refer to Fehw3, non-Fehw3, *TaDreb-B1*, and non-*TaDreb-B1*. (SLW) the sum of leaf witling, (DTW) days to wilting, (RB regrowth biomass), and (DTR) days to recovery. *, ** refer to significance at 0.05, and 0.01, respectively. ns refers to non-significant p values.

2.3.2. Field Experiments

Highly significant genetic variation was found among all genotypes in the winter population for all traits in the two locations in Nebraska, USA [21]. A significant difference was found between the two locations in the rainfall data (Supplementary Table S1), with averages of 3.09 cm and 1.36 cm for Lincoln and North Platte, respectively.

Highly significant variation was found for the same traits among genotypes in spring wheat (Supplementary Figures S1–S4). Soil moisture at 10 and 35 cm under drought stress in the two growing seasons are presented in Supplementary Table S2. Soil moisture (volume %) at 10 cm depth was 19.0 and 19.9% in 2018 and 2019, respectively, while it was 35.4 and 35.5% at 35 cm depth in the two growing seasons, respectively.

For the WP, the marker–trait association between the two KASP markers and the spike-related traits (SL, GNPS, SPS, and TKW) in the two distinct rainfall environments (Lincoln and North Platte) is presented in Table 1. The association between *TaDreb-B1* and spike traits in both locations was non-significant. In Lincoln, the *1-FEH w3* marker was found to be significant for all traits except TKW, with negative target allele effects on SL, GNPS, and SPS. No significant association was found between the *1-FEH w3* marker and spike traits in North Platte.

Table 1. Marker–trait association analysis between the two KASP markers and spike traits under Lincoln and North Platte (low rainfall environment) in the winter population.

Location	Trait	Gene/Target Allele	p-Value	Effect of Target Allele
Lincoln	SL	*1-FEH w3*/C	0.01727 **	−0.38931
		TaDreb-B1/T	0.30918	0.17946
	GNPS	*1-FEH w3*/C	0.00517 **	−3.2963
		TaDreb-B1/T	0.66216	0.56407
	SPS	*1-FEH w3*/C	0.03281 **	−0.53918
		TaDreb-B1/T	0.95199	0.01643
	TKW	*1-FEH w3*/C	0.78477	0.44653
		TaDreb-B1/T	0.44112	−0.17361
North Platte	SL	*1-FEH w3*/C	0.50759	0.08685
		TaDreb-B1/T	0.53501	0.08919
	GNPS	*1-FEH w3*/C	0.60044	0.34391
		TaDreb-B1/T	0.63706	0.33435
	SPS	*1-FEH w3*/C	0.79417	−0.0495
		TaDreb-B1/T	0.89495	0.0268
	TKW	*1-FEH w3*/C	0.48396	−0.3992
		TaDreb-B1/T	0.16237	−0.8338

(SL) Spike length, (SPS) number of spikelets per spike, (GNPS) grain number per spike, and (TKW) thousand-kernel weight. ** refers to significance at $p \leq 0.01$.

In the SP, the association between the two KASP markers and spike traits in two growth seasons (2018 and 2019) under control and drought conditions is shown in Table 2. Under control conditions, a significant association was found between SL and *TaDreb-B1* and between *1-FEH w3* and GNPS in the 2019 growing season. While under drought conditions in two seasons (2018 and 2019), *TaDreb-B1* and *1-FEH W3* were significantly associated with SL and GNPS, respectively.

Table 2. Marker–trait association analysis between the two KASP markers and spike traits under control and drought conditions in two growing seasons (2018 and 2019) in the spring population.

Environment	Trait	Gene/Target Allele	p-Value	Effect of Target Allele
Control 2018				
	SL	1-FEH w3/C	0.647	−0.307
		TaDreb-B1/T	0.105	−0.882
	GNPS	1-FEH w3/C	0.143	−5.190
		TaDreb-B1/T	0.169	4.115
	SPS	1-FEH w3/C	0.941	−0.033
		TaDreb-B1/T	0.788	0.103
	TKW	1-FEH w3/C	0.852	−0.297
		TaDreb-B1/T	0.461	−1.002
Drought 2018				
	SL	1-FEH w3/C	0.371	−0.381
		TaDreb-B1/T	0.000096 **	−1.393
	GNPS	1-FEH w3/C	0.021180 *	3.464
		TaDreb-B1/T	0.191	5.059
	SPS	1-FEH w3/C	0.831	−0.078
		TaDreb-B1/T	0.613	0.152
	TKW	1-FEH w3/C	0.966	−0.073
		TaDreb-B1/T	0.330	−1.382
Control 2019				
	SL	1-FEH w3/C	0.500	−0.358
		TaDreb-B1/T	0.00037 **	−1.540
	GNPS	1-FEH w3/C	0.02732 *	−6.738
		TaDreb-B1/T	0.836	0.519
	SPS	1-FEH w3/C	0.420	−0.374
		TaDreb-B1/T	0.523	0.251
	TKW	1-FEH w3/C	0.503	−0.455
		TaDreb-B1/ T	0.082	−0.942
Drought 2019				
	SL	1-FEH w3/C	0.651	−0.198
		TaDreb-B1/T	0.00051 **	−1.246
	GNPS	1-FEH w3/C	0.00811 **	−8.161
		TaDreb-B1/T	0.524	-d1.588
	SPS	1-FEH w3/C	0.356	−0.377
		TaDreb-B1/T	0.201	0.444
	TKW	1-FEH w3/C	0.802	−0.358
		TaDreb-B1/T	0.546	−0.718

(SL) Spike length, (SPS) number of spikelets per spike, (GNPS) grain number per spike, and (TKW) thousand-kernel weight, *, ** significant at $p \leq 0.05$ and $p \leq 0.01$, respectively.

3. Discussion

3.1. Diversity of KASP Markers

Improving drought tolerance in wheat remains an important task for breeders and geneticists in the face of climate change consequences, including drought stress. DNA markers can be used to accelerate breeding programs. However, they should be validated in different genetic backgrounds to be used for marker-assisted selection to improve target traits. Two KASP markers for *TaDreb-B1* and *1-FEH w3* genes were reported with their association with drought tolerance in wheat [15,16].

In the current study, two wheat populations were used to test the association between the two markers and drought tolerance at seedling and reproductive growth stages. The two populations facilitated reliable investigations on the effectiveness of these two markers in marker-assisted selection because they were produced differently, had different genetic backgrounds, represented two growth types of wheat (winter and spring), and each one had sufficient genotypes (>100 individuals)—recommended to test marker–trait association [22,23].

Both KASP markers were able to separate the genotypes into three clear groups (target allele, non-target allele, and heterozygous/heterogeneous). The diversity features of the two markers were extensively studied. The gene frequency of *1-FEH w3* was 0.76 and 0.81 for the spring and winter populations, respectively, which was higher than the frequency (0.28) reported by Rehman et al. [16] in a set of 153 Pakistani hexaploid wheat cultivars. For *TaDreb-B1*, the frequency of the target allele was 0.54 and 0.84 for the spring and winter populations, respectively, which was also higher than the frequency of the same allele (0.31) reported by Rehman et al. 2021. In the Pakistani hexaploid wheat cultivars, the PIC for *TaDreb-B1* and *1-FEH w3* genes was 0.33 and 0.34 [16], respectively, which were near the PIC values reported in this study for the two genes in both populations. Therefore, both markers can be classified as moderately informative (0.25 < PIC < 0.50), according to Bostein et al. [24]. So, *TaDrebB1* and *1-FEH w3* KSAP markers provided very useful information about the diversity among wheat genotypes and can also be used in diversity studies. The PIC and gene diversity values are an excellent indicator of informative markers that can be used for genetic diversity studies [25–27].

The association between the two target alleles for both markers and drought tolerance at seedling and adult growth stages were investigated. The same traits were recorded in both populations. Unfortunately, very few studies have investigated the effect of these two KASP markers on drought tolerance in the field. Mohamed et al. [17] tested the association between the two KASP markers and germination traits in wheat under 25 and 30% of polyethylene glycol (PEG). The association between yield traits under normal conditions and these two markers was studied by Rehman et al. [16].

3.2. Seedling Experiments

For the seedling experiment, SLW, DTW, RB, and DTR are very important traits that have a direct association with drought tolerance [28]. SLW and DTW are tolerance traits that indicate water deficit on wheat leaves, while RB and DTR are recovery traits indicating the ability of plants to recover and form new shoots after prolonged drought stress [29]. All these traits are highly heritable, so they are very useful for selection to drought tolerance [19,20,29,30]. In the WP, the non-significant association between the two markers and the seedling traits under drought stress indicated that there was no effect on improving drought tolerance for those genotypes that had the target allele. However, it was noted the genotypes that had allele (T) of *TaDreb-B1* had, on average, less leaf wilting symptoms, higher days to wilting, higher regrowth after drought, and lower days to recovery. The target allele (C) of *1-FEH w3* gene, on the other hand, was found to increase SLW, decrease DTW, decrease RB, and increase the DTR. Therefore, this allele can be considered, on average, to decrease drought tolerance. In the SP, the allele C of *1-FEH w3* gene was found to be highly significant and correlated with drought susceptibility for all seedling traits. The target allele of *TaDreb-B1*, on average, was non-significantly associated with seedling traits but, on average, the genotypes possessing this allele had high SLW, high DTW, low DTR, and low RB. Germination is a growth stage before the seedling stage. No significant association was identified between the two KASP markers and germination traits (germination %, germination rate, shoot length, root length, number of roots, fresh and dry weight, and water content) under severe drought stress induced by PEG [17]. Additionally, no clear allele effect, on average, regarding drought tolerance, was found for all traits under both PEG treatments.

3.3. Field Experiments

For the field experiment, both populations were evaluated under normal irrigation (SP) or high-rainfall (WP) and drought or low-rainfall conditions in two different countries (Egypt and USA). The significant differences in the WP between the two locations in the rainfall indicated that the genotypes at North Platte were exposed to greater drought stress compared to those grown in Lincoln. For the spring population, the soil moisture (volume %) was approximately 19% and 35% at 10 and 35 cm depth. In wheat, the optimum volumetric soil moisture content remaining at field capacity was about 45 to 55% (three feet below the soil surface for clay soils and 15 to 20% at the wilting point [31], which is defined as the soil water content when plants growing in that soil wilt and fail to recover their turgor upon rewetting, indicating successful drought stress occurred in spring genotypes).

Under normal (SP) and high-rainfall (WP) conditions, the target allele (C) for the *1-FEH w3* was found to be significantly associated with low GNPS (C19), while in the spring population, it was associated with low SPS, low GNPS and short SL (Lincoln) in the winter population. Additionally, it had a negative effect, on average, on TKW in both populations. These results agreed with previous results reported by Rehman et al. [16], who found that the favorable allele (C) of *1-FEH w3* was significantly associated with decreased grain number per spike (GPS). They also found a non-significant association between the *1-FEH w3* marker and TKW; however, in both populations, in the genotypes that had the C allele, TKW decreased compared to the genotypes that carried the other allele (T). For the *TaDreb-B1* marker, the target allele (T) was significantly associated with decreased spike length. In the study of Rehman et al. [16], the target allele of the *TaDreb-B1* was significantly associated with increased grain number per spike. In the current study, all genotypes in both populations carrying the target allele (T) of the *TaDreb-B1* had, on average, higher GNPS and SPS than the group with the nontarget allele (C). The target allele (T) decreased TKW in both populations. Although Rheman et al. [16] reported a non-significant association between the *TaDreb-B1* and TKW, on average, the genotypes carrying the target allele had higher TKW than the other group.

Under low-rainfall and drought stress conditions, the allele C of the *1-FEH w3* gene was significantly associated with increased GNPS in D18 and decreased GNPS in D19 in the spring population, indicating an environmentally sensitive effect on GNPS across the two years under drought stress. The same markers did not show a significant association in the winter population under low-rainfall environments. The *TaDreb-B1* markers had a significant association with decreased SL in D18 and D19 in the spring population. On average, the target allele of the *TaDreb-B1 (T)* increased SPS and GNPS (except D19) in both populations. Unfortunately, no previous study investigated the marker–trait association between markers and yield traits under drought stress conditions.

To summarize the effects of two markers in both populations in the current study, the allele effect (on average) of the target alleles of each marker for each trait is illustrated in Figure 5. In total, it seemed that *TaDreb-B1* had more positive effects in most of the traits at the seedling stage and adult growth stages than the *1-FEH w3* marker (four traits). Notably, on average, all genotypes carrying the target allele (T) of *TaDreb-B1* had higher GNPS (except D19) and SPS than non-allele genotypes under both conditions in both populations. The findings of Rehman et al. [16] (under normal conditions) support this result. Therefore, the *TaDreb-B1* marker might be more effective in MAS than the *1-FEH w3* marker.

The SNP alleles of *1-FEH w3* were reported by Zhang et al. [32]. The flanking sequence including the target allele (C) was blasted in *EnsemblePlant* (https://plants.ensembl.org/Triticum_aestivum/Info/Index (accessed on 15 May 2023). The blast results revealed that the sequence was located within the TraesCS6B02G080700 gene model, which is known as *1-FEH w3* (https://knetminer.org/Triticum_aestivum/ (accessed on 15 May 2023). Zhang et al. [32] tested the expression of *1-FEH w3* in two contrasting wheat genotypes for drought tolerance. They found that *1-FEH w3* had a higher expression in the drought tolerant genotype than in the drought susceptible genotype. The tolerant genotype had the allele C, while the susceptible genotype had the T allele. This marker segregated very well in the doubled haploid population (DH) derived from the crossing of two genotypes. In the DH population, the *1-FEH w3* allele (C) was found to be significantly associated with high thousand-grain weight under drought and well-watered conditions. Additionally, the target allele (C) of *1-FEH w3* had a non-significant association with kernel number per spike. The DH population only exhibited the variation observed between the two parents. But the diverse populations provide more genetic variation and diversity, making them ideal for identifying alleles and their real effects using marker–trait association analysis [33,34]. In a study by Yáñez et al. [35], the expression of *1-FEH w3* gene was studied in eight contrasting wheat genotypes that were selected based on the drought tolerance index. They found that the *1-FEH w3* gene had a higher expression in the susceptible genotype Fontagro 69 than the tolerant genotype LE2384 at 0, 14, and 21 days after anthesis under water stress. Moreover, the susceptible genotype had a higher thousand kernel weight under normal and drought stress than the tolerant genotype and vice versa for the number of kernels per spike, number of spikelets per spike, and grain yield per plant. This result further supports that *1-FEH w3* as a KASP marker had a non-stable effect on drought tolerance in wheat. As drought tolerance is a complex trait controlled by many genes, it seems that the target allele for the *Fewh3* gene may have only small effects on drought tolerance in wheat. It could be that the other genotypes that did not have the target allele include other drought-tolerant genes. In wheat, the negative and non-significant effect of the C allele of the *1-FEWHw3* gene was found at the germination stage [17] and also at the adult growth stage in a large number of highly diverse wheat genotypes [16].

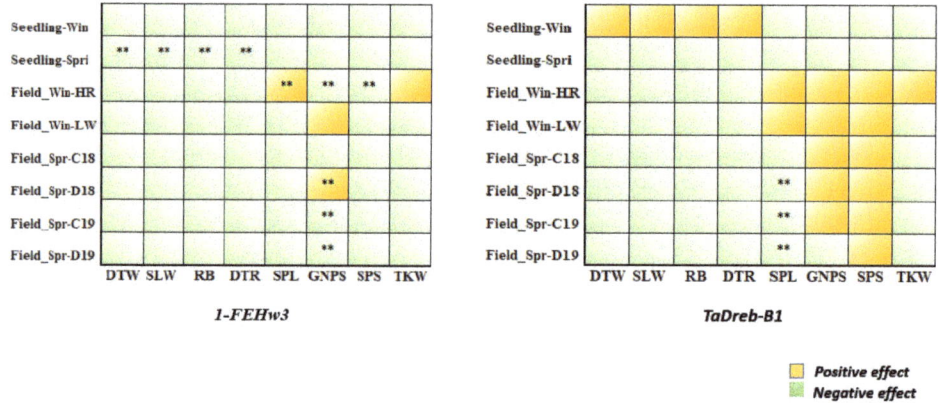

Figure 5. Target allele effect of each gene in all traits scored in spring and winter populations. ** Refers to significant association at $p > 0.01$. (SLW) the sum of leaf witling, (DTW) days to wilting, (RB regrowth biomass), and (DTR) days to recovery, (SL) Spike length, (SPS) number of spikelets per spike, (GNPS) grain number per spike, and (TKW) thousand-kernel weight.

KASP markers are very important for marker-assisted selection to rapidly and genetically target traits [36]. Therefore, KASP markers should be validated in a large number of genotypes with high genetic diversity before use in MAS.

4. Material and Methods

4.1. Plant Material

The plant material used in this study consisted of two wheat populations (winter and spring). The winter population consisted of 157 randomly selected genotypes from the 270 $F_{3:6}$ lines (Nebraska Duplicate Nursery 2017). Details about the production and pedigree of the population were previously published in Eltaher et al. [21]. All genotypes in the winter population (WP) were highly diverse, representing the preliminary yield trial of a standard breeding program [27]. The spring population (SP) consisted of 198 highly diverse wheat genotypes, representing 22 different countries [19]. The list of genotypes in both populations is presented in Supplementary Table S3. The genotypes from SP were collected from the U.S. National Plant Germplasm System database (https://www.ars-grin.gov/ (accessed on 15 May 2023), while the genotypes in WP were from Nebraska Winter Wheat Breeding program (https://agronomy.unl.edu/small-grains (accessed on 15 May 2023).

4.2. Seedlings Experiments

Both populations were previously evaluated under seedling drought stress using the same protocol with very slight modifications in the duration of drought stress. Drought tolerance was evaluated in the spring population by Ahmed et al. [19], while the winter population was evaluated by Eltaher et al. [21].

In both populations, all genotypes were exposed to natural drought stress by water withholding after all plants reached two expanded leaves in the winter population and one expanded leaf in the spring. When 70% of genotypes had full wilting, the drought treatment was ended. Then, all plants were cut and re-irrigated to test their recovery after prolonged drought stress. Four traits were selected from Ahmed et al. [19]. These traits were the sum of leaf witling (SLW), days to wilting (DTW), regrowth biomass (RB), and days to recovery (DTR). Low values of SLW and DTR and high values of RB and DTW indicated tolerance to drought. These traits have a direct relation to seedling drought stress.

The genotypes in spring and winter population were evaluated in seven and three replications, respectively.

4.3. Field Experiments

Important spike traits in the winter population were scored under a higher-rainfall environment (HR) located in Lincoln, NE, and a lower-rainfall environment (LW) located in North Platte, NE, USA by Eltaher et al. [21] in two replications each. The climate data, including temperature and precipitation for the two locations, are presented in Supplementary Table S1.

In this study, the spring population was evaluated under normal and drought stress conditions for two growing seasons (2018 and 2019) at the Experimental Field Station, Assiut University, Egypt, where soil and clay were used in two replications in each of the conditions. In each growing season, the genotypes in the control conditions were irrigated eight times during the growing season, while the irrigation was stopped after two times irrigation at the booting stage in the drought conditions.

Spike length (SL), number of spikelets per spike (SPS), grain number per spike (GNPS), and thousand-kernel weight scored in the spring population, and the data of the same traits were taken from Eltaher et al. [21] for the winter population. Soil water moisture was estimated under drought stress at 10 and 35 cm depth according to method described by Kumar [37].

4.4. DNA Extraction and Kompetitive Allele-Specific PCR (KASP) Genotyping

All genotypes in both populations were sown, and when all plants reached the one leaf stage, the leaves of each genotype were harvested for DNA extraction. All genotypes in the spring and winter populations are highly homozygous. DNA extraction was performed using the BioSprint 96 automatic DNA extractor. For KASP genotyping, the DNA concentrations of all samples were diluted at 50 ng/μL according to LGC Genomics manual instructions. All samples were arrayed in a 384-well plate for KASP PCR. In each well, a 10 μL reaction (5 μL of DNA and 5 μL of the KASP reaction mix, including 0.14 μL of the primer assay mix from LGC-Genomics, Middlesex, UK) was used.

Two KASP markers for *TaDreb-B1* and *1-1-FEH w3* genes were ordered from the Integrated DNA Technologies (IDT) laboratory (Redwood City, CA, USA) (Supplementary Table S4). Thermal cycling conditions lasted 15 min at 94 °C. This was followed by 10 cycles of touchdown PCR as follows: 94 °C for 20 s and 65–57 °C for 60 s (dropping to 0.8 °C per cycle). This was followed by 26 cycles of regular PCR as follows: 94 °C for 20 s and 55 °C for 60 s. The 384-well plate was read via FLUOstar Omega fluorescence. To determine the absence or presence of the *TaDreb-B1* and *1-FEH w3* genes, the target allele (presence of gene) was labelled with FAM (blue), while the non-target allele A (absence of gene) was labelled with HEX (red). The analysis of KASP-PCR products was performed using Klustercaller v2.22.0.5 software, and SNPs of each gene were labelled using SNPviewer software (LGC, Biosearch Technologies, Beverly, MA, USA). The analysis of KASP markers in the two populations was done in 2022.

4.5. Diversity and Single-Marker Analysis

In each population, the polymorphic information content, gene diversity, and gene frequency were estimated for each KASP marker using Power marker software [38].

To test marker–trait association, the SNPs of each gene were analyzed with all traits scored in this study in both populations by single-marker analysis (SMA). The SMA was performed using R software according to the following model [39]:

$$Y = \mu + f\ (marker) + error$$

where Y is the trait value, μ is the nursery mean, and f (marker) is a function of the molecular marker [40].

The significant marker–trait association was determined to be at a significant level of $p < 0.05$. The phenotypic variation explained by marker and gene effect was calculated from the statistical model [39]. Box plots showing the difference between genotypes having the target allele and without the target allele were created using the R package 'ggplot2' [41].

5. Conclusions

In conclusion, two highly diverse populations representing winter and spring types were used to investigate the effectiveness of two KASP markers, *1-FEH w3* and *TaDreb-B1*, on improving drought tolerance in wheat for MAS. The results of this study suggest that the use of *1-FEH w3* as a KASP marker for MAS to improve drought tolerance in wheat is extremely questionable. *TaDreb-B1* provided better consistency in improving drought tolerance than *1-FEH w3*. However, the promising marker allele *TaDreb-B1* should be validated further in different genetic backgrounds.

Supplementary Materials: The following supporting information can be downloaded at: https://www.mdpi.com/article/10.3390/ijms24108986/s1.

Author Contributions: S.E. performed phenotyping and genotyping experiments in winter wheat, analyzed the data, and wrote the manuscript; M.H. and A.A.M.A. performed phenotyping experiment in the spring population; A.B. and P.S.B. provided useful comments during experiments and helped in writing the paper; A.S. designed the whole study, performed KASP genotyping for the spring population, and wrote the manuscript. All authors have read and agreed to the published version of the manuscript.

Funding: Costs for open access publishing were partially funded by the Deutsche Forschungsgemeinschaft (DFG, German Research Foundation, grant 491250510).

Institutional Review Board Statement: Not applicable.

Informed Consent Statement: Not applicable.

Data Availability Statement: All data generated or analyzed during this study are included in this published article and its supplementary information files.

Conflicts of Interest: The authors declare no conflict of interest.

References

1. Guo, J.; Shi, W.; Zhang, Z.; Cheng, J.; Sun, D.; Yu, J.; Li, X.; Guo, P.; Hao, C. Association of Yield-Related Traits in Founder Genotypes and Derivatives of Common Wheat (*Triticum aestivum* L.). *BMC Plant Biol.* **2018**, *18*, 38. [CrossRef] [PubMed]
2. Devate, N.B.; Krishna, H.; Parmeshwarappa, S.K.V.; Manjunath, K.K.; Chauhan, D.; Singh, S.; Singh, J.B.; Kumar, M.; Patil, R.; Khan, H.; et al. Genome-Wide Association Mapping for Component Traits of Drought and Heat Tolerance in Wheat. *Front. Plant Sci.* **2022**, *13*, 2886. [CrossRef] [PubMed]
3. Pingali, P.L. Green Revolution: Impacts, Limits, and the Path Ahead. *Proc. Natl. Acad. Sci. USA* **2012**, *109*, 12302–12308. [CrossRef]
4. Gajghate, R.; Chourasiya, D.; Sharma, R.K. Plant Morphological, Physiological Traits Associated with Adaptation against Heat Stress in Wheat and Maize. In *Plant Stress Biology*; Springer: Berlin/Heidelberg, Germany, 2020; pp. 51–81.
5. Zhang, J.; Zhang, S.; Cheng, M.; Jiang, H.; Zhang, X.; Peng, C.; Lu, X.; Zhang, M.; Jin, J. Effect of Drought on Agronomic Traits of Rice and Wheat: A Meta-Analysis. *Int. J. Environ. Res. Public Health* **2018**, *15*, 839. [CrossRef]
6. Mourad, A.M.I.; Amin, A.E.E.A.Z.; Dawood, M.F.A. Genetic Variation in Kernel Traits under Lead and Tin Stresses in Spring Wheat Diverse Collection. *Environ. Exp. Bot.* **2021**, *192*, 104646. [CrossRef]
7. Esmail, S.M.; Omar, G.E.; Mourad, A.M.I. In-Depth Understanding of the Genetic Control of Stripe Rust Resistance (*Puccinia striiformis* f. Sp. Tritici) Induced in Wheat (*Triticum aestivum*) by Trichoderma Asperellum T34. *Plant Dis.* **2023**, *107*, 457–472. [CrossRef]
8. Mourad, A.M.I.; Abou-Zeid, M.A.; Eltaher, S.; Baenziger, P.S.; Börner, A. Identification of Candidate Genes and Genomic Regions Associated with Adult Plant Resistance to Stripe Rust in Spring Wheat. *Agronomy* **2021**, *11*, 2585. [CrossRef]
9. Yang, H.; Jian, J.; Li, X.; Renshaw, D.; Clements, J.; Sweetingham, M.W.; Tan, C.; Li, C. Application of Whole Genome Re-Sequencing Data in the Development of Diagnostic DNA Markers Tightly Linked to a Disease-Resistance Locus for Marker-Assisted Selection in Lupin (*Lupinus angustifolius*). *BMC Genom.* **2015**, *16*, 660. [CrossRef]
10. Liu, G.; Mullan, D.; Zhang, A.; Liu, H.; Liu, D.; Yan, G. Identification of KASP Markers and Putative Genes for Pre-Harvest Sprouting Resistance in Common Wheat (*Triticum aestivum* L.). *Crop J.* **2022**, *11*, 549–557. [CrossRef]
11. Han, G.; Liu, S.; Jin, Y.; Jia, M.; Ma, P.; Liu, H.; Wang, J.; An, D. Scale Development and Utilization of Universal PCR-Based and High-Throughput KASP Markers Specific for Chromosome Arms of Rye (*Secale cereale* L.). *BMC Genom.* **2020**, *21*, 206. [CrossRef]
12. Makhoul, M.; Rambla, C.; Voss-Fels, K.P.; Hickey, L.T.; Snowdon, R.J.; Obermeier, C. Overcoming Polyploidy Pitfalls: A User Guide for Effective SNP Conversion into KASP Markers in Wheat. *Theor. Appl. Genet.* **2020**, *133*, 2413–2430. [CrossRef] [PubMed]
13. Liu, S.; Sehgal, S.K.; Lin, M.; Li, J.; Trick, H.N.; Gill, B.S.; Bai, G. Independent Mis-Splicing Mutations in Ta PHS 1 Causing Loss of Preharvest Sprouting (PHS) Resistance during Wheat Domestication. *New Phytol.* **2015**, *208*, 928–935. [CrossRef] [PubMed]
14. Semagn, K.; Babu, R.; Hearne, S.; Olsen, M. Single Nucleotide Polymorphism Genotyping Using Kompetitive Allele Specific PCR (KASP): Overview of the Technology and Its Application in Crop Improvement. *Mol. Breed.* **2014**, *33*, 1–14. [CrossRef]
15. Rasheed, A.; Wen, W.; Gao, F.; Zhai, S.; Jin, H.; Liu, J.; Guo, Q.; Zhang, Y.; Dreisigacker, S.; Xia, X.; et al. Development and Validation of KASP Assays for Genes Underpinning Key Economic Traits in Bread Wheat. *Theor. Appl. Genet.* **2016**, *129*, 1843–1860. [CrossRef] [PubMed]
16. Ur Rehman, S.; Ali Sher, M.; Saddique, M.A.B.; Ali, Z.; Khan, M.A.; Mao, X.; Irshad, A.; Sajjad, M.; Ikram, R.M.; Naeem, M.; et al. Development and Exploitation of KASP Assays for Genes Underpinning Drought Tolerance Among Wheat Cultivars From Pakistan. *Front. Genet.* **2021**, *12*, 789. [CrossRef]
17. Mohamed, E.A.; Ahmed, A.A.M.; Schierenbeck, M.; Hussein, M.Y.; Baenziger, P.S.; Börner, A.; Sallam, A. Screening Spring Wheat Genotypes for TaDreb-B1 and Fehw3 Genes under Severe Drought Stress at the Germination Stage Using KASP Technology. *Genes* **2023**, *14*, 373. [CrossRef] [PubMed]
18. Sallam, A.; Alqudah, A.M.; Baenziger, P.S.; Rasheed, A. Editorial: Genetic Validation and Its Role in Crop Improvement. *Front. Genet.* **2023**, *13*, 3705. [CrossRef]
19. Ahmed, A.A.M.; Dawood, M.F.A.; Elfarash, A.; Mohamed, E.A.; Hussein, M.Y.; Börner, A.; Sallam, A. Genetic and Morpho-Physiological Analyses of the Tolerance and Recovery Mechanisms in Seedling Stage Spring Wheat under Drought Stress. *Front. Genet.* **2022**, *13*, 1010272. [CrossRef]
20. Sallam, A.; Eltaher, S.; Alqudah, A.M.; Belamkar, V.; Baenziger, P.S. Combined GWAS and QTL Mapping Revealed Candidate Genes and SNP Network Controlling Recovery and Tolerance Traits Associated with Drought Tolerance in Seedling Winter Wheat. *Genomics* **2022**, *114*, 110358. [CrossRef]
21. Eltaher, S.; Sallam, A.; Emara, H.A.; Nower, A.A.; Salem, K.F.M.; Börner, A.; Stephen Baenziger, P.; Mourad, A.M.I. Genome-Wide Association Mapping Revealed SNP Alleles Associated with Spike Traits in Wheat. *Agronomy* **2022**, *12*, 1469. [CrossRef]
22. Mourad, A.M.I.; Draz, I.S.; Omar, G.E.; Börner, A.; Esmail, S.M. Genome-Wide Screening of Broad-Spectrum Resistance to Leaf Rust (*Puccinia Triticina Eriks*) in Spring Wheat (*Triticum aestivum* L.). *Front. Plant Sci.* **2022**, *13*, 921230. [CrossRef] [PubMed]
23. Alqudah, A.M.; Sallam, A.; Stephen Baenziger, P.; Börner, A. GWAS: Fast-Forwarding Gene Identification and Characterization in Temperate Cereals: Lessons from Barley—A Review. *J. Adv. Res.* **2020**, *22*, 119–135. [CrossRef] [PubMed]
24. Botstein, D.; White, R.L.; Skolnick, M.; Davis, R.W. Construction of a Genetic Linkage Map in Man Using Restriction Fragment Length Polymorphisms. *Am. J. Hum. Genet.* **1980**, *32*, 314. [PubMed]
25. Mourad, A.M.I.; Belamkar, V.; Baenziger, P.S. Molecular Genetic Analysis of Spring Wheat Core Collection Using Genetic Diversity, Population Structure, and Linkage Disequilibrium. *BMC Genom.* **2020**, *21*, 434. [CrossRef]

26. Salem, K.F.M.; Sallam, A. Analysis of Population Structure and Genetic Diversity of Egyptian and Exotic Rice (*Oryza sativa* L.) Genotypes. *Comptes Rendus Biol.* **2016**, *339*, 1–9. [CrossRef]
27. Eltaher, S.; Sallam, A.; Belamkar, V.; Emara, H.A.; Nower, A.A.; Salem, K.F.M.; Poland, J.; Baenziger, P.S. Genetic Diversity and Population Structure of F3:6 Nebraska Winter Wheat Genotypes Using Genotyping-by-Sequencing. *Front. Genet.* **2018**, *9*, 76. [CrossRef]
28. Sallam, A.; Alqudah, A.M.; Dawood, M.F.A.; Baenziger, P.S.; Börner, A. Drought Stress Tolerance in Wheat and Barley: Advances in Physiology, Breeding and Genetics Research. *Int. J. Mol. Sci.* **2019**, *20*, 3137. [CrossRef]
29. Sallam, A.; Mourad, A.M.I.; Hussain, W.; Stephen Baenziger, P. Genetic Variation in Drought Tolerance at Seedling Stage and Grain Yield in Low Rainfall Environments in Wheat (*Triticum aestivum* L.). *Euphytica* **2018**, *214*, 169. [CrossRef]
30. Ahmed, A.A.M.; Mohamed, E.A.; Hussein, M.Y.; Sallam, A. Genomic Regions Associated with Leaf Wilting Traits under Drought Stress in Spring Wheat at the Seedling Stage Revealed by GWAS. *Environ. Exp. Bot.* **2021**, *184*, 104393. [CrossRef]
31. Certified Crop Advisor Study Resources (Northeast Region). Available online: https://nrcca.cals.cornell.edu/soil/CA2/CA0212.1-3.php (accessed on 6 February 2023).
32. Zhang, J.; Xu, Y.; Chen, W.; Dell, B.; Vergauwen, R.; Biddulph, B.; Khan, N.; Luo, H.; Appels, R.; van den Ende, W. A Wheat *1-FEH W3* Variant Underlies Enzyme Activity for Stem WSC Remobilization to Grain under Drought. *New Phytol.* **2015**, *205*, 293–305. [CrossRef]
33. Mondal, S.; Sallam, A.; Sehgal, D.; Sukumaran, S.; Krishnan, J.N.; Kumar, U.; Biswal, A.; Mondal, S.; Sehgal, Á.D.; Sukumaran, Á.S.; et al. Advances in Breeding for Abiotic Stress Tolerance in Wheat. In *Genomic Designing for Abiotic Stress Resistant Cereal Crops*; Springer: Berlin/Heidelberg, Germany, 2021; pp. 71–103. [CrossRef]
34. Mourad, A.M.I.; Alomari, D.Z.; Alqudah, A.M.; Sallam, A.; Salem, K.F.M. Recent Advances in Wheat (*Triticum* Spp.) Breeding. In *Advances in Plant Breeding Strategies: Cereals*; Springer: Berlin/Heidelberg, Germany, 2019; Volume 5.
35. Yáñez, A.; Tapia, G.; Guerra, F.; del Pozo, A. Stem Carbohydrate Dynamics and Expression of Genes Involved in Fructan Accumulation and Remobilization during Grain Growth in Wheat (*Triticum aestivum* L.) Genotypes with Contrasting Tolerance to Water Stress. *PLoS ONE* **2017**, *12*, e0177667. [CrossRef] [PubMed]
36. Grewal, S.; Coombes, B.; Joynson, R.; Hall, A.; Fellers, J.; Yang, C.Y.; Scholefield, D.; Ashling, S.; Isaac, P.; King, I.P.; et al. Chromosome-Specific KASP Markers for Detecting Amblyopyrum Muticum Segments in Wheat Introgression Lines. *Plant Genome* **2022**, *15*, e20193. [CrossRef] [PubMed]
37. Patterson, G.T.; Carter, M.R. *Soil Sampling and Handling*, 2nd ed.; Carter, M.R., Gregorich, E.G., Eds.; CRC Press Taylor & Francis Group: Abingdon, UK, 2006; Volume 44, ISBN 9780849335860.
38. Liu, K.; Muse, S.V. PowerMarker: An Integrated Analysis Environment for Genetic Marker Analysis. *Bioinformatics* **2005**, *21*, 2128–2129. [CrossRef] [PubMed]
39. Collard, B.C.Y.; Jahufer, M.Z.Z.; Brouwer, J.B.; Pang, E.C.K. An Introduction to Markers, Quantitative Trait Loci (QTL) Mapping and Marker-Assisted Selection for Crop Improvement: The Basic Concepts. *Euphytica* **2005**, *142*, 169–196. [CrossRef]
40. Francis, D.M.; Merk, H.L.; Namuth-Covert, D. Introduction to Single Marker Analysis (SMA). Available online: http://www.extension.org/pages/32552/introduction-to-single-marker-analysis-sma (accessed on 25 March 2023).
41. Wickham, H. *Ggplot2: Elegant Graphics for Data Analysis*; Springer: New York, NY, USA, 2016. Available online: https://cran.r-project.org/web/packages/ggplot2/citation.html (accessed on 1 March 2023).

Disclaimer/Publisher's Note: The statements, opinions and data contained in all publications are solely those of the individual author(s) and contributor(s) and not of MDPI and/or the editor(s). MDPI and/or the editor(s) disclaim responsibility for any injury to people or property resulting from any ideas, methods, instructions or products referred to in the content.

Article

Transgenerational Paternal Inheritance of *TaCKX* GFMs Expression Patterns Indicate a Way to Select Wheat Lines with Better Parameters for Yield-Related Traits

Karolina Szala, Marta Dmochowska-Boguta, Joanna Bocian, Waclaw Orczyk and Anna Nadolska-Orczyk *

Department of Functional Genomics, Plant Breeding and Acclimatization Institute—National Research Institute, Radzikow, 05-870 Blonie, Poland
* Correspondence: a.orczyk@ihar.edu.pl

Abstract: Members of the *TaCKX* gene family (GFMs) encode the cytokinin oxygenase/dehydrogenase enzyme (CKX), which irreversibly degrades cytokinins in the organs of wheat plants; therefore, these genes perform a key role in the regulation of yield-related traits. The purpose of the investigation was to determine how expression patterns of these genes, together with the transcription factor-encoding gene *TaNAC2-5A*, and yield-related traits are inherited to apply this knowledge to speed up breeding processes. The traits were tested in 7 days after pollination (DAP) spikes and seedling roots of maternal and paternal parents and their F_2 progeny. The expression levels of most of them and the yield were inherited in F_2 from the paternal parent. Some pairs or groups of genes cooperated, and some showed opposite functions. Models of up- or down-regulation of *TaCKX* GFMs and *TaNAC2-5A* in low-yielding maternal plants crossed with higher-yielding paternal plants and their high-yielding F_2 progeny reproduced gene expression and yield of the paternal parent. The correlation coefficients between *TaCKX* GFMs, *TaNAC2-5A*, and yield-related traits in high-yielding F_2 progeny indicated which of these genes were specifically correlated with individual yield-related traits. The most common was expressed in 7 DAP spikes *TaCKX2.1*, which positively correlated with grain number, grain yield, spike number, and spike length, and seedling root mass. The expression levels of *TaCKX1* or *TaNAC2-5A* in the seedling roots were negatively correlated with these traits. In contrast, the thousand grain weight (TGW) was negatively regulated by *TaCKX2.2.2*, *TaCKX2.1*, and *TaCKX10* in 7 DAP spikes but positively correlated with *TaCKX10* and *TaNAC2-5A* in seedling roots. Transmission of *TaCKX* GFMs and *TaNAC2-5A* expression patterns and yield-related traits from parents to the F_2 generation indicate their paternal imprinting. These newly shown data of nonmendelian epigenetic inheritance shed new light on crossing strategies to obtain a high-yielding F_2 generation.

Keywords: parental imprinting; transgenerational epigenetics; paternal inheritance; wheat; *TaCKX* expression; cytokinin; yield

1. Introduction

Bread wheat (*Triticum aestivum*) is the most important cereal crop in the temperate climate and provides a staple food for more than a third of the world's population [1]. It belongs to the Triticeae tribe, which also includes barley, rye, and triticale. Among the species, wheat has the largest and most complex hexaploid genome (2n = 6x = 42), which consists of the three homoeologous subgenomes A, B, and D. Each gene present as homologues A, B, and D could retain its original function or, as a result of independent evolution, develop heterogeneous expression, and/or one or two copies may be silenced or deleted [2,3].

Cytokinins (CKs) perform a basic role in the growth, development, and productivity of any plant species, including wheat [4]. Their content in developing spikes of wheat is correlated with grain yield, grain number and weight, TGW, chlorophyll content in flag leaves, and seedling root weight [5–9]. There are two types of cytokinins, isoprenoid and aromatic. The most important and widely occurring are isoprenoids, cis-zeatin (cZ), trans-zeatin (tZ), isopentenyl adenine (iP), and dihydrozeatin (DZ), and fewer aromatic forms, e.g., benzylaminopurine (BA). The content of active forms in plant tissues and organs depends on metabolic processes, such as biosynthesis, degradation, inactivation, and reactivation. Active forms and their ribosides can also be transported throughout the plant [10]. Enzymes for all metabolic processes are encoded by members of the gene family (GFMs) [11–15]. One of the most important processes is the irreversible degradation of cytokinins by *TaCKX* GFMs. There are 13 basic *TaCKX* GFMs, 11 of which have homoeologs in subgenomes A, B, and D. Two of them, *TaCKX2.2.2* and *TaCKX2.2.3*, are located only in the D subgenome [11]. The *CKX* GFMs encode the cytokinin oxidase/dehydrogenase enzyme. Their role in the regulation of yield-related traits in wheat [5–7,11] and other species were already shown [4,16–18]. Decreased expression of the *TaCKX1* or *TaCKX2* genes significantly influenced TGW, seed number, and chlorophyl content in flag leaves, and the effect was dependent on the silent gene and/or genotype [5–7]. The *TaCKX4* copy number affected grain yield and chlorophyl content in flag leaves [19]. Haplotype variants of *TaCKX6-D1* (actually *TaCKX2.2.1-3D*) were associated with TGW [20], and the allelic variant of *TaCKX6a02* (annotated as *TACKX2.1*) influenced grain size, filling rate, and weight [21]. In addition to yield, some *CKX* GFMs influence other pleiotropic traits, including root growth, nutrient accumulation, and abiotic stress responses [18,22,23]. These genes could also be regulated at the transcriptional level by transcription factors (TFs), especially those belonging to the NAC family; however, knowledge about their function is still very limited [24]. A promising NAC-encoding candidate with a role in yield-related traits is *TaNAC2-5A* [25,26]. As reported, overexpression of the gene delayed leaf senescence and increased nitrate uptake and concentration, root growth, and grain yield under field conditions. It is interesting to note that in a controlled environment, *TaNAC2-5A* was negatively correlated with the activity of the CKX enzyme in seedling roots and the number of tillers [8].

Crossbreeding and selection are basic steps in crop improvement, and the only potential limitation is too narrow genetic variability. Therefore, it is important to know how yield-related traits are inherited. Moreover, traits, including stably integrated transgenes and edited genes, are inherited according to Mendelian rules [27–30]. An exception to Mendel's principles that encompass both groups of genes is epigenetic inheritance [31] or the polygenic nature of genomic architecture for the linked traits, which can be regulated by transcription factors [32–34] or other gene regulatory networks [35].

The pattern of gene expression could be considered a trait with its own type of inheritance.

This concept, reviewed by Yoo et al. [2], called parental expression additivity, is defined as the arithmetic average of the expression of the parental genes. Expression additivity of parental genes is observed in the offspring of the diploid species. The deviation of additivity called parental non-additive expression is mainly found in the offspring of polyploid species. A bias when the expression of the offspring is similar to that of one of the parents is called expression-level dominance. If total offspring expression is lower or higher than in both parents, the phenomenon is called transgressive expression, and when the contribution of the parental homeologs to the total gene expression is unequal, it is named homeolog expression bias. All this deviation from additivity can be explained as a result of different factors, such as the influence of one of the parental genomes, epigenetic regulation, balance of gene dosage, and *cis-* and/or *trans*-regulatory elements [2].

Expression-level dominance, which is of uniparental origin, is also called genomic imprinting [36,37]. This phenomenon of epigenetic origin is the result of the asymmetries of DNA and histone methylation between maternal and paternal plants. Male and female genotypes are multicellular in origin; therefore, primary gene imprinting can occur in egg cells, central cells, and sperm and, subsequently, in the triploid endosperm or, less frequently, in the diploid embryo [37]. Generally, conservation of imprinting is limited across other crops; however, these genes that show conserved imprinting in cereals showed positive selection and were suggested to perform a dose-dependent function in the regulation of seed development [38]. However, as recently documented by Rodrigues et al. [39], most genes imprinted in the endosperm of seeds were imprinted across cultivars, extending their functions to chromatin and transcriptional regulation, development, and signaling. Only 4% to 11% of the imprinted genes showed divergent imprinting.

Imprinted gene expression affects mostly single genes or groups of genes. Most of them are maternally expressed and inherited [36,40,41]. The best recognized is the maternal effect of genes during embryo development [42]. Early development in Arabidopsis is coordinated by the supply of auxin from the mother integuments of the ovule, which is required for the correct embryo development of embryos [40]. The genomic imprinting of the cereal endosperm influences the timing of endosperm cellularization [43]. An example of imprinted maternally expressed genes in cereals is a polycomb group, which is important for the cellularization of endosperm in rice [44]. Reciprocal crosses between tetraploid and hexaploid wheats showed that imprinted genes were identified in endosperm and embryo tissue, supporting the predominant maternal effect on early grain development [45]. Paternally expressed imprinted genes were associated with hybrid seed lethality in Capsella [46]. In maize, the *Dosage-effect defective1* (*ded1*) locus that contributes to seed size was found to be paternally imprinted [47]. The gene encodes a transcription factor that is specifically expressed during early endosperm development. There is also evidence that small RNAs might determine the paternal methylome by silencing transposons [48]. In addition to these reports, it is very difficult to find examples of paternally inherited genes, especially in cereals. Many studies have indicated dynamic changes in the epigenetic state, including DNA methylation, chromatin modifications, and small RNAs, which are observed during the reproductive development of plants [49,50]. The spatiotemporal pattern of gene expression, imprinting, and seed development in Arabidopsis endosperm is predominantly regulated by small maternal RNAs; however, they also originate from the paternal genome and the seed coat [51]. As reported by Tuteja et al. [52], imprinted paternally expressed genes, but not maternally expressed genes, in Arabidopsis evolve under positive Darwinian selection. These genes were involved in seed development processes, such as auxin biosynthesis and epigenetic regulation. Imprinted paternally expressed genes are mainly associated with hypomethylated maternal DNA alleles, which can be repressed by small genic RNAs and rarer with transposable elements [49,53]. Epigenetic changes can be developmentally regulated (developmental epigenetics). The state in which changes in DNA methylation are stable between generations and heritable is called transgenerational epigenetics [54].

Several *TaCKX* GFMs and *TaNAC2-5A* (*NAC2*) were previously selected as important regulators of yield-related traits. To determine how the expression patterns of selected genes are inherited in the developing spikes and seedling roots of the parents and the F_2 generation, we used a reciprocal crossing strategy. The research hypothesis assumed that knowledge of inheritance of gene expression patterns that regulated yield-related traits indicated the way of selection of genotypes in wheat breeding. There is a research gap in documenting the inheritance of expression patterns for yield-related genes. We found that most of the genes in the F_2 generation were expressed in a pater-of-origin-specific manner, which shed new light on the ways of selecting wheat lines and the breeding strategy.

2. Results

2.1. Reciprocal Crosses Indicate That the Expression Patterns of Most of the TaCKX GFM and Yield-Related Traits Are Mainly Inherited from the Male Parent

Relative values (related to the female parent = 1.0) of the expression profiles of *TaCKX* GFM and *NAC2* in 7 DAP spikes, seedling roots, and phenotypic traits in the female parent, male parent, and their six F_2 progeny from one reciprocal cross of S12B × S6C (C1) and S6C × S12B (C2) are presented in Figure 1. The same data obtained in reciprocal crosses of D16 × KOH7 (C3) and KOH7 × D16 (C4); D19 × D16 (C5) and D16 × D19 (C6); D19 × KOH7 (C7) and KOH7 × D19 (C8) are visualized in Figure S1. S6C, the paternal parent of the S12B × S6C cross (C1), showed higher expression of *TaCKX1* and *NAC2* and lower expression of *TaCKX5* and *TaCKX10* in spikes than the maternal parent (S12B). The expression of *TaCKX1*, *NAC2*, *TaCKX5*, and *TaCKX11* in the spikes of the F_2 progeny of this cross was higher (Figure 1A). In the reverse cross (S6C × S12B), when S12B was a paternal component (Figure 1B), *TaCKX1* and *NAC2* were expressed at low levels, *TaCKX5* and *10* were highly expressed in spikes compared to the maternal parent (S6C), and in F_2, *TaCKX1* and *NAC2* were expressed at low levels, and *TaCKX9*, *10*, *11*, and *5* were upregulated. In seedling roots, the paternal component of S12B × S6C (Figure 1A) showed high expression of *TaCKX5* and *NAC2* and low expression of *TaCKX10* and *11* in the parental parent compared to the maternal parent. In the roots of the F_2 progeny of this cross, *TaCKX5* and *NAC2* were highly expressed, and *TaCKX11* was downregulated. The expression data in the parents of the reverse cross, in which S12B was the paternal component, were opposite. Their F_2 progeny showed a strong upregulation of *TaCKX8*, *10*, and *11* and a downregulation of *TaCKX1*, *TaCKX3*, and *NAC2* in seedling roots. The total grain yield and the number of seeds in F_2 of S12B × S6C were low, similar to the male parent (Figure 1A). The root weight in the F_2 progeny of the same cross (S12B × S6C) was lower than that in the parents. The same yield components in the opposite cross (S6C × S12B) were high in one F_2 sibling, comparable to the maternal parent in two progeny and lower than in the parents in three of them (Figure 1B). Interestingly, the root mass in F_2 was higher than that in both parents.

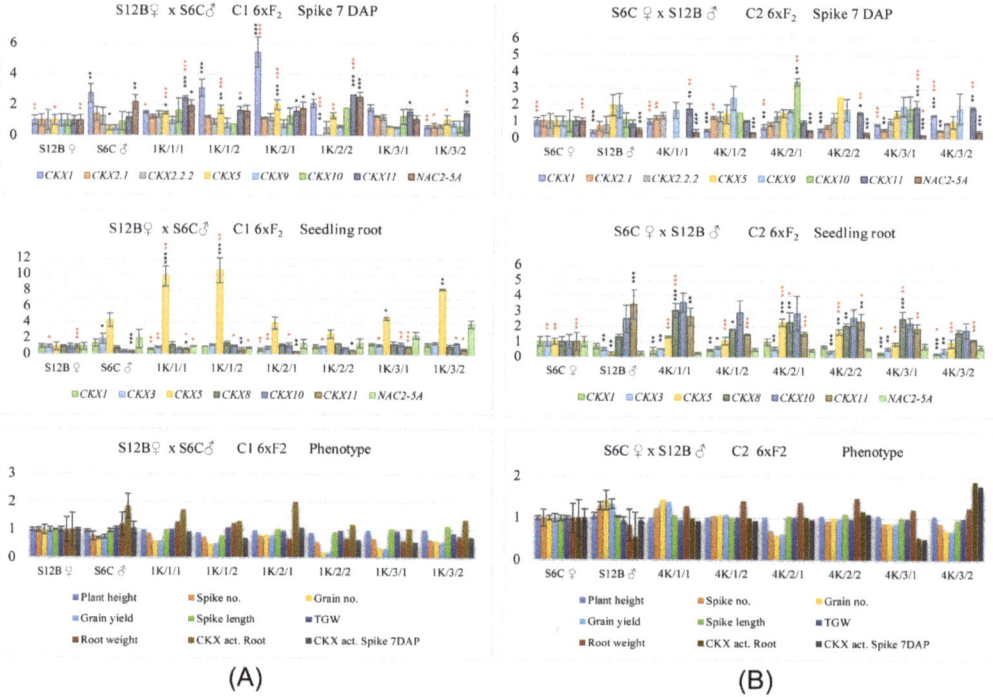

Figure 1. Example of *TaCKX* GFM and *NAC2* expression profiles in 7 DAP spikes, seedling roots, and phenotypic traits in the maternal parent, paternal parent, and their six F_2 progeny, from reciprocal cresses of S12B × S6C, C1 (**A**) and S6C × S12B, C2 (**B**). The data represent mean values with standard deviation. Black and red asterisks indicate statistical significance compared to the maternal parent or paternal parent, respectively (* $0.05 > p \geq 0.01$, ** $0.01 > p \geq 0.001$, *** $p < 0.001$) using the ANOVA test followed by the LSD post hoc test (STATISTICA 10, StatSoft).

As presented in Table 1 by colours, most of the expression patterns tested for *TaCKX* GFM and *NAC2* are inherited from the male parent (red). For example, up-regulated in 7 DAP spikes of the paternal parent *TaCKX1* and *NAC2* compared to the maternal parent is up-regulated in F_2 as well. The upregulated *TaCKX5* and *NAC2* and downregulated *TaCKX11* in seedling roots of the paternal parents are similarly expressed in an F_2. To summarize, the expression levels of all tested *TaCKX* GFMs and *NAC2* in 7 DAP spikes, in addition to being represented in different crosses, showed similar expression patterns to the paternal parents and were independent of the cross path. Among the *TaCKX* GFMs in 7 DAP spikes, which showed the paternal expression patterns were *TaCKX1, 2.1, 2.2.2, 5, 9, 10*, and *11*. In the seedling roots, there were *TaCKX5, 11, NAC2*; *TaCKX10, 11, NAC2*; *TaCKX1, 11, NAC2*; *TaCKX10*; *1, 3, 5, 8, 10, 11, NAC2*; *TaCKX1, 8, 10, NAC2*; and *3, 5, 8, 11*; *NAC2* (all tested but represented in different crosses). The only exceptions are *TaCKX5* in 7 DAP spikes of S12B × S6C (C1) and *TaCKX3* in seedling roots of KOH7 × D16 (C4), whose expression level is similar to that of the maternal parent (green).

Table 1. *TaCKX* GFMs and *NAC2* with high (↑), very high (↑↑), low (↓) or very low (↓↓) expression levels in 7 DAP spikes and seedling roots of the maternal parent (M), the paternal parent (P) and their F_2 progeny from cresses of D16 × KOH7 (C3) and KOH7 × D16 (C4); D19 × D16 (C5) and D16 × D19 (C6); D19 × KOH7 (C7) and KOH7 × D19 (C8), and high (↑) and low (↓) parameters of yield and root mass. Character colours indicate similar patterns of gene expression and yield-related traits in F_2 and paternal parent (red) or in F_2 and maternal parent (green).

	M	P	F_2
	C1 = S12B × S6C		
CKX expression 7 DAP	CKX1↓	CKX1↑	CKX1, 11↑↑
	NAC2↓	NAC2↑	CKX5, NAC2↑
	CKX5, 9↑	CKX5, 9↓	
CKX expression root	CKX5, NAC2↓↓	CKX5↑↑	CKX5↑↑↑, NAC2↑
	CKX3↓	CKX3↑, NAC2↑	
	CKX11, 10↑	CKX11, 10↓	CKX11↓
yield-related traits	yield↑	yield↓	yield↓↓
	CKX act. spike=	CKX act. spike=	CKX act. spike↓↓
	root=↓	root=↑	root=↓
	CKX act. root↓↓	CKX act. root↑↑	CKX act. root↑↑
	C2 = S6C × S12B		
CKX expression 7 DAP	CKX5, 9↓	CKX5, 9↑	CKX9, 10, 11, 5↑
	CKX1, 2.1, NAC2↑	CKX1, 2.1, NAC2↓	CKX1, NAC2↓
CKX expression root	CKX10, 11↓↓	CKX10, 11↑↑	CKX5, 8, 10, 11↑↑
	CKX1, 3, 5, NAC2↑	CKX1, 3, 5, NAC2↓	CKX1, 3, NAC2↓
yield-related traits	yield↓	yield↑	yield=↓
	CKX act. spike=	CKX act. spike=	CKX act. Spike=
	root↑	root=↓	root↑↑
	CKX act. root↑	CKX act. root↓	CKX act. root=
	C3 = D16 × KOH7		
CKX expression 7 DAP	CKX11↓↓	CKX11↑↑	CKX5, 9↑↑
	CKX2.1, 2.2.2↓	CKX2.1, 2.2.2↑	
	CKX10↑	CKX10↓	
CKX expression root	NAC2↓↓	NAC2↑↑	CKX3↓
	CKX1, 5, 10, 11↓↓	CKX1, 5, 10, 11↑↑	CKX1, 8, 11, NAC2↑↑
yield-related traits	yield↑	yield↓	yield↓
	CKX act. spike=↑	CKX act. spike=↓	CKX act. spike↓↓
	root=↑	root=↓	root=
			semi-empty spikes↑↑↑
	C4 = KOH7 × D16		
CKX expression 7 DAP	CKX9, 10↓	CKX9, 10↑	CKX5, 9↑↑
	CKX11↑↑	CKX11↓↓	CKX11, NAC2↓↓
	CKX2.1, 2.2.2↑	CKX2.1, 2.2.2↓	CKX2.1, 2.2.2↓

Table 1. Cont.

	M	P	F$_2$
CKX expression root	CKX3, 8↓	CKX3, 8↑	CKX8↑↑
	CKX1, 5, 10, 11↑	CKX1, 5, 10, 11↓	CKX3, 10↓
	NAC2↑↑	NAC2↓↓	
yield-related traits	yield↓	yield↑	yield↑
	CKX act. spike=↓	CKX act. spike=↑	CKX act. spike=
	root↓	root↑	root=↓
C5 = D19 × D16			
CKX expression 7 DAP	CKX2.2.2↑	CKX2.2.2↓	CKX2.2.2↓
	CKX10↓	CKX10↑	CKX9↑
CKX expression root	CKX 5, 8, NAC2↑	CKX5, 8, NAC2↓↓	CKX3, 5, 8, 11↓
	CKX1, 3, 10, 11↑	CKX1, 3, 10, 11↓	(CKX10, NAC2↑)
yield-related traits	yield↑	yield↓	yield↓↓
	CKX act. spike=↓	CKX act. spike=↑	CKX act. spike=↑
	root=	root=	root↓
	semi-empty spikes↓	semi-empty spikes↑	semi-empty spikes↑↑↑
C6 = D16 × D19			
CKX expression 7 DAP	CKX2.2.2, 5, 9↓	CKX2.2.2, 5, 9↑	
	CKX10↑	CKX10↓	CKX1, 10, 11, NAC2↓
CKX expression root	CKX5, 8, NAC2↓↓	CKX5, 8, NAC2↑↑	CKX1, NAC2↑↑
	CKX1, 3, 10, 11↓	CKX1, 3, 10, 11↑	CKX8, 10↑
yield-related traits	yield↓	yield↑↑	yield↑
	CKX act. spike=↑	CKX act. spike=↓	CKX act. spike=
	root=	root=	root↓
C7 = D19 × KOH7			
CKX expression 7 DAP	CKX2.1, 11, NAC2↓	CKX2.1, 11, NAC2↑	CKX10↑
	CKX9↑	CKX9↓	CKX2.2.2, 9↓
			NAC2↓
CKX expression root	NAC2↓↓	NAC2↑↑	
	CKX1, 10, 11↓	CKX1, 10, 11↑	CKX11↑
	CKX3, 5, 8↑	CKX3, 5, 8↓	CKX3, 5, 8, 10↓
yield-related traits	yield↑	yield↓	yield↓
	CKX act. spike=	CKX act. spike=	CKX act. spike↑
	root=↑	root=↓	root=
		semi-empty spikes↑	semi-empty spikes↑↑
C8 = KOH7 × D19			
CKX expression 7 DAP	CKX5, 9, 10↓	CKX5, 9, 10↑	CKX9, 10↑
	CKX11↑	CKX11↓	CKX11, NAC2↓
	CKX2.1, NAC2=↑	CKX2.1, NAC2=↓	CKX2.1, 2.2.2↓
CKX expression root	CKX8↓↓	CKX8↑↑	CKX3, 8↑↑
	CKX3, 5↓	CKX3, 5↑	CKX1, 10, NAC2↓
	NAC2↑↑	NAC2↓↓	
yield-related traits	yield↓	yield↑↑	yield↑↑
	CKX act. spike=	CKX act. spike=	CKX act. spike↑
	root=↓	root=↑	root=

Bold—cross number and parents.

Yield-related traits are represented by total grain yield and root mass (Table 1). Interestingly, grain yield in 7 out of 8 crosses is inherited from the paternal parent. The exceptions are the F$_2$ progeny of S6C × S12B (C2), which show very large differences in yield, exceeding parental data. The root mass in F$_2$ was lower than that in both parents or higher than that in both parents. In the first case, the root mass in the paternal parent was higher than that in the maternal parent, and in the second, the root mass in the paternal parent was lower than that in the maternal parent.

The results of crossing the low-yielding maternal parent with the higher-yielding paternal parent and their accompanying up- or down-regulated *TaCKX* GFMs and *NAC2* in F₂ generations are presented in Figure 2.

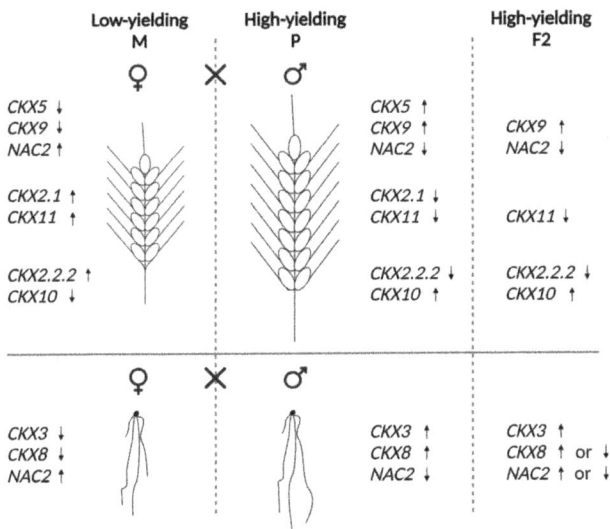

Figure 2. Models of up- (↑) or down-regulation (↓) of *TaCKX* GFMs and *NAC2* in low-yielding maternal parent crossed with higher-yielding paternal parent and their F₂ progeny.

Depending on the crosses, downregulation of *TaCKX5* with *TaCKX9* and upregulation of *NAC2* in spikes of the low-yielding maternal parent and the opposite regulation of these genes in spikes of the higher-yielding paternal parent resulted in high-yielding F₂, characterized, as in the paternal component, by a higher expression level of *TaCKX9* and a lower expression level of *NAC2*. Upregulation of *TaCKX2.1* and *11* in spikes of the maternal parent and downregulation of these genes in the paternal parent were associated with downregulation of *TaCKX11* in high-yielding F₂. Similarly, the upregulation of *TaCKX2.2.2* and the downregulation of *TaCKX10* in the spikes of the low-yielding maternal parent and opposite regulation of these genes, and the yield in the paternal parent, resulted in the downregulation of *TaCKX2.2.2* and the upregulation of *TaCKX10* in the spikes of the high-yielding F₂. The expression of *TaCKX3* in seedling roots of the high-yielding paternal parent and F₂ was upregulated. However, *TaCKX8* expression was upregulated, and *NAC2* was downregulated in the same organ of the paternal parent, but these genes were up- or down-regulated in F₂, depending on the cross.

2.2. Cooperating and Opposite-Functioning Genes

TaCKX5 with *TaCKX9* (yellow) and *TaCKX2.1* with *TaCKX11* (green) showed coordinated up- or downregulation in 7 DAP spikes of the paternal parent of C1, C2, C6, and C8 crosses; and C3, C4, C7, and C8 crosses, respectively (Table 2). Higher expression of *TaCKX5* and *9* in this parent was associated with a higher yield in F₂. However, a higher coordinated expression of *TaCKX2.1* with *TaCKX11* in the paternal parent determined a lower yield in F₂, and, in contrast, a lower expression of these two genes in the paternal parent was associated with a higher yield.

In the 7 DAP spikes, the paternal parent of the C3, C4, C5, and C6 crosses, *TaCKX2.2.2*, showed opposite expression to *TaCKX10* (blue), and upregulation of the first and downregulation of the second were associated with lower yield (but not in C6). In the paternal parent of the C1, C2, C7, and C8 crosses, *NAC2* was oppositely expressed to *TaCKX9*; in these crosses, a high yield was observed when *TaCKX9* was upregulated and *NAC2* was downregulated, and vice versa (only in C7 and C8). Furthermore, upregulated *TaCKX5* and downregulated *TaCKX1* were associated with high root mass in C2 and conversely in the reverse cross (C1).

Table 2. Coordinated expression of *TaCKX* GFMs and *NAC2* genes in 7 DAP spikes and seedling roots of the paternal parent (P) of four reciprocal crosses.

Cross	P Spike P Expr. +/P expr. −	P Root P Expr. +/P Expr. −	F_2 Spike/Root (Yield, Root Mass in F_2)
C1	1, NAC2/5, 9	3, 5, NAC2/10, 11	NAC2+/NAC2+ y−, r = −, As−−
C2	5, 9/1, NAC2	10, 11/3, 5, NAC2	NAC2−/NAC2− y = −, r+, As =
C3	2.1, 11, 2.2.2/10	NAC2, 1, 5, 10, 11/3	10−, 11+/10+,11+, NAC2+ y−, r=, As−, s-e+++
C4	9, 10/2.1, 11, 2.2.2	3, 8/1, 5, 10, 11, NAC2	10+, 11−/10−, 11−, NAC2− y+, r = −, As=
C5	2.2.2/10	?/3, 5, 8, NAC2, 1, 10, 11	10−/10−, NAC2− y−, r−, As = +, s-e+++
C6	2.2.2, 5, 9/10	3, 5, 8, NAC2, 1, 10, 11/none	5+, 10−/ 5+, 10+, NAC2+ y+, r−, As=
C7	2.1, 11, NAC2/9	1, 10, 11, NAC2/3, 5, 8	11+, NAC2+/11+, NAC2+ y−, r=, As+, s-e+++
C8	5, 9, 10/2.1, 11, NAC2	3, 5, 8/NAC2	5+, NAC2−/5+, NAC2− y++, r=, As+

P—paternal parent; expr.+—upregulated; expr.—downregulated; 1, 3, 5, 9 … -*TaCKX* GFMs, As—CKX activity spike; s-e+++—high number of semi-empty spikes, y—yield, r—root mass.

Among the *TaCKX* genes coordinately expressed in the paternal seedling roots were *TaCKX3, 5,* and *8* (green) in C5, C6, C7, and C8 crosses; *TaCKX3* and *8* (green) in C1 and C2 crosses; *TaCKX10, 11,* and *1* (yellow), and *NAC2* in C3 to C7 crosses; and *TaCKX10* and *11* (yellow) in C1 and C2 crosses. However, in one reciprocal cross, C3 and C4, the expression of *TaCKX3* and *TaCKX5* was opposite, and in the case of upregulation of *TaCKX3* and downregulation of *TaCKX5*, the grain yield in F_2 was higher.

In three reciprocal crosses, *NAC2* was downregulated in paternal roots (C2, C4, and C8), and in two of them (C2 and C8), *NAC2* was downregulated in paternal spikes as well. This negative regulation of *NAC2* occurred in the F_2 progeny, which was accompanied by a higher yield and a higher or similar to the parents' mass of the seedling roots. In contrast, in another way crosses, when expression of *NAC2* was increased in paternal roots (C1, C3, and C7) and was upregulated in paternal spikes, the same was observed in F_2 progeny, characterized by lower yield and lower or similar to the parent mass of the roots.

The higher yield in F_2 has been associated with the same or higher CKX activity, as in the paternal parent, in 7 DAP spikes. A higher number of semi-empty spikes, which occurred in low-yielding F_2 of the C3, C5, and C7 crosses, was accompanied by downregulated *TaCKX10* and/or upregulated *TaCKX11* in 7 DAP spikes and upregulated *TaCKX10, 11,* and *NAC2* or downregulated *TaCKX10* and *NAC2* in seedling roots.

2.3. The Correlation Coefficients between TaCKX GFMs and NAC2 Expression, CKX Activity, and Yield-Related Traits Were Significant for Both Parents or the Maternal or Paternal Parent Separately

The correlation coefficients between *TaCKX* GFMs and *NAC2* expression, CKX activity, and yield-related traits in reciprocal crosses were analyzed separately for the maternal parent and F_2, paternal parent, and F_2 for each cross (Table S1).

2.4. Correlations between TaCKX GFM and NAC2 Expression and Yield-Related Traits in the Group of Maternal Plants, and F_2 and Paternal Plants, and F_2 of Reciprocal Crosses

Seed number and spike number were positively correlated (Tables 3 and S1); however, each of these yield-related traits was correlated with different *TaCKX* GFMs.

Table 3. Correlations between *TaCKX GFM* and *NAC2* expression in 7 DAP spikes or seedling roots, and yield-related traits in the group of maternal plants and F_2, and paternal plants and F_2 of reciprocal crosses.

		7 DAP Spike	Seedling Root	Yield-Related Traits	F_2 Phenotype
		Seed Number			
C1 S12B × S6C	M + F_2	CKX1−, CKX5−	CKX11+	spike number+	yield−−, CKX act. −−, root=−, CKX act. root++
	P + F_2	nc	CKX11+	spike number+	
C2 S6C × S12B	M + F_2	nc	nc	CKX act. root−, spike number++	yield−, CKX act.=, root+, CKX act. root=
	P + F_2	nc	CKX3+	CKX act. root−, spike number+	
C3 D16 × KOH7	M + F_2	CKX2.1+,CKX2.2+	CKX3+, CKX8+, CKX11−	plant height+, spike number+	yield−−, CKX act. −−, root=−, semi-empty spikes++
	P + F_2	CKX2.1+, CKX2.2+	CKX1−, CKX3++, CKX8+, CKX11−	plant height+, spike number+	
C4 KOH7 × D16	M + F_2	nc	nc	plant height+, spike number+	yield+, CKX act.=, root=
	P + F_2	nc	nc	plant height+, spike number+	
C5 D19 × D16	M + F_2	CKX1−	CKX3+, CKX5+, NAC2+	plant height+, spike number+	yield−−, CKX act.=, root=−, semi-empty spikes++
	P + F_2	nc	CKX3+, NAC2+	plant height+ spike number+	
C6 D16 × D19	M + F_2	nc	nc	spike number+	yield+, CKX act.=, root=−
	P + F_2	nc	nc	spike number+	
C7 D19 × KOH7	M + F_2	nc	CKX1−	spike number++	yield−, CKX act. +, root=−, semi-empty spikes++
	P + F_2	nc	CKX1−	spike number++	
C8 KOH7 × D19	M + F_2	nc	CKX1−	empty spikes−, spike number+	yield++, CKX act.=+, root=
	P + F_2	CKX2.1+	CKX1−	empty spikes−, spike number+	
		Seed yield			
C1 S12B × S6C	M + F_2	CKX11−, NAC2−	CKX11+	spike number+, seed number++	yield−−, CKX act. −−, root=−, CKX act. root++
	P + F_2	nc	CKX11+	spike number+, seed number++	
C2 S6C × S12B	M + F_2	nc	CKX10+	spike number+, seed number++	yield−, CKX act.=, root+, CKX act. root=
	P + F_2	nc	CKX10+, CKX3+	spike number+, seed number++	
C3 D16 × KOH7	M + F_2	CKX2.1+, CKX2.2+	CKX3+, CKX8+	plant height+, spike number+, seed number++	yield−, CKX act. −−, root=−, semi-empty spikes++
	P + F_2	CKX2.1+, CKX2.2+	CKX3+, CKX8+	plant height+, spike number+, seed number++	
C4 KOH7 × D16	M + F_2	nc	nc	plant height+, spike number+, seed number++	yield+, CKX act.=, root=
	P + F_2	nc	nc	plant height+, spike number+, seed number++	
C5 D19 × D16	M + F_2	CKX1−	CKX3+, CKX5+, NAC2+	plant height+, spike number+, seed number++	yield−, CKX act. +, root=−, semi-empty spikes++
	P + F_2	nc	CKX3+, NAC2+	plant height+, spike number+, seed number++	
C6 D16 × D19	M + F_2	CKX2.1+	nc	plant height+, spike number+, seed number++	yield+, CKX act.=, root=−
	P + F_2	CKX1−	nc	plant height+, spike number+, seed number++	

Table 3. Cont.

Cross		7 DAP Spike	Seedling Root	Yield-Related Traits	F$_2$ Phenotype
C7 D19 × KOH7	M + F$_2$	nc	nc	plant height+, spike number++, seed number++	yield−, CKX act.+, root=−, semi-empty spikes++
	P + F$_2$	nc	nc	plant height+, spike number++, seed number++	
C8 KOH7 × D19	M + F$_2$	nc	CKX1−	spike number+, seed number++	yield++, CKX act.=+, root=
	P + F$_2$	nc	CKX1−	spike number+, seed number++	
Spike number					
C1 S12B × S6C	M + F$_2$	CKX2.1+, CKX2.2+, CKX11−	CKX11+	plant height+	yield−−, CKX act.−−, root=−, CKX act. root++
	P + F$_2$	CKX2.1+, CKX2.2+	nc	CKX act.+, plant height+	
C2 S6C × S12B	M + F$_2$	CKX2.2+	nc	nc	yield−, CKX act.=, root+, CKX act. root=
	P + F$_2$	nc	NAC2−	nc	
C3 D16 × KOH7	M + F$_2$	nc	CKX8+	nc	yield−, CKX act. −−, root=−, semi-empty spikes++
	P + F$_2$	nc	CKX8+	nc	
C4 KOH7 × D16	M + F$_2$	nc	CKX5+	nc	yield+, CKX act.+, root=
	P + F$_2$	nc	CKX5+	nc	
C5 D19 × D16	M + F$_2$	nc	CKX1−, CKX5+, NAC2+	nc	yield−−, CKX act.+, root=−, semi-empty spikes++
	P + F$_2$	nc	CKX1−, CKX5+, NAC2+	nc	
C6 D16 × D19	M + F$_2$	CKX2.1+	nc	nc	yield+, CKX act.=, root=−
	P + F$_2$	nc	nc	nc	
C7 D19 × KOH7	M + F$_2$	nc	CKX1−	nc	yield−, CKX act.+, root=−, semi-empty spikes++
	P + F$_2$	nc	CKX1−, CKX8−	nc	
C8 KOH7 × D19	M + F$_2$	nc	CKX1−, CKX5−, CKX8−	nc	yield++, CKX act.=+, root=
	P + F$_2$	nc	CKX1−, CKX5−	nc	
TGW					
C1 S12B × S6C	M + F$_2$	CKX2.1+, CKX10+, NAC2+	nc	nc	yield−, CKX act. −−, root=−, CKX act. root++
	P + F$_2$	CKX2.2+, NA2C+	nc	nc	
C2 S6C × S12B	M + F$_2$	nc	nc	CKX act. root++, seed number−	yield−, CKX act.=, root+, CKX act. root=
	P + F$_2$	nc	nc	CKX act. root++, seed number−	

Table 3. Cont.

		7 DAP Spike	Seedling Root	Yield-Related Traits	F$_2$ Phenotype
C3 D16 × KOH7	M + F$_2$	nc	nc	plant height+, yield+, semi-empty−, seed number+	yield−, CKX act. −−, root=−, semi-empty spikes++
	P + F$_2$			plant height+, yield+	
C4 KOH7 × D16	M + F$_2$	CKX2.2−	CKX10+, NAC2+	CKX act.−	yield+, CKX act.=, root=
	P + F$_2$	CKX2.2−	CKX10+,NAC2+	CKX act.−	
C5 D19 × D16	M + F$_2$	nc	NAC2+	yield+, seed number+	yield−−, CKX act.+, root=−, semi-empty spikes++
	P + F$_2$	nc	NAC2+	yield+	
C6 D16 × D19	M + F$_2$	nc	nc	plant height+	yield+, CKX act.=, root=−
	P + F$_2$	nc	nc	plant height+	
C7 D19 × KOH7	M + F$_2$	CKX2.1−, CKX2.2−, CKX9−	nc	spike length−	yield−, CKX act.+, root=−, semi-empty spikes++
	P + F$_2$	CKX2.1−, CKX2.2.2−, CKX5+	nc	nc	
C8 KOH7 × D19	M + F$_2$	CKX2.1−, CKX2.2.2−	nc	nc	yield=, CKX act.=+, root=
	P + F$_2$	CKX2.1−, CKX2.2.2−!, CKX10−	nc	nc	
		Root mass			
C1 S12B × S6C	M + F$_2$	CKX5+,CKX11+, NAC2+	CKX1−, CKX3−, CKX10−, CKX11−,	seed number−	yield−, CKX act. −−, root=−, CKX act. root++
	P + F$_2$	nc	NAC2−, CKX1−, CKX10−, CKX11−	seed number−	
C2 S6C × S12B	M + F$_2$	NAC2−	nc	nc	yield−, CKX act.=, root+, CKX act. root=
	P + F$_2$	nc	nc	nc	
C3 D16 × KOH7	M + F$_2$	CKX1+, CKX11+	CKX5+	nc	yield−, CKX act. −−, root=−, semi-empty spikes++
	P + F$_2$	CKX1+, CKX11+, NAC2+	CKX5+, CKX8−	nc	
C4 KOH7 × D16	M + F$_2$	nc	CKX3+, CKX8+, NAC2−	CKX act.+	yield+, CKX act.=, root=
	P + F$_2$	CKX2.1+, CKX2.2+, CKX10+	CKX3+, CKX8+, NAC2−	nc	
C5 D19 × D16	M + F$_2$	NAC2−	CKX1−, CKX11+	nc	yield−−, CKX act.+, root=−, semi-empty spikes++
	P + F$_2$	CKX2.1+		nc	
C6 D16 × D19	M + F$_2$	CKX2.1+, CKX2.2+	nc	yield+ plant height+	yield+, CKX act.=, root=−
	P + F$_2$	CKX2.2+	nc		

Table 3. Cont.

		7 DAP Spike	Seedling Root	Yield-Related Traits	F₂ Phenotype
C7 D19 × KOH7	M + F$_2$	nc	CKX1–	nc	yield–, CKX act.+, root=–, semi-empty spikes+
	P + F$_2$	NAC2–	CKX1–	spike length+	
C8 KOH7 × D19	M + F$_2$	CKX10–	CKX3–	nc	yield++, CKX act.=+, root=
	P + F$_2$	nc	CKX3–	yield+	

Plant height

		7 DAP Spike	Seedling Root	Yield-Related Traits	F₂ Phenotype
C1 S12B × S6C	M + F$_2$	nc	nc	nc	yield–, CKX act. – –, root=–, CKX act.root++
	P + F$_2$	nc	nc	nc	
C2 S6C × S12B	M + F$_2$	CKX11+	CKX1–	nc	yield–, CKX act.=, root+, CKX act.root=
	P + F$_2$	nc	CKX5–, CKX11+	nc	
C3 D16 × KOH7	M + F$_2$	nc	CKX11–	nc	yield–, CKX act.– –, root=–, semi-empty spikes++
	P + F$_2$	nc	CKX11–	nc	
C4 KOH7 × D16	M + F$_2$	nc	nc	nc	yield+, CKX act.=, root=
	P + F$_2$	CKX2.2–	NAC2+	nc	
C5 D19 × D16	M + F$_2$	CKX1–	CKX1–, NAC2+	nc	yield– –, CKX act.+, root=–, semi-empty spikes++
	P + F$_2$	nc	CKX1–, NAC2+	nc	
C6 D16 × D19	M + F$_2$	CKX1–, CKX5–	CKX5–	nc	yield+, CKX act.=, root=–
	P + F$_2$	CKX5–	CKX5–	nc	
C7 D19 × KOH7	M + F$_2$	nc	nc	nc	yield–, CKX act.+, root=–, semi-empty spikes++
	P + F$_2$	CKX2.2–, CKX9–	CKX11+	nc	
C8 KOH7 × D19	M + F$_2$	CKX9–	nc	nc	yield++, CKX act.=+, root=
	P + F$_2$	nc	nc	nc	

Spike length

		7 DAP Spike	Seedling Root	Yield-Related Traits	F₂ Phenotype
C1 S12B × S6C	M + F$_2$	NAC2–	nc	semi-empty spikes–, seed number+, **yield+**	yield–, CKX act.– –, root=–, CKX act.root++
	P + F$_2$	NAC2–	nc	spike number+, seed number+, **yield+**	
C2 S6C × S12B	M + F$_2$	CKX9+, CKX11+	nc	**plant height+**	yield–, CKX act.=, root=+, CKX act.root=
	P + F$_2$	nc	nc	nc	
C3 D16 × KOH7	M + F$_2$	CKX2.1+	CKX3+, CKX11–	**plant height+, seed number+, yield+**	yield–, CKX act.– –, root=–, semi-empty spikes++
	P + F$_2$	CKX2.1+, CKX2.2+	CKX1–, CKX3++, CKX11–	**plant height+, seed number+, yield+**	
C4 KOH7 × D16	M + F$_2$	CKX1+, CKX10+	nc	**plant height+, seed number+, yield+**	yield+, CKX act.=, root=
	P + F$_2$	nc	nc	seed number+, **yield+**	

Table 3. Cont.

		7 DAP Spike	Seedling Root	Yield-Related Traits	F$_2$ Phenotype
C5 D19 × D16	M + F$_2$	**CKX5−**, CKX11+, NAC2−	CKX11+	nc	yield−−, CKX act.+, root=, semi-empty spikes+
	P + F$_2$	nc	nc	seed number+, yield+	
C6 D16 × D19	M + F$_2$	nc	nc	spike number+, seed number+, yield+	yield+, CKX act.=, root=−
	P + F$_2$	nc	nc	seed number+, yield+	
C7 D19 × KOH7	M + F$_2$	CKX1+, CKX9+	nc	seed number+	yield−, CKX act.+, root=−, semi-empty spikes++
	P + F$_2$	CKX9+, NAC2−	nc	nc	
C8 KOH7 × D19	M + F$_2$	nc	nc	nc	yield++, CKX act.=+, root=
	P + F$_2$	**CKX2.1+**	nc	seed number+	
		Semi-empty spikes			
C1 S12B × S6C	M + F$_2$	CKX5+	**CKX5−**, *CKX10−*	nc	yield−, CKX act. −−, root=−, CKX act. root++
	P + F$_2$	CKX5+↓, CKX9+, CKX11+	nc	nc	
C2 S6C × S12B	M + F$_2$	**CKX9+**	nc	nc	yield−, CKX act.=, root+, CKX act. root=
	P + F$_2$	CKX9+	**CKX8−**	empty spikes+	
C3 D16 × KOH7	M + F$_2$	CKX9+	nc	nc	yield−, CKX act. −−, root=−, semi-empty spikes+
	P + F$_2$	CKX11−	nc	nc	
C4 KOH7 × D16	M + F$_2$	nc	nc	spike number+	yield+, CKX act.=, root=
	P + F$_2$	nc	nc	spike number+	
C5 D19 × D16	M + F$_2$	nc	nc	nc	yield−−, CKX act.+, root=−, semi-empty spikes++
	P + F$_2$	CKX2.1−	nc	nc	
C6 D16 × D19	M + F$_2$	CKX5+	**CKX5+**	spike number+	yield+, CKX act.=, root=−
	P + F$_2$	CKX5+	nc	seed number+	
C7 D19 × KOH7	M + F$_2$	CKX10+	nc	nc	yield−, CKX act.+, root=−, semi-empty spikes++
	P + F$_2$	nc	nc	nc	
C8 KOH7 × D19	M + F$_2$	nc	**CKX5−**, *CKX8−*, *NAC2−*	nc	yield++, CKX act.=+, root=
	P + F$_2$	CKX10+	**CKX5−**, *NAC2−*	nc	

All correlation coefficients ≥0.60; bold—significant correlation coefficients; nc—no correlation; +—positive correlation; =+—low positive correlation; ++—strong positive correlation; −—negative correlation; =−—low negative correlation; −−—strong negative correlation.

2.4.1. Seed Number

The decrease in seed number in maternal plants and their F_2 (M and F_2) of the C1 cross (S12B × S6C) was strongly negatively correlated with upregulated *TaCKX1* and *TaCKX5* in spikes and positively correlated with the downregulated *TaCKX11* in the seedling roots of F_2. There was no significant correlation between the expression of *TaCKX* GFM and the yield-related traits in the groups of paternal plants (P) and F_2 in the same cross, and M and F_2, and P and F_2 in the reverse, C2 cross. The F_2 progeny in this reverse cross showed a similar yield and greater root mass compared to the parents.

In both M and F_2, and P and F_2 of C3, the decrease in seed number was strongly positively correlated with *TaCKX2.1* and *TaCKX2.2.2* in spikes and positively correlated with *TaCKX3* and *TaCKX8* but negatively correlated with *TaCKX11* in seedling roots. These correlations were not significant in the reverse, C4 cross, in which the M (KOH7) and F_2 plants showed higher yields.

The decrease in seed number in M × F_2 of C5 was negatively correlated with *TaCKX1* in spikes and positively correlated with downregulated *TaCKX3* and *5*, and upregulated *NAC2* in seedling roots. There were also positive correlations of *TaCKX3* in roots between P and F_2 of the same cross. These correlations were not significant in the reverse C6 cross; however, F_2 of this cross was characterized by higher yield and similar root mass than in the parents.

The increase in seed number was strongly positively correlated with downregulated *TaCKX2.1* in spikes and negatively correlated with downregulated *TaCKX1* in the seedling roots only in F_2 progeny of a C8 cross, 14K (in the case of P × F_2 only for *TaCKX2.1*).

2.4.2. Spike Number

The decrease in the number of spikes in M × F_2 and P × F_2 of the C1 cross was strongly positively correlated with *TaCKX2.1* and *TaCKX2.2.2* in spikes and positively correlated with *TaCKX11* in seedling roots (only in M × F_2). There was also a significant and positive correlation between the expression of *TaCKX2.2.2* and the number of spikes in the reverse C2 cross of M and F_2. Furthermore, in the same cross, the spike number was negatively correlated with downregulated *NAC2*, but only in the P × F_2 group. There were no significant correlations between *TaCKX* GFM expression and spike number in M × F_2 and P × F_2 spikes of C3 and C4. However, there was a strong and positive correlation of the spike number with *TaCKX8* in the roots of C3 and *TaCKX5* in the roots of C4.

The decreased spike number in M × F_2 and P × F_2 of C5 was not correlated with any *TaCKX* expressed in the spikes but was negatively correlated with *TaCKX1*, positively correlated with *TaCKX5*, and positively correlated with *NAC2* expressed in the seedling roots. Conversely, in reverse C6 cross, there was a positive correlation of the spike number with *TaCKX2.1* in a P × F_2, which resulted in a higher yield phenotype in the F_2.

The spike number in C7 and C8 crosses was not correlated with the level of expression of any gene tested in the spikes; however, it was negatively correlated with the expression of *TaCKX1*, *5*, and *8* in seedling roots.

2.4.3. TGW

TGW was positively correlated with *TaCKX2.1*, *10*, and *NAC2* in spikes of M and F_2 of C1 and with *TaCKX2.2.2* of P and F_2 of the same cross. There was no correlation in F_2 between the expression of the genes tested and TGW in spikes of the C2 and roots of the C1 and C2 crosses. There were no correlations between the TGW and *TaCKX* genes in the spikes and roots of C3. However, there was a negative correlation of this trait with *TaCKX2.2.2* in spikes and positive correlations with *TaCKX10* and *NAC2* in roots of the reciprocal C4 cross. Positive correlations of *NAC2* with TGW were also observed in the roots of C5 but not in those of C6. Negative correlations of *TaCKX2.1* and *2.2.2* in spikes with TGW were observed in both reciprocal crosses, C7 and C8. Additionally, *TaCKX9* was negatively correlated with the trait in M + F_2, *TaCKX5* was positively correlated with the trait in P + F_2 of C7, and *TaCKX10* was negatively correlated with TGW in P + F_2 of C8. There was no correlation between TGW and any gene expression in roots.

2.4.4. Root Mass

The mass is positively correlated with the expression of *TaCKX5, 11*, and *NAC2* in spikes of M + F_2 of C1 and negatively correlated with *NAC2* in M + F_2 of C2. A positive correlation between root mass and *TaCKX11*, and *NAC2* was also visible in P + F_2 of C3, and a negative correlation between trait and *NAC2* expression was also observed in spikes of M + F_2 of C5 and P + F_2 of C7. Furthermore, in the C1 cross, this trait was negatively correlated with the expression of *TaCKX1, 3, 10*, and *NAC2* in roots of M + F_2 and with *TaCKX11* in roots of P + F_2. There were also positive correlations between root mass and *TaCKX1* (M + P of C3), root mass and *TaCKX2.1* (P + F_2 of C4; P + F_2 of C_5; M + F_2 of C6), and root mass and *TaCKX2.2.2* (P + F_2 of C4; M + F_2 and P + F_2 of C6). The expression of another gene, *TaCKX10*, in spikes, was positively correlated with root mass in P + F_2 of C4 but negatively correlated in M + F_2 of C8. Correlations between root mass and gene expression tested in roots were dependent on the parent and cross. There were negative correlations with *TaCKX1* in 5 out of 16 combinations tested, negative correlations with *TaCKX3* in 3 combinations, but positive correlations in two combinations, positive correlations with *TaCKX5* in two combinations, and single positive or negative correlations with *TaCKX8, 11*, and *NAC2*. The root mass in single combinations was positively correlated with the yield (twice), height of the plant (once), and length of the spike (once), and negatively correlated with seed number of seeds (twice).

2.4.5. Semi-Empty Spikes

The number of semi-empty spikes was positively correlated with the expression of *TaCKX9* in the P + F_2 C1, C2, and M + F_2 C3 crosses, positively correlated with *TaCKX5* in the M + F_2, and P + F_2 C1 and C6 crosses, and positively correlated with *TaCKX10* in the M + F_2 C7 and P + F_2 C8 crosses, all expressed in 7 DAP spikes. The negative correlation between the number of semi-empty spikes and *TaCKX2.1* was in P + F_2 of C5, and between the same trait and *TaCKX11* was in P + F_2 of C3. In seedling roots of various crosses, this trait was mainly negatively correlated with *TaCKX5, 8, 10*, and *NAC2*.

Generally, negative correlations between the expression of *TaCKX2.1, 2.2.2,* and *10* in spikes and TGW, seed number, seed yield, and spike number were correlated with higher yield, and positive correlations were correlated with lower yield. On the other hand, positive correlations between the expression of these genes and root mass determine a higher yield in F_2. Higher yield in F_2 is also associated with balanced CKX enzyme activity in spikes and seedling roots.

A summary of the regulation of yield-related traits by *TaCKX* GFMs and *NAC2* in the high-yielding progeny of F_2 is presented in Figure 3.

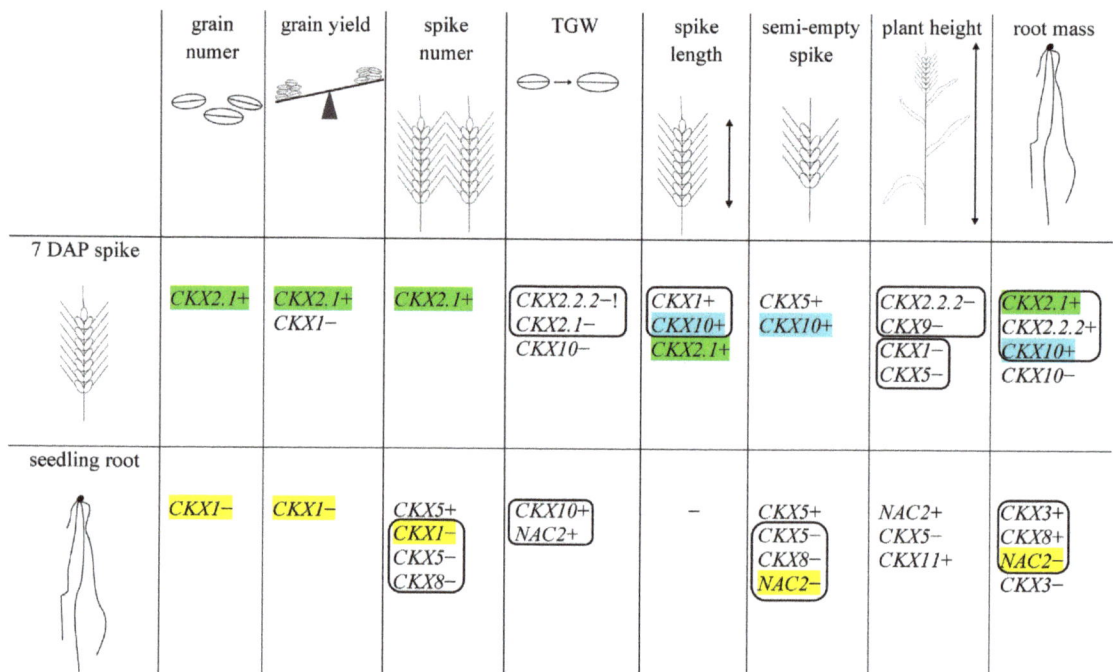

Figure 3. Regulation of yield-related traits by *TaCKX* GFMs and *NAC2* in the high-yielding progeny of F$_2$ based on correlation coefficients.

3. Discussion

Common wheat is a very important cereal crop for feeding the world's population; therefore, continued improvement of the yield of this species is significant. *CKX* GFMs have already been documented to perform a pivotal role in determining yield-related traits in many plant species, including wheat [4,12]. The genes are tissue-specific; they encode cytokinin oxidase/dehydrogenase, the enzyme that irreversibly degrades cytokinins. We have already characterized the role of *TaCKX1* and *TaCKX2* in the regulation of yield traits in awnless and owned-spike cultivars [5–7]. The range of natural variation in the expression levels of most *TaCKX* genes among breeding lines and cultivars was very high, indicating the possibility of selecting beneficial genotypes for breeding purposes [8]. Therefore, we were interested in how the expression of these genes is inherited.

3.1. The Expression Patterns of Most TaCKX GFMs and TaNAC2-5A Are Mainly Inherited from the Paternal Parent

Comparison of the expression patterns of most of the *TaCKX* GFMs and yield-related traits between parents and F$_2$ progeny in all reciprocal crosses tested indicated their inheritance from the paternal parent. This rule includes expression patterns in both tissues tested, 7 DAP spikes, and seedling roots, and all *TaCKX* GFMs and *TaNAC2-5A* tested were represented in different crosses. The exception was *TaCKX5* expressed in 7 DAP spikes, and *TaCKX3* expressed in seedling roots, for which the expression level in single crosses was inherited from the maternal parent. Furthermore, high or low yield was predominantly inherited from the paternal parent, and root mass was inherited from both parents or in one reciprocal cross from the maternal parent. We have not found such examples of inheritance in the literature; however, some deviations from parental additivity of expression in polyploid plants were described [2]. An example of such non-additive gene expression takes place when the gene expression level in progeny is higher than that of one parent. The expression level dominance of one parent, also called genomic imprinting, is epigenetic in origin and was investigated primarily at the molecular level in plants and animals [36,37]. The main regulators of gene imprinting are DNA and histone methylation asymmetries between parental genomes. Most of the imprinted genes in the endosperm of grains of different rice cultivars are imprinted across cultivars, and their functions are associated with the regulation of transcription, development, and signaling [39]. Imprinting might affect a single gene or a group of genes. Genes that showed conserved imprinting in cereals have been shown to reveal positive selection and were suggested to regulate seed development in a dose-dependent manner [38]. The only example of a paternally imprinted locus in maize is *ded1*, which encodes a transcription factor specifically expressed during early embryo development and activates early embryo genes that contribute to grain set and weight [47]. To our knowledge, there are no examples of paternally inherited expression patterns. According to Arabidopsis research, imprinted paternally expressed genes during seed development are mainly related to hypomethylated maternal alleles, repressed by small RNAs or less frequently with transposable elements [48,49]. Contrary to developmental epigenetics, in the case of transgenerational epigenetics, these epigenetic changes do not reset between generations, and this type of inheritance is more related to plants than animals (heritable changes in DNA methylation) [54]. Therefore, we suggest that this paternal inheritance of selected *TaCKX* GFMs is an effect of transgenerational epigenetic changes, not reset between generations. These heritable epigenetic changes might be effects of DNA methylation, repression of maternal alleles by small RNAs, transposable elements, or, most likely, transcription factors. From our in silico analysis and expression analysis (Iqbal et al., not published yet), several NAC transcription factors appear to strongly regulate the expression of *TaCKX* GFMs and *TaIPT* GFMs, influencing yield-related traits [24].

3.2. Cooperation of TaCKX GFMs and TaNAC2-5A in the Determination of Yield-Related Traits

The coordinated high or low level of expression of a few groups of genes in the paternal parent positively or negatively regulates higher or lower yield. In two reciprocal crosses, where both *TaCKX5* and *TACKX9* showed high expression in 7 DAP spikes of the paternal parent, the yield in the F_2 progeny was high and vice versa. In others, the high yield in the F_2 progeny was determined by a low level of expression of *TaCKX2.1* and *TaCKX11* in spikes of the paternal parent and high levels of their expression in the maternal parent. The level of expression of *TaNAC2-5A* in the paternal parent and/or F_2 was in opposition to *TaCKX5* and *9*; however, it was in agreement with *TaCKX11* and *TaCKX2.1*, suggesting their role in the regulation of transcription of these genes. In fact, it was proven by correlation analysis of its expression with yield-related traits [8]. Opposite cooperation of some of the genes in paternal spikes, which resulted in high or low yield in F_2, has also been observed. The high level of expression of *TaCKX2.2.2* and the low level of expression of *TaCKX10* predominantly resulted in the low yield in F_2 progeny and vice versa.

Such common rules of gene expression in the paternal parent associated with yield in the paternal parent and F_2 progeny were also observed in the seedling roots. The high level of expression of *TaCKX3* and *TaCKX8* in the paternal parents of three reciprocal crosses resulted in high yield in the F_2 progeny and vice versa. The expression of *TaNAC2-5A* in spikes and seedling roots of high-yielding paternal parents and F_2 progeny showed a predominantly low expression level and inversely. These principles of paternal inheritance of selected *TaCKX* GFMs and *TaNAC2-5A* expression associated with high yield could be directly involved as molecular markers in high-yielding wheat breeding.

3.3. Regulation of Yield-Related Traits by TaCKX GFMs and TaNAC2-5A in the F_2 Generation

The grain number, grain yield, spike number, and TGW were strongly positively correlated with *TaCKX2.1* and *TaCKX2.2.2* independent of the parent; however, only in the crosses resulted in decreased yield. In contrast, negative correlations were observed between *TaCKX2.1*, *TaCKX2.2.2*, and TGW in a reciprocal cross of C7/C8 and *TaCKX2.2.2* in a one-way cross (C4), in which the F_2 progeny had a higher yield. All these correlations prove our earlier observations [6,7]. Modified wheat lines with 60% decreased expression of *TaCKX2.2.2*, and a slight decrease in the *TaCKX2.2.1* and *2.1* genes exhibit a significantly higher TGW and slightly increased yield [7]. Interestingly, this result was observed in cultivars and breeding lines that represent awnless spikes. In the owned-spike cultivar, silencing of the *TaCKX2* genes co-expressed with other *TaCKX* resulted in decreased yield; however, TGW was at the same level as in non-silent plants [5]. Furthermore, a strong feedback mechanism for regulation of the expression of *TaCKX2* and *TaCKX1* genes was observed in both awnless and owned-spike cultivars [5–7]. Silencing of *TaCKX2* genes upregulated the expression of *TaCKX1* and vice versa. This feedback mechanism could explain the observed positive correlations of the *TaCKX2* genes with yield-related traits in low-yielding F_2 progeny and negative correlations in high-yielding F_2 progeny. A similar mechanism is visible when we analyze individual traits in high-yielding F_2, such as grain number, grain yield, and spike number. These traits are promoted by up-regulated in 7 DAP spikes *TaCKX2.1* and down-regulated *TaCKX1*. Silencing of *HvCKX1* in barley, which is an ortholog of *TaCKX1*, decreased CKX enzyme activity and led to increased seedling root mass and higher plant productivity [55]; however, knock-out of this gene caused a significant decrease in CKX enzyme activity but no changes in grain yield were observed [56]. These differences might be explained by differences in the level of decreased gene expression, which variously coordinate the expression of other genes, regulate phytohormone levels, and determine particular phenotypes, as was already documented in wheat [5,7].

The association of *TaCKX2* genes with yield-related traits has also been reported in different wheat cultivars or genotypes. Zhang et al. [20] showed that *TaCKX6* (renamed by Chen et al. [11] *TaCKX2.2.1-3D*), which is an ortholog of rice *OsCKX2* associated with grain number [57], is related to grain weight. Another allele of *CKX2*, *TaCKX6a02* [21], annotated as *TACKX2.1* [58], significantly correlated with grain size, weight, and grain filling rate. Wheat plants with silenced by RNAi expression of *TaCKX2.2.1-3A* (originally *TaCKX2.4*) showed a strong correlation with the number of grains per spike implied by more filled florets [59]. Since *TaCKX2.2.1-3D* was associated with grain weight, these differences in functions between *TaCKX2.2.1-3A* and *TaCKX2.2.1-3D* were interpreted as subgenome-dependent.

Grain number was also negatively correlated with *TaCKX1* and *TaCKX5*, and grain yield was negatively correlated with *TaCKX1*, predominantly for M and F_2, which were characterized by decreased grain number and lower yield. This observation is also in agreement with previous research. The silencing of *TaCKX1* caused an increase in spike number and grain number but a decrease in TGW because this trait is opposite to grain number [6]. The low-yield progeny of F_2 showed positive correlations between the expression of *TaCKX11*, *3*, *5*, and *8*, and *NAC2* in the seedling roots and the grain number, the spike number and the grain yield; however, the higher-yield progeny of F_2 displayed a negative correlation between *TaCKX1* and these yield-related traits in the seedling roots of some crosses. In summary, high-yielding F_2 was the result of upregulation of *TaCKX2.1* in spikes and downregulation of *TaCKX1* in seedling roots. As documented earlier, *TaCKX11*, *5*, *8* and *TaNAC2-5A* are expressed in all organs, and their expression is correlated with the expression of spike-specific *TaCKX2* and *TaCKX1* [8,58].

Rice *OsCKX11* is an orthologue of wheat *TaCKX11* and is highly expressed in the roots, leaves, and panicles. The gene was shown to coordinate the simultaneous regulation of leaf senescence and grain number by the relationship of source and sink [60]. Since *TaCKX11* is expressed in seedling roots and highly expressed in leaves, inflorescences, and 0, 7, and 14 DAP spikes, it could perform a similar function. This is partly proven by silencing of the *TaCKX2* genes in awnless spikes of cv. Kontesa, which resulted in significant upregulation of *TaCKX11* and growth of TGW, and chlorophyll content in flag leaves [7]. In contrast, *TaCKX11* is significantly negatively regulated by *TaCKX1*, resulting in a higher spike number and grain number [6]. Its orthologue in rice, *OsCKX11*, was found to regulate leaf senescence and grain number by the coordinated source and sink relationship [60].

Based on a summary of the regulation of yield-related traits in high-yielding F_2, it is possible to identify singular genes or groups of genes that are up- or down-regulated in 7 DAP spike or seedling roots and specifically regulate yield-related traits. Upregulated in spikes *TaCKX2.1* and downregulated in seedling roots *TaCKX1* were found to determine grain number, grain yield, spike number, and spike length. Furthermore, upregulated in 7 DAP spikes *TaCKX10* and downregulated *TaNAC2-5A*, together with others, depending on cross, control spike length, semi-empty spikes, root mass, and increased grain yield. As discussed above, high TGW is in contrary to high grain number and partly grain yield and was strongly determined by downregulated *TaCKX2.2.2* together with *TaCKX2.1* in 7 DAP spikes and upregulated *TaCKX10* and *NAC2* in seedling roots. The upregulated in seedling roots *TaNAC2-5A* participates in the determination of TGW and plant height, and the downregulation of *TaNAC2-5A* in seedling roots controls the development of semi-empty spikes and root mass.

In previous research, *TaNAC2-5A* has been documented as a gene encoding a nitrate-inducible wheat transcription factor. Overexpression of the gene improved root growth, grain yield, and grain nitrate concentration [25]. This is in agreement with our observations of growth of TGW but not enhanced roots. The increase was argued to be the consequence of regulation of nitrate concentration and its remobilization in developing grains by direct binding of the *TaNAC2-5A* protein to the promoter of the nitrate transporter, *TaNRT2.5-3A* and positive regulation of its expression [61]. The expression of *TaNAC2-5A* is coregulated by expressed in 7 DAP spikes *TaCKX2* genes and expressed in 7 DAP spikes and seedling roots *TaCKX1* gene [5,7]. Independent of awnless or awned-spike genotype, downregulation of *TaCKX2* genes by RNAi significantly increased *TaNAC2-5A* expression, resulting in higher chlorophyll content in flag leaves and delayed leaf senescence. As discussed above, the strong feedback mechanism between the *TaCKX2* and *TaCKX1* genes implies that downregulation of *TaCKX1* resulted in opposite results. Similar to our observations in wheat, an ortholog of *TaNAC2-5A* in rice, *OsNAC2*, was described as a negative regulator of crown root number and root length [62]. Its expression was positively correlated with cytokinin synthesis genes, *OsIPT3, 5*, the gene determining the formation of active cytokinins, *OsLOG3*, and negatively correlated with *OsCKX4* and *5*. The authors concluded that *OsNAC2* stimulated cytokinin accumulation by suppressing *CKX* expression and stimulating *IPT* expression by binding the OsNAC protein to the promoters of these genes. Therefore, *OsNAC2* functions as an integrator of cytokinin and auxin signals that regulate root growth. In our experiments, orthologous to *OsCKX4*, *TaCKX4* was not tested due to its weak expression in roots. However, the up-regulated expression of highly specific in seedling roots *TaCKX3* and *TaCKX8* [8,58] was antagonistically regulated by *TaNAC2-5A* in these organs, positively influencing seedling growth. Furthermore, our in silico analysis of *TaNACs* with *TaIPTs* and *TaCKXs* showed that the same NAC proteins might join promotor sites of cytokinin synthesis and cytokinin degradation genes in wheat (Iqbal et al., not published yet).

4. Materials and Methods

4.1. Plant Material

Five common wheat breeding lines and cultivars (*Triticum aestivum* L.), named S12B, S6C, D16, KOH7, and D19, which showed differences in the expression levels of *TaCKX* GFMs and *TaNAC2-5A* (*NAC2*) in 7 DAP spikes, seedling roots, and yield-related traits were selected for the study. They were used in four reciprocal crosses: (1) S12B × S6C and S6C × S12B (C1 and C2, respectively); (2) D16 × KOH7 and KOH7 × D16 (C3 and C4, respectively); (3) D19 × D16 and D16 × D19 (C5 and C6, respectively); and (4) D19 × KOH7 and KOH7 × D19 (C7 and C8, respectively) to obtain the F_1 and F_2 generations. The experimental tissue samples were collected from the parental lines and their F_2 progeny growing in a growth chamber during the same period.

Ten seeds of each genotype germinated in Petri dishes for five days at room temperature in the dark. Six out of ten seedlings from each Petri dish were replanted in pots with soil. The plants were grown in a growth chamber under controlled environmental conditions with 20 °C day/18 °C night temperatures and a 16 h light/8 h dark photoperiod. The light intensity was 350 μmol s^{-1}·m^{-2}. The plants were irrigated three times a week and fertilized once a week with Florovit according to the manufacturer's instructions.

The following tissue samples in three biological replicates were collected: 5-day-old seedling roots, which were cut 0.5 cm from the root base before replanting in the pots, and first 7 DAP) spikes from the same plants grown in the growth chamber. All of these samples were collected at 9:00 a.m. The collected material was frozen in liquid nitrogen and kept at −80 °C until use.

4.2. Cross-Breeding

The maternal plant was deprived of its own anthers so that it would not self-fertilize, then pollinated by transferring three anthers from the paternal plant for each ovary of the maternal parent plant and placed in an isolator. The seeds were harvested.

4.3. RNA Extraction and cDNA Synthesis

Total RNA from 7 DAP spikes and roots from 5-day-old seedlings was extracted using TRI Reagent (Invitrogen, Lithuana) according to the manufacturer's protocol. The concentration and purity of the isolated RNA were determined using a NanoDrop spectrophotometer (NanoDrop ND-1000, Thermi Fisher Scientific, Wilmington, DE, USA), and the integrity was checked on 1.5% (*w/v*) agarose gels. To remove residual DNA, RNA samples were treated with DNase I (Thermo Fisher Scientific, Lithuana). Each time, 1 μg of good quality RNA was used for cDNA synthesis using the RevertAid First Strand cDNA Synthesis Kit (Thermo Fisher Scientific, Lithuana) following the manufacturer's instructions. The cDNA was diluted 20 times prior to use in the RT-qPCR assays.

4.4. Quantitative RT-qPCR

RT-qPCR assays were performed for 10 genes: *TaCKX1*, *TaCKX2.1*, *TaCKX2.2*, *TaCKX3*, *TaCKX5*, *TaCKX8*, *TaCKX9*, *TaCKX10*, *TaCKX11*, and *TaNAC2-5A*. The sequences of the primers for each gene are shown in Table S2. All real-time reactions were performed on a Rotor-Gene Q (QIAGEN Hilden, Germany) thermal cycler using 1× HOT FIREPol EvaGreen qPCR Mix Plus (Solis BioDyne, Estonia), 0.2 μM of each primer and 4 μL of cDNA in a total volume of 10 μL. Each reaction was carried out in three biological and three technical replicates in the following temperature profile: initial denaturation and polymerase activation of 95 °C–12 min (95 °C–25 s, 62 °C–25 s, 72 °C–25 s) × 45 cycles, 72 °C–5 min, with melting curve at 72–99 °C 5 s per step. The expression of *TaCKX* genes was calculated according to the two standard curve method using *ADP-ribosylation factor* (*Ref 2*) as a normalizer. The relative expression for each *TaCKX* GFM and *TaNAC2-5A* was calculated in relation to the control female parents, set as 1.00.

4.5. Analysis of CKX Activity

CKX enzyme activity was performed in the same samples subjected to *TaCKX* gene expression analysis according to the procedure developed by Frebort et al. [63] and optimized for wheat tissues. The plant material was powdered with liquid nitrogen using a hand mortar and extracted with a 3-fold excess (*v/w*) of 0.2 M Tris–HCl buffer, pH 8.0, containing 1 mM phenylmethylsulfonyl fluoride (PMSF) and 0.3% Triton X-100 ((St. Louis, MO, USA). Plant samples were incubated in a reaction mixture consisting of 100 mM McIlvaine buffer, 0.25 mM of the electron acceptor dichlorophenolindophenol and 0.1 mM of substrate (N6-isopentenyl adenine). The volume of the enzyme sample used for the assay was adjusted based on the enzyme activity. The incubation temperature was 37 °C for 1–16 h. After incubation, the reaction was stopped by adding 0.3 mL of 40% trichloroacetic acid (TCA) and 0.2 mL of 2% 4-aminophenol (PAF). The product concentration was determined by scanning the absorption spectrum from 230 nm to 550 nm. The total protein concentration was estimated based on the standard curve of bovine serum albumin (BSA) according to the Bradford procedure [64].

4.6. Measurement of Yield-Related Traits

Morphometric measurement of yield-related traits of selected genotypes was performed. The described traits were plant height, spike number, semi-empty spike number, tiller number, spike length, grain yield, grain number, TGW, and 5-day seedling root weight.

4.7. Statistical Analysis

Statistical analysis was performed using Statistica 13 software (StatSoft). The normality of the data distribution was tested using the Shapiro–Wilk test. The significance of the changes was analyzed using ANOVA variance analysis and post hoc tests. The correlation coefficients were determined using parametric correlation matrices (Pearson's test) or a nonparametric correlation (Spearman's test).

5. Conclusions

We indicate, for the first time, that the pattern of expression of selected *TaCKX* GFMs and *TaNAC2-5A*, and grain yield in wheat, is paternally inherited by the F_2 generation. Pater-origin transmission of gene expression levels sheds new light on the method of parent selection and crossing to obtain high-yielding phenotypes. We also showed which genes cooperate together by upregulation or downregulation and which function in the opposite manner in establishing yield-related traits. This knowledge can be applied to select the desirable phenotype in F_2. For example, a high-yielding paternal parent with downregulated, compared to the maternal parent, expression of *TaCKX2.1* and *TaCKX11* in 7 DAP spikes and upregulated expression of *TaCKX3* and *TaCKX8* and downregulated *TaNAC2-5A* in seedling roots is expected to transmit this pattern of expression to F_2, which will result in a high yield. The main problem is the antagonistic expression patterns of genes for some important yield-related traits, such as grain number, grain yield, and spike number, to TGW, which is the result of the feedback mechanism of the regulation of expression between *TaCKX1* and *TaCKX2* genes and others. The expression analysis of *TaNAC2-5A* and the in silico analysis of *TaNAC* GFMs revealed that the encoded proteins participate in the regulation of transcription of selected *TaCKX* genes responsible for cytokinin degradation and *TaIPT* genes responsible for cytokinin biosynthesis. Therefore, *TaNACs* are important additional regulators of yield-related traits in wheat, which should be taken into consideration in wheat breeding.

6. Patents

Nadolska-Orczyk A., Szala K., Dmochowska-Boguta M., Orczyk W. Wzory ekspresji genów jako nowe markery molekularne produktywności zbóż oraz sposób przekazywania wysokiej produktywności I strategia selekcji wysokoplonujących odmian zbóż. (Patterns of gene expression as new molecular markers of cereal productivity and a way of transfer of high yield and the strategy for selecting high-yielding cereal varieties). Patent application filed with the Polish Patent Office (UP RP) 23 January 2023, nr P.443557.

Supplementary Materials: The supporting information can be downloaded at: https://www.mdpi.com/article/10.3390/ijms24098196/s1.

Author Contributions: Conceptualization, A.N.-O., K.S. and W.O.; methodology, K.S. and M.D.-B.; software, K.S. and M.D.-B.; validation, W.O.; formal analysis, K.S., M.D.-B. and A.N.-O.; investigation, K.S. and J.B.; data curation, K.S., M.D.-B. and J.B.; writing—original draft preparation, A.N.-O.; writing—review and editing, A.N.-O. and W.O.; visualization, K.S., M.D.-B., J.B. and A.N.-O.; supervision, A.N.-O.; project administration, A.N.-O.; funding acquisition, A.N.-O. All authors have read and agreed to the published version of the manuscript.

Funding: This research was supported by the Ministry of Agriculture and Rural Development, grant No. 5 PBwPR 4-1-01-4-02, and the Statutory Project of PBAI-NRI. The funding body did not perform a role in the design of the study; the collection, analysis, and interpretation of data; or the writing of the manuscript.

Institutional Review Board Statement: Not applicable.

Informed Consent Statement: Not applicable.

Data Availability Statement: All data generated or analysed during this study are included in this published article [and its Supplementary Materials files].

Acknowledgments: We thank Malgorzata Wojciechowska, Izabela Skuza and Agnieszka Glowacka for excellent technical assistance.

Conflicts of Interest: The authors declare no conflict of interest.

References

1. Shewry, P.R.; Hey, S.J. The contribution of wheat to human diet and health. *Food Energy Secur.* **2015**, *4*, 178–202. [CrossRef] [PubMed]
2. Yoo, M.J.; Liu, X.X.; Pires, J.C.; Soltis, P.S.; Soltis, D.E. Nonadditive Gene Expression in Polyploids. *Annu. Rev. Genet.* **2014**, *48*, 485–517. [CrossRef] [PubMed]
3. Leach, L.J.; Belfield, E.J.; Jiang, C.F.; Brown, C.; Mithani, A.; Harberd, N.P. Patterns of homoeologous gene expression shown by RNA sequencing in hexaploid bread wheat. *BMC Genom.* **2014**, *15*, 276. [CrossRef] [PubMed]
4. Jameson, P.E.; Song, J.C. Cytokinin: A key driver of seed yield. *J. Exp. Bot.* **2016**, *67*, 593–606. [CrossRef]
5. Jablonski, B.; Bajguz, A.; Bocian, J.; Orczyk, W.; Nadolska-Orczyk, A. Genotype-Dependent Effect of Silencing of TaCKX1 and TaCKX2 on Phytohormone Crosstalk and Yield-Related Traits in Wheat. *Int. J. Mol. Sci.* **2021**, *22*, 1494. [CrossRef]
6. Jablonski, B.; Ogonowska, H.; Szala, K.; Bajguz, A.; Orczyk, W.; Nadolska-Orczyk, A. Silencing of TaCKX1 Mediates Expression of Other TaCKX Genes to Increase Yield Parameters in Wheat. *Int. J. Mol. Sci.* **2020**, *21*, 4908. [CrossRef]
7. Jablonski, B.; Szala, K.; Przyborowski, M.; Bajguz, A.; Chmur, M.; Gasparis, S.; Orczyk, W.; Nadolska-Orczyk, A. TaCKX2.2 Genes Coordinate Expression of Other TaCKX Family Members, Regulate Phytohormone Content and Yield-Related Traits of Wheat. *Int. J. Mol. Sci.* **2021**, *22*, 4142. [CrossRef]
8. Szala, K.; Ogonowska, H.; Lugowska, B.; Zmijewska, B.; Wyszynska, R.; Dmochowska-Boguta, M.; Orczyk, W.; Nadolska-Orczyk, A. Different sets of TaCKX genes affect yield-related traits in wheat plants grown in a controlled environment and in field conditions. *BMC Plant Biol.* **2020**, *20*, 1–13. [CrossRef]
9. Nguyen, H.N.; Perry, L.; Kisiala, A.; Olechowski, H.; Emery, R.J.N. Cytokinin activity during early kernel development corresponds positively with yield potential and later stage ABA accumulation in field-grown wheat (*Triticum aestivum* L.). *Planta* **2020**, *252*, 1–16. [CrossRef]
10. Sakakibara, H. Cytokinins: Activity, biosynthesis, and translocation. *Annu. Rev. Plant Biol.* **2006**, *57*, 431–449. [CrossRef]
11. Chen, L.; Zhao, J.Q.; Song, J.C.; Jameson, P.E. Cytokinin dehydrogenase: A genetic target for yield improvement in wheat. *Plant Biotechnol. J.* **2020**, *18*, 614–630. [CrossRef]
12. Chen, L.; Zhao, J.; Song, J.C.; Jameson, P.E. Cytokinin glucosyl transferases, key regulators of cytokinin homeostasis, have potential value for wheat improvement. *Plant Biotechnol. J.* **2021**, *19*, 878–896. [CrossRef]
13. Chen, L.; Jameson, G.B.; Guo, Y.C.; Song, J.C.; Jameson, P.E. The LONELY GUY gene family: From mosses to wheat, the key to the formation of active cytokinins in plants. *Plant Biotechnol. J.* **2022**, *20*, 625–645. [CrossRef]
14. Nguyen, H.N.; Lai, N.; Kisiala, A.B.; Emery, R.J.N. Isopentenyltransferases as master regulators of crop performance: Their function, manipulation, and genetic potential for stress adaptation and yield improvement. *Plant Biotechnol. J.* **2021**, *19*, 1297–1313. [CrossRef]
15. Wang, N.; Chen, J.; Gao, Y.; Zhou, Y.; Chen, M.; Xu, Z.; Fang, Z.; Ma, Y. Genomic analysis of isopentenyltransferase genes and functional characterization of TaIPT8 indicates positive effects of cytokinins on drought tolerance in wheat. *Crop. J.* **2023**, *11*, 46–56. [CrossRef]
16. Bartrina, I.; Otto, E.; Strnad, M.; Werner, T.; Schmulling, T. Cytokinin Regulates the Activity of Reproductive Meristems, Flower Organ Size, Ovule Formation, and Thus Seed Yield in Arabidopsis thaliana. *Plant Cell* **2011**, *23*, 69–80. [CrossRef]
17. Jameson, P.E.; Song, J. Will cytokinins underpin the second 'Green Revolution'? *J. Exp. Bot.* **2020**, *71*, 6872–6875. [CrossRef]
18. Schwarz, I.; Scheirlinck, M.T.; Otto, E.; Bartrina, I.; Schmidt, R.C.; Schmulling, T. Cytokinin regulates the activity of the inflorescence meristem and components of seed yield in oilseed rape. *J. Exp. Bot.* **2020**, *71*, 7146–7159. [CrossRef]
19. Chang, C.; Lu, J.; Zhang, H.P.; Ma, C.X.; Sun, G.L. Copy Number Variation of Cytokinin Oxidase Gene Tackx4 Associated with Grain Weight and Chlorophyll Content of Flag Leaf in Common Wheat. *PLoS ONE* **2015**, *10*, e0145970. [CrossRef]
20. Zhang, L.; Zhao, Y.L.; Gao, L.F.; Zhao, G.Y.; Zhou, R.H.; Zhang, B.S.; Jia, J.Z. TaCKX6-D1, the ortholog of rice OsCKX2, is associated with grain weight in hexaploid wheat. *New Phytol.* **2012**, *195*, 74–584. [CrossRef]
21. Lu, J.; Chang, C.; Zhang, H.P.; Wang, S.X.; Sun, G.; Xiao, S.H.; Ma, C.X. Identification of a Novel Allele of TaCKX6a02 Associated with Grain Size, Filling Rate and Weight of Common Wheat. *PLoS ONE* **2015**, *10*, e0144765. [CrossRef] [PubMed]

32. Ramireddy, E.; Hosseini, S.A.; Eggert, K.; Gillandt, S.; Gnad, H.; von Wiren, N.; Schmulling, T. Root Engineering in Barley: Increasing Cytokinin Degradation Produces a Larger Root System, Mineral Enrichment in the Shoot and Improved Drought Tolerance. *Plant Physiol.* **2018**, *177*, 1078–1095. [CrossRef] [PubMed]
33. Werner, T.; Nehnevajova, E.; Kollmer, I.; Novak, O.; Strnad, M.; Kramer, U.; Schmulling, T. Root-Specific Reduction of Cytokinin Causes Enhanced Root Growth, Drought Tolerance, and Leaf Mineral Enrichment in Arabidopsis and Tobacco. *Plant Cell* **2010**, *22*, 3905–3920. [CrossRef] [PubMed]
34. Iqbal, A.; Bocian, J.; Hameed, A.; Orczyk, W.; Nadolska-Orczyk, A. Cis-Regulation by NACs: A Promising Frontier in Wheat Crop Improvement. *Int. J. Mol. Sci.* **2022**, *23*, 5431. [CrossRef]
35. He, X.; Qu, B.Y.; Li, W.J.; Zhao, X.Q.; Teng, W.; Ma, W.Y.; Ren, Y.Z.; Li, B.; Li, Z.S.; Tong, Y.P. The Nitrate-Inducible NAC Transcription Factor TaNAC2-5A Controls Nitrate Response and Increases Wheat Yield. *Plant Physiol.* **2015**, *169*, 991–2005. [CrossRef]
36. Zhao, D.; Derkx, A.P.; Liu, D.C.; Buchner, P.; Hawkesford, M.J. Overexpression of a NAC transcription factor delays leaf senescence and increases grain nitrogen concentration in wheat. *Plant Biol.* **2015**, *17*, 904–913. [CrossRef]
37. Gimenez, E.; Benavente, E.; Pascual, L.; Garcia-Sampedro, A.; Lopez-Fernandez, M.; Vazquez, J.F.; Giraldo, P. An F-2 Barley Population as a Tool for Teaching Mendelian Genetics. *Plants* **2021**, *10*, 694. [CrossRef]
38. Nadolska-Orczyk, A.; Orczyk, W.; Przetakiewicz, A. Agrobacterium-mediated transformation of cereals—From technique development to its application. *Acta Physiol. Plant.* **2000**, *22*, 77–88. [CrossRef]
39. Howells, R.M.; Craze, M.; Bowden, S.; Wallington, E.J. Efficient generation of stable, heritable gene edits in wheat using CRISPR/Cas9. *BMC Plant Biol.* **2018**, *18*, 1–11. [CrossRef]
40. Luo, M.; Li, H.Y.; Chakraborty, S.; Morbitzer, R.; Rinaldo, A.; Upadhyaya, N.; Bhatt, D.; Louis, S.; Richardson, T.; Lahaye, T.; et al. Efficient TALEN-mediated gene editing in wheat. *Plant Biotechnol. J.* **2019**, *17*, 2026–2028. [CrossRef]
41. Hudzieczek, V.; Hobza, R.; Capal, P.; Safar, J.; Dolezel, J. If Mendel Was Using CRISPR: Genome Editing Meets Non-Mendelian Inheritance. *Adv. Funct. Mater.* **2022**, *32*, 2202585. [CrossRef]
42. Cortes, A.J.; This, D.; Chavarro, C.; Madrinan, S.; Blair, M.W. Nucleotide diversity patterns at the drought-related DREB2 encoding genes in wild and cultivated common bean (*Phaseolus vulgaris* L.). *Theor. Appl. Genet.* **2012**, *125*, 1069–1085. [CrossRef]
43. Blair, M.W.; Cortes, A.J.; This, D. Identification of an ERECTA gene and its drought adaptation associations with wild and cultivated common bean. *Plant Sci.* **2016**, *242*, 250–259. [CrossRef]
44. Zemlyanskaya, E.V.; Dolgikh, V.A.; Levitsky, V.G.; Mironova, V. Transcriptional regulation in plants: Using omics data to crack the cis-regulatory code. *Curr. Opin. Plant Biol.* **2021**, *63*, 102058. [CrossRef]
45. Fagny, M.; Austerlitz, F. Polygenic Adaptation: Integrating Population Genetics and Gene Regulatory Networks. *Trends Genet.* **2021**, *37*, 631–638. [CrossRef]
46. Batista, R.A.; Kohler, C. Genomic imprinting in plants-revisiting existing models. *Gene Dev.* **2020**, *34*, 24–36. [CrossRef]
47. Rodrigues, J.A.; Zilberman, D. Evolution and function of genomic imprinting in plants. *Gene Dev.* **2015**, *29*, 2517–2531. [CrossRef]
48. Waters, A.J.; Bilinski, P.; Eichten, S.R.; Vaughn, M.W.; Ross-Ibarra, J.; Gehring, M.; Springer, N.M. Comprehensive analysis of imprinted genes in maize reveals allelic variation for imprinting and limited conservation with other species. *Proc. Natl. Acad. Sci. USA* **2013**, *110*, 19639–19644. [CrossRef]
49. Rodrigues, J.A.; Hsieh, P.H.; Ruan, D.L.; Nishimura, T.; Sharma, M.K.; Sharma, R.; Ye, X.Y.; Nguyen, N.D.; Nijjar, S.; Ronald, P.C.; et al. Divergence among rice cultivars reveals roles for transposition and epimutation in ongoing evolution of genomic imprinting. *Proc. Natl. Acad. Sci. USA* **2021**, *118*, e2104445118. [CrossRef]
50. Robert, H.S.; Park, C.; Gutierrez, C.L.; Wojcikowska, B.; Pencik, A.; Novak, O.; Chen, J.Y.; Grunewald, W.; Dresselhaus, T.; Friml, J.; et al. Maternal auxin supply contributes to early embryo patterning in Arabidopsis. *Nat. Plants* **2018**, *4*, 548–553. [CrossRef]
51. Zhao, P.; Zhou, X.M.; Shen, K.; Liu, Z.Z.; Cheng, T.H.; Liu, D.N.; Cheng, Y.B.; Peng, X.B.; Sun, M.X. Two-Step Maternal-to-Zygotic Transition with Two-Phase Parental Genome Contributions. *Dev. Cell* **2019**, *49*, 882. [CrossRef] [PubMed]
52. Phillips, A.R.; Evans, M.M.S. Maternal regulation of seed growth and patterning in flowering plants. *Curr. Top Dev. Biol.* **2020**, *140*, 257–282. [PubMed]
53. Olsen, O.A. The Modular Control of Cereal Endosperm Development. *Trends Plant Sci.* **2020**, *25*, 279–290. [CrossRef] [PubMed]
54. Cheng, X.; Pan, M.; Zhiguo, E.; Zhou, Y.; Niu, B.; Chen, C. The maternally expressed polycomb group gene OsEMF2a is essential for endosperm cellularization and imprinting in rice. *Plant Commun.* **2021**, *2*, 100092. [CrossRef]
55. Jia, D.; Chen, L.G.; Yin, G.; Yang, X.; Gao, Z.; Guo, Y.; Sun, Y.; Tang, W. Brassinosteroids regulate outer ovule integument growth in part via the control of inner no outer by brassinozole-resistant family transcription factors. *J. Integr. Plant Biol.* **2020**, *62*, 1093. [CrossRef]
56. Lafon-Placette, C.; Hatorangan, M.R.; Steige, K.A.; Cornille, A.; Lascoux, M.; Slotte, T.; Kohler, C. Paternally expressed imprinted genes associate with hybridization barriers in Capsella. *Nat. Plants* **2018**, *4*, 352. [CrossRef]
57. Dai, D.W.; Mudunkothge, J.S.; Galli, M.; Char, S.N.A.; Davenport, R.; Zhou, X.J.; Gustin, J.L.; Spielbauer, G.; Zhang, J.Y.; Barbazuk, W.B.; et al. Paternal imprinting of dosage-effect defective1 contributes to seed weight xenia in maize. *Nat. Commun.* **2022**, *13*, 1–11. [CrossRef]
58. Long, J.; Walker, J.; She, W.; Aldridge, B.; Gao, H.; Deans, S.; Vickers, M.; Feng, X. Nurse cell–derived small RNAs define paternal epigenetic inheritance in Arabidopsis. *Science* **2021**, *373*, eabh0556. [CrossRef]

49. Gehring, M. Epigenetic dynamics during flowering plant reproduction: Evidence for reprogramming? *New Phytol.* **2019**, *224*, 91–96. [CrossRef]
50. Ono, A.; Kinoshita, T. Epigenetics and plant reproduction: Multiple steps for responsibly handling succession. *Curr. Opin. Plant Biol.* **2021**, *61*, 102032. [CrossRef]
51. Kirkbride, R.C.; Lu, J.; Zhang, C.; Mosher, R.A.; Baulcombe, D.C.; Chen, Z.J. Maternal small RNAs mediate spatial-temporal regulation of gene expression, imprinting, and seed development in Arabidopsis. *Proc. Natl. Acad. Sci. USA* **2019**, *116*, 2761–2766. [CrossRef]
52. Tuteja, R.; McKeown, P.C.; Ryan, P.; Morgan, C.C.; Donoghue, M.T.A.; Downing, T.; O'Connell, M.J.; Spillane, C. Paternally Expressed Imprinted Genes under Positive Darwinian Selection in Arabidopsis thaliana. *Mol. Biol. Evol.* **2019**, *36*, 1239–1253. [CrossRef]
53. Moreno-Romero, J.; Jiang, H.; Santos-Gonzalez, J.; Kohler, C. Parental epigenetic asymmetry of PRC2-mediated histone modifications in the Arabidopsis endosperm. *EMBO J.* **2016**, *35*, 1298–1311. [CrossRef]
54. Quadrana, L.; Colot, V. Plant Transgenerational Epigenetics. *Annu. Rev. Genet.* **2016**, *50*, 467–491. [CrossRef]
55. Zalewski, W.; Galuszka, P.; Gasparis, S.; Orczyk, W.; Nadolska-Orczyk, A. Silencing of the HvCKX1 gene decreases the cytokinin oxidase/dehydrogenase level in barley and leads to higher plant productivity. *J. Exp. Bot.* **2010**, *61*, 1839–1851. [CrossRef]
56. Gasparis, S.; Przyborowski, M.; Kala, M.; Nadolska-Orczyk, A. Knockout of the HvCKX1 or HvCKX3 Gene in Barley (*Hordeum vulgare* L.) by RNA-Guided Cas9 Nuclease Affects the Regulation of Cytokinin Metabolism and Root Morphology. *Cells* **2019**, *8*, 782. [CrossRef]
57. Ashikari, M.; Sakakibara, H.; Lin, S.Y.; Yamamoto, T.; Takashi, T.; Nishimura, A.; Angeles, E.R.; Qian, Q.; Kitano, H.; Matsuoka, M. Cytokinin oxidase regulates rice grain production. *Science* **2005**, *309*, 741–745. [CrossRef]
58. Ogonowska, H.; Barchacka, K.; Gasparis, S.; Jablonski, B.; Orczyk, W.; Dmochowska-Boguta, M.; Nadolska-Orczyk, A. Specificity of expression of TaCKX family genes in developing plants of wheat and their co-operation within and among organs. *PLoS ONE* **2019**, *14*, e0214239. [CrossRef]
59. Li, Y.L.; Song, G.Q.; Gao, J.; Zhang, S.J.; Zhang, R.Z.; Li, W.; Chen, M.L.; Liu, M.; Xia, X.C.; Risacher, T.; et al. Enhancement of grain number per spike by RNA interference of cytokinin oxidase 2 gene in bread wheat. *Hereditas* **2018**, *155*, 1–8. [CrossRef]
60. Zhang, W.; Peng, K.X.; Cui, F.B.; Wang, D.L.; Zhao, J.Z.; Zhang, Y.J.; Yu, N.N.; Wang, Y.Y.; Zeng, D.L.; Wang, Y.H.; et al. Cytokinin oxidase/dehydrogenase OsCKX11 coordinates source and sink relationship in rice by simultaneous regulation of leaf senescence and grain number. *Plant Biotechnol. J.* **2020**, *19*, 335–350. [CrossRef]
61. Li, W.J.; He, X.; Chen, Y.; Jing, Y.F.; Shen, C.C.; Yang, J.B.; Teng, W.; Zhao, X.Q.; Hu, W.J.; Hu, M.Y.; et al. A wheat transcription factor positively sets seed vigour by regulating the grain nitrate signal. *New Phytol.* **2020**, *225*, 1667–1680. [CrossRef] [PubMed]
62. Mao, C.J.; He, J.M.; Liu, L.N.; Deng, Q.M.; Yao, X.F.; Liu, C.M.; Qiao, Y.L.; Li, P.; Ming, F. OsNAC2 integrates auxin and cytokinin pathways to modulate rice root development. *Plant Biotechnol. J.* **2020**, *18*, 429–442. [CrossRef] [PubMed]
63. Frebort, I.; Sebela, M.; Galuszka, P.; Werner, T.; Schmulling, T.; Pec, P. Cytokinin oxidase/cytokinin dehydrogenase assay: Optimized procedures and applications. *Anal. Biochem.* **2002**, *306*, 1–7. [CrossRef] [PubMed]
64. Bradford, M.M.; Williams, W.L. New, Rapid, Sensitive Method for Protein Determination. *Fed. Proc.* **1976**, *35*, 274.

Disclaimer/Publisher's Note: The statements, opinions and data contained in all publications are solely those of the individual author(s) and contributor(s) and not of MDPI and/or the editor(s). MDPI and/or the editor(s) disclaim responsibility for any injury to people or property resulting from any ideas, methods, instructions or products referred to in the content.

Article

OsLPR5 Encoding Ferroxidase Positively Regulates the Tolerance to Salt Stress in Rice

Juan Zhao [1,†], Xin Meng [1,†], Zhaonian Zhang [1], Mei Wang [2], Fanhao Nie [1] and Qingpo Liu [1,*]

[1] The Key Laboratory for Quality Improvement of Agricultural Products of Zhejiang Province, College of Advanced Agricultural Sciences, Zhejiang A&F University, Hangzhou 311300, China; zhaojuan521321@163.com (J.Z.); mx42032@stu.zafu.edu.cn (X.M.); zjnlnxy193@163.com (Z.Z.)
[2] Institute of Horticulture, Zhejiang Academy of Agricultural Sciences, Hangzhou 310021, China
* Correspondence: liuqp@zafu.edu.cn
† These authors contributed equally to this work.

Abstract: Salinity is a major abiotic stress that harms rice growth and productivity. Low phosphate roots (LPRs) play a central role in Pi deficiency-mediated inhibition of primary root growth and have ferroxidase activity. However, the function of LPRs in salt stress response and tolerance in plants remains largely unknown. Here, we reported that the *OsLPR5* was induced by NaCl stress and positively regulates the tolerance to salt stress in rice. Under NaCl stress, overexpression of *OsLPR5* led to increased ferroxidase activity, more green leaves, higher levels of chlorophyll and lower MDA contents compared with the WT. In addition, OsLPR5 could promote the accumulation of cell osmotic adjustment substances and promote ROS-scavenging enzyme activities. Conversely, the mutant *lpr5* had a lower ferroxidase activity and suffered severe damage under salt stress. Moreover, knock out of *OsLPR5* caused excessive Na^+ levels and Na^+/K^+ ratios. Taken together, our results exemplify a new molecular link between ferroxidase and salt stress tolerance in rice.

Keywords: rice; salinity; *OsLPR5*; ferroxidase; stress

Citation: Zhao, J.; Meng, X.; Zhang, Z.; Wang, M.; Nie, F.; Liu, Q. *OsLPR5* Encoding Ferroxidase Positively Regulates the Tolerance to Salt Stress in Rice. *Int. J. Mol. Sci.* **2023**, *24*, 8115. https://doi.org/10.3390/ijms24098115

Academic Editor: Daniela Trono

Received: 17 March 2023
Revised: 14 April 2023
Accepted: 26 April 2023
Published: 30 April 2023

Copyright: © 2023 by the authors. Licensee MDPI, Basel, Switzerland. This article is an open access article distributed under the terms and conditions of the Creative Commons Attribution (CC BY) license (https://creativecommons.org/licenses/by/4.0/).

1. Introduction

Soil salinity is a global environmental problem and an important abiotic stress factor that limits the germination, growth and productivity of plants [1]. High concentrations of salt affect various plant physiological and biochemical processes, causing ionic toxicity, osmotic stress, ROS accumulation and nutritional imbalance [2]. Rice is one of the principal cereal crops for the world's population and is salt-sensitive in the seedling and reproductive stages [3]. To ensure food security, we need to combat the rising threat of soil salinity and develop and grow grain output under saline conditions [4,5]. However, the research on salt tolerance started relatively late in rice compared with in Arabidopsis, and although numerous salt-responsive genes have been identified, very few have been successfully applied in rice [6,7]. Genetic studies aimed at identifying genes underlying salt tolerance and elucidating the corresponding mechanisms are becoming more urgent [8].

Plants have evolved various systems to sense and adapt to a high salinity environment. With salt stress, plants absorb excess Na^+ from soil solution into their roots, which moves sequentially to the shoots and throughout the leaves, ultimately inhibiting the absorption of other nutrients such as K^+ [9,10]. To maintain cellular Na^+/K^+ homeostasis, a series of Na^+/H^+ antiporter, high-affinity K^+ transporters (HKT) and Cl^- channels help the Na^+ efflux, uptake restriction and compartmentalize into the vacuole [2,11–13]. Proline biosynthesis-related genes such as *OsP5CS* and *OsP5CR* and trehalose-6-phosphate synthase/phosphatase-related genes such as *OsTPP1* and *OsTPS1* can promote the accumulation of osmolytes and protect rice from osmotic stress under saline conditions in rice [14,15]. In addition, plants have enzymatic scavengers and nonenzymatic antioxidants to mitigate oxidative stress caused by ROS accumulation. Enzymatic scavengers include superoxide dismutase (SOD), ascorbate peroxidase (APX), catalase (CAT), glutathione peroxidases (GPXs), glutathione peroxidase (GR), glutaredoxin (GRX), glutathione S-transferase (GST) and respiratory burst oxidase homologs (RBOHs) [6,16,17]. Ascorbic acid, glutathione, carotenoids, flavonoids, alkaloids, phenolic compounds and tocopherol are nonenzymatic antioxidants [1]. In addition to the above physiological mechanisms, there are many functional proteins, transcription factors, hormones, etc., involved in signaling and resisting salt stress.

Low phosphate roots (LPRs) encode the multicopper oxidase domain-containing protein, named for its function in Pi deficiency-mediated inhibition of primary root growth [18]. In Arabidopsis, LPR1 interacts genetically with the P5-type ATPase PDR2 in the endoplamic reticulum (ER), and it plays an opposite role in mediating root meristem growth responses to Pi and Fe availability [19]. Furthermore, *LPR1* encodes a cell-wall-targeted ferroxidase and function in iron-dependent callose deposition in low Pi conditions [20]. There are five homologs of LPR1 in rice (OsLPR1-5), and significant increases in the relative expression levels of OsLPR3 and OsLPR5 could be triggered under Pi deficiency [21]. Similar to Arabidopsis, it was reported that OsLPR5 was located in the ER and the cell wall, had ferroxidase activity and was required for normal growth and maintenance of phosphate homeostasis in rice [22]. Moreover, the concentration of Fe (III) and total Fe were increased in the roots and shoots of OsLPR5-overexpressing plants [22]. Therefore, OsLPR5 has a broad spectrum influence on Fe homeostasis and plant development. However, little is known about LPRs functions in abiotic stress response and tolerance.

Although the important roles of OsLPR5 have been reported in plants, its function in salt stress response and tolerance has yet to be defined. In this study, we identified the expression level of OsLPR5 and ferroxidase activity under NaCl stress. The physiological roles in salt stress and tolerance were further identified through overexpression and CRISPR/Cas9-mediated mutation of *OsLPR5*. We have uncovered a previously unknown role that *OsLPR5* positively regulates the salt stress tolerance in rice.

2. Results

2.1. OsLPR5 Is Induced under Salt Stress and Exhibits Higher Expression in Vegetative Organs

To determine the effect of salt stress on the relative transcriptional level of *OsLPR5*, two-week-old wild-type seedlings were treated for 48 h with 120 mM NaCl. The results showed that salt stress triggered a significant induction of *OsLPR5* in the roots and shoots (Figure 1A,B). Furthermore, since *OsLPR5* encodes a ferroxidase, the ferroxidase activity in wild-type seedlings was also measured with or without NaCl treatment, showing that the activity was significantly induced by salt stress (Figure 1C). The above results suggested that *OsLPR5* may play a role in response to salt stress. We examined the expression patterns of *OsLRP5* by qRT-PCR in various wild-type tissues, and it was found that the transcriptional level of *OsLPR5* could be determined both in vegetative and reproductive organs, but was far higher in vegetative organs in the tillering stage (Figure 1D).

2.2. Overexpression of OsLPR5 Increased Tolerance to Salinity Stress in Seedling Stage

To investigate the functional role of *OsLPR5* in regulating the salt stress response, we constructed a knock-out mutant by CRISPR/Cas9 and transgenic lines overexpressing *OsLPR5*. Two independent mutants (*lpr5-1* and *lpr5-2*) were generated and their editing sites were identified (Figure 2A). The expression levels of *OsLPR5* in the overexpressing lines (#1, #2 and #3) were increased by more than 1000-fold compared with the wild type, as determined by qRT-PCR (Figure 2B). Two overexpressing lines (#1 and #3) were then selected for further study. Consistently, the ferroxidase activities extracted from the plant total protein in overexpressing seedlings were significantly elevated compared to the WT. In contrast, *lpr5-1* and *lpr5-2* showed lower ferroxidase activities compared to the WT (Figure 2C). Then, the overexpressing seedlings (#1 and #3) were treated with 100, 120, 150 and 200 mM NaCl for 7 days to investigate whether *OsLPR5* is involved in affecting salt tolerance (Figures S1 and S2). It was found that OE-*OsLPR5* seedlings suffered less damage, had more green leaves and were less sensitive to salt stress compared with the WT at different levels of salt stress (Figures S1 and S2).

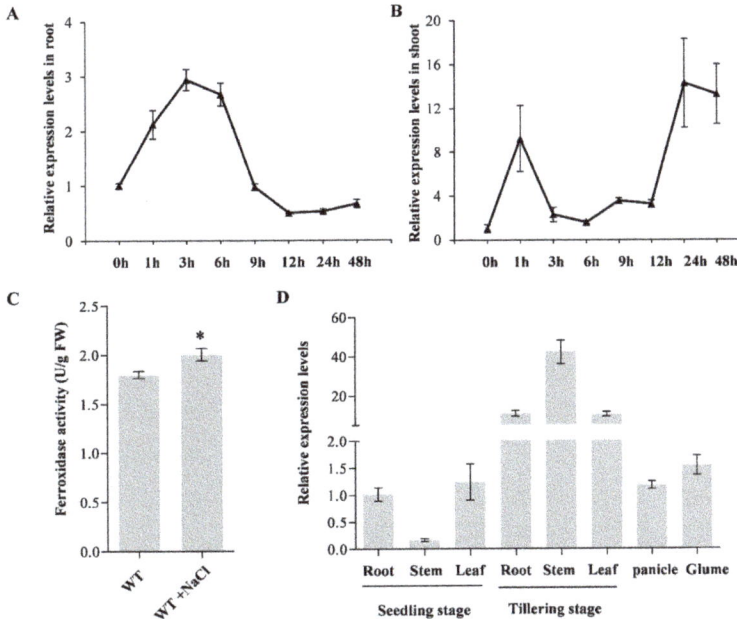

Figure 1. Expression pattern of *OsLPR5*. Time course of *OsLPR5* under salt stress conditions (120 mM NaCl) in roots (**A**) and shoots (**B**) at the seedling stage via qPCR analysis. (**C**) The ferroxidase activity of total protein from wild type (WT) under normal and salt stress conditions. Asterisks represent statistical difference at * $p < 0.05$. (**D**) Expression patterns of *OsLPR5* in different tissues containing root, stem, leaf panicle and glume by RT-qPCR. Data are shown as means ± SD ($n = 3$).

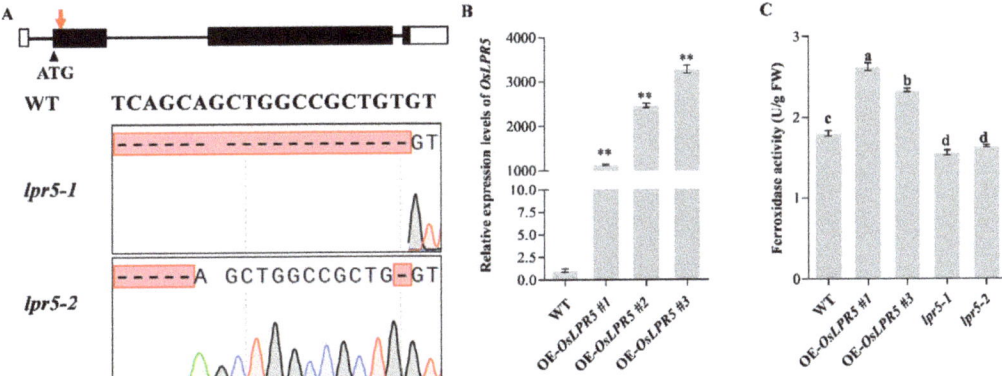

Figure 2. Identification of *lpr5* mutants and overexpressed lines. (**A**) Sequence analysis of *lpr5-1* and *lpr5-2* mutants generated by CRISPR/Cas9. The red arrow indicates the target site. (**B**) Expression analysis of the *OsLPR5* gene in overexpressed transgenic lines. Asterisks represent statistical differences at ** $p < 0.01$. (**C**) The ferroxidase activity assay in WT and *OsLPR5* transgenic seedlings. Equal amounts of total protein from leaves of WT, OE-*OsLPR5* and *lpr5* seedlings were incubated with the substrate $Fe(NH_4)_2(SO_4)_2 \cdot 6H_2O$. Data are shown as the means ± S.D ($n = 3$). Significant differences ($p < 0.05$) are indicated by different letters.

OE-*OsLPR5* (#1 and #3), *lpr5-1*, *lpr5-2* and WT seedlings were subsequently stressed with 120 mM NaCl for further phenotypic observation. Notably, the results showed that *lpr5-1* and *lpr5-2* had significantly reduced fresh weights and seedling heights after salt stress compared with the WT (Figure 3A–C). Moreover, both *lpr5-1* and *lpr5-2* contained extremely lower levels of total chlorophyll and much higher levels of an indicator of membrane lipid peroxidation (malondialdehyde (MDA)) under salt treatment, which indicated that the mutation of *OsLPR5* caused an increased sensitivity to salt stress (Figure 3D,E). In contrast, overexpressing *OsLPR5* in rice increased the tolerance to salt stress, with a lower fresh weight loss, seedling height inhibition and MDA content and higher levels of chlorophyll compared with the WT under NaCl treatment (Figure 3A–E). Collectively, the above results suggest that *OsLPR5* regulates the salinity stress tolerance by acting as a new positive regulator in rice.

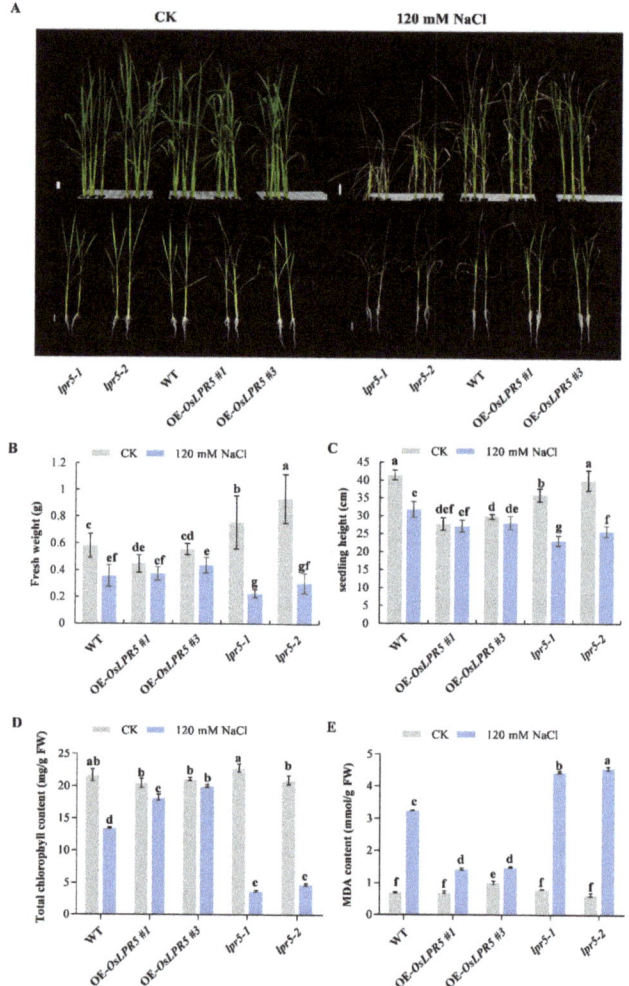

Figure 3. *OsLPR5* contributes to the salinity tolerance of rice. (**A**) Phenotype of wild-type and *OsLPR5* transgenic seedlings exposed to salinity stress for 7 days. Scale bars, 2 cm. (**B–E**) Measurement of fresh weight (**B**), seedling height (**C**), the total chlorophyll (**D**) and MDA (**E**) contents in wild-type and *OsLPR5* transgenic seedlings exposed to salinity stress for 7 days. Data are shown as means ± S.D (n = 3). Different letters indicate significant differences between means as determined using an ANOVA followed by Duncan's test ($p < 0.05$).

2.3. OsLPR5 May Influence Osmotic Adjustment and ROS-Scavenging Enzyme Activities

Upon exposure to salt stress, plants accumulate compatible osmolytes and promote the biosynthesis of a set of ROS-scavenging systems to resist osmotic and oxidative stress [1]. Therefore, we determined the contents of osmolytes, including soluble protein and proline, in plants. Under normal conditions, there were no significant differences among OE-*OsLPR5* (#1 and #3), *lpr5-1*, *lpr5-2* and WT seedlings. Under NaCl treatment, *lpr5-1* and *lpr5-2* contained markedly lower soluble protein and proline than OE-*OsLPR5* and WT (Figure 4). Moreover, the OE-*OsLPR5* (#1 and #3) seedlings accumulated much higher levels of proline content than the WT (Figure 4). Consistent with this, the ROS scavenging enzyme activities in *lpr5-1* and *lpr5-2*, including superoxide dismutase (SOD), ascorbate peroxidase (APX) and catalase (CAT), were also significantly lower than OE-*OsLPR5* and WT, whereas these enzyme activities were higher in OE-*OsLPR5* compared to *lpr5-1*, *lpr5-2* and WT seedlings (Figure 4). Taken together, OsLPR5 appeared to promote the accumulation of cell osmotic adjustment substances and promote ROS-scavenging enzyme activities in response to osmotic and oxidative stress induced by salt stress in rice.

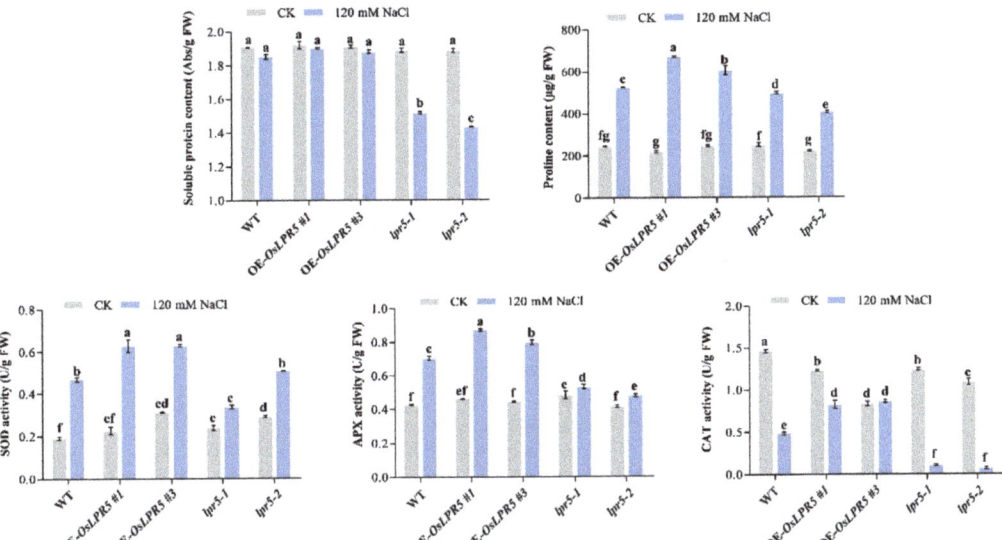

Figure 4. Measurement of the content of soluble protein, proline and ROS-scavenging enzyme activities of SOD, APX and CAT in leaves of wild-type, OE-*OsLPR5* and *lpr5* transgenic seedlings under normal conditions or exposure to salt stress for 7 days. Different letters indicate significant differences between means as determined using an ANOVA followed by Duncan's test ($p < 0.05$).

2.4. Mutation of OsLPR5 Mainly Influence Na$^+$ Levels under NaCl Stress

Excessive Na$^+$ uptake and altered Na$^+$/K$^+$ homeostasis in shoots lead to leaf damage when exposed to high salt concentrations. Thus, the contents of Na$^+$ and K$^+$ were measured in the rice shoots with or without NaCl treatment. Under normal conditions, there were no significant differences in Na$^+$ content among the plants (Figure 5). The Na$^+$ levels of all seedlings tested were dramatically increased after 120 mM NaCl treatment, especially in *lpr5-1* and *lpr5-2* (Figure 5). However, except for the higher K$^+$ level of *lpr5-2*, the contents in other plants were not much different, which suggested that OsLPR5 may not directly regulate K$^+$ homeostasis. Similar to the Na$^+$ levels, the Na$^+$/K$^+$ ratios also showed a sharp increase under salt stress, especially in *lpr5-1* and *lpr5-2* (Figure 5). These results imply that mutation of OsLPR5 mainly causes excessive Na$^+$ levels, Na$^+$/K$^+$ ratios and salt sensitivity.

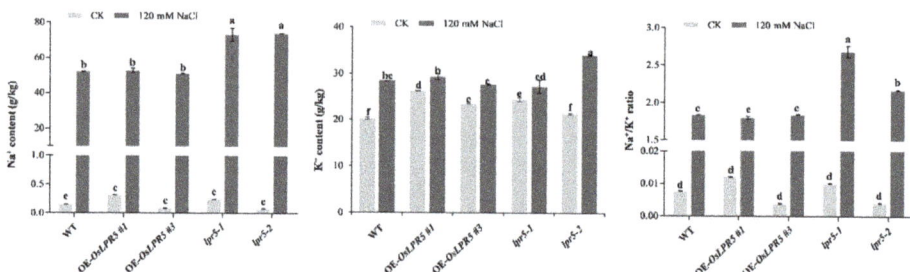

Figure 5. Measurement of the Na$^+$ and K$^+$ contents and Na$^+$/K$^+$ ratio in leaves of wild-type, OE-*OsLPR5* and *lpr5* transgenic seedlings under normal conditions or exposure to salt stress for 7 days. Different letters indicate significant differences between means as determined using an ANOVA followed by Duncan's test ($p < 0.01$).

2.5. OsLPR5 Significantly Changes the Expression Levels of Salt Stress and Iron Homeostasis-Related Genes in Rice

Rice will suffer from osmotic stress, oxidative stress and excessive Na$^+$ uptake when exposed to high salt concentrations. Under NaCl treatment, *lpr5* contained markedly lower proline levels and much higher levels of Na$^+$ than OE-*OsLPR5* and WT (Figures 4 and 5). Therefore, we evaluated the relative expression levels of the key proline-synthesis-related gene (*OsP5CS1*) and high-affinity sodium transporter (*OsHKT2;1*) in WT, OE-*OsLPR5* and *lpr5* seedlings under normal and NaCl treatment conditions by qRT-PCR [23–25]. The results showed that the expression levels of *OsP5CS1* and *OsHKT2;1* were significantly upregulated in WT and OE-*OsLPR5* after NaCl treatment, but downregulated in *lpr5* seedlings (Figure 6). Moreover, the expression of these two genes was most significantly upregulated in the overexpressing seedlings, which is consistent with their increased salt tolerance (Figure 6). Fe is essential for the synthesis of ferredoxins and other redox-related proteins participating in a variety of physiological activities in plants [9,26]. Ai et al. reported that overexpression of *OsLPR5* elevated the concentrations of Fe (III) in the xylem sap and total Fe in rice [22]. In grasses, the acquisition and mobilization of Fe (III) mainly occurs through a chelation-based approach (Strategy II). The YS1-like (YSL) family have been reported to play important roles in this approach. *OsYSL15*, one of the rice YSL genes played a positive role in Fe uptake and distribution during the first stages of growth [27]. Therefore, the relative expression levels of *OsYSL15* was analyzed in WT, OE-*OsLPR5* and *lpr5* seedlings. Under normal conditions, the transcript level in WT, OE-*OsLPR5* and *lpr5* seedlings exhibited no significant difference. However, its level was remarkably increased in *OsLPR5*-overexpressing seedlings in response to salt stress (Figure 6). We speculated that this was related to the increased ferroxidase activity and concentration of Fe (III) in *OsLPR5*-overexpressing seedlings.

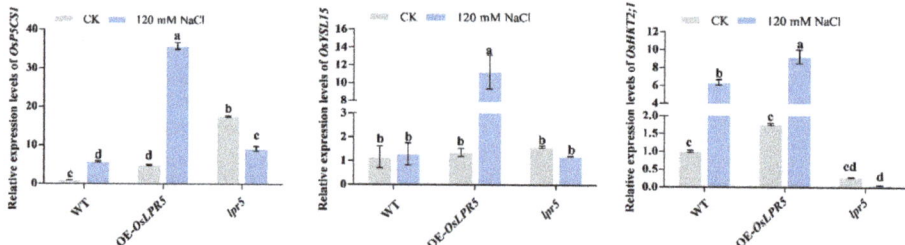

Figure 6. Relative expression levels of stress-related genes in leaves of wild-type, OE-*OsLPR5* and *lpr5* transgenic seedlings under normal conditions or exposure to salt stress for 4 h. Different letters indicate significant differences between means as determined using an ANOVA followed by Duncan's test ($p < 0.05$).

3. Discussion

3.1. OsLPR5 and Ferroxidase Activities Are Induced under Salt Stress

LPRs are highly expressed in roots and play important roles in root meristem development under low Pi in Arabidopsis (*LPR1* and *LPR2*) and rice (*OsLRP5*) [19,22]. Here, we also used qRT-PCR analyses to determine the expression patterns of *OsLRP5* in different tissues, and found consistent higher expression levels in roots and also in the leaves and stems at the tillering stage (Figure 1D). This may be due to the different tissues and periods of the samples tested in our study. The results suggested the broad spectrum influence of *OsLRP5* on plant growth in rice. Therefore, we investigated the relative transcription levels of *OsLRP5* in roots and shoots under NaCl treatment to determine whether this gene was involved in the salt stress response. It showed that salt stress triggered a significant induction of *OsLPR5* both in the roots and shoots, which suggested its possible role in the regulation of salt stress (Figure 1A,B).

Both LPR1 and OsLRP5 have ferroxidase activity in Arabidopsis and rice by catalyzing the oxidation of Fe (II) to Fe (III) using O_2 as a substrate [19,20,22]. In Arabidopsis, the ferroxidase activity in root extracts was up to 5-fold higher in *LPR1*-overexpressing lines compared with the wild type (Col) [20]. By detecting the ferroxidase activity of heterologously expressed purified pGS::GST::OsLPR5 fusion proteins in *E. coli*, OsLPR5 also showed ferroxidase activity [22]. In addition, the activity was significantly higher in the roots of *OsLPR5*-overexpressing lines under Pi-sufficient and Pi-deficient conditions [22]. Consistent with this, our study confirmed that the ferroxidase activity was significantly higher in the shoots of *OsLPR5*-overexpressing lines (#1 and #3) than in the WT (Figure 2C). Moreover, the mutants (*lpr5-1* and *lpr5-2*) showed lower ferroxidase activities compared with the WT (Figure 2C). We further measured the ferroxidase activities in WT at the seedling stage with or without NaCl stress and found a significantly increased activity under NaCl stress (Figure 1C). These results indicated that ferroxidase activity was approximately correlated with OsLPR5 transcription levels.

3.2. OsLPR5 Positively Regulates the Tolerance to Salinity Stress

Salinity stress can cause various physiological and biochemical changes, including ionic toxicity, osmotic stress, ROS accumulation and nutritional imbalance [2]. Enhanced salt-tolerant plants generally have higher chlorophyll and proline contents, lower MDA content and ROS communication and increased ROS-scavenging enzyme activities, for example, in OsR3L1 and OsMADS25 overexpressing seedlings [14,28]. Consistent with this, the *OsLPR5*-overexpressing lines (#1 and #3) had higher fresh weights and chlorophyll contents, lower MDA contents, higher proline contents and increased ROS-scavenging enzyme activities than the WT after NaCl stress. In contrast, the mutants (*lpr5-1* and *lpr5-2*) suffered more severe yellowing, chlorophyll destruction and osmotic stress and had a lower ROS scavenging ability after salt stress (Figures 3 and 4). Consist with this, the transcriptional level of *OsP5CS1* was significantly upregulated in OE-*OsLPR5* after NaCl treatment (Figure 6). These results indicated that OsLPR5 positively regulates the tolerance to salt stress.

An elevated Na^+/K^+ ratio is a characteristic manifestation of ionic toxicity with salt stress. For example, there was a higher K^+/Na^+ ratio and an enhanced salt tolerance in *OsSTAP1* overexpressing lines compared to the WT [29]. In our study, both the Na^+ levels and Na^+/K^+ ratios were elevated in the mutants (*lpr5-1* and *lpr5-2*) compared with the WT, consistent with the salt-hypersensitivity phenotype (Figure 5). Unexpectedly, overexpression of OsLPR5 did not decrease the Na^+ level or Na^+/K^+ ratio after NaCl stress, as they were approximately the same as those in the WT (Figure 5). We speculated that OsLPR5 confers salt tolerance primarily at other physiological levels. The root system is the first tissue to perceive salt stress, and a greater root volume under salt stress in overexpressing *OsAHL1*, *OsHAL3* and *OsMADS25* seedlings exhibited enhanced salt avoidance [1,30]. Similar to this, the primary root length was longer in OsLPR5-overexpressing lines than the WT at the seedling stage (Figure S3). On the other hand, there have been reports that the expression levels of *OsIRO2*, *OsIRT1*, *OsNAS1*, *OsNAS2*, *OsYSL15* and *OsYSL2* were enhanced by a saline-alkaline environment and the plants could acquire Fe more efficiently, thus contributing to a higher accumulation of Fe [6,31]. The increased concentration of Fe (III) and total Fe in the roots and shoots of *OsLPR5*-overexpressing plants may be one of the reasons for its improved salt tolerance [22]. In addition, the transcriptional levels of *OsHKT2;1* and *OsYSL15* were significantly upregulated in OE-*OsLPR5* after NaCl treatment. We speculated that OsLPR5 may influence the salt stress tolerance by regulating the expressions of *OsHKT2;1* and *OsYSL15*. However, the detailed molecular mechanisms and genetic pathways of OsLPR5-mediated salt tolerance remain to be elucidated.

Taking our current results together with those of previous studies, *OsLPR5* is a newly identified gene belong to *LPRs* that significantly improves the salt stress tolerance in rice. We investigated the plant height and thousand-grain weight of OE-*OsLPR5* in the mature stage and found that overexpressing *OsLPR5* causes a slight decrease in both agronomic traits (Figure S4). However, further work is needed to explore the potential use of OsLPR5 in breeding salt-tolerant rice varieties.

4. Materials and Methods

4.1. Experimental Materials and Stress Treatment

The rice variety *japonica* cv. Nipponbare was used as the wild type (WT) and genetic transformation in this study. For salt stress, the sterilized seeds were germinated in water at 32 °C and then grown in culture solution (1.15 mM $(NH_4)_2SO_4$, 0.2 mM NaH_2PO_4, 1 mM $CaCl_2$, 1 mM $MgSO_4$, 0.4 mM K_2SO_4, 0.009 mM $MnCl_2$, 0.075 mM $(NH_4)_6Mo_7O_{24}$, 0.019 mM H_3BO_3, 0.152 mM $ZnSO_4$, 0.155 mM $CuSO_4$ and 0.02 mM Fe-EDTA, pH 5.8) in a growth chamber at 32 °C light/28 °C dark. The 14-day-old seedlings of WT and transgenic lines were treated with 100, 120, 150 and 200 mM NaCl for one week. Seedlings grown in nutrient solution without NaCl served as controls. After treatment, the fresh weight and shoot length of each plant were measured. Each treatment experiment was performed with three replicates with twelve seedlings. Statistically significant differences ($p < 0.05$ or $p < 0.01$) were identified by an ANOVA followed by Duncan's test.

4.2. RNA Isolation and qRT-PCR

Total RNA was extracted from rice tissues using Trizol (Hlingene, Shanghai, China) according to the manufacturer's instructions. For RT-qPCR analyses, 1 μg of RNA sample was depleted of genomic DNA and reverse-transcribed into cDNA using Hifair^{III} 1st Strand cDNA Synthesis SuperMix (YEASEN, Shanghai, China). The Hieff qPCR SYBR Green Master Mix premix (YEASEN, Shanghai, China) was used in qRT-PCR experiments. The reaction was performed with biological triplicates as described previously [32]. The transcription levels were calculated by the $2^{-\Delta\Delta CT}$ values with the expression of the *ubiquitin* gene as the internal control to normalize gene expression data. The primers used in this experiment are listed in Table S1.

4.3. Vector Constructs and Rice Genetic Transformation

To construct the OE-*OsLPR5* plasmid, the coding sequence of *OsLPR5* was cloned and constructed into the pCAMBIA1300-UBI-RBCS vector using the Hieff Clone® Plus One Step Cloning Kit (YEASEN, Shanghai, China). For *lpr5* mutant construction, the CRISPR/Cas9 system was used based on a previous report [33]. The specific CRISPR target site was designed in the protein-encodable region by the CRISPR-GE Software Toolkit (http://crispr.hzau.edu.cn/CRISPR2/, accessed on 16 March 2023) [34]. The annealed 19-bp genomic DNA was ligated into pYLgRNA-OsU3 using T4 ligase (New England Biolabs, Shanghai, China). All constructs were transferred into *Agrobacterium tumefaciens* strain EHA105 and genetically transformed into Nipponbare by the Agrobacterium-mediated transformation method to generate transgenic lines following a previous report [35]. The specific primers are listed in Supplemental Table S1.

4.4. Chlorophyll Content Assay

For total chlorophyll content measurements, 0.1 g of fresh leaves with or without treatment were chopped and put into 10 mL of ethanol/acetone (1:1) mixed solution. Then, pigments were extracted in the dark for 48 h until the leaves turned completely white. Absorbance values at 663 nm and 645 nm were measured with an ultraviolet spectrophotometer. Each sample had three biological replicates. The chlorophyll contents were calculated according to a previously described process [36].

4.5. Physiological Measurements

The free proline content in leaves was measured by the sulfosalicylic acid method according to a previous method [37]. Briefly, 0.2 g of leaves from control and treatment plants were homogenized in solution (3% sulfosalicylic acid/glacial acetic acid/2.5% ninhydrin hydrate 1:1:2), followed by inoculation in a 100 °C water bath for 15 min. After cooling, toluene was added and the UV absorption of the toluene layer at 520 nm was measured. The proline content was calculated according to the standard curve. The soluble protein in leaves was homogenized with a Coomassie brilliant blue solution and measured at 595 nm. For MDA content, 0.2 g of leaves was powdered and homogenized in solution (0.25% TBA dissolved in 10% TCA). After centrifugation, the supernatant was inoculated in a 95 °C water bath for 15 min. After cooling, the absorptions at 532 nm, 600 nm and 450 nm were measured [38].

The antioxidant enzyme activities were measured according to a previous method [14,39]. Briefly, fresh leaves from control and treatment plants were ground by liquid nitrogen and then homogenized with potassium phosphate buffer (pH 7.2). For analyses of APX activity, 5 mM vitamin C (AsA) was added and the absorption at 290 nm was measured. The APX activity was determined by the degradation rate of AsA. The total CAT activity was measured by the rate of decomposition of H_2O_2 at 240 nm. The SOD activity was assayed by monitoring the percentage of inhibition of the pyrogallol autoxidation at 320 nm. Statistically significant differences ($p < 0.05$) were identified by an ANOVA followed by Duncan's test.

4.6. Quantification of Na^+ and K^+ Concentrations

The shoots of rice seedlings from WT and transgenic plants with or without treatment were collected for the determination ion concentrations. All the samples were oven-dried at 105 °C for 30 min and 75 °C for 48 h. The Na^+ and K^+ concentrations of each sample were determined according to a previous report [40]. Statistically significant differences ($p < 0.01$) were identified by an ANOVA followed by Duncan's test.

4.7. Determination of Ferroxidase Activity

To determine the ferroxidase activity of plants, total plant proteins were extracted from seedlings of the wild type and transgenic plants as described previously [41]. Samples were ground with liquid nitrogen and immersed in extracted solution containing a Protease inhibitor cocktail. Then, the total plant proteins were obtained by centrifugating twice at 4 °C and collecting the supernatant. For ferroxidase assays, the reaction contained 1050 µL buffer (450 mM Na-acetate (pH 5.8), 100 mM $CuSO_4$), 225 µL substrate (357 mM $Fe(NH_4)_2(SO_4)_2 \cdot 6H_2O$, 100 mM $CuSO_4$) and 30 µL plant protein. The solution was mixed gently and 16 µL 18 mM ferrozine was added to quench the rection at regular intervals. The reaction mixture without substrate was used as control. The activity of ferroxidase was assayed by the rate of Fe^{2+} oxidation, which was calculated from the decreased absorbance at 560 nm according to the method in [20,22].

Supplementary Materials: The supporting information can be downloaded at: https://www.mdpi.com/article/10.3390/ijms24098115/s1.

Author Contributions: Q.L., J.Z. and M.W. designed the experiments and wrote the manuscript; X.M., J.Z., Z.Z. and F.N. performed the experiments; J.Z., Z.Z. and X.M. analyzed the data. All authors have read and agreed to the published version of the manuscript.

Funding: This work was supported by the National Natural Science Foundation of China (31972959, 32201697 and 32101658), the Zhejiang Provincial Natural Science Foundation of China (LQ21C130001) and the Key Project of Zhejiang Provincial Natural Science Foundation of China (LZ19B070001).

Institutional Review Board Statement: Not applicable.

Informed Consent Statement: Not applicable.

Data Availability Statement: The original data of this present study are available from the corresponding authors.

Conflicts of Interest: The authors declare no conflict of interest.

References

1. Liu, C.; Mao, B.; Yuan, D.; Chu, C.; Duan, M. Salt tolerance in rice: Physiological responses and molecular mechanisms. *Crop J.* **2022**, *10*, 13–25. [CrossRef]
2. Wang, R.; Jing, W.; Xiao, L.; Jin, Y.; Shen, L.; Zhang, W. The Rice High-Affinity Potassium Transporter1;1 Is Involved in Salt Tolerance and Regulated by an MYB-Type Transcription Factor. *Plant Physiol.* **2015**, *168*, 1076–1090. [CrossRef] [PubMed]
3. Razzaq, A.; Ali, A.; Safdar, L.; Zafar, M.M.; Rui, Y.; Shakeel, A.; Shaukat, A.; Ashraf, M.; Gong, W.; Yuan, Y. Salt stress induces physiochemical alterations in rice grain composition and quality. *J. Food Sci.* **2020**, *85*, 14–20. [CrossRef] [PubMed]
4. Ogawa, D.; Abe, K.; Miyao, A.; Kojima, M.; Sakakibara, H.; Mizutani, M.; Morita, H.; Toda, Y.; Hobo, T.; Sato, Y.; et al. RSS1 regulates the cell cycle and maintains meristematic activity under stress conditions in rice. *Nat. Commun.* **2011**, *2*, 278. [CrossRef]
5. Liu, J.; Shabala, S.; Zhang, J.; Ma, G.; Chen, D.; Shabala, L.; Zeng, F.; Chen, Z.; Zhou, M.; Venkataraman, G.; et al. Melatonin improves rice salinity stress tolerance by NADPH oxidase-dependent control of the plasma membrane K(+) transporters and K(+) homeostasis. *Plant Cell Environ.* **2020**, *43*, 2591–2605. [CrossRef]
6. Ganapati, R.K.; Naveed, S.A.; Zafar, S.; Wang, W.; Xu, J. Saline-Alkali Tolerance in Rice: Physiological Response, Molecular Mechanism, and QTL Identification and Application to Breeding. *Rice Sci.* **2022**, *29*, 412–434. [CrossRef]
7. Ganie, S.A.; Molla, K.A.; Henry, R.J.; Bhat, K.V.; Mondal, T.K. Advances in understanding salt tolerance in rice. *Theor. Appl. Genet.* **2019**, *132*, 851–870. [CrossRef]
8. Farhat, S.; Jain, N.; Singh, N.; Sreevathsa, R.; Dash, P.K.; Rai, R.; Yadav, S.; Kumar, P.; Sarkar, A.K.; Jain, A.; et al. CRISPR-Cas9 directed genome engineering for enhancing salt stress tolerance in rice. *Semin. Cell Dev. Biol.* **2019**, *96*, 91–99. [CrossRef]
9. Basu, S.; Kumar, A.; Benazir, I.; Kumar, G. Reassessing the role of ion homeostasis for improving salinity tolerance in crop plants. *Physiol Plant.* **2021**, *171*, 502–519. [CrossRef]
10. Lu, W.; Deng, M.; Guo, F.; Wang, M.; Zeng, Z.; Han, N.; Yang, Y.; Zhu, M.; Bian, H. Suppression of OsVPE3 Enhances Salt Tolerance by Attenuating Vacuole Rupture during Programmed Cell Death and Affects Stomata Development in Rice. *Rice* **2016**, *9*, 65. [CrossRef]
11. Xiao, L.; Shi, Y.; Wang, R.; Feng, Y.; Wang, L.; Zhang, H.; Shi, X.; Jing, G.; Deng, P.; Song, T.; et al. The transcription factor OsMYBc and an E3 ligase regulate expression of a K^+ transporter during salt stress. *Plant Physiol.* **2022**, *190*, 843–859. [CrossRef]
12. Wang, J.; Nan, N.; Li, N.; Liu, Y.; Wang, T.-J.; Hwang, I.; Liu, B.; Xu, Z.-Y. A DNA Methylation Reader-Chaperone Regulator-Transcription Factor Complex Activates OsHKT1;5 Expression during Salinity Stress. *Plant Cell* **2020**, *32*, 3535–3558. [CrossRef]
13. Liu, S.; Zheng, L.; Xue, Y.; Zhang, Q.; Wang, L.; Shou, H. Overexpression of OsVP1 and OsNHX1 Increases Tolerance to Drought and Salinity in Rice. *J. Plant Biol.* **2010**, *53*, 444–452. [CrossRef]
14. Xu, N.; Chu, Y.; Chen, H.; Li, X.; Wu, Q.; Jin, L.; Wang, G.; Huang, J. Rice transcription factor OsMADS25 modulates root growth and confers salinity tolerance via the ABA-mediated regulatory pathway and ROS scavenging. *PLoS Genet.* **2018**, *14*, e1007662. [CrossRef]
15. Joshi, R.; Sahoo, K.K.; Singh, A.K.; Anwar, K.; Pundir, P.; Gautam, R.K.; Krishnamurthy, S.L.; Sopory, S.K.; Pareek, A.; Singla-Pareek, S.L. Enhancing trehalose biosynthesis improves yield potential in marker-free transgenic rice under drought, saline, and sodic conditions. *J. Exp. Bot.* **2020**, *71*, 653–668. [CrossRef]
16. Pandey, M.; Paladi, R.K.; Srivastava, A.K.; Suprasanna, P. Thiourea and hydrogen peroxide priming improved K(+) retention and source-sink relationship for mitigating salt stress in rice. *Sci. Rep.* **2021**, *11*, 3000. [CrossRef]
17. Verma, P.K.; Verma, S.; Tripathi, R.D.; Pandey, N.; Chakrabarty, D. CC-type glutaredoxin, OsGrx_C7 plays a crucial role in enhancing protection against salt stress in rice. *J. Biotechnol.* **2021**, *329*, 192–203. [CrossRef]

18. Wang, X.; Du, G.; Meng, Y.; Li, Y.; Wu, P.; Yi, K. The function of *LPR1* is controlled by an element in the promoter and is independent of SUMO E3 Ligase SIZ1 in response to low Pi stress in Arabidopsis thaliana. *Plant Cell Physiol.* **2010**, *51*, 380–394. [CrossRef]
19. Ticconi, C.A.; Lucero, R.D.; Sakhonwasee, S.; Adamson, A.W.; Creff, A.; Nussaume, L.; Desnos, T.; Abel, S. ER-resident proteins PDR2 and LPR1 mediate the developmental response of root meristems to phosphate availability. *Proc. Natl. Acad. Sci. USA* **2009**, *106*, 14174–14179. [CrossRef]
20. Müller, J.; Toev, T.; Heisters, M.; Teller, J.; Moore, K.L.; Hause, G.; Dinesh, D.C.; Bürstenbinder, K.; Abel, S. Iron-dependent callose deposition adjusts root meristem maintenance to phosphate availability. *Dev. Cell* **2015**, *33*, 216–230. [CrossRef]
21. Cao, Y.; Ai, H.; Jain, A.; Wu, X.; Zhang, L.; Pei, W.; Chen, A.; Xu, G.; Sun, S. Identification and expression analysis of *OsLPR* family revealed the potential roles of *OsLPR3* and *5* in maintaining phosphate homeostasis in rice. *BMC Plant Biol.* **2016**, *16*, 210. [CrossRef] [PubMed]
22. Ai, H.; Cao, Y.; Jain, A.; Wang, X.; Hu, Z.; Zhao, G.; Hu, S.; Shen, X.; Yan, Y.; Liu, X.; et al. The ferroxidase LPR5 functions in the maintenance of phosphate homeostasis and is required for normal growth and development of rice. *J. Exp. Bot.* **2020**, *71*, 4828–4842. [CrossRef] [PubMed]
23. Oomen, R.J.; Benito, B.; Sentenac, H.; Rodríguez-Navarro, A.; Talón, M.; Véry, A.-A.; Domingo, C. HKT2;2/1, a K($^+$)-permeable transporter identified in a salt-tolerant rice cultivar through surveys of natural genetic polymorphism. *Plant J.* **2012**, *71*, 750–762. [CrossRef] [PubMed]
24. Yao, X.; Horie, T.; Xue, S.; Leung, H.-Y.; Katsuhara, M.; Brodsky, D.E.; Wu, Y.; Schroeder, J.I. Differential sodium and potassium transport selectivities of the rice OsHKT2;1 and OsHKT2;2 transporters in plant cells. *Plant Physiol.* **2010**, *152*, 341–355. [CrossRef] [PubMed]
25. Hien, D.T.; Jacobs, M.; Angenon, G.; Hermans, C.; Thu, T.T.; Van Son, L.; Roosens, N.H. Proline accumulation and *Δ1-pyrroline-5-carboxylate synthetase* gene properties in three rice cultivars differing in salinity and drought tolerance. *Plant Sci.* **2003**, *165*, 1059–1068. [CrossRef]
26. Dangol, S.; Chen, Y.; Hwang, B.K.; Jwa, N.S. Iron-and reactive oxygen species-dependent ferroptotic cell death in Rice-*Magnaporthe oryzae* Interactions. *Plant Cell* **2019**, *31*, 189–209. [CrossRef]
27. Lee, S.; Chiecko, J.C.; Kim, S.A.; Walker, E.L.; Lee, Y.; Guerinot, M.L.; An, G. Disruption of *OsYSL15* leads to iron inefficiency in rice plants. *Plant Physiol.* **2009**, *150*, 786–800. [CrossRef]
28. Zhao, W.; Wang, K.; Chang, Y.; Zhang, B.; Li, F.; Meng, Y.; Li, M.; Zhao, Q.; An, S. OsHyPRP06/R3L1 regulates root system development and salt tolerance via apoplastic ROS homeostasis in rice (*Oryza sativa* L.). *Plant Cell Environ.* **2022**, *45*, 900–914. [CrossRef]
29. Wang, Y.; Wang, J.; Zhao, X.; Yang, S.; Huang, L.; Du, F.; Li, Z.; Zhao, X.; Fu, B.; Wang, W. Overexpression of the Transcription Factor Gene *OsSTAP1* Increases Salt Tolerance in Rice. *Rice* **2020**, *13*, 50. [CrossRef]
30. Zhou, L.; Liu, Z.; Liu, Y.; Kong, D.; Li, T.; Yu, S.; Mei, H.; Xu, X.; Liu, H.; Chen, L.; et al. A novel gene *OsAHL1* improves both drought avoidance and drought tolerance in rice. *Sci. Rep.* **2016**, *6*, 30264. [CrossRef]
31. Zhang, H.; Li, Y.; Pu, M.; Xu, P.; Liang, G.; Yu, D. *Oryza sativa* POSITIVE REGULATOR OF IRON DEFICIENCY RESPONSE 2 (*OsPRI2*) and *OsPRI3* are involved in the maintenance of Fe homeostasis. *Plant Cell Environ.* **2020**, *43*, 261–274. [CrossRef]
32. Zhao, J.; Liu, X.; Wang, M.; Xie, L.; Wu, Z.; Yu, J.; Wang, Y.; Zhang, Z.; Jia, Y.; Liu, Q. The miR528-D3 module regulates plant height in rice by modulating the gibberellin and abscisic acid metabolisms. *Rice* **2022**, *15*, 27. [CrossRef]
33. Ma, X.L.; Zhang, Q.Y.; Zhu, Q.L.; Liu, W.; Chen, Y.; Qiu, R.; Wang, B.; Yang, Z.F.; Li, H.; Lin, Y.R.; et al. A robust CRISPR/Cas9 system for convenient, high-efficiency multiplex genome editing in monocot and dicot plants. *Mol. Plant* **2015**, *8*, 1274–1284. [CrossRef] [PubMed]
34. Xie, X.; Ma, X.; Zhu, Q.; Zeng, D.; Li, G.; Liu, Y.-G. CRISPR-GE: A Convenient Software Toolkit for CRISPR-Based Genome Editing. *Mol. Plant* **2017**, *10*, 1246–1249. [CrossRef]
35. Tzfira, T.; Citovsky, V. *Agrobacterium*-mediated genetic transformation of plants: Biology and biotechnology-ScienceDirect. *Curr. Opin. Biotechnol.* **2006**, *17*, 147–154. [CrossRef]
36. Zhao, J.; Qiu, Z.; Ruan, B.; Kang, S.; He, L.; Zhang, S.; Dong, G.; Hu, J.; Zeng, D.; Zhang, G.; et al. Functional Inactivation of Putative *Photosynthetic Electron Acceptor Ferredoxin C2* (*FdC2*) induces delayed heading Date and Decreased Photosynthetic Rate in Rice. *PLoS ONE* **2015**, *10*, e0143361. [CrossRef]
37. Liu, Q.; Dong, G.-R.; Ma, Y.-Q.; Zhao, S.-M.; Liu, X.; Li, X.-K.; Li, Y.-J.; Hou, B.-K. Rice Glycosyltransferase Gene UGT85E1 Is Involved in Drought Stress Tolerance Through Enhancing Abscisic Acid Response. *Front. Plant Sci.* **2021**, *12*, 790195. [CrossRef]
38. Wang, M.; Guo, W.; Li, J.; Pan, X.; Pan, L.; Zhao, J.; Zhang, Y.; Cai, S.; Huang, X.; Wang, A.; et al. The miR528-AO Module Confers Enhanced Salt Tolerance in Rice by Modulating the Ascorbic Acid and Abscisic Acid Metabolism and ROS Scavenging. *J. Agric. Food Chem.* **2021**, *69*, 8634–8648. [CrossRef]
39. Li, J.; Zhang, M.; Yang, L.; Mao, X.; Li, J.; Li, L.; Wang, J.; Liu, H.; Zheng, H.; Li, Z.; et al. OsADR3 increases drought stress tolerance by inducing antioxidant defense mechanisms and regulating *OsGPX1* in rice (*Oryza sativa* L.). *Crop J.* **2021**, *9*, 1003–1017. [CrossRef]

40. Wei, H.; Wang, X.; He, Y.; Xu, H.; Wang, L. Clock component OsPRR73 positively regulates rice salt tolerance by modulating *OsHKT2;1*-mediated sodium homeostasis. *EMBO J.* **2020**, *40*, e105086. [CrossRef]
41. Zhao, J.; Long, T.; Wang, Y.; Tong, X.; Tang, J.; Li, J.; Wang, H.; Tang, L.; Li, Z.; Shu, Y.; et al. *RMS2* Encoding a GDSL Lipase Mediates Lipid Homeostasis in Anthers to Determine Rice Male Fertility. *Plant Physiol.* **2020**, *182*, 2047–2064. [CrossRef] [PubMed]

Disclaimer/Publisher's Note: The statements, opinions and data contained in all publications are solely those of the individual author(s) and contributor(s) and not of MDPI and/or the editor(s). MDPI and/or the editor(s) disclaim responsibility for any injury to people or property resulting from any ideas, methods, instructions or products referred to in the content.

Article

Systematic Analysis of NRAMP Family Genes in *Areca catechu* and Its Response to Zn/Fe Deficiency Stress

Guangzhen Zhou [1], Qiyuan An [1,†], Zheng Liu [2], Yinglang Wan [1] and Wenlong Bao [2,3,*]

[1] Hainan Key Laboratory for Sustainable Utilization of Tropical Bioresources, College of Tropical Crops, Hainan University, Haikou 570228, China
[2] Key Laboratory for Quality Regulation of Tropical Horticultural Crops of Hainan Province, School of Horticulture, Hainan University, Haikou 570228, China
[3] Hainan Yazhou Bay Seed Laboratory, Sanya Nanfan Research Institute, Hainan University, Sanya 572025, China
* Correspondence: wlbao@hainanu.edu.cn
† Current address: Liaoning Provincial Institute of Economic Crops, Liaoyang 111000, China.

Abstract: *Areca catechu* is a commercially important medicinal plant widely cultivated in tropical regions. The natural resistance-associated macrophage protein (NRAMP) is widespread in plants and plays critical roles in transporting metal ions, plant growth, and development. However, the information on *NRAMPs* in *A. catechu* is quite limited. In this study, we identified 12 *NRAMPs* genes in the areca genome, which were classified into five groups by phylogenetic analysis. Subcellular localization analysis reveals that, except for NRAMP2, NRAMP3, and NRAMP11, which are localized in chloroplasts, all other NRAMPs are localized on the plasma membrane. Genomic distribution analysis shows that 12 *NRAMPs* genes are unevenly spread on seven chromosomes. Sequence analysis shows that motif 1 and motif 6 are highly conserved motifs in 12 *NRAMPs*. Synteny analysis provided deep insight into the evolutionary characteristics of *AcNRAMP* genes. Among the *A. catechu* and the other three representative species, we identified a total of 19 syntenic gene pairs. Analysis of Ka/Ks values indicates that *AcNRAMP* genes are subjected to purifying selection in the evolutionary process. Analysis of cis-acting elements reveals that *AcNRAMP* genes promoter sequences contain light-responsive elements, defense- and stress-responsive elements, and plant growth/development-responsive elements. Expression profiling confirms distinct expression patterns of *AcNRAMP* genes in different organs and responses to Zn/Fe deficiency stress in leaves and roots. Taken together, our results lay a foundation for further exploration of the *AcNRAMPs* regulatory function in areca response to Fe and Zn deficiency.

Keywords: *Areca catechu*; NRAMP; Fe and Zn deficiency; stress response; gene expression

1. Introduction

Metal ions play a key role in maintaining the stability of plant physiological and biochemical functions. For example, Fe is an essential element involved in cell respiration, photosynthesis, and the catalytic reaction of metalloproteins, while Zn is a structural cofactor of a variety of enzymes and proteins [1,2]. The balance of metal ions in the plant is also necessary for plant development. A high concentration of free iron (Fe^{3+}/Fe^{2+}) generates oxygen and hydroxyl radicals through the Fenton reaction, which causes a redox reaction and accumulation of superoxide compounds, resulting in intracellular damage to DNA and lipids [3]. Meanwhile, Fe deficiency can also affect plant growth and development. Similarly, excess zinc binds to sulfur, nitrogen, and oxygen-containing functional groups in biological molecules due to its upregulated high affinity, which, in turn, interferes with their biological activities. Zn deficiency leads to the oxidative destruction of chlorophyll, lipids, and proteins [4].

The maintenance of metal ion balance in plants is mainly achieved through the synergies of various metal transporters, including zinc-regulated transporters iron-regulated transporter-like proteins (ZIP), NRAMP, cation diffusion facilitator proteins (CDF), heavy metal ATPase (HMA), yellow stripe-like (YSL), and ATP-binding cassette (ABC) transporters [5,6]. Of these, the NRAMP proteins are a vital membrane transporter family that exist widely in plants, which are mainly involved in the transport of divalent metal cations such as Zn, Fe, and Cu. Moreover, the NRAMP proteins play a principal role in regulating and maintaining the homeostasis of Mn/Fe in plants [7,8]. NRAMP was first found in mice. Subsequently, NRAMP was verified as a metal transporter in diverse organisms including fungi, animals, and plants [9,10]. In the previous study, the roles of NRAMP proteins have been explored in some plant species, including model plants, forest trees, and horticultural plants [10–15]. Different NRAMPs enable the trafficking of different essential or toxic metal ions. For instance, *Arabidopsis* AtNRAMP6 is considered a transporter of cadmium [16], while AtNRAMP1 is a Mn transporter. AtNRAMP3 and AtNRAMP4 are responsible for the transportation of Fe and Mn [17]. Rice OsNRAMP3 is identified as a Mn transporter, while OsNRAMP5 is identified as a Mn/Cd transporter [18,19]. In addition, OsNRAMP4 plays a key role in the trafficking of trivalent Al ions [20]. In summary, NRAMP proteins are critical members of the membrane transporter family involved in metal ion transport, absorption, intracellular transport, and detoxification.

Areca catechu (areca palm) is an evergreen tree of the genus Areca in the palm family, which is native to Malaysia and widely cultivated in tropical Asia. In our previous studies, we treated areca seedlings with Fe and Zn deficiency and found a significant decrease in iron and zinc content in the roots [21]. In this study, 12 *NRAMP* genes were identified from the whole genome of *A. catechu*, and their sequence characteristics, gene structure, phylogeny, promoter sequence, and collinearity were analyzed. Furthermore, the expression levels of these genes in different tissues and under the condition of Fe and Zn deficiency were studied. This work provides a basis for further investigating *NRAMP* function and areca response to Fe and Zn deficiency.

2. Results
2.1. Identification and Characterization of AcNRAMPs

Twelve *AcNRAMP* candidates were identified in the *A. catechu* genome (Figure 1A, Table S2). The amino acid lengths of 12 AcNRAMP proteins altered from 120 (AcNRAMP5) to 971 (AcNRAMP2), while the molecular weight changed from 13.86–106.01 KDa. The theoretical isoelectric point (pI) of AcNRAMP proteins varied from 4.5 (AcNRAMP10)–9.14 (AcNRAMP5). The instability index (Ii) ranged from 25.75 (AcNRAMP10) to 47.77 (AcNRAMP2). Only two AcNRAMP proteins had Ii larger than 40. The aliphatic index (Ai) analysis shows that AcNRAMP2 has the maximum Ai at 82.84, while AcNRAMP12 has the minimum Ai at 132.44. In addition, the grand average of hydropathicity (GRAVY) of all AcNRAMP proteins was positive except for AcNRAMP2. Prediction of subcellular localization reveals that all AcNRAMP proteins are localized to the plasma membrane except for AcNRAMP2, AcNRAMP3, and AcNRAMP11 (Figure 1B).

2.2. Duplication Analysis and Phylogenetic Analysis of AcNRAMPs

To understand the genomic distribution of 12 *AcNRAMPs*, their chromosomal location was analyzed. The twelve genes were located on seven chromosomes, of which chromosome two had a maximum of three genes (Figure 2A). To reveal the *AcNRAMP* collinear gene pairs, the syntenic gene analysis was carried out. The results show that only one collinear *AcNRAMP* (*AcNRAMP3–AcNRAMP4*) is identified in the areca genome (Figure 2A). To shed light on the possible evolutionary process of *AcNRAMPs*, the syntenic relationships of *AcNRAMP* genes with the other plant species were investigated, including dicot *A. thaliana* (belongs to the *Brassicaceae*), monocot *O. sativa* (belongs to the *Poaceae*), and *C. nucifera* (belongs to the *Palmae*). Three *AcNRAMP* homologous genes pairs are detected between *A. catechu* and *A. thaliana* (*AcNRAMP1–AT5G67330.1*;

AcNRAMP4–AT2G23150.1; AcNRAMP4–AT5G67330.1) (Figure 2B, Table S3). Five AcNRAMP homologous gene pairs are identified between *A. catechu* and *O. sativa* (AcNRAMP1–Os12t0581600-01; AcNRAMP1–Os03t0208500-01; AcNRAMP2–Os07t0155600-01; AcNRAMP4–Os03t0208500-01; AcNRAMP6–Os06t0676000-01) (Figure 2C, Table S3). A total of 11 AcNRAMP homologous gene pairs are identified between *A. catechu* and *C. nucifera* (AcNRAMP1–GZ08G0188430.1; AcNRAMP1–GZ09G0194900.1; AcNRAMP1–GZ15G0285060.1; AcNRAMP2–GZ03G0077500.1; AcNRAMP4–GZ08G0188430.1; AcNRAMP4–GZ09G0194900.1; AcNRAMP4–GZ15G0285060.1; AcNRAMP6–GZ01G0002400.1; AcNRAMP7–GZ07G0166110.1; AcNRAMP12–GZ02G0049210.1; AcNRAMP14–GZ08G0185160.1) (Figure 2D, Table S3).

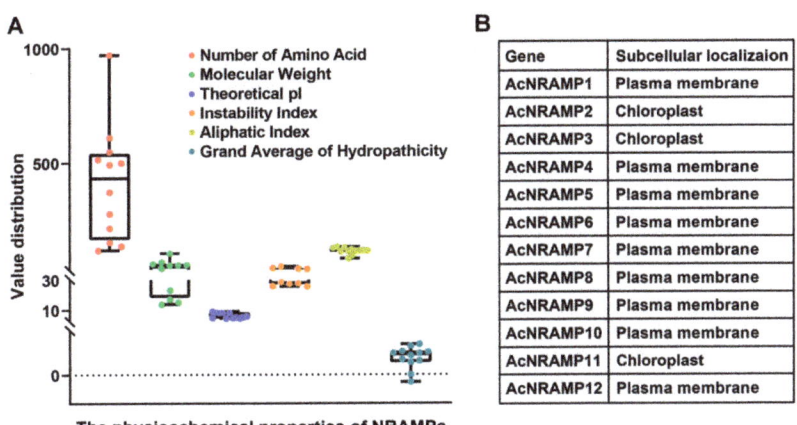

Figure 1. Physicochemical properties of AcNRAMPs (**A**) and prediction of subcellular localization (**B**).

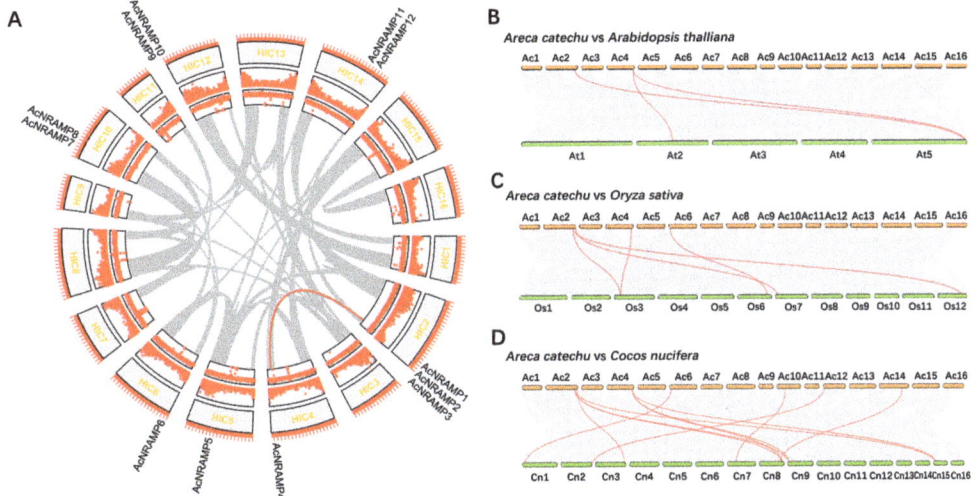

Figure 2. Chromosomal location of *AcNRAMP* genes. (**A**) Chromosomal location of *AcNRAMPs* and their collinear gene pairs in *A. catechu* genome. The outer ring to the inner ring represents 16 chromosomes, gene density, and GC contents, respectively. (**B–D**) Syntenic analysis of *AcNRAMPs* between *A. catechu* and *A. thaliana*, *A. catechu* and *O. sativa*, and *A. catechu* and *C. nucifera*, respectively. The gray lines among each chromosome represent all gene pairs between *A. catechu* and other plant species, while the red lines represent the gene pairs associated with *AcNRAMP* genes.

To uncover the selection pressure on *NRAMP* genes during evolution, we calculated the non-synonymous substitution rate (Ka), the synonymous substitution rate (Ks), and their ratio (Ka/Ks) of homologous *NRAMP* genes in *A. catechu*, *A. thaliana*, *O. sativa*, and *C. nucifera* (Figure 3; Table S4). The results show that the Ka/Ks values vary greatly (from 0.0797 to 0.6172) but are less than 1, suggesting purifying selection on these genes. Generally, the Ka/Ks ratios of *AcNRAMPs* and *OsNRAMPs* gene pairs are larger than those of *AcNRAMPs* and *CnNRAMPs* gene pairs, implying faster evolutionary rates of *AcNRAMP* genes on *O. sativa* than on *C. nucifera* (Table S4).

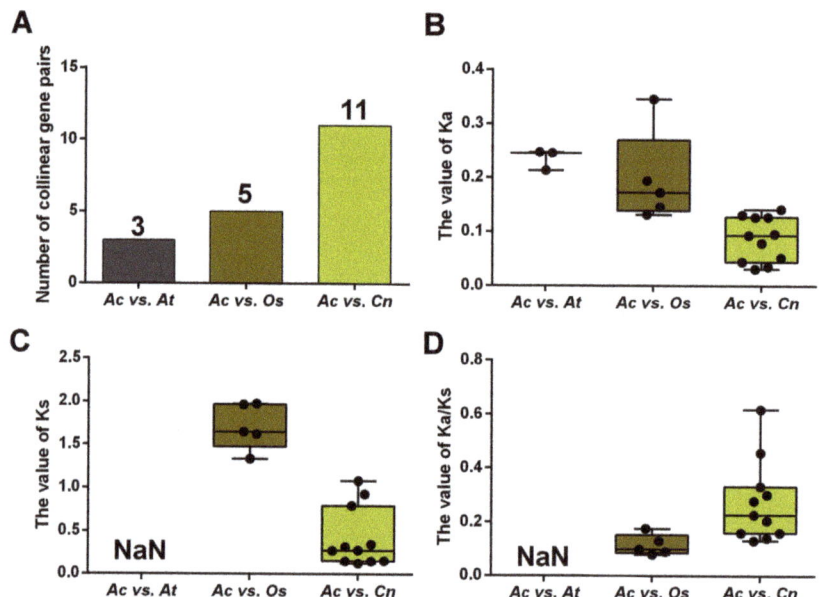

Figure 3. Comparison of *NRAMP* homologous genes substitution rate. (**A**) The number distribution of the collinear gene pairs in three species. Ac: *A. catechu*; At: *A. thaliana*; Os: *O. sativa*; Cn: *C. nucifera*. (**B**–**D**) The Ka, Ks, and Ka/Ks distribution of the *NRAMP* duplicated genes.

To clarify the phylogenetic relationships among *NRAMP* homologs in different plant species, a total of 25 *NRAMP* proteins from *A. thaliana* (6), *O. sativa* (7), and *A. catechu* (12) were used to establish a phylogenetic tree (Figure 4). According to the topology of the phylogenetic tree, 25 NRAMP were subdivided into five subgroups (group 1–group 5). *AcNRAMP* genes are distributed in all five subgroups, among which group 2 has the largest number (Figure 4).

2.3. Sequence Features of AcNRAMPs

To illustrate the sequence features of the *AcNRAMPs*, the MEME program was used to analyze their conserved motifs. In total, six conserved motifs (motif 1–6) are found in 12 AcNRAMP protein sequences. AcNRAMP proteins within the identical subgroup have similar motifs (Figure 5A,B). Motif 1 and motif 6 are present in all proteins. Most of the AcNRAMP proteins have motifs 2 and 3. Only AcNRAMP5 proteins contain two motifs, including motif 1 and motif 6. The analysis of gene structures reveals a great change in introns number within 12 *AcNRAMP* genes (from 3 to 14), while the number distribution of UTR varies from 1 to 3 (Figure 5C). Furthermore, the frequencies of amino acids on the respective position within the sequences of six motifs in the AcNRAMPs are highly different, which are worthy of further elucidation (Figure 5D).

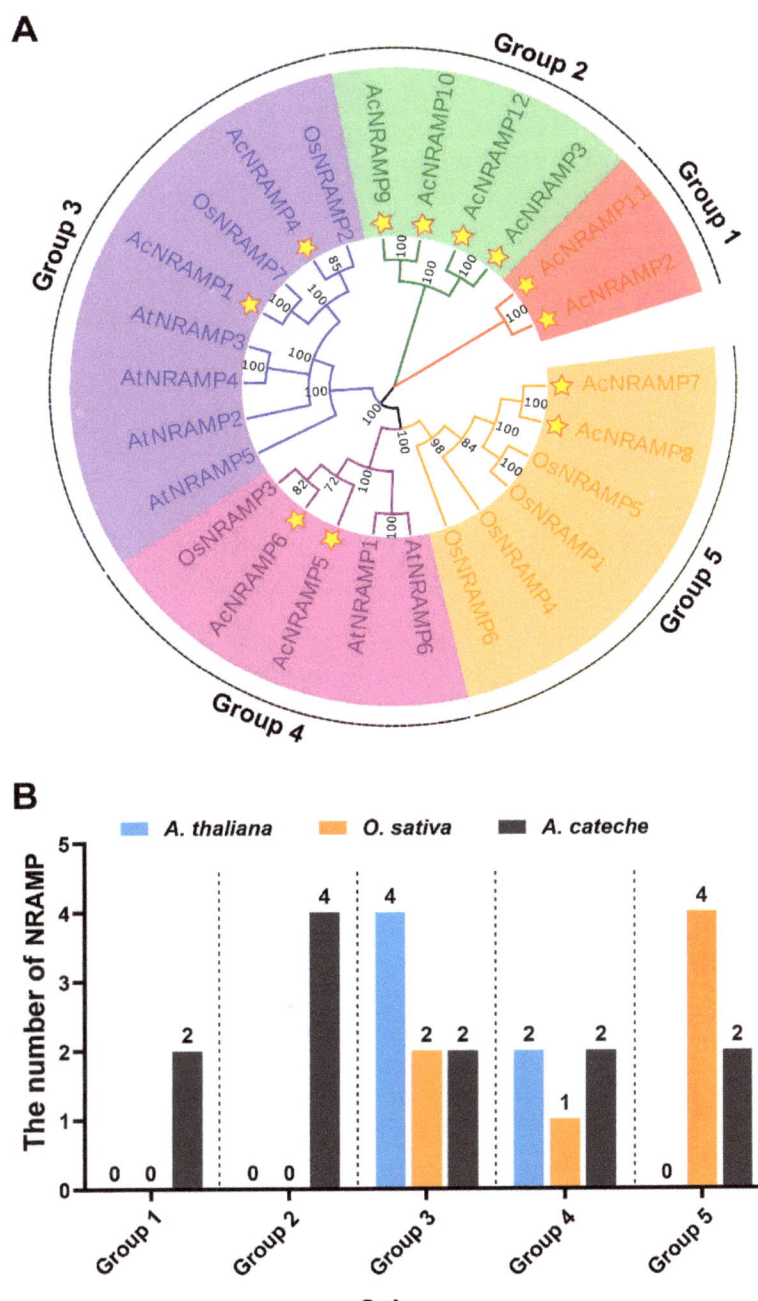

Figure 4. Phylogenetic relationships of NRAMP proteins in *A. catechu*, *O. sativa*, and *A. thaliana*. (**A**) The phylogenetic tree has 25 NRAMP proteins, including 12 AcNRAMP, 7 OsNRAMP, and 6 AtNRAMP). Star-marked 12 NRAMP genes of *A. catechu*. (**B**) The number of AcNRAMPs, Os-NRAMP, and AtNRAMP in the different subgroups.

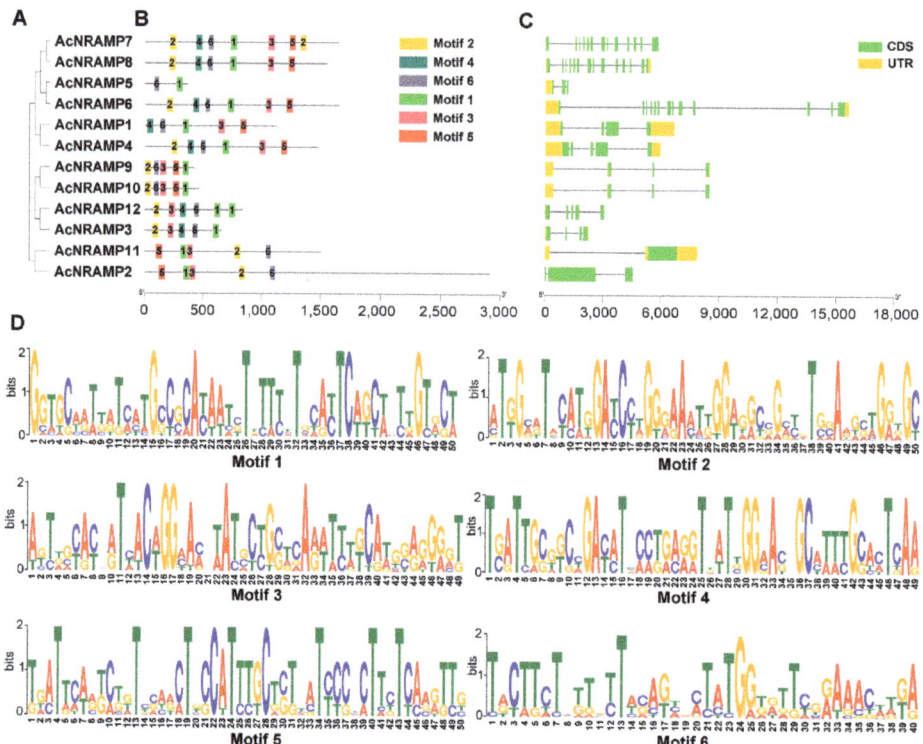

Figure 5. The sequence features of the *AcNRAMP* genes. (**A**) The phylogenetic tree has 12 AcNRAMPs. (**B**) The motif distribution in the 12 AcNRAMPs. The various colored rectangles represent motifs 1 to 6; the black solid lines indicate the sequences outside motifs; the bar scale below represents the number of amino acids. (**C**) The gene structures of the 12 *AcNRAMPs*. The green ellipses and orange ellipses represent exons and untranslated regions (UTR), respectively; the black solid lines linked with the ellipse indicate the introns; the bar scale below represents the gene length. (**D**) Sequence logos of conserved amino acid residues sequences. The X-axis represents the position of the different amino acids in each motif, while the Y-axis represents the bit value of each amino acid.

2.4. Cis-Acting Elements of AcNRAMPs

To gain more insight into the possible regulatory factors of 12 *AcNRAMP* genes, the Plant CARE online program was employed to analyze their cis-acting elements. We identified 25 types of cis-acting elements and classified them into three categories, namely, phytohormone-responsive elements (PREs), defense- and stress-responsive elements (DSREs), and plant growth/development-responsive elements (GDREs) (Figure 6). Of these, the PREs are the most abundant with eight types, containing the ABA-responsive elements such as ABRE, the MeJA-responsive elements such as CGTCA motif and TGACG motif, the auxin-responsive elements such as AuxRR-core, the gibberellin-responsive elements such as P-box and TATC-box, and the other elements such as TCA element and TGA element. The DSREs have 12 types, containing ATC motif, Box 4, GA motif, GARE motif, GATA motif, G-box, I-box, MRE, Sp1, TCCC motif, and TCT motif. However, only five types of GDREs are identified, containing ARE, LTR, MBS, GCN4 motif, and RY element. Moreover, *AcNRAMP9* and *AcNRAMP10* have almost identical cis-acting elements. Box 4 is present in all *AcNRAMPs* as the light-responsive element (Figure 6).

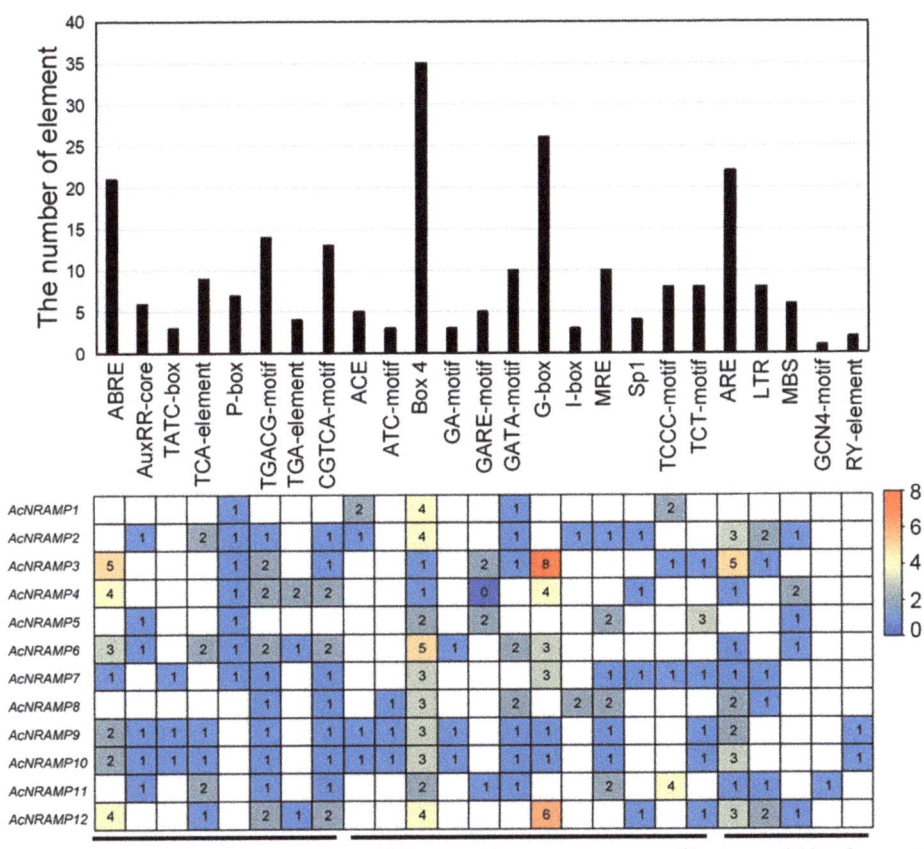

Figure 6. Cis-acting elements of *AcNRAMPs*. The bar chart represents the total number of cis-acting elements within each AcNRAMP. The heatmap shows the number distribution of different cis-acting elements within each *AcNRAMP*.

2.5. Expression Analysis of AcNRAMPs in Different Tissues and under Fe and Zn Deficiency

To elucidate the role of AcNRAMP genes in different tissue and their response to Fe and Zn deficiency, the expression profiles of *AcNRAMP* genes were analyzed by calculating gene FPKM values (Table S5). Tissue-specific expression analysis reveals that *AcNRAMP2*, *AcNRAMP8*, and *AcNRAMP11* are highly expressed in underground roots, and *AcNRAMP4* and *AcNRAMP6* are highly expressed in aerial roots. The expression levels of *AcNRAMP3* are higher in the endosperm than that in the other tissues. *AcNRAMP4* is generally highly expressed in most tissues of areca, especially in aerial roots. *AcNRAMP5* is highly expressed in flowers and pericarp. *AcNRAMP7* is absent from flowers, while the high expression level of *AcNRAMP7* is found in leaves and veins. *AcNRAMP8* is specifically up-regulated in underground roots. *AcNRAMP9* and *AcNRAMP10* are up-regulated in female flowers compared to other tissues. Furthermore, *AcNRAMP11* is only up-regulated in flowers and underground roots, while *AcNRAMP12* shows a higher expression level in leaves and veins (Figure 7A, Table S5). These results suggest that AcNRAMP proteins play vital roles in metal transportation in different areca tissues. Furthermore, the relative expression of six randomly selected *AcNRAMP* genes in female and male flower samples was detected by qRT-PCR. The results show that the gene expression trends determined by qRT-PCR and

RNA-seq are highly consistent, which indicates that transcriptome data are reliable and reproducible (Figure 7B).

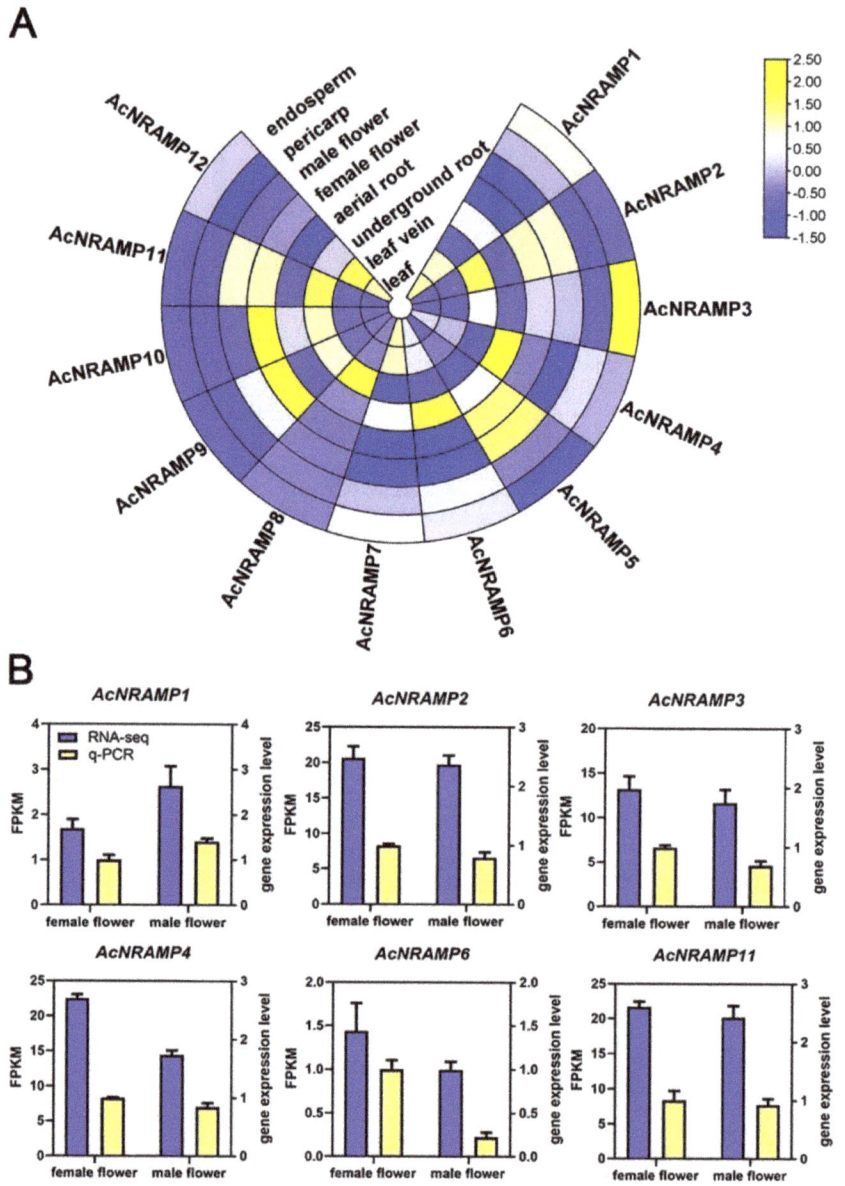

Figure 7. The expression patterns of 12 *AcNRAMP* genes of *A. catechu* in different tissue (**A**) and qRT-PCR validation of RNA-seq (**B**).

To further explore the response of *AcNRAMPs* to Fe and Zn, the expression of *Ac-NRAMP* in leaves and roots of areca seedlings under Fe and Zn deficiency conditions was analyzed (Table S6). In roots, *AcNRAMP1* and *AcNRAMP4* are up-regulated under Zn/Fe deficiency stress, while *AcNRAMP7* is down-regulated. *AcNRAMP3* is up-regulated in Fe-deficient roots, while *AcNRAMP10* is up-regulated in Zn-deficient roots (Figure 8). In leaves, we find that *AcNRAMP1*, *AcNRAMP9*, and *AcNRAMP10* are highly expressed in the first leaf (L1) treated with iron deficiency. Under the condition of zinc deficiency, *AcNRAMP1*, *AcNRAMP2*, *AcNRAMP3*, *AcNRAMP6*, *AcNRAMP8*, *AcNRAMP11*, and *AcNRAMP12* also show a high expression pattern in L1 (Figure 8). In addition, compared with the third leaf (L3), Zn/Fe deficiency treatment greatly affects the expression level of the *AcNRAMPs* gene in L1 (Figure 8). Furthermore, our previous study also suggests that the transcriptome data are reliable and reproducible [21]. These results suggest that the *AcNRAMPs* family plays an important role in coping with Zn/Fe deficiency stress in Areca.

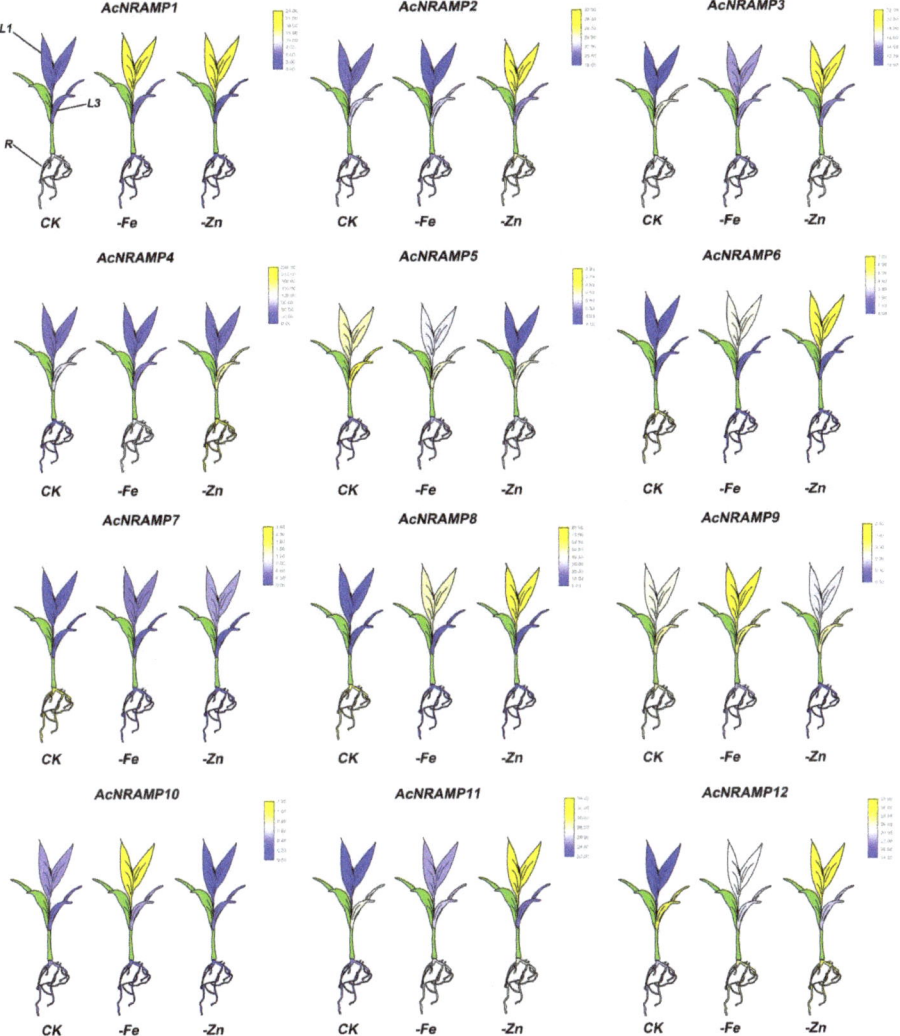

Figure 8. The expression patterns of 12 *AcNRAMP* genes of *A. catechu* in Fe and Zn deficiency. FPKM values were obtained by RNA-seq. Each gene has three cartoon heatmaps that represent areca seedlings in normal, Fe-deficient, and Zn-deficient conditions. The sampling mainly targeted the first leaf (L1), third leaf (L3), and root (R) of areca seedlings.

3. Discussion

The plant *NRAMP* gene family belongs to the conserved metal transport family in natural evolution, which is mainly responsible for the absorption, transport, and intracellular stability of Fe, Mn, and other metal ions [7,22]. In addition, the *NRAMP* gene family also plays a critical regulatory role in photosynthesis, protein activity maintenance, and abiotic stress response [23]. In previous studies, the members of the *NRAMP* gene family have been identified in diverse plant genomes, including *A. thaliana* [11], *O. sativa* [12], *P. alba* [10], *P. vulgaris* [13], *T. cacao* [14], and *B. napus* [15]. In this study, we identified 12 *NRAMP* genes in the *A. catechu* genome. Notably, the *NRAMP* genes in *Arabidopsis* were classified into two subfamilies. However, we identified five subfamilies of *NRAMP* genes in areca, suggesting that *AcNRAMPs* in areca may evolve a series of new functions.

The analysis of physicochemical properties helps decipher the potential functional natures of proteins. For instance, analysis of the protein pI value provides us with an important reference index for protein purification. Similar to other gene family members, 12 AcNRAMP proteins include both basic and acidic proteins [24–26]. All AcNRAMPs are considered thermostable proteins due to their Ai value being higher than 71. In contrast, most of AcNRAMP (10/12) are considered unstable proteins in a test tube due to their Ii values being higher than 40. In addition, except for AcNRAMP2, all AcNRAMP proteins have GRAVY values lower than zero, suggesting their soluble nature. Different organelle localization of NRAMP proteins shows different functions in plants. For example, a previous study shows that AtNRAMP6 is localized in the Golgi/trans-Golgi network and contributes to maintaining intracellular Fe homeostasis [27]. Additionally, AtNRAMP1 is localized on the plasma membrane and participates in Fe/Mn transportation [28]. In rice, most of the NRAMP protein members are also localized on the plasma membrane and are associated with the transport of various intracellular metal ions [17,29]. In the present study, subcellular localization predicts that most *AcNRAMP* genes are localized on the plasma membrane, indicating that most *AcNRAMP* genes transport ions on the membrane, and some *AcNRAMP* genes are located in chloroplast, indicating that they transport ions between organelles.

Numerous studies have found that plant *NRAMP* family genes are associated with the uptake and transport of various divalent metal ions [30]. In *A. thaliana*, *AtNRAMP1* expression can be induced under iron deficiency in the root system, and overexpression lines show high tolerance to iron. These results indicate that *AtNRAMP1* can regulate iron metabolism balance in roots [31]. In addition, *AtNRAMP1* also has the function of absorbing and transporting manganese and iron. Similar to *AtNRAMP1*, *AtNRAMP2* is mainly expressed in the root tip and is involved in manganese ion transport [32]. *AtNRAMP3* and *AtNRAMP4* expression is induced under iron deficiency stress and can regulate the homeostasis of iron and manganese ions in cells [33,34]. Furthermore, the *Atnramp6* mutant shows higher cadmium tolerance than the wild type [27]. Recently, in other plants, an increasing number of *NRAMP* genes have been proven to play a pivotal role in maintaining the balance of iron, manganese, and zinc in cells [35–37]. In our study, we also analyzed the expression profiles of *AcNRAMPs* in different tissues of areca. We found that *AcNRAMP1*, *AcNRAMP4*, *AcNRAMP6*, *AcNRAMP7*, and *AcNRAMP12* are up-regulated in leaves and veins, indicating that they may be involved in the transfer of metal elements. *AcNRAMP3*, *AcNRAMP4*, *AcNRAMP11*, and *AcNRAMP12* are up-regulated in the underground roots, indicating that these four genes may be involved in the accumulation of metal elements in the underground roots. Overall, the expression profile of *AcNRAMP* family genes varies between different organs, suggesting that these genes may exert their functions in specific areca tissues.

Studies have shown that seven *NRAMP* genes are present in *O. sativa*. Of these, *OsNRAMP1* is primarily located on the plasma membrane and is significantly up-regulated in response to Fe deficiency [38]. Aside from transporting Fe, *OsNRAMP1* also facilitates the movement of Cd across the membrane [39]. Moreover, *OsNRMAP5* plays a crucial role in the transportation of Fe, Mn, and Cd in *O. sativa*, thereby significantly contributing to the overall growth and development of the plant [40,41]. In this study, we further analyzed the expression profiles of *AcNRAMP* family genes in areca seedlings under the Zn/Fe deficiency stress. We observe that iron deficiency stress can induce the high expression of *AcNRAMP1*, *AcNRAMP9*, and *AcNRAMP10* in areca L1, suggesting that these genes might be the main iron transporter in areca during the iron deficiency stress. Meanwhile, the expression level of *AcNRAMP12* is higher in iron-deficient roots than in normal areca seedlings, indicating a vital role in iron uptake and transport in roots. Furthermore, the expression levels of *AcNRAMP1*, *AcNRAMP2*, *AcNRAMP3*, *AcNRAMP6*, *AcNRAMP8*, *AcNRAMP11*, and *AcNRAMP12* are significantly higher in the zinc-deficient group than those in the control group, suggesting that these genes may be involved in zinc ion transport. Additionally, we identified three *AcNRAMP* genes (*AcNRAMP2*, *AcNRAMP4*, and *AcNRAMP12*) that are closely related to areca roots' tolerance to zinc deficiency stress.

4. Materials and Methods

4.1. Plant Material and Growth Conditions

An *A. catechu* cultivar "Reyan NO.1", was selected as experimental material in this study. The six-month-old areca seedlings were cultured in Hoagland solution for adaptive cultivation. After two weeks, the seed bulbs of areca seedlings were removed from the base of the stem. Subsequently, the seedlings were cultured in whole nutrient solution (CK), Fe-deficient medium (CK without Fe-$(Na)_2$EDTA), and Zn-deficient medium (CK without Zn^{2+}), respectively.

4.2. Identification of AcNRAMP Genes in A. catechu

The conserved domain of NRAMP protein (PF01566) was obtained from the Pfam database. The AcNRAMP candidates were preliminarily retrieved from the *A. catechu* genome using hidden Markov model (HMM) search with an E-value lower than 10^{-5} based on the methods by Krogh et al. [42]. Pfam Database (http://pfam.xfam.org/) (accessed on 10 January 2023) and SMART (http://smart.embl-heidelberg.de/) (accessed on 10 January 2023) were used to determine the predicted protein as a member of the transporter gene family.

4.3. Analysis of Physicochemical Properties of AcNRAMPs

The physicochemical properties of AcNRAMP proteins were analyzed by using ExPASy6 (https://web.expasy.org/protparam/) (accessed on 12 January 2023) [43], and their subcellular localization was predicted using Plant-mPLoc (http://www.csbio.sjtu.edu.cn/bioinf/plant-multi/) (accessed on 12 January 2023) [44].

4.4. Analysis of Evolutionary Relationships of NRAMP Genes in A. catechu, O. sativa, and A. thaliana

The sequences of AtNRAMPs and OsNRAMPs were obtained from TAIR (https://www.arabidopsis.org/) (accessed on 12 January 2023) and the Rice Annotation Project Database, (https://rapdb.dna.affrc.go.jp/download/irgsp1.html) (accessed on 12 January 2023), respectively. ClustalW was performed to align all sequences with default parameters, and then a phylogenetic tree was constructed using a neighbor-joining (NJ) method of MEGA-X [45]. The bootstrap replications were set as 1000 to test the NJ tree reliability. The Evolview8 program was employed to visualize the phylogenetic tree [46].

4.5. Analysis of Sequence Features, Chromosome Distribution, and Syntenic Relationships of AcNRAMPs

The MEME online (https://meme-suite.org/meme/tools/meme) (accessed on 12 January 2023) was performed to analyze conserved motifs shared among *AcNRAMP* genes [47,48]. The TBtools software (version. x64_1_0987657) was used to visualize the *AcNRAMP* gene structure according to the CDS and genomic DNA sequences of *AcNRAMP* genes [49]. Furthermore, the TBtools software was performed to visualize the information of genome-wide chromosomal density and the distribution of *AcNRAMPs* across all chromosomes of *A. catechu* based on the genome annotation files. The genes distributed on scaffolds were excluded. The TBtools were employed to analyze the syntenic relationships of *AcNRAMPs* genes in interspecies (*A. catechu* vs. *A. thaliana*, *A. catechu* vs. *C. nucifera*, and *A. catechu* vs. *O. sativa*) and intraspecies.

4.6. Analysis of Cis-Acting Elements

Upstream 2,000 bases from the first ATG of CDS of *AcNRAMP* genes were regarded as promoter sequences, which were acquired from the *A. catechu* genome. The cis-acting elements of these sequences were identified using PlantCARE (http://bioinformatics.psb.ugent.be/webtools/plantcare/html/) (accessed on 18 January 2023).

4.7. Transcriptomic Data Analysis of AcNRAMPs

In this experiment, three individual areca seedlings from each group were separately collected as three biological replicates. RNA from a total of 27 samples was extracted and subjected to high-throughput sequencing on the Illumina Hiseq 4000 platform at Biomarker Technologies Co. (Beijing, China). After filtering out low-quality reads, clean reads were aligned to the areca reference genome. Gene expression analysis was performed using BMKCloud (www.biocloud.net) (accessed on 18 January 2023). The expression profiles of *AcNRAMPs* under Fe and Zn deficiency conditions were analyzed by calculating the fragments per kilobase per million mapped reads (FPKM) of *AcNRAMPs* based on the RNA-seq data. The genes with expression levels that met the standards ($|\log 2FC| > 1$, FDR < 0.05, and *p*-value < 0.05) were considered differentially expressed genes (DEGs). The heatmap representing the expression levels of *AcNRAMP* genes was plotted using TBtools. The RNA-seq raw data were deposited at NCBI (accession number: PRJNA767949).

4.8. qRT-PCR Analysis

The RNA quick isolation kit (Tiangen, Beijing, China) was used to extract the high-quality total RNA from the areca samples, and the One-Step gDNA Removal and cDNA Synthesis SuperMix (Tiangen, Beijing, China) was used to synthesize cDNA. The qRT-PCR was conducted to analyze the relative expression of *AcNRAMPs*. Primer Premier 5.0 was used to design the qRT-PCR primers of the *AcNRAMP* genes (Table S1). The qRT-PCR used *β-actin* as an internal reference gene. The 20 µL qRT-PCR reaction mixture contained 10 µL of 2 × SYBR Green Master Mix (Tiangen, Beijing, China), 1 µL of forward and reverse primers, 1 µL of forward and reverse primers, 1 µL cDNA template, and 7 µL ddH$_2$O. The qRT-PCR program follows 94 °C hold for 30 s, 40 cycles at 94 °C, hold for 30 s, and 60 °C hold for 30 s. The final results were analyzed by using the $2^{-\Delta\Delta Ct}$ method. All experiments were repeated with three biological and technical replicates.

5. Conclusions

NRAMP proteins play pivotal roles in plant biological processes, including the trafficking of metal ions, plant growth, and development. However, the information on *NRAMPs* in *A. catechu*, a commercially important medicinal plant of the palm family, is still unclear. In this study, the total members of the *NRAMP* gene family in the areca genome were identified. Subsequently, the sequence characteristics, gene structure, phylogeny, promoter sequence, and collinearity of all *AcNRAMP* genes were comprehensively analyzed. Furthermore, the expression levels of these genes in different tissues and under the condition of Fe and Zn deficiency were studied. The results reveal that *AcNRAMP* genes exert their functions in specific areca tissues and play key roles in areca response to Fe and Zn deficiency. This work provides a valuable reference for the in-depth study of the function of the *AcNRAMP* gene in coping with Fe and Zn deficiency stress in areca.

Supplementary Materials: The supporting information can be downloaded at: https://www.mdpi.com/article/10.3390/ijms24087383/s1.

Author Contributions: W.B., Y.W. and G.Z.: conceptualization. W.B., Y.W. and G.Z.: methodology. G.Z., Z.L. and Q.A.: software, validation, and writing—original draft preparation. Z.L., G.Z. and Q.A.: investigation. W.B. and Y.W.: funding acquisition, writing—review and editing, and supervision. All authors have read and agreed to the published version of the manuscript.

Funding: This research was funded by the Hainan Province Science and Technology Special Fund (ZDYF2021XDNY121), the Hainan Provincial Natural Science Foundation of China (320MS008), and the Initial Funds for the High-level Talents of Hainan University (KYQD(ZR)1935).

Informed Consent Statement: Not applicable.

Data Availability Statement: The datasets presented in this study can be found in online repositories. The names of the repository/repositories and accession number(s) can be found in the article/supplementary material.

Acknowledgments: Special thanks to the reviewers for their precious time and valuable suggestions.

Conflicts of Interest: The authors declare no conflict of interest.

References

1. Cakmak, I. Possible roles of zinc in protecting plant cells from damage by reactive oxygen species. *New Phytol.* **2000**, *146*, 185–205. [CrossRef]
2. Suganya, A.; Saravanan, A.; Manivannan, N. Role of zinc nutrition for increasing zinc availability, uptake, yield, and quality of maize (*Zea mays* L.) grains: An overview. *Commun. Soil Sci. Plant* **2020**, *51*, 2001–2021.
3. Eroglu, S.; Meier, B.; Wirén, N.V.; Peiter, E. The vacuolar manganese transporter MTP8 determines tolerance to iron deficiency-induced chlorosis in *Arabidopsis*. *Plant Physiol.* **2016**, *170*, 1030–1045. [CrossRef]
4. Briat, J.F.; Lebrun, M. Plant responses to metal toxicity. *Comptes Rendus Acad. Sci. III* **1999**, *322*, 43–54. [CrossRef]
5. Yan, A.; Wang, Y.M.; Tan, S.N.; Yusof, M.L.M.; Ghosh, S.; Chen, Z. Phytoremediation: A promising approach for revegetation of heavy metal-polluted land. *Front. Plant Sci.* **2020**, *11*, 359. [CrossRef] [PubMed]
6. Colangelo, E.P.; Guerinot, M.L. Put the metal to the petal: Metal uptake and transport throughout plants. *Curr. Opin. Plant Biol.* **2006**, *9*, 322–330. [CrossRef] [PubMed]
7. Vatansever, R.; Filiz, E.; Ozyigit, I.I. In silico analysis of Mn transporters (NRAMP1) in various plant species. *Mol. Biol. Rep.* **2016**, *43*, 151–163. [CrossRef]
8. Bressler, J.P.; Olivi, L.; Cheong, J.H.; Kim, Y.; Bannona, D. Divalent Metal Transporter 1 in lead and cadmium transport. *Ann. N. Y. Acad. Sci. USA* **2004**, *1012*, 142–152. [CrossRef]
9. Pinner, E.; Gruenheid, S.; Raymond, M.; Gros, P. Functional complementation of the yeast divalent cation transporter family SMF by NRAMP2, a member of the mammalian natural resistance-associated macrophage protein family. *J. Biol. Chem.* **1997**, *272*, 28933–28938. [CrossRef]
10. Gunshin, H.; MacKenzie, B.; Berger, U.V.; Gunshin, Y.; Romero, M.F.; Boron, W.F.; Nussberger, S.; Gollan, J.L.; Hediger, M.A. Cloning and characterization of a mammalian proton-coupled metal-ion transporter. *Nat. Cell Biol.* **1997**, *388*, 482–488. [CrossRef]
11. Mäser, P.; Thomine, S.; Schroeder, J.I.; Ward, J.M.; Hirschi, K.; Sze, H.; Talke, I.N.; Amtmann, A.; Maathuis, F.J.; Sanders, D.; et al. Phylogenetic relationships within cation transporter families of *Arabidopsis*. *Plant Physiol.* **2001**, *126*, 1646–1667. [CrossRef]
12. Chang, J.; Huang, S.; Yamaji, N.; Zhang, W.; Ma, J.F.; Zhao, F. OsNRAMP1 transporter contributes to cadmium and manganese uptake in rice. *Plant Cell Environ.* **2020**, *43*, 2476–2491. [CrossRef]
13. Chen, H.M.; Wang, Y.M.; Yang, H.L.; Zeng, Q.Y.; Liu, Y.J. NRAMP1 promotes iron uptake at the late stage of iron deficiency in poplars. *Tree Physiol.* **2019**, *39*, 1235–1250. [CrossRef]
14. Ishida, J.K.; Caldas, D.G.; Oliveira, L.R.; Frederici, G.C.; Leite, L.M.P.; Mui, T.S. Genome-wide characterization of the *NRAMP* gene family in Phaseolus vulgaris provides insights into functional implications during common bean development. *Genet. Mol. Biol.* **2018**, *41*, 820–833. [CrossRef] [PubMed]
15. Ullah, I.; Wang, Y.; Eide, D.J.; Dunwell, J.M. Evolution, and functional analysis of natural resistance-associated macrophage proteins (NRAMPs) from Theobroma cacao and their role in cadmium accumulation. *Sci. Rep.* **2018**, *8*, 14412. [CrossRef]
16. Zhang, X.D.; Meng, J.G.; Zhao, K.X.; Chen, X.; Yang, Z.M. Annotation and characterization of Cd-responsive metal transporter genes in rapeseed (*Brassica napus*). *BioMetals* **2017**, *31*, 107–121. [CrossRef] [PubMed]
17. Cailliatte, R.; Lapeyre, B.; Briat, J.; Mari, S.; Curie, C. The NRAMP6 metal transporter contributes to cadmium toxicity. *Biochem. J.* **2009**, *422*, 217–228. [CrossRef] [PubMed]
18. Yang, M.; Zhang, W.; Dong, H.X.; Zhang, Y.Y.; Lv, K.; Wang, D.J.; Lian, X.M. OsNRAMP3 Is a vascular bundles-specific manganese transporter that is responsible for manganese distribution in rice. *PLoS ONE* **2013**, *8*, e83990. [CrossRef]
19. Sasaki, A.; Yamaj, N.; Yokosho, K.; Ma, J.F. Nramp5 is a major transporter responsible for manganese and cdmium uptake in rice. *Plant Cell* **2012**, *24*, 2155–2167. [CrossRef]

20. Li, J.Y.; Liu, J.P.; Dong, D.K.; Jia, X.M.; McCouch, S.R.; Kochian, L.V. Natural variation underlies alterations in Nramp aluminum transporter (NRAT1) expression and function that play a key role in rice aluminum tolerance. *Proc. Natl. Acad. Sci.* **2014**, *111*, 6503–6508. [CrossRef]
21. An, Q.Y.; Cui, C.; Muhammad, K.N.; Zhou, G.Z.; Wan, Y.L. Genome-wide investigation of *ZINC-IRON PERMEASE* (*ZIP*) genes in *Areca catechu* and potential roles of ZIPs in Fe and Zn uptake and transport. *Plant Signal. Behav.* **2021**, *16*, 1995647. [CrossRef]
22. Pottier, M.; Oomen, R.; Picco, C.; Giraudat, J.; Scholz-starke, J.; Richaud, P.; Carpaneto, A.; Thomine, S. Identification of mutations allowing natural resistance associated macrophage proteins (NRAMP) to discriminate against cadmium. *Plant J.* **2015**, *83*, 625–637. [CrossRef] [PubMed]
23. Ihnatowicz, A.; Siwinska, J.; Mehar, A.A.; Carey, M.; Koorneef, M.; Reymond, M. Conserved histidine of metal transporter AtNRAMP1 is crucial for optimal plant growth under manganese deficiency at chilling temperatures. *New Phytol.* **2014**, *202*, 1173–1183. [CrossRef]
24. Halligan, B.D. ProMoST: A tool for calculating the pI and molecular mass of phosphorylated and modified proteins on two-dimensional gels. *Methods Mol. Biol.* **2009**, *527*, 283–298. [PubMed]
25. Mohanta, T.K.; Khan, A.; Hashem, A.; Abd-Allah, E.F.; Al-Harrasi, A. The molecular mass and isoelectric point of plant proteomes. *BMC Genom.* **2019**, *20*, 631. [CrossRef]
26. Qing, J.; Dawei, W.; Zhou, J.; Yulan, X.; Bingqi, S.; Fan, Z. Genomewide characterization and expression analyses of the MYB superfamily genes during developmental stages in Chinese jujube. *PeerJ* **2019**, *7*, e6353. [CrossRef]
27. Li, J.; Wang, Y.; Zheng, L.; Li, Y.; Zhou, X.; Li, J.; Gu, D.; Xu, E.; Lu, Y.; Chen, X.; et al. The Intracellular Transporter AtNRAMP6 Is Involved in Fe Homeostasis in *Arabidopsis*. *Front. Plant Sci.* **2019**, *10*, 1124. [CrossRef] [PubMed]
28. Castaings, L.; Caquot, A.; Loubet, S.; Curie, C. The high-affinity metal transporters NRAMP1 and IRT1 team up to take up iron under sufficient metal provision. *Sci. Rep.* **2016**, *6*, 37222. [CrossRef] [PubMed]
29. Yamaji, N.; Sasaki, A.; Xia, J.X.; Yokosho, K.; Ma, J.F. A node-based switch for preferential distribution of manganese in rice. *Nat. Commun.* **2013**, *4*, 2442. [CrossRef] [PubMed]
30. Tiwari, M.; Sharma, D.; Dwivedi, S.; Singh, M.; Tripathi, R.D.; Trivedi, P.K. Expression in *Arabidopsis* and cellular localization reveal involvement of rice NRAMP, OsNRAMP1, in arsenic transport and tolerance. *Plant Cell Environ.* **2014**, *37*, 140–152. [CrossRef] [PubMed]
31. Zha, Q.; Xiao, Z.; Zhang, X.; Han, Z.; Wang, Y. Cloning and functional analysis of MxNRAMP1 and MxNRAMP3, two gens related to high metal tolerance of Malus xiaojinensis. *S Afr. J. Bot.* **2016**, *102*, 75–80. [CrossRef]
32. Alejandro, S.; Cailliatte, R.; Alcon, C.; Dirick, L.; Domergue, F.; Correia, D.; Castaings, L.; Briat, J.F.; Mari, S.; Curie, C. Intracellular distribution of manganese by the Trans-Golgi network transporter NRAMP2 is critical for photosynthesis and cellular redox homeostasis. *Plant Cell* **2017**, *29*, 3068–3084. [CrossRef] [PubMed]
33. Lanquar, V.; Lelièvre, F.; Bolte, S.; Hamès, C.; Alcon, C.; Neumann, D.; Vansuyt, G.; Curie, C.; Schröder, A.; Krämer, U.; et al. Mobilization of vacuolar iron by AtNRAMP3 and AtNRAMP4 is essential for seed germination on low iron. *EMBO J.* **2005**, *24*, 4041–4051. [CrossRef] [PubMed]
34. Mary, V.; Ramos, M.S.; Gillet, C.; Socha, A.L.; Giraudat, J.; Agorio, A.; Merlot, S.; Clairet, C.; Kim, S.A.; Punshon, T.; et al. Bypassing iron storage in endodermal vacuoles rescues the iron mobilization defect in the natural resistance associated-macrophage protein3natural resistance associated-macrophage protein4 double mutant. *Plant Physiol.* **2015**, *169*, 748–759. [CrossRef]
35. Mani, A.; Sankaranarayanan, K. In silico analysis of natural resistance-associated macrophage protein1 (NRAMP) family of transporters in rice. *Protein J.* **2018**, *37*, 237–247. [CrossRef]
36. Zhang, J.; Zhang, M.; Song, H.; Zhao, J.; Shabala, S.; Tian, S.; Yang, X. A novel plasma membrane-based NRAMP transporter contributes to Cd and Zn hyperaccumulation in Sedum alfredii Hance. *Environ. Exp. Bot.* **2020**, *176*, 104121. [CrossRef]
37. Qin, L.; Han, P.P.; Chen, L.Y.; Walk, T.C.; Li, Y.S.; Hu, X.J.; Xie, L.H.; Liao, H.; Liao, X. Genome-Wide identification and expression analysis of NRAMP family genes in soybean (*Glycine max* L.). *Front. Plant Sci.* **2017**, *8*, 1436. [CrossRef]
38. Takahashi, R.; Ishimaru, Y.; Nakanishi, H.; Nishizawa, N.K. Role of the iron transporter OsNRAMP1 in cadmium uptake and accumulation in rice. *Plant Signal. Behav.* **2011**, *6*, 1813–1816. [CrossRef]
39. Takahashi, R.; Ishimaru, Y.; Senoura, T.; Shimo, H.; Ishikawa, S.; Arao, T.; Nakanishi, H.; Nishizawa, N.K. The OsNRAMP1 iron transporter is involved in Cd accumulation in rice. *J. Exp. Bot.* **2011**, *62*, 4843–4850. [CrossRef]
40. Ishimaru, Y.; Bashir, K.; Nakanishi, H.; Nishizawa, N.K. OsNRAMP5, a major player for constitutive iron and manganese uptake in rice. *Plant Signal. Behav.* **2012**, *7*, 763–766. [CrossRef]
41. Yang, M.; Zhang, Y.Y.; Zhang, L.J.; Hu, J.T.; Zhang, X.; Lu, K.; Dong, H.X.; Wang, D.J.; Zhao, F.J.; Huang, C.F.; et al. OsNRAMP5 contributes to manganese translocation and distribution in rice shoots. *J. Exp. Bot.* **2014**, *65*, 4849–4861. [CrossRef]
42. Krogh, A.; Larsson, B.; von Heijne, G.; Sonnhammer, E.L. Predicting transmembrane protein topology with a hidden markov model: Application to complete genomes. *J. Mol. Biol.* **2001**, *305*, 567–580. [CrossRef]
43. Wilkins, M.R.; Gasteiger, E.; Bairoch, A.; Sanchez, J.C.; Williams, K.L.; Appel, R.D.; Hochstrasser, D.F. Protein identification and analysis tools in the ExPASy server. *Methods Mol. Biol.* **1999**, *112*, 531–552. [CrossRef]
44. Xiong, E.H.; Zheng, C.Y.; Wu, X.L.; Wang, W. Protein subcellular location: The gap between prediction and experimentation. *Plant Mol. Biol. Rep.* **2016**, *34*, 52–61. [CrossRef]
45. Kumar, S.; Stecher, G.; Li, M.; Knyaz, C.; Tamura, K.; Mega, X. Molecular evolutionary genetics analysis across computing platforms. *Mol. Biol. Evol.* **2018**, *35*, 1547–1549. [CrossRef] [PubMed]

46. He, Z.; Zhang, H.; Gao, S.; Lercher, M.J.; Chen, W.H.; Hu, S. Evolview v2: An online visualization and management tool for customized and annotated phylogenetic trees. *Nucleic Acids Res.* **2016**, *44*, W236–W241. [CrossRef] [PubMed]
47. Bailey, T.L.; Boden, M.; Busker, F.A.; Frith, M.; Grant, C.E.; Clementi, L.; Ren, J.Y.; Li, W.W.; Noble, W.S. MEME SUITE: Tools for motif discovery and searching. *Nucleic Acids Res.* **2009**, *37*, W202–W208. [CrossRef]
48. Lu, S.N.; Wang, J.Y.; Chitsaz, F.; Derbyshire, M.K.; Geer, R.C.; Gonzales, N.R.; Geadz, M.; Hurwitz, D.I.; Marchler, G.H.; Song, J.S.; et al. CDD/SPARCLE: The conserved domain database in 2020. *Nucleic Acids Res.* **2020**, *48*, D265–D268. [CrossRef] [PubMed]
49. Chen, C.J.; Chen, H.; Zhang, Y.; Thomas, H.R.; Frank, M.H.; He, Y.H.; Xia, R. TBtools: An integrative toolkit developed for interactive analyses of big biological data. *Mol. Plant* **2020**, *13*, 1194–1202. [CrossRef]

Disclaimer/Publisher's Note: The statements, opinions and data contained in all publications are solely those of the individual author(s) and contributor(s) and not of MDPI and/or the editor(s). MDPI and/or the editor(s) disclaim responsibility for any injury to people or property resulting from any ideas, methods, instructions or products referred to in the content.

Article

Comprehensive Analysis of bHLH Transcription Factors in *Ipomoea aquatica* and Its Response to Anthocyanin Biosynthesis

Zheng Liu [1], Xiaoai Fu [1], Hao Xu [1], Yuxin Zhang [1], Zhidi Shi [1], Guangzhen Zhou [2] and Wenlong Bao [1,3,*]

[1] Key Laboratory for Quality Regulation of Tropical Horticultural Crops of Hainan Province, School of Horticulture, Hainan University, Haikou 570228, China
[2] College of Tropical Crops, Hainan University, Haikou 570228, China
[3] Hainan Yazhou Bay Seed Laboratory, Sanya Nanfan Research Institute of Hainan University, Sanya 572025, China
* Correspondence: wlbao@hainanu.edu.cn

Citation: Liu, Z.; Fu, X.; Xu, H.; Zhang, Y.; Shi, Z.; Zhou, G.; Bao, W. Comprehensive Analysis of bHLH Transcription Factors in *Ipomoea aquatica* and Its Response to Anthocyanin Biosynthesis. *Int. J. Mol. Sci.* 2023, 24, 5652. https://doi.org/10.3390/ijms24065652

Academic Editors: Jian Zhang and Zhiyong Li

Received: 7 February 2023
Revised: 11 March 2023
Accepted: 14 March 2023
Published: 15 March 2023

Copyright: © 2023 by the authors. Licensee MDPI, Basel, Switzerland. This article is an open access article distributed under the terms and conditions of the Creative Commons Attribution (CC BY) license (https://creativecommons.org/licenses/by/4.0/).

Abstract: The basic helix-loop-helix (bHLH) proteins compose one of the largest transcription factor (TF) families in plants, which play a vital role in regulating plant biological processes including growth and development, stress response, and secondary metabolite biosynthesis. *Ipomoea aquatica* is one of the most important nutrient-rich vegetables. Compared to the common green-stemmed *I. aquatica*, purple-stemmed *I. aquatica* has extremely high contents of anthocyanins. However, the information on *bHLH* genes in *I. aquatica* and their role in regulating anthocyanin accumulation is still unclear. In this study, we confirmed a total of 157 *bHLH* genes in the *I. aquatica* genome, which were classified into 23 subgroups according to their phylogenetic relationship with the bHLH of *Arabidopsis thaliana* (AtbHLH). Of these, 129 *IabHLH* genes were unevenly distributed across 15 chromosomes, while 28 *IabHLH* genes were spread on the scaffolds. Subcellular localization prediction revealed that most IabHLH proteins were localized in the nucleus, while some were in the chloroplast, extracellular space, and endomembrane system. Sequence analysis revealed conserved motif distribution and similar patterns of gene structure within *IabHLH* genes of the same subfamily. Analysis of gene duplication events indicated that DSD and WGD played a vital role in the *IabHLH* gene family expansion. Transcriptome analysis showed that the expression levels of 13 *IabHLH* genes were significantly different between the two varieties. Of these, the *IabHLH027* had the highest expression fold change, and its expression level was dramatically higher in purple-stemmed *I. aquatica* than that in green-stemmed *I. aquatica*. All upregulated DEGs in purple-stemmed *I. aquatica* exhibited the same expression trends in both qRT-PCR and RNA-seq. Three downregulated genes including *IabHLH142*, *IabHLH057*, and *IabHLH043* determined by RNA-seq had opposite expression trends of those detected by qRT-PCR. Analysis of the cis-acting elements in the promoter region of 13 differentially expressed genes indicated that light-responsive elements were the most, followed by phytohormone-responsive elements and stress-responsive elements, while plant growth and development-responsive elements were the least. Taken together, this work provides valuable clues for further exploring *IabHLH* function and facilitating the breeding of anthocyanin-rich functional varieties of *I. aquatica*.

Keywords: gene family; cis-acting element; gene duplication events; evolutionary relationship; gene expression

1. Introduction

The bHLH proteins comprise one of the largest transcription factor (TF) families in eukaryotes including plants, animals, and fungi. The bHLH TF contains two functional conserved domains with a total of approximately 60 amino acid residues, namely the N-terminal basic region composed of 13–17 amino acids and the C-terminal HLH region with 40–50 amino acids, respectively. Typically, the basic region of bHLH TF in plants possesses a highly conserved HER motif that can specifically bind to the cis-acting element E-box

(CANNTG) or G-box (CACGTG) within the promoter region of the target gene. The HLH region consists of two relatively conserved amphipathic α-helices that are separated by an intervening loop of variable length, which is required for the formation of bHLH dimers through protein-protein interaction to modulate the expression of downstream genes that correlated with diverse signaling pathways [1–4]. With the increasing publications of plant genome data, bHLH TFs in numerous plant species have been identified at the whole genome level. For instance, there are 602 *bHLH* genes in the *Brassica napus* genome [5], 208 in the *Zea mays* genome [6], 162 in the *Arabidopsis thaliana* genome [7], 115 in the *Vitis davidii* genome [8], and 141 in the *Hordeum vulgare* genome [9]. The bHLHs exert a broad range of functions in various biological processes of diverse plant species [3,10,11]. For instance, *Arabidopsis* flowering bHLHs are involved in the regulation of flowering time by activating the gene encoding CONSTANS (CO) protein [12]. Rice *bHLH* gene *INCREASED LAMINA INCLINATION1* (*ILI1*) participates in brassinosteroid-mediated plant development by serving as a downstream target of BRASSINAZOLE-RESISTANT1 (BZR1) TF [13]. In apple, MdbHLH33 facilitates plant cold stress tolerance and anthocyanin accumulation by positively regulating the expressions of the C-repeat binding factor (CBF) TF gene and dihydroflavonol 4-reductase (DFR) gene [14].

Anthocyanins are ubiquitous secondary metabolites generated from the branch of the phenylpropanoid biosynthesis pathway. As water-soluble pigments, anthocyanins are widely distributed in diverse plant organs and tissues and give rise to colors varying with pH value, playing a crucial role in plant pollination and seed dispersal. As natural antioxidants, anthocyanins can help plants defend against various stresses including drought, low temperature, and ultraviolet radiation [15,16]. Consuming a diet high in anthocyanins can enhance immunity and delay the aging process in humans [17]. The key structural genes and regulatory genes involved in the hierarchical regulatory networks governing the anthocyanin biosynthetic pathway are well-studied in many plant species [18,19]. The structural genes involved in the anthocyanin biosynthetic pathway are classified into two categories, namely the early biosynthetic genes (EBGs) and late biosynthetic genes (LBGs). Enzymes encoded by EGBs are responsible for producing large amounts of anthocyanin precursors while those encoded by LBGs are involved in the generation of various anthocyanins [20,21]. Studies showed that the expression of the EBGs is mainly activated or repressed by the MYB TFs, while the LBGs are coordinately modulated by the members of the MYB, bHLH, and WDR TF families [22–24]. Of these TFs, the bHLHs act as an essential cofactor of the MYBs and WDRs and play a central role in the late stage of anthocyanin biosynthesis.

Since the first bHLH protein associated with anthocyanin biosynthesis was discovered in maize, a growing understanding of bHLH functions regulating anthocyanin production in plants was carried out. In *Arabidopsis thaliana*, a bHLH TF TRANSPARENT TESTA8 (TT8) regulates the expression of the LBG *DFR* and *BANYULS* (*BAN*) genes which in turn modulate the seed coat pigmentation [25]. In *Chrysanthemum morifolium*, CmbHLH2 serves as the essential partner of CmMYB6 to facilitate the expression of *CmDFR* and thereby enhance anthocyanin accumulation [26]. In *Malus domestica*, the *MdbHLH3* gene was induced by low-temperature stress and promoted fruit coloration by activating the anthocyanin biosynthetic genes (ABGs) including *MdUFGT* and *MdDFR* [27]. In contrast, the *StbHLH1* gene of *Solanum tuberosum* was down-regulated under high-temperature stress and led to the reduction of anthocyanin contents in flesh by affecting the expression of the ABGs [28]. In *Populus*, the *PdbHLH* (a homolog of TT8 in *Arabidopsis*) acts as an efficient enhancer of PdMYB118 to promote the expression of the ABGs and thereby promotes the anthocyanin production under wound [29]. In *Vitis vinifera*, the VvMYC1 that has been characterized as bHLH TF cooperates with several MYB TFs to participate in the transcriptional cascade involved in the regulation of anthocyanin accumulation in berries [30]. In mulberry fruits, the *bHLH3* was implicated in maintaining flavonoid homeostasis, which regulated the fruit coloration by affecting the anthocyanin compositions [31]. These studies demonstrated that bHLH TFs act as vital regulators of anthocyanin biosynthetic pathways in diverse plant species.

I. aquatica has a wide distribution in tropical and subtropical regions. As a health-promoting leafy vegetable, *I. aquatica* is rich in essential amino acids, flavonoids, and diverse mineral elements such as calcium, potassium, and phosphorus [32,33]. Moreover, *I. aquatica* was used in traditional medicine as a remedy for different diseases including liver disorders, diabetes, and in the treatment of heavy metal intoxication [34–36]. Our previous study revealed that the purple-stemmed *I. aquatica* has extremely high anthocyanin contents relative to the common green-stemmed *I. aquatica* [37]. To understand the possible roles of bHLH TF related to anthocyanin biosynthesis in *I. aquatica*, we comprehensively analyzed the members of the *bHLH* gene family and their expression patterns by combining bioinformatics and experimental methods. Our study laid a foundation for further exploring the roles of *IabHLHs* in regulating anthocyanin biosynthesis in *I. aquatica*.

2. Results

2.1. Identification of IabHLH Genes in the I. aquatica Genome

To identify the members of the *IabHLH* gene family, the HMM profile of bHLH and the AtbHLH protein sequences were used as queries to search against the *I. aquatica* genome. After combining the above results, the redundant sequences and the sequences without the bHLH domain were eliminated based on the running consequences of Pfam, CDD, and SMART programs (Figure 1A). A total of 157 *IabHLH* genes were identified, which were named IabHLH001–IabHLH157 according to their positions on the chromosomes and scaffolds (Table S1). The physicochemical properties of proteins play a crucial role in the bio-function and conformation of proteins. As shown in Figure 1B and Table S2, the length of IabHLHs ranged from 140 aa (IabHLH099) to 701 aa (IabHLH022), and their molecular weight (MW) varied from 15.8 kDa (IabHLH099) to 76.5 kDa (IabHLH014). The IabHLH023 had the highest theoretical isoelectric point (pI), with a value of 9.59, while IabHLH136 had the lowest, with a value of 4.8. Except for IabHLH027, IabHLH071, and IabHLH141, the remaining IabHLHs had instability index (Ii) values greater than 40. In addition, 66 of the 157 IabHLHs had aliphatic index (Ai) values lower than 71. Analysis of the grand average of hydropathicity index (GRAVY) showed that all IabHLHs had negative GRAVY values. Subcellular localization prediction of IabHLHs showed that 139, 13, 4, and one bHLH were localized in the nucleus, chloroplast, extracellular space, and endomembrane system, respectively (Table S2).

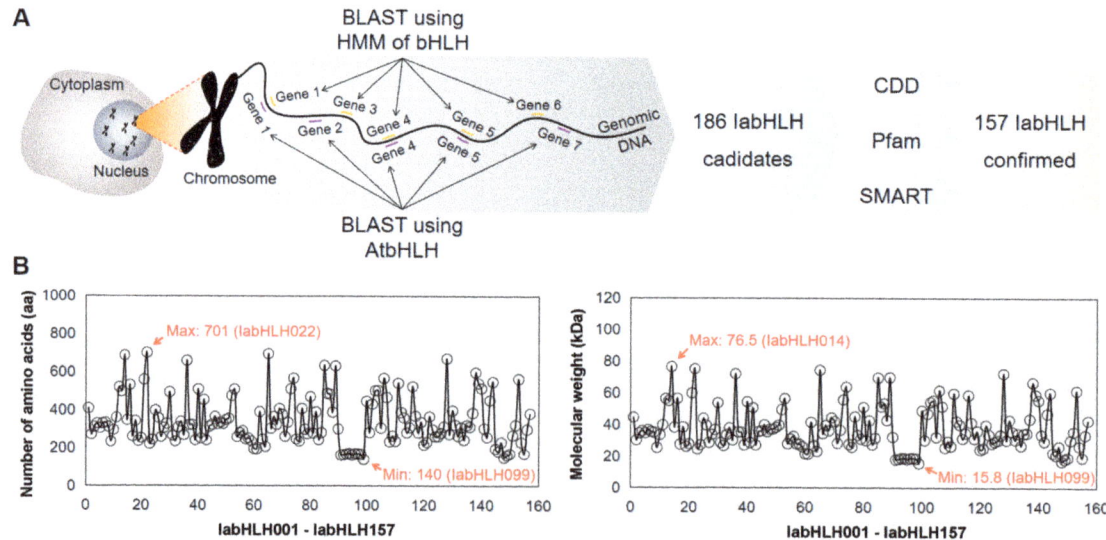

Figure 1. *Cont.*

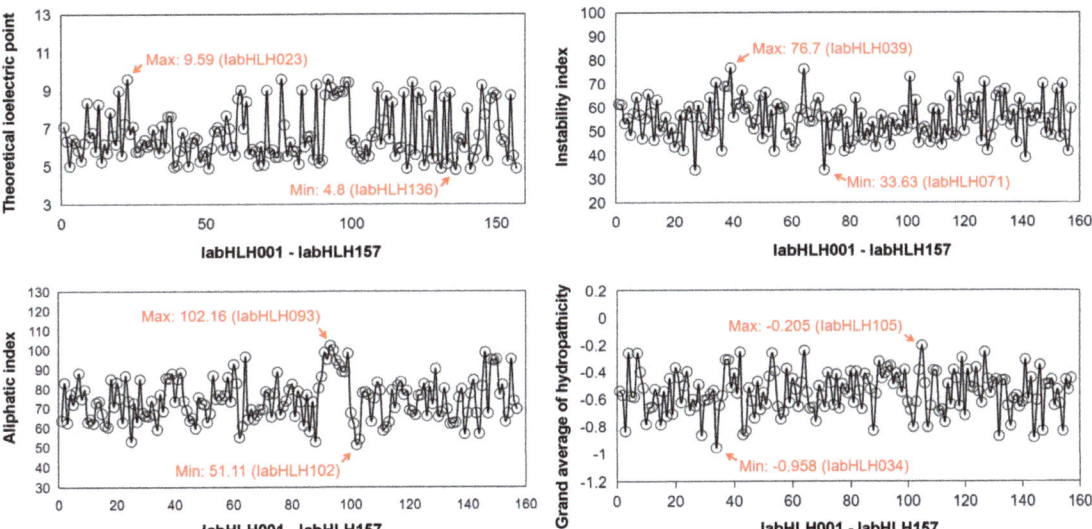

Figure 1. Identification of *IabHLH* genes. (**A**) Schematic diagram of the procedure for retrieving *IabHLH* genes from the *I. aquatica* genome. (**B**) The physicochemical properties of 157 IabHLH proteins.

2.2. Synteny Analysis of IabHLHs

To understand the *IabHLH* gene distribution in the *I. aquatica* genome, their genomic location was analyzed. A total of 129 of the 157 *IabHLH* genes were unevenly distributed on 15 chromosomes, while the remaining genes were spread on the scaffolds (Figure 2A and Table S1).

To reveal the *IabHLH* collinear gene pairs in the *I. aquatica* genome, the analysis of intraspecific collinearity was performed. A total of 61 *IabHLH* colinear gene pairs were identified in the *I. aquatica* genome, of which 44 gene pairs were mapped on the chromosomes (Figure 2A). To explore the potential evolutionary process of *IabHLH* genes, we conducted the analysis of interspecific syntenic relationships among *IabHLH* genes and the other plant species' genomes, including plants of the *Convolvulaceae* family (*Ipomoea batatas*), *Brassicaceae* family (*A. thaliana*), and *Poaceae* family (*Oryza sativa*). As shown in Figure 2B, the highest number of homologous gene pairs were identified between *IabHLH* genes and *I. batatas*, with 239, indicating the closest evolutionary distances between the *IabHLH* gene family and *I. batatas*. As shown in Figure 2C,D, a total of 127 and 43 homologous gene pairs were identified between *I. aquatica* and *A. thaliana*, and between *I. aquatica* and *O. sativa*, respectively, suggesting a closer evolutionary relationship of the *IabHLH* gene family with *A. thaliana* than with *O. sativa*. Moreover, 26 *IabHLH* genes were shared among all homologous gene pairs (Table S3), inferring their relative conservation in the plant evolutionary process.

2.3. Phylogenetic Relationships of IabHLHs and AtbHLHs

To gain further insights into the potential biological functions of IabHLHs, the phylogenetic relationships of IabHLHs and well-characterized AtbHLHs were analyzed. The topology of the phylogenetic tree indicated that the *IabHLH* genes could be clustered into 23 subfamilies, including Ia, Ib(1), Ib(2), II, III(a+c), III(d+e), IIIb, IIIf, IVa, IVb, IVc, IVd, Va, Vb, VII(a+b), VIIIa, VIIIb, VIIIc(1), VIIIc(2), IX, XI, XII, and orphans. The IabHLHs were not clustered together with AtbHLHs in the subfamilies X, XIII, XIV, and XV. In addition, 3 IabHLH orphans were not gathered in the same node (Figure 3A). The subfamily Ib(2) was the largest, with 27 *IabHLH* genes, followed by subfamily XII, with 22 *IabHLH* genes, while the subfamily II and VIIIc(1) were the smallest, with only one *IabHLH* gene (Figure 3A).

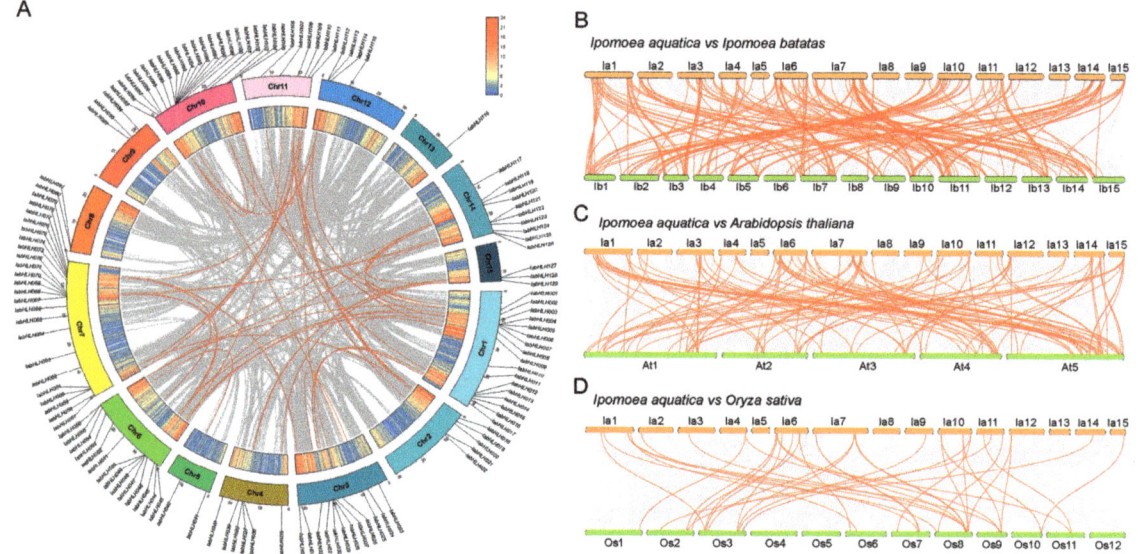

Figure 2. Analysis of *IabHLH* genes at the chromosome level. (**A**) Chromosomal location of the *IabHLH* genes. The gene pairs linked by red lines were the 44 *IabHLH* collinear gene pairs located on chromosomes; the outer circle represents 15 chromosomes of *I. aquatica*, while the inner circle indicates the gene density of each chromosome. (**B**–**D**) Syntenic relationships between *IabHLH* genes and the other plant species' genome. The gray lines among chromosomes indicate all homologous gene pairs at the whole-genome level, while the red lines among chromosomes show homologous gene pairs related to the *IabHLH* genes. More information concerning the homologous gene pairs was shown in Table S3.

2.4. The Sequence Features of the IabHLHs

To investigate the sequence characteristics of the IabHLHs, the multiple sequence alignments of 157 IabHLH protein sequences were carried out. The results showed that the IabHLH proteins contained the typical conserved structure of basic-helix-loop-helix (Figure 4A,B). A total of 16 amino acids had values of frequency greater than 50%. Of these, the highly conserved amino acids with a frequency greater than 85% were identified within the basic region (including Glu-5, Arg-8, and Arg-9), helix 1 region (including Leu-19 and Pro-24), and helix 2 region (Leu-51).

2.5. Conserved Motifs, and Exon-Intron Structures of IabHLH Genes

To further understand the functional and structural features of the IabHLH proteins, the MEME program was performed to analyze the conserved motifs within the IabHLHs. Ten conserved motifs were identified, of which motif 1 and motif 2 were present in all IabHLHs. The IabHLHs within the same group had similar motif distributions, such as the motifs of the IabHLHs in the subfamily Va, IVc, and Ib(1), indicating their similar functions. Motif 5 was only found in the IabHLHs of the subfamily Ib(2), which may impart a specific function to these IabHLHs (Figure 5A,B). The analysis of exon-intron structures within 157 *IabHLH* genes showed that the number of introns ranged from 0 to 11 (Figures 5C and S1, Table S1). The *IabHLH144* in the subfamily Va had the highest number of introns, while a total of 11 *IabHLH* genes in the subfamilies III(d+e) and VIIIb were not disrupted by an intron. Most *IabHLH* genes from the same subfamily had similar exon-intron structures.

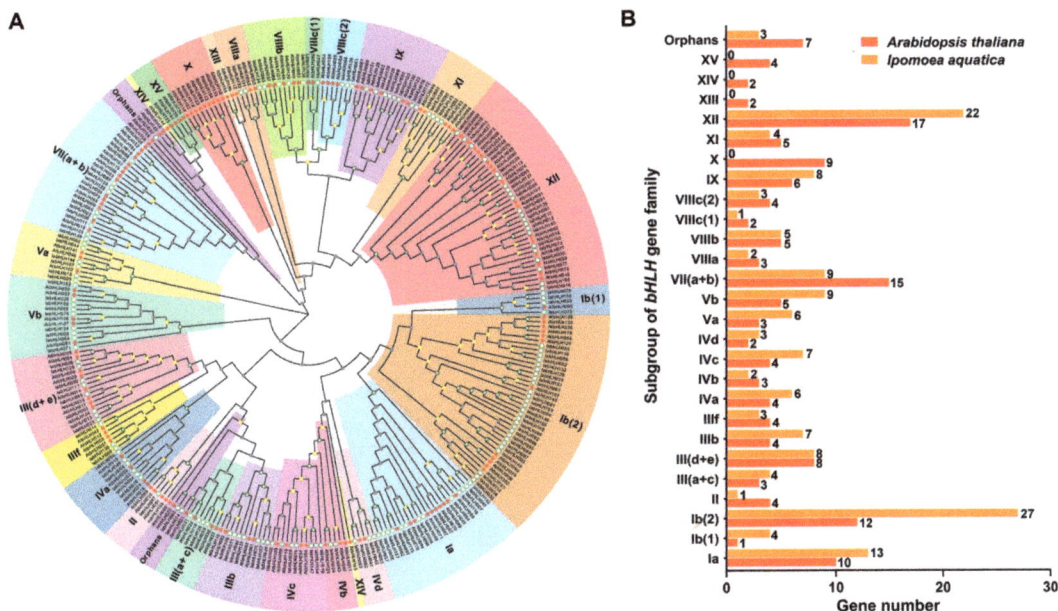

Figure 3. Phylogenetic relationships of IabHLHs and AtbHLHs (**A**) The phylogenetic tree contains 157 IabHLH proteins and 148 AtbHLH proteins. The hollow green circles represent IabHLHs, while the solid red pentagons represent AtbHLHs. The gray, yellow, and green squares represent the bootstrap values varied from 0 to 49, 50 to 79, and 80 to 100, respectively. (**B**) The number of IabHLHs and AtbHLHs in each subfamily.

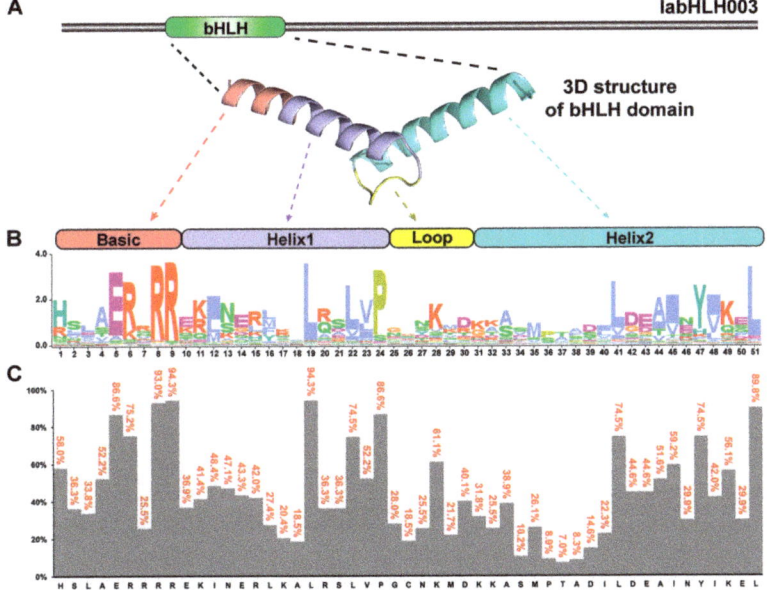

Figure 4. The sequence characteristics of IabHLHs. (**A**) Three-dimensional structure of bHLH domain in the IabHLH003. (**B**) Sequence logos of the IabHLH domain. (**C**) The frequency of the most conserved amino acids at the respective position.

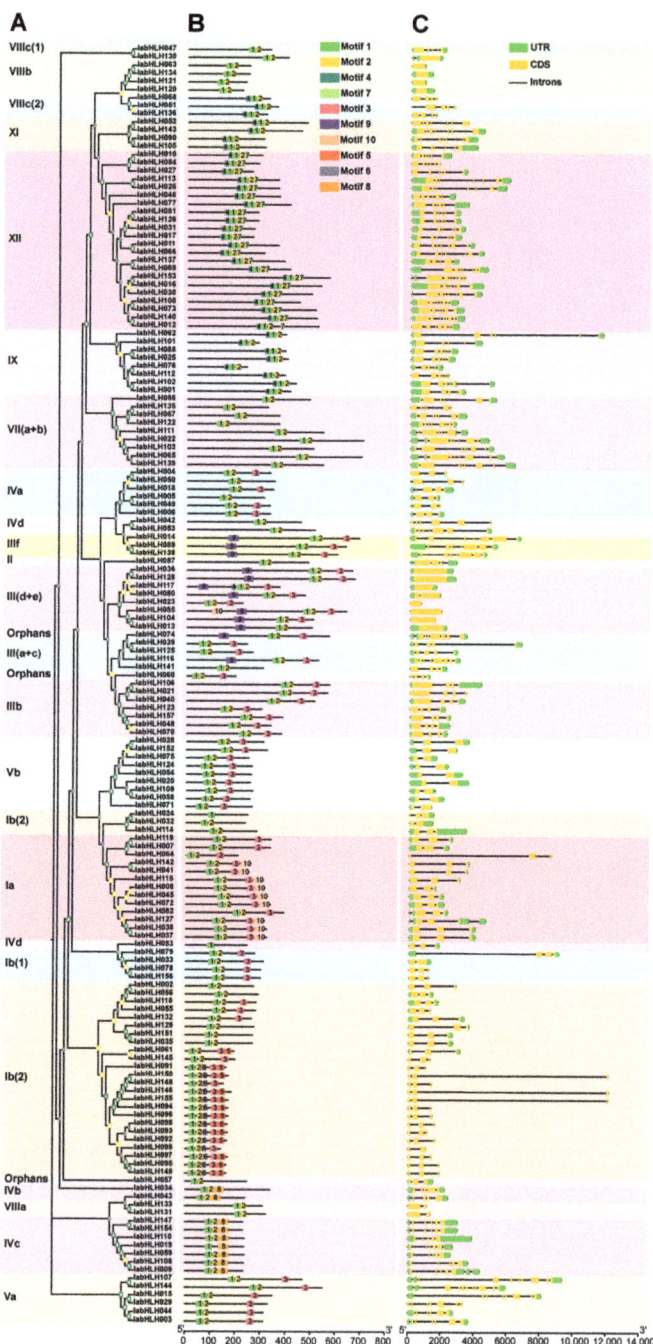

Figure 5. Phylogenetic tree, conserved motif, and exon-intron structure of IabHLHs. (**A**) The phylogenetic tree containing 157 IabHLH protein sequences; (**B**) distribution of conserved motifs within IabHLH proteins. The rectangles with different colors represent different motifs; (**C**) exon-intron structure of the 157 IabHLH genes. The green rectangles indicate exons, while the orange rectangles indicate untranslated regions (UTRs). The black solid lines among the rectangles indicate introns.

2.6. Analysis of Gene Duplication Events of IabHLH Genes

To explore the potential driven force of the *IabHLH* gene family expansion, the Dup-Gen_finder was performed to analyze gene duplication events, including WGD (whole-genome duplication), TD (tandem duplication), PD (proximal duplication), TRD (transposed duplication), and DSD (dispersed duplication). A total of 227 duplicated gene pairs derived from five duplication events were identified, with the maximum number of duplicated gene pairs derived from DSD duplication events (126 duplicated gene pairs), followed by WGD-derived duplicated gene pairs with 61. In contrast, TD and PD duplication events had the fewest number of duplicated gene pairs with only five. Moreover, we identified 30 TRD-derived duplicated gene pairs (Figure 6A and Table S4). These results suggested that the DSD and WGD duplication events played a pivotal role in the *IabHLH* gene family expansion.

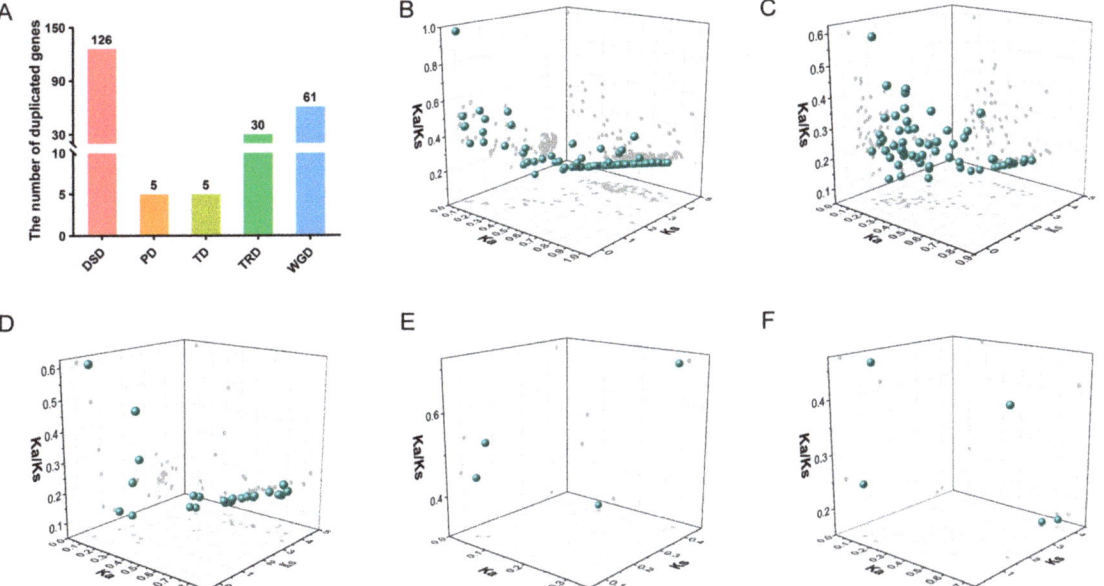

Figure 6. Gene duplication events of *IabHLH* genes. (**A**) The number distribution of *IabHLH* duplicated gene pairs in five duplication events. (**B–F**) The Ka, Ks, and Ka/Ks value distribution of the duplicated gene pairs in five gene duplication events. The DSD, WGD, TRD, PD, and TD events were from (**B–F**), respectively. More information concerning the duplicated genes in five gene duplication events was shown in Table S4.

To investigate the selection pressure on the *IabHLH* genes, the values of the non-synonymous substitution rate (Ka), the synonymous substitution rate (Ks), and Ka/Ks of the duplicated gene pairs derived from five replication events were calculated. As shown in Figure 6B–F, the values of Ka/Ks of all duplicated gene pairs were lower than 1, indicating these genes were subjected to the purifying selection.

2.7. Expression Levels of IabHLHs in Two Varieties

The purple-stemmed *I. aquatica*, which is abundant in anthocyanins, exhibits distinct phenotypic differences relative to the common green-stemmed *I. aquatica* (Figure 7A). To explore the role of the *IabHLH* genes in anthocyanin accumulation in the stems of *I. aquatica*, we analyzed the RNA-seq datasets from the purple-stemmed and green-stemmed varieties. Herein, the expression levels of 157 *IabHLH* genes in two varieties were quantified by calculating gene FPKM values according to the RNA-seq datasets. At this end, the *IabHLHs* with expression levels that met the criteria ($|\log_2 FC| > 1$, FDR < 0.05, and p-value < 0.05) were considered differentially expressed genes (DEGs). A total of 13 DEGs were identified between the two varieties, including six upregulated DEGs (priority of the fold change: *IabHLH027* > *IabHLH130* > *IabHLH084* > *IabHLH022* > *IabHLH112* > *IabHLH085*) and seven downregulated DEGs (priority of the fold change: *IabHLH037* > *IabHLH114* > *IabHLH142* > *IabHLH057* > *IabHLH030* > *IabHLH043* > *IabHLH116*) in the purple-stemmed *I. aquatica* compared to the green-stemmed *I. aquatica* (Table S5). Of these, the log-transformed fold change of *IabHLH027* belonging to subfamily XII between the two varieties was the highest, with 11.3, suggesting its potential positive role in regulating anthocyanin biosynthesis in *I. aquatica*. Contrarily, the *IabHLH037* belonging to subfamily Ia was significantly downregulated in purple-stemmed *I. aquatica*. The expression trends of 13 DEGs determined by RNA-seq were validated using qRT-PCR. As shown in Figure 7B–G, the expression trends of all upregulated DEGs were highly consistent with those detected by qRT-PCR. However, three downregulated DEGs including *IabHLH142* (Figure 7J), *IabHLH057* (Figure 7K), and *IabHLH043* (Figure 7M) determined by RNA-seq had opposite expression trends of those detected by qRT-PCR.

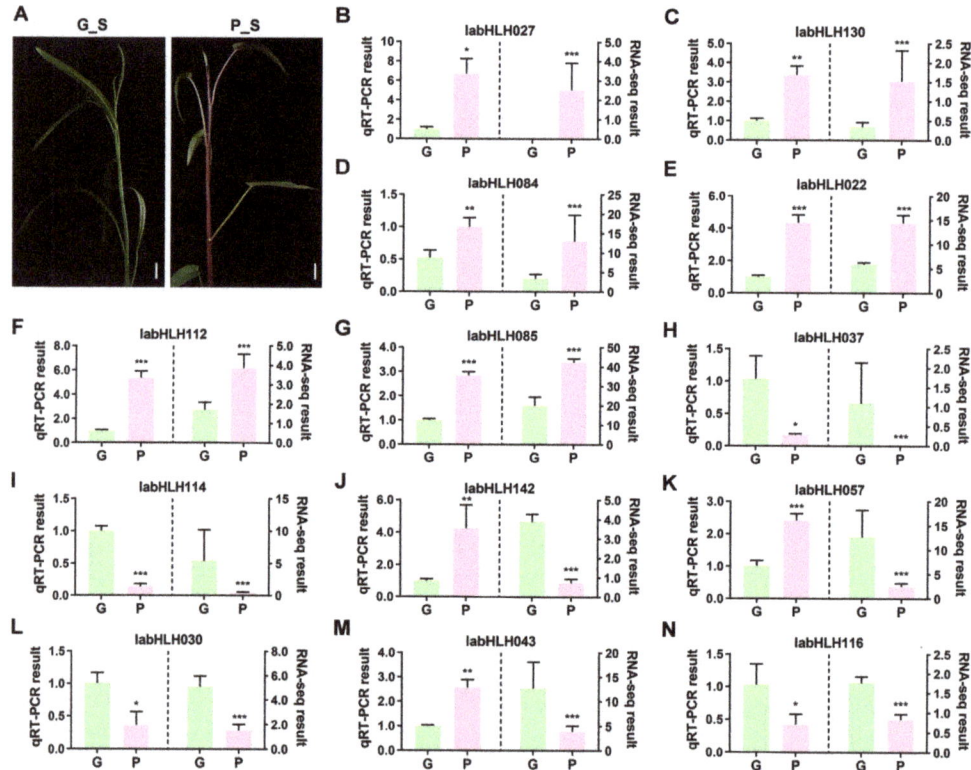

Figure 7. Expression analysis of the 13 DEGs. (**A**) The phenotype of two varieties. The G_S represents green-stemmed *I. aquatica*, while the P_S represents purple-stemmed *I. aquatica*. Scale bars are 2 cm. (**B–N**) The qRT-PCR validation of the expression trends of 13 DEGs was determined by RNA-seq. The left Y-axis and histogram indicated the relative expression levels of the *IabHLH* gene quantified by qRT-PCR. The right Y-axis and histogram indicate the FPKM values of the *IabHLH* gene obtained by RNA-seq datasets. Statistical analysis was performed using Student's *t*-test. Data are shown as mean ± SD of three biological replicates. The *, **, and *** indicate $p < 0.05$, $p < 0.01$, and $p < 0.001$, respectively.

2.8. Analysis of Cis-Acting Elements of DEGs

To gain more information concerning the potential regulatory functions of 13 DEGs, the cis-acting elements within their promoter regions were analyzed by using the PlantCARE program. Except for the common cis-acting elements (TATA-box and CAAT-box) and ones of unknown function, a total of 338 cis-acting elements were detected in 13 DEGs, which were categorized as four classes, including stress-responsive elements, phytohormone-responsive elements, and plant growth and development-responsive elements (Table S6). Of these, the light-responsive elements were the largest, with 20 types including 172 elements, followed by the phytohormone-responsive elements, with 10 types including 104 elements, while the plant growth and development-responsive elements were the least, with only two types including 12 elements. The stress responsive-elements were 50, which were classified into five types. The number distribution of each cis-acting element differs in different DEGs. The light-responsive elements and the phytohormone-responsive elements were detected in all DEGs. Except for the *IabHLH114*, all DEGs had stress-responsive elements (Figure 8A), indicating their importance in plant stress response. The *IabHLH027* with the highest expression fold change between the two varieties was rich in light-responsive elements and phytohormone-responsive elements, suggesting a critical role in regulating light and phytohormone signaling pathways in *I. aquatica*.

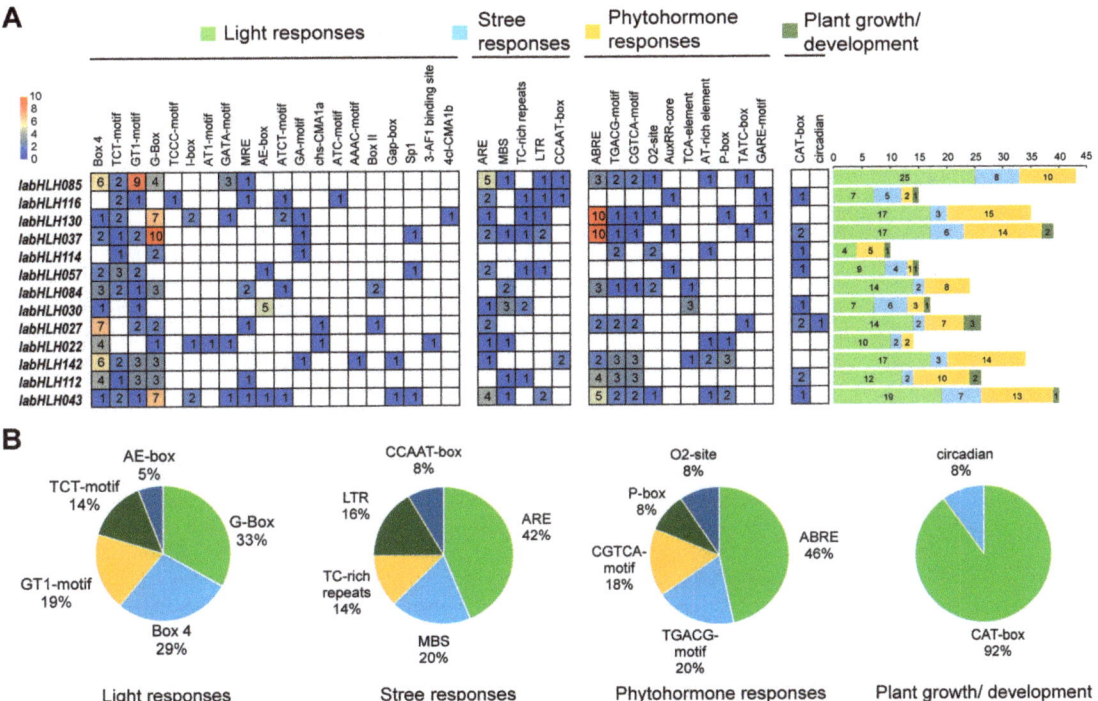

Figure 8. Cis-acting elements within the promoter region of the 13 DEGs. (**A**) Heatmap of the cis-acting elements. The stacked bars on the right side indicate the number distribution of different cis-acting elements in the 13 DEGs; (**B**) percentage of different cis-acting elements in the respective class.

In addition, the number of elements that belonged to the respective class varied greatly. The top five elements of the light-responsive elements were G-Box, Box 4, GT1-motif, TCT-motif, and AE-box. The AREs were the most abundant elements of stress-responsive elements, accounting for 42%. Among the phytohormone-responsive elements, the ABREs were the most numerous elements, accounting for 46%. Only one circadian-related element was detected in the plant growth and development-responsive elements (Figure 8B).

3. Discussion

Basic-helix-loop-helix (bHLH) proteins are widely found in eukaryotic organisms, which compose one of the largest transcription factor families in plants [38]. bHLH proteins play a vital role in regulating plant biological processes, including plant growth and development, metabolic activities, and various stress responses [39,40]. The increasing release of numerous plant genomic databases facilitates genome-wide identification of transcription factors. To date, large amounts of bHLH transcription factors have been identified in diverse plant species at the whole genome level, including model plants such as *Arabidopsis thaliana* [7] and *Nicotiana tabacum* [41], grain crops such as *Oryza sativa* [42] and *Zea mays* [6]; vegetable crops such as *Brassica rapa* [43] and *Solanum lycopersicum* [44], and fruit trees such as *Vitis vinifera* [45] and *Malus×domestica* [46]. Intriguingly, the number of *bHLH* genes varied among different plants, while it was not proportional to their genome sizes. Moreover, studies showed that the number of *bHLH* genes in higher plants is higher than that in lower plants [5,47], suggesting a pivotal role of *bHLHs* in higher plants. In this study, a total of 157 *bHLH* genes were detected in the *I. aquatica* genome based on its high-quality genome data [33]. Most IabHLH proteins were acidic and thermostable. All IabHLH proteins were hydrophilic, a principal property for transcription factors to perform their biological functions. Subcellular localization prediction showed that most IabHLH proteins were localized in the nucleus, while some were localized in the chloroplast, endomembrane system, and extracellular space, inferring that IabHLH proteins might play a distinct role in different plant organelles. These results laid a foundation for further exploring the biological function of *IabHLH* genes.

Studies have shown that the *bHLH* gene family was subdivided into a distinct number of subgroups in plants, probably due to the lack of definite classification criteria for low-conserved sequences outside the bHLH domain. Yang et al. identified 175 *bHLH* genes in *Malus × domestica* and classified them into 23 subgroups [46]. Sun et al. subdivided 159 bHLH genes of *Solanum lycopersicum* into 21 clades [44]. A total of 115 *bHLH* genes in *Vitis davidii* were gathered into 25 branches [8], while 94 *bHLH* genes in *Vitis vinifera* were clustered into 15 subclasses [45]. In the present study, phylogenetic analysis of bHLH sequences from *I. aquatica* and *A. thaliana* indicated that the 157 *IabHLH* genes could be subdivided into 23 subgroups, with three orphan genes. Compared to the *AtbHLH* gene family, although the members of the *IabHLH* gene family were lost in X, XIII, XIV, and XV subfamilies, the *IabHLH* genes in 11 subgroups (including Ia, Ib (1), Ib (2), III (a+c), IIIb, IVa, IVc, IVd, Va, Vb, IX, XII) have expanded, suggesting their key roles in the biological activities of *I. aquatica*. Moreover, three IabHLH proteins (including IabHLH014, IabHLH089, and IabHLH138) were found in the IIIf subfamily. Some reports showed that the *bHLHs* in *O. sativa* belonging to the IIIf subfamily were associated with the regulation of anthocyanin biosynthesis [48,49]. However, the expression levels of these three *IabHLH* genes have no significant difference between purple-stemmed *I. aquatica* and green-stemmed *I. aquatica*, probably other *IabHLH* genes participate in the regulation of purple color formation in the *I. aquatica* stems.

Similar to the previous reports, two adjacent highly conserved motifs and highly conserved gene structural patterns within each subclass were identified [5,50]. Studies have shown that Glu-5 and Arg-8/Arg-9 in the basic region of the bHLH structural domain play an important role in DNA binding, while Leu-19 and Leu-51 in the HLH region may be required for dimerization [1,51]. In the present study, the highly conserved amino acids were identified within the basic region (including Glu-5, Arg-8, and Arg-9), helix 1 region (Leu-19), and helix 2 region (Leu-51) (Figure 4).

Gene duplication events play an indispensable driving role in gene family expansion, which includes WGD, DSD, PD, TRD, and TD [52]. The expansions of the F-box gene family in *Gossypium hirsutum* were mainly impacted by the WGD and TD [53], while the WGD, TRD, and DSD play a dominant role in the expansions of the *P-type ATPase* gene family in *Pyrus bretschneideri* [54]. Similar to previous studies on *BAHD* and *MYB* gene families [52,55], the DSD and WGD made significant contributions to the expansions of the *IabHLH* gene family. The selection pressure could be estimated by calculating the Ka/Ks value. The Ka/Ks value of one, greater than one, and less than one indicated neutral evolution, positive selection (Darwinian selection), and negative selection (purifying selection), respectively [56,57]. The Ka/Ks values of all duplicated gene pairs in five gene duplication events were less than 1, indicating that *IabHLH* genes were subjected to the purifying selection. In addition, the syntenic relationship analysis between *I. aquatica* and other plant species verified closer evolutionary distance between similar plant taxa, suggesting the reliability of the analysis approach.

As transcription factors, the bHLH proteins directly or indirectly regulate the expression level of structural genes and thereby modulating anthocyanin biosynthesis. In grape berries, the expression level of *VdbHLH037* was highly correlated with anthocyanin accumulation in different varieties. Transgenic expression of *VdbHLH037* into *Arabidopsis* resulted in significant upregulation of structural genes involved in anthocyanin biosynthesis and high accumulation of anthocyanins. In strawberry fruits, FvbHLH9 positively regulates light-dependent anthocyanin biosynthesis by forming a heterodimer with HY [58], while in rapeseed, BnbHLH92a impinges on the biosynthesis of anthocyanins by interacting with the BnTTG1 [59]. In the present study, we found that the expression level of 13 *IabHLH* genes exhibited significant differences between purple-stemmed *I. aquatica* and green-stemmed *I. aquatica*. Of these, six DEGs were dramatically upregulated in purple-stemmed *I. aquatica* compared to green-stemmed *I. aquatica*. In particular, the *IabHLH027* belonging to the subgroup XII exhibited approximately 8-fold higher in purple-stemmed *I. aquatica* than that in green-stemmed *I. aquatica* (Figure 7B). The bHLHs of subgroup XII were correlated with positive regulation of the brassinosteroid (BR) signaling pathway [60]. Previous reports showed that BR together with other phytohormone signaling led to the enhancement of anthocyanin accumulation in plants and this process was potentially regulated by ternary MYB-bHLH-WD transcriptional complexes [61,62]. RNA-seq is widely recognized as a highly reliable and accurate method for quantifying gene expression levels, and therefore, additional qRT-PCR validation on RNA-seq results may not always be necessary when using the same samples as those used in RNA-seq. However, given the limited availability of RNA-seq samples and the potential for depletion, qRT-PCR validation of the gene of interest would be valuable when analyzing additional samples grown under the same conditions [63]. In this study, we conducted qRT-PCR analysis on additional samples and observed discrepancies in the expression trends of three downregulated differentially expressed genes (DEGs) between RNA-seq and qRT-PCR. These inconsistencies may be attributed to methodological biases that can affect the accuracy of the results obtained from each method [64–66]. The expression profiles of genes are closely related to their cis-regulatory elements. For instance, some *bHLH* genes in *Hordeum vulgare* containing several drought induction-associated cis-elements within the promoter regions significantly responded to drought stress stimulation [9]. The *Brassica napus bHLH* genes that possess abundant phytohormone-responsive elements were positively regulated upon exposure to more than one phytohormone, such as auxin (IAA), gibberellin (GA3), cytokinin (6-BA), abscisic acid (ABA), and ethylene (ACC) [5]. In the present study, we discovered that all DEGs had plentiful light-responsive elements and phytohormone-responsive elements, suggesting their essential function in *I. aquatica*'s response to light and phytohormone signaling. Taken together, these findings provide us valuable clues for further exploring the molecular function of *IabHLH* in regulating purple color formation in *I. aquatica* stems.

4. Materials and Methods
4.1. Identification of bHLH Genes in the I. aquatica Genome

The whole genome database of *I. aquatica* (BioProject: PRJCA002216) was used as the reference genome. The full-length sequences of AtbHLH proteins were retrieved from the TAIR (https://www.arabidopsis.org/. accessed on 1 August 2022). The BLASTP was performed to search against the *I. aquatica* proteome using AtbHLHs as queries. The HMM (Hidden Markov Model) profile of the bHLH domain (PF00010) was downloaded from the Pfam database (http://pfam-legacy.xfam.org/family/PF00010. accessed on 1 August 2022), which was used to retrieve putative *bHLH* genes from the *I. aquatica* genome. The results of the two searches were integrated to obtain candidate *IabHLH* genes. The online program CDD (https://www.ncbi.nlm.nih.gov/Structure/cdd/wrpsb.cgi. accessed on 8 August 2022), SMART (http://smart.embl-heidelberg.de/. accessed on 7 August 2022), and Pfam (http://pfam-legacy.xfam.org/search#tabview=tab1. accessed on 7 August 2022) were used to verify the structure of candidate *IabHLH* genes, while the redundant sequences and the sequences without bHLH structural domains were manually deleted.

4.2. Analysis of Physicochemical Parameters of IabHLH Proteins

The physicochemical characteristics of the IabHLH proteins were analyzed using the ExPASy program (http://web.expasy.org/protparam/. accessed on 22 November 2022), including the number of amino acids, molecular weight (kDa), theoretical isoelectric point (pI), instability index (Ii), aliphatic index (Ai), and grand average of hydropathicity index (GRAVY). Subcellular localization prediction of IabHLHs was carried out using BUSCA (http://busca.biocomp.unibo.it/. accessed on 13 January 2023).

4.3. Analysis of Phylogenetic Relationships of IabHLHs and AtbHLHs

The AtbHLH protein sequences without conserved bHLH domain were excluded [43,60,67,68]. The ClustalW2.0 program with default parameters was used to perform multiple sequence alignment of IabHLHs and AtbHLHs, and then the phylogenetic tree was constructed using the neighbor-joining (NJ) method within MEGA-X (bootstrap value: 1000) [69]. The phylogenetic tree was visualized using the online tool Evolview-v2 [70] (https://evolgenius.info/. accessed on 13 January 2023). The TBtools (version 1.108) software was used to visualize the sequence logo of the IabHLH domain.

The SWISS-MODEL (https://www.swissmodel.expasy.org/. accessed on 14 January 2023) homology modeling was performed to predict protein tertiary structures. The PyMOL (version 4.6) was used to visualize the protein tertiary structural models [71].

4.4. Analysis of Conserved Motif, Gene Structure, and Cis-Acting Elements

The conserved motif within IabHLH proteins was analyzed using the online program MEME (https://meme-suite.org/meme/tools/meme. accessed on 10 December 2022). The GFF annotation file of *I. aquatica* was used to obtain the intron-exon distributions of the *IabHLHs*.

The promoter sequences of the *IabHLH* genes (the upstream 2000 bp sequences from CDS of *IabHLH* genes) were extracted using TBtools based on the full-length DNA sequences of the *I. aquatica* genomes. The PlantCare online database (http://bioinformatics.psb.ugent.be/webtools/PlantCare/html/, accessed on 1 January 2023) was used to predict the cis-acting elements within promoter regions of *IabHLHs*. The results were visualized using TBtools.

4.5. Analysis of Chromosomal Distribution, Collinearity, and Gene Duplication Events

The chromosomal distribution of *IabHLHs* and the gene density within the chromosome of *I. aquatica* were visualized using TBtools. The collinearity relationships of *I. aquatica*, *A. thaliana*, *Ipomoea batatas*, and *Oryza sativa* were analyzed using Multiple Collinearity Scan Toolkit (MCScanX), The results were visualized using the "Advanced Circos" function within TBtools [72].

Gene duplication events of *IabHLHs* were analyzed using the dupGen_finder pipeline, which includes WGD (whole-genome duplication), TD (tandem duplication), PD (proximal duplication), TRD (transposed duplication), and DSD (dispersed duplication). The

MYN method within KaKs_Calculator 2.0 (https://GitHub.com/qiao-xin/scripts_for_gb. accessed on 23 December 2022) was used to calculate the value of Ka, Ks, and Ka/Ks.

4.6. Plant Materials and Sampling

The purple-stemmed *I. aquatica* and green-stemmed *I. aquatica* were cultivated by cutting in Hoagland nutrient solution at 28 °C with a light cycle of 16 h (light)/8 h (dark). The uniformly colored stems of two varieties were harvested after four weeks of culture and immediately frozen in liquid nitrogen for a moment, which was then stored at −80 °C until use.

4.7. Transcriptome Analysis of IabHLH Genes

The transcript abundance of IabHLH genes was quantified by calculating the value of FPKM (fragments per kilobase per million mapped reads) obtained from the RNA-seq data (BioProject: PRJNA814206). The genes with expression levels that met the criteria ($|\log_2 FC| > 1$, FDR < 0.05, and p-value < 0.05) were considered differentially expressed genes (DEGs).

4.8. qRT-PCR Analysis

Total RNA was extracted from the stems of two varieties using the RNAprep Pure extraction kit (Tiangen, Beijing, China) according to the manufacturer's instructions. The RNA quality was assessed using the microspectrophotometer K5600C (KAIAO, Beijing, China) and agarose gel electrophoresis. The cDNA was synthesized using the FastKing cDNA kit (Tiangen, Beijing, China). The gene-specific primers were designed using Primer Premier 6.0 (Table S7). The reaction mixture of qRT-PCR includes 10 μL 2xSYBR green mix (ChamQ Universal SYBR qPCR Master Mix), 1 μL cDNA, 1 μL primers (0.5 μL forward primer and 0.5 μL reverse primer), and 8 μL PCR-grade water. The qRT-PCR was conducted on the QuantStudio Real-Time Fluorescence PCR System (ThermoFisher, MA, USA). The conditions of two-step qRT-PCR are as follows: 95 °C 3 min followed by 40 cycles of 95 °C 10 s and 55 °C 30 s. The *GAPDH* gene of *I. aquatica* was used as the internal reference gene. The relative expression of the target gene was calculated using the $2^{-\Delta\Delta Ct}$ method. Data were plotted using GraphPad Prism 8.

5. Conclusions

In this study, a total of 157 bHLH genes were identified in *I. aquatica* at the whole genome level, which were subdivided into 23 subgroups. The bHLHs within the identical subgroup had similar motifs and gene structures. The DSD and WGD duplication events made significant contributions to the expansion of the *IabHLH* gene family. All duplicated gene pairs were subjected to the purifying selection. A total of 13 DEGs were identified between the purple-stemmed and green-stemmed varieties. The expression profiling revealed that the *IabHLH027* belonging to subgroup XII was dramatically upregulated in the purple-stemmed *I. aquatica* relative to the green-stemmed *I. aquatica*, suggesting a potential positive role in regulating anthocyanin accumulation in the *I. aquatica* stems. Our study provides important gene candidates for further exploring the regulatory network of anthocyanin biosynthesis of *I. aquatica*.

Supplementary Materials: The following supporting information can be downloaded at: https://www.mdpi.com/article/10.3390/ijms24065652/s1.

Author Contributions: Conceptualization, W.B.; methodology, W.B. and Z.L.; software, Z.L. and X.F.; validation, Z.L., X.F. and H.X.; formal analysis, Z.L., X.F., H.X., Y.Z., Z.S. and G.Z.; investigation, Z.L., X.F., H.X., Y.Z., Z.S. and G.Z.; writing—original draft preparation, Z.L. and X.F.; writing—review and editing, W.B.; funding acquisition, W.B. All authors have read and agreed to the published version of the manuscript.

Funding: This research was funded by the Hainan Provincial Natural Science Foundation of China (2019RC059 and 320MS008) and the Initial Funds for the High-level Talents of Hainan University (KYQD(ZR)1935).

Institutional Review Board Statement: Not applicable.

Informed Consent Statement: Not applicable.

Data Availability Statement: The datasets presented in this study can be found in online repositories. The names of the repository/repositories and accession number(s) can be found in the article/Supplementary Materials.

Conflicts of Interest: The authors declare no conflict of interest.

References

1. Carretero-Paulet, L.; Galstyan, A.; Roig-Villanova, I.; Martinez-Garcia, J.F.; Bilbao-Castro, J.R.; Robertson, D.L. Genome-wide classification and evolutionary analysis of the bHLH family of transcription factors in Arabidopsis, poplar, rice, moss, and algae. *Plant Physiol.* **2010**, *153*, 1398–1412. [CrossRef] [PubMed]
2. Zhang, X.Y.; Qiu, J.Y.; Hui, Q.L.; Xu, Y.Y.; He, Y.Z.; Peng, L.Z.; Fu, X.Z. Systematic analysis of the basic/helix-loop-helix (bHLH) transcription factor family in pummelo (Citrus grandis) and identification of the key members involved in the response to iron deficiency. *BMC Genom.* **2020**, *21*, 233. [CrossRef]
3. Hao, Y.; Zong, X.; Ren, P.; Qian, Y.; Fu, A. Basic Helix-Loop-Helix (bHLH) Transcription Factors Regulate a Wide Range of Functions in Arabidopsis. *Int. J. Mol. Sci.* **2021**, *22*, 7152. [CrossRef] [PubMed]
4. An, F.; Xiao, X.; Chen, T.; Xue, J.; Luo, X.; Ou, W.; Li, K.; Cai, J.; Chen, S. Systematic Analysis of bHLH Transcription Factors in Cassava Uncovers Their Roles in Postharvest Physiological Deterioration and Cyanogenic Glycosides Biosynthesis. *Front. Plant Sci.* **2022**, *13*, 901128. [CrossRef] [PubMed]
5. Ke, Y.Z.; Wu, Y.W.; Zhou, H.J.; Chen, P.; Wang, M.M.; Liu, M.M.; Li, P.F.; Yang, J.; Li, J.N.; Du, H. Genome-wide survey of the bHLH super gene family in Brassica napus. *BMC Plant Biol.* **2020**, *20*, 115. [CrossRef] [PubMed]
6. Zhang, T.; Lv, W.; Zhang, H.; Ma, L.; Li, P.; Ge, L.; Li, G. Genome-wide analysis of the basic Helix-Loop-Helix (bHLH) transcription factor family in maize. *BMC Plant Biol.* **2018**, *18*, 235. [CrossRef]
7. Bailey, P.C.; Martin, C.; Toledo-Ortiz, G.; Quail, P.H.; Huq, E.; Heim, M.A.; Jakoby, M.; Werber, M.; Weisshaar, B. Update on the basic helix-loop-helix transcription factor gene family in Arabidopsis thaliana. *Plant Cell* **2003**, *15*, 2497–2502. [CrossRef]
8. Li, M.; Sun, L.; Gu, H.; Cheng, D.; Guo, X.; Chen, R.; Wu, Z.; Jiang, J.; Fan, X.; Chen, J. Genome-wide characterization and analysis of bHLH transcription factors related to anthocyanin biosynthesis in spine grapes (Vitis davidii). *Sci. Rep.* **2021**, *11*, 6863. [CrossRef]
9. Ke, Q.; Tao, W.; Li, T.; Pan, W.; Chen, X.; Wu, X.; Nie, X.; Cui, L. Genome-wide Identification, Evolution and Expression Analysis of Basic Helix-loop-helix (bHLH) Gene Family in Barley (Hordeum vulgare L.). *Curr. Genom.* **2020**, *21*, 621–644. [CrossRef]
10. Guo, J.; Sun, B.; He, H.; Zhang, Y.; Tian, H.; Wang, B. Current Understanding of bHLH Transcription Factors in Plant Abiotic Stress Tolerance. *Int. J. Mol. Sci.* **2021**, *22*, 4921. [CrossRef]
11. Sun, X.; Wang, Y.; Sui, N. Transcriptional regulation of bHLH during plant response to stress. *Biochem. Biophys. Res. Commun.* **2018**, *503*, 397–401. [CrossRef] [PubMed]
12. Ito, S.; Song, Y.H.; Josephson-Day, A.R.; Miller, R.J.; Breton, G.; Olmstead, R.G.; Imaizumi, T. FLOWERING BHLH transcriptional activators control expression of the photoperiodic flowering regulator CONSTANS in Arabidopsis. *Proc. Natl. Acad. Sci. USA* **2012**, *109*, 3582–3587. [CrossRef] [PubMed]
13. Zhang, L.Y.; Bai, M.Y.; Wu, J.; Zhu, J.Y.; Wang, H.; Zhang, Z.; Wang, W.; Sun, Y.; Zhao, J.; Sun, X.; et al. Antagonistic HLH/bHLH transcription factors mediate brassinosteroid regulation of cell elongation and plant development in rice and Arabidopsis. *Plant Cell* **2009**, *21*, 3767–3780. [CrossRef] [PubMed]
14. An, J.P.; Wang, X.F.; Zhang, X.W.; Xu, H.F.; Bi, S.Q.; You, C.X.; Hao, Y.J. An apple MYB transcription factor regulates cold tolerance and anthocyanin accumulation and undergoes MIEL1-mediated degradation. *Plant Biotechnol. J.* **2020**, *18*, 337–353. [CrossRef]
15. Chen, W.; Miao, Y.; Ayyaz, A.; Hannan, F.; Huang, Q.; Ulhassan, Z.; Zhou, Y.; Islam, F.; Hong, Z.; Farooq, M.A.; et al. Purple stem Brassica napus exhibits higher photosynthetic efficiency, antioxidant potential and anthocyanin biosynthesis related genes expression against drought stress. *Front. Plant Sci.* **2022**, *13*, 936696. [CrossRef] [PubMed]
16. Zhang, Q.; Zhai, J.; Shao, L.; Lin, W.; Peng, C. Accumulation of Anthocyanins: An Adaptation Strategy of Mikania micrantha to Low Temperature in Winter. *Front. Plant Sci.* **2019**, *10*, 1049. [CrossRef] [PubMed]
17. Tena, N.; Martín, J.; Asuero, A.G. State of the Art of Anthocyanins: Antioxidant Activity, Sources, Bioavailability, and Therapeutic Effect in Human Health. *Antioxidants* **2020**, *9*, 451. [CrossRef]
18. Liu, Y.; Ma, K.; Qi, Y.; Lv, G.; Ren, X.; Liu, Z.; Ma, F. Transcriptional Regulation of Anthocyanin Synthesis by MYB-bHLH-WDR Complexes in Kiwifruit (Actinidia chinensis). *J. Agric. Food Chem.* **2021**, *69*, 3677–3691. [CrossRef]
19. Jaakola, L. New insights into the regulation of anthocyanin biosynthesis in fruits. *Trends Plant Sci.* **2013**, *18*, 477–483. [CrossRef]
20. Wan, L.; Li, B.; Lei, Y.; Yan, L.; Huai, D.; Kang, Y.; Jiang, H.; Tan, J.; Liao, B. Transcriptomic profiling reveals pigment regulation during peanut testa development. *Plant Physiol. Biochem.* **2018**, *125*, 116–125. [CrossRef]

21. Zhang, N.; Qi, Y.; Zhang, H.J.; Wang, X.; Li, H.; Shi, Y.; Guo, Y.D. Genistein: A Novel Anthocyanin Synthesis Promoter that Directly Regulates Biosynthetic Genes in Red Cabbage in a Light-Dependent Way. *Front. Plant Sci.* **2016**, *7*, 1804. [CrossRef] [PubMed]
22. Liu, Y.; Tikunov, Y.; Schouten, R.E.; Marcelis, L.F.M.; Visser, R.G.F.; Bovy, A. Anthocyanin Biosynthesis and Degradation Mechanisms in Solanaceous Vegetables: A Review. *Front. Chem.* **2018**, *6*, 52. [CrossRef] [PubMed]
23. Gonzalez, A.; Zhao, M.; Leavitt, J.M.; Lloyd, A.M. Regulation of the anthocyanin biosynthetic pathway by the TTG1/bHLH/Myb transcriptional complex in Arabidopsis seedlings. *Plant J.* **2008**, *53*, 814–827. [CrossRef] [PubMed]
24. Dubos, C.; Stracke, R.; Grotewold, E.; Weisshaar, B.; Martin, C.; Lepiniec, L. MYB transcription factors in Arabidopsis. *Trends Plant Sci.* **2010**, *15*, 573–581. [CrossRef] [PubMed]
25. Nesi, N.; Debeaujon, I.; Jond, C.; Pelletier, G.; Caboche, M.; Lepiniec, L. The TT8 gene encodes a basic helix-loop-helix domain protein required for expression of DFR and BAN genes in Arabidopsis siliques. *Plant Cell* **2000**, *12*, 1863–1878. [CrossRef]
26. Xiang, L.L.; Liu, X.F.; Li, X.; Yin, X.R.; Grierson, D.; Li, F.; Chen, K.S. A Novel bHLH Transcription Factor Involved in Regulating Anthocyanin Biosynthesis in Chrysanthemums (*Chrysanthemum morifolium* Ramat.). *PLoS ONE* **2015**, *10*, e0143892. [CrossRef]
27. Xie, X.B.; Li, S.; Zhang, R.F.; Zhao, J.; Chen, Y.C.; Zhao, Q.; Yao, Y.X.; You, C.X.; Zhang, X.S.; Hao, Y.J. The bHLH transcription factor MdbHLH3 promotes anthocyanin accumulation and fruit colouration in response to low temperature in apples. *Plant Cell Environ.* **2012**, *35*, 1884–1897. [CrossRef]
28. Liu, Y.; Lin-Wang, K.; Espley, R.V.; Wang, L.; Li, Y.; Liu, Z.; Zhou, P.; Zeng, L.; Zhang, X.; Zhang, J.; et al. StMYB44 negatively regulates anthocyanin biosynthesis at high temperatures in tuber flesh of potato. *J. Exp. Bot.* **2019**, *70*, 3809–3824. [CrossRef]
29. Wang, H.; Wang, X.; Yu, C.; Wang, C.; Jin, Y.; Zhang, H. MYB transcription factor PdMYB118 directly interacts with bHLH transcription factor PdTT8 to regulate wound-induced anthocyanin biosynthesis in poplar. *BMC Plant Biol.* **2020**, *20*, 173. [CrossRef]
30. Hichri, I.; Heppel, S.C.; Pillet, J.; Léon, C.; Czemmel, S.; Delrot, S.; Lauvergeat, V.; Bogs, J. The Basic Helix-Loop-Helix Transcription Factor MYC1 Is Involved in the Regulation of the Flavonoid Biosynthesis Pathway in Grapevine. *Mol. Plant* **2010**, *3*, 509–523. [CrossRef]
31. Li, H.; Yang, Z.; Zeng, Q.; Wang, S.; Luo, Y.; Huang, Y.; Xin, Y.; He, N. Abnormal expression of bHLH3 disrupts a flavonoid homeostasis network, causing differences in pigment composition among mulberry fruits. *Hortic. Res.* **2020**, *7*, 83. [CrossRef] [PubMed]
32. Hefny Gad, M.; Tuenter, E.; El-Sawi, N.; Younes, S.; El-Ghadban, E.M.; Demeyer, K.; Pieters, L.; Vander Heyden, Y.; Mangelings, D. Identification of some Bioactive Metabolites in a Fractionated Methanol Extract from Ipomoea aquatica (Aerial Parts) through TLC, HPLC, UPLC-ESI-QTOF-MS and LC-SPE-NMR Fingerprints Analyses. *Phytochem. Anal.* **2018**, *29*, 5–15. [CrossRef] [PubMed]
33. Hao, Y.; Bao, W.; Li, G.; Gagoshidze, Z.; Shu, H.; Yang, Z.; Cheng, S.; Zhu, G.; Wang, Z. The chromosome-based genome provides insights into the evolution in water spinach. *Sci. Hortic.* **2021**, *289*, 110501. [CrossRef]
34. Dua, T.K.; Dewanjee, S.; Khanra, R.; Bhattacharya, N.; Bhaskar, B.; Zia-Ul-Haq, M.; De Feo, V. The effects of two common edible herbs, Ipomoea aquatica and Enhydra fluctuans, on cadmium-induced pathophysiology: A focus on oxidative defence and anti-apoptotic mechanism. *J. Transl. Med.* **2015**, *13*, 245. [CrossRef] [PubMed]
35. Lawal, U.; Maulidiani, M.; Ismail, I.S.; Khatib, A.; Abas, F. Discrimination of Ipomoea aquatica cultivars and bioactivity correlations using NMR-based metabolomics approach. *Plant Biosyst.—Int. J. Deal. All Asp. Plant Biol.* **2016**, *151*, 833–843. [CrossRef]
36. Alkiyumi, S.S.; Abdullah, M.A.; Alrashdi, A.S.; Salama, S.M.; Abdelwahab, S.I.; Hadi, A.H. Ipomoea aquatica extract shows protective action against thioacetamide-induced hepatotoxicity. *Molecules* **2012**, *17*, 6146–6155. [CrossRef]
37. Yuxin, Z.; Zheng, L.; Tangquan, Z.; Yuanyuan, H.; Shanhan, C.; Zhiwei, W.; Jie, Z.; Wuqiang, M.; Wenlong, B. Genome-wide Identification and Expression Analysis of WRKY Gene Family in Water Spinach (Ipomoea aquatica). *Mol. Plant Breed.* **2023**, *21*, 55–66. (In Chinese) [CrossRef]
38. Xiong, Y.; Liu, T.; Tian, C.; Sun, S.; Li, J.; Chen, M. Transcription factors in rice: A genome-wide comparative analysis between monocots and eudicots. *Plant Mol. Biol.* **2005**, *59*, 191–203. [CrossRef]
39. Samira, R.; Li, B.; Kliebenstein, D.; Li, C.; Davis, E.; Gillikin, J.W.; Long, T.A. The bHLH transcription factor ILR3 modulates multiple stress responses in Arabidopsis. *Plant Mol. Biol.* **2018**, *97*, 297–309. [CrossRef]
40. Feller, A.; Machemer, K.; Braun, E.L.; Grotewold, E. Evolutionary and comparative analysis of MYB and bHLH plant transcription factors. *Plant J.* **2011**, *66*, 94–116. [CrossRef]
41. Bai, G.; Yang, D.H.; Chao, P.; Yao, H.; Fei, M.; Zhang, Y.; Chen, X.; Xiao, B.; Li, F.; Wang, Z.Y.; et al. Genome-wide identification and expression analysis of NtbHLH gene family in tobacco (Nicotiana tabacum) and the role of NtbHLH86 in drought adaptation. *Plant Divers.* **2021**, *43*, 510–522. [CrossRef] [PubMed]
42. Wei, K.; Chen, H. Comparative functional genomics analysis of bHLH gene family in rice, maize and wheat. *BMC Plant Biol.* **2018**, *18*, 309. [CrossRef] [PubMed]
43. Song, X.M.; Huang, Z.N.; Duan, W.K.; Ren, J.; Liu, T.K.; Li, Y.; Hou, X.L. Genome-wide analysis of the bHLH transcription factor family in Chinese cabbage (*Brassica rapa* ssp. pekinensis). *Mol. Genet. Genom.* **2014**, *289*, 77–91. [CrossRef] [PubMed]
44. Sun, H.; Fan, H.J.; Ling, H.Q. Genome-wide identification and characterization of the bHLH gene family in tomato. *BMC Genom.* **2015**, *16*, 9. [CrossRef] [PubMed]

45. Wang, P.; Su, L.; Gao, H.; Jiang, X.; Wu, X.; Li, Y.; Zhang, Q.; Wang, Y.; Ren, F. Genome-Wide Characterization of bHLH Genes in Grape and Analysis of their Potential Relevance to Abiotic Stress Tolerance and Secondary Metabolite Biosynthesis. *Front. Plant Sci.* **2018**, *9*, 64. [CrossRef]
46. Yang, J.; Gao, M.; Huang, L.; Wang, Y.; van Nocker, S.; Wan, R.; Guo, C.; Wang, X.; Gao, H. Identification and expression analysis of the apple (Malus × domestica) basic helix-loop-helix transcription factor family. *Sci. Rep.* **2017**, *7*, 28. [CrossRef]
47. Wang, X.J.; Peng, X.Q.; Shu, X.C.; Li, Y.H.; Wang, Z.; Zhuang, W.B. Genome-wide identification and characterization of PdbHLH transcription factors related to anthocyanin biosynthesis in colored-leaf poplar (*Populus deltoids*). *BMC Genom.* **2022**, *23*, 244. [CrossRef]
48. Spelt, C.; Quattrocchio, F.; Mol, J.N.M.; Koes, R. anthocyanin1 of Petunia Encodes a Basic Helix-Loop-Helix Protein That Directly Activates Transcription of Structural Anthocyanin Genes. *Plant Cell* **2000**, *12*, 1619–1631. [CrossRef]
49. Hu, J.; Reddy, V.S.; Wessler, S.R. The rice R gene family: Two distinct subfamilies containing several miniature inverted-repeat transposable elements. *Plant Mol. Biol.* **2000**, *42*, 667–678. [CrossRef]
50. Hong, Y.; Ahmad, N.; Tian, Y.; Liu, J.; Wang, L.; Wang, G.; Liu, X.; Dong, Y.; Wang, F.; Liu, W.; et al. Genome-Wide Identification, Expression Analysis, and Subcellular Localization of Carthamus tinctorius bHLH Transcription Factors. *Int. J. Mol. Sci.* **2019**, *20*, 3044. [CrossRef]
51. Atchley, W.R.; Fitch, W.M. A natural classification of the basic helix-loop-helix class of transcription factors. *Proc. Natl. Acad. Sci. USA* **1997**, *94*, 5172–5176. [CrossRef] [PubMed]
52. Liu, Z.; Zhang, Y.; Altaf, M.A.; Hao, Y.; Zhou, G.; Li, X.; Zhu, J.; Ma, W.; Wang, Z.; Bao, W. Genome-wide identification of myeloblastosis gene family and its response to cadmium stress in Ipomoea aquatica. *Front. Plant Sci.* **2022**, *13*, 979988. [CrossRef] [PubMed]
53. Zhang, S.; Tian, Z.; Li, H.; Guo, Y.; Zhang, Y.; Roberts, J.A.; Zhang, X.; Miao, Y. Genome-wide analysis and characterization of F-box gene family in *Gossypium hirsutum* L. *BMC Genom.* **2019**, *20*, 993. [CrossRef] [PubMed]
54. Zhang, Y.; Li, Q.; Xu, L.; Qiao, X.; Liu, C.; Zhang, S. Comparative analysis of the P-type ATPase gene family in seven Rosaceae species and an expression analysis in pear (*Pyrus bretschneideri* Rehd.). *Genomics* **2020**, *112*, 2550–2563. [CrossRef]
55. Liu, C.; Qiao, X.; Li, Q.; Zeng, W.; Wei, S.; Wang, X.; Chen, Y.; Wu, X.; Wu, J.; Yin, H.; et al. Genome-wide comparative analysis of the BAHD superfamily in seven Rosaceae species and expression analysis in pear (*Pyrus bretschneideri*). *BMC Plant Biol.* **2020**, *20*, 14. [CrossRef]
56. Dong, H.; Chen, Q.; Dai, Y.; Hu, W.; Zhang, S.; Huang, X. Genome-wide identification of PbrbHLH family genes, and expression analysis in response to drought and cold stresses in pear (Pyrus bretschneideri). *BMC Plant Biol.* **2021**, *21*, 86. [CrossRef]
57. Starr, T.K.; Jameson, S.C.; Hogquist, K.A. Positive and negative selection of T cells. *Annu. Rev. Immunol.* **2003**, *21*, 139–176. [CrossRef]
58. Li, Y.; Xu, P.; Chen, G.; Wu, J.; Liu, Z.; Lian, H. FvbHLH9 Functions as a Positive Regulator of Anthocyanin Biosynthesis by Forming a HY5–bHLH9 Transcription Complex in Strawberry Fruits. *Plant Cell Physiol.* **2020**, *61*, 826–837. [CrossRef]
59. Hu, R.; Zhu, M.; Chen, S.; Li, C.; Zhang, Q.; Gao, L.; Liu, X.; Shen, S.; Fu, F.; Xu, X.; et al. BnbHLH92a negatively regulates anthocyanin and proanthocyanidin biosynthesis in *Brassica napus*. *Crop J.* **2022**. [CrossRef]
60. Pires, N.; Dolan, L. Origin and diversification of basic-helix-loop-helix proteins in plants. *Mol. Biol. Evol.* **2010**, *27*, 862–874. [CrossRef]
61. Peng, Z.; Han, C.; Yuan, L.; Zhang, K.; Huang, H.; Ren, C. Brassinosteroid enhances jasmonate-induced anthocyanin accumulation in Arabidopsis seedlings. *J. Integr. Plant Biol.* **2011**, *53*, 632–640. [CrossRef] [PubMed]
62. Yuan, L.B.; Peng, Z.H.; Zhi, T.T.; Zho, Z.; Liu, Y.; Zhu, Q.; Xiong, X.Y.; Ren, C.M. Brassinosteroid enhances cytokinin-induced anthocyanin biosynthesis in Arabidopsis seedlings. *Biol. Plant.* **2015**, *59*, 99–105. [CrossRef]
63. Coenye, T. Do results obtained with RNA-sequencing require independent verification? *Biofilm* **2021**, *3*, 100043. [CrossRef] [PubMed]
64. Everaert, C.; Luypaert, M.; Maag, J.L.V.; Cheng, Q.X.; Dinger, M.E.; Hellemans, J.; Mestdagh, P. Benchmarking of RNA-sequencing analysis workflows using whole-transcriptome RT-qPCR expression data. *Sci. Rep.* **2017**, *7*, 1559. [CrossRef]
65. Love, M.I.; Hogenesch, J.B.; Irizarry, R.A. Modeling of RNA-seq fragment sequence bias reduces systematic errors in transcript abundance estimation. *Nat. Biotechnol.* **2016**, *34*, 1287–1291. [CrossRef]
66. Minshall, N.; Git, A. Enzyme- and gene-specific biases in reverse transcription of RNA raise concerns for evaluating gene expression. *Sci. Rep.* **2020**, *10*, 8151. [CrossRef]
67. Toledo-Ortiz, G.; Huq, E.; Quail, P.H. The Arabidopsis basic/helix-loop-helix transcription factor family. *Plant Cell* **2003**, *15*, 1749–1770. [CrossRef]
68. Heim, M.A.; Jakoby, M.; Werber, M.; Martin, C.; Weisshaar, B.; Bailey, P.C. The basic helix-loop-helix transcription factor family in plants: A genome-wide study of protein structure and functional diversity. *Mol. Biol. Evol.* **2003**, *20*, 735–747. [CrossRef]
69. Kumar, S.; Stecher, G.; Li, M.; Knyaz, C.; Tamura, K. MEGA X: Molecular Evolutionary Genetics Analysis across Computing Platforms. *Mol. Biol. Evol.* **2018**, *35*, 1547–1549. [CrossRef]
70. He, Z.; Zhang, H.; Gao, S.; Lercher, M.J.; Chen, W.H.; Hu, S. Evolview v2: An online visualization and management tool for customized and annotated phylogenetic trees. *Nucleic Acids Res.* **2016**, *44*, W236–W241. [CrossRef]

71. Seeliger, D.; de Groot, B.L. Ligand docking and binding site analysis with PyMOL and Autodock/Vina. *J. Comput. Aided Mol. Des.* **2010**, *24*, 417–422. [CrossRef] [PubMed]
72. Chen, C.; Wu, Y.; Xia, R. A painless way to customize Circos plot: From data preparation to visualization using TBtools. *iMeta* **2022**, *1*, e35. [CrossRef]

Disclaimer/Publisher's Note: The statements, opinions and data contained in all publications are solely those of the individual author(s) and contributor(s) and not of MDPI and/or the editor(s). MDPI and/or the editor(s) disclaim responsibility for any injury to people or property resulting from any ideas, methods, instructions or products referred to in the content.

Article

Development of a Multi-Criteria Decision-Making Approach for Evaluating the Comprehensive Application of Herbaceous Peony at Low Latitudes

Xiaobin Wang [1], Runlong Zhang [1], Kaijing Zhang [1], Lingmei Shao [1], Tong Xu [1], Xiaohua Shi [2], Danqing Li [1], Jiaping Zhang [1,*] and Yiping Xia [1,*]

[1] Genomics and Genetic Engineering Laboratory of Ornamental Plants, Institute of Landscape Architecture, Department of Horticulture, College of Agriculture and Biotechnology, Zhejiang University, Hangzhou 310058, China
[2] Zhejiang Institute of Landscape Plants and Flowers, Hangzhou 311251, China
* Correspondence: zhangjiaping0604@aliyun.com (J.Z.); ypxia@zju.edu.cn (Y.X.)

Citation: Wang, X.; Zhang, R.; Zhang, K.; Shao, L.; Xu, T.; Shi, X.; Li, D.; Zhang, J.; Xia, Y. Development of a Multi-Criteria Decision-Making Approach for Evaluating the Comprehensive Application of Herbaceous Peony at Low Latitudes. *Int. J. Mol. Sci.* **2022**, *23*, 14342. https://doi.org/10.3390/ijms232214342

Academic Editors: Jian Zhang and Zhiyong Li

Received: 31 August 2022
Accepted: 12 November 2022
Published: 18 November 2022

Publisher's Note: MDPI stays neutral with regard to jurisdictional claims in published maps and institutional affiliations.

Copyright: © 2022 by the authors. Licensee MDPI, Basel, Switzerland. This article is an open access article distributed under the terms and conditions of the Creative Commons Attribution (CC BY) license (https://creativecommons.org/licenses/by/4.0/).

Abstract: The growing region of herbaceous peony (*Paeonia lactiflora*) has been severely constrained due to the intensification of global warming and extreme weather events, especially at low latitudes. Assessing and selecting stress-tolerant and high-quality peony germplasm is essential for maintaining the normal growth and application of peonies under adverse conditions. This study proposed a modified multi-criteria decision-making (MCDM) model for assessing peonies adapted to low-latitude climates based on our previous study. This model is low-cost, timesaving and suitable for screening the adapted peony germplasm under hot and humid climates. The evaluation was conducted through the analytic hierarchy process (AHP), three major criteria, including adaptability-related, ornamental feature-related and growth habits-related criteria, and eighteen sub-criteria were proposed and constructed in this study. The model was validated on fifteen herbaceous peonies cultivars from different latitudes. The results showed that 'Meiju', 'Hang Baishao', 'Hongpan Tuojin' and 'Bo Baishao' were assessed as Level I, which have strong growth adaptability and high ornamental values, and were recommended for promotion and application at low latitudes. The reliability and stability of the MCDM model were further confirmed by measuring the chlorophyll fluorescence of the selected adaptive cultivars 'Meiju' and 'Hang Baishao' and one maladaptive cultivar 'Zhuguang'. This study could provide a reference for the introduction, breeding and application of perennials under everchanging unfavorable climatic conditions.

Keywords: multi-criteria decision-making (MCDM); analytic hierarchy process (AHP); herbaceous peony; germplasm resources; breeding; global warming; low latitudes

1. Introduction

The importance of plant breeding for sustainable development is rising rapidly due to extreme weather events and changed climates [1–3]. Utilizing adapted germplasms ensures a sustained yield production and minimizes the negative impacts of climate change on agriculture and landscape ecosystems [4,5]. Breeding a new adapted cultivar is a very time-consuming and tedious process [6]. However, introducing and selecting promising genotypes directly from different latitudes or wild resources could considerably shorten the process [7,8]. Introducing new crops or cultivars—in particular, the introduction between different latitudes—not only leads to the diversification of agricultural production and applications but also has positive effects on biodiversity and ecosystem services [9,10]. In addition, the selected elite germplasms could be used as parental materials to carry out further precise breeding and research work on the molecular mechanism of stress resistance [11]. However, the introduction and selection are very limited to economic plants at low latitudes.

Herbaceous peony (*Paeonia lactiflora*) is a world-renowned economic crop with high ornamental, edible, medicinal and ecological values [3,12,13]. In recent years, herbaceous peonies have gained a new reputation as high-end cut flowers. Up to now, cut peonies have been produced in over 25 countries, with primary markets in Europe, Asia and the United States. Only in Europe, trade in cut peonies has increased 50-fold in the last 30 years, from three million stems produced in the Netherlands at the end of the 1980s to about 140 million stems from 20 countries (data from: Royal Flora, Y. Kohavi). Despite its popularity in the international market, the cultivation area of the peony is gradually limited due to the global warming. At low latitudes (N 30°00′–S 30°00′ areas), the situation is even worse, and the cultivation of the peony has encountered unprecedented challenges [12,14]. High temperatures in the spring (an average of 17–27 °C), especially combined with high precipitation (an average of 127–147 mm, data from https://zh.weatherspark.com (accessed on 10 September 2021)), can cause stem bending, flower bud abortion and severe diseases (Figure 1A–H). High temperatures in the summer, which extreme temperatures usually exceed 40 °C, could lead to severe heat damage and premature withering of the aboveground parts (Figure 1I–K) [12]. Easily formed insufficient chilling requirements due to high temperatures in the autumn and winter affect the establishment and release of dormancy and subsequent normal flowering and vegetative growth (Figure 1L,M) [15,16]. These problems have directly caused the decline of the ornamental value and application of the peony under everchanging unfavorable climatic conditions [17]. Thus, the main goal for improving the herbaceous peony at low latitudes is the screening and breeding of adapted cultivars [18].

Figure 1. Challenges encountered in the cultivation and application of the herbaceous peony at low latitudes. (**A–H**) High spring temperatures and high humidity caused stem bending, flower bud abortion and severe diseases; (**I–K**) High summer temperatures induced heat damage and premature withering of aboveground parts; (**L,M**) High autumn and winter temperatures caused insufficient chilling requirement and induced subsequent abnormal flowering and vegetative growth.

There have been many studies on the introduction, cultivation and comprehensive evaluation of the peony, but most of them were carried out at mid and high latitudes; little work has been undertaken at low latitudes [18,19]. Liu et al. (2019) carried out a comprehensive evaluation of fifteen introduced peony cultivars in the Luoyang alpine region, Henan Province, based on an evaluation model with a global weight of 0.52 for flowering-related traits [19]. Four excellent cultivars were selected by Wu et al. (2014) using the AHP method in Beijing, China, which was established with the numbers of flowers, flower diameter, stem diameter and florescence of every single flower as the four highest weighted indices [20]. Obviously, the introduction and comprehensive evaluation of the peony at these mid and high latitudes were mostly aimed at screening the peony germplasm with high ornamental value rather than for adaptive cultivation [18,19]. The few studies that are available have focused primarily on screen-adapted herbaceous peonies basically conducted at mid-latitudes [21]. We have carried out the resource evaluation of the peony at low latitudes, but due to the limited cultivation experience, the establishment of the evaluation model also focused on the selection of high ornamental resources [18]. With longer practice and scientific research work carried out at low latitudes and continuous consultation with experts and farmers in relevant fields, adaptability, especially heat resistance, has been identified as key factor affecting the cultivation of the peony [12]. Therefore, it is especially important to establish an integrated evaluation methodology with adaptability as a major consideration to screen elite peony germplasms for sustainable growth at low latitudes [7,22].

The development of an integrated evaluation approach frequently encounters with complex multi-criteria situations [23]. A multi-criteria decision-making (MCDM) model is proven to be one of the better tools to address such complex selection issues [24,25]. Construction of a MCDM model typically includes three steps: selection of the criteria indices, weighting of the criteria indices and multi-criteria decision analysis [7,26]. The analytic hierarchy process (AHP) is one of the best subjective weighing methods for obtaining weights of each alternative in an MCDM approach, which was first established by Saaty (1980) [27,28]. The AHP approach helps decision-makers to convert subjective evaluation into objective measures, increasing the validity, efficiency and credibility of the results. It also allows for the combination of qualitative and quantitative factors in the total evaluation [29]. In recent years, MCDM has found its grounding application in various fields and disciplines, such as astronomy, environmental ecology, energy science and artificial intelligence [30–34]. In the agricultural context, MCDM model has been applied to a variety of crops and economic plants [35–38]. In miscanthus (*Miscanthus* spp.), the MCDM model was applied to select high-biomass and high-quality miscanthus varieties for bioenergy production [7]. Similarly, the model has also been used to select the most suitable table grape variety intended for organic viticulture [39]. Continuing under the scope of agriculture, only few studies concerning the selection of species or cultivars for adaptive cultivation under MCDM strategies [40]. Moreover, methods for selecting germplasms have primarily focused on food crops with little to no emphasis on the ornamental crops, such as herbaceous peony [18].

This study creates a multi-criterion integrated decision support framework for selecting an elite herbaceous peony germplasm at low latitudes. The MCDM model established in this study is an extension and improvement of our previous study, aiming to evaluate the comprehensive application of herbaceous peony at low latitudes [18]. The model reconstructed the AHP system, increased adaptability and growth habits-related indices, while reduced reproductive traits and ornamental values-related indices. In addition, the weight of adaptability-related indices was improved via a pairwise comparison. Finally, the model could accurately screen adaptive peony germplasm at low-latitudes and shorten the screening time from six years to three years. The establishment of this model fills a gap in the screening of adapted peonies and greatly facilitates the selection of elite germplasm at low latitudes

The paper structure is organized as follows. Section 2 shows the results associated with the MCDM model, providing an elaborate display of the case study. Section 3 discusses the application of the model and future perspectives. Section 4 describes the methodology, presenting the steps of the development and the application of the MCDM model. The objectives of this research were to: (1) develop a modified MCDM model for evaluating the comprehensive application of herbaceous peony at low latitudes and (2) select representative peony cultivars with strong adaptability for future crossbreeding and studies on the mechanism of stress resistance. This study could promote the cultivation and application of peony at low latitudes and provide a reference for the adaptive cultivation of perennial plants in the context of global warming.

2. Results

2.1. Local Weights of Criteria and Sub-Criteria Indices

Local weights of each criterion were calculated based on expert scoring (Table 1). Consistency index (CI) and consistency ratio (CR) values for each matrix were obtained and all of the CR values < 0.1 (Table 1), which indicated that the consistency of judgment matrix was acceptable.

Table 1. The local weights of criteria and consistency test of the pairwise comparison matrix.

Matrixes	Criteria	Local Weights	Consistency Test		
			λ_{max}	CI	CR
A-B	Adapted peony cultivars at low latitudes (A)				
	Adaptability (B1)	0.540	3.009	0.005	0.009
	Ornamental features (B2)	0.297			
	Growth habits (B3)	0.163			
B1-C	B1				
	Heat damage level (C1)	0.423	4.010	0.003	0.004
	Root rot rate (C2)	0.227			
	Disease rate (C3)	0.227			
	Survival rate (C4)	0.122			
B2-C	B2				
	Flower number per plant (C5)	0.261	7.346	0.058	0.042
	Group blooming period (C6)	0.261			
	Flower type (C7)	0.090			
	Proportion of flowering plant (C8)	0.136			
	Aborted flowers per plant (C9)	0.075			
	Blooming period per flower (C10)	0.118			
	Flower diameter (C11)	0.058			
B3-C	B3				
	Stem bending degree (C12)	0.297	7.037	0.006	0.005
	Plant height (C13)	0.097			
	Plant width (C14)	0.056			
	Stem number (C15)	0.097			
	Stem diameter (C16)	0.178			
	Dates of bud break (C17)	0.178			
	Chlorophyll content (C18)	0.097			

A is the main objective, B1–B3 are the three criteria and C1–C18 are the eighteen sub-criteria in this study; λmax is the largest eigenvalue of the pairwise comparison matrix.

Global weights of criteria and sub-criteria were calculated by using Equation (8) (Table 2). Generally, adaptability was the most important criteria, up to 54%, followed by ornamental features criteria (23%) and growth habits criteria, which is the least important one (16%) (Table 2).

The heat damage level (C1), root rot rate (C2), disease rate (C3), flower number per plant (C5) and group blooming period (C6) were the top five sub-criteria indices of global weight in the whole MCDM model. However, the top three sub-criteria in the global weight ranking were all adaptability-related indices (Table 2). Among the ornamental features-related criteria, the flower number per plant (C5) and group blooming period (C6) were the two sub-criteria with the largest proportion, both of which were 0.078 (Table 2). In growth habits-related sub-criteria, stem bending degree (C12) and dates of bud break (C17) were the most two crucial indices, which were 0.048 and 0.029, respectively (Table 2).

Table 2. Global weights of the three criteria and eighteen sub-criteria.

Criteria (B)	Weights	Sub-Criteria (C)	Global Weights	Ranking
Adaptability (B1)	0.54	Heat damage level (C1)	0.228	1
		Root rot rate (C2)	0.123	2
		Disease rate (C3)	0.123	3
		Survival rate (C4)	0.066	6
Ornamental features (B2)	0.30	Flower number per plant (C5)	0.078	4
		Group blooming period (C6)	0.078	5
		Flower type (C7)	0.027	12
		Proportion of flowering plant (C8)	0.041	8
		Aborted flowers per plant (C9)	0.023	13
		Blooming period per flower (C10)	0.035	9
		Flower diameter (C11)	0.017	14
Growth habits (B3)	0.16	Stem bending degree (C12)	0.048	7
		Plant height (C13)	0.015	15
		Plant width (C14)	0.009	18
		Stem number (C15)	0.015	17
		Stem diameter (C16)	0.029	11
		Dates of bud break (C17)	0.029	10
		Chlorophyll content (C18)	0.015	16

2.2. Observations of Adaptability-Related Traits

We have carried out the introduction, cultivation and breeding of herbaceous peony at low latitudes since 2012 [18]. High temperature and high humidity were found the most important factors restricting the popularization and application of peony under relatively high temperatures at low latitudes (Figure 1). In this study, we observed four adaptability-related sub-indices: heat damage level, root rot rate, disease rate and survival rate. 'Hang Baishao', 'Meiju', 'Bo Baishao' and 'Hongpan Tuojing' performed extremely well in these four sub-criteria indices. Specifically, these four cultivars both showed low disease rates, root rot rates and heat damage levels and maintained almost 100% survival rates after three years of cultivation (Figure 2). On the contrary, some cultivars, such as 'Zhuguang', 'Taohua Feixue' and 'Yangfei Chuyu' showed obvious unsuitability after introduced at low latitudes (Figure 2). These three cultivars observed gradually increased disease rates, even close to 100%, and severe heat damage and root rot, ultimately resulted in a low survival rate (Figure 2). However, most cultivars, such as 'Zaohong', 'Yanzi Xiangyang', 'Zifeng Chaoyang', 'Qing Yunhong' and 'Shanhe Hong' showed moderate performances in these four adaptability-related traits, and the final survival rate was between 50% and 70% after three years of cultivation. In addition, although some of the cultivars have excellent performances in some aspects of the adaptability, it is hard to combine various resistances, such as 'Qihua Lushuang' and 'Chishao', which have strong resistance to disease and root rot but lack heat resistance (Figure 2).

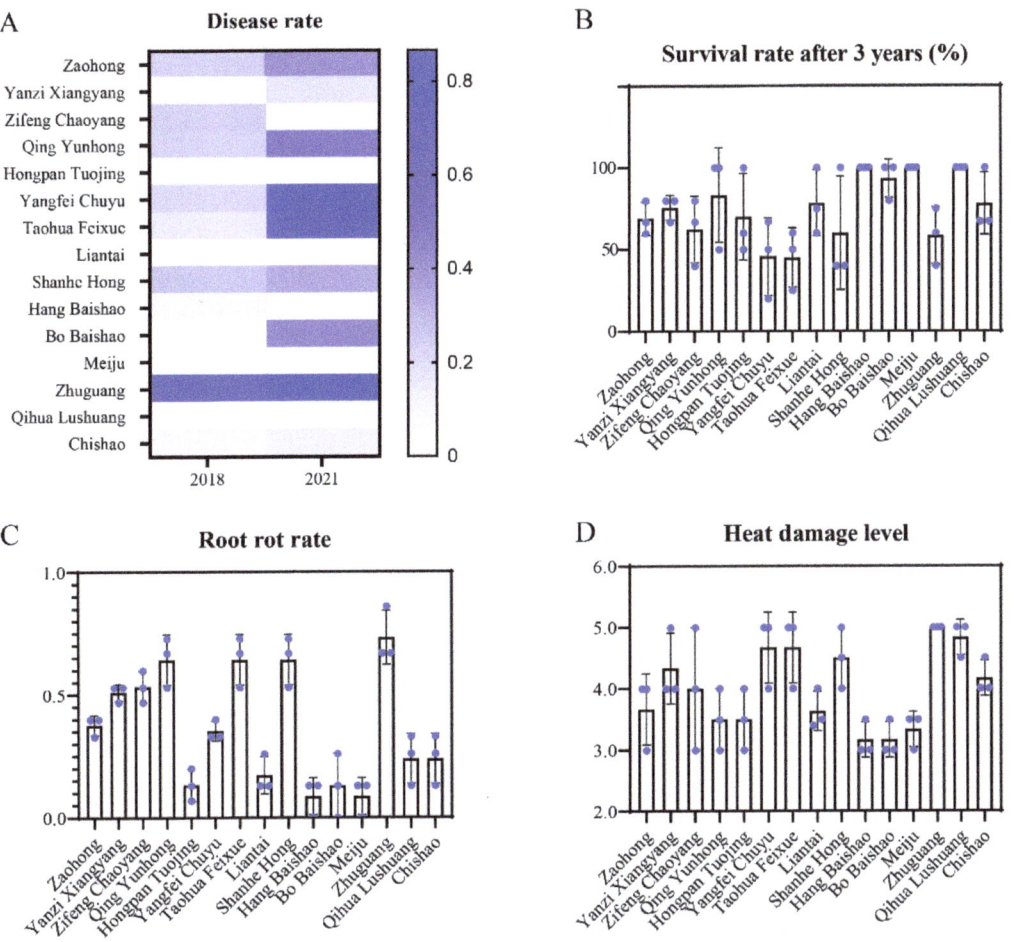

Figure 2. Observations and changes of adaptability-related indices of the fifteen peony cultivars. (**A**) Disease rate; (**B**) Survival rate after three years; (**C**) Root rot rate; (**D**) Heat damage level. Values represent the means± standard deviation of three replicates.

2.3. Observations of Ornamental Features-Related Traits

The ornamental features-related traits of fifteen herbaceous peony cultivars were shown in Figure 3. 'Bo Baishao', 'Qihua Lushuang', 'Liantai' and 'Chishao' performed well in several important ornamental features-related traits, such as flower number per plant, proportion of flowering plant and blooming period per flower, while 'Zaohong', 'Yangfei Chuyu' and 'Zhuguang' performed poorly in these flowering indices (Figure 3A–C). All cultivars showed little difference in the flower diameter indices, within the range of 10–15 cm. Notably, the flower diameters of 'Hangbaishao' and 'Meiju' decreased obviously after three years (Figure 3D). 'Qihua Lushuang' had the highest number of aborted flowers per plant in 2018, but sharply reduced by 2021, contrary to the performance of 'Hang Baishao' (Figure 3E). 'Zaohong' and Chishao' bloom early and have a long group blooming period, while 'Taohua Feixue' and 'Shanhe Hong' have a late and short group flowering period (Figure 3F).

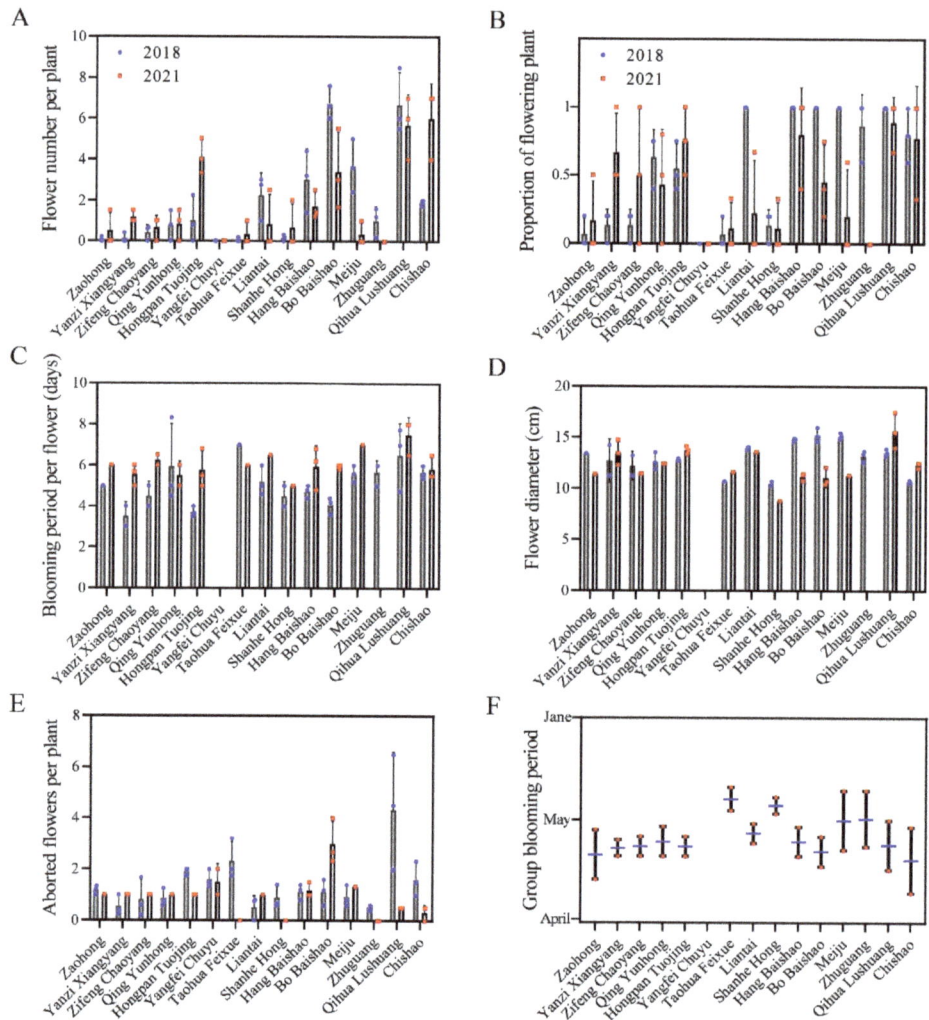

Figure 3. Observations and changes of ornamental features-related indices of the fifteen peony cultivars. (**A**) Flower number per plant; (**B**) Proportion of flowering plant; (**C**) Blooming period per flower; (**D**) Flower diameter; (**E**) Aborted flowers per plant; (**F**) Group blooming period. Values represent the means ± standard deviation of three replicates.

2.4. Observations of Growth Habits-Related Traits

'Bo Baishao', 'Qihua Lushuang', 'Hongpan Tuojing' and 'Hang Baishao' had relatively high plant heights and widths, while 'Taohua Feixue', 'Yangfei Chuyu' and 'Zhuguang' were cultivars with very limited plant height and width (Figure 4A,B). The plant heights and widths decreased in most cultivars, while increased in 'Yanzi Xiangyan', 'Zifeng Chaoyang', 'Zaohong' and 'Hongpan Tuojing' from 2018 to 2021. Besides, the plant heights of 'Shanhe Hong' and 'Qihua Lushuang' increased, but the plant width declined after three years (Figure 4A,B). Similarly, stem numbers of most cultivars decreased after three years, on the contrary, stem diameter of most cultivars increased (Figure 4C,D). The cultivars showed differences in the stem bending degree indices—in particular, 'Hang Baishao', 'Meiju', 'Hongpan Tuojing' and 'Chishao' performed extremely low degrees of stem bending.

However, the rest of the cultivars performed poorly in this crucial index, with stem bending exceeding 20 degrees, and some cultivars, such as 'Yanzi Xiangyan' and 'Qihua Lushuang', even exceeded 40 degrees (Figure 4E). In the observation of bud break, most cultivars bud burst in the middle and early March, while a few cultivars, such as 'Zaohong', 'Yanzi Xiangyang' and 'Zifen Chaoyang', sprouted in the middle and late February (Figure 4F). In addition, most cultivars had chlorophyll content between 50–60 during full blooming, except for 'Zaohong', 'Shanhe Hong' and 'Liantai' (Figure 4G).

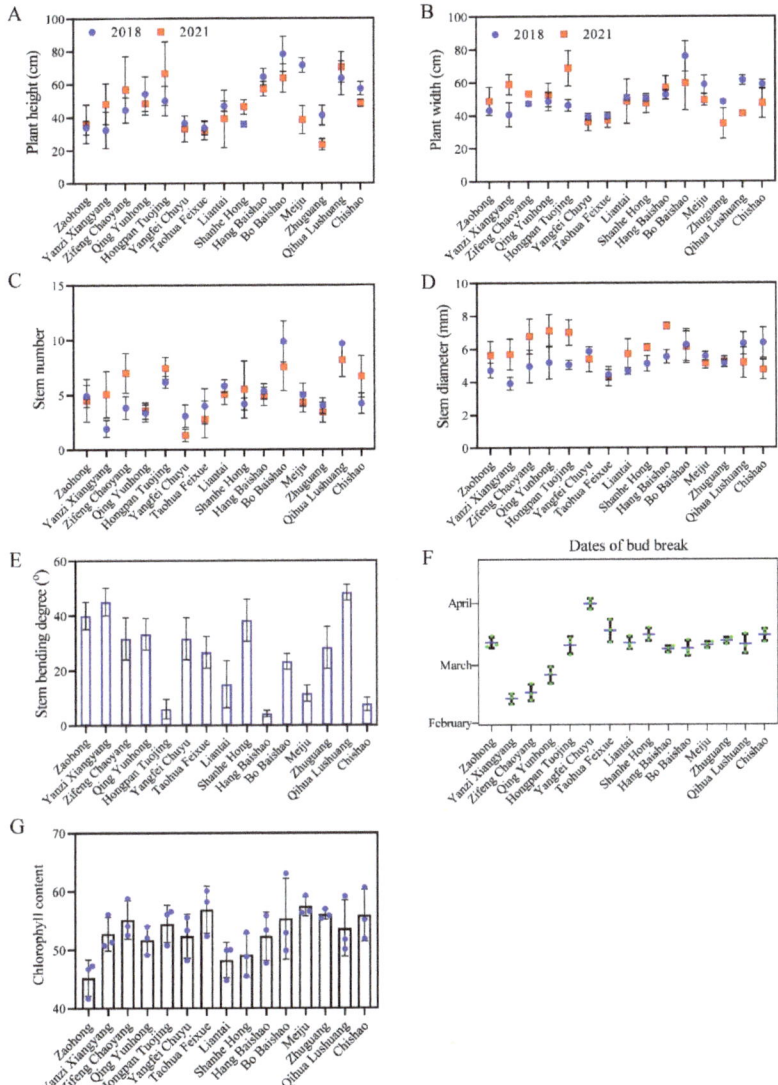

Figure 4. Observations and changes of growth habits-related indices of the fifteen peony cultivars. (**A**) Plant height; (**B**) Plant width; (**C**) Stem number; (**D**) Stem diameter; (**E**) Stem bending degree; (**F**) Dates of bud break; (**G**) Chlorophyll content. Values represent the means ± standard deviation of three replicates.

2.5. Evaluation of Comprehensive Performance of the Fifteen Cultivars by the MCDM Model

Details in comprehensive scores and levels of the fifteen cultivars were presented in Table 3. 'Meiju' acquired the highest points 89.56, and 'Hang Baishao' was ranked second, with 85.05 points, followed by 'Hongpan Tuojing' and 'Bo Baishao', which were the four cultivars with over 80 points. These four cultivars with more than 80 points were classified as level I, indicating they have excellent comprehensive performance and are recommended as germplasms for cultivation and application at low latitudes. The comprehensive scores of 'Qihua Lushuang', 'Liantai', 'Chishao', 'Qing Yunhong', 'Zifeng Chaoyang' and 'Zaohong' were between 60–80 points and classified as level II, indicating that their comprehensive performance is ordinary and could be used as alternative application materials. At the bottom of the ranking, 'Yanzi Xiangyang', 'Shanhe Hong', 'Taohua Feixue', 'Yangfei Chuyu' and 'Zhuguang' were listed. These five cultivars were scored below 60 points and were classified as level III, not recommended for use at low latitudes (Table 3).

Table 3. Statistics of comprehensive evaluation points and rating levels of the fifteen herbaceous peony cultivars.

Cultivars	Adaptability						Ornamental Features							Growth Habits					Points	Levels
	C1	C2	C3	C4	C5	C6	C7	C8	C9	C10	C11	C12	C13	C14	C15	C16	C17	C18		
Meiju	22.85	6.60	12.28	12.28	5.22	7.84	1.80	2.72	1.50	3.55	1.74	3.17	1.03	0.59	1.03	1.90	1.90	1.55	89.56	I
Hang Baishao	22.85	6.60	12.28	12.28	5.22	2.61	0.90	4.08	1.50	2.37	1.16	4.76	0.52	0.59	1.55	2.85	1.90	1.03	85.05	I
Hongpan Tuojing	22.85	4.40	12.28	12.28	5.22	5.22	0.90	2.72	1.50	1.18	1.74	4.76	1.03	0.59	1.55	2.85	1.90	1.03	84.02	I
Bo Baishao	15.23	6.60	12.28	12.28	7.84	7.84	0.90	4.08	0.75	2.37	1.74	3.17	0.52	0.30	1.55	2.85	1.90	1.55	83.73	I
Qihua Lushuang	7.62	6.60	12.28	12.28	7.84	7.84	1.80	4.08	0.75	3.55	1.74	1.59	0.52	0.59	1.55	1.90	1.90	1.03	75.44	II
Liantai	15.23	4.40	12.28	12.28	2.61	5.22	1.80	2.72	2.25	2.37	1.74	3.17	1.55	0.89	1.55	1.90	1.90	0.52	74.39	II
Chishao	7.62	4.40	12.28	12.28	5.22	7.84	0.90	4.08	2.25	2.37	1.16	4.76	1.03	0.59	1.55	1.90	0.95	1.55	72.73	II
Qing Yunhong	22.85	6.60	8.18	4.09	2.61	2.61	2.70	2.72	2.25	2.37	1.16	1.59	1.03	0.59	0.52	2.85	2.85	1.03	68.62	II
Zifeng Chaoyang	15.23	4.40	12.28	8.18	2.61	5.22	2.70	1.36	2.25	2.37	1.16	1.59	1.03	0.59	1.55	1.90	2.85	1.03	68.32	II
Zaohong	15.23	4.40	8.18	8.18	2.61	2.61	2.70	1.36	1.50	2.37	1.16	1.59	0.52	0.89	1.03	1.90	1.90	0.52	61.27	II
Yanzi Xiangyang	7.62	4.40	12.28	8.18	2.61	2.61	0.90	1.36	2.25	1.18	1.74	1.59	1.55	0.89	0.52	0.95	2.85	1.03	54.52	III
Shanhe Hong	7.62	4.40	8.18	4.09	2.61	1.80	1.36	2.25	1.18	0.58	1.59	1.55	0.89	1.03	1.90	1.90	0.52		46.07	III
Taohua Feixue	7.62	2.20	4.09	4.09	2.61	5.22	1.80	1.36	1.50	3.55	1.16	3.17	0.51	0.30	0.52	0.95	1.90	1.55	44.11	III
Yangfei Chuyu	7.62	2.20	4.09	8.18	2.61	1.80	1.36	1.50	2.37	1.16	1.59	0.51	0.30	0.52	1.90	0.95	1.03		42.30	III
Zhuguang	7.62	2.20	4.09	4.09	2.61	2.70	1.36	2.25	2.37	1.16	3.17	0.52	0.89	0.52	0.95	1.90	0.51		41.53	III

The full name of the sub-criteria C1–C18 were listed in Table 1.

2.6. Chlorophyll Fluorescence of the Selected Adaptive and Maladapted Cultivars

'Meiju' and 'Hang Baishao' were screened two adaptive cultivars while 'Zhuguang' was the maladapted cultivar based on the comprehensive evaluation. Chlorophyll fluorescence of the three cultivars was detected in the summer of 2021. The maximum quantum yield of photosystem II (Fv/Fm), nonphotochemical quenching (NPQ), quantum efficiency of photosystem II (YII), photochemical quenching coefficient (qP) values of the three cultivars overall decreased first and then stabilized while apparent electron transfer rate (ETR) behaves in the opposite under high summer temperatures (Figure 5). Among three cultivars, 'Meiju' produced significantly the highest values of the four chlorophyll fluorescence indicators, followed by 'Hangbaishao' and the lowest values were recorded in 'Zhuguang' and were correlated with the chlorophyll fluorescence colors (Figure 5). Additionally, the chlorophyll fluorescence values of the cultivars were consistent with the comprehensive scores calculated by the MCDM model (Table 3).

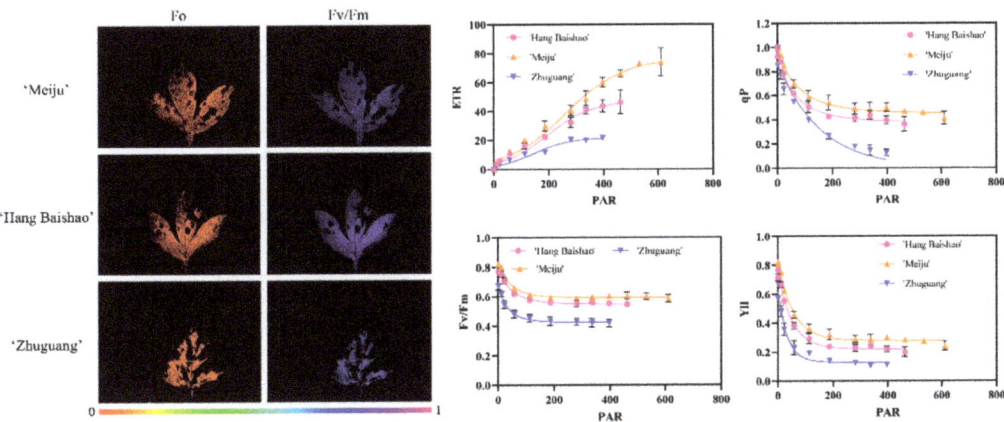

Figure 5. Chlorophyll fluorescence imaging screens of the selected adaptive and maladapted cultivars under summer heat stress in 2021. The fluorescence color indicates the Fo and Fv/Fm values. Data expressed in the figure represents the mean ± standard deviation of three replicates.

3. Discussion

3.1. The Necessity of Introduction and Selection of Adaptive Peony Germplasm at Low Latitudes

Herbaceous peony is a famous ornamental crop worldwide, with the widespread popularity of cut flowers in recent years [17,41]. However, peony is often subjected to multiple abiotic stresses of high temperature and humidity, which seriously and initially affect its normal growth and subsequent flowering under environmental conditions at low latitudes [42,43]. Global warming exacerbates the problem, and extreme temperatures at low latitudes commonly exceed 40 °C in the summer [2,12]. Thus, evaluating germplasm resources, particularly for the purpose of selecting resistant species, is a fundamental work of peony breeding [18]. Numerous introduction and evaluation studies have been carried out in a variety of food crops, but few in ornamental crops, such as herbaceous peony [44]. In fact, herbaceous peony is an ornamental perennial with strong adaptability and could widely distributed in temperate regions of the Northern Hemisphere [13]. Moreover, there are plentiful wild resources, medicinal and ornamental species or cultivars of peonies (both herbaceous peony and tree peony) existed in the mid-latitude region of China, it is completely achievable to directly select excellent existing peony germplasms based on introduction and resource evaluation at low latitudes [18,45]. Therefore, screening or breeding adapted peony germplasm has important theoretical and practical significance in low-latitude areas, especially in the context of global warming [12,42].

3.2. The Development of a Specific Objective-Based Comprehensive Evaluation Model

All models are not omnipotent, their accuracy depends on the specific application purpose [46]. Some common steps in the construction of a completed MCDM evaluation model include: selecting indices, creating criteria and assigning weights (Figure 6). First and foremost, these steps should be completed in accordance with the specific purpose and characteristics of the target plant [18]. When selecting energy-related agronomic traits, dry matter yield is the primary index for screening elite germplasms [7]; when evaluating the rapeseed varieties, economic criterion is the most important index [37]; when establishing a model for selecting herbaceous peony species under protective cultivation conditions, ornamental characteristics are almost all considerations [37]. In this study, the MCDM model was aimed at solving the problem of adaptability in the process of introduction and cultivation of peony at low latitudes and, secondly, considering ornamental values and growth habits. As a consequence, the weights of the three criterion layers of the MCDM model were: adaptability (B1) > ornamental features (B2) > growth habits (B3), and the adaptability (B1) criteria had the highest weight in the criterion layer, up to 0.54 (Table 2). Additionally, the heat damage level (C1), root rot rate (C2) and disease rate (C3) were the top three evaluation sub-criteria of the MCDM model (Table 2), which is in line with our expectations. Specifically, these three adaptability-related sub-criteria have been widely used in studies of peony, rapeseed and grape [19,37,39]. In addition, these sub-criteria could typically reflect the stress resistance to high temperature and high humidity environments based on our cultivation practice at low latitudes. Therefore, the evaluation model established in this study was in accordance with our research purpose.

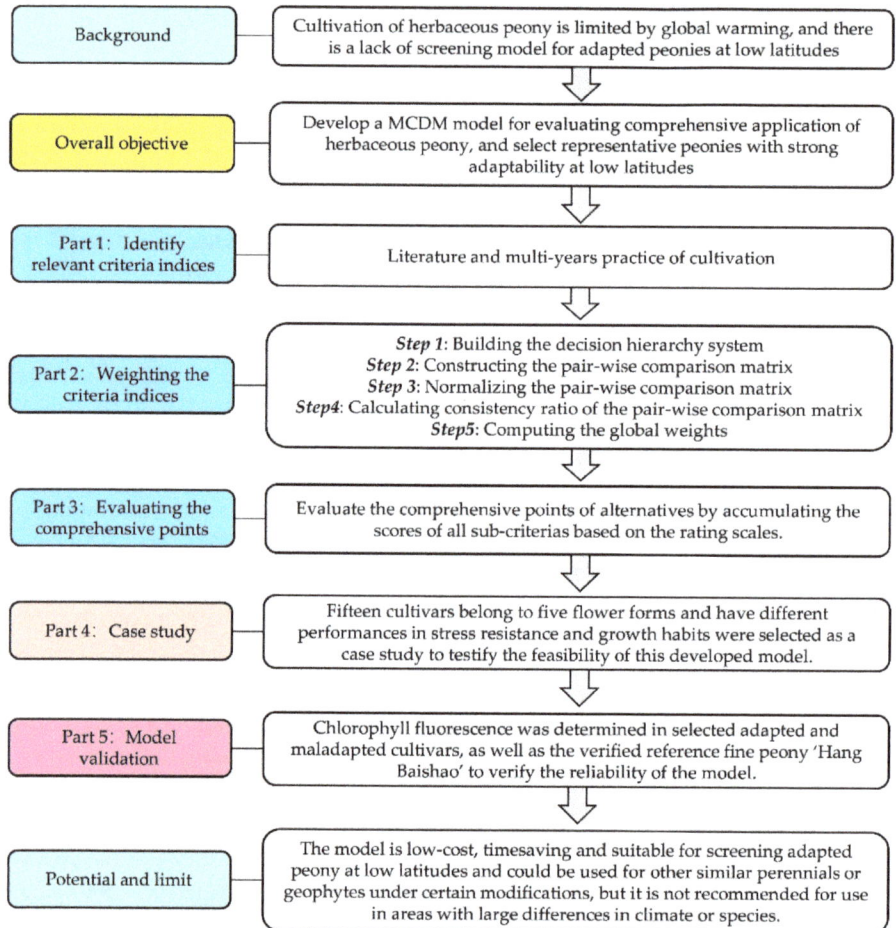

Figure 6. The process to create a multi-criterion integrated decision support framework for selecting elite herbaceous peony germplasm at low latitudes in this study.

3.3. Selection of Elite Peony Germplasms at Low Latitudes

'Meiju', 'Hang Baishao', 'Hongpan Tuojing' and 'Bo Baishao' were the four cultivars that scored over 80 points and classified as level I by the MCDM model (Table 3). These four cultivars both have excellent comprehensive performances in adaptability, ornamental features and growth habits, especially in terms of strong heat and humidity resistance, recommended as elite germplasms for cultivation and application at low latitudes (Figures 2–4). These cultivars could also be valuable for breeding brilliant new germplasms with strong stress resistance [47,48]. In terms of this study, these four cultivars and 'Zhuguang', a representative unsuitable cultivar, could be used as contrasting plant materials to determine the mechanisms governing the differences in the mechanism of stress tolerance in peony [48]. Notably, Liu et al. screened 'Taohua Feixue' and 'Yangfei Chuyu' as excellent cultivars for planting in Guanzhong area of Shaanxi Province by AHP method in 2013, while in this study, these two cultivars were scored below 60 points [21]. This indicates that the performance of the same cultivar varies greatly in different regions, further emphasizing the necessity for comprehensive evaluation of germplasm to reduce economic costs before large-scale introduction of cultivation. In addition, 'Yanzi Xiangyang' showed a tendency to gradually adapt at low latitudes after three years of cultivation, with fewer diseases, improved flowering number and flowering rate and increased plant height and plant width, which deserves a longer time observation (Figures 2–4).

3.4. Accuracy and Reliability of the MCDM Model

The MCDM model constructed in this study aimed to represent an integrated evaluation research strategy based on multi-year cultivation, multi-indices observations and specific application purposes rather than a fixed formula (Figures 6 and 7). The weights of sub-criteria indices will fluctuate once the tested germplasms change since the objective weight value varies with the measured data. The peonies adopted in this study were carefully selected cultivars with different performances in stress resistance and ornamental characteristics at low latitudes based on previous studies [12,18/]. In particular, the native-specific cultivar 'Hangbaishao', which has been proved to be an elite germplasm at low latitudes, was added as a reference for comparing selection results [12,18,49]. In the evaluation of this study, 'Hang Baishao' still performed excellently and ranked second overall (Table 3), which on the one hand showed the rationality and accuracy of this model, on the other hand, it re-emphasized the value of this native germplasm.

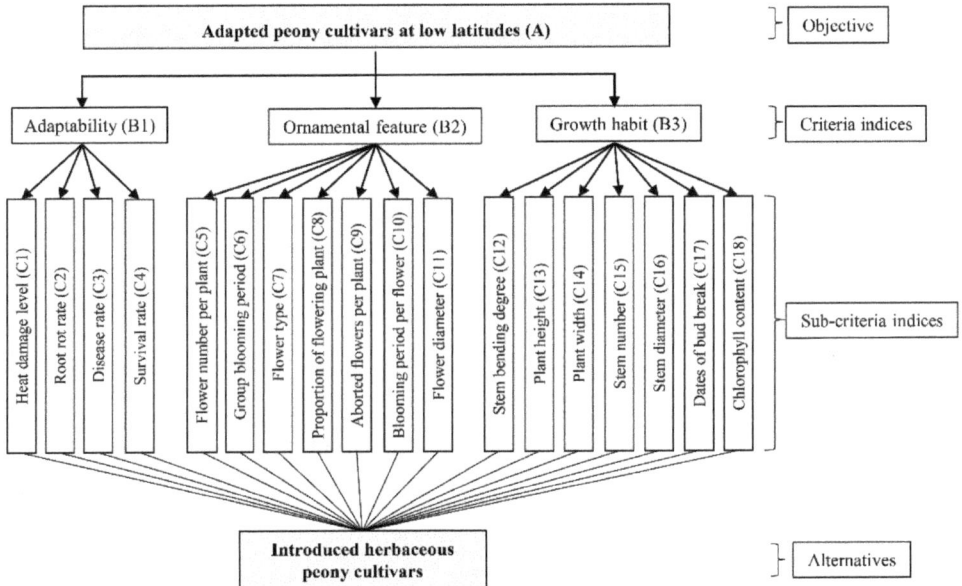

Figure 7. Decision hierarchy system of the MCDM model for evaluating the application potential of herbaceous peony cultivars at low latitudes.

Chlorophyll fluorescence reflects photosynthetic performance and stress in algae and plants, is now widespread in various studies [50–52]. Sharma et al. have successfully identified wheat (*Triticum aestivum* L.) cultivars in tolerance to heat stress by using Fv/Fm [53]. In addition to screening for heat-tolerant wheat, chlorophyll fluorescence has also been used as a method for screening potential wheat cultivars adapted to water deficit environments [54]. In the herbaceous peony, the chlorophyll fluorescence intensity was consistent with the results of comprehensive heat tolerance assessment of different peony cultivars [12]. Thus, in this study, the model was further validated in identified cultivars 'Meiju', 'Hang Baishao' and 'Zhuguang' by the detection of chlorophyll fluorescence. The result showed that the chlorophyll fluorescence values of the cultivars were consistent with the comprehensive scores calculated by the MCDM model (Figure 5 and Table 3). Furthermore, we have previously conducted principal component analysis (PCA) and subordinate function value analysis on six peony cultivars introduced from different latitudes, 'Hang Baishao' and 'Meiju' were determined to be the most heat-tolerant cultivars, while 'Zhuguang' was determined to be the most heat-sensitive one of the six cultivars [12]. Although different evaluation methods were used, the evaluation results were highly consistent, which further demonstrated the reliability and stability of the MCDM model established in this study.

3.5. Limitations, Recommendations and Future Perspectives

The MCDM model is low-cost, timesaving and suitable for screening adapted peony germplasm under hot and humid climates. Fifteen tuberous roots and nine square meters of land surface are sufficient for screening one cultivar. By using this model, it takes only three years to screen the expected cultivars, which is three years less than using our previous model. The model proposed in this study can be directly used at low latitudes, specifically, substituting the observational data of the tested cultivar into the model, then calculate the comprehensive evaluation points, and if it is greater than 80 points, the cultivar is selected. This model requires certain modifications for use in other perennials or geophytes, in that case, due to the specific characteristics of each plant, the major importance is to consider local experts' knowledge and experience, which would possibly add, remove or modify important criteria. Nevertheless, we do not recommend its use in cases where there are large differences in climate and species, as the present model involves a limited number of criteria that may require fundamental modifications.

The model is also a continuous improvement model that will advance in the future. It is well known that with the increasing global warming, the growing environment of plants may face more severe deterioration in the future. This model will be modified accordingly, for example, adding indices such as heat stress oxidation and high light damage, or changing the weights of some indices. In addition to the model itself, this study also aims to provide a research strategy (Figure 6) for other plants with similar situations as peony.

4. Materials and Methods

4.1. Development of a MCDM Model

Figure 6 structurally displayed the main process of developing a complete MCDM model of herbaceous peony in this study. The criteria indices in the current study were selected based on information from the literature and multi-years practice of cultivation (Section 4.1.1). The weights of these criteria and sub-criteria were calculated by the AHP method (Section 4.1.2), and the elite cultivars were selected via the final comprehensive evaluation (Section 4.1.3). The detailed steps are described in the following sections:

4.1.1. Identify Relevant Criteria Indices

The criteria indices were selected through the following two steps in current study: (1) The literature on traits related to comprehensive evaluation of peony cultivation and application, especially when introduced from regions of different latitudes and encountered unfavorable climatic conditions. All reported traits were compared and selected after preliminary screening; (2) Identifying the main limiting factors of peony through years of practice of cultivation at low latitudes. High temperature and high humidity were found the most important factors restricting the popularization and application of peony through years of practice of cultivation at low latitudes (Figure 1). Considering the two main limiting factors, heat damage level, root rot rate, disease rate and survival rate were identified as the corresponding adaptability indices, which have been proven in our previous studies to respond well to the resistance of peonies to humid and hot environments. Based on the above, the core traits with high recommendations were selected for the model establishment (Table 4).

Table 4. Details of evaluation indices and measurements.

Criteria	Sub-Criteria		Definitions and Measurements	Objective	References
Adaptability	Heat damage level	C1	HDL was classified into six categories according to the proportion of leaves that showed signs of discoloration.	Minimize	[12,18]
	Root rot rate	C2	Rot rate of peony root and the rate corresponding to different waterlogging tolerance is 0–30% (strong), 30–60% (medium) and 60–100% (weak).	Minimize	[19]
	Disease rate	C3	Number of diseased plants/number of healthy plants.	Minimize	[18,21]
	Survival rate	C4	Number of surviving plants after three years of cultivation/total number of plants at the beginning.	Maximize	[18,21]
Ornamental features	Flower number per plant	C5	Average number of normally open flowers.	Maximize	[18,19]
	Group blooming period	C6	The day number between the date of the first flower blooming and the date of the last flower falling of a group plants.	Maximize	[18]
	Flower type	C7	According to the shape of the petal, flowers are characterized as different types.	Valved, complex	[17,19]
	Proportion of flowering plant	C8	Number of flowering plants/total number of plants.	Maximize	[21]
	Aborted flowers per plant	C9	Aborted flowers were defined as Flowers could not open and always maintain in the small size and juvenile stage during the full-blooming period.	Minimize	[20]
	Blooming period per flower	C10	Days between the date of the petal blooming and the first petal falling.	Maximize	[18]
	Flower diameter	C11	Average flower diameter in full bloom.	Maximize	[19]
Growth habits	Stem bending degree	C12	The angle between the stem and the vertical direction of the ground.	Minimize	[20,21]
	Plant height	C13	The distance (cm) from the ground to the top of the plant.	Medium	[21]
	Plant width	C14	Maximum width (cm) of aboveground projection of the plant.	Medium	[21]
	Stem number	C15	A mature and normal stem has at least three compound leaves and the height should be more than twenty cm.	Maximize	[21]
	Stem diameter	C16	Diameter of a plant's mature and healthy stem five centimeters above.	Maximize	[20]
	Dates of bud break	C17	Bud break is the opening of bud scales and the emergence of delicate shoots or leaves. emerged.	Earlier	[18,55]
	Chlorophyll content	C18	Average chlorophyll content of peony leaves on May 15.	Maximize	[12]

4.1.2. Weighting the Criteria Indices

AHP is one of the most common subjective methods and was adopted to weight the criteria indices in the present study [56]. Its weighting procedures are described as follows [57]:

Step 1: Building the AHP system

The AHP system is usually defined as a tree, where the main objective is the target layer (A), like the top of the tree; the criteria indices are the second layer (B) for evaluating the target layer, like the trunk; the third layer is the specific sub-criteria (C), and the alternatives are the roots [46,58]. In present study, the AHP system was built in four levels from top to bottom (Figure 7). The selection of adapted peony cultivars at low latitudes (A) is the main objective. The adaptability-related (B1), ornamental features-related (B2) and growth habits-related (B3) traits were selected as the three criteria indices. The eighteen specific traits belonging to the three main criteria indices are: heat damage level (C1), root rot rate (C2), disease rate (C3), survival rate (C4), flower number per plant (C5), group blooming period (C6), flower type (C7), proportion of flowering plant (C8), aborted flowers per plant (C9), blooming period per flower (C10), flower diameter (C11), plant height (C12), plant width (C13), stem number (C14), stem diameter (C15), stem bending degree (C16), dates of bud break (C17) and chlorophyll content (C18), respectively (Figure 7). The fifteen peony cultivars are the alternatives. Observations and measurements of these traits (sub-criteria indices) and their references were shown in Table 4.

Step 2: Constructing the pairwise comparison matrix

In this step, the priority weights of the above-mentioned sub-criteria indices are calculated using a square matrix of pairwise comparisons, as shown in Equation (1) [59].

$$A(a_{ij}) n \times n = \begin{bmatrix} a_{11} & a_{12} & \cdots & a_{1j} & \cdots & a_{1n} \\ a_{21} & a_{22} & \cdots & a_{2j} & \cdots & a_{2n} \\ \cdots & \cdots & \cdots & \cdots & \cdots & \cdots \\ a_{i1} & a_{i2} & \cdots & a_{ij} & \cdots & a_{in} \\ \cdots & \cdots & \cdots & \cdots & \cdots & \cdots \\ a_{n1} & a_{n2} & \cdots & a_{nj} & \cdots & a_{nn} \end{bmatrix}, \quad (1)$$

where i represents the serial number of the former traits, j represents the serial number of the latter traits and n represents the total number of traits in matrix A. a_{ij} is the ratio of the importance of the ith trait compared with the jth trait, and it is scored according to Table 5 [60,61].

Table 5. Scales of pairwise comparison.

Scales	Definition	Explanation
1	Equal importance	Two activities contribute equally to the objective
3	Slight importance	Experience and judgment moderately favor one activity over another
5	Strong importance	Experience and judgement strongly favor one activity over another
7	Very strong importance	An activity is strongly favored and its dominance demonstrated in practice
9	Extreme importance	An activity is extremely favored and its dominance demonstrated in practice
2, 4, 6, 8	Intermediate values	Importance between two corresponding adjacent levels above mentioned

The importance definition between different traits is in accordance with the description proposed by Saaty (1980) [28].

Survey experts and front-line workers in the related field to confirm the importance of these selected traits based on the above steps [62]. A panel of four experts and two workers was contacted and asked about the quantification of the importance of the criteria included in the AHP system (Figure 7). The experts responded to a special questionnaire created for this study, an example of the questionnaire was shown in Table 6.

Table 6. An example of AHP questionnaire used to determine the relative importance of a criteria.

Importance	9	8	More Important 7 6 5 4	3	2	Equal 1	1/2	1/3	Less Important 1/4 1/5 1/6 1/7	1/8	1/9	Criteria
Adaptability				√	√	√						Adaptability Ornamental features Growth habits
Ornamental features					√		√	√				Adaptability Ornamental features Growth habits
Growth habits						√	√	√				Adaptability Ornamental features Growth habits

The ratio of the importance is in accordance with Table 5.

The pairwise comparison matrix was then established after expert judgment in this study (Table 7).

Table 7. The pairwise comparison matrix established in this study.

		Judgment Matrix							
	A	B1	B2	B3					
A-B	B1	1	2	3					
	B2	1/2	1	2					
	B3	1/3	1/2	1					
	B1	C1	C2	C3	C4				
	C1	1	2	2	3				
B1-C	C2	1/2	1	1	2				
	C3	1/2	1	1	2				
	C4	1/3	1/2	1/2	1				
	B2	C5	C6	C7	C8	C9	C10	C11	
	C5	1	1	3	2	4	2	4	
	C6	1	1	3	2	4	2	4	
B2-C	C7	1/3	1/3	1	1/2	2	1/2	2	
	C8	1/2	1/2	2	1	2	1	2	
	C9	1/4	1/4	1/2	1/2	1	2	1	
	C10	1/2	1/2	2	1	1/2	1	3	
	C11	1/4	1/4	1/2	1/2	1	1/3	1	
	B3	C12	C13	C14	C15	C16	C17	C18	
	C12	1	3	4	3	2	2	3	
	C13	1/3	1	2	1	1/2	1/2	1	
B3-C	C14	1/4	1/2	1	1/2	1/3	1/3	1/2	
	C15	1/3	1	2	1	1/2	1/2	1	
	C16	1/2	2	3	2	1	1	2	
	C17	1/2	2	3	2	1	1	2	
	C18	1/3	1	2	1	1/2	1/2	1	

The ratio of the importance is in accordance with Table 5.

Step 3: Normalizing the pairwise comparison matrix

Normalize the pairwise comparison matrix using Equation (2).

$$w'_{ij} = \frac{w_{ij}}{\sum_{i=1}^{n} w_{ij}} \tag{2}$$

where w_{ij} represents the pairwise comparison value, and w_{ij}' is the pairwise comparison value after normalization [63].

Step 4: Calculating consistency ratio of the pairwise comparison matrix

Assess the eigenvalue and the eigenvector using Equations (3)–(5):

$$w_i = \frac{\sum_{j=1}^{n} w'_{ij}}{n} \tag{3}$$

$$W = \begin{bmatrix} w_1 \\ w_2 \\ \ldots \\ w_i \\ \ldots \\ w_n \end{bmatrix} \tag{4}$$

$$\lambda_{max} = \frac{1}{n}\sum_{i=1}^{n}(AW)_i/w_i \tag{5}$$

where w_i is the eigenvalue, W is the eigenvector of the matrix A and λ_{max} is the largest eigenvalue of the pairwise comparison matrix.

Check the consistency index using Equations (6) and (7):

$$CR = \frac{CI}{RI} \quad (6)$$

$$CI = \frac{\lambda\max - n}{n-1} \quad (7)$$

where n denotes the number of criteria, and CR and CI are the consistency ratio and consistency index of the pairwise comparison matrix. RI represents the random consistency index that was introduced by Saaty [58], shown in Table 8.

Table 8. Random Index (RI) values.

Number of Criteria	1	2	3	4	5	6	7	8	9	10
RI	0.0	0.0	0.52	0.89	1.11	1.25	1.35	1.40	1.45	1.49

When CR < 0.1, the consistency degree of judgment matrix A is considered to be within the allowable range, and the eigenvectors of A could be performed to carry out the weight vector calculation [64]; however, if CR ≥ 0.1, the judgment matrix A should be considered for correction [65,66].

Step 5: Computing the global weights

The matrix A in the criteria level contains a series of criteria indices, including ($a_1, a_2, \ldots, a_i, \ldots, a_n$), their eigenvalues should be ($w_1, w_2, \ldots, w_i, \ldots, w_n$), respectively; Matrix B in the sub-criteria level belonging to a_i contains several sub-criteria indices, including ($b_1, b_2, \ldots, b_\alpha, \ldots, b_\beta$), their eigenvalues then should be ($w_1', w_2', \ldots, w_\alpha', \ldots, w_\beta'$), respectively. Finally, the global weights of ($b_1, b_2, \ldots, b_\alpha, \ldots, b_\beta$) should be ($w_i w_1', w_i w_2', \ldots, w_i w_\alpha', \ldots, w_i w_\beta'$), respectively [58].

$$W_g = w_i\, w_\alpha' \quad (8)$$

where W_g is the global weight of the sub-criteria; w_i is the local weight of criteria and w_α' is the local weight of sub-criteria.

4.1.3. Evaluating the Comprehensive Points of Alternatives

The rating scale of all sub-criteria was established based on the literature review, basic knowledge of the various traits of the peony and cultivation practice at low latitudes. The rating scale was divided into three levels, including highly relevant, moderately relevant and slightly relevant with scores of 1, 2/3 and 1/3, respectively (Table 9).

Evaluate the comprehensive points of alternatives by accumulating the scores of all sub-criteria (Equation (9)) [63]. The alternatives are rated according to the comprehensive evaluation score, with points between 80 and 100 as Level I; between 60 and 80 as Level II; between 0 and 60 as Level III.

$$P = 100 \sum_{i=1}^{n} W_{gi} R_i \quad (9)$$

where P is the comprehensive evaluation points; W_{gi} is the global weight of the i-th sub-criteria; R_i is the score of the i-th sub-criteria.

Table 9. Rating scale of all sub-criteria.

Sub Criteria	Score 1/3	Score 2/3	Score 1	References
Heat damage level (C1)	4–5	3.5–4	3–3.5	[12,18]
Root rot rate (C2)	Weak	Medium	Strong	[19]
Disease rate (C3)	40–80%	20–40%	0–20%	[18,21]
Survival rate (C4)	<60%	60–80%	>80%	[18,21]
Flower number per plant (C5)	<2	2–4	>4	[18,19]
Group blooming period (C6)	<10 days	10–15 days	>15 days	[18]
Flower type (C7)	Single or lotus form	Anemone or Crown form	Proliferation form	[19]
Proportion of flowering plant (C8)	<50%	50–80%	>80%	[21]
Aborted flowers per plant (C9)	>2	1–2	<1	[20]
Blooming period per flower (C10)	<5 days	5–6 days	>6 days	[18]
Flower diameter (C11)	<11 cm	11–13 cm	>13 cm	[19]
Stem bending degree (C12)	>30 °C	10–30 °C	<10 °C	[20,21]
Plant height (C13)	>60 or <40 cm	50–60 cm	40–50 cm	[21]
Plant width (C14)	>60 or <40 cm	50–60 cm	40–50 cm	[21]
Stem number (C15)	<4	4–5	>5	[21]
Stem diameter (C16)	<5 mm	5–6 mm	>6 mm	[20]
Dates of bud break (C17)	Later than March 15	March 1–March 15	Earlier than March 1	[18]
Chlorophyll content (C18)	<50	50–55	>55	[12]

4.2. Application of the MCDM Model—A Case Study

To testify the feasibility of this developed model for comprehensive evaluation of herbaceous peony, fifteen representative peony germplasms (*Paeonia lactiflora*) have been selected as a case study (Figure 8). Among these fifteen cultivars, fourteen cultivars were selected midlatitude cultivars introduced from Heze City (E 34°39′-35°52′, N 114°45′-116°25′), Shandong Province, and one cultivar. 'Hang Baishao'. was selected native low-latitude cultivar (Figure 8) [18]. 'Hang Baishao' is a unique traditional Chinese herbaceous peony with strong resistance to heat and humidity and low-chilling requirement trait and can be long cultivated at low latitudes in China [48,49,55]. These fifteen cultivars belong to five flower forms and have different performances in stress resistance and growth habit based on the previous studies [12,18].

In autumn of 2017, four-year-old peony tuberous roots of these fifteen cultivars were cultivated in the low-latitude Perennial Flower Resources Garden of Zhejiang University in Hangzhou (E 118°21′-120°30′, N 29°11′-30°33′), Zhejiang Province, China. Hangzhou has a subtropical monsoon climate, mild and humid. Spring is warm and humid. Summer is hot, with lots of rain and frequently occurring extreme high temperatures (Figure 9). The average annual temperature in Hangzhou is 15–17 °C. The average annual precipitation is 1100–1600 mm, and the average humidity is 76% [67]. The air temperature during the growing seasons in 2018–2021 were shown in Figure 9. Normal field management and fertilization were carried out after introduction. Peony tuberous roots were planted in full sun with rich soil and great drainage. Fertilizer applications were conducted twice times a year, the first time was made during budding and flowering in the spring, and the second time was carried out in the fall to produce roots. Phosphate fertilizer was used in the spring, and compost was used in the fall. In the late summer, when peony leaves lose their luster, turn colors and begin to die back for the winter, cut back peony stems to three to four inches above the ground and throw away the leaves. Additionally, these peony plants were grown under natural sunlight without any shade in summer.

Figure 8. Fifteen *P. lactiflora* cultivars adopted as the material in this study. (**A**) 'Bo Baishao'; (**B**) 'Chishao'; (**C**) 'Hang Baishao'; (**D**) 'Zhuguang'; (**E**) 'Yanzi Xiangyang'; (**F**) 'Hongpan Tuojing'; (**G**) 'Liantai'; (**H**) 'Meiju'; (**I**) 'Qihua Lushuang'; (**J**) 'Shanhe Hong'; (**K**) 'Taohua Feixue'; (**L**) 'Yangfei Chuyu'; (**M**) 'Zaohong'; (**N**) 'Zifeng Chaoyang'; (**O**) 'Qing Yunhong'.

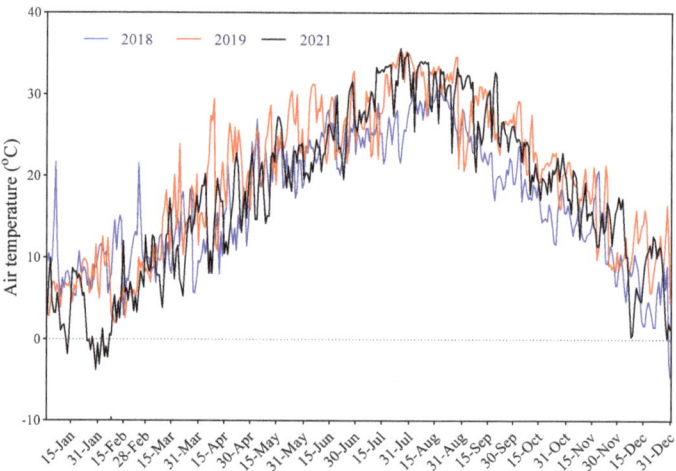

Figure 9. Air temperatures in the garden field in Hangzhou during the growing seasons in 2018–2021. Temperatures were recorded hourly in the field using a GM200-TH temperature and humidity recorder (Zhituo Instruments Limited Company, Hangzhou, China).

The seventeen criteria were observed in the field in 2018 and 2021. Fifteen plants (three biological replicates and five plants per replicate) of each cultivar were monitored, and details of the observations and measurements of evaluation indices were listed in Table 4. The data of the tested cultivars were substituted into the model to calculate the comprehensive evaluation point, and the cultivar was selected when the value was greater than 80 points. The selected brilliant cultivars will continue to be observed for a longer time to confirm whether the model is reliable.

4.3. Chlorophyll Fluorescence Determination of Selected Adapted and Maladapted Cultivars

Chlorophyll fluorescence is now widespread, used to monitor the photosynthetic performance of plants [52], especially the parameter Fv/Fm, which has been widely used to identify heat-tolerant genotypes as a physiological marker [50,68]. We have previously measured Chlorophyll fluorescence of peony cultivars from different latitudes and found that the parameters can accurately reflect the heat resistance of peony [12]. In this study, heat damage level is also a crucial sub-criterion for screening adaptive peony cultivars. Therefore, comparing the evaluation results obtained from the MCDM model with the observation results of Chlorophyll fluorescence indices could confirm whether the model is reliable. For validation, this study only selected the most adapted and maladapted cultivars evaluated by the MCDM model, as well as the verified reference fine peony 'Hang Baishao', for Chlorophyll fluorescence determination. The chlorophyll fluorescence characteristics were observed by an Imaging-PAM Chlorophyll fluorescence system (Hansatech Instruments, Norfolk, England). After a 30-min dark adaptation, the basal fluorescence (Fo) and Fv/Fm were measured on three sun-exposed, fully expanded leaves per cultivar (three biological replicates, one plant per replicate) at 10:00 a.m. on July 15 in 2021. NPQ, YII, qP and ETR parameters were also measured according to Li et al. (2021) [69].

4.4. Data Analysis

The experiments mentioned in this study were following a completely randomized design. Analysis of variance (ANOVA) was adopted to determine the statistical significance of the differences using SPSS 20.0 (IBM Corp., Armonk, NY, USA). Pictures or photos were combined and arranged by the Microsoft Office PowerPoint 2016. GraphPad Prism 8.0 (GraphPad Software, Inc., La Jolla, CA, USA) was used for visualization of the experimental data.

5. Conclusions

Establishing a comprehensive evaluation method to screen elite peony germplasm with adaptation as the main consideration is particularly important in the context of global warming. This study proposed a modified MCDM model for assessing peonies adapted to low-latitude climates, which is an extension and improvement of our previous study. The model reconstructed the AHP system, increased adaptability and growth habit-related indices while reduced reproductive traits and ornamental values-related indices. In addition, the weight of adaptability-related indices was improved via a pairwise comparison which was obtained through expert questionnaires. The model was validated on fifteen herbaceous peonies cultivars from different latitudes. The results showed that 'Meiju', 'Hang Baishao', 'Hongpan Tuojin' and 'Bo Baishao' were assessed as Level I, which have strong growth adaptability and high ornamental values, and were recommended for promotion and application at low latitudes. The reliability and stability of the MCDM model were further confirmed by measuring the Chlorophyll fluorescence of the selected adaptive and maladaptive cultivars. Consequently, the MCDM model developed in this study is low-cost, timesaving and could accurately screen adaptive peony germplasm at low-latitudes with hot and humid climates. This model fills a gap in the screening of adapted peonies and greatly facilitates the selection of elite germplasm at low latitudes. In addition to the establishment of the model; this study also provides a research strategy for other plants with similar situations as peony. In the future, the model will be modified accordingly with the increasing severity of global warming. Specifically, we could improve the applicability of the model by changing the weights of the indices, but there is no doubt that the adaptability-related index is still the most important. Finally, the application and promotion of the peony, as well as other perennial crops at low latitudes, can be addressed by policymakers supporting sustainable development and supporting land use and fundamental research funding.

Author Contributions: Conceptualization, methodology, software, validation and formal analysis, X.W.; investigation, X.W. and K.Z.; resources and data curation, X.W., R.Z., L.S., T.X. and X.S.; writing—original draft preparation, X.W.; writing—review and editing, X.W. and Y.X.; visualization, X.W. and D.L.; supervision, J.Z.; project administration, Y.X. and funding acquisition, Y.X. and J.Z. All authors have read and agreed to the published version of the manuscript.

Funding: This research was funded by the National Key Research and Development Project (grant number 2018YFD1000401) and National Natural Science Foundation of China (grant numbers 32071822 and 31600567).

Institutional Review Board Statement: Not applicable.

Informed Consent Statement: Not applicable.

Data Availability Statement: The data that support the findings of this study are available from the corresponding author, upon reasonable request.

Conflicts of Interest: The authors declare no conflict of interest.

References

1. Sieber, J. Impacts of, and adaptation options to, extreme weather events and climate change concerning thermal power plants. *Clim. Chang.* **2013**, *121*, 55–66. [CrossRef]
2. Johnson, N.C.; Xie, S.P.; Kosaka, Y.; Li, X. Increasing occurrence of cold and warm extremes during the recent global warming slowdown. *Nat. Commun.* **2018**, *9*, 1724. [CrossRef] [PubMed]
3. Zhang, T.; Tang, Y.; Luan, Y.; Cheng, Z.; Wang, X.; Tao, J.; Zhao, D. Herbaceous peony AP2/ERF transcription factor binds the promoter of the tryptophan decarboxylase gene to enhance high-temperature stress tolerance. *Plant Cell Environ.* **2022**, *45*, 2729–2743. [CrossRef] [PubMed]
4. Jiang, R.; He, W.; He, L.; Yang, J.Y.; Qian, B.; Zhou, W.; He, P. Modelling adaptation strategies to reduce adverse impacts of climate change on maize cropping system in Northeast China. *Sci. Rep.* **2021**, *11*, 810. [CrossRef] [PubMed]
5. Ochieng, J.; Kirimi, L.; Mathenge, M. Effects of climate variability and change on agricultural production: The case of small scale farmers in Kenya. *NJAS Wagening. J. Life Sci.* **2021**, *77*, 71–78. [CrossRef]
6. Migicovsky, Z.; Myles, S. Exploiting Wild Relatives for Genomics-assisted Breeding of Perennial Crops. *Front. Plant Sci.* **2017**, *8*, 460. [CrossRef]
7. Xiang, W.; Xue, S.; Qin, S.; Xiao, L.; Liu, F.; Yi, Z. Development of a multi-criteria decision making model for evaluating the energy potential of Miscanthus germplasms for bioenergy production. *Ind. Crops Prod.* **2018**, *125*, 602–615. [CrossRef]
8. Kozak, M.; Bocianowski, J.; Rybinski, W. Selection of promising genotypes based on path and cluster analyses. *J. Agric. Sci.* **2008**, *146*, 85–92. [CrossRef]
9. Meyer, R.S.; Purugganan, M.D. Evolution of crop species: Genetics of domestication and diversification. *Nat. Rev. Genet.* **2013**, *14*, 840–852. [CrossRef]
10. Hufnagel, J.; Reckling, M.; Ewert, F. Diverse approaches to crop diversification in agricultural research. A review. *Agron. Sustain. Dev.* **2020**, *40*, 14. [CrossRef]
11. Yong, H.; Jin, Z.; Gao, L.; Zhang, L.; Liu, X.; Zhang, F.; Zhang, X.; Zhang, D.; Li, M.; Weng, J.; et al. Breeding potential of maize germplasm populations to improve yield and predominant heterotic pattern in Northeast China. *Euphytica* **2017**, *213*, 1–13. [CrossRef]
12. Wang, X.; Shi, X.; Zhang, R.; Zhang, K.; Shao, L.; Xu, T.; Li, D.; Zhang, D.; Zhang, J.; Xia, Y. Impact of summer heat stress inducing physiological and biochemical responses in herbaceous peony cultivars (*Paeonia lactiflora* Pall.) from different latitudes. *Ind. Crops Prod.* **2022**, *184*, 115000. [CrossRef]
13. Wu, G.; Shen, Y.; Nie, R.; Li, P.; Jin, Q.; Zhang, H.; Wang, X. The bioactive compounds and cellular antioxidant activity of Herbaceous peony (Paeonia lactiflora Pall) seed oil from China. *J. Food Sci.* **2020**, *85*, 3815–3822. [CrossRef]
14. Hao, Z.; Wei, M.; Gong, S.; Zhao, D.; Tao, J. Transcriptome and digital gene expression analysis of herbaceous peony (*Paeonia lactiflora* Pall.) to screen thermo-tolerant related differently expressed genes. *Genes Genom.* **2016**, *38*, 1201–1215. [CrossRef]
15. Bogiatzis, K.C.; Wallace, H.M.; Trueman, S.J. Shoot Growth and Flower Bud Production of Peony Plants under Subtropical Conditions. *Horticulturae* **2021**, *7*, 476. [CrossRef]
16. Zhang, R.; Wang, X.; Shi, X.; Shao, L.; Xu, T.; Xia, Y.; Li, D.; Zhang, J. Chilling Requirement Validation and Physiological and Molecular Responses of the Bud Endodormancy Release in *Paeonia lactiflora* 'Meiju'. *Int. J. Mol. Sci.* **2021**, *22*, 8382. [CrossRef]

7. Kamenetsky-Goldstein, R.; Yu, X. Cut peony industry: The first 30 years of research and new horizons. *Hortic. Res.* **2022**, *9*, uhac079. [CrossRef]
8. Zhang, J.; Wang, X.; Zhang, D.; Qiu, S.; Wei, J.; Guo, J.; Li, D.; Xia, Y. Evaluating the Comprehensive Performance of Herbaceous Peonies at low latitudes by the Integration of Long-running Quantitative Observation and Multi-Criteria Decision Making Approach. *Sci. Rep.* **2019**, *9*, 15079. [CrossRef]
9. Liu, J. Comprehensive evaluation of peony traits in Luoyang alpine region. *J. South. Argic.* **2019**, *50*, 809–815.
10. Wu, T.; Gao, J.; Zhao, Z.; Liu, Y.; Zhang, J. Selection of Herbaceous Peony Cultivars under Protected Cultivation Condition. *J. Northwest For. Univ.* **2014**, *29*, 145–150.
11. Si, S.; Meng, Q.; Sun, D.; Zhang, Y. Evaluation on Comprehensive Characteristics of Herbaceous Peonies Introduced in Guanzhong. *J. Northwest For. Univ.* **2021**, *36*, 134–139.
12. Xiang, W.; Xue, S.; Liu, F.; Qin, S.; Xiao, L.; Yi, Z. MGDB: A database for evaluating *Miscanthus* spp. to screen elite germplasm. *Biomass Bioenergy* **2020**, *138*, 105599. [CrossRef]
13. Chen, B.; Wang, Z.; Luo, C. Integrated evaluation approach for node importance of complex networks based on relative entropy. *J. Syst. Eng. Electron.* **2016**, *27*, 1219–1226. [CrossRef]
14. Wang, C.-N.; Tsai, H.-T.; Ho, T.-P.; Nguyen, V.-T.; Huang, Y.-F. Multi-Criteria Decision Making (MCDM) Model for Supplier Evaluation and Selection for Oil Production Projects in Vietnam. *Processes* **2020**, *8*, 134. [CrossRef]
15. Kumar, A.; Sah, B.; Singh, A.R.; Deng, Y.; He, X.; Kumar, P.; Bansal, R.C. A review of multi criteria decision making (MCDM) towards sustainable renewable energy development. *Renew. Sustain. Energy Rev.* **2017**, *69*, 596–609. [CrossRef]
16. Halder, B.; Banik, P.; Almohamad, H.; Al Dughairi, A.A.; Al-Mutiry, M.; Al Shahrani, H.F.; Abdo, H.G. Land Suitability Investigation for Solar Power Plant Using GIS, AHP and Multi-Criteria Decision Approach: A Case of Megacity Kolkata, West Bengal, India. *Sustainability* **2022**, *14*, 11276. [CrossRef]
17. Jovanović, B.; Filipović, J.; Bakić, V. Prioritization of manufacturing sectors in Serbia for energy management improvement – AHP method. *Energy Convers. Manag.* **2015**, *98*, 225–235. [CrossRef]
18. Saaty, T.L. *The Analytic Hierarchy Process*; McGraw-Hill: New York, NY, USA, 1980.
19. van de Water, H.; de Vries, J. Choosing a quality improvement project using the analytic hierarchy process. *Int. J. Qual. Reliab. Manag.* **2006**, *23*, 409–425. [CrossRef]
20. Wątróbski, J.; Jankowski, J.; Ziemba, P.; Karczmarczyk, A.; Zioło, M. Generalised framework for multi-criteria method selection. *Omega* **2019**, *86*, 107–124. [CrossRef]
21. Zeng, S.; Zhang, N.; Zhang, C.; Su, W.; Carlos, L.-A. Social network multiple-criteria decision-making approach for evaluating unmanned ground delivery vehicles under the Pythagorean fuzzy environment. *Technol. Forecast. Soc. Chang.* **2022**, *175*, 121414. [CrossRef]
22. Abdel-Basset, M.; Gamal, A.; Chakrabortty, R.K.; Ryan, M. A new hybrid multi-criteria decision-making approach for location selection of sustainable offshore wind energy stations: A case study. *J. Clean. Prod.* **2021**, *280*, 124462. [CrossRef]
23. Shao, M.; Han, Z.X.; Sun, J.W.; Xiao, C.S.; Zhang, S.L.; Zhao, Y.X. A review of multi-criteria decision making applications for renewable energy site selection. *Renew. Energy* **2020**, *157*, 377–403. [CrossRef]
24. Rahimi, S.; Hafezalkotob, A.; Monavari, S.M.; Hafezalkotob, A.; Rahimi, R. Sustainable landfill site selection for municipal solid waste based on a hybrid decision-making approach: Fuzzy group BWM-MULTIMOORA-GIS. *J. Clean. Prod.* **2020**, *248*, 119186. [CrossRef]
25. Qureshi, M.R.N.; Singh, R.K.; Hasan, M.A. Decision support model to select crop pattern for sustainable agricultural practices using fuzzy MCDM. *Environ. Dev. Sustain.* **2018**, *20*, 641–659. [CrossRef]
26. Pathania, S.; Arora, P.K.; Bhullar, K.S.; Ram, S. Multi criteria decision making analysis to identify micronutrient deficiencies in Kinnow (*Citrus reticulata* Blanco). *J. Plant Nutr.* **2022**, *2022*, 2105718. [CrossRef]
27. Nedeljkovic, M.; Puska, A.; Doljanica, S.; Virijevic Jovanovic, S.; Brzakovic, P.; Stevic, Z.; Marinkovic, D. Evaluation of rapeseed varieties using novel integrated fuzzy PIPRECIA—Fuzzy MABAC model. *PLoS ONE* **2021**, *16*, e0246857. [CrossRef]
28. Ramírez-García, J.; Carrillo, J.M.; Ruiz, M.; Alonso-Ayuso, M.; Quemada, M. Multicriteria decision analysis applied to cover crop species and cultivars selection. *Field Crops Res.* **2015**, *175*, 106–115. [CrossRef]
29. Dragincic, J.; Korac, N.; Blagojevic, B. Group multi-criteria decision making (GMCDM) approach for selecting the most suitable table grape variety intended for organic viticulture. *Comput. Electron. Agric.* **2015**, *111*, 194–202. [CrossRef]
30. Seyedmohammadi, J.; Sarmadian, F.; Jafarzadeh, A.A.; Ghorbani, M.A.; Shahbazi, F. Application of SAW, TOPSIS and fuzzy TOPSIS models in cultivation priority planning for maize, rapeseed and soybean crops. *Geoderma* **2018**, *310*, 178–190. [CrossRef]
31. Yang, Y.; Sun, M.; Li, S.; Chen, Q.; Teixeira da Silva, J.A.; Wang, A.; Yu, X.; Wang, L. Germplasm resources and genetic breeding of *Paeonia*: A systematic review. *Hortic. Res.* **2020**, *7*, 107. [CrossRef]
32. Zhao, D.; Han, C.; Zhou, C.; Tao, J. Shade Ameliorates High Temperature-induced Inhibition of Growth in Herbaceous Peony (*Paeonia lactiflora*). *Int. J. Agric. Biol.* **2015**, *17*, 911–919. [CrossRef]
33. Cohen, M.; Kamenetsky, R.; Yom Din, G. Herbaceous peony in warm climate: Modelling stem elongation and growers profit responses to dormancy conditions. *Inf. Process. Agric.* **2016**, *3*, 175–182. [CrossRef]
34. Weng, J.; Li, P.; Rehman, A.; Wang, L.; Gao, X.; Niu, Q. Physiological response and evaluation of melon (*Cucumis melo* L.) germplasm resources under high temperature and humidity stress at seedling stage. *Sci. Hortic.* **2021**, *288*, 110317. [CrossRef]

45. Xue, Y.; Liu, R.; Xue, J.; Wang, S.; Zhang, X. Genetic diversity and relatedness analysis of nine wild species of tree peony based on simple sequence repeats markers. *Hortic. Plant J.* **2021**, *7*, 579–588. [CrossRef]
46. Russo, R.d.F.S.M.; Camanho, R. Criteria in AHP: A Systematic Review of Literature. *Procedia Comput. Sci.* **2015**, *55*, 1123–1132. [CrossRef]
47. Rossi, S.; Burgess, P.; Jespersen, D.; Huang, B. Heat-Induced Leaf Senescence Associated with Chlorophyll Metabolism in Bentgrass Lines Differing in Heat Tolerance. *Crop Sci.* **2017**, *57*, S-169. [CrossRef]
48. Wang, X.; Li, D.; Zhang, D.; Shi, X.; Wu, Y.; Qi, Z.; Ding, H.; Zhu, K.; Xia, Y.; Zhang, J. Improving crucial details and selecting the optimal model for evaluating the chilling requirement of *Paeonia lactiflora* Pall. at low latitudes during four winters. *Sci. Hortic.* **2020**, *265*, 109175. [CrossRef]
49. Zhang, J.; Zhang, D.; Wei, J.; Shi, X.; Ding, H.; Qiu, S.; Guo, J.; Li, D.; Zhu, K.; Horvath, D.P.; et al. Annual growth cycle observation, hybridization and forcing culture for improving the ornamental application of *Paeonia lactiflora* Pall. in the low-latitude regions. *PLoS ONE* **2019**, *14*, e0218164. [CrossRef]
50. Sharma, D.K.; Andersen, S.B.; Ottosen, C.O.; Rosenqvist, E. Wheat cultivars selected for high Fv/Fm under heat stress maintain high photosynthesis, total chlorophyll, stomatal conductance, transpiration and dry matter. *Physiol. Plant.* **2015**, *153*, 284–298. [CrossRef]
51. Ferguson, J.N.; McAusland, L.; Smith, K.E.; Price, A.H.; Wilson, Z.A.; Murchie, E.H. Rapid temperature responses of photosystem II efficiency forecast genotypic variation in rice vegetative heat tolerance. *Plant J.* **2020**, *104*, 839–855. [CrossRef]
52. Baker, N.R. Chlorophyll fluorescence: A probe of photosynthesis in vivo. *Annu. Rev. Plant Biol.* **2008**, *59*, 89–113. [CrossRef] [PubMed]
53. Sharma, D.K.; Andersen, S.B.; Ottosen, C.O.; Rosenqvist, E. Phenotyping of wheat cultivars for heat tolerance using chlorophyll a fluorescence. *Funct. Plant Biol.* **2012**, *39*, 936–947. [CrossRef] [PubMed]
54. Nyachiro, J.M.; Briggs, K.G.; Hoddinott, J.; Johnson-Flanagan, A.M. Chlorophyll content, chlorophyll fluorescence and water deficit in spring wheat. *Cereal Res. Commun.* **2001**, *29*, 135–142. [CrossRef]
55. Wang, X.; Zhang, R.; Huang, Q.; Shi, X.; Li, D.; Shao, L.; Xu, T.; Horvath, D.P.; Xia, Y.; Zhang, J. Comparative Study on Physiological Responses and Gene Expression of Bud Endodormancy Release Between Two Herbaceous Peony Cultivars (*Paeonia lactiflora* Pall.) With Contrasting Chilling Requirements. *Front. Plant Sci.* **2022**, *12*, 772285. [CrossRef] [PubMed]
56. Li, X.; Li, J.; Sui, H.; He, L.; Cao, X.; Li, Y. Evaluation and determination of soil remediation schemes using a modified AHP model and its application in a contaminated coking plant. *J. Hazard. Mater.* **2018**, *353*, 300–311. [CrossRef]
57. Ilangkumaran, M.; Kumanan, S. Selection of maintenance policy for textile industry using hybrid multi-criteria decision making approach. *J. Manuf. Technol. Manag.* **2009**, *20*, 1009–1022. [CrossRef]
58. Saaty, T.L. The Modern Science of Multicriteria Decision Making and Its Practical Applications: The AHP/ANP Approach. *Oper. Res.* **2013**, *61*, 1101–1118. [CrossRef]
59. Singh, R.P.; Nachtnebel, H.P. Analytical hierarchy process (AHP) application for reinforcement of hydropower strategy in Nepal. *Renew. Sustain. Energy Rev.* **2016**, *55*, 43–58. [CrossRef]
60. Saaty, R.W. The analytic hierarchy process—What it is and how it is used. *Math. Model.* **1987**, *9*, 161–176. [CrossRef]
61. Misran, M.F.R.; Roslin, E.N.; Mohd Nur, N. AHP-Consensus Judgement on Transitional Decision-Making: With a Discussion on the Relation towards Open Innovation. *J. Open Innov. Technol. Mark. Complex.* **2020**, *6*, 63. [CrossRef]
62. Stirn, L.Z.; Groselj, P. Multiple critera methods with focus on analytic hierarchy process and group decision making. *Croat. Oper. Res. Rev.* **2010**, *1*, 2–11.
63. Thanki, S.; Govindan, K.; Thakkar, J. An investigation on lean-green implementation practices in Indian SMEs using analytical hierarchy process (AHP) approach. *J. Clean. Prod.* **2016**, *135*, 284–298. [CrossRef]
64. Alonso, J.A.; Lamata, M.T. Consistency in the analytic hierarchy process: A new approach. *Int. J. Uncertain. Fuzziness Knowl.-Based Syst.* **2006**, *14*, 445–459. [CrossRef]
65. Díaz, H.; Guedes Soares, C. A novel multi-criteria decision-making model to evaluate floating wind farm locations. *Renew. Energy* **2022**, *185*, 431–454. [CrossRef]
66. Zhang, B.; Pedrycz, W.; Fayek, A.R.; Dong, Y. A Differential Evolution-Based Consistency Improvement Method in AHP with an Optimal Allocation of Information Granularity. *IEEE Trans. Cybern.* **2022**, *52*, 6733–6744. [CrossRef]
67. Sheng, L.; Tang, X.; You, H.; Gu, Q.; Hu, H. Comparison of the urban heat island intensity quantified by using air temperature and Landsat land surface temperature in Hangzhou, China. *Ecol. Indic.* **2017**, *72*, 738–746. [CrossRef]
68. Camejo, D.; Jimenez, A.; Alarcon, J.J.; Torres, W.; Gomez, J.M.; Sevilla, F. Changes in photosynthetic parameters and antioxidant activities following heat-shock treatment in tomato plants. *Funct. Plant Biol.* **2006**, *33*, 177–187. [CrossRef]
69. Li, N.N.; Shi, F.; Gao, H.Y.; Khan, A.; Wang, F.Y.; Kong, X.H.; Luo, H.H. Improving photosynthetic characteristics and antioxidant enzyme activity of capsule wall and subtending leaves increases cotton biomass under limited irrigation system. *Photosynthetica* **2021**, *59*, 215–227. [CrossRef]

Article

Identification of Genetic Variations and Candidate Genes Responsible for Stalk Sugar Content and Agronomic Traits in Fresh Corn via GWAS across Multiple Environments

Jianjian Chen [1], Jinming Cao [2], Yunlong Bian [2], Hui Zhang [3], Xiangnan Li [1], Zhenxing Wu [1], Guojin Guo [1] and Guihua Lv [1,*]

1. Institute of Maize and Featured Upland Crops, Zhejiang Academy of Agricultural Sciences, Hangzhou 310004, China
2. Jiangsu Key Laboratory of Crop Genomics and Molecular Breeding, Co-Innovation Center for Modern Production Technology of Grain Crops, Key Laboratory of Plant Functional Genomics of the Ministry of Education, Jiangsu Key Laboratory of Crop Genetics and Physiology, Yangzhou University, Yangzhou 225009, China
3. Zhejiang Agricultural Technology Extension Center, Hangzhou 310004, China
* Correspondence: lvgh@zaas.ac.cn; Tel.: +86-013454997051

Citation: Chen, J.; Cao, J.; Bian, Y.; Zhang, H.; Li, X.; Wu, Z.; Guo, G.; Lv, G. Identification of Genetic Variations and Candidate Genes Responsible for Stalk Sugar Content and Agronomic Traits in Fresh Corn via GWAS across Multiple Environments. *Int. J. Mol. Sci.* **2022**, *23*, 13490. https://doi.org/10.3390/ijms232113490

Academic Editors: Jian Zhang and Zhiyong Li

Received: 11 September 2022
Accepted: 28 October 2022
Published: 4 November 2022

Publisher's Note: MDPI stays neutral with regard to jurisdictional claims in published maps and institutional affiliations.

Copyright: © 2022 by the authors. Licensee MDPI, Basel, Switzerland. This article is an open access article distributed under the terms and conditions of the Creative Commons Attribution (CC BY) license (https://creativecommons.org/licenses/by/4.0/).

Abstract: The stem and leaves of fresh corn plants can be used as green silage or can be converted to biofuels, and the stalk sugar content and yield directly determine the application value of fresh corn. To identify the genetic variations and candidate genes responsible for the related traits in fresh corn, the genome-wide scan and genome-wide association analysis (GWAS) were performed. A total of 32 selective regions containing 172 genes were detected between sweet and waxy corns. Using the stalk sugar content and seven other agronomic traits measured in four seasons over two years, the GWAS identified ninety-two significant single nucleotide polymorphisms (SNPs). Most importantly, seven SNPs associated with the stalk sugar content were detected across multiple environments, which could explain 13.68–17.82% of the phenotypic variation. Accessions differing in genotype for certain significant SNPs showed significant variation in the stalk sugar content and other agronomic traits, and the expression levels of six important candidate genes were significantly different between two materials with different stalk sugar content. The genetic variations and candidate genes provide valuable resources for future studies of the molecular mechanism of the stalk sugar content and establish the foundation for molecular marker-assisted breeding of fresh corn.

Keywords: fresh corn; stalk sugar content; agronomic traits; genome-wide association analysis; selective sweep; single nucleotide polymorphisms

1. Introduction

The stem and leaves of fresh corn plants are often applied to green silage and converted to biofuels. Fresh corn, mainly comprising sweet corn and waxy corn types, is an important staple food consumed as a vegetable or fruit, and is also used for livestock feed and fuel [1]. The stems and leaves of fresh corn plants, after harvesting of the ears, are rich in nutrients and can be used as green silage, thus realizing the utilization of multiple corn resources [2]. Corn silage is widely used in animal husbandry because of its high energy value [3]. Therefore, improvement of the quality of silage corn feed is important for the development of animal husbandry. Fresh corn must go through a suitable fermentation process to become high-quality silage, which produces a large amount of lactic acid, and sugar is the crucial component in this process [4,5]. However, an insufficient amount of lactic acid not only leads to the reduction in silage quality but also might favor silage mildew development. Therefore, the sugar content of green silage is an important factor that affects silage quality, and a higher stalk sugar content can improve the feed quality and palatability of silage corn [6,7]. In addition, plant height, plant weight, and stalk sugar content are important factors that affect the utility of fresh corn; increasing the plant weight and stalk sugar content of fresh corn can promote the application of corn green silage and promote the development of animal husbandry. Most importantly, the stalk sugar content is also an important attribute affecting the overall feasibility and profitability of biofuel production, such as bioethanol. Given the current challenges faced in the world, identifying biomass resources with superior properties is important to balance food safety and biofuel production [8]. Increasing the stalk sugar content of fresh corn will facilitate a more economically viable production of higher-generation biofuels, which are derived from biomass rather than from edible resources.

Sugar supply is crucial for plant growth and development. Calcium signaling-mediated transcriptional regulation might play an important role in glucose metabolism. Calcium-dependent protein kinases can phosphorylate and regulate sucrose synthase and sucrose phosphate synthase, thus influencing sucrose accumulation [9,10]. Receptor protein kinases may regulate sucrose metabolism in sugarcane [11], and MYB transcription factors are involved in the regulation of water-soluble polysaccharide biosynthesis [12,13]. Pyruvate dehydrogenase plays an important role in energy production by controlling the entry of carbon into the tricarboxylic acid cycle [14,15]. Mutation of Sh2 (shrunken2), which encodes the large subunit of adenosine diphosphate glucose pyrophosphorylases, directly affects starch synthesis and leads to an accumulation of abundant sugars in corn [16,17]. Therefore, studies on the regulation of glucose metabolism-related genes provide an important foundation for the improvement of the stalk sugar content.

The stalk sugar content of corn is a complex quantitative trait controlled by hereditable factors despite its complex phenotype [18]. The stalk sugar content is strongly associated with kernel filling rate [19]. In sweet sorghum, a reduction in sugar storage in the grain promotes stalk sugar accumulation and increases stem biomass [20]. Elucidation of the molecular mechanism of stalk sugar accumulation in corn is important for breeding new cultivars with an enhanced stalk sugar content. In previous studies, using 202 recombinant inbred lines, the major quantitative trait locus (QTL) qSSC-2.1 (bnlg1909-umc1635) was identified, which explained 13.8% of the phenotypic variance and exhibited high stability. The average stem sugar content of lines carrying the main-effect QTL was 12.5%, and as much as 15% in several lines [3,21]. In addition, QTLs for sugar content has been mapped in sweet sorghum [22]. Two QTLs for stalk sugar content were identified in sweet sorghum and were useful to improve the sugar content by marker-assisted selection [23]. Using a diverse panel of one hundred and twenty-five sorghum accessions, three significant associations for plant height and stem sugar (Brix) traits were detected through association mapping in sweet sorghum [24]. A QTL for stem sugar content on chromosome 3 was identified and explained 25% of the genetic variance [25].

Linkage analysis is limited by parental differences and population size, and the number of localized QTLs is limited [26]. In recent years, genome-wide association analysis (GWAS) has proven to be an effective method to study complex traits [27]. Important progress has been achieved in corn and a series of candidate genes has been detected. To date, GWAS has been applied to corn in studies of resistance to head smut [28], kernel oil concentration and fatty acid composition [29], drought tolerance [30], cold tolerance [31], and other phenotypic traits [32]. However, GWAS of stalk sugar content-related traits in corn has not been reported previously.

To uncover the molecular regulation mechanism of the stalk sugar content and yield in fresh corn, the identification of the genetic variations and candidate genes responsible for the stalk sugar content and agronomic traits is the key problem. In this study, based on one hundred and eighty-eight sweet, waxy, and hybrid corn accessions differing in genetic background, the stalk sugar content and other agronomic traits (e.g., plant height, whole weight per plant, and ear weight with bract) were investigated in four seasons of two years during the spring and autumn sowing. A GWAS analysis was performed to detect the significant single-nucleotide polymorphisms (SNPs) associated with stalk sugar content and the agronomic traits, and the candidate genes were screened, which provides significant insights into the genomic footprints of the stalk sugar content in fresh corn and facilitates the breeding of corn cultivars with higher stalk sugar content.

2. Results

2.1. Identification of SNPs

In this study, 188 sweet corn, waxy corn, and hybrid accessions were genotyped. Among them, 41 were sweet corn inbred lines, 74 were waxy corn inbred lines, and 73 were hybrid accessions (Table S1). After comparison with the reference genome (B73 RefGen_v4) and filtering, a total of 36,069 high-quality SNP markers (minimum allele frequency > 0.05 and missing data < 30%) were retained. Among these SNPs, 35,824 markers were distributed on 10 chromosomes (Figure 1). After annotation with ANNOVAR, 5590 (14.99%) SNPs were determined to be located in coding regions, comprising 2133 (5.72%) nonsynonymous SNPs and 3457 (9.27%) synonymous SNPs (Figure S1). In addition, 5698 (15.28%) and 4418 (11.85%) SNPs were located in intronic and intergenic regions, respectively. Examination of the SNP mutation types (Figure S2) revealed that the transition/transversion ratio was 2.81 (Table S2).

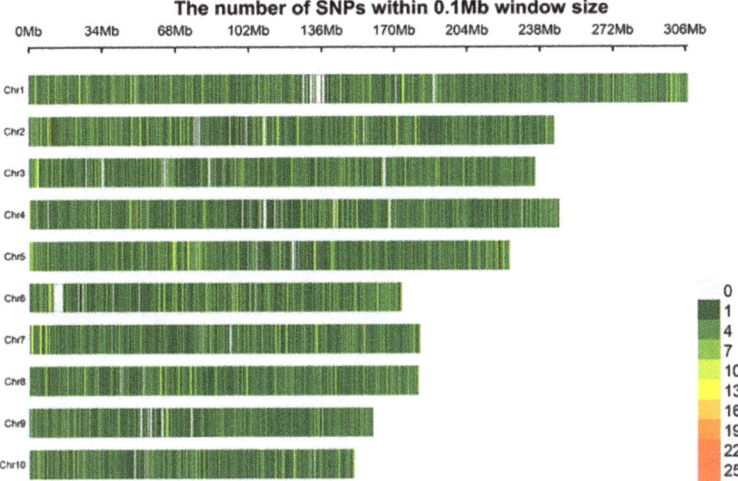

Figure 1. Density of SNPs on the 10 chromosomes of corn within a 0.1 Mb window size. The spectrum column indicates the densities by different colors.

2.2. Population Structure, Relative Kinship, and LD Decay

To understand the population structure of the 188 fresh corn accessions, we first constructed a phylogenetic tree through the maximum likelihood method using 36,069 high-quality SNPs. The sweet corn and waxy corn accessions were mainly clustered into two subclasses, and most hybrid accessions were classified together with sweet corn accessions (Figure 2a), which indicated that sweet corn constituted a higher proportion of the genetic background of the hybrid accessions. Then, a two-dimensional principal component analysis (PCA) was performed using the GCTA software (Figures 2b and S3). The first and second principal components (PC1 and PC2) explained 8.8% and 3.6% of the total variations, respectively. The PC1 separated the sweet corn (group A) and waxy corn (group B) accessions, and the hybrid accessions (group C) were placed near the sweet corn accessions, which was consistent with the relationships represented in the phylogenetic tree. To further analyze the genetic background of the sweet corn (group A), waxy corn (group B), and hybrid (group C) accessions, a population structure analysis was conducted through ADMIXTURE software (Figures 2c and S4, Table S3), which also supported the relationships suggested from the phylogenetic analysis and PCA. Thus, fresh corn was indicated to be the core breeding material among the study population, and the genetic relationships among the germplasm were complicated likely as a result of long-term crossing and selection. To further clarify the relationships among the accessions used in this study, we evaluated the relative kinship of the 188 corn accessions using the TASSEL software (Figure 2d). Most accessions were weakly related to each other except for a small number of samples, which did not affect the subsequent association analysis. The estimated linkage disequilibrium (LD) decay in the three groups was similar and decayed to $r^2 = 0.2$ at about 50 kb (Figure 2e).

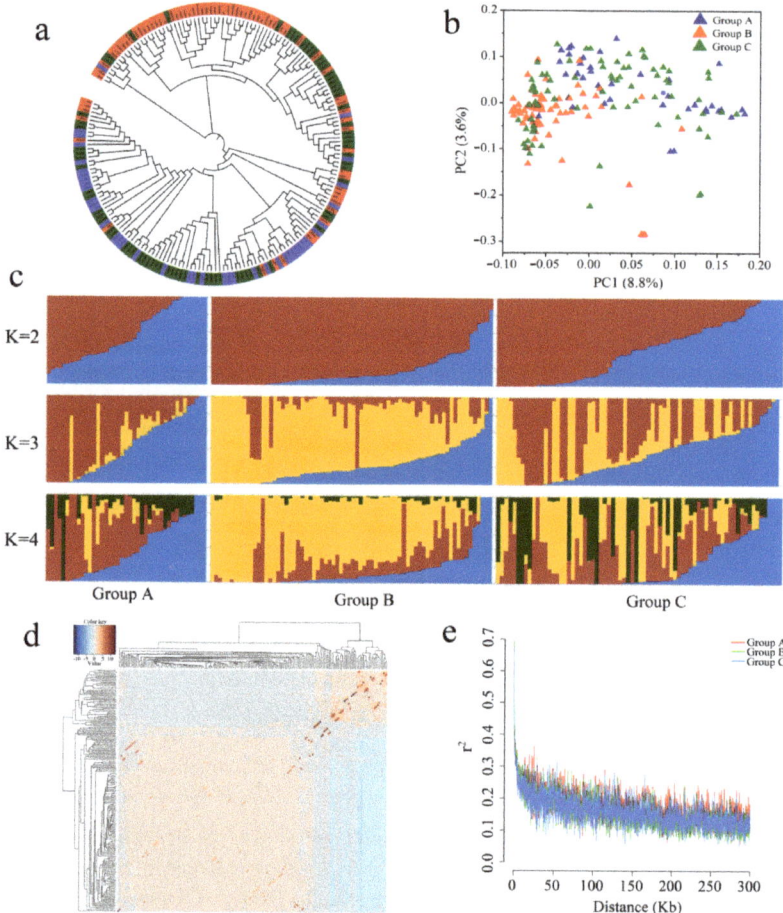

Figure 2. Population structure, principal component analysis, and linkage disequilibrium decay of the 188 fresh corn accessions. (**a**) Phylogenetic tree inferred from the 36,069 high-quality SNPs. Blue, red, and green colors represent sweet corn, waxy corn, and hybrid accessions, respectively. (**b**) Principal component analysis plot of the 188 fresh corn accessions on the first two principal components (PC1 and PC2). Groups A, B, and C include sweet corn, waxy corn, and hybrid accessions, respectively. (**c**) Population structure analysis of the corn accessions performed with ADMIXTURE software. Different results are displayed when $K = 2, 3,$ or 4. (**d**) Relative kinship heatmap of the corn accessions. (**e**) Genome-wide linkage disequilibrium (LD) decay (measured by r^2) of the different groups.

2.3. Selective Sweep Analysis

To determine the possible selective signals during the breeding history of the study population, we performed a selective sweep analysis with the genetic differentiation coefficient (F_{ST}) and nucleotide diversity between the sweet corn (group A) and waxy corn (group B) accessions (π_A/π_B). In total, 18 and 20 significant selective regions were detected from the F_{ST} and π_A/π_B values, respectively (Figure 3). A total of 85 and 110 genes were located in these selective regions, respectively (Tables S4 and S5). Among the selective regions, six were detected both from the F_{ST} and π_A/π_B, and 23 genes were located in these regions (Table 1), including a cation transport protein (*Zm00001d040627*), proline-rich receptor-like protein kinase (*Zm00001d033104*), and zinc finger protein (*Zm00001d052883*). A gene ontology (GO) analysis of all selective genes revealed that 53 GO terms were enriched (Figures S5 and S7). A Kyoto Encyclopedia of Genes and Genomes (KEGG) pathway analysis indicated that "ribosome" and "terpenoid backbone biosynthesis" were the top two enriched pathways (Figures S6 and S8). These results indicated that specific genes might have played important roles in the natural and artificial selection of the studied sweet corn and waxy corn accessions.

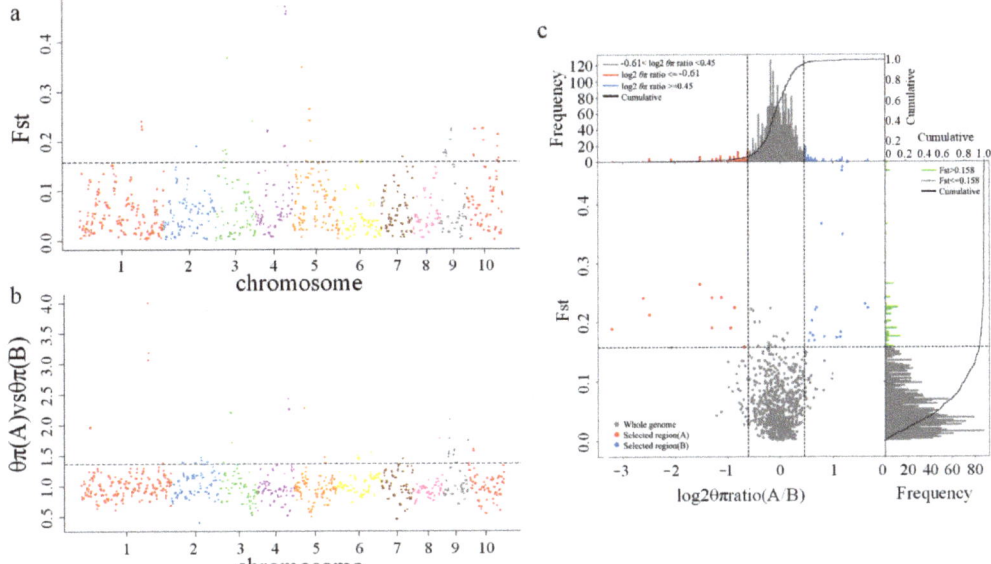

Figure 3. Selective sweeps between sweet corn and waxy corn accessions. (**a**) Distribution of the selective sweeps identified by genetic differentiation coefficient (F_{ST}). The threshold is represented by a dashed line (top 5% of F_{ST} values). (**b**) Distribution of the selective sweeps identified by nucleotide diversity (π_A/π_B). The threshold is represented by a dashed line (top 5% of π_A/π_B values). (**c**) Selective sweep regions with the top 5% of F_{ST} and π_A/π_B values.

2.4. Phenotyping Results

Stalk sugar content and related agronomic traits from four seasons in two years were investigated (Table S6). The eight traits (summarized in Tables S7–S10) each exhibited broad continuous variation in all environments (Figures S9–S16). Significant differences in sugar content traits were observed among the inbred lines (Table S10), including the sugar content of the stalk (SCS), sugar content in the lower ear parts (SCLEP), three-node sugar content in the ear (TNSCE), and sugar content in the upper ear parts (SCUEP). The mean SCS in the spring and autumn of 2020 was 9.14% and 8.15%, with ranges between 3.6–18.1% and 2.7–17.6%, respectively. The mean SCS in the spring and autumn of 2021 was 8.73% and 7.48%, with ranges between 1.9–16.1% and 2.7–17.6%, respectively (Table S11). Thus, considerable genetic variation in SCS was indicated among the accessions. This genetic variation provided a genetic basis for the improvement of SCS in corn. In addition, all agronomic traits showed continuous and approximately normal distributions (Figures S9–S16) and thus were suitable for GWAS.

Table 1. Summary of the genes located in the selective sweep regions both with the top 5% of F_{ST} and π_A/π_B between sweet and waxy corn accessions.

Gene ID	Chromosome	Start	End	Gene Function
Zm00001d033104	1	250291343	250304767	Proline-rich receptor-like protein kinase PERK
Zm00001d033105	1	250320688	250322547	Ribosomal protein
Zm00001d033106	1	250323656	250324358	Unknown
Zm00001d033107	1	250328690	250338437	Sulfated surface glycoprotein
Zm00001d033108	1	250339356	250347246	Lysine–tRNA ligase
Zm00001d033109	1	250346963	250360412	P-loop containing nucleoside triphosphate hydrolase superfamily protein
Zm00001d040502	3	47172433	47173432	Early nodulin-like protein
Zm00001d040503	3	47223813	47241740	Retrovirus-related Pol polyprotein
Zm00001d040504	3	47292181	47293995	Lamin-like protein
Zm00001d040627	3	54948300	54950704	Cation transport protein
Zm00001d040628	3	55023821	55026070	Unknown
Zm00001d052883	4	203324761	203327583	Zinc finger protein
Zm00001d052885	4	203405706	203407600	40S ribosomal protein
Zm00001d052886	4	203407872	203412281	Sulfotransferase
Zm00001d052888	4	203494690	203511433	Polyadenylate-binding protein
Zm00001d045596	9	28028596	28031674	BAG family molecular chaperone regulator
Zm00001d045597	9	28031820	28047755	Rho GTPase-activating protein
Zm00001d045598	9	28060153	28063330	Unknown
Zm00001d045599	9	28082328	28086696	Monocopper oxidase-like protein
Zm00001d045600	9	28126449	28127152	NAC domain-containing protein
Zm00001d046109	9	63900291	63900708	60S ribosomal protein
Zm00001d046111	9	63970902	63971348	Unknown
Zm00001d046112	9	63973033	63977492	Spermidine synthase

Based on the phenotypic data measured in the four environments, the effects of genotype on SCS and the other traits were strongly significant, but also were affected by seeding year and location (Table S12). The broad-sense heritability was greater than 90% for the whole weight per plant (WWP), SCS, SCLEP, TNSCE, and SCUEP (Table S12). These results indicated that the stalk sugar content of fresh corn was mainly controlled by genetic factors and thus was suitable for further association analysis.

2.5. Genome-Wide Association Analysis

In this study, the generalized linear model (GLM) and mixed linear model (MLM) models were constructed for the association analysis, but the GLM model showed poor correlations for most traits except the ear node (EN). Therefore, the more reliable MLM model was adopted for most traits and the GLM model was used only for EN (Table S13). Significant signals consistently detected in the different environments were considered to represent high-confidence associations (Table 2). For the agronomic traits, eight significant SNPs associated with plant height (PH) were identified (Figure S17). Among them, three and two significant SNPs were detected in the spring and autumn of 2020, respectively, and the association analysis of the traits from the spring and autumn of 2021 identified two and two significant SNPs, respectively (Table S13). Among these SNPs, one (AX-86251807) was detected in two environments and explained 15.4% and 14.3% of the phenotypic variance.

The relevant candidate gene (*Zm00001d020297*) encoded a calcium-dependent protein kinase (CDPK). Fifteen significant SNPs associated with WWP were identified (Figure 4). Among them, five and one significant SNPs were detected in the spring and autumn of 2020, respectively, and the association analysis of the traits from the spring and autumn of 2021 identified ten and six significant SNPs, respectively (Table S13). Importantly, six SNPs were detected in at least two environments. The candidate genes encoded ATP-binding cassette (ABC) transporter, leghemoglobin reductase, and plant homeobox domain (PHD) finger protein, among others. Most importantly, SNP AX-86252871 was identified in three environments and explained 15.0–15.4% of the phenotypic variance. This SNP was located within the exon of the candidate gene *Zm00001d009167*, which encoded a polygalacturonase belonging to the glycosyl hydrolase family. The accessions carrying AX-86252871-AA had higher WWP than the accessions carrying AX-86252871-GG, and the genotypes carrying the other five significant SNPs (AX-91358539, AX-86263111, AX-90573879, AX-86252872, and AX-86296303) differed in WWP (Figure 4E). Twenty-four significant SNPs associated with EWB were detected (Figure S18, Table S13). The relevant candidate genes encoded lipid phosphate phosphatase, premnaspirodiene oxygenase, E3 ubiquitin-protein ligase, acyl transferase, and zinc finger protein, among others. With regard to EN, 11 significant SNPs were detected with the GLM model (Figure S19, Table S13). The relevant candidate genes encoded serine/threonine protein kinase, glycoprotein 3-alpha-L-fucosyltransferase, and a MADS-box transcription factor, among others.

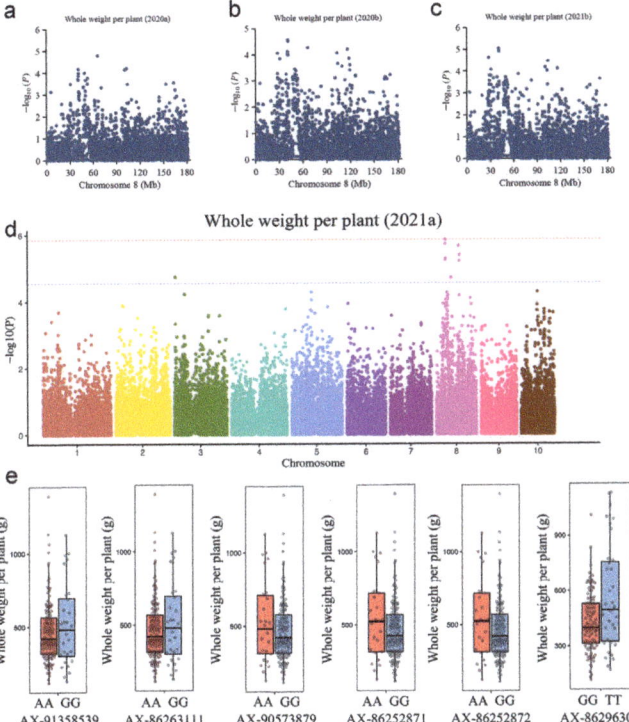

Figure 4. Genome-wide association analysis of whole weight per plant (WWP) using the mixed linear model. The dashed line indicates the significance threshold. (**a**) Local Manhattan plot of WWP from spring 2020 (2020a) on chromosome 8. (**b**) Local Manhattan plot of WWP from autumn 2020 (2020b) on chromosome 8. (**c**) Local Manhattan plot of WWP from autumn 2021 (2021b) on chromosome 8. (**d**) Local Manhattan plot of WWP from spring 2021 (2021a). (**e**) Box plots for WWP based on the different genotypes of important significant SNPs detected in multiple environments.

Table 2. Significant SNPs and candidate genes simultaneously identified in different environments.

Trait	Chromosome	SNP	Position	−log10 (P)	Marker R^2	Location	Gene ID	Gene Function	Locus Tag	Literature
PH	7	AX-86251807	104761239	4.71	15.4%	intronic	Zm00001d020297	Calcium-dependent protein kinase	LOC100194392	None
WWP	8	AX-86296303	65780194	4.80	15.3%	intergenic	Zm00001d009466	PHD finger protein	None	None
WWP	8	AX-91358539	39945148	5.72	17.2%	intronic	Zm00001d009156	Protein ABC transporter	LOC103635101	None
WWP	8	AX-86263111	40371639	5.73	16.9%	exonic	Zm00001d009163	Leghemoglobin reductase	LOC100501719	None
WWP	8	AX-90573879	40619812	5.85	17.5%	intergenic	Zm00001d009165	Unknown	None	None
WWP	8	AX-86252871	40632157	4.96	15.0%	exonic	Zm00001d009167	Polygalacturonase	LOC103635108	None
WWP	8	AX-86252872	40752630	4.92	15.0%	exonic	Zm00001d009171	Mitochondrial carrier protein	LOC100274176	None
SCS	1	AX-116875096	301500710	4.72	15.2%	5′UTR	Zm00001d034759	Acyl-CoA-binding domain-containing protein	LOC100384139	[33]
SCS	4	AX-86312905	71985807	4.90	15.7%	intergenic	Zm00001d050190	Vegetative cell wall protein	LOC100191378	None
SCS	4	AX-86312908	71985862	4.84	15.6%	intergenic	Zm00001d050190	Vegetative cell wall protein	LOC100191378	None
SCS	4	AX-86246392	75672064	5.44	17.8%	3′UTR	Zm00001d050244	ATP-dependent DEAD-box RNA helicase	LOC100279228	None
SCS	4	AX-95657025	18609 0916	5.49	15.3%	exonic	Zm00001d052289	Pyruvate dehydrogenase 1	LOC732791	[34]
SCS	8	AX-86279495	109191809	5.02	15.0%	3′UTR	Zm00001d010314	Unknown	LOC100272847	None
SCS	10	AX-86257654	149652762	5.12	15.9%	3′UTR	Zm00001d026668	Receptor protein kinase TMK1	LOC100192365	None
SCLEP	1	AX-116875096	301500710	4.65	15.2%	5′UTR	Zm00001d034759	Acyl-CoA-binding domain-containing protein	LOC100384139	[33]
SCLEP	4	AX-95657025	18609 0916	4.87	13.4%	exonic	Zm00001d052289	Pyruvate dehydrogenase 1	LOC732791	[34]
SCLEP	8	AX-86279495	109191809	5.63	17.1%	3′UTR	Zm00001d010314	Unknown	LOC100272847	None
SCLEP	10	AX-86257654	149652762	5.04	15.7%	3′UTR	Zm00001d026668	Receptor protein kinase TMK1	LOC100192365	None
SCUEP	4	AX-86246392	75672064	5.03	16.3%	3′UTR	Zm00001d050244	ATP-dependent DEAD-box RNA helicase	LOC100279228	None

Table 2. Cont.

Trait	Chromosome	SNP	Position	−log10 (P)	Marker R^2	Location	Gene ID	Gene Function	Locus Tag	Literature
SCUEP	8	AX-86319101	5579363	5.08	16.9%	intronic	Zm00001d008334	Pectinesterase precursor	LOC100191701	[35]
SCUEP	4	AX-86300656	181467257	5.69	18.6%	exonic	Zm00001d052157	Unknown	LOC100280272	None
SCUEP	4	AX-86283479	181472159	4.95	16.0%	3′UTR	Zm00001d052157	Unknown	LOC100280272	None
SCUEP	4	AX-95657025	186090916	5.27	14.6%	exonic	Zm00001d052289	Pyruvate dehydrogenase 1	LOC732791	[34]
SCUEP	5	AX-86293407	203302677	5.42	17.9%	downstream	Zm00001d017656	Pigment biosynthesis protein	LOC100282483	None
SCUEP	8	AX-86279495	109191809	5.09	16.7%	3′UTR	Zm00001d010314	Unknown	LOC100272847	None
SCUEP	1	AX-86328029	300979520	4.63	14.3%	3′UTR	Zm00001d034736	Cysteine–tRNA ligase CPS1	LOC103644288	None
SCUEP	10	AX-86257654	149652762	4.75	15.3%	3′UTR	Zm00001d026668	Receptor protein kinase TMK1	LOC100192365	None
SCUEP	1	AX-86308222	278512731	5.22	15.8%	upstream	Zm00001d033915	Glycerol-3-phosphate 2-O-acyltransferase	LOC103643973	[36]
SCUEP	4	AX-86313840	185513108	5.54	16.6%	3′UTR	Zm00001d052271	Protease Do-like 9 CBS	LOC103655753	None
SCUEP	4	AX-91346570	68058771	5.00	16.9%	downstream	Zm00001d050130	domain-containing protein CBSX1	LOC100282782	None
TNSCE	4	AX-86312905	71985807	5.20	16.9%	intergenic	Zm00001d050190	Vegetative cell wall protein	LOC100191378	None
TNSCE	4	AX-86312908	71985862	5.11	16.7%	intergenic	Zm00001d050190	Vegetative cell wall protein	LOC100191378	None
TNSCE	4	AX-86246392	75672064	5.22	17.0%	3′UTR	Zm00001d050244	DEAD-box ATP-dependent RNA helicase	LOC100279228	None
TNSCE	1	AX-116875096	301500710	5.21	17.1%	5′UTR	Zm00001d034759	Acyl-CoA-binding domain-containing protein	LOC100279228	[33]
TNSCE	3	AX-86263525	159083718	5.02	16.6%	exonic	Zm00001d042287	MYB transcription factor	LOC541962	None
TNSCE	4	AX-95657025	186090916	5.20	14.4%	exonic	Zm00001d052289	Pyruvate dehydrogenase 1	LOC732791	[34]
TNSCE	10	AX-86258646	135575124	5.02	16.4%	exonic	Zm00001d025992	Unknown	LOC100276746	[37]

To explore the candidate genes associated with stalk sugar content, we conducted a GWAS on the sugar content of the different stalk parts. For SCS, 18 significant SNPs were identified, which were distributed on chromosomes 1, 3, 4, 6, 7, 8, and 10 (Figure 5). Additionally, six and twelve significant SNPs were detected in the spring and autumn of 2020, respectively, and the association analyses of the SCS from the spring and autumn of 2021 identified five and three significant SNPs, respectively (Table S13). Among these SNPs, seven were detected in multiple environments and eleven were detected in a single environment. The seven significant SNPs explained 13.68% to 17.82% of the phenotypic variation and could be regarded to be stable genetic markers. Most importantly, the significant SNP AX-116875096 on chromosome 1 was detected in three environments in both years and explained 13.68–15.15% of the phenotypic variance. The significant SNP AX-116875096 was located in the 5′ untranslated region (UTR) of the candidate gene *Zm00001d034759*, which encoded an acyl-CoA-binding domain-containing protein. Through phenotypic analysis of the accessions with different genotypes, the accessions carrying AX-116875096-CC were shown to have higher stalk sugar content than the accessions carrying AX-116875096-GG (Figure 5g). In addition, the association analysis of SCLEP, TNSCE, and SCUEP also detected the significant SNP AX-116875096 (Figure 5e, Table S13), which demonstrated that the significant SNP AX-116875096 and candidate gene *Zm00001d034759* were associated with high credibility. An additional important significant SNP, AX-86257654, was also detected in the association analysis of SCLEP, TNSCE, and SCUEP (Figure 5f, Table S13). The sugar content of the accessions carrying AX-86257654-AA was higher than that of the accessions harboring AX-86257654-GG (Figure 5f,g). The relevant candidate gene (*Zm00001d026668*) encoded the receptor protein kinase TMK1. In addition, the sugar content of the accessions with different genotypes of the significant SNPs AX-86246392, AX-86279495, AX-86312905, and AX-86312908 also be analyzed, which shows that the accessions carrying AX-86246392-CC, AX-86279495-GG, AX-86312905-GG, and AX-86312908-CC had higher stalk sugar content (Figure 5g).

In the association analysis for SCLEP, three, twelve, four, and three significant SNPs were detected in the four environments, respectively (Table S13). Among these SNPs, four significant SNPs were detected in two environments. AX-116875096 and AX-95657025 were associated with SCLEP in 2020, whereas AX-86279495 and AX-86257654 were associated with SCLEP in 2021 (Figure S20, Table S13). The top four significant SNPs were also detected in the GWAS of SCS and SCUEP (Table S13). The significant SNP AX-95657025 was located in the exon of the candidate gene *Zm00001d052289*, which encoded pyruvate dehydrogenase, and AX-86279495 was located in the 3′ UTR of the candidate gene *Zm00001d010314*, which encoded an unknown protein. For SCUEP, twenty-five significant SNPs were identified (Figure S21, Table S13), of which eleven SNPs were detected in at least two environments, including AX-86308222 and AX-86319101, which were detected in three environments. AX-86308222 was located upstream of the candidate gene *Zm00001d033915*, which encoded a glycerol-3-phosphate 2-O-acyltransferase, and AX-86319101 was located in the intron of the candidate gene *Zm00001d008334*, which encoded a pectinesterase precursor. The accessions carrying AX-86308222-TT and AX-86319101-CC had a higher sugar content in the upper ear parts than the accessions carrying AX-86308222-GG and AX-86319101-TT (Figure S21). The GWAS analysis of TNSCE identified 15 significant SNPs (Table S13). Among these SNPs, eight significant SNPs were detected in two environments, of which five SNPs were also identified in the GWAS of SCS, and AX-86263525, AX-91346570, and AX-86258646 were the other three significant SNPs (Figure S22, Table S13). The SNP AX-86263525 was located in the exon of the candidate gene *Zm00001d042287*, which encoded an MYB transcription factor, and AX-86258646 was located in the exon of the candidate gene *Zm00001d025992*, which encoded an unknown protein. Analysis of the sugar content of the accessions with different genotypes revealed that the accessions carrying AX-86263525-TT and AX-86258646-CC had a higher sugar content in the upper ear parts than the accessions carrying AX-86263525-CC and AX-86258646-TT (Figure S22). In this study, the Q-Q plots were shown in the Figures S17–S20 and S23–S26.

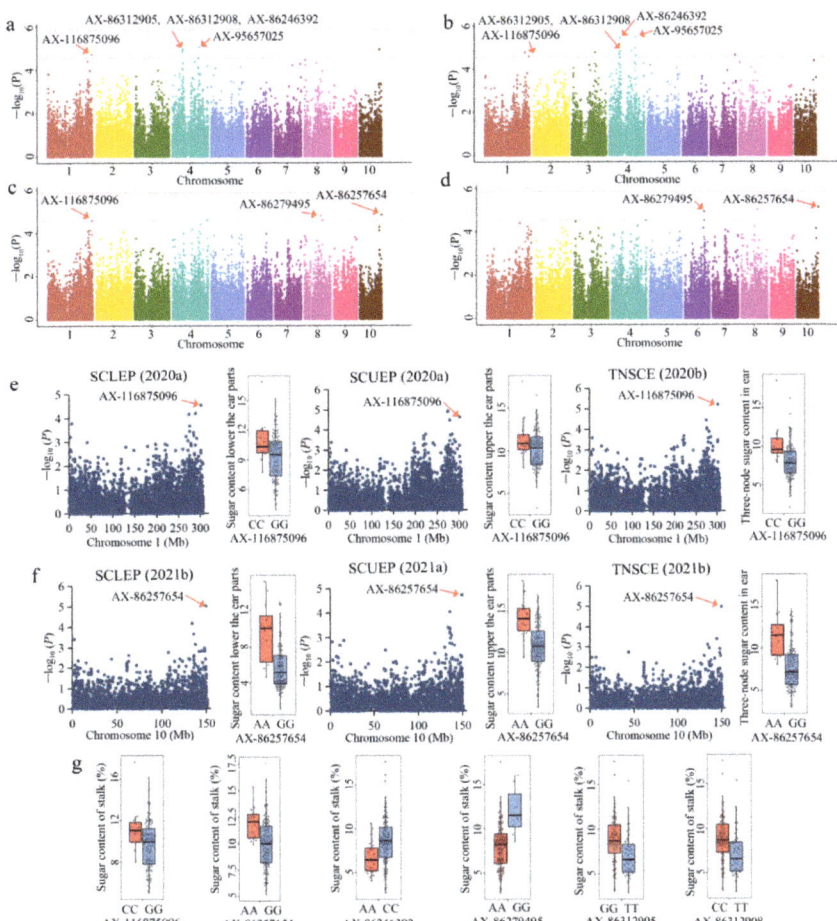

Figure 5. Identification of significant SNPs associated with sugar content of the stalk (SCS) by genome-wide association analysis. (**a–d**) Manhattan plots of SCS from spring 2020, autumn 2020, spring 2021, and autumn 2021, respectively. (**e**) Important significant SNP AX-116875096 detected by association analysis of sugar content in the lower ear parts (SCLEP), three-node sugar content in the ear (TNSCE), and sugar content in the upper ear parts (SCUEP). Box plots of the traits are based on the different genotypes of AX-116875096. (**f**) Important significant SNP AX-86257654 detected by association analysis of SCLEP, TNSCE, and SCUEP. Box plots of the traits are based on the different genotypes of AX-86257654. (**g**) Box plots for sugar content of the stalk based on the different genotypes of important significant SNPs detected in multiple environments.

2.6. qRT-PCR Analysis of Candidate Genes

To better understand the function of candidate genes and genetic variants, the expression levels of sixteen important candidate genes were verified by a quantitative real-time PCR (qRT-PCR) in two fresh corn materials with significantly different stalk sugar content. The average stalk sugar content in multiple environments of the high-sugar content material was 15.5%, while the average stalk sugar content of the low-sugar content material was 3.7%. Based on the qRT-PCR results, we found that the expression levels of *Zm00001d009466, Zm00001d026668, Zm00001d034759, Zm00001d050190, Zm00001d050244, Zm00001d052271, Zm00001d052289* in the high-sugar content material were significantly higher than the low-sugar content material, especially *Zm00001d026668, Zm00001d050190* and *Zm00001d050244* (Figure 6). Furthermore, the significant SNP AX-86257654 was associated with *Zm00001d026668*. We found that the high-sugar content material carried AX-86257654-AA and the low-sugar content material carried AX-86257654-GG. Additionally, the expression levels of *Zm00001d050244* showed an obvious change between the high-sugar content material and the low-sugar content material. The high-sugar content material carried AX-86246392-CC and the low-sugar content material carried AX-86246392-AA. Moreover, the expression level of *Zm00001d050190* in the high-sugar content material carrying AX-86312905-GG was higher than the low-sugar content material carrying AX-86312905-TT. Additionally, *Zm00001d009163* had a significantly higher expression level in low-sugar content materials, and *Zm00001d009167* did not express in the two materials (Figure 6).

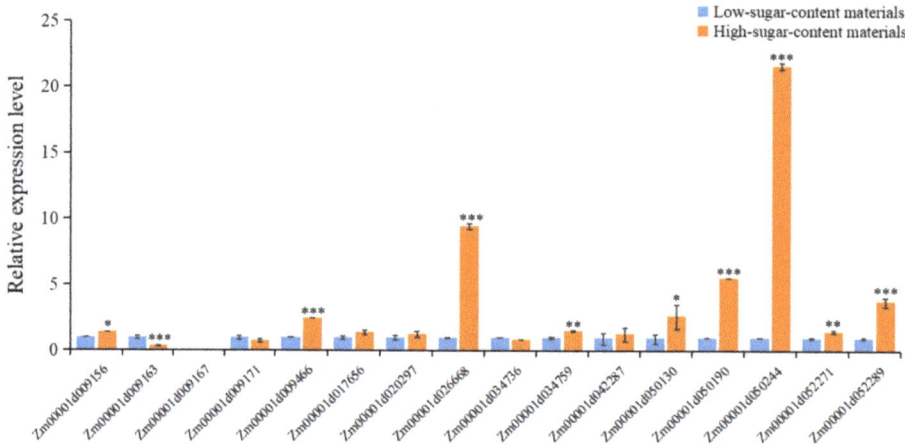

Figure 6. The expression levels of important candidate genes in two fresh corn materials with significantly different stalk sugar content by quantitative RT-PCR. Error bars were used to express the standard deviation of three biological replications. * Represents $p < 0.05$, ** represents $p < 0.01$, *** represents $p < 0.001$.

3. Discussion

Sugars are the primary products of photosynthesis and accumulate in sinks, including the stalk. Studying the sugar content and yield-related traits in corn is helpful to improve the yield and quality of fresh corn, which is beneficial to the breeding of crosses with higher stalk sugar content and yield. The stalk sugar content also directly affects the application value of fresh corn as a biofuel feedstock. Research on the stalk sugar content of fresh corn may improve the efficiency of corn utilization and expand its application for green silage and biofuel production. The stalk sugar content may also influence the sustainability features of the processes of green silage and biofuels, using advanced sustainability assessment tools is beneficial to evaluate the application value of fresh corn as biofuel production [38]. In recent years, GWAS has become an important method for the discovery of crucial candidate genes associated with complex traits [39]. Previously, QTL mapping was used to locate genes associated with stalk sugar content [3,21]. In the present study, to detect candidate genes associated with stalk sugar content and related agronomic traits in fresh corn, we performed GWAS based on GLM and MLM models. The GLM model led to many false positives, whereas the MLM model effectively reduced the number of false positives, as reported in previous studies [40,41]. Numerous significant SNPs were detected in more than one environment, and thus these genetic variations and candidate genes are of substantial research and practical value.

The QTLs that contribute to the stalk sugar content also affect the leaf area of corn [21]. In addition, yield is critical for the use of fresh corn in silage. In the association analysis of PH and WWP, reliable significant SNPs were simultaneously detected in multiple environments. Calcium-dependent protein kinase (CDPK) plays an important role in plant growth, development, and stress response [42]. Previous studies in sugarcane have shown that CDPK may phosphorylate and regulate sucrose synthase and sucrose phosphate synthase, and therefore is important for sucrose accumulation [9,10]. In the current study, the candidate gene *Zm00001d020297* encodes a CDPK, which was identified in two environments and might be a crucial candidate gene regulating plant height. Moreover, the alleles of AX-86252871 had differential effects on WWP, the accessions carrying the allele AX-86252871-AA had a higher WWP (Figure 4e). AX-86252871 is located in the exon region of candidate gene *Zm00001d009167*, which encoded a polygalacturonase and belonged to the glycosyl hydrolase family. Polygalacturonases are ubiquitous in higher plant cells and are strongly associated with plant growth and development owing to their functions in cell wall degradation, promotion of cell division, and fruit ripening [43]. Ethylene can promote the expression of polygalacturonase (PG) and plays a crucial role in sucrose metabolism in blueberries [44]. A previous study reported that the PHD finger protein MePHD1 negatively regulates the transcript level of the *MeAGPS1a* gene, which encods an ADP-glucose pyrophosphorylase (AGPase) [45]. Our study found that the accessions carrying AX-86296303-TT had a higher WWP than the accessions carrying AX-86296303-GG (Figure 4e), which was significantly associated with candidate gene *Zm00001d009466* encoding a PHD finger protein. Additionally, we also identified a candidate gene (*Zm00001d009156*) encoding an ATP-binding cassette (ABC) transporter through significant SNP AX-91358539. The study in *Arabidopsis* has shown that ABC transporters are involved in the transport of lipids, glycosides, hormones, and so on [46]. The results showed that the whole weight per plant (WWP) of the accessions carrying AX-91358539-GG was higher than the accessions carrying AX-91358539-AA (Figure 4e), which might directly influence the function of the ABC transporter and ultimately affect the whole weight per plant.

Sugar is an important photosynthate and is the main product of photosynthesis transported in the plant body [47]. In this study, a suite of significant SNPs and candidate genes associated with SCS (sugar content of the stalk) was detected with high confidence. Among these SNPs, the significant SNP AX-95657025 was located in the first exon of the candidate gene *Zm00001d052289*, which encodes a pyruvate dehydrogenase involved in energy production and conversion. Pyruvate dehydrogenase is a crucial enzyme required for glucose metabolism, which converts pyruvate to acetyl CoA to complete the tricarboxylic acid cycle [15]. Mutation of genes encoding components of the pyruvate dehydrogenase complex affects enzymatic activity and plant growth [48]. We found that the expression level of *Zm00001d052289* in the high-sugar content material was significantly higher than in the low-sugar content material (Figure 6), and future research should be focused on whether the variation of the SNP AX-95657025 affects the function of the candidate gene *Zm00001d052289*. Additionally, previous studies have reported that Acyl-CoA-binding domain-containing protein (ACBD) is required for the transferal of glucosylceramide [49], and plays an important role in plant growth, development, and stress response [50,51]. In the current study, four traits related to the stalk sugar content (SCS, SCLEP, TNSCE, and SCUEP) were associated with the significant SNP AX-116875096, which was also detected in multiple environments. The SNP AX-116875096 was located in the 5' UTR of *Zm00001d034759*, which encodes an ACBD protein. The accessions carrying AX-116875096-CC had higher sugar content in the stalk (Figure 5e), and the expression level of *Zm00001d034759* was higher in the high-sugar content material (Figure 6), which reflected that the genetic variation might influence the expression of the candidate gene *Zm00001d034759*.

Analysis of the expression profile of signal transduction components in sugarcane has shown that the receptor protein kinase may regulate sucrose metabolism [11]. In the present study, the candidate gene *Zm00001d026668* encoding a receptor protein kinase TMK was identified through the significant SNP AX-86257654. AX-86257654 was detected in four traits (SCS, SCLEP, TNSCE, and SCUEP) and multiple environments, which had higher credibility. We found that the accessions carrying AX-86257654-AA had higher talk sugar content than the accessions carrying AX-86257654-GG (Figure 5g). The expression level of *Zm00001d026668* in the high-sugar content material was significantly higher than the low-sugar content material (Figure 6), while the high-sugar content material carrying AX-86257654-AA and the low-sugar content material carrying AX-86257654-GG. Moreover, a candidate gene (*Zm00001d042287*) encoding an MYB transcription factor was identified in multiple environments, and the SNP AX-86263525 was located in the exon of *Zm00001d042287*. The alleles of AX-86263525 had different effects on stalk sugar content and explained 15.16%–16.57% of the phenotypic variance. The accessions carrying AX-86263525-TT had higher stalk sugar content than the accessions carrying AX-86263525-CC (Figure S22), and the expression level of *Zm00001d042287* was significantly higher in the high-sugar content material, reflecting that the mutation of AX-86263525 might affect the function of *Zm00001d042287*. The studies have reported that MYB transcription factors are involved in the biosynthesis of the water-soluble polysaccharide and MYB46 directly regulates the three *CESA* genes, which are essential for cellulose production [12,13]. We also found that *Zm00001d050244* had a significantly higher expression level in the high-sugar content material (Figure 6), which encoded a DEAD-box ATP-dependent RNA helicase, while the high-sugar content material carried AX-86246392-CC and the high-sugar content material carried AX-86246392-AA (Figure 5g). The DEAD-box ATP-dependent RNA helicase plays an important role in regulating translation initiation and growth in plants [52]. Thus, we inferred that *Zm00001d050244* might take part in sugar metabolism by regulating the translation of other genes. In addition, candidate genes encoding the pectinesterase precursor and cell wall protein were also detected in this study.

Differing from the QTL mapping in previous studies, our study detected the candidate gene responsible for the stalk sugar content of fresh corn through the genome-wide association analysis. Despite many significant genetic variations and candidate genes detected in the present study, which have a potential value for the future breeding of fresh corn, there is still room to further validate the true phenotypic effects of the corresponding two haplotypes through transgenic experiments. Hence, further investigation of the molecular mechanism of the genetic variations and candidate genes associated with the stalk sugar content in fresh corn is required.

4. Materials and Methods

4.1. Plant Materials

The plant seed materials included 188 sweet corn, waxy corn, and hybrid accessions. Among them, 41 were sweet corn inbred lines, 74 were waxy corn inbred lines, and 73 were hybrid accessions. Seeds were obtained from southeastern China, northern China, and the Huang-Huai-Hai region of eastern China.

4.2. Trait Measurements and Analysis

Seeds of the experimental materials were sown in an experimental field at Dongyang, Zhejiang, China (29°16′ N, 120°19′ E) in the spring and autumn of 2020 and 2021. A completely randomized design with three replications was used in this study. Each replicate comprised two rows of each accession of 3 m in length, and the plant density was 0.75 m × 0.3 m. Field management (e.g., watering, fertilization, and weed management) was applied uniformly throughout the growing period.

Before tasseling, 10 plants of each accession exhibiting normal growth and development were selected for measurement of stalk sugar content and agronomic traits. Artificial pollination was performed to prevent pollen mixing. In the harvest season for fresh corn, eight plants per growth period were selected for measurement of plant height (PH), whole weight per plant (WWP), ear weight with bract (EWB), ear node (EN), sugar content of the stalk (SCS), sugar content in the lower ear parts (SCLEP), three-node sugar content in the ear (TNSCE), and sugar content in the upper ear parts (SCUEP). For estimation of the sugar content, we extracted the whole stem juice from each plant, and after thorough mixing, the sugar content was determined with a hand-held sugar meter (PAL-1, Atago, Tokyo, Japan). To minimize measurement error, the stem juice of each plant was measured three times, and the average was taken as the sugar content of each individual. Frequency distribution analysis, correlation analysis, and variance analysis of the data were performed with SPSS software (version 22.0).

4.3. Genotypic Data

The 188 sweet corn, waxy corn, and hybrid accessions were genotyped with the Axiom® Maize56K SNP Array, which includes 56,000 SNPs in total. After comparison with the corn genome (B73 RefGen_v4), SNP markers with minimum allele frequency >0.05 and missing data <30% were filtered using TASSEL software [53]. Annotation of the SNPs was performed with ANNOVAR [54]. A total of 36,069 SNPs were used for subsequent analysis.

4.4. Population Structure Analysis, Linkage Disequilibrium, and Relative Kinship Estimation

A phylogenetic tree was constructed from a data set of 36,069 SNPs using the maximum likelihood method in FastTree 2.1.9 [55]. The tree was colored using iTQL (https://itol.embl.de/ (accessed on 27 October 2022)). Principal component analysis (PCA) was conducted using GCTA software (version 1.92.0) [56]. The first two eigenvectors were selected to generate a scatterplot of the principal component scores to visualize similarities among the accessions in two dimensions. The ADMIXTURE software was used to investigate the population structure [57]. The relative kinship matrix (K) was calculated using TASSEL software [53]. Linkage disequilibrium (LD) was calculated using PopLDdecay software [58].

4.5. Detection of Selective Sweeps

The selective sweeps between sweet corn (group A) and waxy corn (group B) were detected using VCFtools [59]. First, the nucleotide diversity of the sweet corn and waxy corn accessions (π_A and π_B, respectively) was calculated using a 100 kb window and a step size of 10 kb. Then, π_A/π_B was calculated and the windows, with the top 5% largest π_A/π_B values were identified as selective sweeps. In addition, the genetic differentiation (F_{ST}) among sweet corn (group A) and waxy corn (group B) was determined using a 100 kb window and a step size of 10 kb, and the windows with the top 5% largest F_{ST} values were identified as selective sweeps. To explore the function of selected genes located in the selective regions, the AgriGO analysis toolkit [60] and the Kyoto Encyclopedia of Genes and Genomes (KEGG) database were used to perform gene ontology (GO) and KEGG pathway analyses.

4.6. Genome-Wide Association Analysis

Plant height (PH), whole weight per plant (WWP), ear weight with bract (EWB), ear node (EN), sugar content of the stalk (SCS), sugar content in the lower ear parts (SCLEP), three-node sugar content in the ear (TNSCE), and sugar content in the upper ear parts (SCUEP) recorded in four environments were used for GWAS analysis. For the association analysis, a generalized linear model (GLM) and mixed linear model (MLM) were generated with TASSEL [53]. Manhattan plots and quantile-quantile plots were generated with R software based on the p-values from the association analysis [61]. The significant association threshold was defined as $0.05/n$ and $1/n$, where n is the number of SNPs [62].

4.7. RNA Extraction and Quantitative RT-PCR Analysis

To verify the expression levels of important candidate genes, total RNA was extracted from two fresh corn materials with significantly different stalk sugar content using the RNAprep Pure Plant Kit (TIANGEN, Beijing, China): the high-sugar content material LQ5 from a domestic hybrid second cycle line and the high-sugar content material SH23 from a domestic inbred line. A total of 1 ug RNA was used to synthesize cDNA according to the instructions of a PrimeScript RT Reagent Kit (TaKaRa, China). The qRT-PCR reactions were performed through the SYBR-Green PrimeScript RT-PCR Kit (Takara) following the manufacturer's instructions. *GAPDH* was used as the reference gene for the qRT-PCR. The original data were processed by the $2^{-\Delta\Delta Ct}$ method, where Ct is the threshold cycle [63], and the primers for candidate genes were shown in Table S14.

5. Conclusions

In this study, genome-wide association analysis was used to detect significant SNPs and candidate genes associated with stalk sugar content and agronomic traits. A total of 92 significant associated SNPs were identified. Among them, 24 significant SNPs and 22 candidate genes were identified in multiple environments, including CDPK, MYB transcription factor, pyruvate dehydrogenase, and so on. Most importantly, accessions differing in genotype for several significant SNPs showed significant phenotypic differences, indicating that the genetic variations in candidate genes might be closely related to the phenotypic traits. The expression levels of six important candidate genes were significantly different between the high-sugar content material and the low-sugar content material. In general, the significant genetic variations might be useful in molecular marker-assisted breeding, and the investigation of candidate genes may provide important insights into the mechanism of stalk sugar content in fresh corn, which establishes the foundation for fresh corn breeding. Future research should focus on illustrating how the genetic variations and candidate genes regulate the stalk sugar content in fresh corn, and concentrate on breeding new varieties with higher stalk sugar content.

Supplementary Materials: The following are available online at https://www.mdpi.com/article/10.3390/ijms232113490/s1.

Author Contributions: Conceptualization, Y.B. and G.L.; methodology, J.C. (Jianjian Chen), Z.W., G.G., X.L. and H.Z.; software, J.C. (Jianjian Chen); validation, J.C. (Jinming Cao); formal analysis, J.C. (Jinming Cao), Z.W., G.G., X.L. and H.Z.; writing—original draft preparation, J.C. (Jianjian Chen); supervision, Y.B. and G.L.; funding acquisition, G.L. All authors have read and agreed to the published version of the manuscript.

Funding: This research was supported by grants from the Public Service Technology Application Research of Zhejiang Province (GN20C020002), the Key Research and Development Plan of Zhejiang Province (2022C04024), and the Zhejiang Science and Technology Major Program on Agricultural New Variety Breeding (2021C02064-4).

Institutional Review Board Statement: Not applicable.

Informed Consent Statement: Not applicable.

Data Availability Statement: The original data presented in this study can be accessed at https://ngdc.cncb.ac.cn/gsub/submit/bioproject/list (project ID: PRJCA009494).

Conflicts of Interest: The authors declare no conflict of interest.

Abbreviations

GWAS	Genome-wide association analysis
SNPs	Single nucleotide polymorphisms
QTL	Quantitative trait locus
MYB	V-myb avian myeloblastosis viral oncogene homolog
LD	Linkage disequilibrium
PCA	Principal components analysis
F_{ST}	Genetic differentiation coefficient
π_A/π_B	Nucleotide diversity between sweet corns (group A) and waxy corns (group B)
KEGG	Kyoto encyclopedia of genes and genomes
GO	Gene ontology

SCS	Sugar content of the stalk
SCLEP	Sugar content in the lower ear parts
TNSCE	Three-node sugar content in the ear
SCUEP	Sugar content in the upper ear parts
WWP	Whole weight per plant
EN	Ear node
PH	Plant height
EWB	Ear weight with bract
GLM	Generalized linear model
MLM	Mixed linear model
UTR	Untranslated region
qRT-PCR	Quantitative real-time PCR
CDPK	Calcium-dependent protein kinase
ABC	ATP-binding cassette
PHD	Plant homeobox domain
ACBD	Acyl-CoA-binding domain-containing protein
CESA	Cellulose synthase

References

1. Shi, Z.S.; Zhong, X.M. Breeding principle and technical skill of fresh corn varieties. *J. Maize Sci.* **2016**, *24*, 1–5.
2. Bian, Y.-L.; Du, K.; Wang, Y.-J.; Deng, D.-X. Distribution of Sugar Content in Corn Stalk. *Acta Agron. Sin.* **2009**, *35*, 2252–2257. [CrossRef]
3. Bian, Y.; Gu, X.; Sun, D.; Wang, Y.; Yin, Z.; Deng, D.; Wang, Y.; Li, G. Mapping dynamic QTL of stalk sugar content at different growth stages in maize. *Euphytica* **2015**, *205*, 85–94. [CrossRef]
4. Kleinschmit, D.H.; Schmidt, R.J.; Jr, L.K. The effects of various antifungal additives on the fermentation and aerobic stability of corn silage. *J. Dairy Sci.* **2005**, *88*, 2130–2139. [CrossRef]
5. Zhang, D.Y.; Li, Z.Q.; Liu, C.L. Progress in the study of infection factors on silage quality. *Acta Ecol. Anim. Domastici* **2007**, *28*, 109–112.
6. Froetschel, M.; Ely, L.; Amos, H. Effects of Additives and Growth Environment on Preservation and Digestibility of Wheat Silage Fed to Holstein Heifers. *J. Dairy Sci.* **1991**, *74*, 546–556. [CrossRef]
7. Bai, Q.L.; Chen, S.J.; Dai, J.R. Stalk quality traits and their correlations of maize inbred lines in China. *Acta Agron. Sin.* **2007**, *33*, 1777–1781.
8. Esfandabadi, Z.S.; Ranjbari, M.; Scagnelli, S.D. The imbalance of food and biofuel markets amid Ukraine-Russia crisis: A systems thinking perspective. *Biofuel Res. J.* **2022**, *9*, 1640–1647. [CrossRef]
9. Hardin, S.C.; Winter, H.; Huber, S.C. Phosphorylation of the Amino Terminus of Maize Sucrose Synthase in Relation to Membrane Association and Enzyme Activity. *Plant Physiol.* **2004**, *134*, 1427–1438. [CrossRef]
10. Papini-Terzi, F.S.; Rocha, F.R.; Vêncio, R.Z.; Felix, J.M.; Branco, D.S.; Waclawovsky, A.J.; Del Bem, L.E.; Lembke, C.G.; Costa, M.D.; Nishiyama, M.Y.; et al. Sugarcane genes associated with sucrose content. *BMC Genom.* **2009**, *10*, 120. [CrossRef]
11. Felix, J.D.M.; Papini-Terzi, F.S.; Rocha, F.R.; Vêncio, R.Z.N.; Vicentini, R.; Nishiyama, M.Y.; Ulian, E.C.; Souza, G.M.; Menossi, M. Expression Profile of Signal Transduction Components in a Sugarcane Population Segregating for Sugar Content. *Trop. Plant Biol.* **2009**, *2*, 98–109. [CrossRef]
12. Kim, W.-C.; Ko, J.-H.; Kim, J.-Y.; Kim, J.; Bae, H.-J.; Han, K.-H. MYB46 directly regulates the gene expression of secondary wall-associated cellulose synthases in Arabidopsis. *Plant J.* **2012**, *73*, 26–36. [CrossRef] [PubMed]
13. He, C.; Da Silva, J.A.T.; Wang, H.; Si, C.; Zhang, M.; Zhang, X.; Li, M.; Tan, J.; Duan, J. Mining MYB transcription factors from the genomes of orchids (Phalaenopsis and Dendrobium) and characterization of an orchid R2R3-MYB gene involved in water-soluble polysaccharide biosynthesis. *Sci. Rep.* **2019**, *9*, 13818. [CrossRef] [PubMed]
14. Yu, H.; Du, X.; Zhang, F.; Zhang, F.; Hu, Y.; Liu, S.; Jiang, X.; Wang, G.; Liu, D. A mutation in the E2 subunit of the mitochondrial pyruvate dehydrogenase complex in Arabidopsis reduces plant organ size and enhances the accumulation of amino acids and intermediate products of the TCA Cycle. *Planta* **2012**, *236*, 387–399. [CrossRef] [PubMed]
15. Ohbayashi, I.; Huang, S.; Fukaki, H.; Song, X.; Sun, S.; Morita, M.; Tasaka, M.; Millar, A.H.; Furutani, M. Mitochondrial Pyruvate Dehydrogenase Contributes to Auxin-Regulated Organ Development. *Plant Physiol.* **2019**, *180*, 896–909. [CrossRef]
16. Kramer, V.; Shaw, J.R.; Senior, M.L.; Hannah, L.C. The sh2-R allele of the maize shrunken-2 locus was caused by a complex chromosomal rearrangement. *Theor. Appl. Genet.* **2014**, *128*, 445–452. [CrossRef]
17. Chhabra, R.; Hossain, F.; Muthusamy, V.; Baveja, A.; Mehta, B.; Zunjare, R.U. Mapping and validation of Anthocyanin1 pigmentation gene for its effectiveness in early selection of shrunken2 gene governing kernel sweetness in maize. *J. Cereal Sci.* **2019**, *87*, 258–265. [CrossRef]
18. Reen, R.V.; Singleton, W.R. Sucrose content in the stalks of maize inbreds. *Agron. J.* **1952**, *44*, 610–614. [CrossRef]
19. Hua, H.L.; Zhao, Q.; Jie, W.; Liu, Q.G.; Bian, Y.L. Kernel filling characteristics at different grain positions on an ear and their relationship with stalk sugar content in maize. *Int. J. Agric. Biol.* **2019**, *21*, 184–190.

20. Jebril, J.; Wang, D.; Rozeboom, K.; Tesso, T. Grain sink removal increases stalk juice yield, sugar accumulation, and biomass in sweet sorghum [*Sorghum bicolor* (L.) Moench]. *Ind. Crops Prod.* **2021**, *173*, 114089. [CrossRef]
21. Bian, Y.; Sun, D.; Gu, X.; Wang, Y.; Yin, Z.; Deng, D.; Wang, Y.; Wu, F.; Li, G. Identification of QTL for stalk sugar-related traits in a population of recombinant inbred lines of maize. *Euphytica* **2014**, *198*, 79–89. [CrossRef]
22. Shiringani, A.L.; Frisch, M.; Friedt, W. Genetic mapping of QTLs for sugar-related traits in a RIL population of Sorghum bicolor L. Moench. *Theor. Appl. Genet.* **2010**, *121*, 323–336. [CrossRef]
23. Bian, Y.L.; Seiji, Y.; Maiko, I.; Cai, H.W. QTLs for sugar content of stalk in sweet sorghum (*Sorghum bicolor* L. Moench). *Agric. Sci. China* **2006**, *5*, 736–744. [CrossRef]
24. Murray, S.C.; Rooney, W.L.; Hamblin, M.T.; Mitchell, S.E.; Kresovich, S. Sweet Sorghum Genetic Diversity and Association Mapping for Brix and Height. *Plant Genome* **2009**, *2*, 48–62. [CrossRef]
25. Murray, S.C.; Sharma, A.; Rooney, W.L.; Klein, P.E.; Mullet, J.E.; Mitchell, S.E.; Kresovich, S. Genetic Improvement of Sorghum as a Biofuel Feedstock: I. QTL for Stem Sugar and Grain Nonstructural Carbohydrates. *Crop Sci.* **2008**, *48*, 2165–2179. [CrossRef]
26. Zhang, G.; Gao, M.; Zhang, G.; Sun, J.; Jin, X.; Wang, C.; Zhao, Y.; Li, S. Association Analysis of Yield Traits with Molecular Markers in Huang-Huai River Valley Winter Wheat Region, China. *Acta Agron. Sin.* **2013**, *39*, 1187–1199. [CrossRef]
27. Flint-Garcia, S.A. Genetics and Consequences of Crop Domestication. *J. Agric. Food Chem.* **2013**, *61*, 8267–8276. [CrossRef] [PubMed]
28. Wang, M.; Yan, J.; Zhao, J.; Song, W.; Zhang, X.; Xiao, Y.; Zheng, Y. Genome-wide association study (GWAS) of resistance to head smut in maize. *Plant Sci.* **2012**, *196*, 125–131. [CrossRef]
29. Li, H.; Peng, Z.; Yang, X.; Wang, W.; Fu, J.; Wang, J.; Han, Y.; Chai, Y.; Guo, T.; Yang, N.; et al. Genome-wide association study dissects the genetic architecture of oil biosynthesis in maize kernels. *Nat. Genet.* **2012**, *45*, 43–50. [CrossRef] [PubMed]
30. Wang, X.; Wang, H.; Liu, S.; Ferjani, A.; Li, J.; Yan, J.; Yang, X.; Qin, F. Genetic variation in ZmVPP1 contributes to drought tolerance in maize seedlings. *Nat. Genet.* **2016**, *48*, 1233–1241. [CrossRef]
31. Hu, G.; Li, Z.; Lu, Y.; Li, C.; Gong, S.; Yan, S.; Li, G.; Wang, M.; Ren, H.; Guan, H.; et al. Genome-wide association study Identified multiple Genetic Loci on Chilling Resistance During Germination in Maize. *Sci. Rep.* **2017**, *7*, 10840. [CrossRef] [PubMed]
32. Tian, F.; Bradbury, P.J.; Brown, P.J.; Hung, H.; Sun, Q.; Flint-Garcia, S.; Rocheford, T.R.; McMullen, M.D.; Holland, J.; Buckler, E. Genome-wide association study of leaf architecture in the maize nested association mapping population. *Nat. Genet.* **2011**, *43*, 159–162. [CrossRef] [PubMed]
33. Zhu, J.; Li, W.; Zhou, Y.; Pei, L.; Liu, J.; Xia, X.; Che, R.; Li, H. Molecular characterization, expression and functional analysis of acyl-CoA-binding protein gene family in maize (*Zea mays*). *BMC Plant Biol.* **2021**, *21*, 94. [CrossRef]
34. Thelen, J.J.; Miernyk, J.A.; Randall, D.D. Molecular cloning and expression analysis of the mitochondrial pyruvate dehydro-genase from maize. *Plant Physiol.* **1999**, *119*, 635–644. [CrossRef]
35. Zhang, P.; Wang, H.; Qin, X.; Chen, K.; Zhao, J.; Zhao, Y.; Yue, B. Genome-wide identification, phylogeny and expression analysis of the PME and PMEI gene families in maize. *Sci. Rep.* **2019**, *9*, 19918. [CrossRef] [PubMed]
36. Zhu, T.; Wu, S.; Zhang, D.; Li, Z.; Xie, K.; An, X.; Ma, B.; Hou, Q.; Dong, Z.; Tian, Y.; et al. Genome-wide analysis of maize GPAT gene family and cytological characterization and breeding application of ZmMs33/ZmGPAT6 gene. *Theor. Appl. Genet.* **2019**, *132*, 2137–2154. [CrossRef]
37. Li, C.; Guan, H.; Jing, X.; Li, Y.; Wang, B.; Li, Y.; Liu, X.; Zhang, D.; Liu, C.; Xie, X.; et al. Genomic insights into historical improvement of heterotic groups during modern hybrid maize breeding. *Nat. Plants* **2022**, *8*, 750–763. [CrossRef]
38. Aghbashlo, M.; Hosseinzadeh-Bandbafha, H.; Shahbeik, H.; Tabatabaei, M. The role of sustainability assessment tools in realizing bioenergy and bioproduct systems. *Biofuel Res. J.* **2022**, *9*, 1697–1706. [CrossRef]
39. Alqudah, A.M.; Sallam, A.; Baenziger, P.S.; Börner, A. GWAS: Fast-forwarding gene identification and characterization in temperate Cereals: Lessons from Barley–A review. *J. Adv. Res.* **2019**, *22*, 119–135. [CrossRef]
40. Yang, X.; Gao, S.; Xu, S.; Zhang, Z.; Prasanna, B.M.; Li, L.; Li, J.; Yan, J. Characterization of a global germplasm collection and its potential utilization for analysis of complex quantitative traits in maize. *Mol. Breed.* **2010**, *28*, 511–526. [CrossRef]
41. Cai, D.; Xiao, Y.; Yang, W.; Ye, W.; Wang, B.; Younas, M.; Wu, J.; Liu, K. Association mapping of six yield-related traits in rapeseed (*Brassica napus* L.). *Theor. Appl. Genet.* **2013**, *127*, 85–96. [CrossRef] [PubMed]
42. Zhao, P.; Liu, Y.; Kong, W.; Ji, J.; Cai, T.; Guo, Z. Genome-wide identification and characterization of calcium-dependent protein kinase (CDPK) and CDPK-related kinase (CRK) gene families in Medicago truncatula. *Int. J. Mol. Sci.* **2021**, *22*, 1044. [CrossRef]
43. Wang, F.; Sun, X.; Liu, B.; Kong, F.; Pan, X.; Zhang, H. A polygalacturonase gene PG031 regulates seed coat permeability with a pleiotropic effect on seed weight in soybean. *Theor. Appl. Genet.* **2022**, *135*, 1603–1618. [CrossRef] [PubMed]
44. Wang, S.; Zhou, Q.; Zhou, X.; Zhang, F.; Ji, S. Ethylene plays an important role in the softening and sucrose metabolism of blueberries postharvest. *Food Chem.* **2019**, *310*, 125965. [CrossRef]
45. Ma, P.; Chen, X.; Liu, C.; Xia, Z.; Song, Y.; Zeng, C.; Li, Y.; Wang, W. MePHD1 as a PHD-Finger Protein Negatively Regulates ADP-Glucose Pyrophosphorylase Small Subunit1a Gene in Cassava. *Int. J. Mol. Sci.* **2018**, *19*, 2831. [CrossRef] [PubMed]
46. Li, J.; Peng, Z.; Liu, Y.; Lang, M.; Chen, Y.; Wang, H.; Li, Y.; Shi, B.; Huang, W.; Han, L.; et al. Over-expression of peroxisome-localized GmABCA7 promotes seed germination in Arabidopsis thaliana. *Int. J. Mol. Sci.* **2022**, *23*, 2389. [CrossRef] [PubMed]
47. Shi, J.G.; Cui, H.Y.; Zhao, B.; Dong, S.T.; Liu, P.; Zhang, J.W. Effect of light on yield and characteristics of grain-filling of summer maize from flowering to maturity. *Sci. Agric. Sin.* **2013**, *46*, 4427–4434.

48. Song, L.; Liu, D. Mutations in the three Arabidopsis genes that encode the E2 subunit of the mitochondrial pyruvate dehydrogenase complex differentially affect enzymatic activity and plant growth. *Plant Cell Rep.* **2015**, *34*, 1919–1926. [CrossRef]
49. Liao, J.; Guan, Y.; Chen, W.; Shi, C.; Yao, D.; Wang, F.; Lam, S.M.; Shui, G.; Cao, X. ACBD3 is required for FAPP2 trans-ferring glucosylceramide through maintaining the Golgi integrity. *J. Mol. Cell Biol.* **2019**, *11*, 107–117. [CrossRef]
50. Hsiao, A.-S.; Haslam, R.P.; Michaelson, L.V.; Liao, P.; Chen, Q.-F.; Sooriyaarachchi, S.; Mowbray, S.L.; Napier, J.A.; Tanner, J.A.; Chye, M.-L. Arabidopsis cytosolic acyl-CoA-binding proteins ACBP4, ACBP5 and ACBP6 have overlapping but distinct roles in seed development. *Biosci. Rep.* **2014**, *34*, e00165. [CrossRef]
51. Xie, L.J.; Yu, L.J.; Chen, Q.F.; Wang, F.Z.; Huang, L.; Xia, F.N.; Zhu, T.R.; Wu, J.X.; Yin, J.; Liao, B.; et al. Arabidopsis acyl-CoA-binding protein ACBP3 participates in plant response to hypoxia by modulating very-long-chain fatty acid metabolism. *Plant J.* **2014**, *81*, 53–67. [CrossRef] [PubMed]
52. Tyagi, V.; Parihar, V.; Singh, D.; Kapoor, S.; Kapoor, M. The DEAD-box RNA helicase eIF4A1 interacts with the SWI2/SNF2-related chromatin remodelling ATPase DDM1 in the moss Physcomitrella. *Biochim. Biophys. Acta Proteins Proteom.* **2021**, *1869*, 140592. [CrossRef] [PubMed]
53. Bradbury, P.J.; Zhang, Z.; Kroon, D.E.; Casstevens, T.M.; Ramdoss, Y.; Buckler, E.S. TASSEL: Software for association mapping of complex traits in diverse samples. *Bioinformatics* **2007**, *23*, 2633–2635. [CrossRef]
54. Yang, H.; Wang, K. Genomic variant annotation and prioritization with ANNOVAR and wANNOVAR. *Nat. Protoc.* **2015**, *10*, 1556–1566. [CrossRef]
55. Price, M.N.; Dehal, P.S.; Arkin, A.P. FastTree 2—Approximately Maximum-Likelihood Trees for Large Alignments. *PLoS ONE* **2010**, *5*, e9490. [CrossRef]
56. Yang, J.; Lee, S.H.; Goddard, M.E.; Visscher, P.M. GCTA: A Tool for Genome-wide Complex Trait Analysis. *Am. J. Hum. Genet.* **2011**, *88*, 76–82. [CrossRef]
57. Alexander, D.H.; Novembre, J.; Lange, K. Fast model-based estimation of ancestry in unrelated individuals. *Genome Res.* **2009**, *19*, 1655–1664. [CrossRef] [PubMed]
58. Zhang, C.; Dong, S.-S.; Xu, J.-Y.; He, W.-M.; Yang, T.-L. PopLDdecay: A fast and effective tool for linkage disequilibrium decay analysis based on variant call format files. *Bioinformatics* **2019**, *35*, 1786–1788. [CrossRef]
59. Danecek, P.; Auton, A.; Abecasis, G.; Albers, C.A.; Banks, E.; DePristo, M.A.; Handsaker, R.E.; Lunter, G.; Marth, G.T.; Sherry, S.T.; et al. The variant call format and VCFtools. *Bioinformatics* **2011**, *27*, 2156–2158. [CrossRef]
60. Du, Z.; Zhou, X.; Ling, Y.; Zhang, Z.; Su, Z. agriGO: A GO analysis toolkit for the agricultural community. *Nucleic Acids Res.* **2010**, *38*, W64–W70. [CrossRef]
61. Turner, S.D. qqman: An R package for visualizing GWAS results using Q-Q and manhattan plots. *J. Open Source Softw.* **2018**, *3*, 731. [CrossRef]
62. Benjamini, Y.; Hochberg, Y. Controlling the False Discovery Rate: A Practical and Powerful Approach to Multiple Testing. *J. R. Stat. Soc. Ser. B* **1995**, *57*, 289–300. [CrossRef]
63. Livak, K.J.; Schmittgen, T.D. Analysis of relative gene expression data using real-time quantitative PCR and the $2^{-\Delta\Delta CT}$ method. *Methods* **2001**, *25*, 402–408. [CrossRef] [PubMed]

Sucrose Facilitates Rhizome Development of Perennial Rice (*Oryza longistaminata*)

Zhiquan Fan [†], Guanwen Huang [†], Yourong Fan [*] and Jiangyi Yang [*]

State Key Laboratory for Conservation and Utilization of Subtropical Agro-Bioresources, College of Life Science and Technology, Guangxi University, Nanning 530004, China
* Correspondence: yrfan@gxu.edu.cn or fanyourred@163.com (Y.F.); yangjy@gxu.edu.cn or yangjy598@163.com (J.Y.)
† These authors contributed equally to this work.

Abstract: Compared with annual crops, perennial crops with longer growing seasons and deeper root systems can fix more sunlight energy, and have advantages in reducing soil erosion and saving water, fertilizer and pesticide inputs. Rice is one of the most important food crops in the world. Perennial rice can be of great significance for protecting the ecological environment and coping with the shortage of young farmers due to urbanization. *Oryza longistaminata* (*OL*) is a rhizomatous wild rice with an AA genome and has strong biotic and abiotic resistances. The AA genome makes *OL* easy to cross with cultivated rice, thus making it an ideal donor material for perennial rice breeding. Sucrose plays an important role in the development and growth of plants. In this study, *OL* seedlings were cultured in medium with different concentrations of sucrose, and it was found that sucrose of appropriate concentrations can promote the sprout of basal axillary buds and the subsequent development of rhizomes. In order to explore the molecular mechanism, comparative transcriptome analysis was carried out with *OL* cultured under two concentrations of sucrose, 20 g/L and 100 g/L, respectively. The results showed that the boost of sucrose to rhizome elongation may be due to the glucose and fructose, hydrolyzed from the absorbed sucrose by vacuolar acid invertase. In addition, the consequent increased osmotic pressure of the cells would promote water absorption, which is benefit for the cell elongation, eventually causing the rhizome elongation. These results may provide a reference for elucidating the regulatory mechanism of sucrose on the rhizome development of *OL*.

Keywords: *Oryza longistaminata*; rhizome; sucrose; RNA-Seq; transcriptome

1. Introduction

O. longistaminata (*OL*) is a perennial wild rice with well-developed rhizomes. Rhizome is the key organ for vegetative reproduction and perenniality of *OL* [1–5]. The rhizome of *OL* is a horizontally growing underground stem that can grow out of the soil and become a new plant (ramet). Axillary buds on the nodes of the rhizome can produce secondary rhizomes or new ramets (new plants genetically identical to the parent). In addition, the rhizome is also an important storage organ of water and nutrients, such as starch [6], which is of great significance for *OL* to survive through harsh environments.

The development of rhizome in *OL* is a complex trait, maybe controlled by single, dual or multiple genes. A single dominant allele *Rhz* on chromosome 4 was firstly proposed for the rhizomatous growth habit in *OL*, and modifying genes have also been suggested to be involved for the various *Rhz* phenotypes among F_2 plants [7]. A pair of dominant complementary genes, *Rhz2* and *Rhz3*, have subsequently been identified for the regulation of rhizomes in *OL* [4]. However, only *Rhz2* and *Rhz3* did not guarantee the presence of rhizomes in the hybrid progeny between cultivated rice and *OL* (Hu, 2015, presentation entitled "Progress in Perennial Rice Breeding and Genetics", accessible from http://pwheat.anr.msu.edu/2015/02/ (accessed on 25 September 2022)). Thirteen QTLs (quantitative trait loci) regulating rhizome development have recently been mapped by Fan, et al. [8]. Further study from Hu's group identified 13 major-effect loci that jointly control rhizomatousness in *OL*, and as many as 51 QTLs for the rhizome abundance [9]. On the whole, the development of rhizome is regulated by multiple genes [8–10].

In addition to the genetic factor, environmental factors, such as sugars, also play an important role in rhizome development [6,11]. Sucrose can promote secondary rhizome development from the axillary buds of isolated rhizomes [6,11]. The rhizome usually grows horizontally below the ground for a certain distance, and then grows upward due to the negative gravitropism, and finally grows out of the soil and becomes a new ramet. It has been found that sucrose can retard the upward growth of rhizomes [6,11], causing speculation that sucrose is the key stimulus inducing the development and transformation of rhizomes [6]. However, the molecular mechanism by which sucrose affects rhizome development in *OL* remains unclear. In the previous studies, the isolated rhizomes were all cultured with sucrose in vitro [6,11], and the valid experiment about the effect of sucrose in rhizome development should be carried out in vivo. In this study, *OL* seedling was cultured in solid medium with different concentrations of sucrose to observe the effects of sucrose on the formation and development of rhizomes. In addition, the regulatory network of rhizome development in *OL* was probed by comparative transcriptome analysis.

2. Results

2.1. Sucrose Is Important for Sprouting of the Axillary Buds and Rhizome Growth in O. longistaminata (OL)

In order to explore the effect of sucrose on rhizome development, the aseptic *OL* seedlings at three–four leaf stage were cultured with different concentration of sucrose. It was found that when the sucrose concentration was 20 g/L, there are no axillary bud sprouted (Figure 1A). With the increase of sucrose concentration, such as to 40 g/L, a few sprouted axillary buds were found to form tillers or rhizomes (Figure 1B). When sucrose concentration rose to 60 g/L, the axillary buds of all seedlings sprouted, with many rhizomes formed (Figure 1C). At 80 g/L and 100 g/L, a large number of axillary buds sprouted to form rhizomes (Figure 1D,E). Compared with other concentrations, longer and more robust rhizomes were found at these two sucrose concentrations, with a more pronounced effect observed at the 100 g/L group. With further increase of sucrose concentration, the positive effect of sucrose on rhizome development became less prominent. At 120 g/L sucrose, even though the plant growth was slightly inhibited, rhizomes still could be found (Figure 1F). When the sucrose concentration increased to 140 g/L or 160 g/L, a harmful effect appeared and the sprouted axillary buds would turn up quickly (Figure 1G,H), with the leaves withered, maybe due to the high-water stress (Figure 1I). The results showed that a certain concentration of sucrose promotes the sprouting of axillary buds and the formation of rhizomes in *OL*.

2.2. Seedling Culture for Transcriptome Sequencing

In order to explore the molecular mechanism of sucrose affecting the development of *OL* rhizomes, crowns (the shortened internodes at stem base, also called shortened basal internodes), together with attached axillary buds, were collected for transcriptome sequencing. Usually, when cultivated rice seedlings are grown in a medium containing 20 g/L sucrose, axillary buds can sprout. However, when 20 g/L sucrose was used to culture *OL*, there was no axillary bud sprouted (Figure 1A). Therefore, a certain concentration of uniconazole, which can promote the sprouting of axillary buds [12], was added to the sucrose medium to promote sprouting of axillary buds. According to the effects of different uniconazole concentrations on the plant height, axillary bud sprouting and rhizome development, the optimum concentration of uniconazole was determined. The optimum concentration of uniconazole was 0.2 mg/L. If the concentration of uniconazole was too low, the sprouting of axillary buds in 20 g/L sucrose group was rare, which is not appropriate for the observation of the development of axillary buds. When uniconazole concentration was too high, although it could fully induce the sprouting of axillary buds, the plants were too short and this was able to surpass the promoting effect of 100 g/L sucrose on rhizome growth, resulting in the development of all axillary buds into tillers in both the 20 g/L and 100 g/L sucrose groups. When the concentration of uniconazole was 0.2 mg/L, the sprouted axillary buds in the 100 g/L sucrose group could develop into both tillers and rhizomes (Figure 2A); however, compared with the same concentration of sucrose without uniconazole (Figure 1E), the rhizomes of plants with uniconazole (Figure 2A) were shorter. However, at the concentration of 0.2 mg/L uniconazole, although the axillary buds of the 20 g/L sucrose group could sprout, they almost all developed into tillers (Figure 2B). Therefore, the crowns, together with attached axillary buds in the 20 g/L sucrose plus 0.2 mg/L uniconazole, were taken as the control group, denoted S20, and the same parts in the 100 g/L sucrose plus 0.2 mg/L uniconazole were the treatment group, denoted S100. RNA samples of both S20 and S100 were committed to transcriptome sequencing.

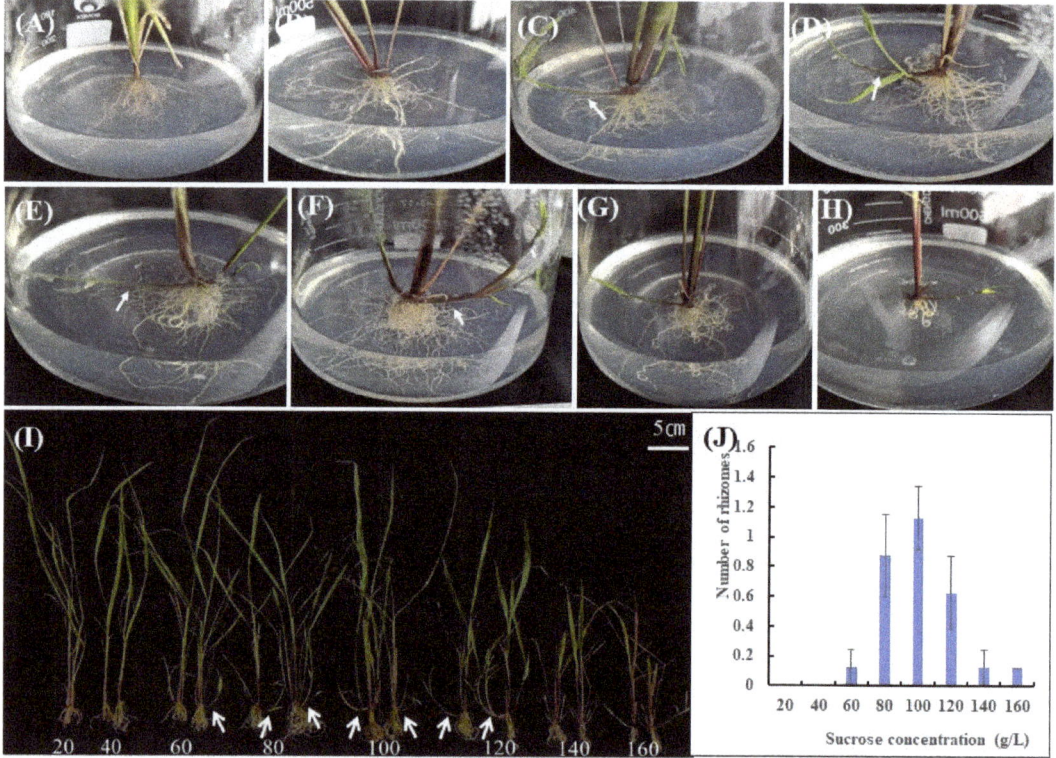

Figure 1. Effects of different sucrose concentrations on rhizomes of *Oryza longistaminata*. The corresponding sucrose concentrations in (**A–H**) are 20 g/L (**A**), 40 g/L (**B**), 60 g/L (**C**), 80 g/L (**D**), 100 g/L (**E**), 120 g/L (**F**), 140 g/L (**G**) and 160 g/L (**H**), respectively. Rhizomes are indicated with white arrows. (**I**) Phenotypes of the whole plants cultivated with various concentrations of sucrose. In (**J**), the abscissa represents different sucrose concentrations, and the ordinate represents the average number of rhizomes per plant under the corresponding sucrose concentration. The error bars represent standard deviation.

Figure 2. Branch formation in mediums with different sucrose concentrations and additional uniconazole. The working concentration of uniconazole is 0.2 mg/L. The sucrose concentration in (**A**) is 100 g/L, and that in (**B**) is 20 g/L. Arrows indicate rhizomes.

2.3. Sample Correlation Analysis

Based on the Illumina Nova seq 6000 sequencing platform, all samples were sequenced according to the instructions. The original sequencing data (raw data) were trimmed to remove low-quality and short sequence data, resulting in a total of 47.94 Gb of clean data in six samples; the clean data of each sample totaled over 7.66 Gb, and the percentage of Q30 bases was over 94.81% (Table S3). The clean reads of each sample were aligned with the reference genome, and the mapped rates ranged from 81.39% to 82.56% (Table S3).

In order to test whether the variation between biological repeats was in line with the expectations of the experimental design, and to provide a basic reference for differential gene analysis, the TPM (transcripts per million reads) value was used as the gene expression level, the correlation analysis of each biological repeat was carried out based on the Pearson correlation coefficient and the method of complete linkage was used for clustering. The results showed that the correlation coefficients between biological replicates in this study were all greater than 0.97 (Table S4), and the samples of biological replicates were clustered together (Figure S1). The above results show that the sampling was reasonable, and all sample data can be used for the subsequent analysis. In the study, a total of 38,866 genes were detected. These genes were aligned with six major databases (NR, Swiss-Prot, Pfam, EggNOG, GO and KEGG) to obtain gene functional information (Table S5). Given that the genome of *O. sativa* was used as the reference genome, all interpretations of gene functions are based on the annotation of *O. sativa* gene.

2.4. Functional Analysis of Differentially Expressed Genes

TPM (transcripts per million reads) was used as a quantitative index to analyze the quantitative results of gene expression. Based on the relative expression of genes, with

$|\log_2 FC| \geq 1$ and *P-adjust* < 0.05 as the threshold, 1255 differentially expressed genes (DEGs) with significance were obtained between S20 and S100 groups (Table S5). Compared with the S20 group, 529 were up-regulated and 726 were down-regulated in the S100 group (Figure S2). DEGs were classified and analyzed by COG (cluster of orthologous groups of proteins), GO (gene ontology) and KEGG (Kyoto Encyclopedia of Genes and Genomes).

It was found that, except for genes with unknown COG functions, the most functional one was G type (carbohydrate transport and metabolism) (Figure 3), and this functional type contains 88 genes (Table S6), of which *Os10g0555651* and *Os10g0555700* (*OsEXPB2*) were described as being able to cause loosening and extension of plant cell walls by disrupting non-covalent bonding between cellulose microfibrils and matrix glucans. *Os10g0577500* and *Os10g0545500* were annotated as xyloglucan endotransglucosylase hydrolase protein gene, and *Os06g0725300* as an expansin-like gene. *Os01g0249100* and *Os01g0249050* were plant-type cell wall organization genes. *Os09g0530250* was endoglucanase. *Os01g0813800*, *Os05g0365700* and *Os01g0940800* were beta-glucosidase. *Os03g0300600* was pectinesterase. In addition, nine genes, including *Os02g0106100* (*OsINV3*), were described as hydrolases. This indicated that some genes involved in cell wall loosening and extension, hydrolysis and remodeling of cell wall components were induced by sucrose. It means that sucrose treatment induced the loosening and extension of plant cell walls, and the hydrolysis and remodeling of cell wall components. In addition, these 88 genes included some genes related to aquaporin proteins, such as *Os07g0448100* (*OsPIP2;4*), *Os01g0975900* (*OsTIP1;2*), *Os07g0448150*, *Os05g0231700*, *Os02g0658100* (*OsTIP2;1*), *Os04g0521100*, *Os06g0552900*, *Os06g0553001*, *Os04g0233851*, *Os07g0448200* and *Os07g0448600*, which were described as channels that permit osmotically driven movement of water in both directions and are involved in the osmoregulation and in the maintenance of cell turgor during volume expansion in rapidly growing cells. It indicates that sucrose treatment induces the change of osmotic pressure of plant cells.

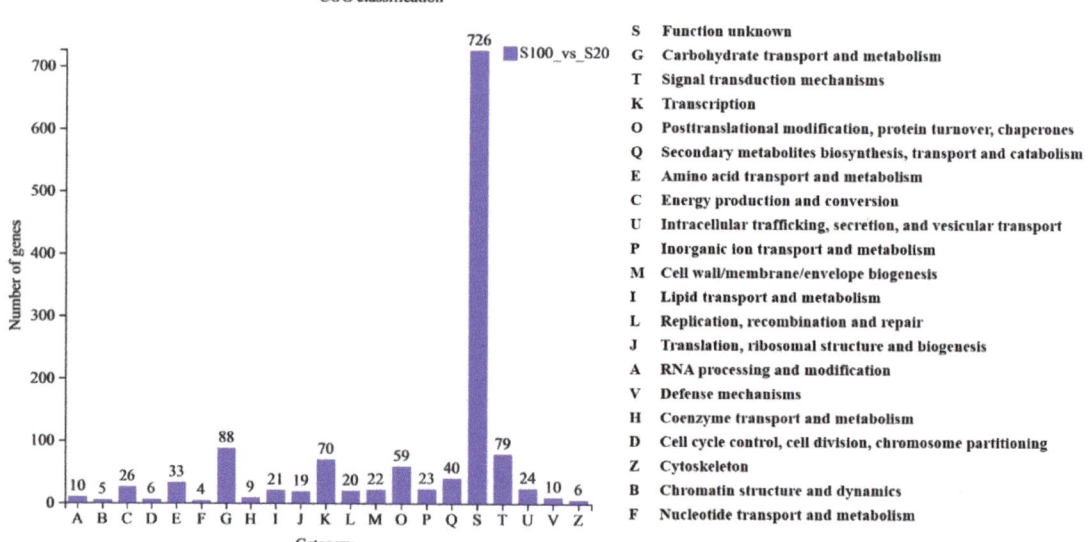

Figure 3. Histogram of COG classification statistics of differentially expressed genes. The abscissa represents the functional type of COG, and the ordinate represents the number of genes with this type of function.

GO annotation was performed on these 1255 DEGs and these genes were classified into 49 GO items. These included 21 biological_process entries, 15 cellular_component entries and 13 molecular_function entries (Figure 4). Biological_process had the largest number of DEGs in the metabolic process (GO:0008152), followed by cellular process (GO:0009987). Cellular_component had the largest number of DEGs in the cell (GO:0005623), followed by cell part (GO:0044464). Molecular_function had the largest number of DEGs in binding (GO:0005488), followed by catalytic activity (GO:0003824) (Figure 4).

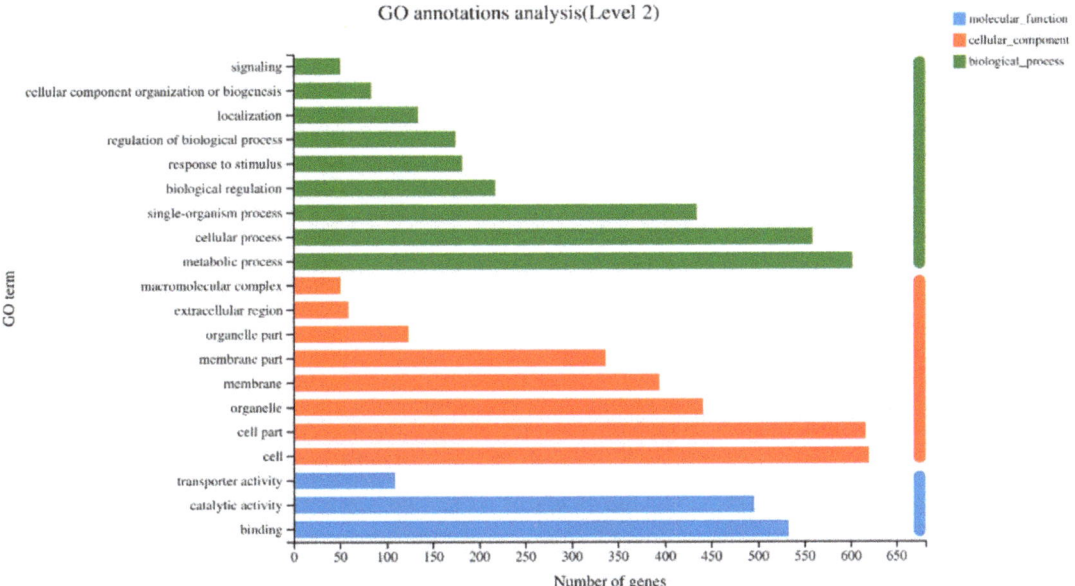

Figure 4. Histogram of GO classification statistics of differentially expressed genes. The vertical axis in the figure represents the secondary classification term of GO, and the horizontal axis represents the number of genes in the secondary classification.

The DEGs were further analyzed for GO functional enrichment to obtain the main GO functions of the genes. When *P-adjust* < 0.05, this GO item was considered to be significantly enriched. It was found that a total of 125 GO items were significantly enriched, including circadian rhythm (GO:0007623), energy storage and metabolism (GO:0006112), carbohydrate transport and metabolism (GO:0008643, GO: 0044262, GO: 0005975), responses to water (GO: 0009414, GO: 0009415), etc. (Figure 5). Among them, there were 130 genes involved in carbohydrate transport and metabolism (Table S7). Many of these genes were the same as the genes in the G functional classification of COG, including vacuolar acid invertase gene *Os02g0106100* (*OsINV3*) and xyloglucan endotransglucosylase hydrolase protein genes *Os10g0545500*, *Os06g0697000* and *Os10g0577500*. There were also some other cell wall component-related genes, such as pectinesterase (*Os11g0172100*), polygalacturonase (*Os01g0636500*, *Os05g0279900*, *Os05g0279850*, *Os05g0578750*, *Os05g0542900*, *Os01g0296200*, *Os05g0578600* (*PSL1*), *Os01g0329300*) and endoglucanase (*Os09g0530250*, *Os08g0387400*). In addition, there were some glycosyltransferase genes, sugar transporter genes, amylase genes, sucrose synthase genes and starch synthase genes among these 130 genes, such as *Os03g0170900* (*OsSUT1*), *Os07g0616800* (*RSUS3*; *SUS3*), *Os07g0106200* (*OsMST3*), *Os02g0513100* (*OsSWEET15*) and *Os06g0160700* (*OsSSI*). There were 17 genes that responded to water (Table S8), mainly related to drought stress, such as catalase gene *Os02g0115700* (*OsCATA*) [13], aquaporin protein gene *Os07g0448100* (*OsPIP2;4*), *Os04g0521100*, *Os02g0823100* (*OsPIP1;3*), *Os07g0448200*, ABA-related genes *Os05g0213500*

(*OsPYL/RCAR5*; *OsPYL5*), *Os11g0167800* (*OsASR1*), *Os11g0454300* (*RAB21*; *Rab16A*) and transcription factor *Os03g0741100* (*OsbHLH148*), which regulates drought tolerance in rice [14].

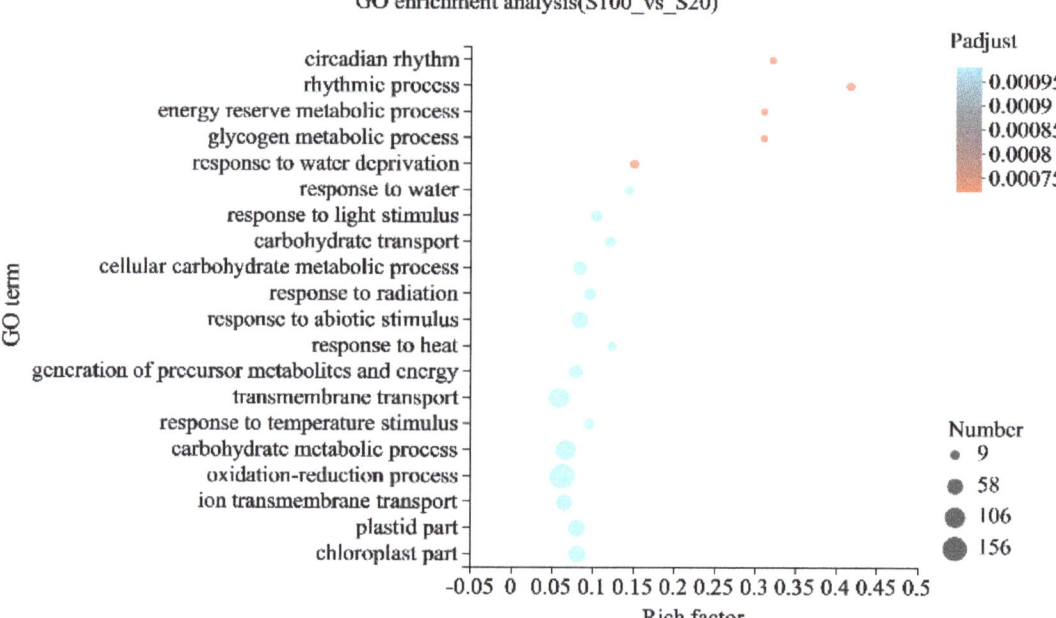

Figure 5. GO enrichment analysis. The vertical axis is the GO entry and the horizontal axis is the ratio of the enriched gene number (sample number) to the annotated gene number (background number) in the entry. The larger the ratio, the higher the enrichment degree. The size of the dot represents the number of genes in the entry; the larger the dot, the more genes there are. The color of the dot shows different *P−adjust* values; the smaller the *P−adjust*, the redder the dot. The figure shows the enrichment results of the top 20 entries of *P−adjust* from small to large.

Genes in the GO enrichment analysis results were used for the GO directed acyclic graph analysis, which shows the relationship between the upper and lower levels of the GO items and the degree of enrichment. It was found that these GO items were mainly divided into two branches, namely carbohydrates metabolism and response to water deprivation (Figure 6). In addition, there were three minor branches, namely carbohydrate transport, dioxygenase activity and response to light stimulation (Figure 6). Carbohydrate transport and metabolism were involved in the utilization of high concentrations of sucrose and the response to water deprivation, suggesting that high concentrations of sucrose may lead to increased intracellular osmotic pressure.

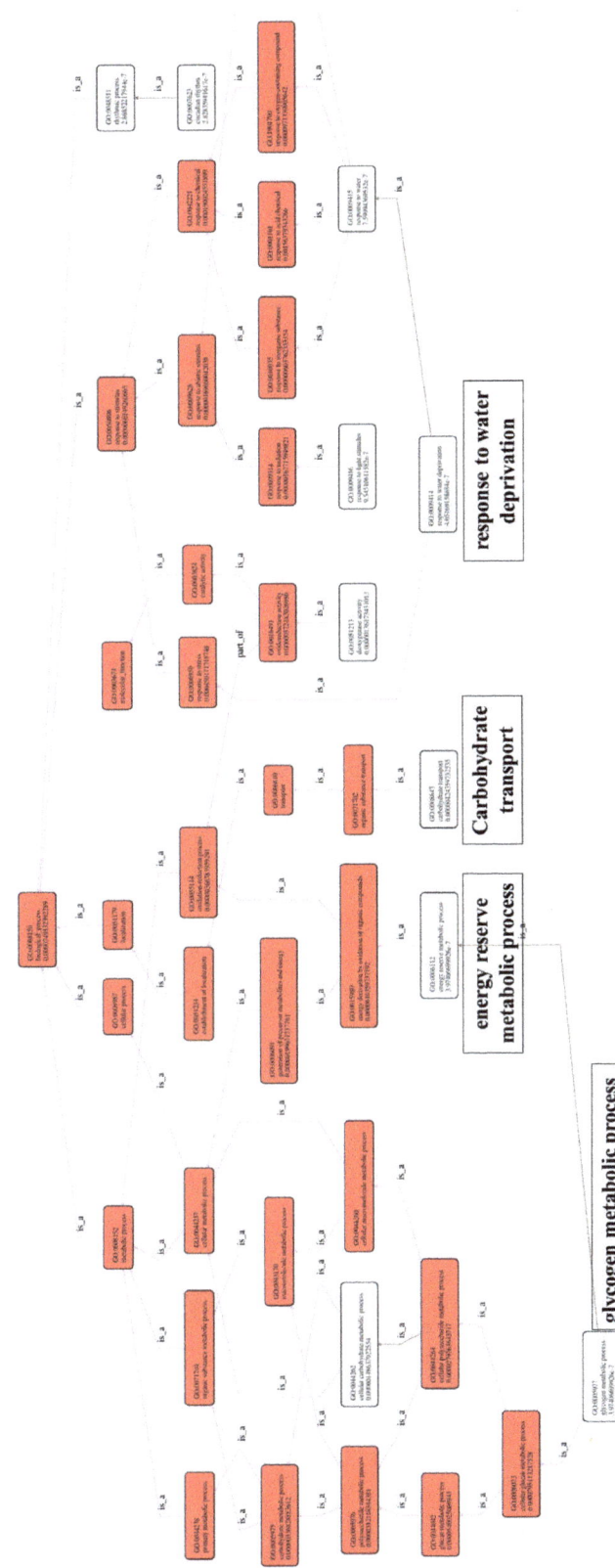

Figure 6. GO directed acyclic graph. The boxes show the GO IDs and annotations of the GO items. Red indicates the GO items with significant enrichment; the redder the color, the more significant the enrichment. The connecting line indicates the relationship between the two GOs. Please refer to http://geneontology.org/docs/ontology-relations/ (accessed on 25 September 2022) for the relationship explanation.

KEGG annotation analysis of DEGs found that there were 5 first-level KEGG pathway categories and 21 s-level KEGG pathway categories. Among them, the DEGs in the classification of carbohydrate metabolism were the most numerous ones, followed by biosynthesis of other secondary metabolites and energy metabolism (Figure 7).

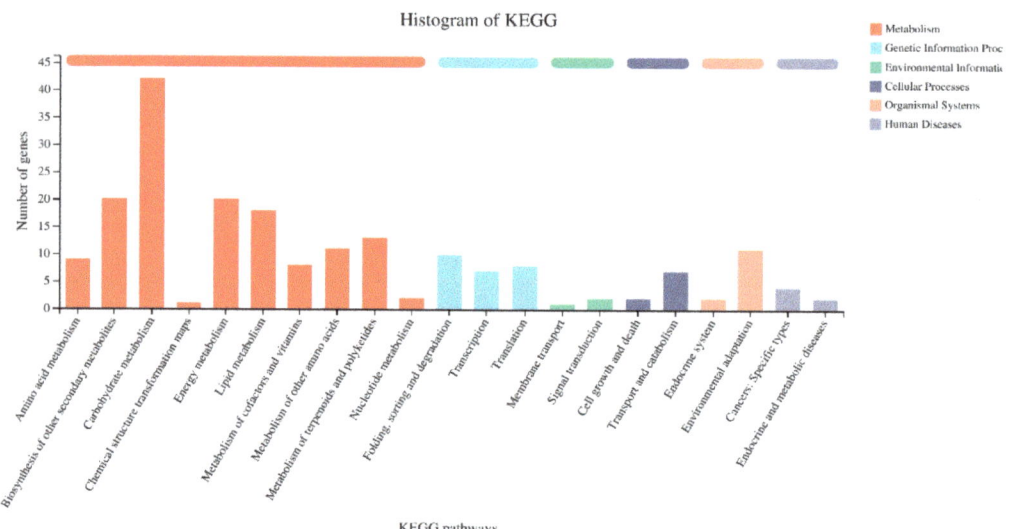

Figure 7. Classification statistics of differentially expressed genes in the KEGG pathway. The abscissa is the name of the KEGG metabolic pathway; the ordinate is the number of genes annotated to this pathway.

Taking *P-adjust* < 0.05 as the threshold, the DEGs were analyzed for KEGG pathways with significant enrichment. The results showed that two pathways were enriched, namely starch and sucrose metabolism (map00500) and plant circadian rhythm (map04712) (Figure 8). Combined with the results of GO enrichment analysis, it is speculated that the related processes of carbohydrate metabolism and response to water may be involved in the sucrose-induced formation and development of *OL* rhizomes. Among them, 32 genes were enriched in the carbohydrate transport metabolism KEGG pathway (Table S9), and these genes were mainly sucrose and starch synthase genes, such as *Os06g0160700* (*OsSSI*), *Os01g0919400* (*OsSPS1*), and *Os10g0189100* (*OspPGM*), indicating that high-concentration sucrose can be converted into starch for storage. In addition, there were a few genes related to cell wall components, such as endoglucanase (*Os09g0530250*).

2.5. Quantitative Real-Time PCR (qPCR) Validation

Among the differentially expressed genes, 14 genes were selected for qPCR. The expression level of both the qPCR and the RNA-Seq results of the control group S20 were assigned as 1, and the relative expression level of the treatment group S100 was calculated. The results showed that the up- and down-regulation of gene expression detected by qPCR was consistent with the results of RNA-Seq (Figure S3).

According to the above results, it is inferred that with the increase of sucrose concentration in the medium, the absorption of sucrose by *OL* increases. After sucrose enters the cell vacuole, it is decomposed into glucose and fructose by vacuolar acid invertase, which increases the cell osmotic pressure, and promotes the cell to absorb water and swell, causing the expression of genes involved in remodeling of plant cell walls, such as expansin, and then cell elongation. The final result is to promote the elongation of rhizomes in *OL* (Figure 9).

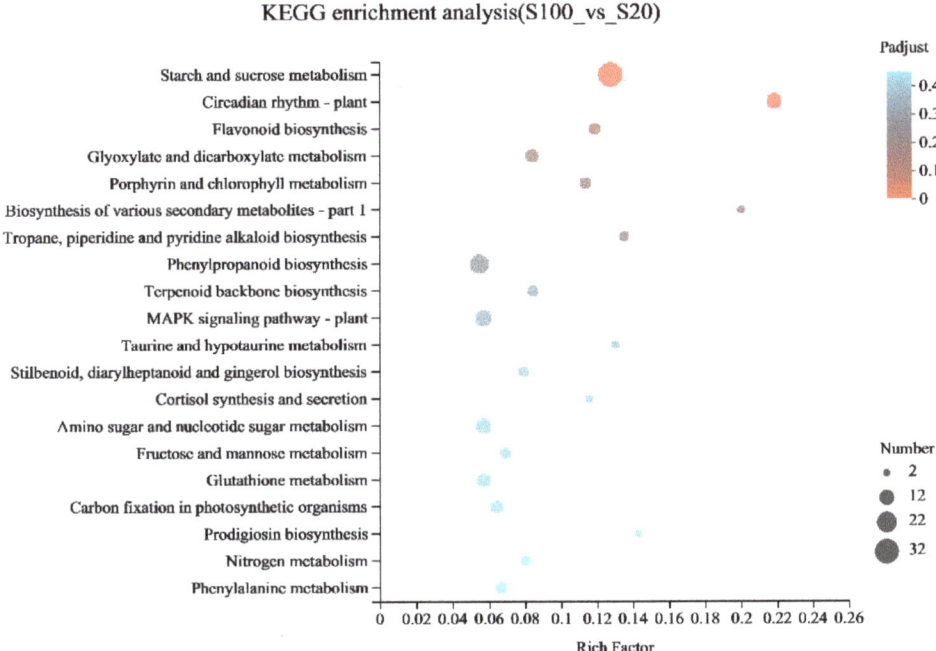

Figure 8. KEGG enrichment analysis. The vertical axis is the pathway entry, and the horizontal axis represents the ratio of the enriched gene number (sample number) to the annotated gene number (background number) in the entry. The greater the rich factor, the greater the degree of enrichment. The size and color of the dots indicate the number of genes in this pathway and the corresponding *P-adjust* range, respectively. The figure shows the enrichment results of the top 20 pathways of *P-adjust* from small to large.

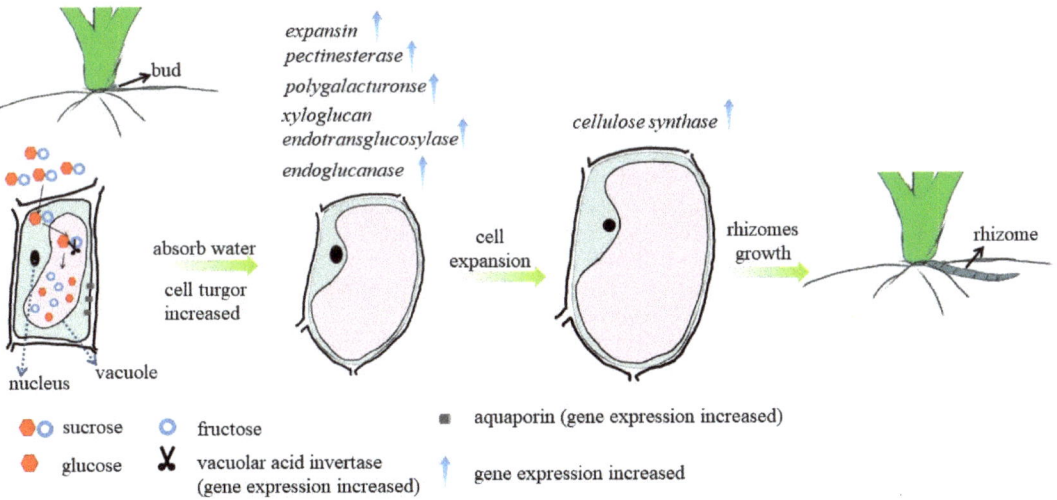

Figure 9. The regulatory mechanism of sucrose on rhizome elongation of *Oryza longistaminata*.

3. Discussion

Sucrose is the main product of photosynthesis and the main transport form of carbohydrates in plants, and it plays an important role in the entire life cycle of plants. Sugar is the main source of nutrients required for the growth of plant axillary buds [15], and elevated levels of sucrose promote the growth of buds [16]. An adequate supply of sucrose can reduce the competition of apical buds for sucrose, breaking the apical dominance, and facilitate the sprouting of lateral buds [17]. In addition, sucrose can also act as an osmotic agent, which can induce a certain degree of osmotic stress [18].

In this study, it was observed that when plants were transferred to a sucrose gradient medium, with the increase of culture time to about one month, phenotypic differences appeared in plants cultured at different concentration gradients. The 20 g/L and 40 g/L sucrose treatments had no rhizome formation, 60–100 g/L sucrose treatment had rhizome formation, and 120–160 g/L sucrose treatment had unhealthy plant growth. With the further extension of the culture time, even up to 3 months, the axillary buds of the plants in the 20 g/L and 40 g/L sucrose treatment groups still did not germinate or germinated individually, and no rhizomes were formed. The rhizomes in the 60–100 g/L sucrose treatment group grew and developed to the wall of the triangular flask, and began to grow up along the bottle wall, growing true leaves, and developing into ramets. In the 120–160 g/L sucrose treatment group, the leaves of the plants turned yellow, and some plants died. It shows that a certain concentration of sucrose is beneficial to the development of rhizome. However, too high sucrose concentration formed stress on the plant, resulting in a plant that could not grow healthily.

In *OL* rhizomes, the content of sucrose is similar to the concentration pattern of starch, decreasing gradually from the base to the tip of the rhizome [6], which indicates that the accumulation of sugar in the rhizome increases with the rhizome growth and development. By adding sucrose to the culture medium, the negative gravity of the *OL* rhizome is inhibited and the horizontal growth is promoted [6,11]. Horizontal growth of the stolon of *Cynodon dactylon* and *Paspalum vaginatum Swartz* is also promoted when sucrose supply is sufficient [19,20]. It was shown that sucrose could weaken the negative gravity response of both stolons and rhizomes, and high levels of sucrose are beneficial to the horizontal growth of rhizomes. It is suggested that sucrose plays an important role in the formation and development of rhizomes.

In this study, the expression of the vacuolar invertase gene *OsINV3* was up-regulated in the high-sucrose group (Tables S6 and S7), and the action of vacuolar acid invertase promoted the degradation of sucrose into glucose and fructose [21,22], resulting in an increase in the content of osmotic regulators and an increase in vacuolar osmotic potential, and also an up-regulation of cell aquaporin gene expression. High water channel activity of aquaporin [23], which promoted the water absorption and expansion of cells, thereby promotes cell elongation [24]. During the process of cell swelling and water absorption, the cell wall relaxed and remodeled, and the related genes involved in cell wall relaxation and remodeling were differently expressed (Table S6). Among them, the expression of expansin gene, polygalacturonase/pectin degrading enzyme genes and xyloglucan endoglucan glycosylase/hydrolase genes increased, and the expression of pectin methylesterase inhibitor gene decreased (Table S6).

Expansin, also known as a cell wall relaxation protein, is involved in various functions, such as plant stem elongation and water stress [25–28]. Expansins are able to weaken or break the non-covalent bonds between cell wall polysaccharides, causing hemicellulose to move, thereby causing cell wall loosening and, ultimately, cell wall stretching [29]. Both xyloglucan polygalacturonase/pectin degrading enzymes [30] are able to degrade pectin, which is the main polysaccharide in the primary cell wall of plants and is involved in cell wall plasticity [31,32]. The balance between pectin methylesterase (PME) activity and inhibition of post-transcriptional PME by pectin methylesterase inhibitors determines pectin methylesterification in plant cells [33]. *OsPMEI28* functions as a key component in regulating the level of methyl esterification of pectin, and the impaired methyl esterification level of pectin affects the biochemical properties of cell wall composition and leads to abnormal cell elongation in rice stem tissue and dwarf plants [33]. Xyloglucan endoglucan glycosylase/hydrolase catalyzes the hydrolysis or transfer of xyloglucan molecules and plays an important role in plant cell wall extensibility [34]. The action of these proteins leads to cell wall remodeling and elongation, cell elongation and, ultimately, rhizome elongation.

RNA-Seq also detected that the sucrose phosphate synthase gene *OsSPS1*, the starch synthase genes *OsSSI* were up-regulated in the high sucrose concentration group. Sucrose phosphate synthase is one of the key enzymes controlling sucrose synthesis in higher plants [35]. Starch synthases play a role in plant starch synthesis and accumulation [36,37]. Cytoplasmic fructose-1,6-bisphosphatase is an important enzyme in the sucrose synthesis pathway, and its lack of activity leads to insufficient sucrose supply, resulting in the inhibition of tiller growth in rice mutants of this gene [38]. It is speculated that the sucrose entering the *OL* was metabolized and used as a carbon source and energy for the sprouting of axillary buds and the growth and development of rhizomes (Figure 9).

4. Materials and Methods

4.1. Plant Material

The plant material used was the aseptic seedlings from medium-germinated seeds of *OL* (No. WYD-108). WYD-108 was introduced from Senegal, Africa, and preserved in the National Germplasm Resource Nursery (wild rice) of China.

4.2. Method

4.2.1. Culture of *OL* Seedlings

The seeds of *OL* were collected by bagged selfing to avoid outcrossing, and were treated at 50 °C for 5 days to break dormancy. The dehulled seeds were disinfected with 0.15% mercuric chloride for 15–20 min, and then rinsed with sterilized distilled water 5 times [39]. The seeds were then transferred to the rooting medium until the 3–4 leaf stage at 26 °C, with 12 h light/12 h dark cycle [39]. The light intensity was about 140 μmol/m^2/s. It took about 18 days for the seeds to grow to the 3–4 leaf stage on the rooting medium. The rooting medium was prepared based on 1/2 MS basic medium with 20 g/L sucrose added [40]. The basic medium formula is shown in Table S1.

4.2.2. Culture of *OL* with Different Concentrations of Sucrose

Under aseptic conditions, the 3–4 leaf-stage *OL* seedlings were pulled out from the rooting medium and most leaves and roots were cut off, keeping the youngest ones. Then, the seedlings were transplanted to solid basic medium with different sucrose concentrations (sucrose concentrations of 20 g/L, 40 g/L, 60 g/L, 80 g/L, 100 g/L, 120 g/L, 140 g/L and 160 g/L), and cultured under 26 °C, 12 h light/12 h dark cycle, with light intensity 140 μmol/m^2/s. Although rhizomes and tillers can develop from axillary buds, the morphology of rhizome buds is completely different from that of tiller buds. Rhizome bud in *OL* is onion-like-shaped and develops in an oblique direction [41]. In addition, the leaf morphology of tillers and rhizomes is also different. The tiller leaves are composed of a distal leaf and a proximal leaf sheath, while the formation of rhizome leaves is suppressed, and most rhizome leaves are wrapped by leaf sheaths [42]. These phenotypes can help us distinguish between tillers and rhizomes.

4.2.3. Sample Collection for Transcriptome Analysis

Based on our initial results, the axillary buds of *OL* did not sprout under 20 g/L sucrose, and then 0.2 mg/L uniconazole was added into the medium to ensure sprouting of the axillary bud [12]. Although the axillary bud of *OL* can sprout spontaneously under 100 g/L sucrose, 0.2 mg/L uniconazole was also added to control the experimental parameters. The *OL* seedlings at the 3–4 leaf stage were transplanted into these two sucrose–uniconazole media with only the sucrose differing (100 g/L vs. 20 g/L). After the sprouting of axillary buds, the crowns (which are the shortened internodes at the base of the stem of grass species, also called shortened basal internodes), together with attached axillary buds, were quickly cut and placed into liquid nitrogen, numbered S20 and S100, respectively. Three biological replicates were sampled for both S20 and S100. All collected samples were stored at −80 °C. RNA isolation, library construction and sequencing were performed by Shanghai Majorbio Co., Ltd. (Shanghai, China). The sequenced data were uploaded in NCBI (the accession numbers is PRJNA894815).

4.3. Data Processing and Reads Alignment

After removing the adapter sequence, low-quality reads, sequences with high N rate (N represents uncertain base) and those too short raw reads, the quality control sequences (clean reads) were obtained. The clean reads were aligned with the reference genome (*Oryza sativa Japonica* Group (IRGSP-1.0), http://plants.ensembl.org/Oryza_sativa/Info/Index (accessed on 25 September 2022)). Mapped reads were assembled and spliced using the software StringTie (http://ccb.jhu.edu/software/stringtie/ (accessed on 25 September 2022)) and compared with known transcripts.

4.4. Abundance Estimation and Correlation Analysis

RSEM software (http://deweylab.github.io/RSEM/ (accessed on 25 September 2022)) [43] was used to quantify gene abundances. The expression level of each transcript was calculated according to the transcripts per million reads (TPM) method. Based on the normalized expression levels, the Pearson correlation coefficient between any two biological replicates was calculated.

4.5. GO and KEGG Ontology Enrichment Analysis

Using TPM as a quantitative indicator of gene/transcript expression, the genes with significant differential expression (DEGs) between groups were identified using DESeq2 software [44], with $|\log_2 FC| \geq 1$ and *p-adj* < 0.05 as the threshold. The DEGs were functionally annotated using COG (cluster of orthologous groups), GO (gene ontology) and KEGG (Kyoto Encyclopedia of Genes and Genomes). Taking *P-adj* < 0.05 as the threshold, the software Goatools (https://github.com/tanghaibao/GOatools (accessed on 25 September 2022)) was used to perform GO enrichment analysis. KEGG enrichment was performed using R-Package.

4.6. Quantitative Real-Time PCR Validation

The expression abundance of 14 DEGs randomly selected was evaluated by quantitative real-time PCR (qPCR). Total RNAs were reverse transcribed into cDNA using RT kit (PrimeScript™ RT reagent Kit with gDNA Eraser, TaKaRa). qPCR analysis was performed on CFX96™ Real-Time System (Bio-Rad, Hercules, CA, USA) with SYBR Mix (ChamQ Universal SYBR qPCR Master Mix, Vazyme, Nanjing, China). *Ubiquitin* gene (*LOC_Os03g13170*) was used as the reference gene, and the relative expression level of target genes was calculated according to the $2^{-\Delta\Delta Ct}$ method [45]. Primers are listed in Table S2.

Supplementary Materials: The following supporting information can be downloaded at: https://www.mdpi.com/article/10.3390/ijms232113396/s1.

Author Contributions: J.Y. and Y.F. designed the experiments; Z.F. and G.H. conducted experiments; Z.F., J.Y. and Y.F. analyzed the data; Z.F. wrote the manuscript; J.Y. and Y.F. reviewed and edited the manuscript. All authors have read and agreed to the published version of the manuscript.

Funding: This work was supported by grants from the Guangxi Natural Science Foundation of China (2018GXNSFDA050010), Guangxi Science and Technology Development Program (AD19110145), the Scientific Research Foundation of Guangxi University (XTZ131548, XMPZ160942) and the State Key Laboratory for Conservation and Utilization of Subtropical Agro-bioresources (SKLCUSA-a201916, SKLCUSA-a202004).

Institutional Review Board Statement: Not applicable.

Informed Consent Statement: Not applicable.

Data Availability Statement: The datasets supporting the conclusions of this article are included within the article (and its Supplementary Materials).

Acknowledgments: We thank Ruiyang Zhou of Agricultural College of Guangxi University for providing wild rice.

Conflicts of Interest: The authors declare no conflict of interest.

References

1. Chen, Z.; Hu, F.; Xu, P.; Li, J.; Deng, X. QTL analysis for hybrid sterility and plant height in interspecific populations derived from a wild rice relative, *Oryza longistaminata*. *Breed. Sci.* **2009**, *59*, 441–445. [CrossRef]
2. Ghesquiere, A. Evolution of *Oryza longistaminata*. In *Rice Genetics*; IRRI, Ed.; World Scientific Publishing Company: Los Banos, Philippines, 1985; pp. 15–27.
3. Hu, F.; Wang, D.; Zhao, X.; Zhang, T.; Sun, H.; Zhu, L.; Zhang, F.; Li, L.; Li, Q.; Tao, D.; et al. Identification of rhizome-specific genes by genome-wide differential expression analysis in *Oryza longistaminata*. *BMC Plant Biol.* **2011**, *11*, 18. [CrossRef] [PubMed]
4. Hu, F.Y.; Tao, D.Y.; Sacks, E.; Fu, B.Y.; Xu, P.; Li, J.; Yang, Y.; McNally, K.; Khush, G.S.; Paterson, A.H.; et al. Convergent evolution of perenniality in rice and sorghum. *Proc. Natl. Acad. Sci. USA* **2003**, *100*, 4050–4054. [CrossRef] [PubMed]
5. Sacks, E.J.; Dhanapala, M.P.; Sta. Cruz, M.T.; Sallan, R. Breeding for perennial growth and fertility in an *Oryza sativa*/*O. longistaminata* population. *Field Crop. Res.* **2006**, *95*, 39–48. [CrossRef]
6. Bessho-Uehara, K.; Nugroho, J.E.; Kondo, H.; Angeles-Shim, R.B.; Ashikari, M. Sucrose affects the developmental transition of rhizomes in *Oryza longistaminata*. *J. Plant Res.* **2018**, *131*, 693–707. [CrossRef]
7. Maekawa, M.; Inukai, T.; Rikiishi, K.; Matsuura, T.; Govidaraj, K. Inheritance of the rhizomatous trait in hybrid of *Oryza longistaminata* Chev. et Roehr. and *O. sativa* L. *SABRAO J. Breed. Genet.* **1998**, *30*, 69–72.
8. Fan, Z.; Wang, K.; Rao, J.; Cai, Z.; Yang, J. Interactions among multiple quantitative trait loci underlie rhizome development of perennial rice. *Front. Plant Sci.* **2020**, *11*, 591157. [CrossRef]
9. Li, W.; Zhang, S.; Huang, G.; Huang, L.; Zhang, J.; Li, Z.; Hu, F. A genetic network underlying rhizome development in *Oryza longistaminata*. *Front. Plant Sci.* **2022**, *13*, 866165. [CrossRef]
10. Fan, Z.; Wang, K.; Fan, Y.; Tao, L.-Z.; Yang, J. Combining ability analysis on rhizomatousness via incomplete diallel crosses between perennial wild relative of rice and Asian cultivated rice. *Euphytica* **2020**, *216*, 140. [CrossRef]
11. Fan, Z.; Cai, Z.; Shan, J.; Yang, J. Bud position and carbohydrate play a more significant role than light condition in the developmental transition between rhizome buds and aerial shoot buds of *Oryza longistaminata*. *Plant Cell Physiol.* **2017**, *58*, 1281–1282. [CrossRef]
12. Ma, Y. *Studies on Anther Culture of Rice Hybrid F_1 and the Technique of Rapid Propagation of Oryza Longistaminata by Tissue Culture*; Guangxi University: Nanning, China, 2019.
13. Ye, N.; Zhu, G.; Liu, Y.; Li, Y.; Zhang, J. ABA controls H_2O_2 accumulation through the induction of *OsCATB* in rice leaves under water stress. *Plant Cell Physiol.* **2011**, *52*, 689–698. [CrossRef] [PubMed]
14. Seo, J.-S.; Joo, J.; Kim, M.-J.; Kim, Y.-K.; Nahm, B.H.; Song, S.I.; Cheong, J.-J.; Lee, J.S.; Kim, J.-K.; Choi, Y.D. OsbHLH148, a basic helix-loop-helix protein, interacts with OsJAZ proteins in a jasmonate signaling pathway leading to drought tolerance in rice. *Plant J.* **2011**, *65*, 907–921. [CrossRef] [PubMed]
15. Fichtner, F.; Barbier, F.F.; Feil, R.; Watanabe, M.; Annunziata, M.G.; Chabikwa, T.G.; Hoefgen, R.; Stitt, M.; Beveridge, C.A.; Lunn, J.E. Trehalose 6-phosphate is involved in triggering axillary bud outgrowth in garden pea (*Pisum sativum* L.). *Plant J.* **2017**, *92*, 611–623. [CrossRef] [PubMed]
16. Kebrom, T.H.; Mullet, J.E. Photosynthetic leaf area modulates tiller bud outgrowth in sorghum. *Plant Cell Environ.* **2015**, *38*, 1471–1478. [CrossRef]
17. Mason, M.G.; Ross, J.J.; Babst, B.A.; Wienclaw, B.N.; Beveridge, C.A. Sugar demand, not auxin, is the initial regulator of apical dominance. *Proc. Natl. Acad. Sci. USA* **2014**, *111*, 6092–6097. [CrossRef]
18. Cui, X.-H.; Murthy, H.N.; Wu, C.-H.; Paek, K.-Y. Sucrose-induced osmotic stress affects biomass, metabolite, and antioxidant levels in root suspension cultures of *Hypericum perforatum* L. *Plant Cell Tissue Organ Cult.* **2010**, *103*, 7–14. [CrossRef]
19. Montaldi, E.R. Gibberellin-sugar interaction regulating the growth habit of bermudagrass (*Cynodon dactylon* (L.) Pers). *Experientia* **1969**, *25*, 91–92. [CrossRef]
20. Willemoës, J.; Beltrano, J.; Montaldi, E.R. Diagravitropic growth promoted by high sucrose contents in *Paspalum vaginatum*, and its reversion by gibberellic acid. *Can. J. Bot.* **1988**, *66*, 2035–2037. [CrossRef]
21. Koch, K. Sucrose metabolism: Regulatory mechanisms and pivotal roles in sugar sensing and plant development. *Curr. Opin. Plant Biol.* **2004**, *7*, 235–246. [CrossRef]
22. Ruan, Y.L.; Jin, Y.; Yang, Y.J.; Li, G.J.; Boyer, J.S. Sugar input, metabolism, and signaling mediated by invertase: Roles in development, yield potential, and response to drought and heat. *Mol. Plant* **2010**, *3*, 942–955. [CrossRef]

23. Sakurai, J.; Ishikawa, F.; Yamaguchi, T.; Uemura, M.; Maeshima, M. Identification of 33 rice aquaporin genes and analysis of their expression and function. *Plant Cell Physiol.* **2005**, *46*, 1568–1577. [CrossRef] [PubMed]
24. Kutschera, U. Cell expansion in plant development. *Rev. Bras. Fisiol. Veg.* **2000**, *12*, 65–95.
25. Cho, H.T.; Kende, H. Expression of expansin genes is correlated with growth in deepwater rice. *Plant Cell* **1997**, *9*, 1661–1671. [CrossRef] [PubMed]
26. Lee, Y. Expression of β-expansins is correlated with internodal elongation in deepwater rice. *Plant Physiol.* **2001**, *127*, 645–654. [CrossRef]
27. Reidy, B.; McQueen-Mason, S.; Nosberger, J.; Fleming, A. Differential expression of alpha- and beta-expansin genes in the elongating leaf of *Festuca pratensis*. *Plant Mol. Biol.* **2001**, *46*, 491–504. [CrossRef] [PubMed]
28. Wu, Y.J.; Meeley, R.B.; Cosgrove, D.J. Analysis and expression of the alpha-expansin and beta-expansin gene families in maize. *Plant Physiol.* **2001**, *126*, 222–232. [CrossRef]
29. Simon, M.-M.; Le, N.T.; Brocklehurst, D. *Expansins*; Springer: Berlin/Heidelberg, Germany, 2006.
30. Zhang, G.; Hou, X.; Wang, L.; Xu, J.; Chen, J.; Fu, X.; Shen, N.; Nian, J.; Jiang, Z.; Hu, J.; et al. PHOTO-SENSITIVE LEAF ROLLING 1 encodes a polygalacturonase that modifies cell wall structure and drought tolerance in rice. *New Phytol.* **2021**, *229*, 890–901. [CrossRef]
31. Mohnen, D. Pectin structure and biosynthesis. *Curr. Opin. Plant Biol.* **2008**, *11*, 266–277. [CrossRef]
32. Oh, C.S.; Kim, H.; Lee, C. Rice cell wall polysaccharides: Structure and biosynthesis. *J. Plant Biol.* **2013**, *56*, 407. [CrossRef]
33. Nguyen, H.P.; Jeong, H.Y.; Jeon, S.H.; Kim, D.; Lee, C. Rice pectin methylesterase inhibitor28 (*OsPMEI28*) encodes a functional PMEI and its overexpression results in a dwarf phenotype through increased pectin methylesterification levels. *J. Plant Physiol.* **2017**, *208*, 17–25. [CrossRef]
34. Atkinson, R.G.; Johnston, S.L.; Yauk, Y.K.; Sharma, N.N. Analysis of xyloglucan endotransglucosylase/hydrolase (XTH) gene families in kiwifruit and apple. *Postharvest Biol. Technol.* **2009**, *51*, 149–157. [CrossRef]
35. Hirose, T.; Hashida, Y.; Aoki, N.; Okamura, M.; Yonekura, M.; Ohto, C.; Terao, T.; Ohsugi, R. Analysis of gene-disruption mutants of a sucrose phosphate synthase gene in rice, *OsSPS1*, shows the importance of sucrose synthesis in pollen germination. *Plant Sci.* **2014**, *225*, 102–106. [CrossRef] [PubMed]
36. Fujita, N.; Yoshida, M.; Asakura, N.; Ohdan, T.; Miyao, A.; Hirochika, H.; Nakamura, Y. Function and characterization of starch synthase I using mutants in rice. *Plant Physiol.* **2006**, *140*, 1070–1084. [CrossRef] [PubMed]
37. Hirose, T.; Terao, T. A comprehensive expression analysis of the starch synthase gene family in rice (*Oryza sativa* L.). *Planta* **2004**, *220*, 9–16. [CrossRef] [PubMed]
38. Koumoto, T.; Shimada, H.; Kusano, H.; She, K.C.; Iwamoto, M.; Takano, M. Rice monoculm mutation moc2, which inhibits outgrowth of the second tillers, is ascribed to lack of a fructose-1,6-bisphosphatase. *Plant Biotechnol.* **2013**, *30*, 47–56. [CrossRef]
39. Yang, J.; Fan, Z.; Zhang, Y.; Fan, Y. Tissue Culture Method of Rice Rhizome. China ZL202110344638.7. 31 March 2021.
40. Sivakumar, P.; Law, Y.; Ho, C.; Harikrishna, J. High frequency plant regeneration from mature seed of elite, recalcitrant malaysian indica rice (*Oryza Sativa* L.) CV. MR 219. *Acta Biol. Hung.* **2010**, *61*, 313–321. [CrossRef]
41. Yoshida, A.; Terada, Y.; Toriba, T.; Kose, K.; Ashikari, M.; Kyozuka, J. Analysis of rhizome development in *Oryza longistaminata*, a wild rice species. *Plant Cell Physiol.* **2016**, *57*, 2213–2220. [CrossRef]
42. Toriba, T.; Tokunaga, H.; Nagasawa, K.; Nie, F.; Yoshida, A.; Kyozuka, J. Suppression of leaf blade development by BLADE-ON-PETIOLE orthologs is a common strategy for underground rhizome growth. *Curr. Biol.* **2020**, *30*, 509–516. [CrossRef]
43. Li, B.; Dewey, C.N. RSEM: Accurate transcript quantification from RNA-Seq data with or without a reference genome. *BMC Bioinf.* **2011**, *12*, 323. [CrossRef]
44. Love, M.I.; Huber, W.; Anders, S. Moderated estimation of fold change and dispersion for RNA-seq data with DESeq2. *Genome Biol.* **2014**, *15*, 550. [CrossRef]
45. Livak, K.J.; Schmittgen, T.D. Analysis of relative gene expression data using real-time quantitative PCR and the $2^{-\Delta\Delta Ct}$ method. *Methods* **2001**, *25*, 402–408. [CrossRef] [PubMed]

Article

The Shift in Synonymous Codon Usage Reveals Similar Genomic Variation during Domestication of Asian and African Rice

Guilian Xiao [1], Junzhi Zhou [1], Zhiheng Huo [1], Tong Wu [1], Yingchun Li [1], Yajing Li [1], Yanxia Wang [2] and Mengcheng Wang [1,*]

[1] The Key Laboratory of Plant Development and Environment Adaptation Biology, Ministry of Education, School of Life Science, Shandong University, Qingdao 266237, China
[2] Shijiazhuang Academy of Agriculture and Forestry Sciences, Shijiazhuang 050041, China
* Correspondence: wangmc@sdu.edu.cn

Abstract: The domestication of wild rice occurred together with genomic variation, including the synonymous nucleotide substitutions that result in synonymous codon usage bias (SCUB). SCUB mirrors the evolutionary specialization of plants, but its characteristics during domestication were not yet addressed. Here, we found cytosine- and guanidine-ending (NNC and NNG) synonymous codons (SCs) were more pronounced than adenosine- and thymine-ending SCs (NNA and NNT) in both wild and cultivated species of Asian and African rice. The ratios of NNC/G to NNA/T codons gradually decreased following the rise in the number of introns, and the preference for NNA/T codons became more obvious in genes with more introns in cultivated rice when compared with those in wild rice. SCUB frequencies were heterogeneous across the exons, with a higher preference for NNA/T in internal exons than in terminal exons. The preference for NNA/T in internal but not terminal exons was more predominant in cultivated rice than in wild rice, with the difference between wild and cultivated rice becoming more remarkable with the rise in exon numbers. The difference in the ratios of codon combinations representing DNA methylation-mediated conversion from cytosine to thymine between wild and cultivated rice coincided with their difference in SCUB frequencies, suggesting that SCUB reveals the possible association between genetic and epigenetic variation during the domestication of rice. Similar patterns of SCUB shift in Asian and African rice indicate that genomic variation occurs in the same non-random manner. SCUB representing non-neutral synonymous mutations can provide insight into the mechanism of genomic variation in domestication and can be used for the genetic dissection of agricultural traits in rice and other crops.

Keywords: rice; domestication; genomic variation; synonymous codon usage bias; DNA methylation

1. Introduction

Cultivated rice is an ancient and widely consumed staple food crop. Two representative cultivated rice species, Asian *Oryza sativa* and African *Oryza glaberrima*, were domesticated from the sympatric Asian wild *O. rufipogon* and African wild *Oryza barthii*, respectively [1,2]. During the domestication process, genome-wide genetic variations occurred, including single nucleotide polymorphisms (SNPs), small insertions and deletions (indels), large size structural variants and so on [3–5]. These genome-scale variations provide the genetic basis for the differences in a wide range of morphological and physiological traits between wild and cultivated rice [2,6].

SNPs are the most common and plentiful genetic variations in genomes. SNPs in protein-coding sequences are classified into synonymous and non-synonymous, with much attention paid to the latter; this is because these change peptide sequences and may affect phenotypes. The synonymous SNPs result from the shift in synonymous codons (SCs) that encode the amino acids, except for methionine and tryptophan. The frequencies of SCs encoding a given amino acid are heterogeneous in the genome of a species, resulting in synonymous codon usage bias (SCUB). Nucleotide substitution between SCs does not change the corresponding amino acid residue and is, therefore, often believed to be functionally neutral [7,8]. However, SCUB affects recombination rate, splicing regulation, transcription efficiency, RNA secondary structure, mRNA stability, translational efficiency and accuracy in the regulation of gene expression and protein folding [9–13]. Especially, synonymous mutations in representative yeast genes proved to be strongly non-neutral [14]. Based on these, SCUB may influence mutation rates and the extent of genetic drift and natural selection [15–18], and is, therefore, an important genetic force for plant evolution. Although genome-wide nucleotide substitutions and other genetic variations occurred during the domestication of rice [3–5], whether SCUB shifting played a role in cultivated rice was not studied.

As forms of insertion and deletion (indel), intron gain and loss induce genetic variation and are key evolutionary forces [19–21]. The process of indels, such as intron gain and loss, comprises DNA break and repair, which could lead to genomic shock [22,23] and, therefore, induce local single-nucleotide polymorphisms [24–26]. As a part of nucleotide substitutions, SCUB in exons is related to adjacent introns in the nuclear genomes [27]. The occurrence of intron gain or loss is associated with both the number of introns and exon position within the gene body [28], so SCUB is reasonably related to these factors. Although the contribution of intron gain and loss in genomic variation during the domestication of rice is still unclear, the number of introns and the exon position were proved to be associated with SCUB shift during plant evolution and somatic hybridization [29,30]. Thus, the relationship between SCUB and the number of introns or exon position following the domestication of rice is worthy of being addressed as it could provide genetic clues for the further understanding of the role of introns in genetic variation.

Apart from genetic variation, the profiles of DNA methylation as a major epigenetic variation altered widely during the domestication of rice [31]. DNA methylation is also a source of genetic variation, because methylated cytosine (5^mC) can be converted into thymine [32]. Thus, DNA methylation-mediated conversion from cytosine to thymine affects SCUB in plants [29,30]. However, the issues concerning the contribution of DNA methylation to SCUB, as well as the role of SCUB in DNA methylation alteration, were not reported thus far.

In this study, we used Asian and African cultivated rice and their wild species to analyze the characteristics of SCUB during domestication, with the aim of knowing whether domestication can affect SCUB and how DNA methylation contributes to SCUB during domestication. We found that SCUB was obviously affected during the domestication of both Asian and African rice in the same manner, and the dual associations between SCUB and DNA methylation-mediated conversion from cytosine to thymine implies a close link between genetic and epigenetic variations during the domestication process. Our work provides novel data indicating that the SCUB shift possibly provides DNA methylation sites to promote epigenetic variation in the genome during rice domestication; it also demonstrates the bidirectional orchestration between genetic and epigenetic variation.

2. Results

2.1. C/G-Ending Synonymous Codons Are Preferred in Cultivated Rice

The frequencies of 61 amino acid encoding codons ranged from 0.048 (CGA in *O. glaberrima*) to 0.385 (GAG in *O. glaberrima*) and showed similar patterns in wild and cultivated rice (Figure S1). The frequencies in African wild rice *O. barthii* were slightly distinct from those of the other species, among which A/T-ending codons had higher frequencies but C/G-ending codons lower. Fifty-nine synonymous codons (SCs) encoding eighteen amino acids, except for start codon ATG and AGG encoding Trp, were used for a more detailed analysis. Generally, C/G-ending codons (NNCs and NNGs) were more frequent than A/T-ending codons (NNAs and NNTs) (Figure S2). The patterns of SC frequencies were extremely correlative to RSCU values (Figure S2).

To directly compare the SCUB of SCs, the SCUB frequency of a given amino acid encoded by SCs, defined as the ratio of the number of C/G-ending SCs (NNCs/Gs) to that of the A/T-ending SCs (NNAs/Ts), was used for analysis. The SCUB frequencies of the 18 amino acids ranged from 0.805 (Ile in *O. barthii*) to 2.335 (Leu in *O. sativa*) (Figure 1A). The SCUB frequencies of the amino acids except for Ile were all higher than 1 ($p < 1.26 \times 10^{-45}$, Table S1), showing the bias to C/G-ending SCs in both wild and cultivated rice. Moreover, the SCUB frequencies of SCs showed significant differences between cultivated and wild rice, and all 18 amino acids had higher SCUB frequencies in cultivated rice compared with those of wild rice (Figure 1A; Table S2).

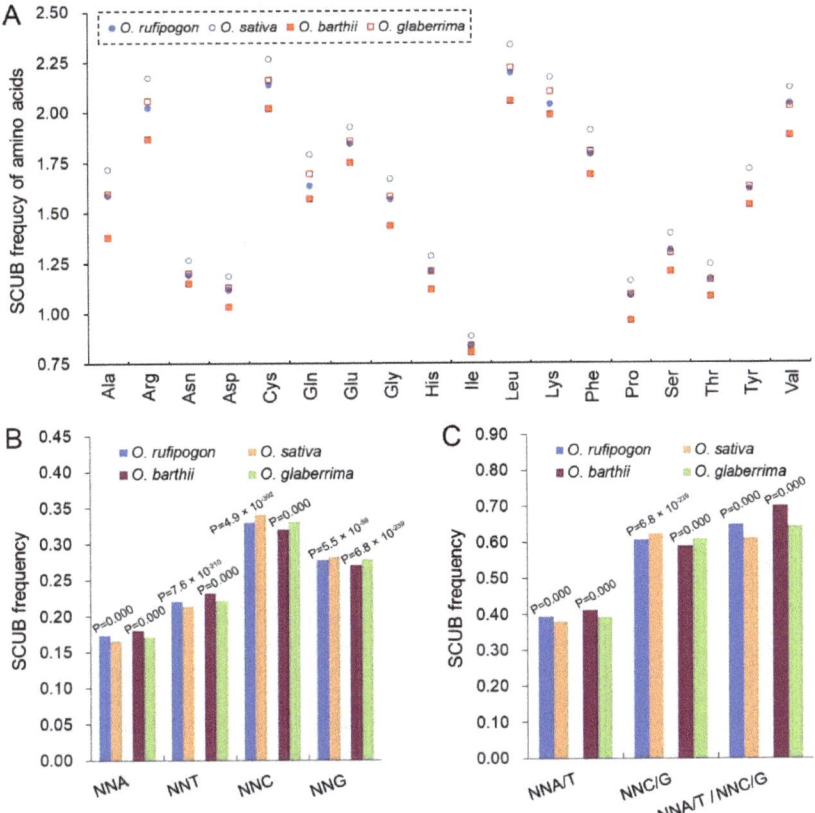

Figure 1. SCUB is heterogeneous between wild and cultivated rice. (**A**): The SCUB frequencies of 18 amino acids, except for Met and Trp, that were defined as the ratios between the number of C/G-ending SCs and A/T-ending SCs. (**B**): The frequencies of NNA, NNT, NNC and NNG codons. NNA, NNT, NNC and NNG: SCs with A, T, C and G as their final base, respectively, N denotes any base. The frequency was calculated as the ratio of the number of all SCs ending with A, T, C or G to the number of all SCs. (**C**): The frequencies of NNA/T and NNC/G codons. NNA/T and NNC/G: SCs with A and T, as well as C and G, as their final base, respectively, N denotes any base. The statistical comparison between wild and cultivated rice was conducted using the Chi-square (χ^2) test of cross-table analysis.

SCUB was further evaluated by the total SCUB frequencies of NNA, NNT, NNC and NNG, which were, respectively, defined as the ratios of the number of all NNAs, NNTs, NNCs and NNGs of 59 SCs to the number of the 59 SCs in all the CDS in a genome. In all four species, NNC and NNG were more pronounced than NNA and NNT; NNC frequency was obviously higher than NNG frequency, while NNA frequency was lower than NNT frequency ($p = 0.000$, Figure 1B). The frequencies of NNC/G were higher than those of NNA/T ($p = 0.000$), and the ratios of NNA/T to NNC/G ranged from 0.609 to 0.699 (Figure 1C). The frequencies of NNA, NNT, NNC or NNG codons differed between wild and cultivated rice ($p = 0.000$, χ^2 test; Figure 1B,C). The frequencies of NNA, NNT and NNA/T were higher in wild rice than in cultivated rice, but the frequencies of NNC, NNG and NNC/G were the converse ($p = 0.000 \sim 5.5 \times 10^{-56}$, χ^2 test). Moreover, SCUB frequencies were different in Asian and African rice (Figure 1B,C; Table S3). NNA, NNT and NNA/T frequencies in both wild and cultivated Asian rice were correspondingly lower than those in African rice, while NNC, NNG and NNC/G frequencies were the converse. Consistent with SCUB frequencies, the indices such as CAI, CBI and Nc were higher in cultivated rice than in wild rice (Table S4), showing the SCUB frequency used here can reflect the alteration of SCUB during rice domestication. These results indicate that although SCUB is different in Asian and African rice, the bias to NNC/Gs became stronger during the domestication in both Asian and African rice.

We further compared the SCUB frequencies among 12 chromosomes (Figure S3). In both wild and cultivated rice, NNA, NNT and NNA/T frequencies were lower than NNC, NNG and NNC/G frequencies across the 12 chromosomes, similar to the difference based on the whole genome. The frequencies were similar from the first to tenth chromosomes; in the eleventh and twelfth chromosomes, the NNA and NNT frequencies were higher, but the NNC and NNG frequencies were lower, and the difference between NNA/T and NNC/G frequencies decreased. The differences in the frequencies among the 12 chromosomes were similar among all the species (Figure S4). Compared with wild rice, cultivated rice had lower NNA and NNT frequencies but higher NNC and NNG frequencies across the chromosomes. Compared with African rice, Asian rice had higher NNC/G frequencies but lower NNA/T frequencies across the chromosomes, as was consistent with the results based on the whole genome, further confirming the bias to C/G-ending codons during the domestication of rice.

2.2. Cultivated Rice Exhibits Stronger Bias to A/T-Ending Synonymous Codons Following the Rise in Intron Number

Intron evolution is a common event in the course of plant evolution and affects SCUB [29,30], so we compared the relation between the SCUB shift frequency and intron number during the domestication of rice. In both wild and cultivated rice, NNC had higher frequencies than NNG in genes harboring no to nine introns, and NNT had higher frequencies than NNA (Figure S5). The frequencies of NNA and NNT gradually increased with the rise in intron number, while the frequencies of NNC and NNG gradually decreased (Figures 2A–D and S5). NNC and NNG frequencies were obviously higher than NNA and NNT frequencies in genes with no or few introns, and the difference became weaker linearly following the rise in intron number. NNT frequencies were even higher than NNG frequencies in genes with more than six introns, and became the highest in genes with nine introns. Consistently, in genes with fewer than nine introns, the frequencies of NNC/G were higher than those of NNA/T ($p = 1.27 \times 10^{-4} \sim 0.001$, t-test); for genes with nine introns, NNA/T and NNC/G were comparable ($p = 0.377$) (Figure 2E,F). Following the rise in intron number, the increase in NNT frequencies was stronger than that of NNA frequencies, and the difference between NNA and NNT frequencies grew; the decrease in NNC frequencies was greater than that of NNG frequencies, and the difference between NNC and NNG frequencies decreased (Figure S5; Table S5).

Compared with wild rice, NNA and NNT frequencies of cultivated rice were lower in genes with no or few introns (except for intronless genes in Asian rice), but higher in genes with more introns (Figure 2A,B; Table S6). For example, the ratios of NNA frequencies between cultivated to wild increased from 0.962 (one intron) to 1.031 (nine introns) in Asian rice, and from 0.917 (no introns) to 1.050 (nine introns) in African rice (Table S6). The difference in NNA frequencies between cultivated and wild rice became more obvious in genes following the rise in intron number (*p* values became smaller, χ^2 test). Conversely, compared with wild rice, NNC and NNG frequencies in cultivated rice were higher in genes with no or few introns (except for intronless genes in Asian rice), but lower in genes with more introns (Figure 2C,D; Table S6). NNA/T and NNC/G frequencies exhibited similar alteration trends to NNA/NNT and NNC/NNG frequencies, respectively, in genes with no to nine introns between cultivated and wild rice, so that the ratios of NNC/G to NNA/T were larger in genes with no or few introns in cultivated rice than in wild rice (except for intronless genes in Asian rice), but were smaller in genes with more introns (Figure 2E,F; Table S6). These results indicate that following the rise in intron number, the bias to NNA/T became stronger, and the bias appeared to be more drastic in cultivated rice than in wild rice.

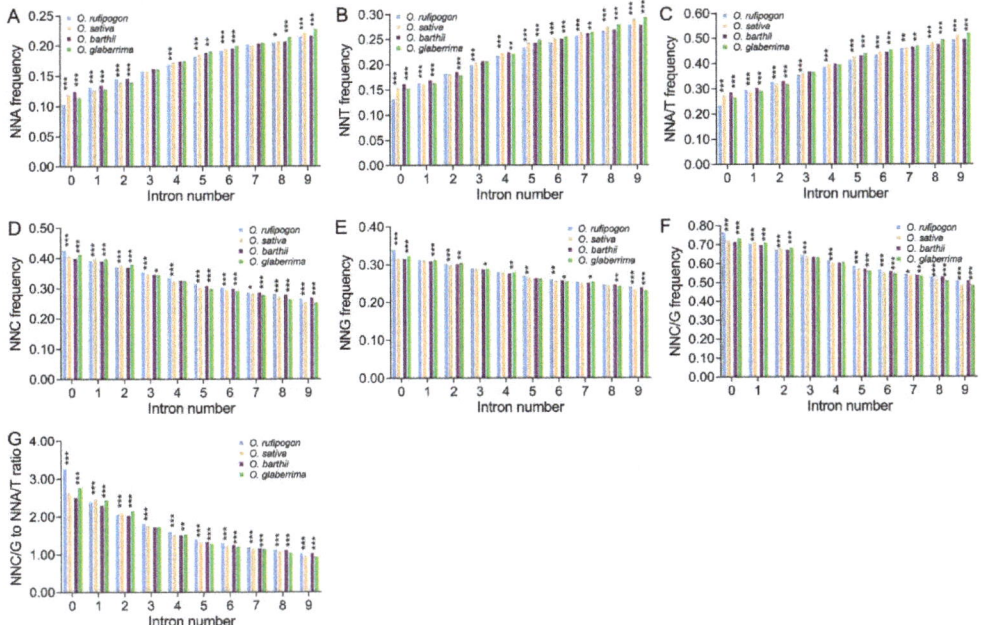

Figure 2. The difference in SCUB frequencies between wild and cultivated rice is associated with intron number. (**A**): The frequencies of A-ending SCs (NNA) in genes with no to nine introns. (**B**): The frequencies of T-ending SCs (NNT) in genes with no to nine introns. (**C**): The frequencies of A- and T-ending SCs (NNA/T) in genes with no to nine introns. (**D**): The frequencies of C-ending SCs (NNC) in genes with no to nine introns. (**E**): The frequencies of G-ending SCs (NNG) in genes with no to nine introns. (**F**): The frequencies of C- and G-ending SCs (NNC/G) in genes with no to nine introns. (**G**): The ratios of C/G-ending SCs to A/T-ending SCs (NNC/G to NNA/T ratios) in genes with no to nine introns. N denotes any base. The difference between wild and cultivated rice was calculated with chi square (χ^2) test of cross-table analysis (*: $p < 0.05$; **: $p < 0.01$; ***: $p < 0.001$); the results are presented in Table S7.

2.3. Cultivated Rice Has Stronger Bias to A/T-Ending Codons in Internal Exons

Given the SCUB patterns based on intron number were different in wild and cultivated rice, the association between SCUB frequency and exon position along the genes was further analyzed. In both wild and cultivated rice, for genes with two to ten exons, the first exons had lower NNA, NNT and NNA/T frequencies but higher NNC, NNG and NNC/G frequencies than the other exons (Figures 3A,B, S6 and S7), resulting in the lowest NNA/T to NNC/G ratios in the first exons (Figure 3C). The frequencies of NNA, NNT, NNC, NNG, NNA/T and NNC/G, as well as the ratios between NNA/T and NNC/G frequencies, were almost comparable in the first exons (CV = 0.010~0.104) (Table S7), showing the SCUB of the first exon remained constant in the genomes of wild and cultivated rice. In the last exons, the frequencies of NNA, NNT and NNA/T were also lower than those of NNC, NNG and NNC/G, but the difference was not as large as that in the first exons. Furthermore, unlike the first exons, the SCUB frequencies did not remain constant across the last exons in genes with two to ten exons (CV = 0.055~0.128). NNA, NNT and NNA/T frequencies gradually increased but NNC, NNG and NNC/G frequencies gradually decreased with the rise in the exon number up to six (CV = 0.046~0.121), and they were comparable in genes with six to ten exons (CV = 0.007~0.042). Thus, the ratios between NNA/T and NNC/G frequencies gradually increased in genes with two to six exons and then remained constant in genes with six to ten exons (Figure 3C). Moreover, in genes with three to ten exons, the frequencies and ratios of the second exons were correspondingly similar to those of the last exons.

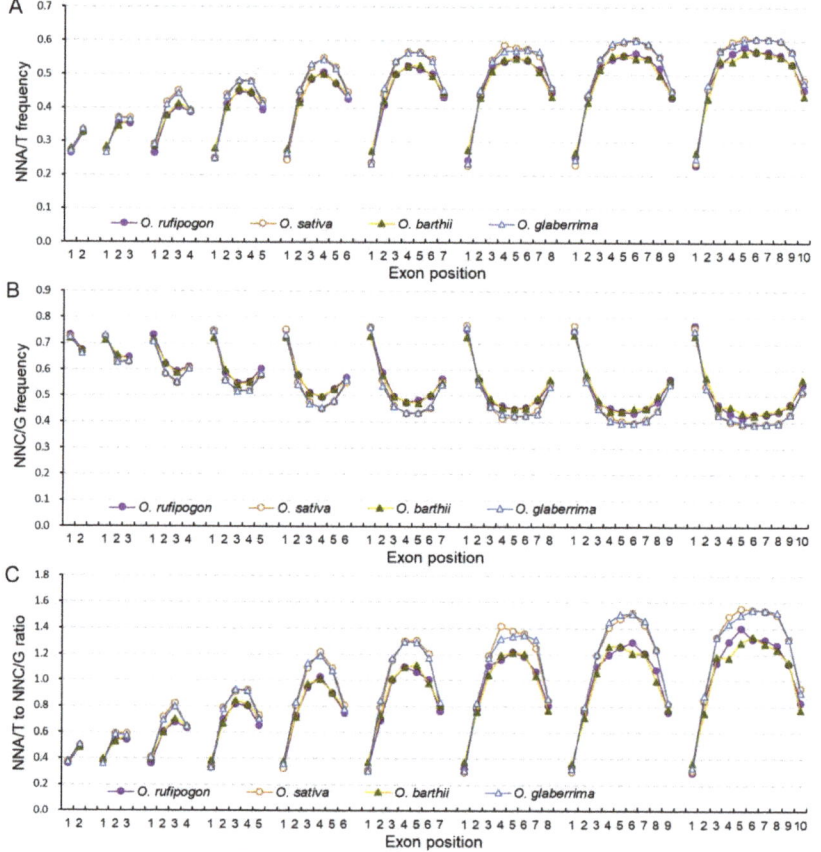

Figure 3. The heterogeneity of SCUB among exons differentiates wild and cultivated rice. (**A**): The frequencies of A- and T-ending SCs (NNA/T) across the exons in genes with two to ten exons. (**B**): The frequencies of C- and G-ending SCs (NNC/G) across the exons in genes with two to ten exons. (**C**): The ratios of A- and T-ending SCs to C- and G-ending SCs (NNA/T to NNC/G ratios) across the exons in genes with two to ten introns. N denotes any base. The difference between wild and cultivated rice was calculated with the chi square (χ^2) test of cross-table analysis; the results are presented in Table S9.

In genes harboring four to ten exons, internal exons had higher NNA, NNT and NNA/T frequencies but lower NNC, NNG and NNC/G frequencies compared with terminal (the first, second and last) exons (Figures 3, S6 and S7). In internal exons, NNA, NNT and NNA/T frequencies increased but NNC, NNG and NNC/G frequencies decreased close to the middle exons. Thus, NNC, NNG and NNC/G frequencies formed in the shape of concave curves ("∪") across the exons, but NNA, NNT and NNA/T frequencies, as well as the ratio between NNA/T and NNC/G frequencies, formed convex curves ("∩"). The curves appeared to be symmetric from the second to the last exons. Moreover, the curve peaks of the NNA, NNT and NNA/T frequencies increased gradually following the rise in exon number, and the increase became quite weak in genes with eight to ten exons; NNC, NNG and NNC/G frequencies showed contrasting patterns (Figures S6 and S7). Across the third to the last-but-one exons, NNA, NNT and NNA/T frequencies were lower than NNC, NNG and NNC/G frequencies in genes with fewer exons, while they became higher in genes with more exons. For internal exons, the shift in NNC and NNT frequencies was more obvious with the rise in exon numbers than in NNG or NNA frequencies. Thus, the differences between NNA and NNT frequencies became larger, while those of NNCs and NNGs became smaller. The profiles of SCUB frequencies based on exon position were similar in wild and cultivated rice, indicating that the stronger bias to A/T-ending codons in the middle exons was maintained during the domestication of rice.

An obvious difference was present in wild and cultivated rice (Figures 3 and S8; Table S8). In the first exons, wild rice had lower NNC, NNG and NNC/G frequencies but higher NNA, NNT and NNA/T frequencies and NNA/T to NNC/G ratios than cultivated rice, while the trend was the converse in the other exons. In both Asian and African rice, the difference in NNA, NNT or NNA/T, as well as in NNC, NNG or NNC/G frequencies, between wild and cultivated rice, was almost constant in the first, second and last exons of the genes with two to ten exons; the difference, however, became larger in the internal exons following the rise in exon number, being much closer to that in the middle exons. On the other hand, the SCUB frequencies across the exons in genes with two to ten exons were almost the same in Asian and African wild rice, and the curves of SCUB frequencies appeared to almost coincide with each other, as was also found between Asian and African cultivated rice. These results indicate that the heterogeneity of SCUB frequencies across exons, as well as the stronger preference for A/T-ending codons in internal exons after domestication, is the same in both Asian and African rice.

2.4. SCUB Shift in Cultivated Rice Is Associated with DNA Methylation-Mediated Conversion of Cytosine to Thymine

DNA methylation serves as a source of nucleotide substitution because methylated cytosine (5^mC) can be converted into thymine [33]. To investigate whether SUCB shift during rice domestication is associated with DNA methylation-mediated nucleotide substitution, we evaluated the frequencies of NNA and NNG with different nucleotides in the second position (conversion of C to T in the antisense strand causes conversion G to A), as well as the frequencies of NNT and NNC with different nucleotides in the first position of the downstream codon (NT|N and NC|N) (conversion of C to T in the sense strand).

Generally, NAA, NCA, NGA and NTA frequencies were slightly lower than NAG, NCG, NGG and NTG frequencies ($p = 0.055 \sim 0.130$, t-test) (Figure 4A). NCA frequencies were higher than other NNA frequencies; NCG frequencies were lower than NAG, higher than NGG frequencies, but similar to NGG frequencies. NCA/NCG ratios mirroring the methylation-mediated conversion of C to T in the antisense strand were significantly higher than NAA/NAG, NGA/NGG and NTA/NTG ratios (Figure 4C). NNT|G frequencies were drastically higher than NT|A, NT|C and NT|T frequencies and NC|G frequencies were also higher than NC|A, NC|C and NC|G frequencies; this resulted in drastically higher NT|G/NC|G ratios mirroring the methylation-mediated conversion of C to T in the sense strand than for the NT|A/NC|A, NT|C/NC|G and NT|T/NC|T ratios (Figure 4B,D). These data show that C in the second position of the codons and G in the first position of the next codons had a stronger effect in increasing the bias of A and T in the third position of the codons, which indicates the association between methylation-mediated nucleotide conversion and SCUB.

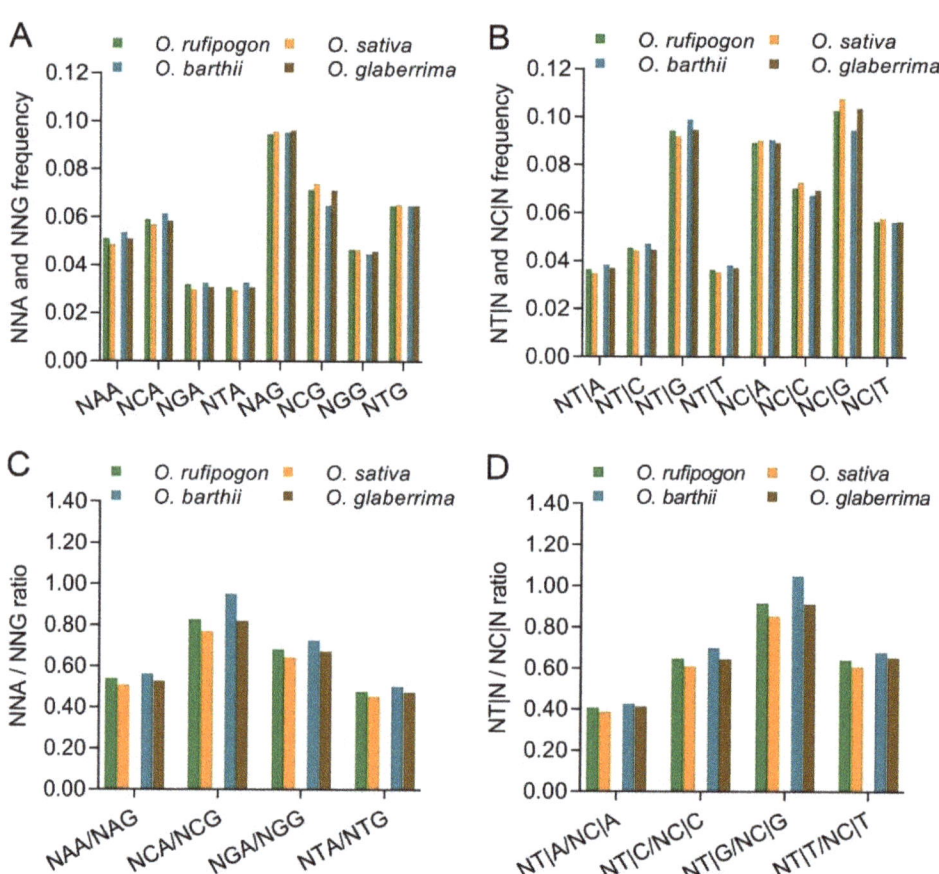

Figure 4. DNA methylation-driven conversion of cytosine to thymine is associated with the difference in SCUB frequencies between wild and cultivated rice. (**A**): The frequencies of NNA and NNG indicating the effect of the second codon nucleotide on the conversion of C to T at the third position in the antisense strand. (**B**): The frequencies of NT|N and NC|N indicating the effect of the first nucleotide of next codons on the conversion of C to T at the third position in the sense strand. (**C**): The ratios of NNA to NNG. (**D**): The ratios of NT|N to NC|N. NNA and NNG: SCs with A and G as the final bases and A, T, C and G at the second position, N denotes any base. NT|N and NC|N: the triple nucleotide combinations with C and T as the final bases and A, T, C and G at the first position of the next codons. The difference in wild and cultivated rice was calculated with chi square (χ^2) test of cross-table analysis, and the results are presented in Table S10.

In both Asian and African rice, the frequencies of the four types of NNA codons were lower in cultivated rice than in wild rice; in contrast, among the four types of NNG codons, NCG frequencies were obviously higher in cultivated rice compared with wild rice, but the frequencies of the other three types of NNG codons were substantially less different in the wild and cultivated rice (Figure 4A; Table S9). Thus, the ratios of the four NNA/NNG combinations were lower in cultivated rice than in wild rice, among which the ratios of NCA/NCG had the most pronounced difference between the wild and cultivated rice.

Consistently, NT|A, NT|T, NT|C and NT|G frequencies were higher in wild rice than in cultivated rice; for NC|N combinations, NC|G frequencies were significantly lower in wild rice than in cultivated rice, and the difference in the frequencies of NC|A, NC|C and NC|T in wild and cultivated rice was not as remarkable as the NC|G frequencies.

To further analyze the effect of the second nucleotide on DNA methylation-mediated SCUB, the frequencies of C/G-ending SC pairs of amino acids with the same nucleotides in the first and second positions were calculated (Figure 5A). The ratios of NCA/NCG (encoding alanine, proline, serine and threonine) varied from 0.595 to 1.193, significantly higher than those of N(A/G/T)A/N(A/G/T)G (encoding arginine, glycine, glutamic acid, glutamine, leucine, lysine and valine) (0.264 to 0.636 except for glycine (0.858~0.976)) (p = 0.009~0.023 and 0.001~0.004 without glycine; t-test). Moreover, the ratios of both NCA/NCG and N(A/G/T)A/N(A/G/T)G combinations of these amino acids were lower in cultivated rice than in wild rice (Table S10), and the difference between wild and cultivated rice was more significant in African than in Asian rice (p = 0.002 in African rice and 0.501 in Asian rice, t-test). On the other hand, the first nucleotide G of the adjacent codons also caused lower ratios of NNT|G/NNC|G in cultivated rice than in wild rice (Figure 5B).

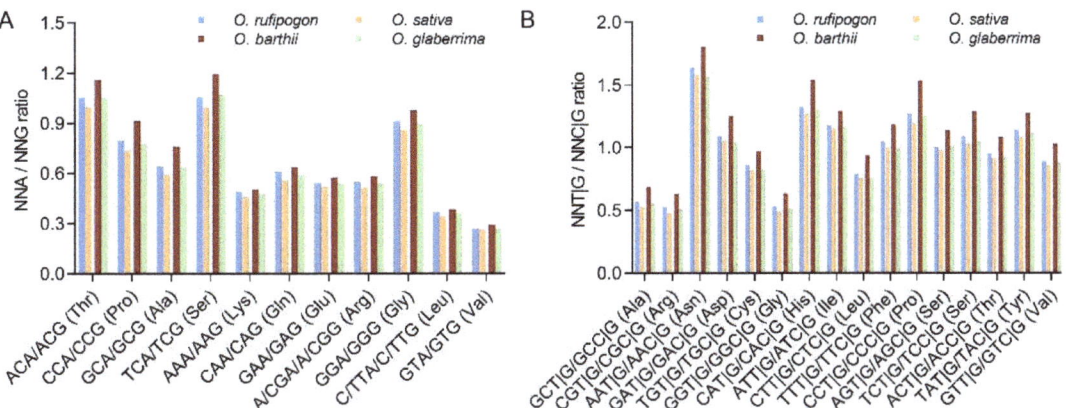

Figure 5. The effect of the adjacent nucleotides on the ratios of A-ending SCs to G-ending SCs, as well as T-ending SCs to C-ending SCs, encoding the given amino acids. (**A**): The effect of the second nucleotide on the codons. (**B**): The effect of the first nucleotide on the next codons. The statistical comparison was conducted using chi square (χ^2) test, and the results are presented in Table S11. The difference between the ratios of Ala, Pro, Ser, Thr and the ratios of Arg, Gln, Glu, Gly, Leu, Lys and Val in a species was calculated with two-sample t-test (p < 0.05).

The association between DNA methylation and the heterogeneity of SCUB based on introns was further analyzed. The ratios of the four NNA/NNG combinations and the four NT|N/NC|N combinations improved following the increase in intron number (Figures 6A,B and S9). The increase in NCA/NCG and NT|G/NC|G ratios was much sharper than that in the ratios of the other NNA/NNG and NT|N/NC|N combinations (Figure S9), showing DNA methylation-associated SCUB was more preferential in genes containing more introns. Moreover, when compared to wild and cultivated rice, the ratios of both NCA/NCG and NT|G/NC|G were comparable with genes with fewer introns, but were higher in cultivated rice than in wild rice for genes with more introns (Figure 6A,B; Table S11).

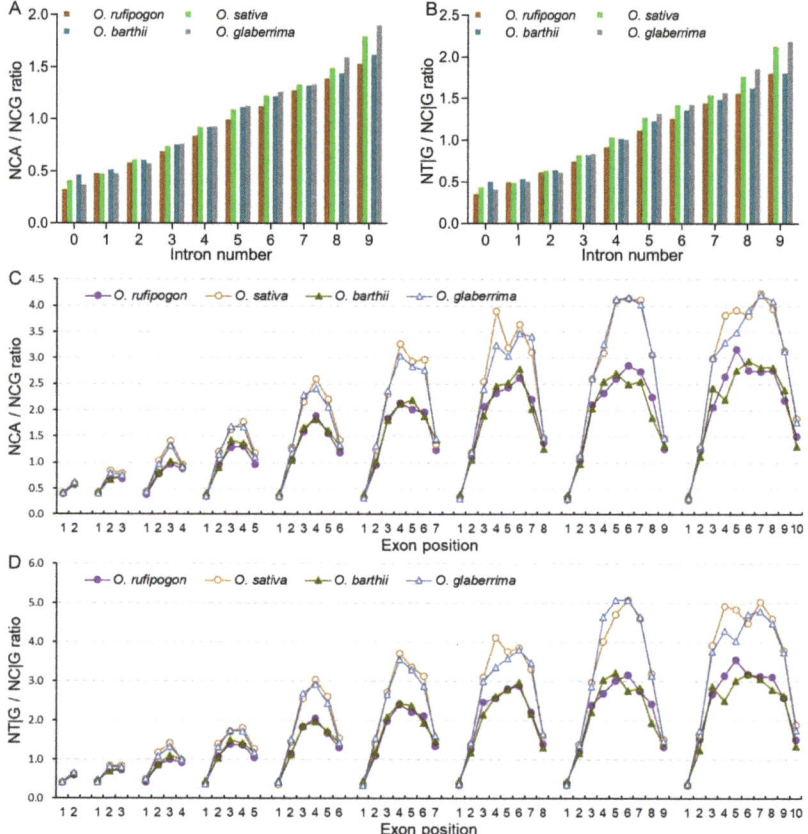

Figure 6. The association between DNA methylation and SCUB heterogeneity based on intron number and exon position reveals the difference between wild and cultivated rice. (**A**): The ratios of NCA to NCG in genes with no to nine introns. (**B**): The ratios of NT|G to NC|G in the genes with no to nine introns. (**C**): The ratios of NCA to NCG across the exons in genes with two to ten exons. (**D**): The ratios of NT|N to NC|N across the exons in genes with two to ten introns. The difference was calculated via chi square (χ^2) test of cross-table analysis, and the results are presented in Table S12.

The ratios of the four NAA/NNG combinations were almost the same in the first exons of genes with two to ten exons (Figures 6C and S10). In the last exons, the NCA/NCG ratios gradually improved following the increase in the exon number, but the ratios of the other three NNA/NNG combinations were comparable to those in the first exons (Figure S10). The second exons of genes with three to ten exons exhibited similar patterns to the last exons. As for genes with four to ten exons, the ratios of the four NNA/NNG combinations in the internal (the third to the last but one) exons were higher than those in the terminal exons, and gradually increased to be closer to the middle exons, resulting in "∩" curves. The ratios of the four NNA/NNG combinations in the internal exons gradually increased following the rise in exon number, of which the NCA/NCG ratios drastically increased up to more than three in genes with eight to ten exons, but the ratios of NAA/NAG, NGA/NGG and NTA/NTG weakly increased to approximately one (Figure S10). The ratios of NT|A/NC|A, NT|C/NC|C, NT|G/NC|G and NT|T/NC|T among the exons exhibited similar "∩" profiles, and the NT|G/NC|G ratios in the internal exons were more predominant than the other NT|N/NC/N combinations; the increase in the ratios of NT|N/NC|N combinations in the internal exons was more obvious than that of NNA/NNG combinations following the rise in exon number (Figure S11). In both African and Asian rice, the ratios of NAA/NAG, NGA/NGG and NTA/NTG, as well as NT|A/NC|A, NT|C/NC|C and NT|T/NC|T, across the exons in genes with two to ten exons were either comparable to wild and cultivated rice or slightly higher in cultivated rice than in wild rice (Figures S10 and S11). However, the NCA/NCG and NT|G/NC|G ratios in the internal exons were remarkably higher in cultivated rice compared with those in wild rice, and the difference became larger close to the middle exons and more significant following the rise in exon number; the ratios in the first two exons and the last exons were similar in wild and cultivated rice (Figure 6C,D). Moreover, in both wild and cultivated rice, the ratios of the four NNA/NNG and the four NT|N/NC|N combinations across the exons in genes with two to ten exons were similar in African and Asian rice, showing the similar SCUB patterns in wild and cultivated rice.

The association between DNA methylation and SCUB based on exons was further confirmed by C- and G-ending SC pairs of amino acids sharing the same nucleotides in their first and second positions (Figures S12 and S13). The ratios of NCA/NCG combinations were higher than those of N(A/G/T)A/N(A/G/T)G combinations, and the difference became more remarkable following the increase in intron number (Figure S12A–D). Compared with wild rice, cultivated rice had higher ratios of NCA/NCG combinations in genes with more exons (Figure S12E,F). On the other hand, the ratios of NCA/NCG and N(A/G/T)A/N(A/G/T)G combinations among exons exhibited similar "∩" patterns (Figure S13; Table S12). The ratios of the NCA/NCG combinations in internal exons were significantly higher than those of the N(A/G/T)A/N(A/G/T)G combinations; among the N(A/G/T)A/N(A/G/T)G combinations, the ratios of GGA/GGG of glycine in the internal exons were higher than those of the other combinations. The ratios of NCA/NCG combinations in internal exons were obviously higher in cultivated rice than in wild rice, and the difference became more drastic close to the middle exons, as well as following the rise in exon number (Figure S14; Table S13).

2.5. SCUB Mirrors the Effect of Domestication

Phylogenetic analysis was conducted to outline the association between SCUB and the domestication of rice. The cluster based on both SCUB frequencies and RUSC values of the 59 SCs indicate that African wild rice *O. barthii* and the other three are clustered into two distinct clades, and in the latter clade, African cultivated rice *O. glaberrima* and Asian wild rice *O. rufipogon* are grouped into a sub-clade, differentiated from Asian cultivated rice *O. sativa* (Figures 7A and S15A). This cladistic analysis was confirmed by PCA based on SCUB frequencies and RUSC values (Figures 7C and S15B). The scatter plots of the first and second principal components (PC1 and PC2) distinguish *O. glaberrima* and *O. rufipogon* from *O. barthii* and *O. sativa*. Moreover, in both African and Asian rice, the scatter points of cultivated rice positioned at the top right corner of wild rice, show that SCUB altered in a similar manner during the domestication of Asian and African rice. The cluster using the SCUB frequencies based on exon position and intron number differentiates wild and cultivated rice in a different manner (Figures 7E and S16A). Wild and cultivated rice are clustered into two groups in the PC1–PC2 plot (Figures 7G and S16C). Especially, the PCA using SCUB frequencies based on exon position show that both Asian and African wild rice were close to each other, as were both Asian and African cultivated rice (Figure 7G). These data show that a similar alteration of SCUB in Asian and African rice during their domestication was closely associated with intron. The cluster and PCA using the frequencies of methylation-associated codon combinations obtained similar results to the SCUB frequencies of the 59 SCs (Figure 7B,D), and the analysis using methylation-associated frequencies based on exon position and intron number also differentiated wild and cultivated rice (Figures 7F,H and S16B). Correlation analysis indicates that there are similar correlations with wild and cultivated rice, as well as with Asian and African rice based on the SCUB frequencies of the 59 SCs (Figure S17A–D). As for the SCUB frequencies based on exon position, the correlation of wild and cultivated species was similar between Asian and African rice, and weaker than the correlation between Asian and African wild rice and between Asian and African cultivated rice (Figure S17E–H). Together with the data from the phylogenic tree and the PCA, SCUB appears to reflect the domestication of rice and the association of DNA methylation to SCUB alteration.

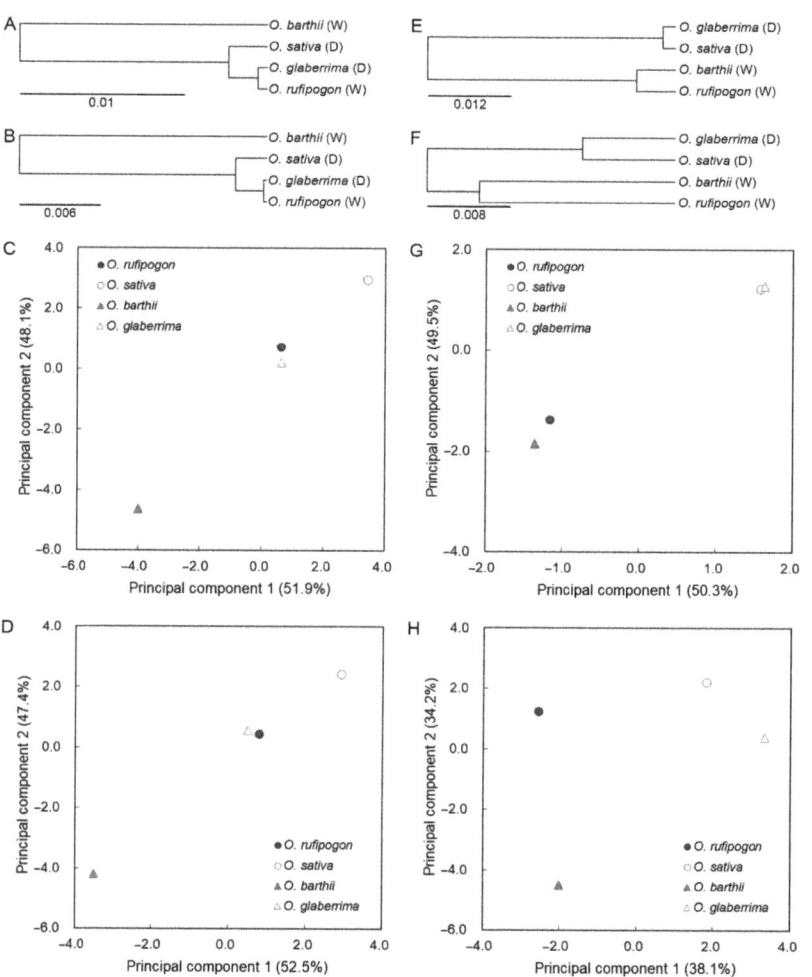

Figure 7. The cluster and principal component analyses of SCUB reflect the domestication of rice. (**A**): A cluster tree based on SCUB frequencies of 59 SCs encoding 18 amino acids. (**B**): A cluster tree based on the ratios of DNA methylation-associated codon combinations. (**C**): A scatter plot of PC1 and PC2 score coefficients from PCA based on SCUB frequencies in 59 SCs encoding 18 amino acids. (**D**): A scatter plot of PC1 and PC2 score coefficients from PCA based on the ratios of DNA methylation-associated codon combinations. (**E**): A cluster tree based on SCUB frequencies across the exons in genes with two to ten exons. (**F**): A cluster tree based on the ratios based on the ratios of DNA methylation-associated codon combinations. (**G**): A scatter plot of PC1 and PC2 score coefficients from PCA based on SCUB frequencies across the exons in genes with two to ten exons. (**H**): A scatter plot of PC1 and PC2 score coefficients from PCA based on the ratios of DNA methylation-associated codon combinations.

3. Discussion

As a type of genetic variation, SCUB exhibits diverse profiles in the nuclear genomes of land plants, and can mirror the evolution of plants [29]. Here, both wild and cultivated rice show a preference for C/G-ending SCs (Figure 1). The domestication of wild rice caused a genome-scale genetic variation including nucleotide substitution [3–5], and the nucleotide substitution may have changed the SCUB. We found that in both Asian and African rice, the cultivated rice showed more preference for C/G-ending SCs than did wild rice (Figure 1). This demonstrates that domestication indeed affects SCUB and promotes the preference for C/G-ending SCs in cultivated rice. Codon usage bias correlates with the trend of GC content variations [34], so it was proposed that codon usage bias may be driven by GC content changes [35,36]. On other hand, GC-rich regions appear to be prone to homologous recombination, a force of genetic variation, leading to biased gene conversion [37]; this increases the GC content across transcripts [38] and affects codon bias because GC-rich codons tend to be over-represented in ORFs, especially in higher organisms [39]. Thus, homoeologous recombination may partially account for the bias to C/G-ending SCs, and this bias in turn could have promoted homoeologous recombination and genomic variation during the domestication of rice.

Intron gain and loss is a typical genetic event in eukaryotic genomes [40] and causes nuclear substitution in exon sequences; the process is commonly preferential to a lower GC content [41]. Intron-rich genes suffer from stronger selection pressure, so they tend to retain A/T-ending codons [42,43]. Consistent with this, in both wild and cultivated rice, the bias to A/T-ending codons appears to be more pronounced with the rise in intron number (Figure 2). On the other hand, as a type of sequence insertions and deletions (indels), intron evolution could induce nucleotide substitution in adjacent exons, because indels cause nucleotide substitution in several hundred bases of flanking sequences [25,44]. Moreover, a higher bias to A/T-ending codons in internal exons, compared with that in terminal exons, is present in the genes of rice, and the bias in internal exons is more distinguishable in genes with more introns (Figure 3), consistent with the increase in the bias to A/T-ending codons following the increase in the number of introns (Figure 2). These data indicate that the internal exons may be the key point of genetic variation in gene sequences, and are largely responsible for the effect of introns on SCUB. In comparison with wild rice, cultivated rice has higher frequencies of A/T-ending codons in genes with more introns and in the internal exons of the genes (Figures 2 and 3). Especially, the patterns of SCUB frequencies across the exons are almost the same, both between Asian and African wild rice and between Asian and African cultivated rice. Although the association between intron evolution and domestication was not addressed, our findings find new characteristics of genetic variation and suggest that intron evolution may have played an important role in SCUB during the domestication of rice.

Genome-scale DNA methylation was found in both wild and cultivated rice [31]. The methylated cytosine can be converted to thymine [32], so DNA methylation is a source of SNP formation [45]. Consistent with this, the ratios of NCA/NCG and NT|G/NC|G are higher than those of the other NXA/NXG (X = A, G and T) and NT|X/NC|X (X = A, C and T) (Figure 4). Furthermore, the difference in the ratios of both NCA/NCG and NT|G/NC|G in wild and cultivated rice is more pronounced than that of the other NXA/NXG or NT|X/NC|X (Figures 4–6), and coincides with the difference in SCUB frequencies in wild and cultivated rice. This indicates that DNA methylation-driven nucleotide substitution is associated with the SCUB shift in the domestication of rice. In this respect, the ratios of both NCA/NCG and NT|G/NC|G should be higher in cultivated rice than in wild rice. However, these ratios were lower in cultivated than in wild rice (Figure 4). Given that the methylation density and average methylation level of all cytosines in the genome of cultivated rice are higher than those in wild rice [31], it could be speculated that the decrease in DNA methylation-mediated SCUB alteration may result from the conversion of T to C, so as to produce methylation sites, thereby increasing genome-wide DNA methylation levels. On the other hand, the ratios of both NCA/NCG and NT ǀ G/NC ǀ G in the internal exons and intron-rich genes of cultivated rice were higher than those in wild rice (Figure 6). In the genes with body methylation, the internal exons have higher DNA methylation levels than do the terminal exons [46], so introns have a positive effect on DNA methylation-mediated bias to A/T-ending SCs. Therefore, there may be two effects on the association between SCUB and DNA methylation during the domestication of rice: (i) the bias to C- and G-ending SCs contributes to a higher DNA methylation level, and (ii) the higher DNA methylation level results in the bias to A and T-ending SCs by DNA methylation, driving C to T conversion. The synonymous variation seems to be a nonrandom event to orchestrate the domestication and evolution of plants. This is an interesting point to be investigated in the future. Moreover, epigenetic variation such as DNA methylation governs the balance of gene expression [47]. Given the role of SCs in transcription efficiency, mRNA stability, translational efficiency and accuracy [9–13], a shift in SCUB may be detrimental to the phenotype of cultivated rice. Thus, the substitution between SCs can also be used for mining genes and excellent allelic variation governing agricultural traits in rice and other crops.

In summary, our work found that SCUB shifted during the domestication of rice, and that the shift in SCUB exhibits similar characteristics in Asian and African rice, as illustrated by cluster analysis and PCA (Figure 7); this indicates that SCUB and genetic variation is not a random event and provides a new insight into the genomic variation during domestication. Nucleotide substitution polymorphism is an important genetic force in plant evolution and crop improvement. Given the non-neutral effect of synonymous codons within the cells [9–14], SCUB may have a detrimental effect on the improvement of agricultural traits in crops; it is, therefore, necessary to focus more attention on the genetic dissection of agricultural traits in the future.

4. Materials and Methods

4.1. Genome Sequences and Codon Count

Asian wild rice *Oryza rufipogon* and cultivated rice *Oryza sativa*, and African wild rice *Oryza barthii* and cultivated rice *Oryza glaberrima*, were used for analysis. Their genome sequences were downloaded from the EnsemblPlants database (http://plants.ensembl.org/info/data/ftp/index.html (accessed on 16 April 2021)). The coding sequences (CDS) of annotated genes were extracted according to the GFF3 gene-annotation files that were also downloaded from the EnsemblPlants genome database. For genes with more than one transcript type, the first transcript sequence was used for analysis. Any extracted CDS without a length that was a multiple of three, containing N, with start codon not ATG, stop codons not TAA, TAG and TGA were excluded. Codons interrupted by an intron between the first and the second nucleotide were treated as belonging to the downstream exon, while those interrupted between the second and the third nucleotides were deemed to belong to the upstream exon.

4.2. Calculation of SCUB Indices

Using CodonW 1.4.2 software (https://sourceforge.net/projects/codonw/ (accessed on 27 February 2013)), all filtered coding sequences (CDS) in the genome of a species were used to calculate the relative synonymous codon usage (RSCU), codon adaptation index (CAI) and other indices of SCUB.

4.3. Calculation of SCUB Frequency

We adopted SCUB frequencies to measure the bias of SCs. The frequency of each of the 61 amino acid-encoding codons was calculated using the ratio of the number of this codon to the number of all codons of the filtered CDS in a species. In total, 59 SCs encoding 18 amino acids, except for Met and Trp, were used to calculate SCUB frequency. The SCUB frequency of an amino acid encoded by SCs was defined as the ratio of the number of C- and G-ending SCs to the number of A- and T-ending SCs of this amino acid. Total SCUB frequency was defined as the ratio of the number of all SCs with A, T, C or G at the third position (abbreviated as NNA, NNT, NNC or NNG) to the number of all codons represented in the filtered CDS, except for start codon, stop codons and TGG.

Methylated cytosine (5^mC) can be converted into thymine [33]; methylation is mainly present in the C of CpG, so the conversion of 5^mC results in TpG in the sense strand and CpA in the antisense strand. The conversion of NCG to NCA (the second to third position) and NC|G to NT|G (the third-next codon's first position) can lead to the bias to A- and T-ending codons. Thus, the ratios of the NXA number to the NXG number (X = A, T, C, or G) can indicate the effect of the second nucleotide on the conversion from G and C to A and T at the third position, respectively; in addition, the ratios of the NG|X number to the NC|X number (X = A, T, C, or G) can indicate the effect of the first nucleotide of the next codon on the conversion from G and C to A and T at the third position, respectively. Based on this, the association between DNA methylation and SCUB was evaluated by comparing the difference in the NCA/NCG ratio with the NAA/NAG, NGA/NGG and NTA/NTG ratios, and the difference in the NT|G/NC|G ratio with the NT|A/NC|A, NT|C/NC|C and NT|T/NC|T ratios.

4.4. Cluster Analysis and Principal Component Analysis

Employing the average linkage method and the distance measurement of correlation in Minitab 17 statistical software, cluster analysis was performed using the SC frequencies and RSCU values of 59 SCs, the SCUB frequencies based on exon position, the number of introns and the frequencies of codon combinations associated with DNA methylation. The dendrogram was generated on the basis of similarity. The data for cluster analysis were also subjected to principal component analysis in JMP 13 software with default parameters. The factor score coefficients given by the first two principal components were used to generate the scatter plot diagrams.

4.5. Statistical Analysis

The difference in the frequencies of NNA, NNT, NNC and NNG of a species and their difference in wild and cultivated rice were calculated using the chi square (χ^2) test of the cross-table analysis, and the NNA, NNT, NNC and NNG numbers were used for the calculations. The same statistical analysis was performed to compare the difference in the frequencies of NNA/T (A- and T-ending SCs) and NNC/G (C- and G-ending SCs). The χ^2 test of the cross-table analysis was conducted to evaluate the difference in SCUB frequency related to the third nucleotide position concerning DNA methylation, where the difference between NCA/NCG and NXA/NXG ratios (X = A, G or T) was analyzed using the numbers of NCA, NCG, NXA and NXG; the difference between NC|G/NG|G and NC|X/NG|X ratios (X: A, C, or T, respectively) was analyzed using the numbers of NC|G, NG|G, NC|X and NG|X. The difference between NXC and NXG (X: A, C, G or T, respectively) SCs of an amino acid encoding by G- and C-ending SCs was measured with the χ^2 test using the numbers of NXC and NXG. The difference in the SCUB frequencies, based on intron number and exon position in wild and cultivated rice, was calculated via the two-sample t-test. Fluctuation was assessed by the coefficient of variation (CV), which was calculated as the ratio of standard deviation to mean.

Supplementary Materials: The following supporting information can be downloaded at: https://www.mdpi.com/article/10.3390/ijms232112860/s1.

Author Contributions: M.W. designed and conceived the work. G.X., J.Z., Z.H., T.W., Y.L. (Yingchun Li), Y.L. (Yajing Li) and Y.W. analyzed the data. G.X. and M.W. conducted the statistical analysis. M.W. and G.X. wrote the paper. All authors have read and agreed to the published version of the manuscript.

Funding: This work was supported by the National Natural Science Foundation of China (31870242, 32170297), the National Transgenic Project (2020ZX08009-11B), and the Key Project of Natural Science Foundation of Shandong (ZR2021ZD32).

Institutional Review Board Statement: Not applicable.

Informed Consent Statement: Not applicable.

Data Availability Statement: Not applicable.

Conflicts of Interest: The authors declare no conflict of interest.

References

1. Fuller, D.; Sato, Y.; Castillo, C.; Qin, L.; Weisskopf, A.; Kingwell Banham, E.; Song, J.; Ahn, S.; Van Etten, J. Consilience of genetics and archaeobotany in the entangled history of rice. *Archaeol. Anthropol. Sci.* **2010**, *2*, 115–131. [CrossRef]
2. Wing, R.A.; Purugganan, M.D.; Zhang, Q. The rice genome revolution: From an ancient grain to Green Super Rice. *Nat. Rev. Genet.* **2018**, *19*, 505–517. [CrossRef] [PubMed]
3. Huang, X.; Kurata, N.; Wei, X.; Wang, Z.-X.; Wang, A.; Zhao, Q.; Zhao, Y.; Liu, K.; Lu, H.; Li, W.; et al. A map of rice genome variation reveals the origin of cultivated rice. *Nature* **2012**, *490*, 497–501. [CrossRef] [PubMed]
4. Zhao, Q.; Feng, Q.; Lu, H.; Li, Y.; Wang, A.; Tian, Q.; Zhan, Q.; Lu, Y.; Zhang, L.; Huang, T.; et al. Pan-genome analysis highlights the extent of genomic variation in cultivated and wild rice. *Nat. Genet.* **2018**, *50*, 278–284. [CrossRef] [PubMed]
5. Kou, Y.; Liao, Y.; Toivainen, T.; Lv, Y.; Tian, X.; Emerson, J.J.; Gaut, B.S.; Zhou, Y. Evolutionary Genomics of Structural Variation in Asian Rice (*Oryza sativa*) Domestication. *Mol. Biol. Evol.* **2020**, *37*, 3507–3524. [CrossRef] [PubMed]
6. Li, C.; Zhou, A.; Sang, T. Rice domestication by reducing shattering. *Science* **2006**, *311*, 1936–1939. [CrossRef]
7. Nei, M.; Gojobori, T. Simple methods for estimating the numbers of synonymous and nonsynonymous nucleotide substitutions. *Mol. Biol. Evol.* **1986**, *3*, 418–426.
8. King, J.; Jukes, T. Non-Darwinian evolution. *Science* **1969**, *165*, 788–798. [CrossRef]
9. Marais, G.; Mouchiroud, D.; Duret, L. Does recombination improve selection on codon usage? Lessons from nematode and fly complete genomes. *Proc. Natl. Acad. Sci. USA* **2001**, *98*, 5688–5692. [CrossRef]
10. Warnecke, T.; Hurst, L. Evidence for a trade-off between translational efficiency and splicing regulation in determining synonymous codon usage in Drosophila melanogaster. *Mol. Biol. Evol.* **2007**, *24*, 2755–2762. [CrossRef]
11. Zhang, G.; Hubalewska, M.; Ignatova, Z. Transient ribosomal attenuation coordinates protein synthesis and co-translational folding. *Nat. Struct. Mol. Biol.* **2009**, *16*, 274–280. [CrossRef] [PubMed]
12. Presnyak, V.; Alhusaini, N.; Chen, Y.; Martin, S.; Morris, N.; Kline, N.; Olson, S.; Weinberg, D.; Baker, K.; Graveley, B.; et al. Codon optimality is a major determinant of mRNA stability. *Cell* **2015**, *160*, 1111–1124. [CrossRef] [PubMed]
13. Tuller, T.; Carmi, A.; Vestsigian, K.; Navon, S.; Dorfan, Y.; Zaborske, J.; Pan, T.; Dahan, O.; Furman, I.; Pilpel, Y. An evolutionarily conserved mechanism for controlling the efficiency of protein translation. *Cell* **2010**, *141*, 344–354. [CrossRef] [PubMed]
14. Shen, X.; Song, S.; Li, C.; Zhang, J. Synonymous mutations in representative yeast genes are mostly strongly non-neutral. *Nature* **2022**, *606*, 725–731. [CrossRef] [PubMed]
15. Akashi, H.; Eyre-Walker, A. Translational selection and molecular evolution. *Curr. Opin. Genet. Dev.* **1998**, *8*, 688–893. [CrossRef]
16. Akashi, H. Gene expression and molecular evolution. *Curr. Opin. Genet. Dev.* **2001**, *11*, 660–666. [CrossRef]
17. Guo, F.B.; Yuan, J.B. Codon usages of genes on chromosome, and surprisingly, genes in plasmid are primarily affected by strand-specific mutational biases in Lawsonia intracellularis. *DNA Res.* **2009**, *16*, 91–104. [CrossRef]
18. Wang, Z.; Lucas, F.; Qiu, P.; Liu, Y. Improving the sensitivity of sample clustering by leveraging gene co-expression networks in variable selection. *BMC Bioinform.* **2014**, *15*, 153. [CrossRef]
19. Knowles, D.G.; McLysaght, A. High Rate of Recent Intron Gain and Loss in Simultaneously Duplicated *Arabidopsis* Genes. *Mol. Biol. Evol.* **2006**, *23*, 1548–1557. [CrossRef]
20. Sharpton, T.J.; Neafsey, D.E.; Galagan, J.E.; Taylor, J.W. Mechanisms of intron gain and loss in Cryptococcus. *Genome Biol.* **2008**, *9*, R24. [CrossRef]
21. Tarrío, R.; Ayala, F.J.; Rodríguez-Trelles, F. Alternative splicing: A missing piece in the puzzle of intron gain. *Proc. Natl. Acad. Sci. USA* **2008**, *105*, 7223–7228. [CrossRef] [PubMed]
22. Stoltzfus, A. Molecular Evolution: Introns Fall into Place. *Curr. Biol.* **2004**, *14*, R351–R352. [CrossRef] [PubMed]
23. Rodríguez-Trelles, F.; Tarrío, R.; Ayala, F.J. Origins and Evolution of Spliceosomal Introns. *Annu. Rev. Genet.* **2006**, *40*, 47–76. [CrossRef] [PubMed]
24. Choi, K.; Weng, M.-L.; Ruhlman, T.A.; Jansen, R.K. Extensive variation in nucleotide substitution rate and gene/intron loss in mitochondrial genomes of Pelargonium. *Mol. Phylogenet. Evol.* **2021**, *155*, 106986. [CrossRef]
25. Tian, D.; Wang, Q.; Zhang, P.; Araki, H.; Yang, S.; Kreitman, M.; Nagylaki, T.; Hudson, R.; Bergelson, J.; Chen, J.-Q. Single-nucleotide mutation rate increases close to insertions/deletions in eukaryotes. *Nature* **2008**, *455*, 105–108. [CrossRef] [PubMed]
26. Chen, J.-Q.; Wu, Y.; Yang, H.; Bergelson, J.; Kreitman, M.; Tian, D. Variation in the Ratio of Nucleotide Substitution and Indel Rates across Genomes in Mammals and Bacteria. *Mol. Biol. Evol.* **2009**, *26*, 1523–1531. [CrossRef]
27. Hershberg, R.; Petrov, D.A. Selection on Codon Bias. *Annu. Rev. Genet.* **2008**, *42*, 287–299. [CrossRef]
28. Coulombe-Huntington, J.; Majewski, J. Characterization of intron loss events in mammals. *Genome Res.* **2007**, *17*, 23–32. [CrossRef]
29. Qin, Z.; Cai, Z.; Xia, G.; Wang, M. Synonymous codon usage bias is correlative to intron number and shows disequilibrium among introns in plants. *BMC Genom.* **2013**, *14*, 56. [CrossRef] [PubMed]
30. Xu, W.; Li, Y.; Li, Y.; Liu, C.; Wang, Y.; Xia, G.; Wang, M. Asymmetric Somatic Hybridization Affects Synonymous Codon Usage Bias in Wheat. *Front. Genet.* **2021**, *12*, 682324. [CrossRef]

31. Li, X.; Zhu, J.; Hu, F.; Ge, S.; Ye, M.; Xiang, H.; Zhang, G.; Zheng, X.; Zhang, H.; Zhang, S.; et al. Single-base resolution maps of cultivated and wild rice methylomes and regulatory roles of DNA methylation in plant gene expression. *BMC Genom.* **2012**, *13*, 300. [CrossRef] [PubMed]
32. Ossowski, S.; Schneeberger, K.; Lucas-Lledó, J.I.; Warthmann, N.; Clark, R.M.; Shaw, R.G.; Weigel, D.; Lynch, M. The Rate and Molecular Spectrum of Spontaneous Mutations in Arabidopsis thaliana. *Science* **2010**, *327*, 92–94. [CrossRef]
33. Nabel, C.S.; Manning, S.A.; Kohli, R.M. The Curious Chemical Biology of Cytosine: Deamination, Methylation, and Oxidation as Modulators of Genomic Potential. *ACS Chem. Biol.* **2012**, *7*, 20–30. [CrossRef]
34. Bernardi, G. Codon usage and genome composition. *J. Mol. Evol.* **1985**, *22*, 363–365. [CrossRef]
35. Knight, R.D.; Freeland, S.J.; Landweber, L.F. A simple model based on mutation and selection explains trends in codon and amino-acid usage and GC composition within and across genomes. *Genome Biol.* **2001**, *2*, RESEARCH0010. [PubMed]
36. Zhang, Z.; Yu, J. Modeling compositional dynamics based on GC and purine contents of protein-coding sequences. *Biol. Direct.* **2010**, *5*, 63. [CrossRef]
37. Hanson, G.; Coller, J. Codon optimality, bias and usage in translation and mRNA decay. *Nat. Rev. Mol. Cell Biol.* **2018**, *19*, 20–30. [CrossRef]
38. Galtier, N.; Piganeau, G.; Mouchiroud, D.; Duret, L. GC-content evolution in mammalian genomes: The biased gene conversion hypothesis. *Genetics* **2001**, *159*, 907–911. [CrossRef] [PubMed]
39. Plotkin, J.B.; Kudla, G. Synonymous but not the same: The causes and consequences of codon bias. *Nat. Rev. Genet.* **2011**, *12*, 32–42. [CrossRef]
40. Fawcett, J.A.; Rouzé, P.; Van de Peer, Y. Higher intron loss rate in *Arabidopsis thaliana* than *A. lyrata* is consistent with stronger selection for a smaller genome. *Mol. Biol. Evol.* **2012**, *29*, 849–859. [CrossRef]
41. Singh, N.D.; Arndt, P.F.; Petrov, D.A. Genomic Heterogeneity of Background Substitutional Patterns in Drosophila melanogaster. *Genetics* **2005**, *169*, 709–722. [CrossRef]
42. Xing, Y.; Lee, C. Alternative splicing and RNA selection pressure—Evolutionary consequences for eukaryotic genomes. *Nat. Rev. Genet.* **2006**, *7*, 499–509. [CrossRef]
43. Bernardi, G. Isochores and the evolutionary genomics of vertebrates. *Gene* **2000**, *241*, 3–17. [CrossRef]
44. Zhang, W.; Sun, X.; Yuan, H.; Araki, H.; Wang, J.; Tian, D. The pattern of insertion/deletion polymorphism in Arabidopsis thaliana. *Mol. Genet. Genom.* **2008**, *280*, 351–361. [CrossRef]
45. Laird, P.W. Principles and challenges of genome-wide DNA methylation analysis. *Nat. Rev. Genet.* **2010**, *11*, 191–203. [CrossRef] [PubMed]
46. Bewick, A.J.; Schmitz, R.J. Gene body DNA methylation in plants. *Curr. Opin. Plant Biol.* **2017**, *36*, 103–110. [CrossRef] [PubMed]
47. Mutti, J.S.; Bhullar, R.K.; Gill, K.S. Evolution of Gene Expression Balance Among Homeologs of Natural Polyploids. *G3 Genes Genomes Genet.* **2017**, *7*, 1225–1237. [CrossRef]

Article

Genome-Wide Identification of Maize Protein Arginine Methyltransferase Genes and Functional Analysis of *ZmPRMT1* Reveal Essential Roles in *Arabidopsis* Flowering Regulation and Abiotic Stress Tolerance

Qiqi Ling [†], Jiayao Liao [†], Xiang Liu, Yue Zhou and Yexiong Qian *

Anhui Provincial Key Laboratory of Conservation and Exploitation of Important Biological Resources, College of Life Sciences, Anhui Normal University, Wuhu 241000, China
* Correspondence: qyx2011@ahnu.edu.cn; Tel.: +86-55-3386-9297
† These authors contributed equally to this work.

Citation: Ling, Q.; Liao, J.; Liu, X.; Zhou, Y.; Qian, Y. Genome-Wide Identification of Maize Protein Arginine Methyltransferase Genes and Functional Analysis of *ZmPRMT1* Reveal Essential Roles in *Arabidopsis* Flowering Regulation and Abiotic Stress Tolerance. *Int. J. Mol. Sci.* **2022**, *23*, 12781. https://doi.org/10.3390/ijms232112781

Academic Editors: Zhiyong Li and Jian Zhang

Received: 27 September 2022
Accepted: 20 October 2022
Published: 24 October 2022

Publisher's Note: MDPI stays neutral with regard to jurisdictional claims in published maps and institutional affiliations.

Copyright: © 2022 by the authors. Licensee MDPI, Basel, Switzerland. This article is an open access article distributed under the terms and conditions of the Creative Commons Attribution (CC BY) license (https://creativecommons.org/licenses/by/4.0/).

Abstract: Histone methylation, as one of the important epigenetic regulatory mechanisms, plays a significant role in growth and developmental processes and stress responses of plants, via altering the methylation status or ratio of arginine and lysine residues of histone tails, which can affect the regulation of gene expression. Protein arginine methyltransferases (PRMTs) have been revealed to be responsible for histone methylation of specific arginine residues in plants, which is important for maintaining pleiotropic development and adaptation to abiotic stresses in plants. Here, for the first time, a total of eight *PRMT* genes in maize have been identified and characterized in this study, named as *ZmPRMT1-8*. According to comparative analyses of phylogenetic relationship and structural characteristics among *PRMT* gene family members from several representative species, all maize 8 PRMT proteins were categorized into three distinct subfamilies. Further, schematic structure and chromosome location analyses displayed evolutionarily conserved structure features and an unevenly distribution on maize chromosomes of *ZmPRMT* genes, respectively. The expression patterns of *ZmPRMT* genes in different tissues and under various abiotic stresses (heat, drought, and salt) were determined. The expression patterns of *ZmPRMT* genes indicated that they play a role in regulating growth and development and responses to abiotic stress. Eventually, to verify the biological roles of *ZmPRMT* genes, the transgenic *Arabidopsis* plants overexpressing *ZmPRMT1* gene was constructed as a typical representative. The results demonstrated that overexpression of *ZmPRMT1* can promote earlier flowering time and confer enhanced heat tolerance in transgenic *Arabidopsis*. Taken together, our results are the first to report the roles of *ZmPRMT1* gene in regulating flowering time and resisting heat stress response in plants and will provide a vital theoretical basis for further unraveling the functional roles and epigenetic regulatory mechanism of *ZmPRMT* genes in maize growth, development and responses to abiotic stresses.

Keywords: histone methylation; protein arginine methyltransferase; *Zea mays* L.; abiotic stress; functional analysis

1. Introduction

In eukaryotic, nucleosome is largely comprised of 146–147 base pairs of DNA and a histone octamer, including four types of histones (namely H2A, H2B, H3 and H4) [1]. The N-terminal tails of these histones can be subjected to post-translational and covalent modifications, including methylation, acetylation, phosphorylation, glycosylation, ADP-ribosylation, sumoylation and ubiquitination, designated as histone codes that regulate gene expression epigenetically through various mechanisms [2]. These histone codes have been demonstrated to not only directly affect and change the structure of chromatin but also extensively involve the regulation of gene expression [3]. It has been revealed that the levels of histone methylation can be altered in abiotic and biotic stress responses of plants, and the methylation of some specific residues at the N-terminal tails of histone is closely related to the upregulated or downregulated expression of stress response genes [4]. Therefore, histone methylation has become one of the hot topics in epigenetic regulation research in recent years.

Various arginine (R) and lysine (K) residues at the N-terminal tails of histone can be methylated via protein arginine and lysine methyltransferases, respectively. Protein arginine methyltransferases (PRMTs) can methylate histone H3 at R2 (H3R2), R8 (H3R8), R17 (H3R17), R26 (H3R26) and H4 at R3 (H4R3) through transferring the methyl group from S-adenosylmeth ionine (Ado-Met) to the nitrogen atom of arginine side chain [5]. Generally, there are three main forms of methylated arginine: monomethylarginines (MMA), asymmetric dimethylarginines (ADMA) and symmetric dimethylarginines (SDMA). Based on various methylated arginine forms, PRMTs can be categorized into four major types, namely types I, II, III or IV enzymes [6,7]. The type I and II PRMTs can regulate gene expression through the methylation of histone tails. Among them, the type I PRMTs include PRMT1, PRMT2, PRMT3, PRMT4, PRMT6 and PRMT10 and usually methylate H3R2 and H4R3 residues to generate ADMA, leading to transcriptional activation and ribosomal biosynthesis [8]. In contrast, the type II PRMTs, including PRMT5 and PRMT9, are required for the formation of SDMA at H3R8 and H4R3 residues, resulting in transcriptional repression [9,10]. However, the type III PRMTs (mainly PRMT7) only catalyze the generation of MMA [11]. In addition, the type IV PRMTs can methylate secondary amine on arginine residues, which has only been revealed in yeast [11–13]. Since PRMTs are involved in regulating diverse biological processes in animals and yeast, their significance of PRMTs in the model plant *Arabidopsis* has been paid great attention to in recent years [14].

In plants, genes encoding PRMTs have been identified and analyzed in several species, including *Arabidopsis thaliana* [15], *Oryza sativa* [14], *Eucalyptus grandis* [16] and *Glycine max* [17]. Studies have revealed that PRMTs share rather conserved features in eukaryotic cells and play crucial roles in chromatin structure, RNA processing, altered gene transcription, transport and translation, DNA repair, cellular differentiation and signal transduction [18,19]. Previous studies have demonstrated that the absence of AtPRMT3 can lead to multiple developmental defects in *Arabidopsis*, including unbalanced polyribosome spectra and abnormal rRNA preprocessing, where rRNA precursor processing is required for ribosomal biogenesis [20]. In *Arabidopsis*, AtPRMT4a and AtPRMT4b, two orthologs of human PRMT4/CARM1 protein, can methylate histone H3R2, H3R17 and H3R26 in vitro and are required for the methylation at H3R17 in vivo. The double mutant of *AtPRMT4a* and *AtPRMT4b* genes exhibited an FLC-dependent late flowering phenotype [21]. In addition, AtPRMT5/Skb1 (Shk1binding protein 1) belongs to a type II PRMT and can result in the generation of SDMA at H4R3 residue in vitro, and the *atprmt5* mutant exhibited pleiotropic phenotypes, such as growth retardation, curly and dark green leaves and late flowering in FLC-dependent manner in *Arabidopsis* [22,23]. Increasing evidence has suggested that the PRMT5-mediated arginine methylation plays a crucial role in alterative splicing of normal pre-mRNA in plants and animals [24,25] and that the late-flowering phenotype shown in *atprmt5-1* and *atprmt5-2* mutants may be resulted from alterative splicing of flowering time regulatory genes associated with RNA processing [24]. Previous studies have demonstrated that PRMT5 acts as a key determinant of circadian period in *Arabidopsis* and Drosophila, which may link circadian cycle with alternative splicing [25]. It has also been revealed that the expression level of *FLOWERING LOCUS C (FLC)* gene is upregulated, which might be resulted from alternative splicing of *FLK (Flowering Locus C)* in *atprmt5* mutant [26]. In addition, AtPRMT5/Skb1 has been revealed to regulate gene transcription and involve alterative splicing of pre-mRNA through symmetrically dimethylating H4R3 residues of histone and small nuclear ribonucleoprotein LSM4 and thereby confer a high salt stress tolerance in *Arabidopsis* [27]. Furthermore, AtPRMT10, a plant-specific type I PRMT, has been revealed to play divergent roles in flowering time control [15]. A mutation in the *AtPRMT10* gene resulted in late flowering by upregulating the transcript level of

FLC gene. Moreover, the correlation between the *FLC* expression and flowering time and vernalization has been determined, which indicates that the *FLC* gene acts as an important determinant in natural variation of flowering time [28,29]. Further, the *Arabidopsis* FLC protein has been demonstrated to function as a flowering repressor, which may be involved in regulating related genes in autonomous or vernalization pathways. Previous studies have demonstrated that the *FLC* gene can function in repressing expression of the flowering regulatory genes *SOC1 (Super of overexpression COI)* and *FT (Flowering Locus T)* through genetic and transgenic methods [30].

Furthermore, recent studies have revealed that many epigenetic factors are involved in various abiotic stress responses, and distinct chromatin modifications can be altered when plants are exposed to adverse environmental conditions, resulting in a dynamic chromatin environment to regulate gene expression [31]. At present, the cross-talk between diverse abiotic stress response pathways and epigenetic regulatory pathways has been thoroughly studied in plants [31]. However, the underlying mechanisms of epigenetic regulation of plant responses to heat stress remain to be elucidated, especially in regulating the dynamic histone arginine methylation patterns of stress-responsive genes, which partly depends on the catalytic function of PRMTs in plants. Moreover, proline is an essential multifunctional amino acid, which plays an important role in abiotic stress tolerance of plants. However, little is known about the biological functions of proline metabolism in plant responses to heat stress. Previous studies have revealed that short-term heat shock at 42 °C can result in proline accumulation in plant seedlings, and the external application of proline can induce the level of endogenous free proline and activities of antioxidant enzymes, followed by enhancing heat tolerance of plant seedlings [32]. Moreover, proline accumulation is mainly caused by the following three aspects: increasing the synthesis of proline, reducing the oxidation and degradation of proline and reducing the utilization of protein synthesis [33].

So far, much is well-known regarding proline metabolism in plants. The metabolic pathway of proline includes synthetic pathway and catabolic pathway. The synthetic pathway involves two enzymes, P5CS (pyrroline-5-carboxylate synthetase) and P5CR (P5C reductase). The degradation pathway involves other two enzymes, ProDH (proline dehydrogenase) and P5CDH (pyrroline-5-carboxylate dehydrogenase) [34]. Proline accumulation has been demonstrated to play adaptive roles in abiotic stress tolerance of plants [35]. Accumulated free proline could be implicated in adjusting cytosolic osmotic potential in order to save water [36], protect the membrane structure, sustain the structures of soluble proteins and activities of enzymes [37], act as reactive oxygen species (ROS) scavenger [38] and retain storage of carbon and nitrogen [39]. For example, overexpression of bean *P5CS1* gene in tobacco can slow down the decline of osmotic potential of transgenic plants under water stress [36].

Maize (*Zea mays* L.), as one of the most important crop species in the world and is vulnerable to environmental factors with climate change. Thus, how to use the methods of molecular biology to improve stress resistance and yield of maize is one of the hot topics of current biological research. Moreover, it has been demonstrated that plant *PRMT* genes are extensively involved in the regulation of growth and development and responses to various abiotic stresses. However, little is known regarding identification and function analysis of *PRMT* gene family in maize. In the present study, we carried out a comprehensive identification and functional analysis of *PRMT* genes in maize, including their phylogenetic relationships, gene and protein structures, conserved domain and motif architecture, chromosome location, gene duplication events and diverse expression profiles, which will facilitate further studies to unravel the exact biological roles of *ZmPRMT* genes in maize. Furthermore, the roles of *ZmPRMT1* in regulating flowering time and conferring heat stress tolerance were further clarified in transgenic *Arabidopsis*.

In this study, we selected transgenic *Arabidopsis* lines that overexpress maize *ZmPRMT1* gene to further reveal its functional roles in plants based on the conserved evolutionary relationship among ZmPRMT1 protein and some orthologous PRMT proteins in *Arabidopsis*, rice and sorghum. Although the direct use of *Arabidopsis* as a model plant for maize still has limitations, the current performance of *Arabidopsis* in plants is better than all other models. For example, when the purpose of the research is to gain a fundamental understanding of the specific growth process of plants, such as the regulation of flowering time, the *Arabidopsis* model system is still of reference significance [40]. Of course, it is also necessary to further verify the function of genes in maize through the overexpression and latest CRISPR/Cas9 gene editing techniques. In the case of flowering time regulation, the factors that advance the flowering time of *Arabidopsis* may inhibit the flowering time of rice (*Oryza sativa*) [41], and the network variation that affects the flowering time is significant even in gramineous plants [42,43]. Taken together, this study may contribute to an in-depth comprehension of the evolution of *ZmPRMT* genes and their crucial roles in maize growth, development and responses to abiotic stresses.

2. Results
2.1. Identification of the Members of PRMT Gene Family in Maize

To identify the total possible orthologs of *PRMT* gene family in maize, the PRMT protein sequences of *Arabidopsis*, rice and sorghum and their PRMT domains (Pfam: PF05185) were used as queries in the maize genome database. After the redundant sequences were removed, a total of eight maize PRMT proteins (namely ZmPRMT1–ZmPRMT8) were obtained. Further, these eight ZmPRMT sequences were scanned using Pfam and SMART databases to confirm the presence of the PRMT domain, respectively. The sequences of eight AtPRMTs, two OsPRMTs and two SbPRMTs were obtained from NCBI database for further investigation. The phylogenetic tree containing these representative AtPRMTs, OsPRMTs and SbPRMTs was constructed with ZmPRMT proteins, which was classified into three distinct subfamilies (Figure 1). Moreover, a total of eight *ZmPRMT* genes were uniformly named as *ZmPRMT1–ZmPRMT8* referring to their corresponding encoded protein in the above method section. All the conserved domains in PRMTs in maize were very similar to those PRMT proteins in *Arabidopsis*. Then the basic information such as the protein molecular weight, the isoelectric point and the number of amino acid residues was analyzed by ExPASY (https://web.expasy.org/protparam/) (accessed on 10 September 2021) (Table 1). The length of the *ZmPRMT* coding sequence varied from 798 bp from *ZmPRMT1* to 3201 bp for *ZmPRMT3*, with the respective coding potential of 306 and 1066 amino acids. The protein isoelectric point was between 5.27 (ZmPRMT4) and 8.36 (ZmPRMT3). Furthermore, the molecular weight varied from 29.86 KDa (ZmPRMT1) to 117.06 KDa (ZmPRMT3), indicating that ZmPRMT proteins share a large molecular weight range.

Table 1. Basic information about ZmPRMT in maize.

Gene Name	Accession Number Ensemble Transcript	Genome Location Coordinates (5'-3')	CDS (bp)	Protein Length (a.a)	Mol. Wt (kDa)	PI	Chr No.
ZmPRMT1	Zm00001d015228	80075042–80104640	798	265	29.86	5.761	5
ZmPRMT2	Zm00001d054001	244881916–244901087	2070	689	77.33	5.618	4
ZmPRMT3	Zm00001d022469	178246469–178259200	3201	1066	115.42	7.413	7
ZmPRMT4	Zm00001d007133	223062728–223070243	1647	548	60.44	5.127	2
ZmPRMT5	Zm00001d026614	148995156–149000603	1164	387	43.65	5.274	10
ZmPRMT6	Zm00001d032633	232602048–232615275	1212	403	44.97	5.971	1
ZmPRMT7	Zm00001d036131	73847683–73858041	1134	377	42.37	5.423	6
ZmPRMT8	Zm00001d020188	98545533–98549295	921	306	34.52	6.862	7

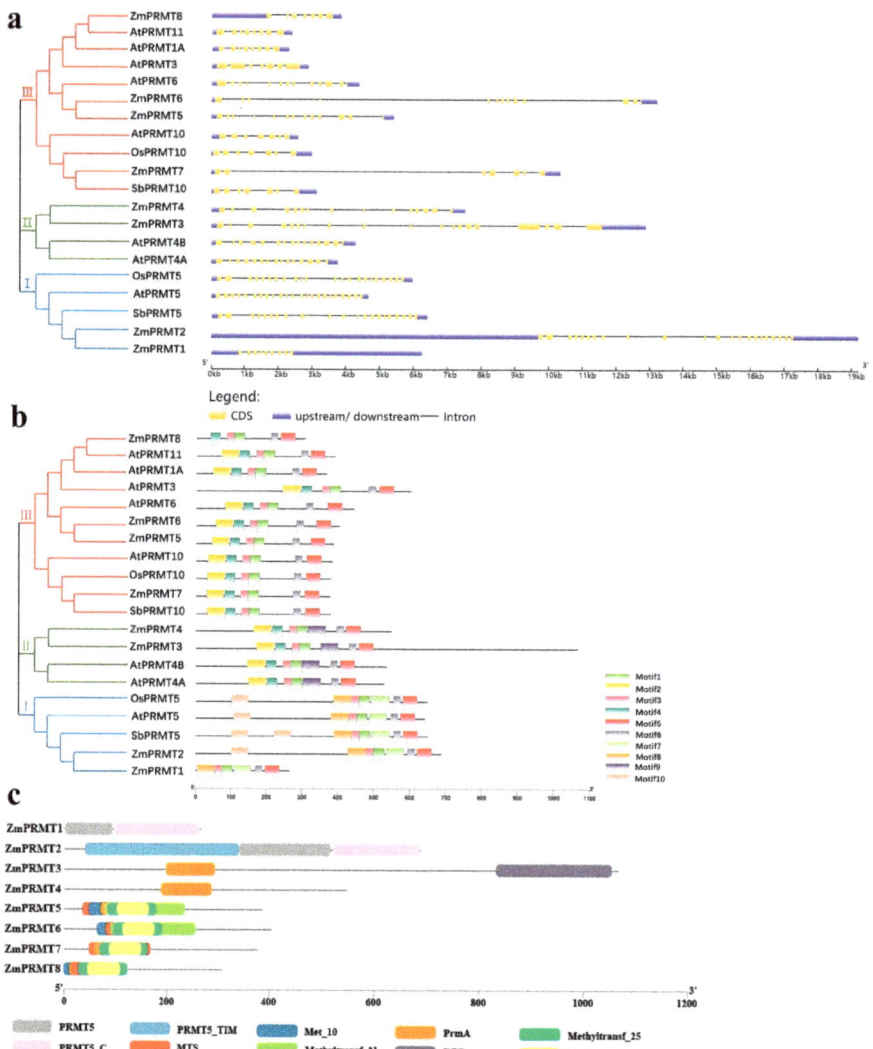

Figure 1. Phylogenetic relationship, gene structure, conserved motif and conserved domain analysis of maize PRMT family from maize, rice, sorghum and *Arabidopsis*. (**a**) Phylogenetic tree and the corresponding exon–intron structures of PRMT proteins. The maximum-likelihood phylogenetic tree was constructed using MEGA 7.0 with 1000 replicates. Three main clades are marked: I, II and III with different colored ranges. In the exon–intron structures, black lines represent introns, yellow boxes represent exons and the blue boxes represent upstream/downstream regions of PRMT genes. Protein sequences were downloaded from the Maize genome database and NCBI database. (**b**) Phylogenetic tree and the corresponding conserved motifs of PRMT proteins. Ten different colored boxes with numbers are used to represent different conserved motifs. The motifs identified in each group of PRMT proteins were schematically represented using the MEME motif search tool. (**c**) Phylogenetic tree and the corresponding conserved domains of maize PRMT proteins. The 8 PRMT domain sequences of maize were downloaded from the Smart database. Protein is represented by gray line. The position to the left of the gray line represents the N-terminal of each protein. Different colored rectangles are used to represent the domains contained in proteins.

2.2. Phylogenetic and Conserved Domain Analysis of Maize PRMT Proteins

To investigate the phylogenetic relationships among PRMTs from maize, *Arabidopsis*, rice and sorghum, the multiple sequence alignment with MEGA7.0 was performed using the protein sequences of PRMTs from these plants and the unrooted phylogenetic tree was constructed from the alignment full-length protein sequences of eight AtPRMTs, two OsPRMTs, two SbPRMTs and eight ZmPRMTs through the Maximum-Likelihood method. Based on the comparison and analysis of evolutionary relationships among PRMT proteins, these proteins were categorized into three major subfamilies (Figure 1). In the first subfamily, maize ZmPRMT1 and ZmPRMT2 exhibit high homology with OsPRMT5, SbPRMT5 and AtPRMT5. In the second subfamily, maize ZmPRMT3, ZmPRMT4 and AtPRMT4A, AtPRMT4B belong to orthologous proteins. In the third subfamily, ZmPRMT7, OsPRMT10, AtPRMT10 and SbPRMT10 belong to orthologous proteins, and ZmPRMT5, ZmPRMT6 and ZmPRMT8 share high homology with AtPRMT1A, AtPRMT3, AtPRMT6 and AtPRMT11. Further, the genetic structure of maize *PRMT* genes is highly conserved, and the structure is simpler than that of other plants *PRMTs* (Figure 1a), which is consistent with their motif distribution (Figure 1b).

In order to understand the difference of the domain architecture, the SMART database was used to determine the structure types of eight maize PRMT proteins with default parameters (Figure 1c). According to their relationships with *Arabidopsis*, rice and sorghum PRMT proteins, the eight ZmPRMT proteins were classified into three different subfamilies, including the subfamily I (two members), the subfamily II (two members) and the subfamily III (four members). In subfamily I, ZmPRMT1 and ZmPRMT2 share two conserved domains including PRMT5 (PF05185) and PRMT5_C. The PRMT5 domain is the conserved functional domain of arginine methyltransferase. ZmPRMT2 has a complete PRMT_Tim domain at the N-terminal, which may play a role in the quaternary structure of this protein. In subfamily II, ZmPRMT4 has only one conserved PrmA domain, while ZmPRMT3 has another REF domain at the C-terminal. Among the members of subfamily III, all these proteins share several common conserved domains: MTS, PrmA, Methyltransf_11 and Methyltransf_25. Among them, ZmPRMT5, ZmPRMT6 and ZmPRMT7 contain more than six domains, which are concentrated in the anterior middle segment of the protein sequence, and ZmPRMT5 and ZmPRMT7 have only one Met_10 domain at the C-terminal. The conserved domains of ZmPRMT8 are concentrated at the N-terminal. In conclusion, the distribution of conserved domains of each protein is related to the evolution of each other.

2.3. Chromosomal Localization and Gene Duplication of PRMT Gene Family in Maize

The physical locations of *ZmPRMT* genes on chromosomes were investigated to generate the chromosomal position graphics of *ZmPRMT* genes. The graphics showed that all eight *ZmPRMT* genes were distributed unevenly across seven of all the ten chromosomes in the maize genome (Figure 2). These newly identified genes were distributed individually in various regions of these chromosomes (i.e., telomere, near centromere and other regions). Chromosome 7 has the highest number of *ZmPRMT* genes, i.e., two, while Chromosomes 1, 2, 4, 5, 6 and 10 have only one in each, respectively. In detail, *ZmPRMT3* and *ZmPRMT8* were located on Chromosome 7, whereas the *ZmPRMT6*, *ZmPRMT4*, *ZmPRMT2*, *ZmPRMT1*, *ZmPRMT7* and *ZmPRMT5* were located on Chromosomes 1, 2, 4, 5, 6 and 10, respectively. Furthermore, *ZmPRMT2*, *ZmPRMT3* and *ZmPRMT5* are distributed near the telomeres of chromosomes. Considering the importance of telomere structure for maintaining chromosome stability and ensuring the complete replication of genes on chromosomes, the above three genes are considered to be very conserved in the process of species evolution and may play an irreplaceable role in normal growth and developmental process of maize. Furthermore, gene duplication events were investigated to determine the evolutionary patterns of the maize *PRMT* gene family. Based on the analysis of sequence alignment, it was revealed that a pair of genes (*ZmPRMT3*/*ZmPRMT4*) was involved in the segmental duplication of maize for they share 88.67% homology in the sequences (Supplementary Figure S1). This result indicates that gene duplication events may have occurred during the evolution process of PRMT genes to preserve the function of PRMT proteins.

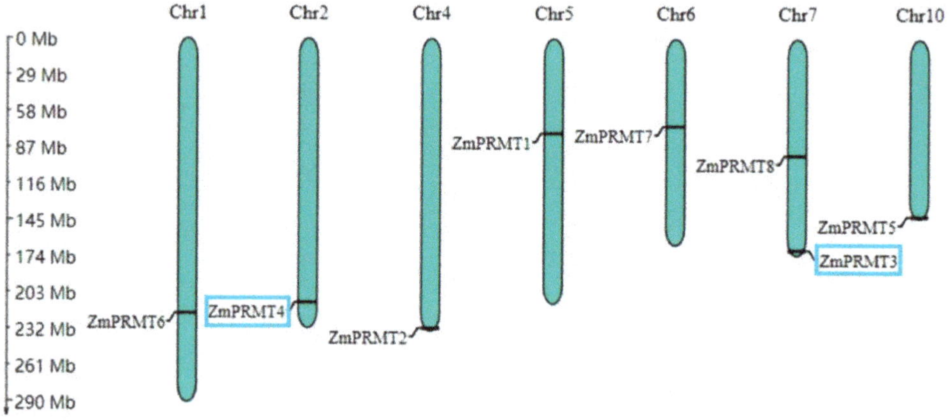

Figure 2. Chromosomal location of PRMT family genes in maize. Eight maize *PRMT* genes were located on 7 of all the 10 maize chromosomes. The number of chromosomes is marked at the top of each blue column bar. The approximate position of each *ZmPRMT* gene on the chromosome corresponds to the gene name on the left of the blue bar in the figure. The scale on the left in the figure is in megabytes. The blue box indicates that there is a replication relationship between genes.

2.4. The Cis-Acting Regulatory Elements in the Promoter of Maize PRMT Genes

To further investigate the potential regulatory mechanism in biotic or abiotic stress responses, the cis-acting elements were detected in 2000 bp upstream of promoter of *ZmPRMT* genes via using the PlantCARE database. All of the 16 cis-acting regulatory elements associated with stress and hormones were detected in the promoter regions of the *PRMT* gene in maize (Figure 3). Most *ZmPRMT* genes contain ARE that is related to the anaerobic reaction, except *ZmPRMT6*. Further, most *ZmPRMT* genes share the CGTCA-motif (MeJA reactive element), TGACG-motif (MeJA reactive element), ABRE (abscisic acid reactive element) and LTR (low-temperature relative element). The results indicated that these genes not only respond to hormone but also may respond to abiotic stress. In addition, MBS (drought induction element), TC-rich repeats (defense and stress response element), GARE motif (gibberellin response element), P-box (gibberellin response element), TGA element (auxin response element) and TCA element (salicylic acid response element) were detected in the 2000 bp upstream region of the promoter of these genes. Taken together, these results revealed that expression of maize *PRMT* genes could be modulated by various hormones and adversity stress, which may participate in the regulation of biotic or abiotic stress responses and hormone signal transduction.

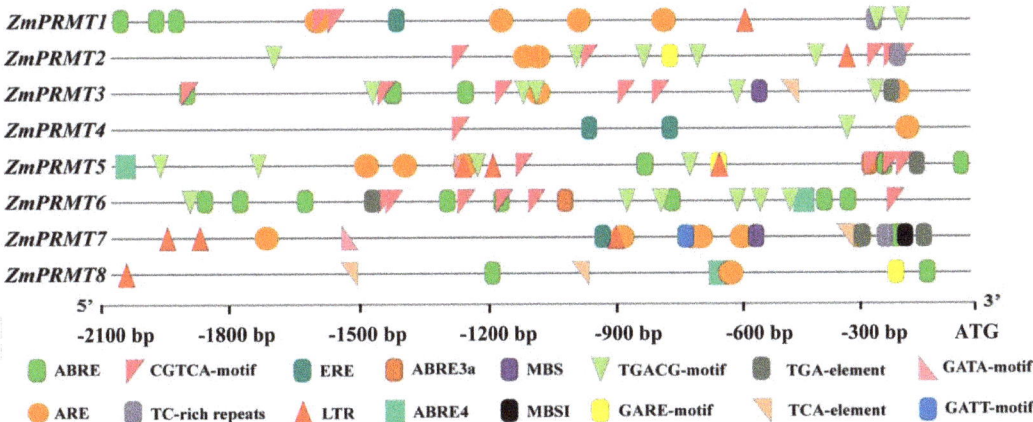

Figure 3. Distribution of major stress-related cis-elements in the promoter sequences of the 8 *ZmPRMT* genes. Putative ABRE, CGTCA-motif, ERE, MBS, TGACG-motif, GATA-motif, TGA-element, ARE, TC-rich repeats, LTR, GARE-motif, TCA-element and GATT-motif core sequences are represented by different symbols as shown in the symbols at the bottom. ABRE: cis-acting element involved in the abscisic acid responsiveness; CGTCA-motif: cis-acting regulatory element involved in the MeJA responsiveness; ERE: ethylene-responsive element; MBS: drought -responsive element; TGACG-motif: cis-acting regulatory element involved in the MeJA-responsiveness; GATA-motif: zinc-finger transcription factor; TGA element: auxin-responsive element; ARE: anaerobic-responsive element; TC-rich repeats: cis-acting element involved in defense and stress responsiveness; LTR: low-temperature relative element; GARE-motif: gibberellin-responsive element; TCA-element: cis-acting element involved in salicylic acid responsiveness. The position of each cis-element in the 2 kb sequence upstream of the initiation codon (ATG) of the *ZmPRMT* genes was measured at a scale of 300 bp.

2.5. Analysis of Microarray Expression Profile of Maize PRMT Genes in Different Tissues

To further explore the transcription patterns of maize *PRMT* genes, the expression profile of eight *ZmPRMT* genes in 60 different developmental periods was characterized with microarray data (Figure 4). Heat map shows that *ZmPRMT* genes have a trend of differential expression in different growth periods of all these tissues. The signal values for all these *ZmPRMT* genes were shown in Supplementary Table S2. According to the heat map, some specifically expressed *ZmPRMT* genes in tissues or organs were discovered at 60 diverse developmental stages. All eight *ZmPRMT* transcripts investigated were expressed in the whole growth process, although these members were expressed at low levels in some tissues at different growth and development stages. In addition, the results showed that some *ZmPRMT* genes begin to express at a certain time during plant growth and development, indicating that these *ZmPRMT* genes are very important in maize growth and development processes. Moreover, we found that the expression patterns of *ZmPRMT* genes can be roughly classified into two periods with the R2_Outer Husk and the R2_Innermost Husk as the boundaries. In the first period, the expression levels of most *ZmPRMT* genes were generally very low at various tissues. However, some *ZmPRMT* genes were highly expressed in some specific tissues such as 6DAS_GH_WTeoptile, 6DAS_GH_Primary Root, VE_Primary Root, V1_Stem and SAM, V3_Stem and SAM, V13_Immature Tassel and V18_Immature Cob, implying that these genes might play important roles in these specific tissues at different periods of maize growth process. Furthermore, according to the heat map, several *ZmPRMT* genes such as *ZmPRMT2*, *ZmPRMT3* and *ZmPRMT4* were also revealed to be highly or specifically expressed in some tissues. In the second period, the majority of *ZmPRMT* genes were expressed highly or specifically at different developmental stages of seed and endosperm after pollination. To summarize, the expression pattern showed that the transcriptional levels of *ZmPRMT* genes in seed and endosperm were obviously higher than those in roots, stems and leaves. Taken together, the results demonstrated that these identified *ZmPRMT* genes exhibited differential expression patterns at diverse growth and development stages of maize, suggesting that these genes may function in multiple tissues.

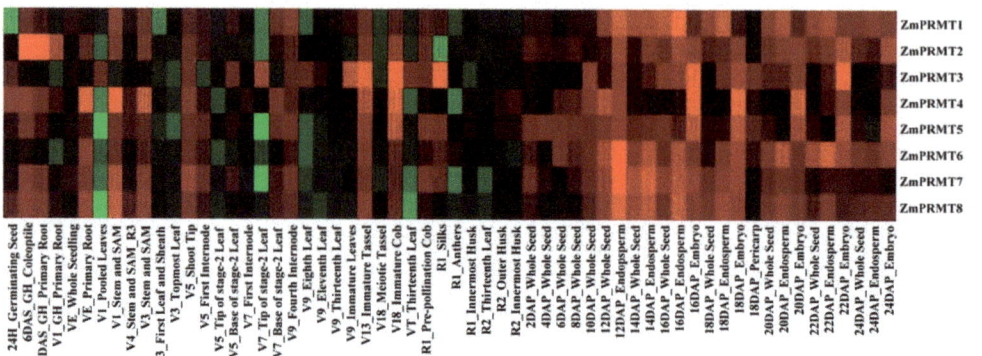

Figure 4. Hierarchical cluster analysis of expression profiles of 8 maize *ZmPRMT* genes. Hierarchical cluster analysis of expression profiles of *ZmPRMT* genes family in all 60 tissues that including the whole growth and development process of maize. The color code on the right indicates the log2 signal value. The gene name of each *ZmPRMT* is shown on the left of each line. Red represents a high level and green indicates a low level of transcript abundance. The tissues and/or organs in different development periods are noted on the bottom of each lane.

2.6. Expression Profile Analysis of Maize PRMT Genes under Abiotic Stress Treatment

To further confirm the responsiveness of *ZmPRMT* genes to abiotic stresses, the expression profile of these eight genes was further explored by qRT-PCR analysis and with at least three biological repeats to make sure the reliability of the qRT-PCR results. The experiment was accomplished by using the cDNA of leaves, stems and roots of maize seedlings, which were exposed to three diverse abiotic stress treatments, including heat, drought and salt. Firstly, the different maize tissues (including leaves, stems and roots) at five different time points (0 h, 1 h, 2 h, 4 h and 8 h) under heat stress treatment were collected to investigate the transcription levels of these eight *ZmPRMT* genes by using qRT-PCR, respectively (Figure 5). Firstly, the expression levels of these eight *ZmPRMT* genes in maize leaves, stems and roots at five different time points (0 h, 1 h, 2 h, 4 h and 8 h) after heat treatment were detected by qRT-PCR analysis. After heat treatment, these eight *ZmPRMT* genes displayed differential accumulation of expression levels. One of the most obvious results we can observe is that the majority of *ZmPRMT* genes exhibited differential upregulated expression levels in leaves after heat treatment, whereas only one *ZmPRMT* gene (*ZmPRMT5*) in stems and two *ZmPRMT* genes (*ZmPRMT1*, *ZmPRMT5*) in roots showed the significantly upregulated expression levels after heat treatment compared with those in leaves. Only two *ZmPRMT* genes (*ZmPRMT4* and *ZmPRMT6*) in leaves didn't show upregulation after heat stress at different time points, while the others exhibited upregulation in leaves at distinct time hours, indicating that most of these *ZmPRMT* genes may play potential roles in heat stress responses. Moreover, under heat treatment, the majority of these *ZmPRMT* genes showed more obvious responses in leaves than those in stems or roots. Based on the above analysis, it can be speculated that the *ZmPRMT* genes in leaves might be involved in heat stress responses of plants and more sensitive than those in steams and roots. Therefore, these results indicate that plant leaves rather than stems and roots may be the main organs of plants in heat stress responses. Moreover, it is noticeable that the upregulated expression patterns of *ZmPRMT1* and *ZmPRMT5* in all the three different tissues revealed that these two genes might play crucial roles in maize in response to heat stress.

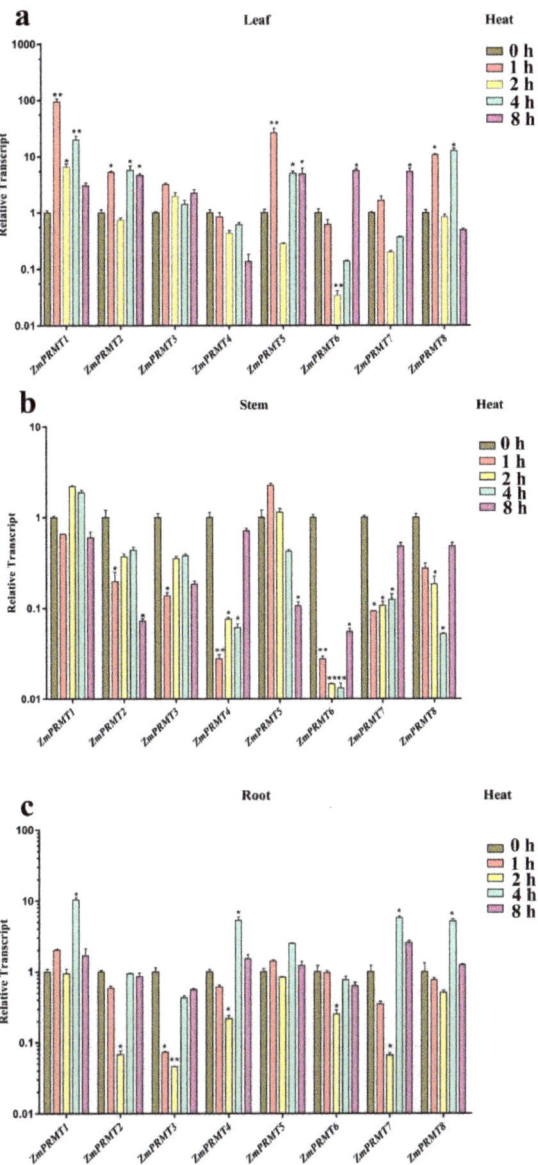

Figure 5. Expression profiling of the 8 *ZmPRMT* genes exposed to heat stress treatment in maize tissues (roots, stems and leaves). (**a**) The expression profiling of the 8 *ZmPRMT* genes exposed to heat stress treatment in maize leaves. (**b**) The expression profiling of the 8 *ZmPRMT* genes exposed to heat stress treatment in maize stems. (**c**) The expression profiling of the 8 *ZmPRMT* genes exposed to heat stress treatment in maize roots. The qRT-PCR data was standardized with the maize *ZmActin1* gene. The control condition was represented as CK (0 h), and the heat stress conditions were represented as Heat (1 h, 2 h, 4 h and 8 h). X-axes represent the genes under different treatments, while y-axes represent the scale of the relative expression level of genes. Error bar is generated by three biological repetitions. The significance level is indicated by the asterisk at the top of the error bar. $p < 0.05$ is significantly different *; $p < 0.01$ is significantly different **.

Further, we performed the drought stress treatment. The majority of *ZmPRMT* genes showed differential downregulated expression levels in leaves after drought treatments, whereas the most of *ZmPRMT* genes in stems and roots showed significantly upregulated expression levels after drought treatments compared with those under normal conditions (CK) (Figure 6). For example, there were only two *ZmPRMT* genes (*ZmPRMT5* and *ZmPRMT6*) in leaves exhibiting upregulation under drought treatment, whereas only two *ZmPRMT* genes (*ZmPRMT1* and *ZmPRMT2*) in the stems and only one *ZmPRMT* gene (*ZmPRMT6*) in the roots showed lower expression levels under drought treatment. It should be noted that the expression profiles of these *ZmPRMT* genes in maize under drought treatment were different from those under heat treatment. Especially in roots, most of the *ZmPRMT* genes expressed highly after drought treatment, and most of them were upregulated significantly and responded rapidly at 1 h after drought treatment in roots, indicating that these genes showed faster and stronger responses in roots than those in stems under drought stress. Similarly, it is also possible that the plant stems and roots rather than leaves are the main organs of plants in response to drought stress.

Finally, the salt stress treatment was completed. The majority of *ZmPRMT* genes showed differential downregulated expression patterns in all three different tissues after salt treatment, and there were only one *ZmPRMT* gene (*ZmPRMT1*) in stems and three *ZmPRMT* genes (*ZmPRMT1*, *ZmPRMT2* and *ZmPRMT4*) in roots exhibited higher expression levels under salt treatment than that under normal conditions (CK) (Figure 7). Therefore, these results show that salt stress can significantly inhibit the expression of these genes at all time points. Taken together, these results suggest that *ZmPRMT* genes may play an important role in responding abiotic stress and most of them exhibited immediate response to abiotic stress.

2.7. Generation of Overexpressed ZmPRMT1 Transgenic Arabidopsis Plants

To evaluate the function of *ZmPRMT* genes, we firstly constructed the overexpression vector of *ZmPRMT1* gene under 35 s promoter through genetic engineering method. The *ZmPRMT1* clone map and the double enzyme (Hind3/Xba1) digestion map of the recombinant plasmid were shown in Supplementary Figure S2A and S2B, respectively. Then, the recombinant vector with the *ZmPRMT1* gene was transformed into the *Arabidopsis* line using *Agrobacterium*-mediated method by dipping the *Arabidopsis* floral to obtain the overexpressed *Arabidopsis* plants. Consequently, a total of 14 transgenic lines were generated in this study (Supplementary Figure S3). Among them, three representative stable homozygous lines 2, 6 and 7 were further selected for functional analysis. The expression of *ZmPRMT1* gene in three transgenic *Arabidopsis* lines was validated through qRT-PCR. In addition, according to the analysis of phylogenetic tree, the *ZmPRMT1* gene was located in the Group I and its orthologous gene (*AtPRMT5*) (Supplementary Figure S4) has been confirmed to regulate flowering time in *Arabidopsis*. Thus, the *ZmPRMT1* gene may share similar functions with the orthologous *AtPRMT5* gene, by which the protein encoded encompasses high homology, locates in nucleus and functions in repressing target genes by methylating H4R3 and H3R8 residues of histone, as well as transcription factors/regulators. Thus, this result implies that the *ZmPRMT1* gene may be involved in the modulation of vegetative growth and control of flowering time in plants.

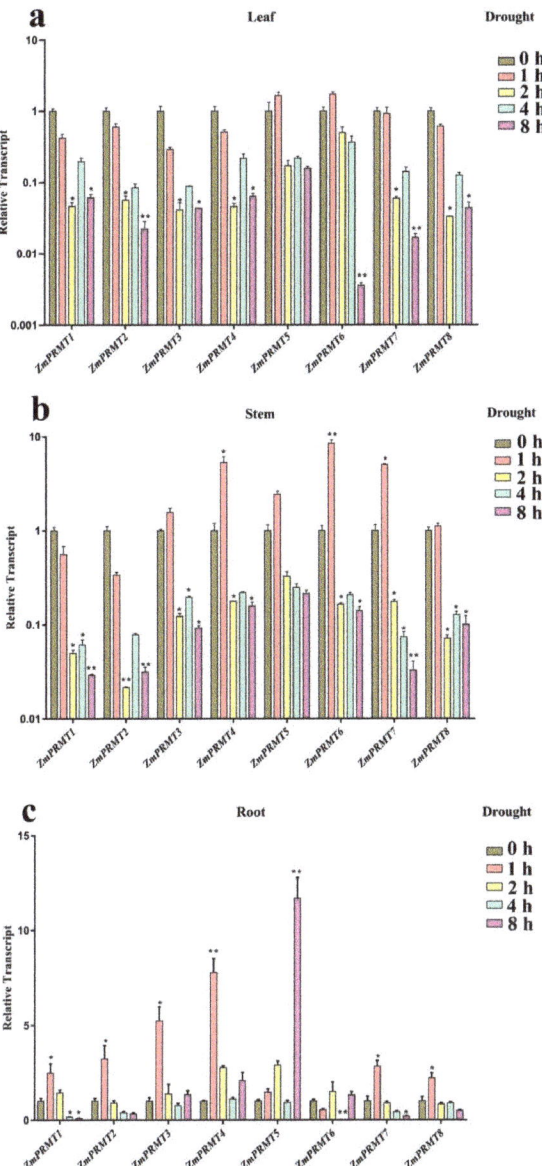

Figure 6. Expression profiling of the 8 *ZmPRMT* genes exposed to drought stress treatment in maize tissues (roots, stems and leaves). (**a**) The expression profiling of the 8 *ZmPRMT* genes exposed to drought stress treatment in maize leaves. (**b**) The expression profiling of the 8 *ZmPRMT* genes exposed to drought stress treatment in maize stems. (**c**) The expression profiling of the 8 *ZmPRMT* genes exposed to drought stress treatment in maize roots. The qRT-PCR data was standardized with the maize *ZmActin1* gene. The control condition was represented as CK (0 h), and the heat stress conditions were represented as Drought (1 h, 2 h, 4 h and 8 h). X-axes represent the genes under different treatments, while the y-axes represent the scale of the relative expression level of genes. Error bar is generated by three biological repetitions. The significance level is indicated by the asterisk at the top of the error bar. $p < 0.05$ is significantly different *; $p < 0.01$ is significantly different **.

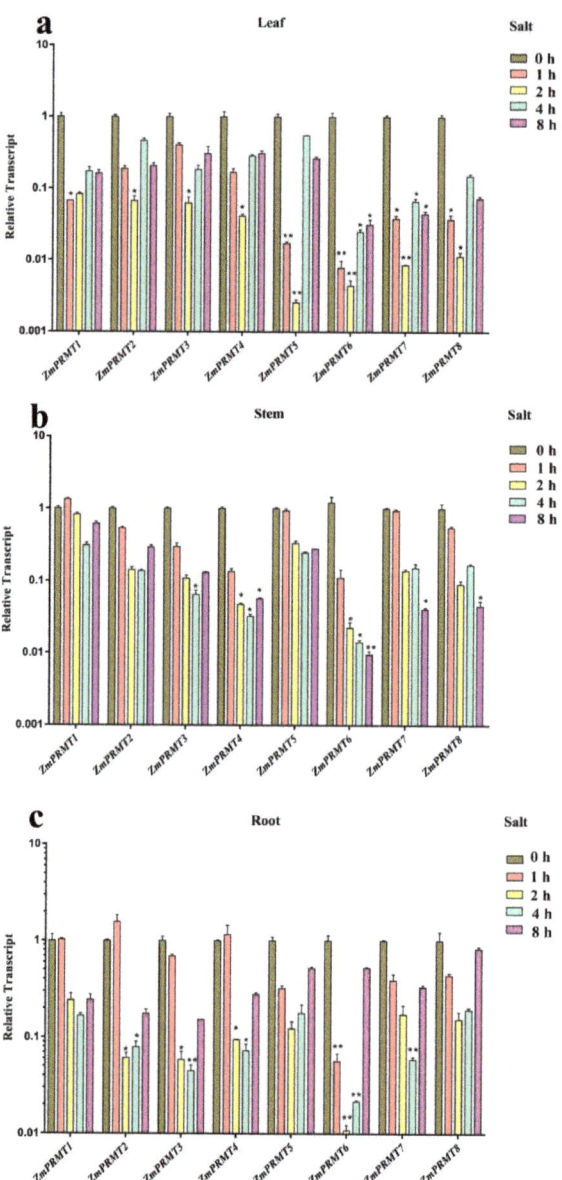

Figure 7. Expression profiling of the 8 *ZmPRMT* genes exposed to salt stress treatment in maize tissues (roots, stems and leaves). (**a**) The expression profiling of the 8 *ZmPRMT* genes exposed to salt stress treatment in maize leaves. (**b**) The expression profiling of the 8 *ZmPRMT* genes exposed to salt stress treatment in maize stems. (**c**) The expression profiling of the 8 *ZmPRMT* genes exposed to salt stress treatment in maize roots. The qRT-PCR data was standardized with maize *ZmActin1* gene. The control condition was represented as CK (0 h), and the heat stress conditions were represented as Salt (1 h, 2 h, 4 h and 8 h). X-axes represent the genes under different treatments, while y-axes represent the scale of the relative expression level of genes. Error bar is generated by three biological repetitions. The significance level is indicated by the asterisk at the top of the error bar. $p < 0.05$ is significantly different *; $p < 0.01$ is significantly different **.

2.8. Overexpression of ZmPRMT1 Gene Advances the Flowering Time of Transgenic Arabidopsis

Previous studies have demonstrated that *AtPRMT5* in *Arabidopsis* plays a critical role in regulating flowering time by virtue of FLC determinative factor [30], which indicates that the *ZmPRMT1* gene, the orthologous gene of *AtPRMT5* in maize, may have similar physiological function to *AtPRMT5*—that is, to advance the flowering time of plants in an FLC-dependent manner. Therefore, in this study, *ZmPRMT1* gene was selected from the *ZmPRMT* gene family to explore its expression pattern and genetic interaction with some flowering-related regulatory genes in transgenic lines. Firstly, the phenotype of these four different lines including WT, transgenic lines 2, 6 and 7 were surveyed in growth progress around flowering. These results suggest that there is no obvious difference between the transgenic *Arabidopsis* and WT lines in seedlings. However, the transgenic *Arabidopsis* lines were flowered earlier than WT plants, as well as the leaf numbers of them were slightly lower than those of WT plants when these *Arabidopsis* plants have been cultivated for about 30 days (Figure 8a,b), which suggested that overexpression of *ZmPRMT1* gene might result in early flowering in transgenic *Arabidopsis* through advancing the flowering time of these plants.

Figure 8. Regulation of flowering time and leaf number by *ZmPRMT1* in transgenic *Arabidopsis* lines. (**a**) The early-flowering phenotype of transgenic *Arabidopsis* lines was observed. The left figure shows the phenotype of the WT and transgenic *Arabidopsis* lines 2, 6 and 7. Seedlings of these 4 different *Arabidopsis* lines were grown in the temperature incubator at 22 °C under long-day conditions (16 h light and 8 h dark) and 60% relative humidity. After 27 days, 4 different *Arabidopsis* lines flowering successively. (**b**) Total leaf number of transgenic *Arabidopsis* lines was calculated. The right figure shows leaf number of the corresponding plants left. Total leaf number of WT, transgenic *Arabidopsis* 2, 6 and 7 was calculated under same conditions.

To further confirm the potential regulatory pathways of overexpressed *ZmPRMT1* gene in regulating flowering time of transgenic *Arabidopsis*, the *Arabidopsis* flowering-related regulatory genes, including *FLC*, *SOC1*, *FT*, MADS Affecting Flowering genes (*MAF1*, *MAF2*, *MAF3*, *MAF4* and *MAF5*) were investigated through qRT-PCR, respectively (Figure 9). The experiment was repeated biologically for at least three times to ensure the accuracy of qRT-PCR analysis. The *Arabidopsis FLC* gene has been reported to encode a MADS domain protein, which acts as an inhibitive factor of flowering promoting genes: *SOC1* and *FT* [28,44–47]. The results showed that the *ZmPRMT1* gene exhibited differential upregulated expression levels in all three transgenic lines compared with those in WT after flowering. The expression levels of *FLC* in transgenic and WT lines were significantly downregulated after flowering, whereas the expression levels of *FT* and *SOC1* in all four lines were upregulated after flowering. The *SOC1* gene in transgenic line 2 was shown to exhibit higher expression level than two other transgenic lines and WT after flowering. The *FT* gene was significantly upregulated in transgenic lines 2, 6 and 7 than that in WT after flowering. Furthermore, the expression levels of *MAF* genes, which shared highly conversed MADS domain with *FLC*, were also be investigated in these *Arabidopsis* lines after flowering. The transcript levels of *MAF* genes (*MAF1-4*) showed a consistent decrease with the *FLC* expression level after flowering in all investigated *Arabidopsis* lines, whereas *MAF5* exhibited high transcript level after flowering in all investigated *Arabidopsis* lines. Taken together, the early flowering phenotype of these transgenic *Arabidopsis* may be often closely linked to the expression of *ZmPRMT1*. Furthermore, *ZmPRMT1* gene may share the similar physiological function as *AtPRMT5* to promote plant flowering time in FLC-dependent manner. Given the expression pattern of flowering-related regulatory genes, it can be speculated that *ZmPRMT1* gene may be required for the promotion of flowering in maize by modulating the expression of flowering-related regulatory genes under natural environmental conditions.

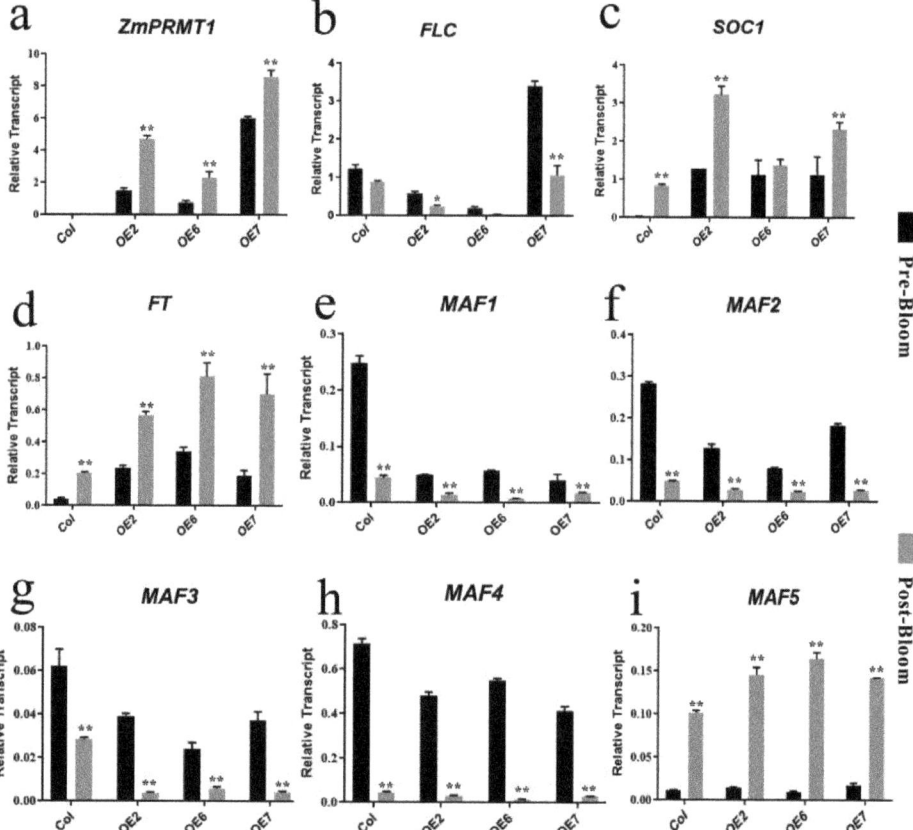

Figure 9. Expression of flowering-related genes by qRT-PCR analysis. The flowering-related genes, including (**b**) FLC (AT5G10140), MADS BOX PROTEIN FLOWERING LOCUS F; (**c**) SOC1 (AT2G45660), SUPPRESSOR OF OVEREXPRESSION OF CONSTANS 1; (**d**) FT (AT1G65480), FLOWERING LOCUS T; (**e**) MAF1 (AT1G77080), MADS AFFECTING FLOWERING 1; (**f**) MAF2 (AT5G65050) MADS AFFECTING FLOWERING 2; (**g**) MAF3 (AT5G65060) MADS AFFECTING FLOWERING 3; (**h**) MAF4 (AT5G65070) MADS AFFECTING FLOWERING 4; (**i**) MAF5 (AT5G65080) MADS AFFECTING FLOWERING 5. The mRNA levels were determined using qRT-PCR, and the values are reported relative to *ZmPRMT1* mRNA levels in each line (**a**). The expression levels of related genes before flowering are represented by black columns, and the expression levels of related genes after flowering is represented by gray columns. Error bar is generated by three biological repetitions. The significance level is indicated by the asterisk at the top of the error bar. $p < 0.05$ is significantly different *; $p < 0.01$ is significantly different **.

2.9. Overexpression of ZmPRMT1 Gene Enhances Heat Tolerance in Transgenic Arabidopsis

To further explore heat stress-responsiveness of *ZmPRMT1* gene in transgenic *Arabidopsis*, the leaves were sampled at two different time points (0 h and 8 h) under heat stress treatment and performed for the qRT-PCR analysis, respectively. The experiment was repeated biologically at least three times to ensure the accuracy of qRT-PCR analysis. Firstly, the phenotype of these four different lines: WT, transgenic *Arabidopsis* lines 2, 6 and 7 were surveyed under 0 h and 8 h heat stress. Under the control condition, no significant phenotypic difference was observed between transgenic *Arabidopsis* and WT lines. However, under heat stress condition (42 °C), the transgenic *Arabidopsis* lines 2, 6 and 7 were respectively more resistant to heat stress than WT. Especially, the transgenic *Arabidopsis* lines exhibited higher vitality than the WT plants (Figure 10). Overall, these results indicate that the overexpressed *ZmPRMT1* gene in *Arabidopsis* can increase thermotolerance.

Figure 10. Heat tolerance analysis of transgenic *Arabidopsis* lines. The phenotypes of the WT and transgenic *Arabidopsis* seedling lines 2, 6 and 7 are shown following their treatment with different heat stress. CK (0 h), the control condition and Heat (8 h), the heat stress conditions. The 3-week-old seedlings were directly exposed to 42 °C to detect heat tolerance in 0 h and 8 h. The *Arabidopsis* were photographed in 3 d after the 0 h and 8 h heat stress treatment. The wilting degree of leaves of both the WT and transgenic lines was observed at 3 days after the 8 h heat stress treatment.

Furthermore, to explore molecular mechanism of heat tolerance of transgenic *Arabidopsis* in response to heat stress, we compared the proline contents of WT and transgenic *Arabidopsis* lines under 42 °C treatment at 0 h and 8 h. The proline contents of these 4 *Arabidopsis* lines exposed to heat stress treatment have been displayed to increase significantly compared with those under the normal conditions (Figure 11). For example, the proline contents in the transgenic *Arabidopsis* lines 2 and 6 were revealed to exhibit increased accumulation compared with that in WT plants after heat treatment, indicating that the *ZmPRMT1* gene could be involved in modulating heat-induced proline accumulation in transgenic *Arabidopsis*. To further explore heat stress tolerance mechanism of *ZmPRMT1* gene in proline metabolic pathway, the *Arabidopsis* leaves were further sampled after heat treatments at 0 h and 8 h and carried out for the qRT-PCR analyses to further discern the transcription patterns of some genes associated with proline metabolic pathway (Figure 11). The results showed that the related genes of proline synthesis pathway showed differential upregulated expression levels, whereas the related genes of proline degradation pathway exhibited differential downregulated expression levels after heat treatment. The *P5CS1* and *P5CR* genes in proline synthesis pathway showed significantly upregulated expression levels in all three different transgenic *Arabidopsis* lines compared with in WT plants after heat treatment, while the *P5CDH* and *PDH* genes of proline degradation pathway showed significantly downregulated expression levels in all three different over-expression lines compared with in WT plants under heat stress. Taken together, these results reveal that the overexpression of *ZmPRMT1* gene can result in greater proline accumulation by increasing the expression levels of proline synthesis pathway genes and decreasing the expression levels of proline degradation pathway genes, which is likely to lead to enhance of water potential and osmotic potential necessary for holding photosynthetic activity to alleviate heat stress and thereby enhance the heat stress tolerance of transgenic *Arabidopsis*.

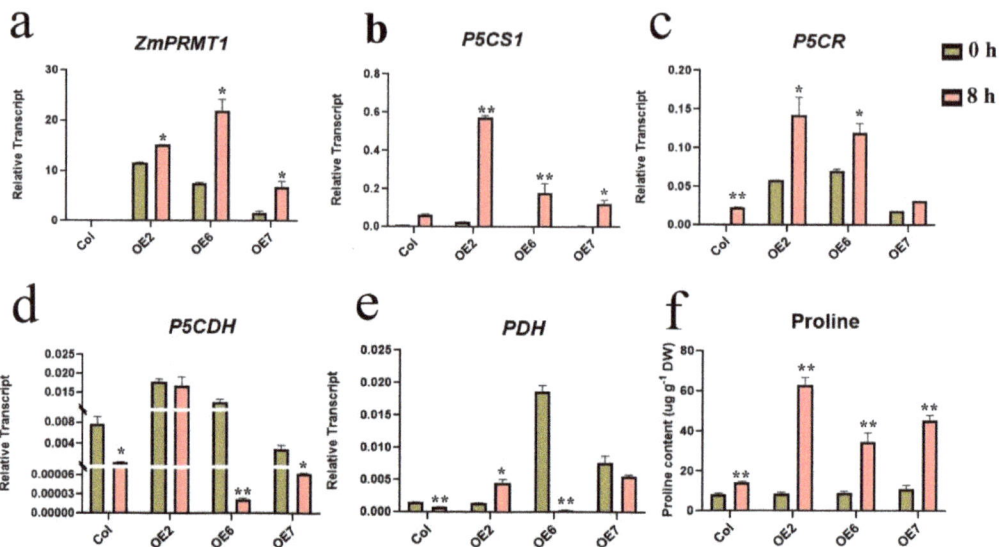

Figure 11. Analysis of heat tolerance-related genes expression and proline accumulation under heat stress by qRT-PCR. The plant abiotic stress tolerance-related genes, including genes in the proline synthesis way such as (**b**) P5CS1 (AT2G39800), delta1-pyrroline-5-carboxylate synthase 1; (**c**) P5CR (AT5G14800), pyrroline-5-carboxylate (P5C) reductase and genes in the proline degradation way such as (**d**) P5CDH (AT5G62530), DELTA1-PYROLINE-5-CARBOXYLATE DEHYDROGENASE; (**e**) PDH (AT3G30775) and PROLINE DEHYDROGENASE. The mRNA levels were determined using qRT-PCR, and the values are reported relative to *ZmPRMT1* mRNA levels in each line (**a**). The proline accumulation contents were shown in (**f**). Error bar is generated by three biological repetitions. The significance level is indicated by the asterisk at the top of the error bar. $p < 0.05$ is significantly different *; $p < 0.01$ is significantly different **.

3. Discussion

In plants, histone methylation is an important epigenetic modification involved in various biological processes by adjusting the homeostasis of histone methylation and demethylation [48,49]. Histone methyltransferase and demethylase are indispensable of accommodating the homeostasis of histone methylation in growth, development and responses to biotic and abiotic stresses in plants. Protein arginine methyltransferase (PRMT) can regulate the methylation status of arginine residues at histone tails and thereby activate or repress the transcription of target genes, which may play vital roles in plant growth and development and responses to abiotic stresses. In this study, a comprehensive set of eight nonredundant *ZmPRMT* genes were identified and analyzed from the AGPv4 version of maize (B73 inbred line) genome, including their phylogenetic relationships, gene and protein structures, conserved domain and motif architecture, chromosome location, gene duplication events and diverse expression profiles in maize developmental process and responses to various abiotic stresses. In addition, we also explored the effect of *ZmPRMT1* gene on heat stress and flowering time control via transgenic *Arabidopsis*.

Firstly, the analysis of phylogenetic relationship was carried out to provide insights into the evolution of *PRMT* gene family members and gene multiplicity in maize. In the present study, an unrooted phylogenetic tree was constructed through multiple sequence alignments of conserved PRMT domain-containing proteins from these putative PRMT proteins in maize and their representative orthologs from *Arabidopsis*, rice and sorghum. In our investigation and exploration of PRMT proteins, we discovered that maize *PRMT* genes were mainly clustered into three distinct subfamilies. In addition, the type IV PRMT has only been discovered in yeast so far [11–13]. Furthermore, although maize possesses a larger genome size (2300 Mbp) compared with *Arabidopsis* (125 Mbp) and rice (389 Mbp), only eight *PRMT* genes were identified and analyzed in maize. The number of *ZmPRMT* genes is similar to that in *Arabidopsis* (nine) or rice (eight). The unique phenomenon may be resulted from the occurrence of less gene duplication events in maize genome or undergoing a large gene loss during the process of maize genome duplication.

Furthermore, the analysis of conserved domain and motif architecture revealed evolutionarily conserved structure features and highly similarity among members of PRMT protein family, indicating that the PRMT proteins are highly conserved during the process of evolution. In addition, the gene structure map further reveals that the position and phase of intron/exon in the same group or subgroup exhibit considerable difference between maize and other plants, which may be resulted from too many nonfunctional sequences in the huge genome of maize. The genetic structure of maize *PRMT* genes is simpler than that of *PRMT* genes in other plants, but they share similar motif or domain distribution, implying that the protein structure in the same group or subgroup may be quite conserved during the process of evolution. For example, all of the ZmPRMT proteins investigated are mainly clustered into three distinct subclasses and each ZmPRMT protein contains the arginine methyltransferase-related conserved domains. It is noteworthy that the ZmPRMT1 and ZmPRMT2 proteins share two conserved domains, including PRMT5 (PF05185) and PRMT5_C, which act as the functional domain of arginine methyltransferase. In addition, the ZmPRMT2 protein has a complete PRMT_Tim domain at the N-terminal, which may play a crucial role in the quaternary structure of protein. In the subfamily II, ZmPRMT4 has only one conserved PrmA domain, while ZmPRMT3 has another REF domain at the C-terminal. Among the members of subfamily III, the majority of these proteins share several common conserved domains: MTS, PrmA, Methyltransf_11 and Methyltransf_25. Other proteins have different conserved arginine methyltransferase-related domains in different positions. In conclusion, the distribution of conserved domains of each protein may be closely linked to the evolutionary functions of each other. For example, it has been reported that the PrmA domain from *Arabidopsis* is dually targeted to chloroplasts and mitochondria. In addition, the conserved PrmA domain in photosynthetic eukaryotes accompanied by the methylated binding sites of translation factor to ribosome indicates that the PrmA domain in plants may participate in binding extra post-translational modifications or enhancing the function of ribosome [50]. Moreover, the REF domain (Apurinic/apyrimidinic endonuclease/redox factor) contained in ZmPRMT3 has been revealed to function in cellular responses to DNA damage or oxidative stress [51].

To further confirm potential functions of *ZmPRMT* genes in developmental processes and stress-responsiveness of them to abiotic stresses, the expression patterns were conducted by applying the available transcriptomic data at different developmental stages of maize [52]. The result demonstrated the *ZmPRMT* genes showed apparent differential expression patterns in the main tissues of different growth periods, indicating that these genes may function in growth and development processes of maize. Moreover, among 60 developmental periods, some genes were revealed to express in a tissue-specific manner, whereas the other genes were shown to express in a time-specific manner. For example, if we had divided the expression profile of these *PRMT* genes into two stages, we could found that some tissue/organ specific genes such as *ZmPRMT2*, *ZmPRMT3* and *ZmPRMT4* were revealed to express highly or specifically in some tissues of the first stage such as 6DAS_GH_WTeoptile, 6DAS_GH_Primary Root, VE_Primary Root, V1_Stem and SAM,

V3_Stem and SAM, V13_Immature Tassel and V18_Immature Cob, whereas the time-specific genes like *ZmPRMT1* were expressed highly in the whole second stage. Overall, the majority of *ZmPRMT* genes show obviously differential expression levels, suggesting that *ZmPRMT* genes may play crucial roles in the growth and development of maize.

Furthermore, previous studies have showed that plant PRMTs participate extensively in diverse regulation of transcriptional and post-transcriptional levels, which are involved in gene expression, mRNA processing, translation and intracellular signaling during the growth and development of plants [53]. For example, the PRMT1 can mediate asymmetric dimethylation of histone H4 arginine 3 (H4R3me2a), as well as a number of nonhistone proteins, such as C/EBPα12, Twist113 and Gli114, and thereby promote transcription [54]. Moreover, it has been reported that EgPRMT1 plays a critical role in the initiation and elongation of root hair, which is resulted from the methylation of β-tubulin that is associated with cytoskeleton formation [16]. Also, AtPRMT3 has been revealed to participate in RNA processing and ribosomal biogenesis in *Arabidopsis*. Moreover, AtPRMT5 belongs to the Type II PRMT and is extensively involved in pre-mRNA splicing [25], flowering time [24], salt stress tolerance [27], primary root length [22], root stem cell maintenance during DNA damage [26] and circadian rhythms [26]. In this study, to confirm the expression patterns and their stress-responsiveness of these newly-identified *ZmPRMT* genes in response to various stresses including heat, drought and salt treatments in three different tissues, the expression patterns of *ZmPRMTs* under three abiotic stresses were performed by means of qRT-PCR analysis in three different tissues. The results demonstrated that all *ZmPRMT* genes showed obviously differential expression levels in three distinct tissues under heat, drought and salt stress treatments. For instance, the majority of *ZmPRMT* genes were significantly upregulated in leaves at different time points after heat treatment, whereas most of *ZmPRMT* genes were downregulated in stems and roots after heat treatment. After drought treatment, the majority of *ZmPRMT* genes showed differential downregulated expression levels in leaves, whereas most of *ZmPRMT* genes in stems and roots showed significantly upregulated expression levels. Furthermore, the majority of *ZmPRMT* genes showed differential downregulated expression levels in all three tissues after salt treatment, and there was only 1 *ZmPRMT* gene (*ZmPRMT1*) in stems and three *ZmPRMT* genes (*ZmPRMT1*, *ZmPRMT2* and *ZmPRMT4*) in roots exhibited upregulated expression levels under salt treatment. Based on the above analysis, these results suggest that the majority of *ZmPRMT* genes may play potential roles in response to diverse abiotic stresses. One possible explanation is that this phenomenon may be resulted from the specifically temporal and spatial expression regulation when responding to abiotic stresses. Additionally, the promoter analysis further demonstrated that many stress- and hormone-related cis-elements in promoter regions of *ZmPRMT* genes might be involved in transcriptional regulation of diverse abiotic stress responses.

In plants, the timing of transition from vegetative to reproductive development has been revealed to become a critical adaptive trait, which is necessary for plants to accomplish flower development, pollination, and seed production in favorable conditions. Previous studies have showed that the continuous growth under low temperature conditions can accelerate flowering in most plant species [45]. This phenomenon is referred to vernalization, which is a key determinant involving the switch from vegetative to reproductive development. In *Arabidopsis*, the *FLOWERING LOCUS C (FLC)* is required for most vernalization-requiring *Arabidopsis* accessions transcriptionally. The *Arabidopsis FLC* gene has been revealed to encode a MADS domain protein that functions both in leaves and in the apical meristem as a repressor of flowering promoting genes: *SOC1* and *FT* [15,28,30,46,47]. Similar to *FLC*, the expression levels of *MAF* genes are usually dependent on vernalization regulation. Vernalization represses the expression of *MAF1*, *MAF2*, and *MAF3*, but it induces the expression of *MAF5* and does not significantly affect the expression of *MAF4* [55]. According to the previous studies, *AtPRMT5* and *AtPRMT10* showed critical effects on flowering time and *FLC* mRNA levels. In this study, *ZmPRMT1*, an orthologous gene of *AtPRMT5*, was assumed to function in regulating flowering time in plants. Thus, to further explore the role of *ZmPRMT1* in flowering time regulation, we constructed transgenic lines overexpressing *ZmPRMT1* gene in *Arabidopsis* for preliminary functional verification. Firstly, the results of phenotypic analyses revealed that the most common effect of transgenic *Arabidopsis* overexpressing *ZmPRMT1* was early flowering. Then, the flowering-related regulatory genes including *FLC*, *SOC1*, *FT*, and *MAF1* to *MAF5* were investigated through quantitative real-time PCR analysis, respectively. Consistent with our observations, the expression levels of *FLC* in transgenic *Arabidopsis* lines 2 and 6 decreased compared with that in WT plants after flowering, whereas the expression levels of *FT* and *SOC1* in transgenic *Arabidopsis* lines 2, 6 and 7 increased after flowering. In addition, the expression levels of *MAF1-4* genes showed a consistent decrease with the *FLC* expression level after flowering in all investigated *Arabidopsis* lines, whereas *MAF5* exhibited high transcript level after flowering in all investigated *Arabidopsis* lines. Furthermore, previous studies showed that some other *FLC* clade members are required for FLC protein binding to the chromatin associated with the *FT* and *SOC1* genes, indicating that these proteins may be involved in regulating flowering via MADS-domain complexes [56].

Therefore, we conclude that *ZmPRMT1* is required for the promotion of flowering in plants, which may control the floral transition in an FLC-dependent manner based on the above analysis. Moreover, histone arginine methylation has been elucidated to be a conserved epigenetic mechanism involving dynamics regulation of eukaryotic chromatin in three different methylation manners. Among them, monomethylation (MMA) and asymmetric dimethylation (ADMA) are generally involved in transcriptional activation, whereas symmetric dimethylation (SDMA) is associated with transcriptional silencing [9]. For instance, *Arabidopsis PRMT5* can symmetrically methylate some arginine residues of relative proteins involving RNA processing and histones, specifically histone 4 [10]. In this study, it can be speculated that *ZmPRMT1* might promote flowering time by repressing *FLC* expression level, as well as regulating other related *MAF* genes. Therefore, a schematic model of *ZmPRMT1*-mediated flowering time regulation in transgenic *Arabidopsis* was proposed in this study (Figure 12).

Plants are sessile organisms that are subject to constantly endure a variety of adverse environmental conditions, some of which can result in abiotic stress responses. High temperatures causing heat stress responses generally bring about the disturbance of cellular homeostasis and the impedance of growth and development in plants. One of the major stresses in crop plants is heat stress, which is usually accompanied by other stresses that are resulted from extra environmental conditions such as drought or salinity [57]. Accumulating studies have revealed that proline, as an essential multifunctional amino acid, as well as a stress signal, is extensively involved in multiple physiological pathways such as adjusting osmotic potential, scavenging reactive oxygen species and buffering redox reactions, and functions as a small molecular chaperone, as well as a plant development signal [58–61]. Moreover, previous studies have also revealed that short-term heat shock at 42 °C can result in proline accumulation in maize seedling and the exogenous application of proline can improve the level of endogenous free proline and thereby enhance the activities of antioxidant enzymes, followed by an increased heat tolerance of maize seedling [32]. Thus, we proposed a schematic model of the *ZmPRMT1* gene involved in proline catabolism for improving heat tolerance in transgenic *Arabidopsis* (Figure 13). In this study, our results demonstrated that the proline accumulation of WT and transgenic *Arabidopsis* lines under heat stress treatment all increased. However, the proline accumulation in transgenic *Arabidopsis* lines increased more significantly than that in WT plants. Furthermore, the analysis of qRT-PCR showed that *P5CS1* and *P5CR* gene in proline synthesis pathway exhibited upregulated expression levels in transgenic *Arabidopsis* lines compared with in WT plants after heat stress treatment, whereas *P5CDH* and *PDH* gene in proline degradation pathway exhibited significantly downregulated expression levels in transgenic *Arabidopsis* lines compared with in WT plants under heat stress treatment. Based on previous studies, *Arabidopsis AtPRMT5* usually functions in repressing target genes by symmetrically demethylating histone H4 Arginine 3 (H4R3me2s) in vitro [22]. If *ZmPRMT1* is involved in regulating proline-related genes directly in this study, it might be expected to reduce the expression levels of proline-related regulatory genes including *P5CS1*, *P5CR*, *P5CDH* and *PDH* in transgenic *Arabidopsis*. However, overexpression of the *ZmPRMT1* gene in transgenic *Arabidopsis* lines was revealed to result in increased expression levels of *P5CS1* and *P5CR* genes, as well as decreased expression levels of *P5CDH* and *PDH* genes. Together with these results, it can be speculated that the protein encoded by *ZmPRMT1* may be involved in both positive and negative regulatory effects on gene expression through modifying diverse histone methylation of specific arginine residues. Overall, the results suggested that the maize *ZmPRMT1* gene might play a crucial role in resisting heat stress response in plants. However, the underlying regulatory mechanism of this gene still remains to be further elucidated.

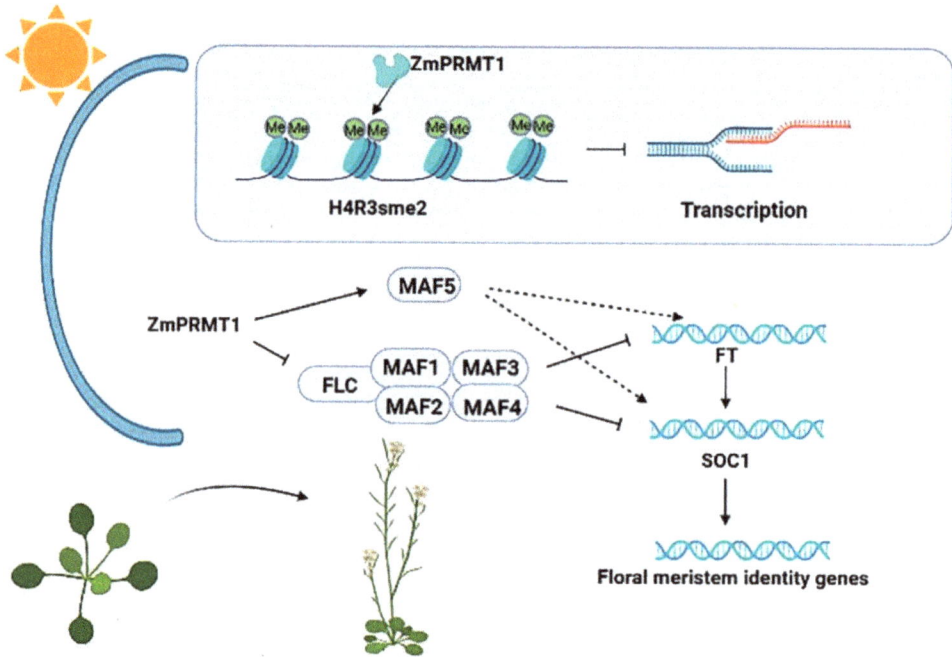

Figure 12. Hypothetical model for the regulation of flowering time controlled by *ZmPRMT1* in *Arabidopsis*. Overexpression of *ZmPRMT1* in *Arabidopsis* leads to downregulation of the expression levels of *FLC* and *MAF* genes (*MAF1-MAF4*), and its products assembled into MADS-box repressor complexes. However, the expression levels of *MAF5* were upregulated with an unclear manner in plant growth process. FLC and MADS domain complex forms large protein complexes at the *FT* and *SOC1* sites to repress the expression of these two genes. H4R3sme2 is one of the possible targets for *ZmPRMT1* to function. The arrows indicate positive modulation, and the "T" bars indicate negative modulation. The dashed lines represent assumed interactions.

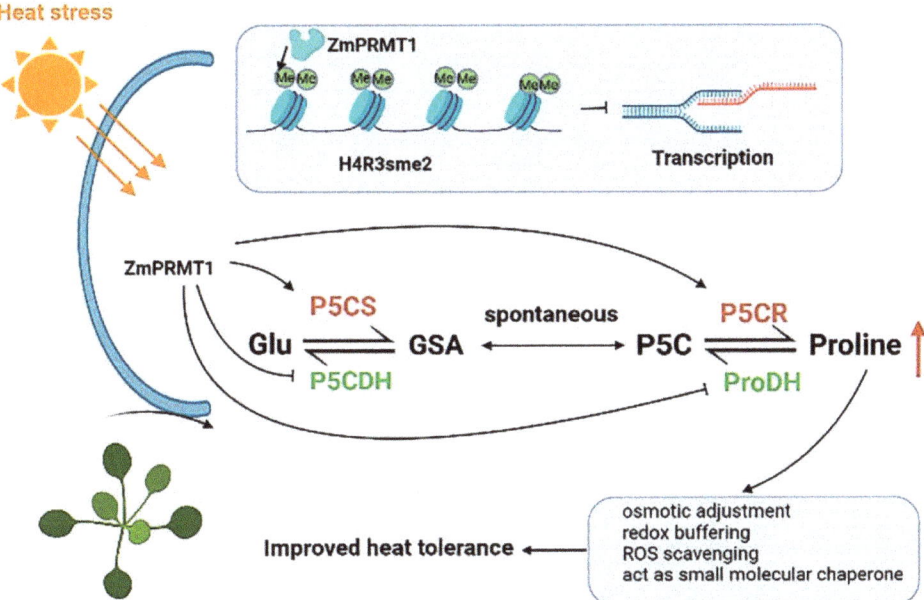

Figure 13. Hypothetical model showing the overexpression of *ZmPRMT1* results in improved resistance under heat stress in *Arabidopsis* involving proline metabolism pathway. The pathway of proline metabolism. Glu glutamic acid, GSA glutamate-1-semi-aldehyde, P5C pyrroline-5-carboxylate, P5CS1 delta-1-pyrroline-5-carboxylate synthase1, P5CR pyrroline-5-carboxylate reductase, P5CDH delta-1-pyrroline-5-carboxylate dehydrogenase, ProDH proline dehydrogenase. Overexpression of *ZmPRMT1* in *Arabidopsis* leads to upregulation of the expression levels of *P5CS1* and *P5CR* gene in proline synthesis pathway while downregulation of the expression levels of *P5CDH* and *PDH* gene in the proline degradation pathway. Finally, proline accumulation increased. H4R3sme2 is one of the possible targets for *ZmPRMT1* to function. The upregulation of gene expression levels is represented by red font, whereas downregulation of gene expression levels is represented by green font. The arrows indicate positive modulation, and the "T" bars indicate negative modulation. The dashed lines represent assumed interactions.

4. Materials and Methods

In this study, a comprehensive strategy combining bioinformatics and expression profiling analysis were used to identify all *ZmPRMT* genes and explore their function in response to abiotic stress. Furthermore, through the *Arabidopsis* model overexpressing maize *ZmPRMT1* gene, the molecular mechanism of *ZmPRMT1* gene responding to heat stress and promoting early flowering time was clarified in this study. The schematic flowchart of the study is shown in Figure 14.

4.1. Plant Materials and Growth Conditions

The plants of maize B73 inbred line were grown in the temperature incubator at 28 °C under long-day conditions (15 h light and 9 h dark) and 60% relative humidity. In order to explore the expression pattern of maize *ZmPRMT* genes under three abiotic stresses (heat, drought and salt), we set up three stress treatment methods. At 21 days after emergence, heat stress was induced by 42 °C. The maize seedlings grown in the dark incubator at 28 °C with enough water were used as the control. Maize seedlings in the same growth period were used as treat materials under salt stress. The seedlings of the treat group were cultured in the nutrient solution containing 200 mM NaCl, while the seedlings of the control group were grown in the nutrient solution lacking NaCl. After being treated with

NaCl, the maize seedlings were washed with distilled water for subsequent experiments. In the drought treatment, the three-week-old seedlings were gently pulled out of the soil and placed on clean white paper in a dark incubator as the treat group, while the seedlings of the control group were grown under normal conditions. The root, stem and leaf tissues of maize seedlings were collected at 0 h, 1 h, 2 h, 4 h and 8 h treated with heat, salt and drought stress and stored in liquid nitrogen to extract RNA.

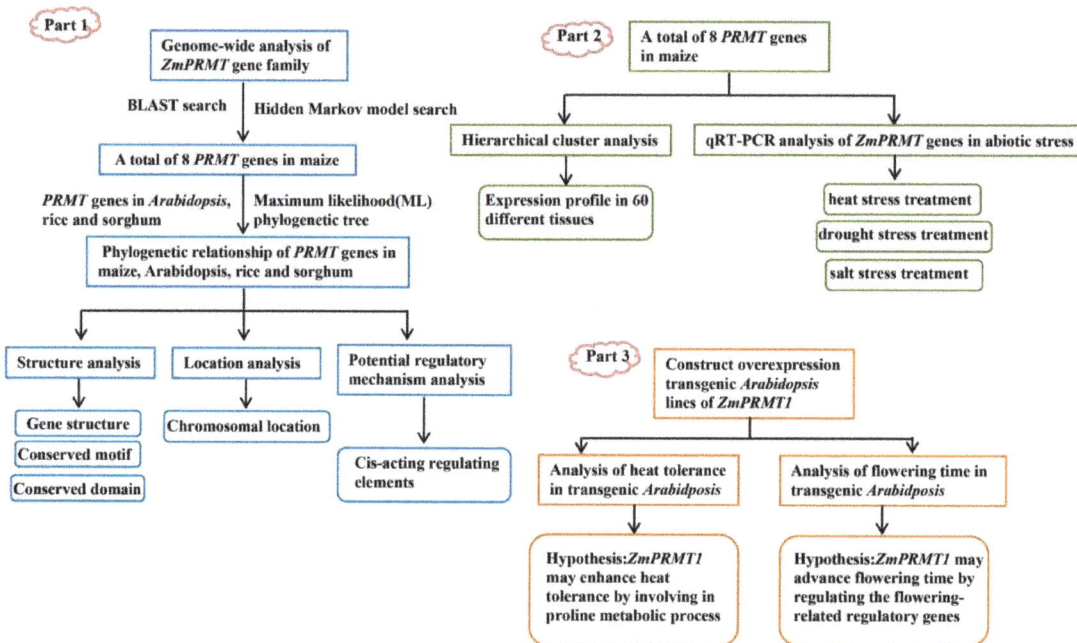

Figure 14. The schematic flowchart of the study.

4.2. Identification of the Members of PRMT Gene Family in Maize

To identify all possible orthologs of *PRMT* gene family in maize, we took the *PRMT* genes in *Arabidopsis*, rice and sorghum as homologous reference genes and carried out the following analysis. The whole-genome sequences of maize were downloaded from the maize genomic database (http://www.maizesequence.org/index.html) (accessed on 5 September 2021) by referring to our previous studies [49]. The BLASTP program (p-value $< 1 \times 10^{-5}$) was used to confirm proteins containing PRMT domain from the maize genomic database using the published eight *Arabidopsis* PRMTs (AtPRMTs), two rice PRMTs (OsPRMTs) and two sorghum PRMTs (SbPRMTs), with sequences of their proteins and PRMT domains as queries, respectively. The information of PRMT genes in these plants were listed in Supplementary Table S1. The Hidden Markov Model (HMM) program was used to identify putative PRMT protein sequences in maize genome database, which was downloaded from the SMART database (http://smart.embl-heidelberg.de/) (accessed on 3 September 2021) and Pfam database (http://pfam.xfam.org/) (accessed on 3 September 2021) [62,63]. Redundant protein sequences were removed manually by searching against the SMART database and the NCBI database (https://www.ncbi.nlm.nih.gov) (accessed on 3 September 2021). The newly identified genes were assigned by referring to the subfamily classification order of this gene family, combined with their phylogenetic relatedness to orthologous proteins of PRMTs in *Arabidopsis*, rice and sorghum. Then, the genome-wide files of maize B73 genome were downloaded to obtain the basic information of each *PRMT* gene from the maize genomic databases, including the length of CDS, the number of amino acids and the location of chromosomes. The physicochemical parameters of these putative proteins, including molecular weight (kDa) and isoelectric point (pI), were calculated using the online calculation pI/Mw tool ExPASy (http://www.expasy.org/tools/) (accessed on 10 September 2021), and the parameter was set to average [64]. Other relevant information was collected from the NCBI database. Similarly, the protein sequences of corresponding orthologs in *Arabidopsis*, rice and sorghum were also retrieved, and a dataset was created for bioinformatics analysis.

4.3. Phylogenetic Relationship, Gene Structure and Conserved Domain Analyses of the PRMT Gene Family in Maize

To clarify the phylogenetic relationship and structural characteristics among *PRMT* genes, we constructed an evolutionary tree, a gene structure map and a conservative structure map. All predicted PRMT protein sequences of maize and their orthologous sequences from *Arabidopsis*, rice and sorghum were aligned by Clustal-W software using default parameters [65,66]. All protein sequences were downloaded from the NCBI database. Then, MEGA7.0 program was used to analyze the phylogenetic relationship among these different species via the Maximum-Likelihood method with default parameters. The genetic structures of maize *PRMT* genes were obtained using online website GSDS (http://gsds.gao-lab.org/) (accessed on 12 September 2021), according to the alignment of the cDNA sequences with their corresponding genomic sequences. The Pfam database was used to investigate the PRMT conserved domains with default parameters. The conserved motifs of maize PRMT proteins encoded by putative *ZmPRMT* genes were analyzed by using MEME (https://meme-suite.org/meme/tools/meme) (accessed on 12 September 2021) online website with full-length protein sequences [67].

4.4. Chromosome Localization and Prediction of Cis-Acting Elements in Maize PRMT Genes

To investigate the position of genes on chromosomes and their possible functions, we completed the localization and cis-acting element analysis. The chromosome location of maize *PRMT* genes was obtained from the maize genomic database. The online mapping tool MG2C (http://mg2c.iask.in/mg2c_v2.0/) (accessed on 13 September 2021) is used to generate chromosome localization image of maize *PRMT* genes. In addition, the cis-acting elements in the promoter sequences can be involved in regulating gene transcription in abiotic stress responses, which depends on binding to different transcription factors. The 2 kb sequences upstream of the initiation codon (ATG) of each *ZmPRMT* gene were downloaded from the phytozome database (https://phytozome.jgi.doe.gov/pz/portal.html) (accessed on 13 September 2021) to further investigate the potential regulatory elements. The stress- and hormone-related cis-acting regulatory elements in promoter regions of *PRMT* genes in maize were analyzed by using the PlantCARE database (http://bioinformatics.psb.ugent.be/webtools/plantcare/html) (accessed on 13 September 2021) [68].

4.5. Expression Profile Analysis of Maize PRMT Genes in Developmental Tissues and Response to Abiotic Stress

To investigate the spatiotemporal expression patterns of *ZmPRMT* genes, the previously reported whole genome gene expression data of maize inbred line B73 was used to analyze the tissue-specific expression patterns of *ZmPRMT* genes as shown in Supplementary Table S2 [52]. Using Treeview software (http://jtreeview.sourceforge.net/) (accessed on 15 September 2021) and Cluster 3.0 (http://bonsai.hgc.jp/mdehoon/software/cluster/software.htm) (accessed on 15 September 2021), we performed the hierarchical clustering analysis to investigate the corresponding gene expression patterns at different developmental stages of maize, according to the Pearson coefficients with average linkage. The result was further visualized to obtain the heat map. To further confirm the influence of external environmental factors on the expression of *ZmPRMT* genes in maize, the qRT-PCR expression profiles of leaves, stems and roots after abiotic stress treatment were analyzed. The several four-week B73 seedlings under similar growth status were transferred to the incubator of 42 °C and sampled the roots, stem and leaf of 0 h, 1 h, 2 h, 4 h and 8 h (0 h as the control groups). To ensure the reliability of the quantitative analysis, at least 3 biological repeats were completed. Total RNA was extracted from the collected tissues via Trizol reagent (Invitrogen, USA) and was purified to remove genomic DNA contamination using DNase I. Reverse transcription reactions were performed using 1μg of total RNA by the QuantiTectRev. Transcription Kit (Qiangen, Germany). The qRT-PCR reverse transcription reactions were carried out using a CFX96™ Real-Time PCR Detection System (Bio-Rad Laboratories, Inc., USA). Each cDNA sample was biologically replicated at least three times. The maize *Actin1* gene was used as a standardized internal control. The relative mRNA levels of PRMT genes were calculated by using $2^{-\Delta\Delta CT}$ method to compare the fold changes of related gene expression levels. The total primers in this study were listed in Supplementary Table S3.

4.6. Vector Construction and Arabidopsis Genetic Transformation

To verify the biological roles of *ZmPRMT* genes, the transgenic *Arabidopsis* plants overexpressing *ZmPRMT1* gene was constructed as a typical representative to explore functional roles in plants. The plasmid was constructed by amplifying the *ZmPRMT1* gene using PCR primers carrying proper restriction enzyme sites. The amplified *ZmPRMT1* cDNA was cloned into PHB vector driven by 35S promoter. The primers with enzyme sites are shown in Supplementary Table S4. The vectors were transformed into the Agrobacterium strain for transformation into *Arabidopsis* by the floral dip method. Seeds of the T0 and T1 generations were selected on 1/2 MS plates supplemented with hygromycin (30 ug/mL) and confirmed by PCR with primers (Hyg-F: GGTCGCGGAGGCTATGGATGC; Hyg-R: GCTTCTGCGGGCGATTTGTGT). The presence of *ZmPRMT1* was verified by qRT-PCR using specific primers. Three independent homozygous overexpression lines 2, 6 and 7 were selected, and the expression level of *ZmPRMT1* gene was quantified. At least 3 biological repeats were taken for all samples to ensure the reliability of the results of qRT-PCR analysis.

4.7. Flowering Time Assessment of Transgenic Arabidopsis

To verify the effect of *ZmPRMT1* gene on flowering time of *Arabidopsis*, we observed the phenotypic changes of *Arabidopsis* before and after flowering. Seedlings of WT and transgenic *Arabidopsis* were grown in the temperature incubator at 22 °C under day-long conditions (16 h light and 8 h dark) and 60% relative humidity. Flowering time of *Arabidopsis* was calculated by counting the number of rosette leaves after flowering. At least 3 plants were counted for each line.

4.8. Heat Tolerance Assay of Transgenic Arabidopsis

To verify the function of *ZmPRMT1* gene on heat tolerance of *Arabidopsis*, we conducted heat stress treatment on transgenic *Arabidopsis*. The seeds of *Arabidopsis* were sterilized with 75% alcohol and then vernalized at low temperature on the plate. Then the sprouted Arabidopsis seedlings are transplanted into the soil. Seedlings of *Arabidopsis* were also grown in the temperature incubator at 22 °C with a growth photoperiod of 16 h light and 8 h dark and 60% relative humidity. After 21 days, seedlings were exposed to heat stress (42 °C) for 8 h. Seedlings grown in the dark incubator at 22 °C with enough water were used as the control, and seedling leaves were collected after heat stress treatment and stored at −80 °C with liquid nitrogen to isolate RNA.

4.9. RNA Extraction and Quantitative Real-Time PCR Analysis of Transgenic Arabidopsis

To verify the biological function of *ZmPRMT1* gene in the transgenic *Arabidopsis*, we conducted qRT-PCR experiments to verify the expression profile of related genes. Based on the manufacturer's instruction, total RNA was isolated from the collected samples using Trizol RNA isolation (USA). For quantitative real-time PCR, cDNA from three distinct biological samples was used for analysis. The PCR condition was carried out as follows: pre-denaturation for 15 min at 95 °C, 40 cycles at 95 °C for 10 s, 55 °C for 30 s and

72 °C for 30 s. The gene expression levels were calculated using the $2^{-\triangle\triangle Ct}$ method [69]. Each experiment was conducted in the form of at least three technologies and biological replication. The total primers in this study were listed in Supplementary Table S3.

4.10. Physiological Parameter Determination

In order to explore the function of *ZmPRMT1* gene on the change of proline accumulation under heat stress, we conducted physiological experiments to determine the change of proline accumulation. The content of proline was measured using the proline assay kit (Nanjing Jiancheng, Nanjing, China). The principle of determination is that in plants, only proline can react with acid ninhydrin to produce stable red compounds. The product has a maximum absorption peak at 520 nm, and its absorption value is linearly related to the content of proline. The testing tube, contrast tube and blank tube were boiled about 30 min, and the absorbance was measured on a spectrophotometer at 520 nm using double-distilled water as a standard. The experiment was performed for repeating at least three times.

5. Conclusions

In conclusion, we identified eight *ZmPRMT* genes encoding protein arginine methyltransferases in this study. The classification, evolutionary relationship, conserved motifs and domains, gene structure and stress-responsive cis-regulatory elements in promoter regions of *ZmPRMT* genes were determined to provide insights into their potential functions in this study. The comprehensive expression profiles of all maize *PRMT* genes in different tissues revealed that these genes may play functional roles at diverse developmental stages of maize. Furthermore, the expression levels of *ZmPRMT* genes were upregulated or downregulated under three various abiotic stress treatments, implying their potential roles in abiotic stress responses. What's more, the construction of transgenic *Arabidopsis* lines further demonstrated the *ZmPRMT1* gene may play a crucial role in flowering time regulation and heat resistance. Compared with the WT plants, the transgenic *Arabidopsis* plants exhibited early flowering by modulating the expression of flowering-related regulatory genes, which might result in alteration in expression levels of floral meristem identity genes. In addition, the transgenic *Arabidopsis* exhibited enhanced heat tolerance through modulating proline metabolism related genes, which might result in changes in proline accumulation. Finally, via analyzing the expression levels of *ZmPRMT1* gene in flowering process and under heat stress treatment, we conclude that *ZmPRMT1* may play a critical role in growth and development processes, as well as abiotic stress responses. Taken together, the exploration of the maize *PRMT* gene family in organization, structure, evolution, expression profiling and the preliminary functional verification of *ZmPRMT1* will facilitate future functional verification of *ZmPRMT* genes and provide an important theoretical basis for a better comprehension of molecular epigenetic mechanism of regulating flowering and adaptation to abiotic stresses in maize. What's more important, this study provides new insights and ideas for applying epigenetic methods to crop breeding.

Supplementary Materials: The supporting information can be downloaded at: https://www.mdpi.com/article/10.3390/ijms232112793/s1.

Author Contributions: Project administration, Y.Q. and J.L.; funding acquisition, Y.Q.; supervision, Y.Q.; writing—original draft preparation, Q.L. and J.L.; investigation, X.L. and Y.Z.; data curation, Q.L. and J.L.; methodology, Q.L., J.L., X.L. and Y.Z. All authors have read and agreed to the published version of the manuscript.

Funding: This study was supported by grants from the National Natural Science Foundation of China (NSFC) (Grant NO. 31571673) and the open fundings of National Engineering Laboratory of Crop Stress Resistance Breeding (Grant NO. KNZJ1023) and Anhui Provincial Key Laboratory of the Conservation and Exploitation of Biological Resources (Grant NO. Swzy202003) and Anhui Provincial Academic Funding Project for Top Talents in Disciplines (Majors) (Grant NO. gxbjZD 2021044).

Institutional Review Board Statement: Not applicable.

Informed Consent Statement: Not applicable.

Data Availability Statement: Not applicable.

Acknowledgments: The authors are grateful to the editor and reviewers for critically evaluating the manuscript and providing constructive comments for its improvement.

Conflicts of Interest: The authors declare no conflict of interest. The funders had no role in the study design, collection, analysis and interpretation of data, or in the writing of the report or decision to submit the article for publication.

References

1. Egger, G.; Liang, G.N.; Aparicio, A.; Jones, P.A. Epigenetics in human disease and prospects for epigenetic therapy. *Nature* **2004**, *429*, 457–463. [CrossRef]
2. Holliday, R. DNA methylation and epigenetic defects in carcinogenesis. *Mutat. Res.* **1987**, *181*, 215–217. [CrossRef]
3. Boyko, A.; Kovalchuk, I. Genetic and Epigenetic Effects of Plant-Pathogen Interactions: An Evolutionary Perspective. *Mol. Plant* **2011**, *4*, 1014–1023. [CrossRef] [PubMed]
4. Liu, C.; Lu, F.; Cui, X.; Cao, X. Histone methylation in higher plants. *Annu. Rev. Plant Biol.* **2010**, *61*, 395–420. [CrossRef] [PubMed]
5. Turner, B.M. Cellular memory and the histone code. *Cell* **2002**, *111*, 285–291. [CrossRef]
6. Di Lorenzo, A.; Bedford, M.T. Histone arginine methylation. *FEBS Lett.* **2011**, *585*, 2024–2031. [CrossRef]
7. Gary, J.D.; Clarke, S. RNA and protein interactions modulated by protein arginine methylation. *Prog. Nucleic Acid Res. Mol. Biol.* **1998**, *61*, 65–131.
8. Lee, J.H.; Cook, J.R.; Yang, Z.H.; Mirochnitchenko, O.; Gunderson, S.I.; Felix, A.M.; Herth, N.; Hoffmann, R.; Pestka, S. PRMT7, a new protein arginine methyltransferase that synthesizes symmetric dimethylarginine. *J. Biol. Chem.* **2005**, *280*, 3656–3664. [CrossRef]
9. Nishioka, K.; Reinberg, D. Methods and tips for the purification of human histone methyltransferases. *Methods* **2003**, *31*, 49–58. [CrossRef]
10. Gonsalvez, G.B.; Tian, L.; Ospina, J.K.; Boisvert, F.M.; Lamond, A.I.; Matera, A.G. Two distinct arginine methyltransferases are required for biogenesis of Sm-class ribonucleoproteins. *J. Cell Biol.* **2007**, *178*, 733–740. [CrossRef]
11. Shen, E.C.; Henry, M.F.; Weiss, V.H.; Valentini, S.R.; Silver, P.A.; Lee, M.S. Arginine methylation facilitates the nuclear export of hnRNP proteins. *Genes Dev.* **1998**, *12*, 679–691. [CrossRef] [PubMed]
12. McBride, A.E.; Zurita-Lopez, C.; Regis, A.; Blum, E.; Conboy, A.; Elf, S.; Clarke, S. Protein arginine methylation in Candida albicans: Role in nuclear transportv. *Eukaryot. Cell* **2007**, *6*, 1119–1129. [CrossRef] [PubMed]
13. Margueron, R.; Trojer, P.; Reinberg, D. The key to development: Interpreting the histone code? *Curr. Opin. Genet. Dev.* **2005**, *15*, 163–176. [CrossRef] [PubMed]
14. Ahmad, A.; Dong, Y.Z.; Cao, X.F. Characterization of the *PRMT* Gene Family in Rice Reveals Conservation of Arginine Methylation. *PLoS ONE* **2011**, *6*, e22664. [CrossRef]
15. Niu, L.F.; Lu, F.L.; Pei, Y.X.; Liu, C.Y.; Cao, X.F. Regulation of flowering time by the protein arginine methyltransferase AtPRMT10. *Embo Rep.* **2007**, *8*, 1190–1195. [CrossRef]
16. Plett, K.L.; Raposo, A.E.; Bullivant, S.; Anderson, I.C.; Piller, S.C.; Plett, J.M. Root morphogenic pathways in *Eucalyptus grandis* are modified by the activity of protein arginine methyltransferases. *BMC Plant Biol.* **2017**, *17*, 62. [CrossRef]
17. Liew, L.C.; Singh, M.B.; Bhalla, P.L. An RNA-seq transcriptome analysis of histone modifiers and RNA silencing genes in soybean during floral initiation process. *PLoS ONE* **2013**, *8*, e77502.
18. Ahmad, A.; Cao, X. Plant PRMTs broaden the scope of arginine methylation. *J. Genet. Genom.* **2012**, *39*, 195–208. [CrossRef]
19. Bedford, M.T. Arginine methylation at a glance. *J. Cell Sci.* **2007**, *120*, 4243–4246. [CrossRef]
20. Hang, R.L.; Liu, C.Y.; Ahmad, A.; Zhang, Y.; Lu, F.L.; Cao, X.F. *Arabidopsis* protein arginine methyltransferase 3 is required for ribosome biogenesis by affecting precursor ribosomal RNA processing. *Proc. Natl. Acad. Sci. USA* **2014**, *111*, 16190–16195. [CrossRef]
21. Niu, L.F.; Zhang, Y.; Pei, Y.X.; Liu, C.Y.; Cao, X.F. Redundant requirement for a pair of PROTEIN ARGININE METHYLTRANS-FERASE4 homologs for the proper regulation of *Arabidopsis* flowering time. *Plant Physiol.* **2008**, *148*, 490–503. [CrossRef]
22. Pei, Y.X.; Niu, L.F.; Lu, F.L.; Liu, C.Y.; Zhai, J.X.; Kong, X.F.; Cao, X.F. Mutations in the type II protein arginine methyltransferase AtPRMT5 result in pleiotropic developmental defects in *Arabidopsis*. *Plant Physiol.* **2007**, *144*, 1913–1923. [CrossRef] [PubMed]
23. Wang, X.; Zhang, Y.; Ma, Q.B.; Zhang, Z.L.; Xue, Y.B.; Bao, S.L.; Chong, K. SKB1-mediated symmetric dimethylation of histone H4R3 controls flowering time in *Arabidopsis*. *EMBO J.* **2007**, *26*, 1934–1941. [CrossRef] [PubMed]
24. Deng, X.A.; Gu, L.F.; Liu, C.Y.; Lu, T.C.; Lu, F.L.; Lu, Z.K.; Cui, P.; Pei, Y.X.; Wang, B.C.; Hu, S.N.; et al. Arginine methylation mediated by the *Arabidopsis* homolog of PRMT5 is essential for proper pre-mRNA splicing. *Proc. Natl. Acad. Sci. USA* **2010**, *107*, 19114–19119. [CrossRef]
25. Sanchez, S.E.; Petrillo, E.; Beckwith, E.J.; Zhang, X.; Rugnone, M.L.; Hernando, C.E.; Cuevas, J.C.; Herz, M.A.G.; Depetris-Chauvin, A.; Simpson, C.G.; et al. A methyl transferase links the circadian clock to the regulation of alternative splicing. *Nature* **2010**, *468*, 112–116. [CrossRef]

26. Hong, S.; Song, H.R.; Lutz, K.; Kerstetter, R.A.; Michael, T.P.; McClung, C.R. Type II protein arginine methyltransferase 5 (PRMT5) is required for circadian period determination in *Arabidopsis thaliana*. *Proc. Natl. Acad. Sci. USA* **2010**, *107*, 21211–21216. [CrossRef]
27. Zhang, Z.L.; Zhang, S.P.; Zhang, Y.; Wang, X.; Li, D.; Li, Q.L.; Yue, M.H.; Li, Q.; Zhang, Y.E.; Xu, Y.Y.; et al. *Arabidopsis* Floral Initiator SKB1 Confers High Salt Tolerance by Regulating Transcription and Pre-mRNA Splicing through Altering Histone H4R3 and Small Nuclear Ribonucleoprotein LSM4 Methylation. *Plant Cell* **2011**, *23*, 396–411. [CrossRef]
28. Coupland, G. Genetic and environmental-control of flowering time *Arabidopsis*. *Trends Genet.* **1995**, *11*, 393–397. [CrossRef]
29. Shindo, C.; Aranzana, M.J.; Lister, C.; Baxter, C.; Nicholls, C.; Nordborg, M.; Dean, C. Role of FRIGIDA and FLOWERING LOCUS C in determining variation in flowering time of *Arabidopsis*. *Plant Physiol.* **2005**, *138*, 1163–1173. [CrossRef] [PubMed]
30. Helliwell, C.A.; Wood, C.C.; Robertson, M.; Peacock, W.J.; Dennis, E.S. The *Arabidopsis* FLC protein interacts directly in vivo with SOC1 and FT chromatin and is part of a high-molecular-weight protein complex. *Plant J.* **2006**, *46*, 183–192. [CrossRef] [PubMed]
31. Chang, Y.N.; Zhu, C.; Jiang, J.; Zhang, H.M.; Zhu, J.K.; Duan, C.G. Epigenetic regulation in plant abiotic stress responses. *J. Integr. Plant Biol.* **2020**, *62*, 563–580. [CrossRef]
32. Li, Z.G.; Ding, X.J.; Du, P.F. Hydrogen sulfide donor sodium hydrosulfide-improved heat tolerance in maize and involvement of proline. *J. Plant Physiol.* **2013**, *170*, 741–747. [CrossRef] [PubMed]
33. Liu, J.; Wang, Y.S. Proline metabolism and molecular cloning of *AmP5CS* in the mangrove Avicennia marina under heat stress. *Ecotoxicology* **2020**, *29*, 698–706. [CrossRef]
34. Ter Kuile, B.H.; Opperdoes, F.R. A chemostat study on proline uptake and metabolism of Leishmania donovani. *J. Protozool.* **1992**, *39*, 555–558. [CrossRef] [PubMed]
35. Verbruggen, N.; Hermans, C. Proline accumulation in plants: A review. *Amino Acids* **2008**, *35*, 753–759. [CrossRef]
36. Kishor, P.B.K.; Hong, Z.L.; Miao, G.H.; Hu, C.A.A.; Verma, D.P.S. Overexpression of [delta]-pyrroline-5-carboxylate synthetase increases proline production and confers osmotolerance in transgenic plants. *Plant Physiol.* **1995**, *108*, 1387–1394. [CrossRef]
37. Solomon, A.; Beer, S.; Waisel, Y.; Jones, G.P.; Paleg, L.G. Effects of NaCl on the carboxylating activity of Rubisco from Tamarix jordanis in the presence and absence of proline-related compatible solutes. *Physiol. Plant.* **1994**, *90*, 198–204. [CrossRef]
38. Brands, S.; Schein, P.; Castro-Ochoa, K.F.; Galinski, E.A. Hydroxyl radical scavenging of the compatible solute ectoine generates two N-acetimides. *Arch. Biochem. Biophys.* **2019**, *674*, 108097. [CrossRef] [PubMed]
39. Hare, P.D.; Cress, W.A. Metabolic implications of stress-induced proline accumulation in plants. *Plant Growth Regul.* **1997**, *21*, 79–102. [CrossRef]
40. Hayama, R.; Coupland, G. The molecular basis of diversity in the photoperiodic flowering responses of *Arabidopsis* and rice. *Plant Physiol.* **2004**, *135*, 677–684. [CrossRef]
41. Brutnell, T.P.; Bennetzen, J.L.; Vogel, J.P. *Brachypodium distachyon* and Setaria viridis: Model Genetic Systems for the Grasses. *Annu. Rev. Plant Biol.* **2015**, *66*, 465–485. [CrossRef] [PubMed]
42. Cockram, J.; Jones, H.; Leigh, F.J.; O'Sullivan, D.; Powell, W.; Laurie, D.A.; Greenland, A.J. Control of flowering time in temperate cereals: Genes, domestication, and sustainable productivity. *J. Exp. Bot.* **2007**, *58*, 1231–1244. [CrossRef]
43. Higgins, J.A.; Bailey, P.C.; Laurie, D.A. Comparative genomics of flowering time pathways using *Brachypodium distachyon* as a model for the temperate grasses. *PLoS ONE* **2010**, *5*, e10065. [CrossRef]
44. Ratcliffe, O.J.; Nadzan, G.C.; Reuber, T.L.; Riechmann, J.L. Regulation of flowering in *Arabidopsis* by an FLC homologue. *Plant Physiol.* **2001**, *126*, 122–132. [CrossRef] [PubMed]
45. Alexandre, C.M.; Hennig, L. FLC or not FLC: The other side of vernalization. *J. Exp. Bot.* **2008**, *59*, 1127–1135. [CrossRef]
46. Zhao, X.Y.; Li, J.R.; Lian, B.; Gu, H.Q.; Li, Y.; Qi, Y.J. Global identification of *Arabidopsis* lncRNAs reveals the regulation of MAF4 by a natural antisense RNA. *Nat. Commun.* **2018**, *9*, 5056. [CrossRef]
47. Blazquez, M.A.; Ahn, J.H.; Weigel, D. A thermosensory pathway controlling flowering time in *Arabidopsis thaliana*. *Nat. Genet.* **2003**, *33*, 168–171. [CrossRef]
48. Qian, Y.; Xi, Y.; Cheng, B.; Zhu, S. Genome-wide identification and expression profiling of DNA methyltransferase gene family in maize. *Plant Cell Rep.* **2014**, *33*, 1661–1672. [CrossRef]
49. Qian, Y.X.; Chen, C.L.; Jiang, L.Y.; Zhang, J.; Ren, Q.Y. Genome-wide identification, classification and expression analysis of the JmjC domain-containing histone demethylase gene family in maize. *BMC Genom.* **2019**, *20*, 256. [CrossRef]
50. Mazzoleni, M.; Figuet, S.; Martin-Laffon, J.; Mininno, M.; Gilgen, A.; Leroux, M.; Brugiere, S.; Tardif, M.; Alban, C.; Ravanel, S. Dual Targeting of the Protein Methyltransferase PrmA Contributes to Both Chloroplastic and Mitochondrial Ribosomal Protein L11 Methylation in *Arabidopsis*. *Plant Cell Physiol.* **2015**, *56*, 1697–1710. [CrossRef] [PubMed]
51. Choi, S.; Joo, H.K.; Jeon, B.H. Dynamic Regulation of APE1/Ref-1 as a Therapeutic Target Protein. *Chonnam Med. J.* **2016**, *52*, 75–80. [CrossRef] [PubMed]
52. Sekhon, R.S.; Lin, H.; Childs, K.L.; Hansey, C.N.; Buell, C.R.; de Leon, N.; Kaeppler, S.M. Genome-wide atlas of transcription during maize development. *Plant J.* **2011**, *66*, 553–563. [CrossRef]
53. Guccione, E.; Richard, S. The regulation, functions and clinical relevance of arginine methylation. *Nat. Rev. Mol. Cell Biol.* **2019**, *20*, 642–657. [CrossRef] [PubMed]
54. Wang, G.H.; Wang, C.F.; Hou, R.; Zhou, X.Y.; Li, G.T.; Zhang, S.J.; Xu, J.R. The AMT1 Arginine Methyltransferase Gene Is Important for Plant Infection and Normal Hyphal Growth in *Fusarium graminearum*. *PLoS ONE* **2012**, *7*, e38324. [CrossRef] [PubMed]

55. Ratcliffe, O.J.; Kumimoto, R.W.; Wong, B.J.; Riechmann, J.L. Analysis of the *Arabidopsis* MADS AFFECTING FLOWERING gene family: MAF2 prevents vernalization by short periods of cold. *Plant Cell* **2003**, *15*, 1159–1169. [CrossRef]
56. Gu, X.; Le, C.; Wang, Y.; Li, Z.; Jiang, D.; Wang, Y.; He, Y. *Arabidopsis* FLC clade members form flowering-repressor complexes coordinating responses to endogenous and environmental cues. *Nat. Commun.* **2013**, *4*, 1947. [CrossRef]
57. Prasad, P.V.V.; Bheemanahalli, R.; Jagadish, S.V.K. Field crops and the fear of heat stress-Opportunities, challenges and future directions. *Field Crops Res.* **2017**, *200*, 114–121. [CrossRef]
58. Hare, P.D.; Cress, W.A.; van Staden, J. Proline synthesis and degradation: A model system for elucidating stress-related signal transduction. *J. Exp. Bot.* **1999**, *50*, 413–434. [CrossRef]
59. Fracheboud, Y.; Haldimann, P.; Leipner, J.; Stamp, P. Chlorophyll fluorescence as a selection tool for cold tolerance of photosynthesis in maize (*Zea mays* L.). *J. Exp. Bot.* **1999**, *50*, 1533–1540. [CrossRef]
60. Mattioli, R.; Falasca, G.; Sabatini, S.; Altamura, M.M.; Costantino, P.; Trovato, M. The proline biosynthetic genes *P5CS1* and *P5CS2* play overlapping roles in *Arabidopsis* flower transition but not in embryo development. *Physiol. Plant.* **2009**, *137*, 72–85. [CrossRef]
61. Verslues, P.E.; Sharma, S. Proline metabolism and its implications for plant-environment interaction. *Arab. Book* **2010**, *8*, e0140. [CrossRef] [PubMed]
62. Finn, R.D.; Mistry, J.; Schuster-Bockler, B.; Griffiths-Jones, S.; Hollich, V.; Lassmann, T.; Moxon, S.; Marshall, M.; Khanna, A.; Durbin, R.; et al. Pfam: Clans, web tools and services. *Nucleic Acids Res.* **2006**, *34*, D247–D251. [CrossRef] [PubMed]
63. Letunic, I.; Copley, R.R.; Schmidt, S.; Ciccarelli, F.D.; Doerks, T.; Schultz, J.; Ponting, C.P.; Bork, P. SMART 4.0: Towards genomic data integration. *Nucleic Acids Res.* **2004**, *32*, D142–D144. [CrossRef] [PubMed]
64. Gasteiger, E.; Gattiker, A.; Hoogland, C.; Ivanyi, I.; Appel, R.D.; Bairoch, A. ExPASy: The proteomics server for in-depth protein knowledge and analysis. *Nucleic Acids Res.* **2003**, *31*, 3784–3788. [CrossRef] [PubMed]
65. Larkin, M.A.; Blackshields, G.; Brown, N.P.; Chenna, R.; McGettigan, P.A.; McWilliam, H.; Valentin, F.; Wallace, I.M.; Wilm, A.; Lopez, R.; et al. Clustal W and clustal X version 2.0. *Bioinformatics* **2007**, *23*, 2947–2948. [CrossRef]
66. Zhu, X.Z.; Zhang, Y.X.; Liu, X.; Hou, D.Y.; Gao, T. Authentication of commercial processed *Glehniae radix* (*Beishashen*) by DNA barcodes. *Chin. Med.* **2015**, *10*, 35. [CrossRef]
67. Bailey, T.L.; Johnson, J.; Grant, C.E.; Noble, W.S. The MEME Suite. *Nucleic Acids Res.* **2015**, *43*, W39–W49. [CrossRef]
68. Lescot, M.; Dehais, P.; Thijs, G.; Marchal, K.; Moreau, Y.; Van de Peer, Y.; Rouze, P.; Rombauts, S. PlantCARE, a database of plant cis-acting regulatory elements and a portal to tools for in silico analysis of promoter sequences. *Nucleic Acids Res.* **2002**, *30*, 325–327. [CrossRef]
69. Rao, X.; Huang, X.; Zhou, Z.; Lin, X. An improvement of the 2^(−delta delta CT) method for quantitative real-time polymerase chain reaction data analysis. *Biostat. Bioinform. Biomath.* **2013**, *3*, 71–85.

Article

Transcriptome Analysis Reveals the Genes Related to Pollen Abortion in a Cytoplasmic Male-Sterile Soybean (*Glycine max* (L.) Merr.)

Zhiyuan Bai [1,2,†], Xianlong Ding [1,†], Ruijun Zhang [2], Yuhua Yang [2], Baoguo Wei [2], Shouping Yang [1,*] and Junyi Gai [1,*]

1. Soybean Research Institute, National Center for Soybean Improvement, Key Laboratory of Biology and Genetic Improvement of Soybean (General, Ministry of Agriculture and Rural Affairs of the People's Republic of China), State Key Laboratory of Crop Genetics and Germplasm Enhancement, Jiangsu Collaborative Innovation Center for Modern Crop Production, College of Agriculture, Nanjing Agricultural University, Nanjing 210095, China
2. Center for Agricultural Genetic Resources Research, Shanxi Agricultural University, Taiyuan 030031, China
* Correspondence: spyang@njau.edu.cn (S.Y.); sri@njau.edu.cn (J.G.)
† These authors contributed equally to this work.

Abstract: Cytoplasmic male sterility (CMS) lays a foundation for the utilization of heterosis in soybean. The soybean CMS line SXCMS5A is an excellent CMS line exhibiting 100% male sterility. Cytological analysis revealed that in SXCMS5A compared to its maintainer SXCMS5B, its tapetum was vacuolated and abnormally developed. To identify the genes and metabolic pathways involving in pollen abortion of SXCMS5A, a comparative transcriptome analysis was conducted between SXCMS5A and SXCMS5B using flower buds. A total of 372,973,796 high quality clean reads were obtained from 6 samples (3 replicates for each material), and 840 differentially expressed genes (DEGs) were identified, including 658 downregulated and 182 upregulated ones in SXCMS5A compared to SXCMS5B. Among them, 13 DEGs, i.e., 12 open reading frames (ORFs) and 1 *COX2*, were mitochondrial genome genes in which *ORF178* and *ORF103c* were upregulated in CMS lines and had transmembrane domain(s), therefore, identified as CMS candidate mitochondrial genes of SXCMS5A. Furthermore, numerous DEGs were associated with pollen wall development, carbohydrate metabolism, sugar transport, reactive oxygen species (ROS) metabolism and transcription factor. Some of them were further confirmed by quantitative real time PCR analysis between CMS lines with the same cytoplasmic source as SXCMS5A and their respective maintainer lines. The amount of soluble sugar and adenosine triphosphate and the activity of catalase and ascorbic acid oxidase showed that energy supply and ROS scavenging decreased in SXCMS5A compared to SXCMS5B. These findings provide valuable information for further understanding the molecular mechanism regulating the pollen abortion of soybean CMS.

Keywords: soybean (*Glycine max* (L.) Merr.); cytoplasmic male sterility; pollen abortion; gene expression; RNA sequencing

Citation: Bai, Z.; Ding, X.; Zhang, R.; Yang, Y.; Wei, B.; Yang, S.; Gai, J. Transcriptome Analysis Reveals the Genes Related to Pollen Abortion in a Cytoplasmic Male-Sterile Soybean (*Glycine max* (L.) Merr.). *Int. J. Mol. Sci.* **2022**, *23*, 12227. https://doi.org/10.3390/ijms232012227

Academic Editors: Jian Zhang and Zhiyong Li

Received: 22 September 2022
Accepted: 11 October 2022
Published: 13 October 2022

Publisher's Note: MDPI stays neutral with regard to jurisdictional claims in published maps and institutional affiliations.

Copyright: © 2022 by the authors. Licensee MDPI, Basel, Switzerland. This article is an open access article distributed under the terms and conditions of the Creative Commons Attribution (CC BY) license (https://creativecommons.org/licenses/by/4.0/).

1. Introduction

A common biological phenomenon in nature, heterosis serves as an efficient agricultural approach for increasing crop yield. Utilization of heterosis in rice and corn could increase crop yield by 15% to 50% [1,2]. Research on utilization of soybean heterosis started late and progressed slowly. The discovery of soybean cytoplasmic male sterility (CMS) laid a foundation for the utilization of soybean heterosis [3,4]. At present, a number of research institutions in China had realized a three-line support system for hybrid soybean production [5,6].

Plant male sterility refers to the phenomenon that plants cannot produce functional pollen. Plant male sterility could be used not only as an important tool for heterosis utilization [7] but also as an ideal material for studying plant reproductive development [8,9]. Research indicated that male sterility was an extremely complex process, with diverse abortion forms and degrees [10,11]. Male sterility in plants was caused mainly by abnormal function of genes in nucleus or cytoplasm involving in pollen development, while toxic proteins, insufficient energy supply, abnormal programmed cell death (PCD), and other factors might lead to abnormal plant fertility [12–16]. In view of the complexity of male sterility in plants, it was very difficult to analyze the genetic mechanism from the perspective of individual genes by conventional methods. The transcriptome refers to the sum of all RNA transcribed by a specific cell under a certain functional state, and it thus can provide information on gene expression, gene regulation, and amino acid content [17]. The study of transcriptomics could screen and find the target genes regulating biological traits, infer the function of corresponding unknown genes, and reveal the action and molecular mechanism of genes in biological processes, which had been widely used in the study of plant male sterility [18–20]. However, there were still few reports on transcriptomics between soybean cytoplasmic male sterile lines and maintainer lines. Soybean cytoplasmic male sterile line NJCMS1A had been studied, but its sequencing library construction only involved nuclear genome [21]. The underlying molecular mechanism of CMS and the genes related to pollen abortion in soybean remains unclear.

The soybean CMS line SXCMS5A is a new male-sterile line successfully transferred from the variety JY20 with H3A cytoplasm. In the present study, we performed transcriptomic analyses of SXCMS5A vs. SXCMS5B, combined with quantitative real time PCR (qRT-PCR) analysis, cyto-morphological characteristic and enzyme activity assay, and substance content analysis in order to reveal the male sterility mechanism of SXCMS5A. We aimed to identify differences between CMS line SXCMS5A and its maintainer SXCMS5B at the transcriptional level and to find important differentially expressed genes (DEGs) and metabolic pathways related to pollen abortion. These findings might contribute to greater understanding of the molecular mechanism underlying CMS and provide useful information to facilitate progress in hybrid breeding in soybean.

2. Results

2.1. Comparison of the Cyto-Morphological Characteristics between Soybean CMS Line SXCMS5A and Its Maintainer SXCMS5B

In order to describe the cyto-morphological characteristics of pollen abortion of soybean CMS line SXCMS5A, the flower buds of SXCMS5A and SXCMS5B were observed and compared by paraffin sections. As shown in Figure 1A, at the tetrad stage, the tapetum cells of SXCMS5A were closely arranged, vacuolated, expanded inward, and tended to squeeze microspores. Subsequently, the tapetum was gradually broken and disintegrated, and clear contours and disintegrated fragments could be observed (Figure 1B,C). After that, the diaphragm between the two chambers did not open. And there were signs of vacuolization (Figure 1D). The pollen grains were abnormally developed and could not be stained by I_2-KI (Figure 1E). In contrast, the tapetum of SXCMS5B normally initiated PCD (Figure 1F). Subsequently, the tapetum continued to degrade (Figure 1G). After that, microspores gradually developed and matured, the diaphragm between the two chambers opened normally (Figure 1H,I). The pollen grains developed normally and could be stained by I_2-KI (Figure 1J). It was speculated that the tapetum of soybean CMS line SXCMS5A was vacuolated and abnormally developed, which could not provide necessary nutrients for microspore development, resulting in abnormal pollen development.

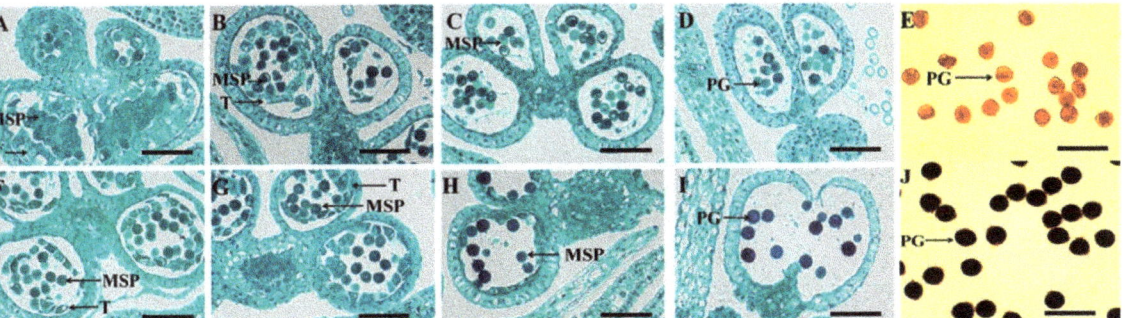

Figure 1. Microscopic observations of anthers from the soybean cytoplasmic male sterility (CMS) line SXCMS5A and its maintainer SXCMS5B. (**A–D**) Transverse sections of sterile anthers; abnormal tapetum and abnormal anthers developed in SXCMS5A. (**E**) Mature pollen grains stained by I_2-KI in SXCMS5A. (**F–I**) Transverse sections of fertile anthers; normal tapetum and normal anthers developed in SXCMS5B. (**J**) Mature pollen grains stained by I_2-KI in SXCMS5B. MSP, microspore; T, tapetum; PG, pollen grain; Bars = 20 µm.

2.2. Transcriptome Sequencing, Sequence Alignment and Quality Evaluation

To further understand the molecular mechanism of CMS in soybean, RNA sequencing (RNA-Seq) analysis of flower buds of SXCMS5A and SXCMS5B was conducted using Illumina technology. As shown in Table S1, 58.97 Gb clean data and 393,126,454 clean reads were obtained from 6 samples. After strict filtering of the original data, 54.15 GB high quality clean data and 372,973,796 high quality clean reads were obtained. The average percentages of Q20 base, Q30 base, and GC content of all samples were 97.52%, 93.37% and 43.93%, respectively, indicating that the sequencing data was high quality to meet the standards for subsequent gene function analysis. Next, Tophat2 software was used to compare the filtered ribosomal reads to the reference genome. A total of 331,374,194 mapped reads were obtained, with an average matching rate of 89.98%. Pearson correlation coefficient analysis revealed that the correlation coefficients (R^2) value between samples was greater than 0.97, showing that the expression mode of SXCMS5A was very close to SXCMS5B (Figure S1).

2.3. Identification and Confirmation of DEGs

To identify putative DEGs between SXCMS5A and SXCMS5B, the thresholds of "False Discovery Rate (FDR) < 0.05 and | $\log_2$ Fold Change (FC) | > 1" was used to screen for DEGs. There were 840 DEGs between SXCMS5A and SXCMS5B, among which 658 downregulated and 182 upregulated in SXCMS5A compared to SXCMS5B (Figure 2A; Table S2). The expression of most DEGs in SXCMS5A was downregulated compared to SXCMS5B (Figure 2B).

To validate the results of RNA-Seq, 13 DEGs (3 upregulated and 10 downregulated genes) were randomly selected and assayed by qRT-PCR. In Figure 2C, 12 DEGs showed the same trend in both RNA-Seq analysis and qRT-PCR; the coincidence rate between qRT-PCR and RNA-Seq data was 92.31%, suggesting that transcriptome analysis was accurate and reliable.

2.4. Functional Classification of DEGs between SXCMS5A and SXCMS5B

Through analysis of gene ontology (GO) function, with Q value $\leq$ 0.05 as the threshold, 324 DEGs were annotated to 479 GO terms in the biological process, 28 of which were significantly enriched, and the first 5 GO terms were external encapsulation structure organization, cell wall organization, cell wall organization or biogenesis, carbohydrate metabolic process and cell wall modification (Table S3). A total of 418 DEGs were annotated to 335 GO terms in molecular functions, 25 of which were significantly enriched. The first 5 GO terms were enzyme inhibitor activity, molecular function regulator, enzyme regulator activity, catalytic activity and pectinesterase activity (Table S4). In addition, 131 DEGs were annotated to 111 GO terms in the cell components, 7 of which were significantly enriched, and the first 5 GO terms were cell wall, external encapsulation structure, membrane, cell peripheral, and intrinsic component of membrane (Table S5).

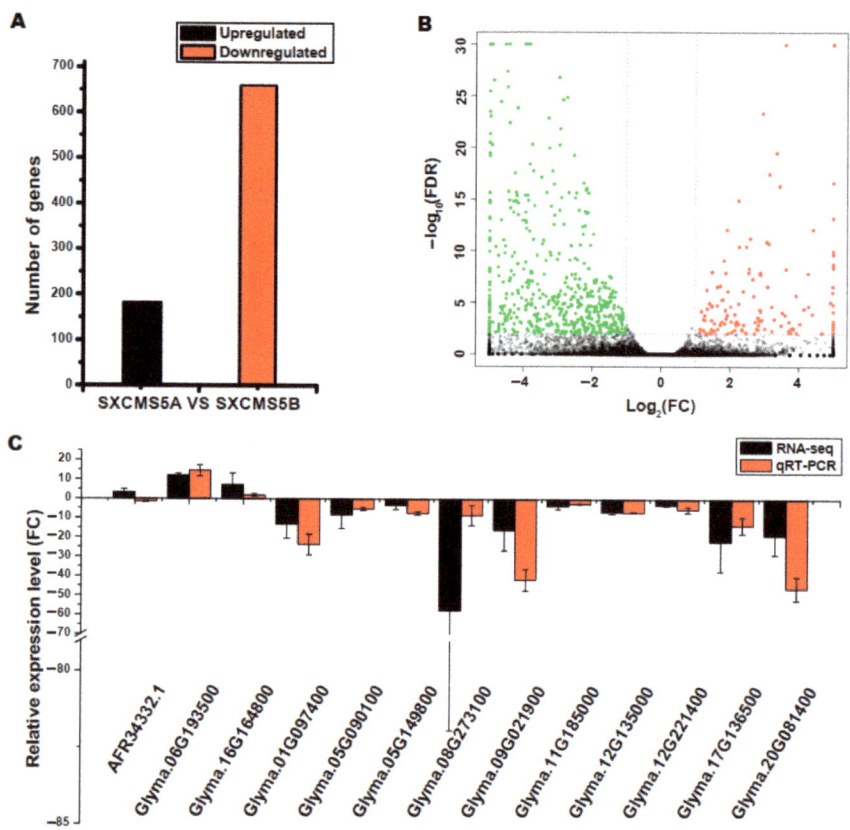

Figure 2. Analysis of differentially expressed genes (DEGs) between SXCMS5A and SXCMS5B. (**A**) Number of upregulated and downregulated DEGs. (**B**) Volcano plot comparing DEGs. Red dots, green dots, and black dots indicated DEGs that were significantly upregulated, significantly downregulated, or showed no significant difference in expression, respectively. (**C**) Relative expression level of selected DEGs. The y-axis indicated relative mRNA expression level, determined by RNA sequencing (RNA-seq) and quantitative real time PCR (qRT-PCR) analysis. The results were obtained from three biological replicates. FC, fold change; FDR, false discovery rate.

To identify the metabolic pathways in which the DEGs were involved and enriched, pathway analysis was performed using the Kyoto encyclopedia of genes and genomes (KEGG) pathway database. The results showed that starch and sucrose metabolism, pentose and glucuronate interconversions, thiamine metabolism, glycolysis/gluconeogenesis, biosynthesis of amino acids, and selenocompound metabolism were the main metabolic pathways (Figure 3).

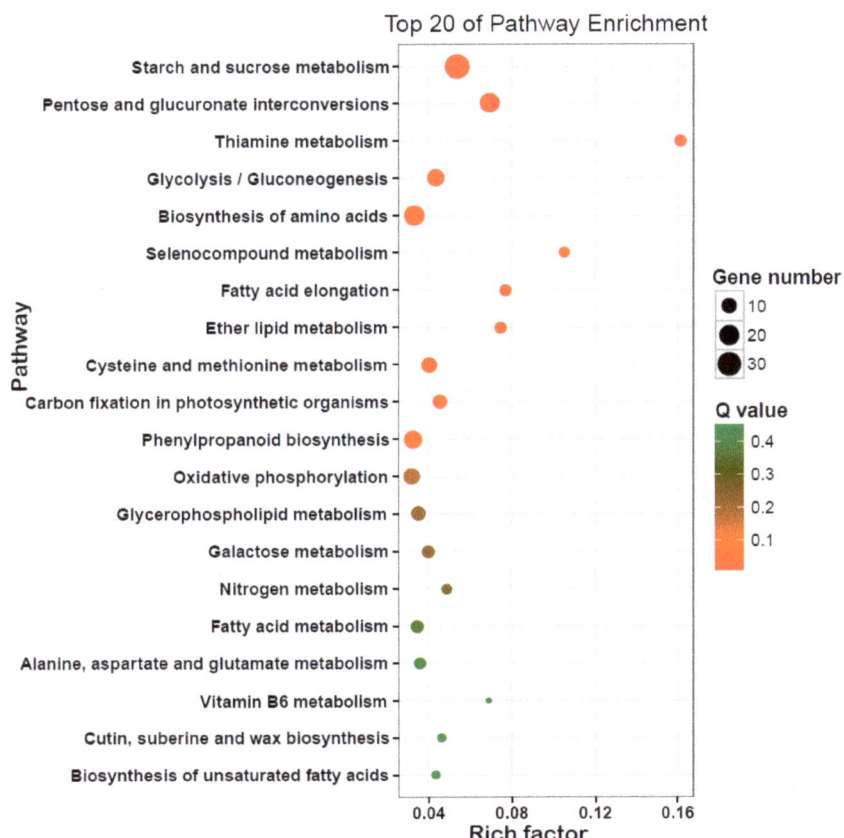

Figure 3. Top 20 Kyoto encyclopedia of genes and genome (KEGG) pathway analysis of DEGs between SXCMS5A and SXCMS5B. The x-axis indicated the rich factor corresponding to each pathway and the y-axis indicated name of the KEGG pathway. The dot color represented the Q values of the enrichment analysis. The size and color of bubbles represented the number and degree of enrichment of DEGs, respectively.

2.5. Identification of DEGs Associated with Mitochondrial Genome

Numerous open reading frames (ORFs) in the mitochondrial genome were closely correlated with plant CMS [14]. In this study, 13 DEGs, i.e., 12 ORFs and 1 *COX2* in the soybean mitochondrial genome were differentially expressed between SXCMS5A and SXCMS5B (Table S2; Figure 4A), 10 out of the 13 genes were upregulated in SXCMS5A compared to SXCMS5B. Interestingly, 3 ORFs (*ORF151*, *ORF103c* and *ORF178*) were expressed at very low levels or not expressed in SXCMS5B. Especially, qRT-PCR analysis confirmed that these 3 ORFs genes were upregulated in SXCMS5A, SXCMS6A and SXCMS7A (the latter two CMS lines having a same cytoplasm source as SXCMS5A) compared to their respective maintainer SXCMS5B, SXCMS6B and SXCMS7B (Figure 4B–D). Furthermore, He et al. [22] found that *ORF178* was formed during the process of genome recombination in a soybean CMS line NJCMS1A. This indicated that *ORF178* might be a CMS gene of soybean. Since ORF103c also contains a transmembrane domain like ORF178 (Figure 4E,F and Figure S2) [22], *ORF103c* might also be a CMS gene of soybean.

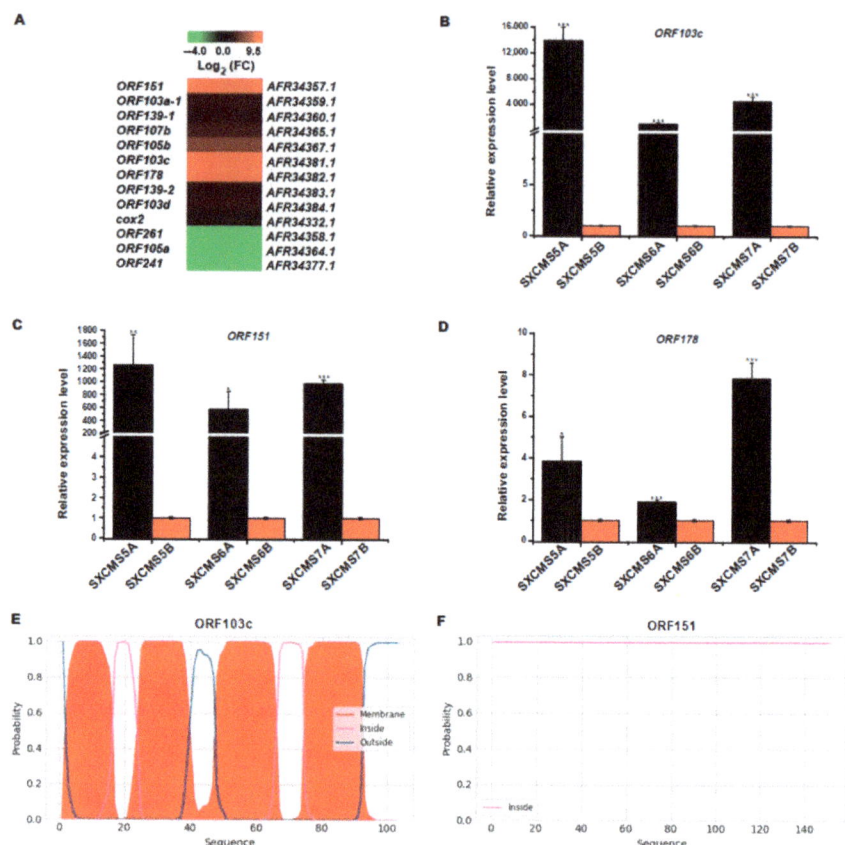

Figure 4. Analysis of DEGs in mitochondrial genome between soybean CMS lines and their maintainer lines. (**A**) Heat map of DEGs in mitochondrial genome between SXCMS5A and SXCMS5B. The heat map was conducted using MeV 4.9 software. Log$_2$(FC) values were obtained from the RNA-seq data. (**B**–**D**) Relative expression level of *ORF103c*, *ORF151* and *ORF178* between soybean CMS lines and their maintainer lines. by qRT-PCR analysis. Asterisk indicated statistical significance: *, $p < 0.05$; **, $p < 0.01$; ***, $p < 0.001$. (**E**,**F**) Transmembrane domain analysis of ORF103c and ORF151. The abscissa indicated the amino acid length of ORFs. The ordinate represented the probability of the predicted transmembrane domain.

2.6. Identification of DEGs Associated with Pollen Development

Pectin methylesterase (PME, also named pectinesterase) and pectate lyase (PL) were two key enzymes involved in the degradation of plant pectin, and played important roles in the regulation of pollen development [23,24]. As noted above, 11 DEGs and 3 DEGs were found associated with pectinesterase activity (GO:0030599) and pectate lyase activity (GO:0030570) GO terms, respectively (Table S4). As shown in Figure 5A, all the 14 DEGs were downregulated in SXCMS5A compared to SXCMS5B. Most importantly, RNA-seq data in Phytozome v12.0 indicated all these transcripts were enriched in soybean flowers (Figure 5B). *GmPME* (*Glyma.02G008300*) and *GmPL* (*Glyma.13G064700*) were selected for qRT-PCR analysis, which were all downregulated in SXCMS5A, SXCMS6A and SXCMS7A compared to their respective maintainer SXCMS5B, SXCMS6B and SXCMS7B (Figure 5C,D). These findings suggested that the two gene types might involve in pollen development processes and that their reduced expression in soybean CMS might lead to pollen abortion.

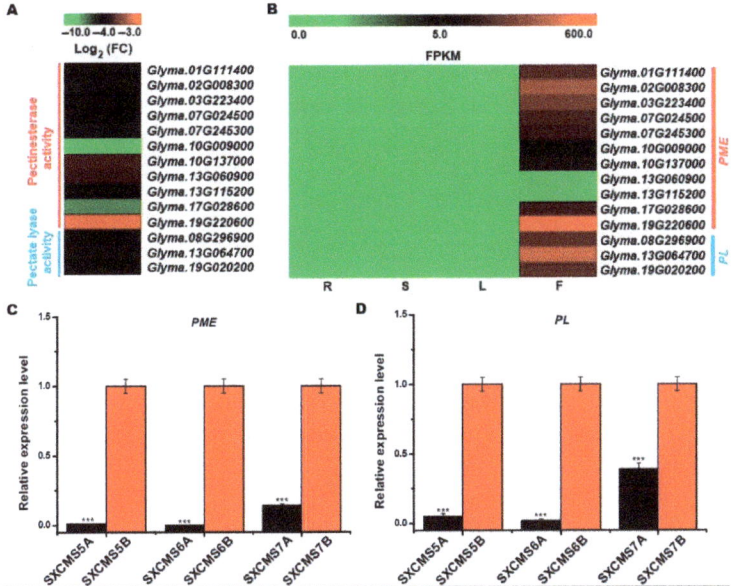

Figure 5. Analysis of DEGs related to pollen wall development between soybean CMS lines and their maintainer lines. (**A**) Heat map of the DEGs related to pollen wall development between SXCMS5A and SXCMS5B. The heat map was created using MeV 4.9 software. Log$_2$(FC) values were obtained from RNA-seq data in this study. (**B**) Heat map of the DEGs related to pollen wall development in four different tissues. The color scale represented the relative transcript abundance of the DEGs in four soybean tissues. The heat map was created using MeV 4.9 software. Fragments per kilobase of transcript per million mapped reads (FPKM) values were obtained from RNA-seq data in Phytozome v12.0. R, root; S, stem; L, leaf; F, flower. (**C,D**) Relative expression level of *GmPME* (*Glyma.02G008300*) and *GmPL* (*Glyma.13G064700*) between soybean CMS lines and their maintainer lines by qRT-PCR analysis. PME, Pectin methylesterase; PL, pectate lyase. Asterisk indicated statistical significance: ***, $p < 0.001$.

2.7. Identification of DEGs Associated with Carbohydrate Metabolism and Sugar Transport

Many DEGs between SXCMS5A and SXCMS5B involved in carbohydrate metabolism during flower bud development. Among these DEGs, 29, 17, and 12 were associated with starch and sucrose metabolism, pentose and glucuronate interconversions, and glycolysis/gluconeogenesis pathways, respectively (Figure 6A). Especially, most of these genes (55/58) were downregulated in SXCMS5A compared to SXCMS5B. *UDP-glucuronic acid decarboxylase 2-like* (*UDP-GAD2, Glyma.07G246600*), *exopolygalacturonase* (*exoPG, Glyma.07G245100*) were selected for qRT-PCR analysis, which were all downregulated in SXCMS5A, SXCMS6A and SXCMS7A compared to their respective maintainer SXCMS5B, SXCMS6B and SXCMS7B (Figure 6B,C). In addition, 16 DEGs were involved in sugar transport, and 14 of these were downregulated in SXCMS5A compared to SXCMS5B (Figure 6A). *Sugar transport protein 11* (*STP11, Glyma.20G103900*) was selected for qRT-PCR analysis, which was downregulated in SXCMS5A, SXCMS6A, and SXCMS7A, compared to their respective maintainer SXCMS5B, SXCMS6B, and SXCMS7B (Figure 6D). Furthermore, we measured soluble sugar, starch, and adenosine triphosphate (ATP) amounts in flower buds of SXCMS5A and SXCMS5B. The results showed that soluble sugar and ATP amounts decreased in SXCMS5A, while the starch amount decreased slightly in SXCMS5A, relative to SXCMS5B (Figure 6E–G). All these findings suggested that inhibition of carbohydrate metabolism and sugar transport might be two of the causes of soybean CMS.

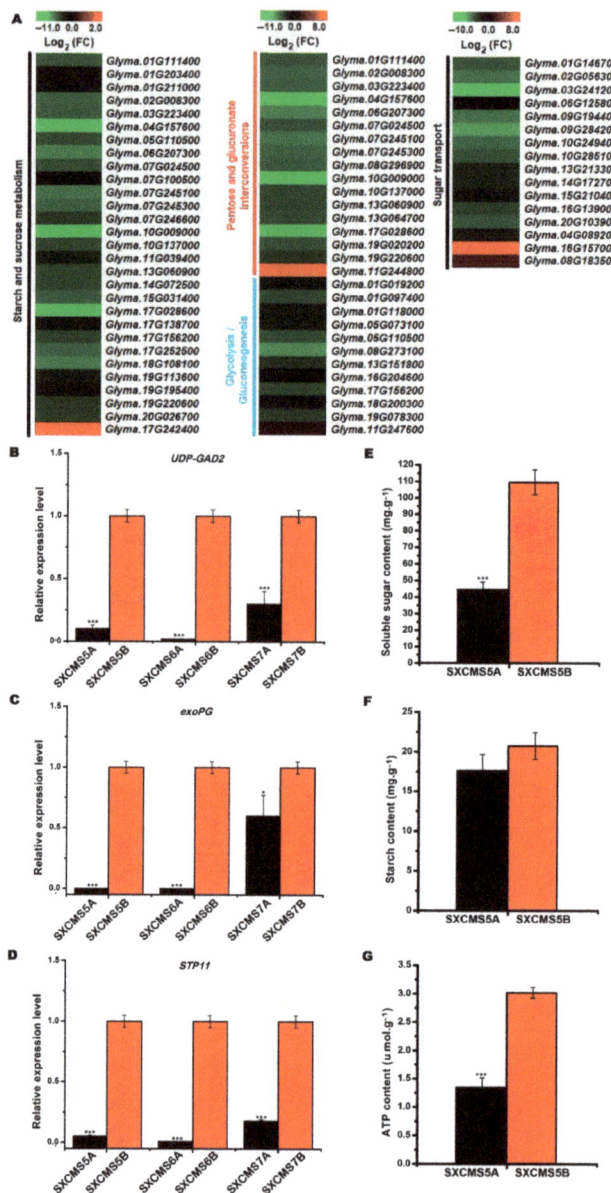

Figure 6. Analysis of DEGs related to carbohydrate metabolism and sugar transport between soybean CMS lines and their maintainers. (**A**) Heat map of the DEGs related to starch and sucrose metabolism, pentose and glucuronate interconversions, glycolysis/gluconeogenesis, and sugar transport between SXCMS5A and SXCMS5B. The heat map was created using MeV 4.9 software. Log$_2$(FC) values were obtained from RNA-seq data. (**B–D**) Relative expression level of *UDP-glucuronic acid decarboxylase 2-like* (*UDP-GAD2*, *Glyma.07G246600*), *exopolygalacturonase* (*exoPG*, *Glyma.07G245100*) and *sugar transport protein 11* (*STP11*, *Glyma.20G103900*) between soybean CMS lines and their maintainers by qRT-PCR analysis. (**E–G**) Soluble sugar, starch, and adenosine triphosphate (ATP) contents analysis between SXCMS5A and SXCMS5B. Asterisk indicated statistical significance: *, $p < 0.05$; ***, $p < 0.001$.

2.8. Identification of DEGs Associated with Reactive Oxygen Species (ROS) Metabolism

Several DEGs were found involved in ROS metabolism, including glutathione metabolism and ascorbate and aldarate metabolism (Figure 7A). Among these DEGs, 4 DEGs were associated with glutathione metabolism, and exactly half number of these genes were downregulated or upregulated in SXCMS5A compared to SXCMS5B. In addition, 2 DEGs were found associated with ascorbate and aldarate metabolism, and the RNA-seq showed that they were downregulated in SXCMS5A compared to SXCMS5B. *Glutathione S-transferase-like* (*GST, Glyma.02G154400*) and *L-ascorbate oxidase homolog* (*L-AO, Glyma.07G225400*) were selected for qRT-PCR analysis, which were downregulated in SXCMS5A, SXCMS6A, and SXCMS7A compared to their respective maintainer SXCMS5B, SXCMS6B, and SXCMS7B (Figure 7B,C). In addition, we measured catalase (CAT), ascorbic acid oxidase (AAO), and glutathione peroxidase (GPX) activities in flower buds of SXCMS5A and SXCMS5B. The results showed that CAT and AAO activities decreased in SXCMS5A, while the GPX activity was slightly decreased in SXCMS5A, relative to SXCMS5B (Figure 7D–F). All these findings suggested that downregulation of genes associated with ROS metabolism might be one of the causes of soybean CMS.

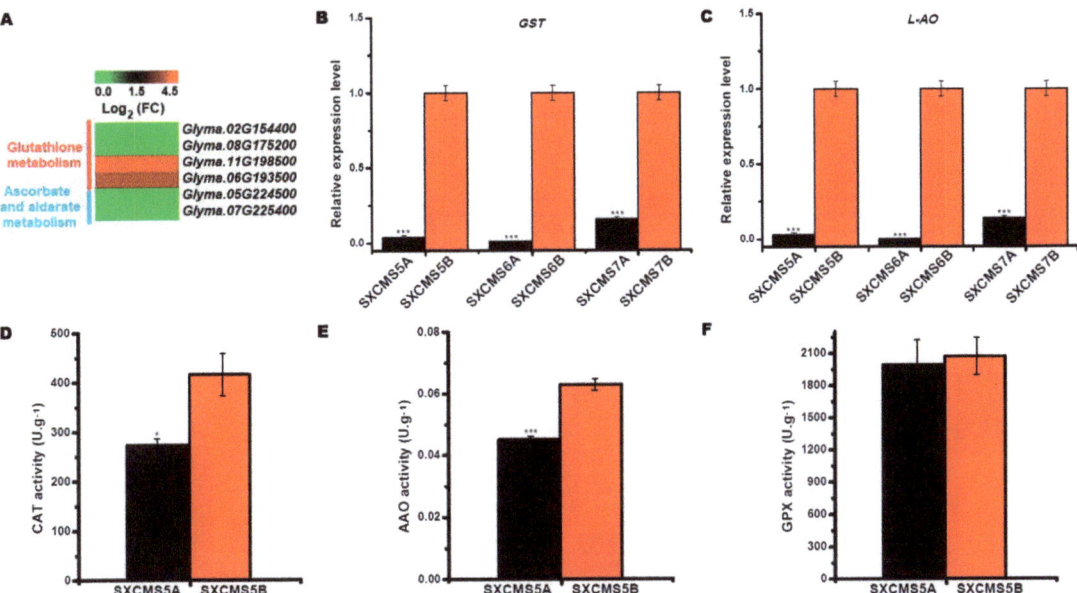

Figure 7. Analysis of DEGs related to reactive oxygen species (ROS) metabolism between soybean CMS lines and their maintainers. (**A**) Heat map of the DEGs related to glutathione metabolism and ascorbate and aldarate metabolism between SXCMS5A and SXCMS5B. The heat map was created using MeV 4.9 software. Log$_2$(FC) values were obtained from RNA-seq data. (**B,C**) Relative expression level of *Glutathione S-transferase-like* (*GST, Glyma.02G154400*) and *L-ascorbate oxidase homolog* (*L-AO, Glyma.07G225400*) between soybean CMS lines and their maintainers by qRT-PCR analysis. (**D–F**) Activity assays of catalase (CAT), ascorbic acid oxidase (AAO) and glutathione peroxidase (GPX) between SXCMS5A and SXCMS5B. Asterisk indicated statistical significance: *, $p < 0.05$; ***, $p < 0.001$.

2.9. Identification of DEGs Associated with Transcription Factor

Many transcription factors (TFs) were differentially expressed between SXCMS5A and SXCMS5B. As shown in Figure 8A, a total of 20 differentially expressed transcription factors were found, including 15 downregulated and 5 upregulated TFs in SXCMS5A compared to SXCMS5B. The 15 downregulated DEGs were 3 MYB family TFs,

3 NAC family TFs, 2 bHLH family TFs, 2 nuclear family transcription factor Y subunit, 1 heat stress transcription factor B-3-like isoform X2, and 4 other TFs. The 5 upregulated DEGs were 3 MYB family TFs, 1 WRKY TF and 1 ethylene-responsive TF. Furthermore, *GmMYB35* (*Glyma.06G188400*), *GmMYB* (*Glyma.16G218900*), *GmbHLH118* (*Glyma.09G150000*) and *GmWRKY43* (*Glyma.18G238600*) were selected for qRT-PCR analysis. Among them, the expression trends of *GmMYB35* (*Glyma.06G188400*), *GmbHLH118* (*Glyma.09G150000*) and *GmWRKY43* (*Glyma.18G238600*) were consistent in SXCMS5A, SXCMS6A and SXCMS7A, compared to their respective maintainer SXCMS5B, SXCMS6B, and SXCMS7B (Figure 8B,D,E). However, the expression trend of *GmMYB* (*Glyma.16G218900*) was not consistent in SXCMS5A, SXCMS6A, and SXCMS7A compared to their respective maintainer SXCMS5B, SXCMS6B, and SXCMS7B (Figure 8C). This inconsistency suggested that *GmMYB* (*Glyma.16G218900*) might play different roles in different CMS-maintainer pairs. All these findings suggested that these TFs might involve in regulation of pollen development in soybean CMS.

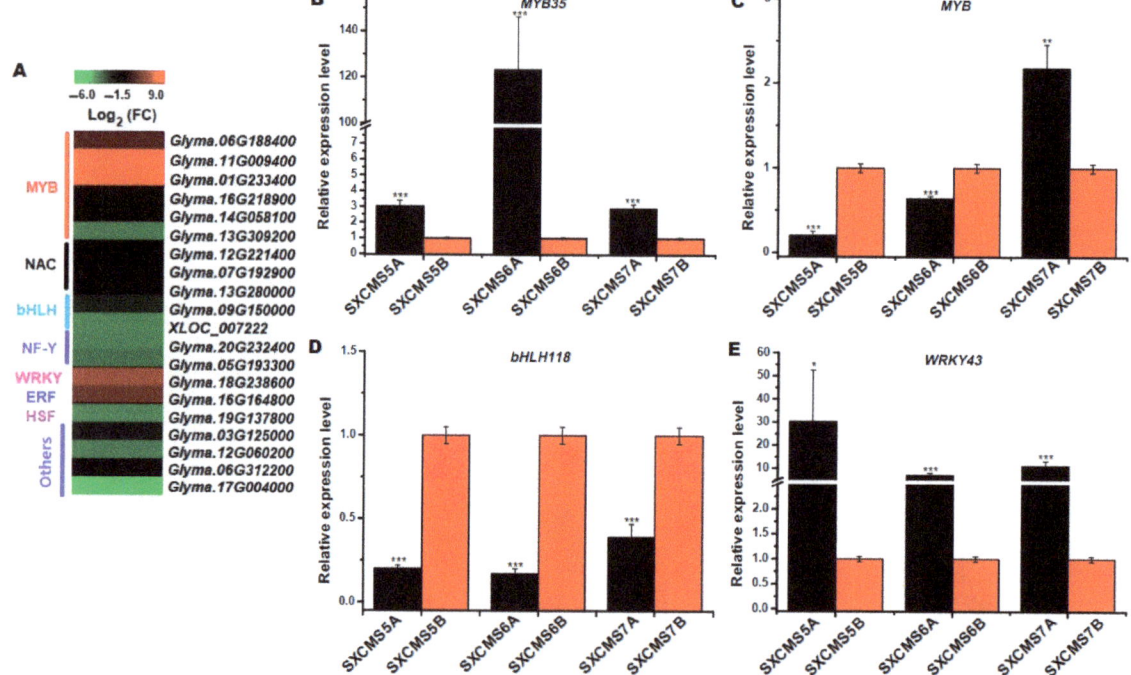

Figure 8. Analysis of DEGs related to transcription factors (TFs) between soybean CMS lines and their maintainers. (**A**) Heat map of the DEGs related to TFs between SXCMS5A and SXCMS5B. The heat map was created using MeV 4.9 software. Log$_2$(FC) values were obtained from RNA-seq data. (**B–E**) Relative expression level of *GmMYB35* (*Glyma.06G188400*), *GmMYB* (*Glyma.16G218900*), *GmbHLH118* (*Glyma.09G150000*) and *GmWRKY43* (*Glyma.18G238600*) between soybean CMS lines and their maintainers by qRT-PCR analysis. Asterisk indicated statistical significance: *, $p < 0.05$; **, $p < 0.01$; ***, $p < 0.001$.

3. Discussion

In plants, CMS, i.e., cytoplasmic nuclear interaction male sterility, is controlled by cytoplasmic and nuclear male sterility genes in a coordinated manner. It is generally believed that CMS is caused by the coupling of mitochondrial genes and nuclear genes [14]. The mitochondrial genome and nuclear genome locate on different parts in a cell, but they relate and influence each other in functions [25]. The nuclear genes encode various protein factors and enzymes needed for mitochondrial gene replication, transcription and translation, while mitochondrial gene mutation leads to the abnormality of its coding protein polypeptide, which can reverse-regulate the replication and expression of a series of nuclear genes through the signal pathway, leading to pollen abortion. In general, nuclear genes related to pollen wall development, carbohydrate metabolism and sugar transport, ROS metabolism and TFs play important roles in male fertility regulation. The relationship between DEGs and pollen abortion of SXCMS5A were discussed as follows.

3.1. ORF178 and ORF103c Identified as CMS Candidate Mitochondrial Genes of SXCMS5A

In plants, the production of most CMS was closely related to the variation, recombination and rearrangement of mitochondrial genome, resulting in a large number of new chimeric ORFs, which changed the transcription and translation products of genes, affected the expression and loss of function of genes, and led to male sterility in plants [12,26,27]. In rice, the CMS of the wild abortion type and Honglian type had been intensively investigated. Among the genes involved in these CMS variants, *WA352*, a CMS sterility gene, interacted with mitochondrial protein COX11 to stimulate the degradation of tapetum in anther, which in turn led to pollen abortion [28]. ORFH79, a protein expressed by a CMS sterility gene of the rice Honglian type, interacted with p61, a small subunit of mitochondrial respiratory electron transport chain complex III, resulting in ATP concentration decreased and ROS amount increased, which eventually led to cytoplasmic male sterility [29]. In this study, 13 DEGs included 12 ORFs and 1 *COX2* were mitochondrial genome genes between SXCMS5A and SXCMS5B of which three (*ORF103c*, *ORF151* and *ORF178*) were expressed almost exclusively in differential soybean CMS lines. Furthermore, both *ORF103c* and *ORF178* encode transmembrane proteins, which is one of the main characteristics of CMS genes [14,22]. Interestingly, *ORF178* was formed during the process of genome recombination in a soybean CMS line NJCMS1A, and it was also found that *ORF178* was expressed in NJCMS1A, NJCMS4A and NJCMS5A [22]. However, *ORF103c* was found in SXCMS5A, SXCMS6A and SXCMS7A, but not in NJCMS1A. In addition, *ORF261*, another CMS candidate gene in NJCMS1A, was downregulated in SXCMS5A compared to SXCMS5B, which might be caused by cytoplasmic differences. Thus, the upregulated expression of *ORF178* and *ORF103c* might change the transcription and translation products of genes and affect gene expression and loss of function, which was related to pollen abortion and male sterility in SXCMS5A. Since the role of these DEGs in pollen development had not been previously reported, these findings offered a new direction for investigations of the molecular mechanisms underlying soybean CMS.

3.2. Down-Regulation of DEGs Associated with Pollen Wall Development Is One of the Key Factors of Pollen Development Defect in SXCMS5A

The development of the pollen wall in pollen grains was a requirement for plant sexual reproduction, and most of the characters associated with male sterility were related to abnormal development of the pollen wall [30]. Pectin metabolism played an important role in pollen development; thus, inhibition of pectin metabolism during pollen development would lead to delayed pollen development, male sterility, and a lower seed setting rate [31,32]. In this study, we identified 11 PMEs, all of which were downregulated in the CMS line SXCMS5A compared to SXCMS5B. These genes were predicted to be correlated with pectinesterase activity. In addition, three PLs were also downregulated in the CMS line SXCMS5A compared to SXCMS5B. PL and PME were two important enzymes involved in the degradation of plant pectin and the formation of pollen walls in plants [23,24]. PL, Exo-PG, and PME were associated with male fertility restoration of the CMS line in pepper, and PL and PME played an important role in pollen development [33]. Thus, the downregulated expression of PL and PME genes might result in abnormal pollen wall development, which was related to pollen abortion and male sterility in SXCMS5A.

3.3. Blocked Carbohydrate Metabolism and Sugar Transport Leads to Abnormal Pollen Development in SXCMS5A

Carbohydrate metabolism and sugar transport were the most basic metabolic processes in plant, providing energy and carbon for anther development, and starch and sucrose serve as energy reserves for pollen maturation [34,35]. In this study, starch and sucrose metabolism, pentose and glucuronate interconversions, and glycolysis/gluconeogenesis were three enriched pathways for carbohydrate metabolism. These 3 key pathways contained 58 DEGs of which 55 were downregulated in SXCMS5A. Male sterility in the soybean CMS line NJCMS1A was associated with these three pathways [21]. Similarly, we found that many sugar transport-related DEGs, such as *STP11* and *sucrose transport protein SUC4-like* (*SUC4*), were downregulated in SXCMS5A. In cucumber, Sun et al. [16] found that downregulation of the sugar transporter *CsSUT1* inhibited pollen germination and caused male sterility. In soybean CMS-based F_1, Ding et al. [36] had found downregulation of sugar transport-related DEGs and reduction of sugar accompanied by the decrease of male fertility under heat stress. Furthermore, substance amount analysis also showed that energy supply was decreased in SXCMS5A compared to SXCMS5B. Thus, the downregulated expression of carbohydrate metabolism and sugar transport related genes might lead to insufficient energy supply, which was related to pollen abortion and male sterility in SXCMS5A.

3.4. Abnormal ROS Metabolism Leads to Pollen Abortion in SXCMS5A

PCD was a common phenomenon in the development of animals and plants, regulated by genes under specific physiological or pathological conditions [37]. Tapetum provides nutrients for pollen development, and its abnormal PCD process is one of the direct causes of plant male sterility [14,15,38,39]. In this study, cytological analysis showed that the tapetum was vacuolated and abnormally developed in SXCMS5A compared to SXCMS5B, which had typical morphological characteristics of abnormal PCD. Although the male-sterile lines formed microspores through meiosis, they could not provide related materials for the development of pollen and could not ultimately form functional pollen. Most importantly, there was a close relationship between PCD and ROS metabolism [40,41]. Studies had shown that abnormal ROS metabolism during anther or spikelet development was related to male sterility [42–44]. Ascorbic acid and glutathione had important physiological functions in plants, with special roles in maintaining the redox balance of cells in the plant antioxidant system [45,46]. In this study, *L-AO* and *GST* genes (components of ascorbic acid and glutathione metabolism, respectively) were downregulated in flower buds of the soybean CMS lines compared to their maintainer lines. Furthermore, enzyme activity analysis also showed that ROS scavenging were decreased in SXCMS5A compared to SXCMS5B. Thus, the downregulated expression of ROS metabolism genes might lead to abnormal PCD, which was related to pollen abortion and male sterility in SXCMS5A.

3.5. Abnormal Expression of TF Related DEGs Causes Pollen Abortion in SXCMS5A

TFs are important regulators of gene expression, and their expression changes may have an important impact on plant growth and development [47,48]. For example, MYB, bHLH and WRKY participated in regulation of rates of gene transcription and regulation of meiosis, which was very important for stamen development and maturation [49–52]. Most of these TFs played a key role in the process of tapetum PCD and pollen formation, and their abnormal functioning often caused male sterility [13,53,54]. In this study, 20 coding TF DEGs were found between SXCMS5A and SXCMS5B among which 15 downregulated and 5 upregulated in SXCMS5A compared to SXCMS5B. Among these DEGs, 6 were related to MYB TFs, 2 were related to bHLH TFs, and 1 was related to WRKY TFs. In addition, effective activation of the ethylene signaling pathway was required for plant responses to growth and environmental signals, but continuous over activation of the ethylene signaling pathway had obvious inhibitory and toxic effects on plant growth and reproduction [55]. Previous research had shown that increasing ethylene concentration could lead to male sterility in wheat [55]. Up-regulation of the *Glyma.16G164800* encoding ethylene response transcription factor was also consistent with this result. Thus, the abnormal expression of these TFs might cause disturbance in expression of genes related to tapetum and pollen development, which was related to pollen abortion and male sterility in SXCMS5A.

3.6. Proposed Model for the Mechanism of Male Sterility in Soybean CMS Line SXCMS5A

According to previous reports and the data presented in this study, we made the following speculation on the mechanism of male sterility in soybean CMS line SXCMS5A (Figure 9). First, the rearrangement of soybean CMS line SXCMS5A mitochondrial genome generated CMS genes, including *ORF178* and *ORF103c*. The production of ORF178 or ORF103c leads to mitochondrial dysfunction, such as blocked energy synthesis and massive production of ROS. Subsequently, mitochondrial defects directly/indirectly lead to the down-regulation of genes related to carbohydrate metabolism, sugar transport and pollen wall development in the nucleus, leading to further energy shortage and abnormal pollen development during pollen development. In addition, the downregulation of enzymatic ROS scavenging related genes in the nucleus leads to the dysfunction of the enzymatic ROS scavenging system, resulting in the inability of effective ROS clearance and the accumulation of ROS, which affects the normal PCD process of anther tapetum. The combination of these processes eventually leads to male sterility in soybean CMS line SXCMS5A. Further studies are needed to validate this proposed model.

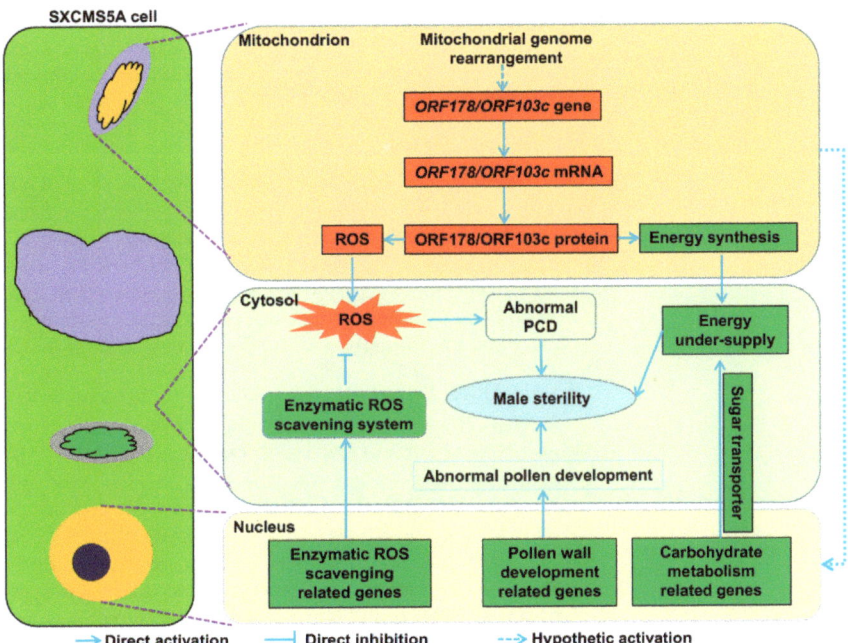

Figure 9. A proposed model for the mechanism of male sterility in soybean CMS line SXCMS5A. The upregulated and downregulated genes or metabolite contents are in red and green backgrounds, respectively.

4. Materials and Methods

4.1. Plant Materials and Sample Collection

Three soybean CMS lines, SXCMS5A, SXCMS6A, and SXCMS7A with their respective maintainer lines were used in the present study. SXCMS5A was developed by continuous backcross with the CMS line H3A as the donor parent and the variety JY20 (designated as SXCMS5B afterwards) as the recurrent parent; SXCMS6A was developed by continuous backcross with the CMS line H3A as the donor parent and the strain LX11 (designated as

SXCMS6B afterwards) as the recurrent parent; and SXCMS7A was developed by continuous backcross with the CMS line H3A as the donor parent and the strain JDX (designated as SXCMS7B afterwards) as the recurrent parent. Here, the CMS line H3A was developed by continuous backcross with the CMS line JLCMS1A as the donor parent and the strain H3 as the recurrent parent, whereas JLCMS1A was introduced from Jilin Academy of Agricultural Sciences.

SXCMS5A and SXCMS5B were planted in the summer of 2017 at Dangtu Experimental Station of Nanjing Agricultural University. Because it was difficult to judge the precise pollen development stage from flower bud appearance in soybean, the mixture of flower buds with different sizes were collected from three individual plants in the afternoon as three biological replicates for SXCMS5A and SXCMS5B. The samples were immediately placed in liquid nitrogen and then stored at $-80\ °C$ for RNA-Seq.

SXCMS5A and SXCMS5B, SXCMS6A and SXCMS6B, and SXCMS7A and SXCMS7B were planted in the summer of 2019 at Dangtu Experimental Station. The mixture of flower buds with different sizes were collected in the afternoon during flowering period, immediately placed in liquid nitrogen, then stored at $-80\ °C$ for qRT-PCR. All qRT-PCR reactions were performed with three biological replicates.

SXCMS5A and SXCMS5B were planted in the spring of 2019 at Dongyang Experimental Station of Shanxi Agricultural University. Flower buds with different sizes were collected in the afternoon at the flowering stage and fixed in formaldehyte-alcohol-acetic acid (FAA) for cytological examination. The mixture of flower buds of different sizes was collected in the afternoon at the flowering stage, immediately placed in liquid nitrogen, then stored in at $-80\ °C$ for enzyme activity assay and substance content analysis. All enzyme activity assay and substance content analysis were performed with three biological replicates.

4.2. Cytological Examination

To observe the cyto-morphological characteristics of pollen development of SXCMS5A and SXCMS5B, flower buds with different sizes were fixed, dehydrated, embedded, sectioned and stained according to a previous report [56]. To observe the pollen fertility of SXCMS5A and SXCMS5B, the anthers of unopened flowers (the flowers that will open in the morning of next day) in the afternoon were taken and stained with a 1% I_2-KI solution [57]. All samples were observed using a light microscope (Nikon Eclipse CI, Tokyo, Japan), and photographed under the imaging system (Nikon DS-U3, Tokyo, Japan).

4.3. Total RNA Extraction, Library Construction, and Sequencing

Trizol (Invitrogen, Carlsbad, CA, USA) was used to extract total RNA from the flower buds of SXCMS5A and SXCMS5B. The construction of cDNA library referred to prokaryote, considering that the plant mitochondrial genomes were similar to its ring genome. So, after total RNA was extracted, sample mRNA was enriched by removing rRNA by a Ribo-ZeroTM Magnetic Kit (Epicenter, Madison, WI, USA). Next, the enriched mRNA was fragmented into short fragments using fragmentation buffer and reverse-transcribed into cDNA with random primers. Second-strand cDNA was synthesized with DNA polymerase I, RNase H, dNTPs, and buffer. The cDNA fragments were then purified with a QiaQuick PCR extraction kit, end-repaired, poly(A) was added, and ligated to Illumina sequencing adapters. The ligation products were size selected by agarose gel electrophoresis, PCR amplified, and sequenced using Illumina HiSeqTM 2500 by Gene Denovo Biotechnology Co. (Guangzhou, China).

4.4. Raw Sequencing Data Analysis and Bioinformatics Analysis

The raw data from the sequencing machines were initially filtered to get clean data. The short-reads alignment tool Bowtie2 [58] was used to compare and remove reads containing rRNA. Tophat2 software [59] was used to compare the reads of the filtered rRNA to the reference genome (Nucleus, Wm82.a2.v1; mitochondria, JX463295.1; chloroplast, DQ317523.1). Next, the transcripts of a group of different repeats were fused into comprehensive transcripts with Cufflinks software [60]; transcripts of multiple groups were then merged into a group of final comprehensive transcripts for further analysis of downstream differential expression. Transmembrane domain of ORF was predicted using DeepTMHMM (V.1.0.12, https://dtu.biolib.com/DeepTMHMM, accessed on 15 September 2022) [61]. The FPKM values of DEGs between SXCMS5A and SXCMS5B in soybean root, stem, leaf and flower tissues were obtained from the RNA-seq data in Phytozome v12.0 (https://phytozome.jgi.doe.gov/pz/portal.html#, accessed on 15 August 2020), and the heat map was conducted using MeV 4.9 software.

4.5. Quantification of Gene Abundance and DEG-Analysis

Gene abundance was quantified with RSEM software [62]. The gene expression level was normalized with fragments per kilobase of transcript per million mapped reads (FPKM) method. To identify differentially expressed genes, the edgeR package was used. A standard of "FDR < 0.05 and $|\log_2 FC| > 1$" was used as the threshold to screen for significant DEGs. DEGs were then subjected to enrichment analysis of GO functions and KEGG pathways.

4.6. GO and KEGG Pathway Enrichment Analysis

GO enrichment analysis identified all GO terms that were significantly enriched in DEGs comparing to the genome background, and filtered the DEGs that correspond to biological functions. All DEGs were mapped to GO terms in the gene ontology database (http://www.geneontology.org/, accessed on 15 December 2017); gene numbers were calculated for every term, and significantly enriched GO terms in DEGs compared to the genome background were defined by hypergeometric test. The calculated p values were run through FDR correction, taking a Q value ≤ 0.05 as a threshold. GO terms meeting this condition were defined as significantly enriched GO terms in DEGs. This analysis enabled identification of the main biological functions correlated with the DEGs in question.

KEGG was the major public pathway-related database [63]. Pathway enrichment analysis identified significantly enriched metabolic pathways or signal transduction pathways in DEGs compared to the whole genome background. The calculated p value was run through FDR correction, taking a Q value ≤ 0.05 as a threshold value. Pathways meeting this condition were defined as significantly enriched pathways in DEGs.

4.7. qRT-PCR Analysis

qRT-PCR was used to validate the gene expression levels of DEGs detected by RNA-Seq. All primers (Table S6) were designed based on the mRNA sequences, and synthesized commercially (General Biosystems, Chuzhou, China). Total RNAs from the flower buds of SXCMS5A and SXCMS5B, SXCMS6A and SXCMS6B, SXCMS7A and SXCMS7B were used for the validation of RNA-Seq. Using the protocol provided in the HiScript Q RT SuperMix for qPCR kit (+gDNA wiper, Vazyme, Nanjing, China), 1 μg of total RNA was reverse-transcribed using oligo (dT) primers. qPCR analysis was carried out using the AceQ qPCR SYBR Green Master Mix (Vazyme, Nanjing, China) on a Bio-Rad CFX96 instrument (CFX96 Touch, BIO-RAD, USA). *GmTubulin* (accession number: *NM_001252709.2*) was used as the internal control gene [36]. The maintainer lines were used as the control of their male sterile lines. Relative expression levels of the genes were quantified using the $2^{-\Delta\Delta Ct}$ method [64].

4.8. Substance Contents and Enzyme Activity Assay

Soluble sugar and starch contents were measured by visible light spectrophotometer according to the operation procedure of soluble sugar content detection kit (Solarbio, Beijing, China) and starch content detection kit (Solarbio, Beijing, China), respectively. ATP content was measured by UV spectrophotometer according to the operation procedures of ATP content detection kit (Solarbio, Beijing, China).

CAT and AAO activities were measured by UV spectrophotometer according to the operation procedure of CAT activity test kit (Solarbio, Beijing, China) and AAO activity test kit (Solarbio, Beijing, China), respectively. GPX activity was measured by visible light spectrophotometer according to the operation procedure of GPX activity test kit (Solarbio, Beijing, China).

5. Conclusions

In this study, two ORFs in mitochondria, including *ORF178* and *ORF103c*, were upregulated in sterile lines and had transmembrane domains, which were identified as two candidate CMS genes of soybean CMS line SXCMS5A as well as its two half-sib sister lines with a same cytoplasm source (SXCMS6A and SXCMS7A). Our study showed that pollen wall development, carbohydrate metabolism, sugar transport, ROS metabolism related genes and TFs were involved in the process of pollen abortion and male sterility. The male sterility mechanism of SXCMS5A might be the rearrangement of soybean mitochondrial genome to produce CMS gene, which directly or indirectly affected a series of biological processes, such as the decrease of energy supply and the outbreak of ROS, leading to the abnormal development of anther tapetum and finally pollen abortion. Future research will focus on cloning CMS related candidate genes in soybean.

Supplementary Materials: The following supporting information can be downloaded at: https://www.mdpi.com/article/10.3390/ijms232012227/s1.

Author Contributions: J.G. and S.Y. designed the experiments. Z.B., X.D., R.Z., Y.Y. and B.W. performed the experiments. Z.B. and X.D. wrote the manuscript. Z.B., X.D., S.Y. and J.G. revised the manuscript. All authors have read and agreed to the published version of the manuscript.

Funding: This work was supported by grants from the National Key R&D Program of China (2016YFD0101500, 2016YFD0101504).

Institutional Review Board Statement: Not applicable.

Informed Consent Statement: Not applicable.

Data Availability Statement: The datasets generated by this study can be found in the NCBI using accession number PRJNA887481.

Acknowledgments: We thank Huan Sun (Jilin Academy of Agricultural Sciences, China) for kindly providing seeds of the soybean CMS line JLCMS1A, and Baoguo Wei (Center for Agricultural Genetic Resources Research, Shanxi Agricultural University, China) for providing seeds of the soybean CMS line H3A derived from JLCMS1A for this study.

Conflicts of Interest: The authors declare no conflict of interest.

References

1. Li, S.Q.; Yang, D.C.; Zhu, Y.G. Characterization and use of male sterility in hybrid rice breeding. *J. Integr. Plant Biol.* **2007**, *49*, 791–804. [CrossRef]
2. Tester, M.; Langridge, P. Breeding technologies to increase crop production in a changing world. *Science* **2010**, *327*, 818–822. [CrossRef] [PubMed]
3. Sun, H.; Zhao, L.M.; Huang, M. Study on soybean cytoplasmic nuclear interaction sterile line. *Sci. Bull. China* **1993**, *38*, 1535–1536.
4. Gai, J.Y.; Cui, Z.L.; Ji, D.F.; Ren, Z.J.; Ding, D.R. A report on the nuclear cytoplasmic male sterility from a cross between two soybean cultivars. *Soybean Genet. Newsl.* **1995**, *22*, 55–58.
5. Zhao, L.M.; Sun, H.; Wang, S.M.; Wang, Y.Q.; Huang, M.; Li, J.P. Breeding of hybrid soybean HybSoy 1. *Chin. J. Oil Crop Sci.* **2004**, *26*, 15–17.
6. Wei, B.G.; Wei, Y.C.; Bai, Z.Y.; Lei, M.L.; Zhang, H.P.; Zhang, R.J. Breeding and seed production technology of new hybrid soybean variety Jindou 48. *China Seed Ind.* **2015**, *9*, 65–66.
7. Perez-Prat, E.; Van Lookeren Campagne, M.M. Hybrid seed production and the challenge of propagating male-sterile plants. *Trends Plant Sci.* **2002**, *7*, 199–203. [CrossRef]
8. Guo, J.X.; Liu, Y.G. Molecular control of male reproductive development and pollen fertility in rice. *J. Integr. Plant Biol.* **2012**, *54*, 967–978. [CrossRef]
9. Gómez, J.F.; Talle, B.; Wilson, Z.A. Anther and pollen development: A conserved developmental pathway. *J. Integr. Plant Biol.* **2015**, *57*, 876–891. [CrossRef]

10. Cai, H.; Shamsi, I.H.; Zhao, H.J.; Meng, H.B.; Jilani, G.; Zou, Q.; Xu, X.F.; Zhu, L.X. Cytological evidences of pollen abortion in *Ornithogalum caudatum* Ait. *Afr. J. Biotechnol.* **2011**, *10*, 14061–14066. [CrossRef]
11. Sheng, Z.H.; Tang, L.Q.; Shao, G.N.; Xie, L.H.; Jiao, G.A.; Tang, S.Q.; Hu, P.S. The rice thermo-sensitive genic male sterility gene *tms9*: Pollen abortion and gene isolation. *Euphytica* **2015**, *203*, 145–152. [CrossRef]
12. Hu, J.; Wang, K.; Huang, W.C.; Liu, G.; Gao, Y.; Wang, J.M.; Huang, Q.; Ji, Y.X.; Qin, X.J.; Wan, L.; et al. The rice pentatricopeptide repeat protein RF5 restores fertility in Hong-Lian cytoplasmic male-sterile lines via a complex with the glycine-rich protein GRP162. *Plant Cell* **2012**, *24*, 109–122. [CrossRef] [PubMed]
13. Niu, N.N.; Liang, W.Q.; Yang, X.J.; Jin, W.L.; Wilson, Z.A.; Hu, J.P.; Zhang, D.B. EAT1 promotes tapetal cell death by regulating aspartic proteases during male reproductive development in rice. *Nat. Commun.* **2013**, *4*, 167–191. [CrossRef] [PubMed]
14. Chen, L.T.; Liu, Y.G. Male sterility and fertility restoration in crops. *Annu. Rev. Plant Biol.* **2014**, *65*, 579–606. [CrossRef]
15. Song, Y.L.; Wang, J.W.; Zhang, G.S.; Zhang, P.F.; Zhao, X.L.; Niu, N.; Ma, S.C. Microspore abortion of abnormal tapetal degeneration in a male-sterile wheat line induced by chemical hybridizing agent SQ-1. *Crop Sci.* **2015**, *55*, 1117–1128. [CrossRef]
16. Sun, L.L.; Sui, X.L.; Lucas, W.J.; Li, Y.X.; Feng, S.; Ma, S.; Fan, J.W.; Gao, L.H.; Zhang, Z.X. Down-regulation of the sucrose transporter *CsSUT1* causes male sterility by altering carbohydrate supply. *Plant Physiol.* **2019**, *180*, 986–997. [CrossRef]
17. Wei, W.L.; Qi, X.Q.; Wang, L.H.; Zhang, Y.X.; Hua, W.; Li, D.H.; Lv, H.X.; Zhang, X.R. Characterization of the sesame (*Sesamum indicum* L.) global transcriptome using Illumina paired-end sequencing and development of EST-SSR markers. *BMC Genom.* **2011**, *12*, 451. [CrossRef]
18. Zheng, B.B.; Wu, X.M.; Ge, X.X.; Deng, X.X.; Grosser, J.W.; Guo, W.W. Comparative transcript profiling of a male sterile cybrid pummelo and its fertile type revealed altered gene expression related to flower development. *PLoS ONE* **2012**, *7*, e43758. [CrossRef]
19. Yan, X.H.; Dong, C.H.; Yu, J.Y.; Liu, W.H.; Jiang, C.H.; Liu, J.; Hu, Q.; Fang, X.P.; Wei, W.H. Transcriptome profile analysis of young floral buds of fertile and sterile plants from the self-pollinated offspring of the hybrid between novel restorer line NR1 and Nsa CMS line in *Brassica napus*. *BMC Genomics* **2013**, *14*, 26. [CrossRef]
20. Qiu, Y.L.; Liao, L.J.; Jin, X.R.; Mao, D.D.; Liu, R.S. Analysis of the meiotic transcriptome reveals the genes related to the regulation of pollen abortion in cytoplasmic male-sterile pepper (*Capsicum annuum* L.). *Gene* **2018**, *641*, 8–17. [CrossRef]
21. Li, J.J.; Han, S.H.; Ding, X.L.; He, T.T.; Dai, J.Y.; Yang, S.P.; Gai, J.Y. Comparative transcriptome analysis between the cytoplasmic male sterile line NJCMS1A and its maintainer NJCMS1B in soybean (*Glycine max* (L.) Merr.). *PLoS ONE* **2015**, *10*, e0126771. [CrossRef] [PubMed]
22. He, T.T.; Ding, X.L.; Zhang, H.; Li, Y.W.; Chen, L.F.; Wang, T.L.; Yang, L.S.; Nie, Z.X.; Song, Q.J.; Gai, J.Y.; et al. Comparative analysis of mitochondrial genomes of soybean cytoplasmic male-sterile lines and their maintainer lines. *Funct. Integr. Genomics.* **2021**, *21*, 43–57. [CrossRef] [PubMed]
23. Micheli, F. Pectin methylesterases: Cell wall enzymes with important roles in plant physiology. *Trends Plant Sci.* **2001**, *6*, 414–419. [CrossRef]
24. Corral-Martínez, P.; García-Fortea, E.; Bernard, S.; Driouich, A.; Seguí-Simarro, J.M. Ultrastructural immunolocalization of arabinogalactan protein, pectin and hemicellulose epitopes through anther development in *Brassica napus*. *Plant Cell Physiol.* **2016**, *57*, 2161–2174. [CrossRef]
25. Yang, J.H.; Zhang, M.F. Mechanism of cytoplasmic male-sterility modulated by mitochondrial retrograde regulation in higher plants. *Hereditas* **2007**, *29*, 1173–1181. [CrossRef]
26. Rathburn, H.B.; Hedgcoth, C. A chimeric open reading frame in the 5′ flanking region of coxI mitochondrial DNA from cytoplasmic male-sterile wheat. *Plant Mol. Biol.* **1991**, *16*, 909–912. [CrossRef]
27. Singh, M.; Brown, G.G. Characterization of expression of a mitochondrial gene region associated with the *Brassica* "Polima"CMS: Developmental influences. *Curr. Genet.* **1993**, *24*, 316–322. [CrossRef]
28. Luo, D.P.; Xu, H.; Liu, Z.L.; Guo, J.X.; Liu, Y.G.; Li, H.; Chen, L.; Fang, C.; Zhang, Q.Y.; Bai, M.; et al. A detrimental mitochondrial-nuclear interaction causes cytoplasmic male sterility in rice. *Nat. Genet.* **2013**, *45*, 573–577. [CrossRef]
29. Wang, K.; Gao, F.; Ji, Y.X.; Liu, Y.; Dan, Z.W.; Yang, P.F.; Zhu, Y.G.; Li, S.Q. ORFH79 impairs mitochondrial function via interaction with a subunit of electron transport chain complex III in Honglian cytoplasmic male sterile rice. *New Phytol.* **2013**, *198*, 408–418. [CrossRef]
30. Zhou, Q.; Zhu, J.; Cui, Y.L.; Yang, Z.N. Ultrastructure analysis reveals sporopollenin deposition and nexine formation at early stage of pollen wall development in *Arabidopsis*. *Sci. Bull.* **2015**, *60*, 273–276. [CrossRef]
31. Zhang, G.Y.; Feng, J.; Wu, J.; Wang, X.W. BoPMEI1, a pollen specific pectin methylesterase inhibitor, has an essential role in pollen tube growth. *Planta* **2010**, *231*, 1323–1334. [CrossRef] [PubMed]
32. Marín-Rodríguez, M.C.; Orchard, J.; Seymour, G.B. Pectate lyases, cell wall degradation and fruit softening. *J. Exp. Bot.* **2002**, *53*, 2115–2119. [CrossRef] [PubMed]
33. Wei, B.Q.; Wang, L.L.; Bosland, P.W.; Zhang, G.Y.; Zhang, P. Comparative transcriptional analysis of *Capsicum* flower buds between a sterile flower pool and a restorer flower pool provides insight into the regulation of fertility restoration. *BMC Genom.* **2019**, *20*, 837. [CrossRef]

34. Wei, M.M.; Song, M.Z.; Fan, S.L.; Yu, S.X. Transcriptomic analysis of differentially expressed genes during anther development in genetic male sterile and wild type cotton by digital gene-expression profiling. *BMC Genom.* **2013**, *14*, 97. [CrossRef] [PubMed]
35. Wu, Z.M.; Cheng, J.W.; Qin, C.; Hu, Z.Q.; Xin, C.X.; Hu, K.L. Differential proteomic analysis of anthers between cytoplasmic male sterile and maintainer lines in *Capsicum annuum* L. *Int. J. Mol. Sci.* **2013**, *14*, 22982–22996. [CrossRef]
36. Ding, X.L.; Guo, Q.L.; Li, Q.; Gai, J.Y.; Yang, Y.P. Comparative transcriptomics analysis and functional study reveal important role of high temperature stress response gene *GmHSFA2* during flower bud development of CMS-based F_1 in soybean. *Front. Plant Sci.* **2020**, *11*, 600217. [CrossRef]
37. Bialik, S.; Zalckvar, E.; Ber, Y.; Rubinstein, A.D.; Kimchi, A. Systems biology analysis of programmed cell death. *Trends Biochem. Sci.* **2010**, *10*, 556–564. [CrossRef]
38. Goldberg, R.B.; Beals, T.P.; Sanders, P.M. Anther development: Basic principles and practical applications. *Plant Cell* **1993**, *5*, 1217–1229. [CrossRef]
39. Li, H.; Yuan, Z.; Vizcay-Barrena, G.; Yang, C.Y.; Liang, W.Q.; Zong, J.; Wilson, Z.A.; Zhang, D.B. Persistent tapetal cell1 encodes a phd-finger protein that is required for tapetal cell death and pollen development in rice. *Plant Physiol.* **2011**, *156*, 615–630. [CrossRef]
40. Farrugia, G.; Balzan, R. Oxidative stress and programmed cell death in yeast. *Front. Oncol.* **2012**, *2*, 64. [CrossRef]
41. Murik, O.; Elboher, A.; Kaplan, A. Dehydroascorbate: A possible surveillance molecule of oxidative stress and programmed cell death in the green alga *Chlamydomonas reinhardtii*. *New Phytol.* **2014**, *202*, 471–484. [CrossRef]
42. Jiang, P.D.; Zhang, X.Q.; Zhu, Y.G.; Zhu, W.; Xie, H.Y.; Wang, X.D. Metabolism of reactive oxygen species in cotton cytoplasmic male sterility and its restoration. *Plant Cell Rep.* **2007**, *26*, 1627–1634. [CrossRef]
43. Liu, G.B.; Xu, H.; Zhang, L.; Zheng, Y.Z. Fe binding properties of two soybean (*Glycine max* (L.) Merr.) LEA4 proteins associated with antioxidant activity. *Plant Cell Physiol.* **2011**, *52*, 994–1002. [CrossRef]
44. Ding, X.L.; Wang, X.; Li, Q.; Yu, L.F.; Song, Q.J.; Gai, J.Y.; Yang, S.P. Metabolomics studies on cytoplasmic male sterility during flower bud development in soybean. *Int. J. Mol. Sci.* **2019**, *20*, 2869. [CrossRef]
45. Yu, T.; Li, Y.S.; Chen, X.F.; Hu, J.; Chang, X.; Zhu, Y.G. Transgenic tobacco plants overexpressing cotton glutathione S-transferase (GST) show enhanced resistance to methyl viologen. *J. Plant Physiol.* **2003**, *160*, 1305–1311. [CrossRef]
46. Huang, C.H.; He, W.L.; Guo, J.K.; Chang, X.X.; Su, P.X.; Zhang, L.X. Increased sensitivity to salt stress in an ascorbate-deficient *Arabidopsis* mutant. *J. Exp. Bot.* **2005**, *56*, 3041–3049. [CrossRef]
47. Kater, M.M.; Colombo, L.; Franken, J.; Busscher, M.; Masiero, S.; Van Lookeren Campagne, M.M.; Angenent, G.C. Multiple AGAMOUS homologs from cucumber and petunia differ in their ability to induce reproductive organ fate. *Plant Cell* **1998**, *10*, 171–182. [CrossRef] [PubMed]
48. Hao, Q.N.; Zhao, X.A.; Sha, A.H.; Wang, C.; Zhou, R.; Chen, S.L. Identification of genes associated with nitrogen-use efficiency by genome-wide transcriptional analysis of two soybean genotypes. *BMC Genom.* **2011**, *12*, 525. [CrossRef]
49. Wilson, Z.A.; Zhang, D.B. From *Arabidopsis* to rice: Path-ways in pollen development. *J. Exp. Bot.* **2009**, *60*, 1479–1492. [CrossRef]
50. Jiang, Y.; Zeng, B.; Zhao, H.N.; Zhang, M.; Xie, S.J.; Lai, J.S. Genome-wide transcription factor gene prediction and their expressional tissue-specificities in maize. *J. Integr. Plant Biol.* **2012**, *54*, 616–630. [CrossRef]
51. Dukowic-Schulze, S.; Harris, A.; Li, J.H.; Sundararajan, A.; Mudge, J.; Retzel, E.F.; Pawlowski, W.P.; Chen, C.B. Comparative transcriptomics of early meiosis in *Arabidopsis* and maize. *J. Genet. Genomics.* **2014**, *41*, 139–152. [CrossRef] [PubMed]
52. Xu, J.; Ding, Z.W.; Vizcay-Barrena, G.; Shi, J.X.; Liang, W.Q.; Yuan, Z.; Werck-Reichhart, D.; Schreiber, L.; Wilson, Z.A.; Zhang, D.B. *ABORTED MICROSPORES* acts as a master regulator of pollen wall formation in *Arabidopsis*. *Plant Cell* **2014**, *26*, 1544–1556. [CrossRef] [PubMed]
53. Li, N.; Zhang, D.S.; Liu, H.S.; Yin, C.S.; Li, X.X.; Liang, W.Q.; Yuan, Z.; Xu, B.; Chu, H.W.; Wang, J.; et al. The rice *tapetum degeneration retardation* gene is required for tapetum degradation and anther development. *Plant Cell* **2006**, *18*, 2999–3014. [CrossRef] [PubMed]
54. Feng, B.M.; Lu, D.H.; Ma, X.; Peng, Y.B.; Sun, Y.J.; Ning, G.; Ma, H. Regulation of the *Arabidopsis* anther transcriptome by *DYT1* for pollen development. *Plant J.* **2012**, *72*, 612–624. [CrossRef]
55. Veselova, T.D.; Il'ina, G.M.; Levinskikh, M.A.; Sychev, V.N. Ethylene is responsible for a disturbed development of plant reproductive system under conditions of space flight. *Russ. J. Plant Physiol.* **2003**, *50*, 339–354. [CrossRef]
56. Yang, S.P.; Gai, J.Y.; Xu, H.Q. A genetical and cytomorphological study on the male sterile mutant *nj89-1* in soybeans. *Soybean Sci.* **1998**, *17*, 32–37.
57. Nie, Z.X.; Zhao, T.J.; Liu, M.F.; Dai, J.Y.; He, T.T.; Lyu, D.; Zhao, J.M.; Yang, Y.P.; Gai, J.Y. Molecular mapping of a novel male-sterile gene *msNJ* in soybean [*Glycine max* (L.) Merr.]. *Plant Reprod.* **2019**, *32*, 371–380. [CrossRef]
58. Langmead, B.; Salzberg, S.L. Fast gapped-read alignment with Bowtie 2. *Nat. Methods* **2012**, *9*, 357–359. [CrossRef]
59. Kim, D.; Pertea, G.; Trapnell, C.; Pimentel, H.; Kelley, R.; Salzberg, S.L. TopHat2: Accurate alignment of transcriptomes in the presence of insertions, deletions and gene fusions. *Genome Biol.* **2013**, *14*, 621–628. [CrossRef]
60. Trapnell, C.; Roberts, A.; Goff, L.; Pertea, G.; Kim, D.; Kelley, D.R.; Pimentel, H.; Salzberg, S.L.; Rinn, J.L.; Pachter, L. Differential gene and transcript expression analysis of RNA-seq experiments with TopHat and Cufflinks. *Nat. Protoc.* **2012**, *7*, 562–578. [CrossRef] [PubMed]

1. Hallgren, J.; Tsirigos, K.D.; Pedersen, M.D.; Armenteros, J.J.A.; Marcatili, P.; Nielsen, H.; Krogh, A.; Winther, O. DeepTMHMM predicts alpha and beta transmembrane proteins using deep neural networks. *bioRxiv* **2022**. [CrossRef]
2. Li, B.; Dewey, C.N. RSEM: Accurate transcript quantification from RNA-Seq data with or without a reference genome. *BMC Bioinf.* **2011**, *12*, 323. [CrossRef]
3. Kanehisa, M.; Araki, M.; Goto, S.; Hattori, M.; Hirakawa, M.; Itoh, M.; Katayama, T.; Kawashima, S.; Okuda, S.; Tokimatsu, T.; et al. KEGG for linking genomes to life and the environment. *Nucleic Acids Res.* **2008**, *36*, 480–484. [CrossRef]
4. Livak, K.J.; Schmittgen, T.D. Analysis of relative gene expression data using realtime quantitative PCR and the 2(-Delta Delta C(T)) method. *Methods* **2001**, *25*, 402–408. [CrossRef]

Article

A Mutation in the *MYBL2-1* Gene Is Associated with Purple Pigmentation in *Brassica oleracea*

Emil Khusnutdinov, Alexander Artyukhin, Yuliya Sharifyanova and Elena V. Mikhaylova *

Institute of Biochemistry and Genetics Ufa Federal Research Center RAS, Prospekt Oktyabrya 71, 450054 Ufa, Russia
* Correspondence: mikhele@list.ru

Abstract: Anthocyanins are well-known antioxidants that are beneficial for plants and consumers. Dihydroflavonol-4-reductase (*DFR*) is a key gene of anthocyanin biosynthesis, controlled by multiple transcription factors. Its expression can be enhanced by mutations in the negative regulator of anthocyanin biosynthesis myeloblastosis family transcription factor-like 2 (*MYBL2*). The expression profiles of the *DFR* gene were examined in 43 purple and green varieties of *Brassica oleracea* L., *Brassica napus* L., *Brassica juncea* L., and *Brassica rapa* L. *MYBL2* gene expression was significantly reduced in purple varieties of *B. oleracea*, and green varieties of *B. juncea*. The *MYBL2* gene sequences were screened for mutations that can affect pigmentation. Expression of the *DFR* gene was cultivar-specific, but in general it correlated with anthocyanin content and was higher in purple plants. Two single nucleotide polymorphisms (SNPs) were found at the beginning of the DNA-binding domain of *MYBL2* gene in all purple varieties of *B. oleracea*. This mutation, leading to an amino acid substitution and the formation of a mononucleotide repeat $(A)_8$, significantly affects RNA structure. No other noteworthy mutations were found in the *MYBL2* gene in green varieties of *B. oleracea* and other studied species. These results bring new insights into the regulation of anthocyanin biosynthesis in genus *Brassica* and provide opportunities for generation of new purple varieties with precise mutations introduced via genetic engineering and CRISPR/Cas.

Keywords: Brassicaceae; Brassica oleracea; DFR; *MYBL2*; SNP; RNA; CRISPR

Citation: Khusnutdinov, E.; Artyukhin, A.; Sharifyanova, Y.; Mikhaylova, E.V. A Mutation in the *MYBL2-1* Gene Is Associated with Purple Pigmentation in *Brassica oleracea*. *Int. J. Mol. Sci.* **2022**, *23*, 11865. https://doi.org/10.3390/ijms231911865

Academic Editors: Zhiyong Li and Jian Zhang

Received: 25 August 2022
Accepted: 3 October 2022
Published: 6 October 2022

Publisher's Note: MDPI stays neutral with regard to jurisdictional claims in published maps and institutional affiliations.

Copyright: © 2022 by the authors. Licensee MDPI, Basel, Switzerland. This article is an open access article distributed under the terms and conditions of the Creative Commons Attribution (CC BY) license (https://creativecommons.org/licenses/by/4.0/).

1. Introduction

The genus *Brassica* includes species distributed worldwide, such as *B. oleracea*, *B. napus*, *B. juncea*, and *B. rapa*. *Brassica* vegetables are economically important dietary products rich in phenolic compounds, carotenoids, vitamins C, and E [1]. Purple varieties of *Brassica* sp. are the most rare and beneficial for health due to the presence of anthocyanins. These phenolic compounds can be used as natural dyes and colorants in food due to the high thermal stability [2]. Anthocyanins are also associated with increased stress tolerance [3].

The anthocyanin biosynthesis pathway is well-studied; however, the specific genes responsible for purple pigmentation in *Brassica* sp. have not been discovered yet. The genetic regulation of anthocyanin accumulation is traditionally studied in ornamental plants, such as petunia, phalaenopsis, and snapdragon [4–7].

The key gene of anthocyanin biosynthesis, encoding dihydroflavonol-4-reductase (*DFR*), is controlled by multiple transcription factors and micro RNAs. Low expression of this gene, which catalyzes the synthesis of leucoanthocyanidin, is suggested to be the bottleneck, preventing anthocyanin synthesis in plants [8]. PAP1 (AN2), PAP2, TT8, TTG1, MYB2, and Delila are known as its positive regulators. GLABRA2, HAT1, CPC, MYBL2, MYB1, and MYB57 were reported to downregulate the expression of the *DFR* gene. Three types of transcription factors: MYB, MYC/basic Helix-Loop-Helix (bHLH), and WD40-repeat (WDR) proteins usually act in MBW complexes. These complexes bind to the DFR promoter and regulate the transcription of *DFR* gene to alter anthocyanin biosynthesis [9].

In phalaenopsis, *DFR* and *MYB* transcripts were present in the purple rather than the white sectors of the flower; however, the expression level of other structural and regulatory genes was the same in both sectors [4]. Zoysiagrass cultivars with green and purple spikes and stolons differed by the expression level of *Anthocyanidin synthase* and *DFR* genes [10]. In black-skinned pomegranate, the *DFR* gene expression was higher, and *AN2* gene expression was lower than in samples with white skin [11]. In the white petunia, the introduction of an exogenous *DFR* gene resulted in a pink-flowered phenotype [12]. In general, in white flowers the expression of structural anthocyanin genes is undetectable [4,13]. However green varieties of *Brassica* sp. produce some anthocyanins, their content is by several orders of magnitude lower than in purple varieties [14]. Unlike petunia, which has three different *DFR* genes to control pigmentation in different organs, most *Brassicaceae* have only one functional gene and usually accumulate anthocyanins in all tissues [15]. Overexpression of the exogenous *AtDFR* gene in *B. napus* increased *DFR* transcript levels, anthocyanin accumulation in the shoots, and salt tolerance [3]. Purple pigmentation may appear or decrease in response to stress, lighting intensity, changes in nitrogen concentration and other factors. This suggests that in green *Brassica* sp. *DFR* expression is not blocked absolutely, but is controlled via a quantitative regulatory system [16].

Some structural and regulatory genes of *Brassicaceae* can be solely responsible for purple pigmentation or its absence [17]. Mutations such as indels can affect anthocyanin pigmentation. A mutation in *GLABRA3* gene contributed to a 98% decrease in the *DFR* transcript content and anthocyaninless phenotype in *Arabidopsis thaliana* [8]. Insertion of a natural transposon in a *TT8* gene of *B. rapa* contributed to the loss of pigmentation in the seeds [14]. Mutation of the promoter activated the expression of the *BoMYB2* gene in purple cauliflower [18].

Mutations identified in the coding sequence of the *DFR* gene itself also can disrupt the functions of the protein [19,20]. Deletion of the *DFR* gene in purple ornamental kale resulted in a loss of pigmentation [21].

The negative regulator, *MYBL2*, which encodes a R3-MYB-related protein, appears to make the greatest contribution to changes in *DFR* gene expression and anthocyanin pigmentation in many plant species. *MYBL2* expression appears to be dependent on light intensity and temperature [22,23]. The MYBL2 transcription factor has a DNA-binding MYB domain and a repressive domain. In *A. thaliana* the minimal repression domain of MYBL2 consists of six amino acids (TLLLFR) at the carboxyl terminus [24]. Possible inhibition mechanisms of this negative regulator have not yet been clarified. It either binds to the bHLH protein and prevents the formation of MBW complexes or replaces it in the complex [25–27].

It has been confirmed that silencing of the *MYBL2* gene in *A. thaliana* promoted *DFR* expression [26,27]. Unlike *DFR* gene, *MYBL2* might have several orthologs in *Brassica* sp. [28]. For example, *MYBL2-1* mRNA was not detectable in purple *B. oleracea*; however *MYBL2-2* was expressed. This suggests that only *MYBL2-1* is closely associated with anthocyanin production in *B. oleracea*; however, the second ortholog probably did not retain this function. Anthocyanin hyperaccumulation in these plants resulted from either the promoter substitution or deletion of the *MYBL2-1* gene [25]. Data on the role of *MYBL2-1* gene in *B. rapa* are contradictory. It has been suggested that a 100-bp insertion in the third exon of *MYBL2-1* gene is associated with purple pigmentation in this species [17,29]. However, *MYBL2-1* gene is highly expressed in another purple variety of *B. rapa* and therefore may serve as a positive regulator [30].

However it is clear that the *MYBL2* gene plays a key role in the regulation of anthocyanin biosynthesis, data on mutations that reliably induce anthocyanin accumulation in genus *Brassica* are very limited. Most of the experiments were carried out on the model plant *A. thaliana*, which has a single *MYBL2* gene copy. The studied *B. oleracea* and *B. rapa* varieties were of Asian origin; however, purple *Brassica* vegetables are supposed to be native to the Mediterranean region of Europe [29]. In *B. napus* and *B. juncea* this gene has never been studied. *MYBL2* has not yet been subjected to CRISPR/Cas editing, although it appears to be a perfect target gene, responsible for phenotypic changes [9].

To figure out if any mutations in the *MYBL2* gene can be associated with pigmentation in 43 purple and green varieties of *B. oleracea* L., *B. napus* L., *B. juncea* L., and *B. rapa* L., expression profiles of *DFR* and *MYBL2* genes and sequencing of the *MYBL2* gene were performed. Anthocyanin content was measured in green and purple varieties. The possibilities to introduce precise mutations to the *MYBL2* gene via CRISPR/Cas editing were evaluated.

2. Results

2.1. Expression Profiles of the DFR Gene in Brassicaceae

Visible purple pigmentation in *B. oleracea* was not directly related to the expression of the *DFR* gene (Figure 1). In general, purple varieties of the same cultivar produced more mRNA of the *DFR* gene than green ones, but there was a great difference between cultivars. Purple kohlrabi (Figure 1(20–22)) did not differ significantly from several green varieties of headed cabbage by the expression of *DFR* gene (Figure 1(6,7)); however, the phenotypic differences were clearly visible. The primary leaf veins, petioles, and stems were the most pigmented organs in *B. oleracea*, and not the leaf lamina.

In several varieties of headed cabbage and Brussels sprout (Figure 1(2,3,8)) the *DFR* gene was expressed at the level of the reference gene (70–90%); however, in cauliflower and kohlrabi it only reached 20–30% of the reference gene expression. Our results indicate that the *DFR* gene is not decisive for the formation of anthocyanin pigmentation in all varieties of cabbage; therefore, other genes may be involved.

B. rapa was the only species where mRNA content of the *DFR* gene was higher than those of the reference gene (Figure 2(38–40)). The cotyledons of *B. rapa* were also the most pigmented among all studied *Brassica* species. On the contrary, *B. napus* with pigmented roots had green leaves with unremarkable expression level of the *DFR* gene (Figure 2(33,34)). One of the *B. juncea* varieties was supposed to accumulate anthocyanins (Figure 2(28)); however, its leaves were not visibly pigmented and demonstrated low expression level of the *DFR* gene, but not as low as in green varieties. In green cabbage this gene was expressed at a certain level (Figure 1); however, in green *B. juncea* and *B. rapa* its mRNA content was close to zero (Figure 2(31,32,42,43)).

Our results indicate that expression profiles of the *DFR* gene in *Brassicaceae* are species-specific and even cultivar-specific.

2.2. Expression Profiles of the MYBL2 Gene in Brassicaceae

However *DFR* gene expression in several purple varieties of *B. oleracea* was no higher than in green ones, *MYBL2* mRNA content was very low and did not exceed 10% in all purple varieties (Figure 1(1,4,21). In green varieties of *B. oleracea MYBL2* gene expression was also cultivar-specific. In headed cabbage and kohlrabi it was the highest (40–50% of the reference gene), but in Brussels sprout and cauliflower it was almost as low as in purple plants (Figure 1(8–19)). Nevertheless, in these two cultivars a strong negative correlation was observed between the expression levels of *DFR* and *MYBL2* genes (correlation coefficients = −0.78 and −0.8, respectively); however, in headed cabbage and kohlrabi coefficients were lower (−0.5 and −0.65).

Interestingly, in three purple varieties of *B. juncea*, the expression of *MYBL2* gene was significantly higher than in green varieties (Figure 2(26–28)), which was unexpected of a putative negative regulator of anthocyanin biosynthesis. No correlation was found between the expression levels of *DFR* and *MYBL2* genes in this species.

The content of *MYBL2* mRNA in purple *B. rapa* and *B. oleracea* was comparable, except for the Mizuna Red cultivar (Figure 2(39)). The level of *DFR* expression in this cultivar was also lower than in Ruby little mermaid and Mizuna purple (Figure 2(38,40)). The correlation coefficient between the expression levels of the two studied genes in *B. rapa* was −0.62.

Unexpectedly, the correlation between the expression levels of *DFR* and *MYBL2* genes appeared to be the strongest in *B. napus* (−0.99), which were not visibly pigmented.

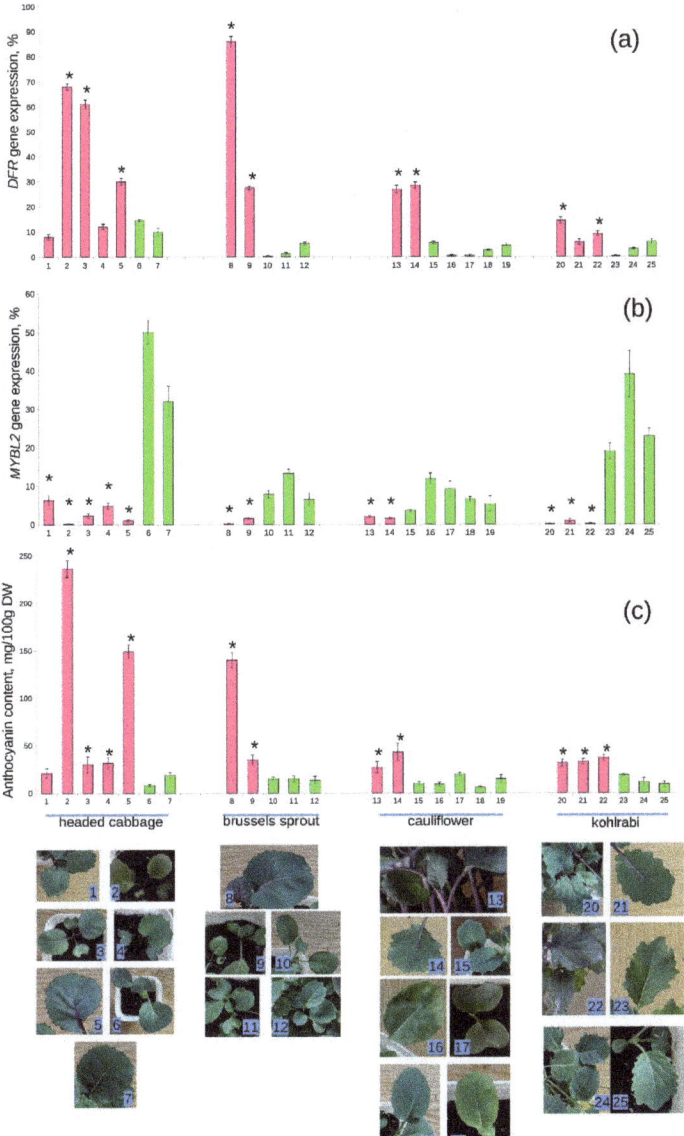

Figure 1. Relative expression level of the *DFR* gene (**a**), *MYBL2* gene (**b**), and anthocyanin content (**c**) in purple and green cultivars of *B. oleracea*. Visible pigmentation is shown in the photographs. Cultivars are numbered as follows: 1—Kalibos; 2—Ludmila; 3—Ruby; 4—Mars; 5—Firebird; 6—Royal Vantage; 7—Moscow Late; 8—Garnet bracelet; 9—Bunch of grapes; 10—Rosella; 11—Hercules; 12—Sapphire; 13—Gardener's dream; 14—Purple ball; 15—Bird's milk; 16—Baby; 17—Flame star; 18—Green bunch; 19—Alpha; 20—Madonna; 21—Violetta; 22—Vienna purple; 23—Vienna white; 24—Gulliver; 25—Picante. Varieties declared as purple are highlighted in pink. Asterisk (*) indicates a significant difference from unpigmented plants of the same cultivar.

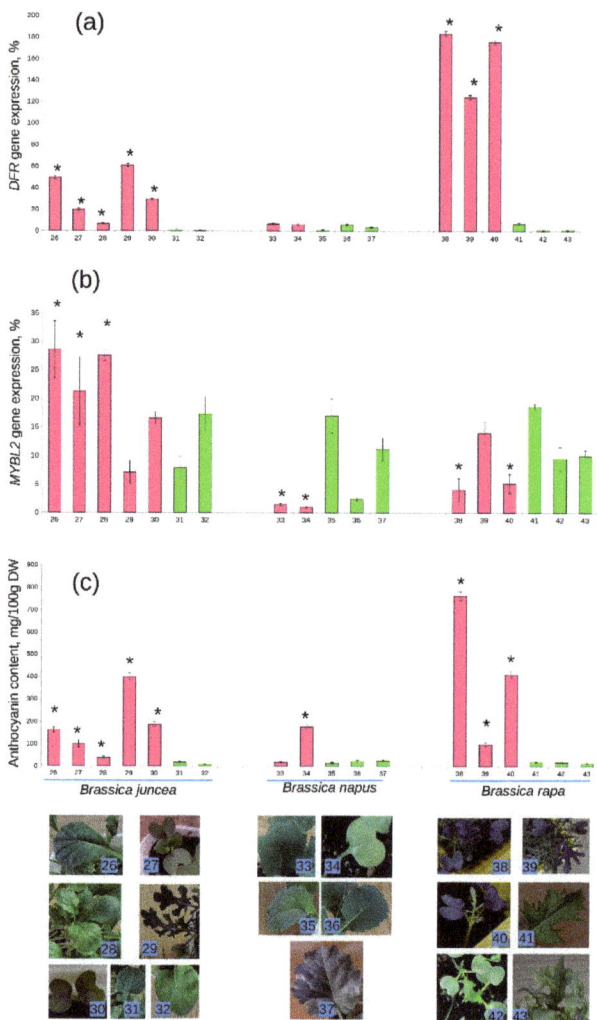

Figure 2. Relative expression level of the *DFR* gene (**a**), *MYBL2* gene (**b**) and anthocyanin content (**c**) in purple and green cultivars of *B. juncea*, *B. napus*, and *B. rapa*. Visible pigmentation is shown in the photographs. Cultivars are numbered as follows: 26—Vitamin; 27—Red velvet; 28—Freckle; 29—Miracles in the sieve; 30—Red giant; 31—Wavelet; 32—Vigorous; 33—rutabaga Krasnoselskaya; 34—rutabaga Gera; 35—rutabaga Novgorodskaya; 36—rutabaga Child love; 37—fodder cabbage Veha; 38—Ruby little mermaid; 39—Mizuna Red; 40—Mizuna purple; 41—Impulse; 42—The little mermaid; 43—Mizuna green. Varieties declared as purple are highlighted in pink. Asterisk (*) indicates a significant difference from unpigmented plants of the same species.

2.3. Total Anthocyanin Content

To reveal how *DFR* and *MYBL2* gene expression levels affect the accumulation of anthocyanins, their content was measured in green and purple varieties.

The content of anthocyanins in the studied plants ranged from 6 to 411 mg/100 g of dry weight (DW). In general, this parameter positively correlated with the *DFR* gene expression level (correlation coefficient = 0.83). Among groups, only in *B. napus* and headed cabbage correlation between *DFR* mRNA and anthocyanin content was weaker (0.4 and

0.66). Correlation with *MYBL2* gene expression was negative, but very weak. It was notable only in kohlrabi (−0.92), Brussels sprout (−0.69), and *B. rapa* (−0.75). These results are indicative of the direct involvement of the *DFR* gene in anthocyanin biosynthesis and regulatory role of the *MYBL2* gene, which may differ depending on the plant variety and exogenous factors.

Green varieties contained on average 14 mg/100 g of anthocyanins, while in purple varieties anthocyanin content was 10 times higher. These results are consistent with the literature data [10]. However, there were some exceptions. For example, in the rutabaga Gera (Figure 2(34)), which was supposed to have pigmentation only in roots, the leaves contained no less anthocyanins than the purple varieties of *B. oleracea*. The expression of the *DFR* gene in *B. napus* was also higher than in most of the anthocyaninless varieties of other species. This indicates that the regulation of anthocyanin biosynthesis in *B. napus* may be more complex due to ploidy and multiple copies of the gene.

In purple-headed cabbages Kalibos and Mars (Figure 1(1,4)) *DFR* mRNA content was the same as in green varieties. Indeed, anthocyanin content in these plants was also low, and *MYBL2* expression was higher than in other purple varieties of *B. oleracea*, but much lower than in green-headed cabbages.

The highest anthocyanin content as well as the highest *DFR* gene expression was observed in *B. rapa*. As far as these species have thin, delicate leaves, they may contain less debris and more anthocyanins per unit weight. The involvement of the *MYBL2* gene in anthocyanin biosynthesis in these species is very likely. On the contrary, no noticeable effect of the *MYBL2* expression on the accumulation of anthocyanins was observed in *B. juncea*.

To determine if the negative regulator of anthocyanin biosynthesis MYBL2 is related to the *DFR* expression profiles and anthocyanin content, we subjected the samples of each species and cultivar to sequencing.

2.4. Sequencing of the MYBL2 Gene

However it was earlier reported that purple varieties of *B. oleracea* lacked *BoMYBL2-1* coding sequences [25], amplification was successful in all studied samples. A 100-bp insertion, associated with purple pigmentation in *B. rapa* [29], also was not detected.

We discovered two completely new SNPs in a *MYBL2* sequence of all analyzed purple varieties of *B. oleracea*, which affected the beginning of the DNA-binding MYB domain, resulting in the replacement of two amino acids (from KESN to KKNN), compared with *B. rapa* (Figure 3). The mutation was common for all cultivars of *B. oleracea*; however, it was not detected in any of the green varieties, or in *B. juncea*, *B. napus*, and *B. rapa*. In green varieties of *B. oleracea* there was also one SNP in this site, compared with the *MYBL2* gene of *B. rapa*, resulting in an amino acid change (GAA into CAA – KESN into KQSN); however, it was outside of the DNA-binding domain. Therefore, the mutation probably did not affect DNA binding. Other species did not have any mutations within this site. Close to the carboxyl terminus there was also one SNP, specific for purple cabbage (GTG into GTT); however, it did not result in an amino acid substitution.

It should be noted that there were occasional heterozygous mutations in green varieties of cabbage (Figure 3C); however, in purple varieties there were almost none which means they have homozygous alleles of *MYBL2* gene.

There were other differences between the varieties, which cannot be directly associated with anthocyanin pigmentation. For example, a 30 bp long insertion (GTAATGTGAATGT-TATTTTTTTTGCAAAAA) was observed in the noncoding sequence of the *MYBL2* gene in purple and green *B. oleracea*, compared with *B. rapa*.

Therefore, we were unable to detect any noteworthy mutations in purple varieties of *B. juncea*, *B. napus*, and *B. rapa*, which can be associated with hyperpigmentation.

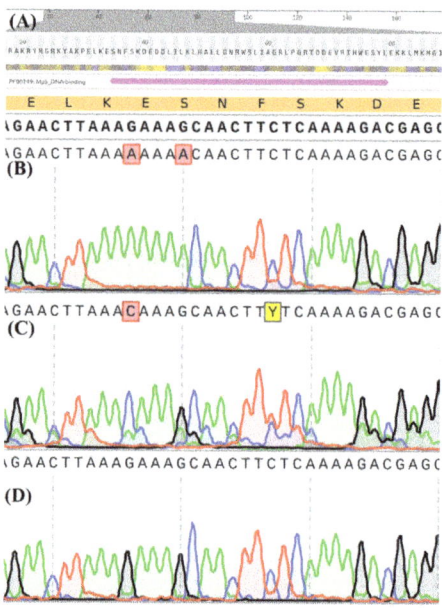

Figure 3. The localization of a DNA-binding domain in a *MYBL2* gene (**A**) and the sequence of the beginning of this domain in *B. oleracea* varieties Gardener's dream (**B**), Gulliver (**C**), and green *B. juncea* variety Vigorous (**D**).

3. Discussion

As a result of a present study, we report a mutation in the beginning of a DNA-binding site of a *MYBL2* gene of purple *B. oleracea* for the first time. According to the NCBI database, AAAGAAAGCAAC into AAAAAAAACAAC (KESN into KKNN) mutation is unique and has never previously been found. The presence of the same mutation in all purple varieties of *B. oleracea* strongly suggests that it is directly associated with anthocyanin pigmentation.

There are several possible mechanisms by which this mutation can result in phenotypic changes. First of all, it is a mononucleotide $(A)_8$ repeat (SSR) in the coding region, which can lead to gene inactivation, changes in function and phenotype [31]. SSRs can serve as binding sites for regulatory elements, may form altered DNA secondary structures, and affect gene translation. For example, translational repression of the *MYBL2* gene by MiR858 was shown to enhance anthocyanin biosynthesis in seedlings of *A. thaliana* [32]. SSRs often cause DNA slippage during DNA replication [33]. This mutation also creates the KKXX dilysine motif, which promotes the relocation of proteins from the Golgi apparatus to the endoplasmic reticulum [34].

The mutation affects the secondary structure of RNA (Figure 4), which plays a central role in post-transcriptional regulation of gene expression and affect recognition by ribosomes. U- and A-rich sites may correlate with the recognition of endonucleases for regulating alternative polyadenylation [35].

In all purple varieties of *B. oleracea* expression level of the *MYBL2* gene was lower than 10%; however, in green varieties it can be as high as 50% (Figure 1). Variability in the mRNA content of this gene was not so pronounced in other plant species (Figure 2). It is interesting that strong correlation between the expression of *MYBL2* and *DFR* genes was also observed in *B. napus*, which is indicative of the special role of *MYBL2* gene in the C genome. In *B. oleracea* and *B. napus* this gene negatively affected the expression of *DFR*, such as in *A. thaliana* [26,27].

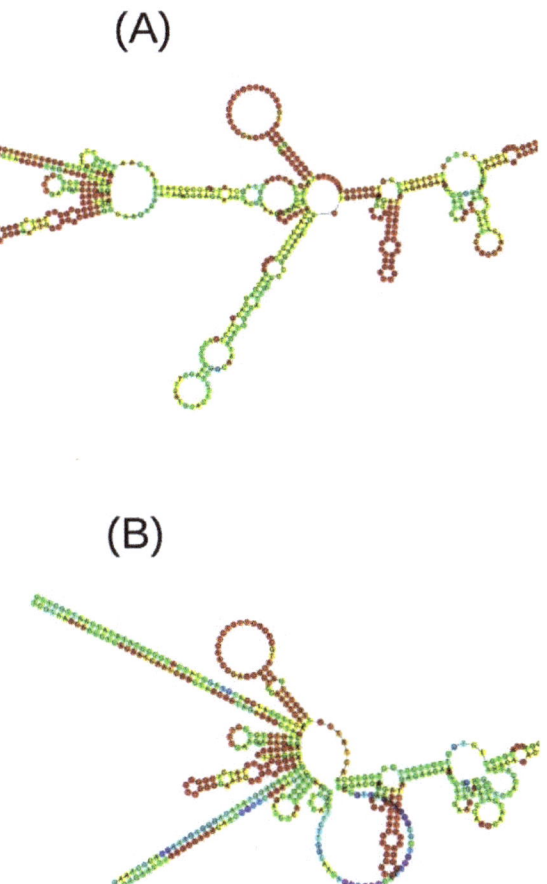

Figure 4. Part of a predicted RNA structure of the normal (**A**) and mutant (**B**) *MYBL2* gene.

No significant mutations were detected in the *MYBL2-1* gene sequence in purple varieties of *B. juncea*, *B. napus*, and *B. rapa*, which suggests that *DFR* gene expression in these species can be regulated differently. It has been shown that six amino acids (TLLLFR) at the carboxyl terminus are obligatory for the repression activity of the *MYBL2* transcription factor in *A. thaliana* [24]. However, in *B. rapa* there is a substitution of the last amino acid from R to Q (CAG). In *B. juncea MYBL2* expression was even higher in purple varieties, which suggests that, such as in *B. rapa*, this gene can also serve as a positive regulator of anthocyanin biosynthesis [30]. Therefore, in these species *MYBL2* may work through a different mechanism of repression activity or have none at all. In our study the reverse primer for gene amplification covered half of the sequence encoding this repression domain. Successful amplification indicated that in purple varieties of *B. rapa* and *B. juncea* it was present.

This also suggests that the absence of the PCR product of *MYBL2-1* gene in purple varieties of cabbage [25] may be due to primer design or contamination of the DNA with polyphenols, which can inhibit PCR reaction. There can be significant differences between European and Asian varieties of cabbage. All cultivars of *B. oleracea* involved in our study may have the same European (possibly Mediterranean) origin, because none of them were of Asian breeding. Some varieties, such as Kalibos, have been known in Europe for more than 200 years.

Interestingly, varieties of *B. oleracea*, such as Brussels sprout and kohlrabi, had different levels of *MYBL2* and *DFR* genes expression and anthocyanin content. In purple and green kohlrabi the difference between *MYBL2* mRNA content was dramatic; however, in green Brussels sprout *MYBL2* mRNA content was rather low. On the contrary, there was a great difference in *DFR* expression between green and purple cultivars of Brussels sprout. This may be explained by the fact that *DFR* is not the only gene controlled by MYBL2, and MYBL2 is not the only anthocyanin biosynthesis repressor in *Brassica*. In *B. oleracea* the content of anthocyanins increases with maturation, which indicates the important role of regulatory genes such as *MYBL2* in this process. The loss of *MYBL2* activity in *A. thaliana* resulted in an increase in mRNA content of leucoanthocyanidin dioxygenase *(LDOX)* and, to a lesser extent - flavanone 3β-hydroxylase *(F3H)* and chalcone isomerase *(CHI)* structural genes [22]. Similar to *DFR*, *LDOX* is regulated by MBW complexes [36]. LDOX is involved in anthocyanin biosynthesis pathway downstream of the DFR, promoting the transformation of leucocyanidin to cyanidin [37]. F3H and CHI are involved upstream and are able to provide more substrate for DFR.

It has been demonstrated that an insertion in the coding region of *LDOX* gene resulted in a complete loss of pigmentation in pomegranate [38]. In *A. thaliana* a single aminoacid substitution in LDOX resulted in a pale brown seed color and lack in anthocyanin accumulation [36]. *LDOX* gene of *Reaumuria trigyna* complemented reduced proanthocyanin and anthocyanin levels in the Arabidopsis loss of function mutant [39]. Transgenic *A. thaliana* overexpressing *RtLDOX2* accumulated more anthocyanins and flavonols only under abiotic stress, but also increased primary root length, biomass accumulation, and stress resistance [40]. There are several functional copies of endogenous *LDOX* gene in the genomes of *Brassica* sp., which complicates the investigation of their impact on the anthocyanin accumulation. Therefore, they remain understudied in these species. Consequently, *LDOX* may be a gene of interest in purple cabbage cultivars with low expression of *DFR* gene, such as kohlrabi.

We were unable to detect any significant mutations in *MYBL2-1* gene of *B. juncea*, *B. napus*, and *B. rapa*, including those previously described [25,28]. This suggests that other genes may be responsible for anthocyanin accumulation in studied varieties. There are a lot of candidate genes, including not only transcriptional repressors *CPC*, *LBD*, and *GLABRA2*, but also activators such as *PAP1*, *PAP2*, *MYB1*, and *MYB2* [9]. For example, a mutation in the promoter region activated the expression of *BoMYB2* gene in purple *B. oleracea* [18]. Whole genome sequencing might be a more suitable method for further studies of *B. juncea*, *B. napus*, and *B. rapa* purple varieties.

Knowledge of beneficial mutations can allow fast generation of new purple varieties via traditional selection, genetic engineering, and CRISPR/Cas. Precise editing in the mutation site of the *MYBL2-1* gene can allow to prove if two SNPs in the beginning of the DNA-binding domain are enough to induce athocyanin hyperaccumulation not only in *B. oleracea*, but also in other *Brassica* species.

To evaluate the possibility of CRISPR/Cas editing, we examined the mutation site in the beginning of the DNA-binding domain for the presence of the PAM sequences. There are several types of Cas nucleases, which recognize different PAM sequences [41]. The site of interest is AT-rich; therefore, the most common nuclease SpCas9 is poorly applicable (Figure 5a). Two of the three possible gRNAs close to the target site are inefficient, according to CRISPOR software calculations. The second most used nuclease Cas12a also is not appropriate for introducing the desired mutation, since the predicted double strand breaks are quite far from the target site (Figure 5b).

Modified nucleases such as iSpyMacCas9 [42], recognizing A-rich PAM sequence 5′-NAAR, can be a better choice to introduce desired mutations in the beginning of the DNA-binding domain of a *MYBL2* gene (Figure 5c).

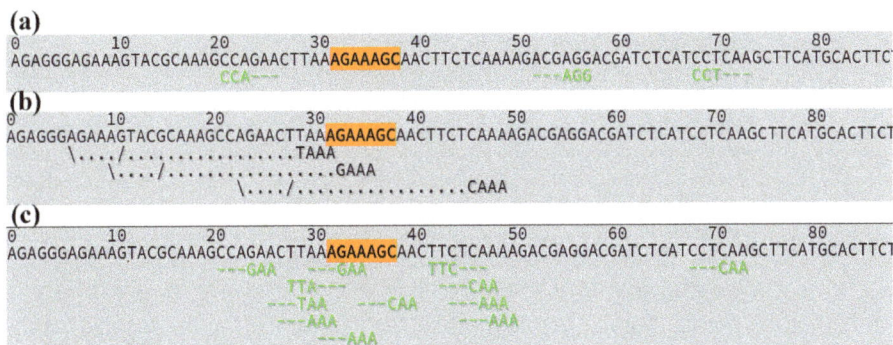

Figure 5. PAM sites in the beginning of the DNA-binding domain of a *MYBL2* gene, recognized by (**a**) Cas9, (**b**) Cas12a, (**c**) iSpyMacCas9.

Prime editing [43] might be the most suitable method to check if this particular mutation cause anthocyanin pigmentation in B. oleracea. With the use of elongated guide RNA, containing a donor sequence, it is possible to introduce the precise point mutations, identical to those in naturally bred purple cabbage. Nevertheless, according to the previous reports, many types of mutations in the MYBL2 gene, which are much easier to obtain, are applicable for generation of new purple varieties [25–27,29]. Any of the described nucleases can be used with multiple gRNAs to ensure complete excision of the target region or the whole gene. However complete excision may not be recommended as far as it can affect other functions of MYBL2-1 gene, such as trichome initiation and brassinosteroid signaling [26,44], absence of this gene copy did not seem to affect fitness of purple *B. oleracea* [25]. Therefore, attempts for introduction of the two desired SNPs are of little value to agriculture.

The discovery and introduction of point mutations, ensuring anthocyanin hyperaccumulation, can help to produce new purple varieties of economically important *Brassicaceae* crops. *Brassica* vegetables can be considered a functional food that can be used directly or as a dietary supplement [45]. Altering competition for substrate between flavonol synthase and dihydroflavonol-4-reductase by creating mutations in the *MYBL2* gene may also increase stress tolerance and nutritional value of new purple varieties.

4. Materials and Methods

4.1. Plant Materials

We analyzed 43 varieties of four economically important species of *Brassica*, including 22 varieties that were supposed to be purple, according to the manufacturer. *B. oleracea* was represented by 25 varieties, including purple-headed cabbage (Kalibos, Ludmila, Ruby, Mars, and Firebird), purple Brussels sprout (Garnet bracelet and Bunch of grapes), purple cauliflower (Gardener's dream and Purple ball), and purple kohlrabi (Madonna, Violetta, and Vienna purple), as well as green varieties of headed cabbage (Royal Vantage and Moscow Late), Brussels sprout (Rosella, Hercules, and Sapphire), cauliflower (Bird's milk, Baby, Flame star, Green bunch, and Alpha), and kohlrabi (Vienna white, Gulliver, and Picante). Anthocyanin pigmentation was observed in all varieties that were supposed to be purple, according to the manufacturer.

Five purple varieties of *B. juncea* (Vitamin, Red velvet, Freckle, Miracles in the sieve, and Red giant) and two green varieties (Wavelet and Vigorous) were used in this study. However, Freckle lacked anthocyanin pigmentation. Among five varieties of *B. napus*, rutabaga Krasnoselskaya and Gera were supposed to accumulate anthocyanins in roots; however rutabaga Novgorodskaya and Child love, as well as fodder cabbage Veha, were supposed to be anthocyaninless. Visible purple pigmentation was not detected in the leaves of any of these varieties. We also studied three purple varieties of *B. rapa* (Ruby little mermaid, Mizuna Red, and Mizuna purple) and three green varieties (Impulse, The little mermaid, and Mizuna green).

Green and purple varieties of each cultivar of *B. oleracea* and other studied species were subjected to Sanger sequencing of the *MYBL2* gene. Namely, green varieties Moscow late, Sapphire, Flame star, Gulliver, Vigorous, Novgorodskaya, Child love, and Impulse as well as purple varieties Ludmila, Garnet bracelet, Gardener's dream, Purple ball, Madonna, Vienna purple, Vitamin, Krasnoselskaya, Gera, and Mizuna red were examined.

Plants were grown at 20 °C under 10,000 lux and a 16:8 h day:night photoperiod, generated by LED grow light. True leaves were used for anthocyanin content measurement. Other experiments were carried out on cotyledons, homogenized using FastPrep-24 bead beating grinder and lysis system (MP Biomedicals, Santa Ana, CA, USA).

4.2. Real-Time Quantitative PCR

RNA was extracted from 50 mg of the plant tissue using Lira reagent (Biolabmix, Novosibirsk, Russia). cDNA was synthesized with OT-M-MuLV-RH kit (Biolabmix, Novosibirsk, Russia) using oligo(dT) primer.

Real-time quantitative PCR (RT-qPCR) was performed on Quant studio 5 Real-Time PCR System (Thermo Fisher Scientific, USA) using BioMaster HS-qPCR Lo-ROX SYBR mix (Biolabmix, Novosibirsk, Russia) with the primers 5′-CGGATCTGCAGGTTTAACTG-3′ and 5′-GCCGACATGTGACTATCGG-3′ for the *DFR* gene and two primer pairs 5′-AGAGGGAGAAAGTACGCAAAG-3′ and 5′-CGGTTCGTCCAGGCAATCTT-3′; 5′-TTCATGCACTTCTTGGCAAT-3′ and 5′-CGGTTCGTCCAGGCAATCTT-3′ for the *MYBL2* gene. Primers were designed using NCBI Primer BLAST tool [46]. Actin7 (primers 5′-AGGAATCGCTGACCGTATGAG-3′ and 5′-GCTGAGGGATGCAAGGATGGA-3′) was used as a reference gene. This gene is stably expressed in normal conditions and under biotic and abiotic stress [47].

4.3. DNA Sequencing

DNA was extracted from plant tissue using CTAB method [48].

MYBL2-1 gene was amplified using primers 5′-AGAGGGAGAAAGTACGCAAAG-3′ and 5′-AGAAGTGTTTCTTGACTCGTTGA-3′. PCR products were purified with ExoSAP-IT™ PCR Product Cleanup Reagent (Thermo Fisher Scientific, Waltham, MA, USA), prepared using BigDye™ Terminator v3.1 Cycle Sequencing Kit and subjected to Sanger sequencing via Applied Biosystems 3500 genetic analyzer (Thermo Fisher Scientific, Waltham, MA, USA).

4.4. Total Anthocyanin Content

Anthocyanin pigment concentration was measured by the spectrophotometric pH differential method using a Perkin Elmer LS 55 Luminescence Spectrometer (Perkin Elmer, Waltham, MA, USA) and expressed as cyanidin-3-glucoside equivalents. Anthocyanins in *Brassicaceae* are known to be cyanidin derivatives. Extracts from 15 mg of dried leaf tissue in pH 1.0 and pH 4.5 buffers were prepared and analyzed according to Lee et al. [49]. Measurements were performed in three technical and three biological replicates in 96-well plates.

4.5. Statistical Analysis

Three samples of each variety were used for DNA and RNA extraction. Primers were designed using NCBI Primer BLAST tool [46]. RT-qPCR was carried out for three replicates of each sample. Results of RT PCR were assessed by $2^{-\Delta\Delta CT}$ method. Sequencing was performed using forward and reverse primers for each sample. Sequences were aligned to the reference sequence of the *MYBL2-1* gene of *B. rapa* (JN379102) and analyzed using SnapGene software. Functional domains were analyzed via InterPro [50]. Search for gRNAs was performed in CRISPOR web tool [51]. RNA structure was predicted by RNAfold web tool [52]. For all experiments, means and standard deviation ($p < 0.05$) were compared by analysis of variance (ANOVA), and Pearson correlation coefficient was calculated using LibreOffice v.

Author Contributions: DNA and RNA extraction and PCR, E.K.; sequencing, Y.S. and A.A.; methodology, analysis, and supervision, E.V.M. All authors have read and agreed to the published version of the manuscript.

Funding: This research was funded by Russian Science Foundation, grant number 20-74-10053.

Institutional Review Board Statement: Not applicable.

Informed Consent Statement: Not applicable.

Data Availability Statement: The sequence of the mutant genotype was submitted to the NCBI database (ON464161).

Conflicts of Interest: The authors declare no conflict of interest.

References

1. Podsedek, A. Natural antioxidants and antioxidant capacity of Brassica vegetables: A review. *LWT* **2007**, *40*, 1–11. [CrossRef]
2. Azad, M.O.K.; Adnan, M.; Kang, W.S.; Lim, J.D.; Lim, Y.S. A Technical Strategy to Prolong Anthocyanins Thermal Stability in Formulated Purple Potato (*Solanum Tuberosum* L. Cv Bora Valley) Processed by Hot-melt Extrusion. *Int. J. Food Sci. Technol.* **2021**, 15485. [CrossRef]
3. Kim, J.; Lee, W.J.; Vu, T.T.; Jeong, C.Y.; Hong, S.W.; Lee, H. High accumulation of anthocyanins via the ectopic expression of AtDFR confers significant salt stress tolerance in *Brassica napus* L. *Plant Cell Rep.* **2017**, *36*, 1215–1224. [CrossRef]
4. Ma, H.; Pooler, M.; Griesbach, R. Anthocyanin Regulatory/Structural Gene Expression in *Phalaenopsis*. *J. Am. Soc. Hort. Sci.* **2009**, *134*, 88–96. [CrossRef]
5. Almeida, J.; Carpenter, R.; Robbins, T.P.; Martin, C.; Coen, E.S. Genetic Interactions Underlying Flower Color Patterns in Antirrhinum Majus. *Genes Dev.* **1989**, *3*, 1758–1767. [CrossRef] [PubMed]
6. Holton, T.A.; Cornish, E.C. Genetics and Biochemistry of Anthocyanin Biosynthesis. *Plant Cell* **1995**, *7*, 1071–1083. [CrossRef] [PubMed]
7. Chen, G.; He, W.; Guo, X.; Pan, J. Genome-Wide Identification, Classification and Expression Analysis of the MYB Transcription Factor Family in *Petunia*. *Int. J. Mol. Sci.* **2021**, *22*, 4838. [CrossRef]
8. Dugassa, N.; Feyissa, T.L.; Kristine, M.O.; Rune, S.; Cathrine, L. The endogenous *GL3*, but not *EGL3*, gene is necessary for anthocyanin accumulation as induced by nitrogen depletion in *Arabidopsis* rosette stage leaves. *Planta* **2009**, *230*, 747–754. [CrossRef]
9. Khusnutdinov, E.; Sukhareva, A.; Panfilova, M.; Mikhaylova, E. Anthocyanin Biosynthesis Genes as Model Genes for Genome Editing in Plants. *Int. J. Mol. Sci.* **2021**, *22*, 8752. [CrossRef] [PubMed]
10. Ahn, J.H.; Kim, J.-S.; Kim, S.; Soh, H.Y.; Shin, H.; Jang, H.; Ryu, J.H.; Kim, A.; Yun, K.-Y.; Kim, S.; et al. De Novo Transcriptome Analysis to Identify Anthocyanin Biosynthesis Genes Responsible for Tissue-Specific Pigmentation in Zoysiagrass (*Zoysia Japonica* Steud.). *PLoS ONE* **2015**, *10*, e0124497. [CrossRef]
11. Rouholamin, S.; Zahedi, B.; Nazarian-Firouzabadi, F.; Saei, A. Expression Analysis of Anthocyanin Biosynthesis Key Regulatory Genes Involved in Pomegranate (*Punica Granatum* L.). *Sci. Horticult.* **2015**, *186*, 84–88. [CrossRef]
12. Davies, K.M.; Schwinn, K.E.; Deroles, S.C.; Manson, D.G.; Bloor, S.J.; Bradley, J.M. Enhancing Anthocyanin Production by Altering Competition for Substrate between Flavonol Synthase and Dihydroflavonol 4-Reductase. *Euphytica* **2003**, *131*, 259–268. [CrossRef]
13. Tsuda, T.; Yamaguchi, M.; Honda, C.; Moriguchi, T. Expression of Anthocyanin Biosynthesis Genes in the Skin of Peach and Nectarine Fruit. *J. Am. Soc. Hortic. Sci.* **2004**, *129*, 857–862. [CrossRef]
14. Zhuang, H.; Lou, Q.; Liu, H.; Han, H.; Wang, Q.; Tang, Z.; Ma, Y.; Wang, H. Differential Regulation of Anthocyanins in Green and Purple Turnips Revealed by Combined De Novo Transcriptome and Metabolome Analysis. *Int. J. Mol. Sci.* **2019**, *20*, 4387. [CrossRef] [PubMed]
15. Chen, Q.-F.; Dai, L.-Y.; Xiao, S.; Wang, Y.-S.; Liu, X.-L.; Wang, G.-L. The COI1 and DFR Genes Are Essential for Regulation of Jasmonate-Induced Anthocyanin Accumulation in Arabidopsis. *J. Integr. Plant Biol.* **2007**, *49*, 1370–1377. [CrossRef]
16. Li, X.; Chen, L.; Hong, M.; Zhang, Y.; Zu, F.; Wen, J.; Yi, B.; Ma, C.; Shen, J.; Tu, J.; et al. A Large Insertion in BHLH Transcription Factor BrTT8 Resulting in Yellow Seed Coat in *Brassica Rapa*. *PLoS ONE* **2012**, *7*, e44145. [CrossRef]
17. Kim, J.; Kim, D.-H.; Lee, J.-Y.; Lim, S.-H. The R3-Type MYB Transcription Factor BrMYBL2.1 Negatively Regulates Anthocyanin Biosynthesis in Chinese Cabbage (*Brassica rapa* L.) by Repressing MYB–bHLH–WD40 Complex Activity. *Int. J. Mol. Sci.* **2022**, *23*, 3382. [CrossRef] [PubMed]
18. Chiu, L.W.; Zhou, X.; Burke, S.; Wu, X.; Prior, R.L.; Li, L. The purple cauliflower arises from activation of a MYB transcription factor. *Plant Physiol.* **2010**, *154*, 1470–1480. [CrossRef]
19. Song, S.; Kim, C.W.; Moon, J.S.; Kim, S. At least nine independent natural mutations of the *DFR-A* gene are responsible for appearance of yellow onions (*Allium cepa* L.) from red progenitors. *Mol. Breed.* **2014**, *33*, 173–186. [CrossRef]

20. Kim, S.; Yoo, K.S.; Pike, L.M. Development of a PCR-Based Marker Utilizing a Deletion Mutation in the Dihydroflavonol 4-Reductase (DFR) Gene Responsible for the Lack of Anthocyanin Production in Yellow Onions (*Allium Cepa*). *Theor. Appl. Genet.* 2005, *110*, 588–595. [CrossRef]
21. Liu, X.P.; Gao, B.Z.; Han, F.Q.; Fang, Z.Y.; Yang, L.M.; Zhuang, M.; Lv, H.H.; Liu, Y.M.; Li, Z.S.; Cai, C.C.; et al. Genetics and fine mapping of a purple leaf gene, BoPr, in ornamental kale (*Brassica oleracea* L. var. *acephala*). *BMC Genom.* 2017, *18*, 230. [CrossRef]
22. Dubos, C.; Le Gourrierec, J.; Baudry, A.; Huep, G.; Lanet, E.; Debeaujon, I.; Routaboul, J.-M.; Alboresi, A.; Weisshaar, B.; Lepiniec, L. MYBL2 Is a New Regulator of Flavonoid Biosynthesis in *Arabidopsis Thaliana*. *Plant J.* 2008, *55*, 940–953. [CrossRef]
23. Rowan, D.D.; Cao, M.; Lin-Wang, K.; Cooney, J.M.; Jensen, D.J.; Austin, P.T.; Hunt, M.B.; Norling, C.; Hellens, R.P.; Schaffer, R.J.; et al. Environmental Regulation of Leaf Colour in Red 35S:PAP1 Arabidopsis Thaliana. *New Phytol.* 2009, *182*, 102–115. [CrossRef] [PubMed]
24. Guo, N.; Cheng, F.; Wu, J.; Liu, B.; Zheng, S.; Liang, J.; Wang, X. Anthocyanin Biosynthetic Genes in *Brassica Rapa*. *BMC Genom.* 2014, *15*, 426. [CrossRef] [PubMed]
25. Song, H.; Yi, H.; Lee, M.; Han, C.-T.; Lee, J.; Kim, H.; Park, J.-I.; Nou, I.-S.; Kim, S.-J.; Hur, Y. Purple *Brassica Oleracea Var. Capitata F. Rubra* Is Due to the Loss of BoMYBL2-1 Expression. *BMC Plant Biol.* 2018, *18*, 82. [CrossRef]
26. Nemie-Feyissa, D.; Olafsdottir, S.M.; Heidari, B.; Lillo, C. Nitrogen Depletion and Small R3-MYB Transcription Factors Affecting Anthocyanin Accumulation in Arabidopsis Leaves. *Phytochemistry* 2014, *98*, 34–40. [CrossRef]
27. Matsui, K.; Umemura, Y.; Ohme-Takagi, M. AtMYBL2, a Protein with a Single MYB Domain, Acts as a Negative Regulator of Anthocyanin Biosynthesis in Arabidopsis. *Plant J.* 2008, *55*, 954–967. [CrossRef] [PubMed]
28. Goswami, G.; Nath, U.K.; Park, J.-I.; Hossain, M.R.; Biswas, M.K.; Kim, H.-T.; Kim, H.R.; Nou, I.-S. Transcriptional Regulation of Anthocyanin Biosynthesis in a High-Anthocyanin Resynthesized *Brassica Napus* Cultivar. *J. Biol. Res. Thessalon.* 2018, *25*, 19. [CrossRef] [PubMed]
29. Zhang, X.; Zhang, K.; Wu, J.; Guo, N.; Liang, J.; Wang, X.; Cheng, F. QTL-Seq and Sequence Assembly Rapidly Mapped the Gene BrMYBL2.1 for the Purple Trait in *Brassica Rapa*. *Sci. Rep.* 2020, *10*, 2328. [CrossRef]
30. Rameneni, J.J.; Choi, S.R.; Chhapekar, S.S.; Kim, M.-S.; Singh, S.; Yi, S.Y.; Oh, S.H.; Kim, H.; Lee, C.Y.; Oh, M.-H.; et al. Red Chinese Cabbage Transcriptome Analysis Reveals Structural Genes and Multiple Transcription Factors Regulating Reddish Purple Color. *Int. J. Mol. Sci.* 2020, *21*, 2901. [CrossRef] [PubMed]
31. Li, Y.-C. Microsatellites Within Genes: Structure, Function, and Evolution. *Mol. Biol. Evol.* 2004, *21*, 991–1007. [CrossRef] [PubMed]
32. Wang, Y.; Wang, Y.; Song, Z.; Zhang, H. Repression of MYBL2 by Both MicroRNA858a and HY5 Leads to the Activation of Anthocyanin Biosynthetic Pathway in *Arabidopsis*. *Mol. Plant* 2016, *9*, 1395–1405. [CrossRef] [PubMed]
33. Li, Y.-C.; Korol, A.B.; Fahima, T.; Beiles, A.; Nevo, E. Microsatellites: Genomic Distribution, Putative Functions and Mutational Mechanisms: A Review. *Mol. Ecol.* 2002, *11*, 2453–2465. [CrossRef] [PubMed]
34. Benghezal, M.; Wasteneys, G.O.; Jones, D.A. The C-Terminal Dilysine Motif Confers Endoplasmic Reticulum Localization to Type I Membrane Proteins in Plants. *Plant Cell* 2000, *12*, 1179–1201. [CrossRef]
35. Yang, X.; Yang, M.; Deng, H.; Ding, Y. New Era of Studying RNA Secondary Structure and Its Influence on Gene Regulation in Plants. *Front. Plant Sci.* 2018, *9*, 671. [CrossRef] [PubMed]
36. Appelhagen, I.; Jahns, O.; Bartelniewoehner, L.; Sagasser, M.; Weisshaar, B.; Stracke, R. Leucoanthocyanidin Dioxygenase in *Arabidopsis Thaliana*: Characterization of Mutant Alleles and Regulation by MYB-BHLH-TTG1 Transcription Factor Complexes. *Gene* 2011, *484*, 61–68. [CrossRef] [PubMed]
37. Abrahams, S.; Lee, E.; Walker, A.R.; Tanner, G.J.; Larkin, P.J.; Ashton, A.R. The *Arabidopsis TDS4* Gene Encodes Leucoanthocyanidin Dioxygenase (LDOX) and Is Essential for Proanthocyanidin Synthesis and Vacuole Development: The Role of LDOX in PA Biosynthesis. *Plant J.* 2003, *35*, 624–636. [CrossRef] [PubMed]
38. Ben-Simhon, Z.; Judeinstein, S.; Trainin, T.; Harel-Beja, R. A "White" Anthocyanin-less Pomegranate (*Punica granatum* L.) Caused by an Insertion in the Coding Region of the Leucoanthocyanidin Dioxygenase (LDOX; ANS) Gene. *PLoS ONE* 2015, *10*, e0142777. [CrossRef] [PubMed]
39. Zhang, H.; Du, C.; Wang, Y.; Wang, J.; Zheng, L.; Wang, Y. The *Reaumuria Trigyna* Leucoanthocyanidin Dioxygenase (RtLDOX) Gene Complements Anthocyanidin Synthesis and Increases the Salt Tolerance Potential of a Transgenic Arabidopsis *LDOX* Mutant. *Plant Physiol. Biochem.* 2016, *106*, 278–287. [CrossRef] [PubMed]
40. Li, N.; Wang, X.; Ma, B.; Wu, Z.; Zheng, L.; Qi, Z.; Wang, Y. A Leucoanthocyanidin Dioxygenase Gene (RtLDOX2) from the Feral Forage Plant *Reaumuria Trigyna* Promotes the Accumulation of Flavonoids and Improves Tolerance to Abiotic Stresses. *J. Plant Res.* 2021, *134*, 1121–1138. [CrossRef]
41. Mikhaylova, E.V.; Khusnutdinov, E.A.; Chemeris, A.V.; Kuluev, B.R. Available Toolkits for CRISPR/CAS Genome Editing in Plants. *Russ. J. Plant Physiol.* 2022, *69*, 3. [CrossRef]
42. Sretenovic, S.; Yin, D.; Levav, A.; Selengut, J.D.; Mount, S.M.; Qi, Y. Expanding Plant Genome-Editing Scope by an Engineered iSpyMacCas9 System That Targets A-Rich PAM Sequences. *Plant Commun.* 2020, *2*, 100–101. [CrossRef] [PubMed]
43. Anzalone, A.V.; Randolph, P.B.; Davis, J.R.; Sousa, A.A.; Koblan, L.W.; Levy, J.M.; Chen, P.J.; Wilson, C.; Newby, G.A.; Raguram, A.; et al. Search-and-Replace Genome Editing without Double-Strand Breaks or Donor DNA. *Nature* 2019, *576*, 149–157. [CrossRef]
44. Sawa, S. Overexpression of the *AtmybL2* Gene Represses Trichome Development in *Arabidopsis*. *DNA Res.* 2002, *9*, 31–34. [CrossRef] [PubMed]

45. Azad, M.O.K.; Adnan, M.; Sung, I.J.; Lim, J.D.; Baek, J.; Lim, Y.S.; Park, C.H. Development of Value-added Functional Food by Fusion of Colored Potato and Buckwheat Flour through Hot-melt Extrusion. *Food Process. Preserv.* **2022**, *46*, e15312. [CrossRef]
46. Ye, J.; Coulouris, G.; Zaretskaya, I.; Cutcutache, I.; Rozen, S.; Madden, T.L. Primer-BLAST: A Tool to Design Target-Specific Primers for Polymerase Chain Reaction. *BMC Bioinform.* **2012**, *13*, 134. [CrossRef]
47. Mikhaylova, E.; Khusnutdinov, E.; Shein, M.Y.; Alekseev, V.Y.; Nikonorov, Y.; Kuluev, B. The Role of the *GSTF11* Gene in Resistance to Powdery Mildew Infection and Cold Stress. *Plants* **2021**, *10*, 2729. [CrossRef]
48. Lee, J.; Durst, R.W.; Wrolstad, R.E. Determination of Total Monomeric Anthocyanin Pigment Content of Fruit Juices, Beverages, Natural Colorants, and Wines by the pH Differential Method: Collaborative Study. *J. AOAC Int.* **2005**, *88*, 1269–1278. [CrossRef]
49. Porebski, S.; Bailey, L.G.; Baum, B.R. Modification of a CTAB DNA Extraction Protocol for Plants Containing High Polysaccharide and Polyphenol Components. *Plant Mol. Biol. Rep.* **1997**, *15*, 8–15. [CrossRef]
50. Blum, M.; Chang, H.-Y.; Chuguransky, S.; Grego, T.; Kandasaamy, S.; Mitchell, A.; Nuka, G.; Paysan-Lafosse, T.; Qureshi, M.; Raj, S.; et al. The InterPro Protein Families and Domains Database: 20 Years On. *Nucleic Acids Res.* **2021**, *49*, 344–354. [CrossRef]
51. Concordet, J.-P.; Haeussler, M. CRISPOR: Intuitive Guide Selection for CRISPR/Cas9 Genome Editing Experiments and Screens. *Nucleic Acids Res.* **2018**, *46*, 242–245. [CrossRef] [PubMed]
52. Gruber, A.R.; Lorenz, R.; Bernhart, S.H.; Neubock, R.; Hofacker, I.L. The Vienna RNA Websuite. *Nucleic Acids Res.* **2008**, *36*, W70–W74. [CrossRef] [PubMed]

Article

Physiological and Biochemical Regulation Mechanism of Exogenous Hydrogen Peroxide in Alleviating NaCl Stress Toxicity in Tartary Buckwheat (*Fagopyrum tataricum* (L.) Gaertn)

Xin Yao [1], Meiliang Zhou [2], Jingjun Ruan [1,*], Yan Peng [1], Chao Ma [1], Weijiao Wu [1], Anjing Gao [1], Wenfeng Weng [1] and Jianping Cheng [1,*]

1. College of Agronomy, Guizhou University, Guiyang 550025, China
2. Institute of Crop Science, Chinese Academy of Agriculture Science, Beijing 100081, China
* Correspondence: jjruan@gzu.edu.cn (J.R.); jpcheng@gzu.edu.cn (J.C.)

Citation: Yao, X.; Guo, H.; Zhou, M.; Ruan, J.; Peng, Y.; Ma, C.; Wu, W.; Gao, A.; Weng, W.; Cheng, J. Physiological and Biochemical Regulation Mechanism of Exogenous Hydrogen Peroxide in Alleviating NaCl Stress Toxicity in Tartary Buckwheat (*Fagopyrum tataricum* (L.) Gaertn). *Int. J. Mol. Sci.* **2022**, *23*, 10698. https://doi.org/10.3390/ijms231810698

Academic Editors: Zhiyong Li and Jian Zhang

Received: 23 August 2022
Accepted: 8 September 2022
Published: 14 September 2022

Publisher's Note: MDPI stays neutral with regard to jurisdictional claims in published maps and institutional affiliations.

Copyright: © 2022 by the authors. Licensee MDPI, Basel, Switzerland. This article is an open access article distributed under the terms and conditions of the Creative Commons Attribution (CC BY) license (https://creativecommons.org/licenses/by/4.0/).

Abstract: We aimed to elucidate the physiological and biochemical mechanism by which exogenous hydrogen peroxide (H_2O_2) alleviates salt stress toxicity in Tartary buckwheat (*Fagopyrum tataricum* (L.) Gaertn). Tartary buckwheat "Chuanqiao-2" under 150 mmol·L^{-1} salt (NaCl) stress was treated with 5 or 10 mmol·L^{-1} H_2O_2, and seedling growth, physiology and biochemistry, and related gene expression were studied. Treatment with 5 mmol·L^{-1} H_2O_2 significantly increased plant height (PH), fresh and dry weights of shoots (SFWs/SDWs) and roots (RFWs/RDWs), leaf length (LL) and area (LA), and relative water content (LRWC); increased chlorophyll a (Chl a) and b (Chl b) contents; improved fluorescence parameters; enhanced antioxidant enzyme activity and content; and reduced malondialdehyde (MDA) content. Expressions of all stress-related and enzyme-related genes were up-regulated. The *F3'H* gene (flavonoid synthesis pathway) exhibited similar up-regulation under 10 mmol·L^{-1} H_2O_2 treatment. Correlation and principal component analyses showed that 5 mmol·L^{-1} H_2O_2 could significantly alleviate the toxic effect of salt stress on Tartary buckwheat. Our results show that exogenous 5 mmol·L^{-1} H_2O_2 can alleviate the inhibitory or toxic effects of 150 mmol·L^{-1} NaCl stress on Tartary buckwheat by promoting growth, enhancing photosynthesis, improving enzymatic reactions, reducing membrane lipid peroxidation, and inducing the expression of related genes.

Keywords: Tartary buckwheat; H_2O_2; NaCl; physiology and biochemistry

1. Introduction

Buckwheat (*Fagopyrum esculentum* Moench, 2n = 8) is an annual or perennial grain crop used as food and forage [1]. In China, buckwheat is divided into three main cultivars: Tartary buckwheat (*F. tataricum*), sweet buckwheat (*F. esculentum*), and golden buckwheat (*F. cymosum*), in addition to wild species [2]. Tartary buckwheat is a dicotyledonous plant in the family Polygonaceae [3]. China is not only the largest producer of Tartary buckwheat, but also the global center of buckwheat diversity. Tartary buckwheat is suitable for growing in high-altitude southwestern regions of China, such as Yunnan, Guizhou, and Sichuan Provinces, which have a cold climate and short frost-free period [4]. The southwest region of China is internationally recognized as the center of origin for buckwheat [5]. China ranks among the highest in the world in terms of planting area and output, with a current output second only to Russia. Tartary buckwheat has been promoted, grown, and eaten in many countries in the world [6]. Tartary buckwheat is rich in flavonoids and other biologically active substances, including rutin, quercetin, kaempferol, morin, and other natural compounds, and the rutin is the main essential ingredient in Tartary buckwheat. The content of rutin in the shoots is significantly higher than that in the roots [7]. Tartary buckwheat also contains a variety of nutrients, such as protein, starch, and vitamins, which can relieve cardiovascular sclerosis, diabetes, and other diseases [4,8]. Therefore, with the rising demand for Tartary buckwheat products, its development and utilization have gained increasing attention.

In natural biological populations, interactions between genotype and environment affect the phenotype of plants. In plants with the same genotype, different environments can lead to various phenotypic differences [9,10]. With increasingly serious global warming and rising land evaporation, the problem of soil salinization has become prominent in countries at middle and low latitudes [11]. Salt stress refers to the adverse effect of excess soil salt on plants [12], and is a major abiotic stress that seriously affects the growth and development of plants, inhibits their physiological and biochemical metabolic processes, even causing plant death, greatly reduces agricultural yields, and restricts agricultural development potential [13,14]. When plants are subjected to salt stress, with increasing salt concentration, the growth trend of plants decreases [15], the content of reactive oxygen species (ROS) increases [16], the activity of antioxidant enzymes reduces [17], photosynthesis weakens [18], and the homeostasis of sodium (Na^+) and potassium (K^+) ions is unbalanced [19]. Therefore, the improvement, development, and utilization of soil salinization is an effective way to improve the growth environment of plants, and improving plant growth and yield under salt stress has become a hot research topic globally.

Reactive oxygen species (ROS) are biochemical substances that play a vital role in seed dormancy and germination. In particular, the intracellular homeostasis of hydrogen peroxide (H_2O_2), superoxide anion (O_2^-), and hydroxyl radical (OH^-) is involved in signaling cascades, and this determines the growth and development process and stress response [20]. H_2O_2 is an important regulatory component of plant signal transduction. It is not only a free radical produced by oxidative stress products, but also helps to maintain plant cell homeostasis [21]. Concurrently, H_2O_2 can regulate the expression of various genes in physiological metabolism, including the genes encoding antioxidant enzymes, regulating biotic and abiotic stress response proteins [22,23]. H_2O_2 has a concentration-dependent effect on physiological and biochemical processes. That is, high concentrations of H_2O_2 produce an oxidative stress reaction in plants, leading to cell damage [24]; whereas at low concentrations, H_2O_2 acts as a signal molecule participating in tolerance to various abiotic [25] and biological stresses [26], and also plays a regulatory role. A number of studies have reported that H_2O_2 is involved in the regulation of physiological activities such as seed germination [27] and photosynthesis [28].

As a multifunctional signaling molecule, H_2O_2 plays an important role in a series of physiological and biochemical processes. However, the involvement of H_2O_2 in the anti-stress metabolism of Tartary buckwheat, and particularly the mechanism of salt tolerance, is rarely studied. In previous research by the research group, it was found that 1–5 mmol·L^{-1} H_2O_2 (without NaCl treatment) could significantly increase the root growth of Tartary buckwheat at the germination stage [29], and at the germination stage 5–10 mmol·L^{-1}, H_2O_2 can significantly promote the germination of Tartary buckwheat under 50 mmol·L^{-1} NaCl treatment [1]. Therefore, in this study, different concentrations of exogenous H_2O_2 were applied to the leaves of Tartary buckwheat to explore the growth, photosynthesis, antioxidant enzyme activity, membrane lipid peroxidation, and expression of related genes of Tartary buckwheat under NaCl stress. The physiological metabolic mechanism by which H_2O_2 regulates the salt tolerance of Tartary buckwheat was evaluated through correlation and principal component analyses to provide a scientific theoretical basis for exogenous H_2O_2 treatment to improve the salt tolerance of Tartary buckwheat.

2. Results

2.1. Effect of Exogenous H_2O_2 on Plant Height and Biomass under NaCl Stress

Figure 1 shows that the growth of Tartary buckwheat under the CK treatment ($H_2O + H_2O$) and 5H + N (5 mmol·L^{-1} H_2O_2 + 150 mmol·L^{-1} NaCl) treatments was significantly better than that under NaCl (150 mmol·L^{-1} NaCl + H_2O) and 10H + N (10 mmol·L^{-1} H_2O_2 + 150 mmol·L^{-1} NaCl) treatments, and the root growth with 5H + N was better than that with the other three treatments.

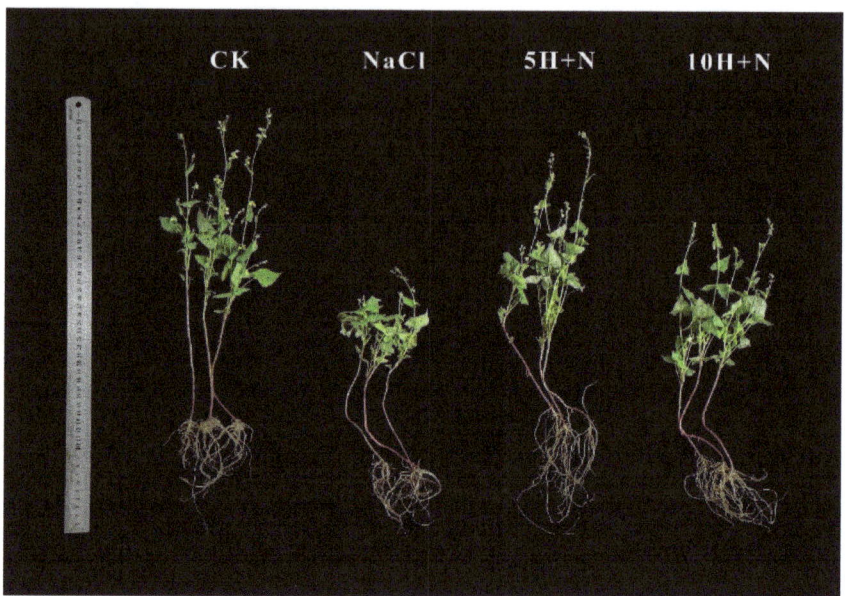

Figure 1. Effects of exogenous H_2O_2 on growth of Tartary buckwheat under NaCl stress. The photos were taken on the 40th day of transplanting. CK: control, H_2O foliar spray + H_2O irrigation, NaCl: H_2O foliar spray + 150 mmol·L^{-1} NaCl irrigation, 5H + N: 5 mmol·L^{-1} H_2O_2 foliar spray + 150 mmol·L^{-1} NaCl irrigation, 10H + N: 10 mmol·L^{-1} H_2O_2 foliar spray + 150 mmol·L^{-1} NaCl irrigation.

Under NaCl stress, plant height (PH), stem fresh weight (SFW), stem dry weight (SDW), root fresh weight (RFW), and root dry weight (RDW) of Tartary buckwheat were significantly decreased by 28.22%, 51.90%, 44.63%, 59.60%, and 48.19%, respectively, compared with what they were under the CK treatment (Figure 2A–E). Under the 10H + N treatment, PH, SFW, SDW, RFW, and RDW were significantly decreased by 0.78-fold, 0.43-fold, 0.63-fold, 0.58-fold, and 0.61-fold, respectively, relative to what they were under the CK treatment (Figure 2A–E). The SFW of Tartary buckwheat decreased by 9.80% under the 10H + N treatment compared with what it was under the NaCl treatment (Figure 2B). Under the 5H + N treatment, the PH, SFW, SDW, RFW, and RDW of Tartary buckwheat were significantly higher than they were under the NaCl treatment (Figure 2A–E); SFW and RFW increased by 124.60% and 141.58%, respectively (Figure 2B,D), and SFW was significantly decreased by 2.49 times under 10H + N treatment (Figure 2B).

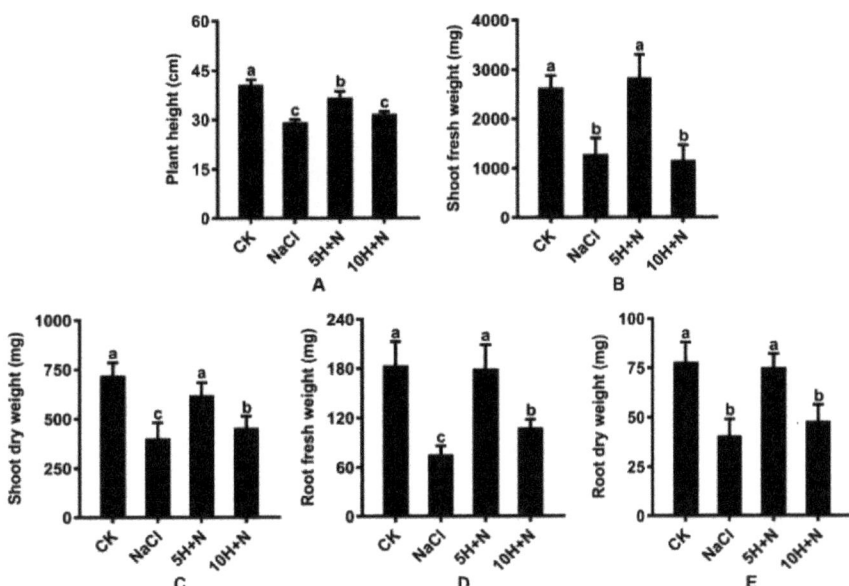

Figure 2. Effects of exogenous H_2O_2 on PH (**A**), SFW (**B**), SDW (**C**), RFW (**D**), and RDW (**E**) of Tartary buckwheat under NaCl stress. CK: control, H_2O foliar spray + H_2O irrigation, NaCl: H_2O foliar spray + 150 mmol·L^{-1} NaCl irrigation, 5H + N: 5 mmol·L^{-1} H_2O_2 foliar spray + 150 mmol·L^{-1} NaCl irrigation, 10H + N: 10 mmol·L^{-1} H_2O_2 foliar spray + 150 mmol·L^{-1} NaCl irrigation. PH: plant height (cm), SFW: shoot fresh weight (mg), SDW: shoot dry weight (mg), RFW: root fresh weight (mg), RDW: root dry weight (mg). All values are expressed as mean ± SD. According to Duncan's multiple comparisons, different letters represent significant differences among different treatments ($p < 0.05$).

2.2. Effect of Exogenous H_2O_2 and NaCl Stress on Leaf Growth and Relative Water Content

Compared with the CK, the leaf area (LA) and leaf length (LL) of Tartary buckwheat increased significantly by 24.75% and 18.60%, respectively, under 5H + N treatment (Figure 3A,B), and the leaf relative water content (LRWC) decreased slightly (Figure 3C); under 10H + N treatment, LA was significantly decreased by 11.88% (Figure 3A), and LL and LRWC were significantly decreased by 23.97% and 22.84%, respectively (Figure 3B,C). Compared with NaCl treatment, the LA, LL, and LRWC of Tartary buckwheat were significantly increased under the 5H + N treatment (Figure 3A–C), among which LL increased the most (91.33%) (Figure 3C). LA, LL, and LRWC were significantly increased under 10H + N treatment; however, the increasing trend of LRWC was not significant, and all values were significantly lower under 10H + N treatment than under 5H + N treatment (Figure 3A–C).

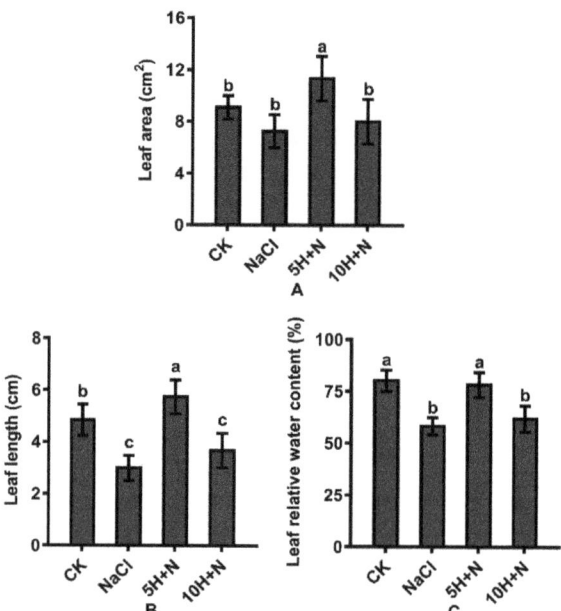

Figure 3. Effects of exogenous H_2O_2 on LA (**A**), LL (**B**), and LRWC (**C**) of Tartary buckwheat under NaCl stress. CK: control, H_2O foliar spray + H_2O irrigation, NaCl: H_2O foliar spray + 150 mmol·L^{-1} NaCl irrigation, 5H + N: 5 mmol·L^{-1} H_2O_2 foliar spray + 150 mmol·L^{-1} NaCl irrigation, 10H + N: 10 mmol·L^{-1} H_2O_2 foliar spray + 150 mmol·L^{-1} NaCl irrigation. LA: leaf area (cm^2), LL: leaf length (cm), LRWC: leaf relative water content (%). All values are expressed as mean ± SD. According to Duncan's multiple comparisons, different letters represent significant differences among different treatments ($p < 0.05$).

2.3. Effect of Exogenous H_2O_2 and NaCl Stress on Leaf Photosynthetic Pigments

Compared with CK, Chl a, Chl b, and Car were significantly reduced by 40.63–69.33% and 31.69–55.11%, respectively, under the NaCl stress and 10H + N treatments (Figure 4A). Under 5H + N treatment, Chl a, Chl b, and Car showed an insignificant downward trend (Figure 4A). Compared with NaCl stress, Chl a, Chl b, and Car were significantly increased in Tartary buckwheat under 5H + N treatment (Figure 4A), among which Car showed the largest increase (191.30%) (Figure 4A). Under 10H + N treatment, Chl a was significantly decreased, while Chl b and Car were significantly increased (Figure 4A).

The total chlorophyll content included Chl a and Chl b. As shown in Figure 4B,C, the maximum observed value of Chl a + b was in the CK treatment (1.147), and the maximum Chl a/b was in the NaCl treatment (2.639). Under the 10H + N treatment, Chl a + b and Chl a/b of Tartary buckwheat were significantly decreased compared with what they were under CK treatment (Figure 4B,C). Under NaCl stress, Chl a + b was significantly decreased (Figure 4B), while Chl a/b showed no significant increase compared to under CK treatment (Figure 4C). Under 5H + N treatment, Chl a + b of Tartary buckwheat was significantly increased (56.61%) compared with under NaCl stress (Figure 4B).

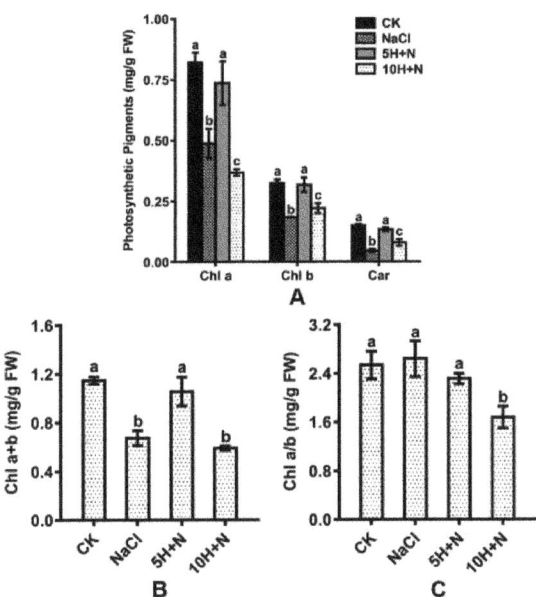

Figure 4. Effects of exogenous H_2O_2 on Chl a, Chl b, and Car (**A**), Chl a + b (**B**), and Chl a/b (**C**) of Tartary buckwheat under NaCl stress. CK: control, H_2O foliar spray + H_2O irrigation, NaCl: H_2O foliar spray + 150 mmol·L^{-1} NaCl irrigation, 5H + N: 5 mmol·L^{-1} H_2O_2 foliar spray + 150 mmol·L^{-1} NaCl irrigation, 10H + N: 10 mmol·L^{-1} H_2O_2 foliar spray + 150 mmol·L^{-1} NaCl irrigation. Chl a: chlorophyll a (mg/g FW), chlorophyll b (mg/g FW), Car (mg/g FW). All values are expressed as mean ± SD. According to Duncan's multiple comparisons, different letters represent significant differences among different treatments ($p < 0.05$).

2.4. Effect of Exogenous H_2O_2 and NaCl Stress on Leaf Fluorescence Parameters

The change trends of PSII maximum photochemical efficiency (Fv/Fm), effective quantum yield (Fv'/Fm'), non-photochemical quenching coefficient (NPQ), and effective electron transfer rate (ETR) of Tartary buckwheat were similar (Figure 5A,B,E,F). Compared with CK, the Fv/Fm, Fv'/Fm', NPQ, and ETR of Tartary buckwheat under NaCl stress were significantly decreased by 25.74%, 27.69%, 30.86%, and 36.26%, respectively (Figure 5A,B,E,F). The Fv/Fm, Fv'/Fm', NPQ, and ETR of Tartary buckwheat under 10H + N treatment were all reduced (Figure 5A,E,F), with the largest reduction observed in NPQ (29.14%) (Figure 5E); this reduction was significant for all except Fv'/Fm'. Compared with NaCl stress, the Fv/Fm, Fv'/Fm', NPQ, and ETR were significantly increased by 22.70–82.66% (Figure 5A,B,E,F), with the largest increase observed in the ETR (Figure 5F).

The change trends of Φ_{PSII} and qP of Tartary buckwheat were similar (Figure 5C,D). Compared with CK, Φ_{PSII} and qP were significantly increased under NaCl stress and the 10H + N treatment. The largest increases in Φ_{PSII} and qP (1.36 and 1.45 times, respectively) were observed under 10H + N treatment (Figure 5C,D), while Φ_{PSII} and qP did not significantly change with 5H + N treatment (Figure 5C,D). Compared with 5H + N treatment, Φ_{PSII} and qP were significantly increased under NaCl stress and 10H + N treatment, while Φ_{PSII} and qP were not significantly increased under either NaCl stress or 10H + N treatment (Figure 5C,D).

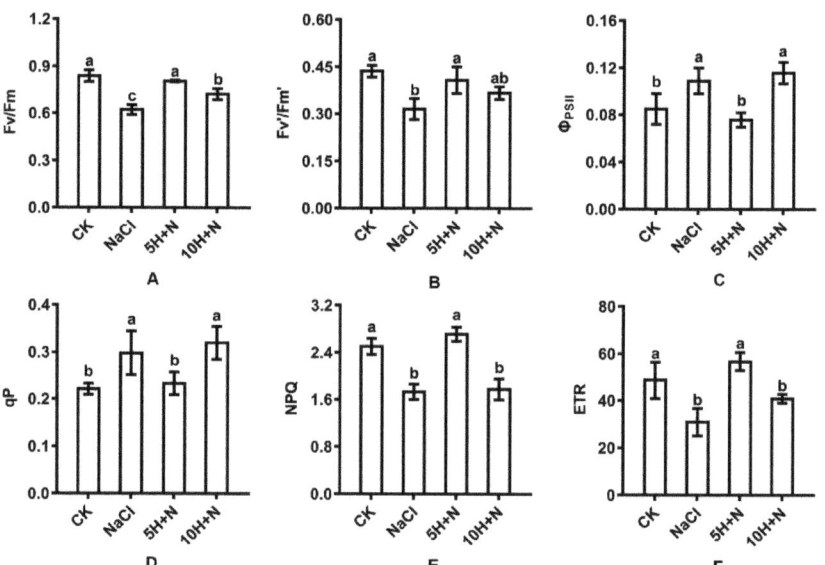

Figure 5. Effects of exogenous H_2O_2 on Fv/Fm (**A**), Fv'/Fm' (**B**), Φ_{PSII} (**C**), qP (**D**), NPQ (**E**), and ETR (**F**) of Tartary buckwheat under NaCl stress. CK: control, H_2O foliar spray + H_2O irrigation, NaCl: H_2O foliar spray + 150 mmol·L^{-1} NaCl irrigation, 5H + N: 5 mmol·L^{-1} H_2O_2 foliar spray + 150 mmol·L^{-1} NaCl irrigation, 10H + N: 10 mmol·L^{-1} H_2O_2 foliar spray + 150 mmol·L^{-1} NaCl irrigation. Fv/Fm: PSII maximum photochemical efficiency, Fv'/Fm': effective quantum yield, Φ_{PSII}: actual photochemical efficiency of PSII, qP: photochemical quenching coefficient, NPQ: non-photochemical quenching coefficient, ETR: effective electron transfer rate. All values are expressed as mean ± SD. According to Duncan's multiple comparisons, different letters represent significant differences among different treatments ($p < 0.05$).

2.5. Effect of Exogenous H_2O_2 and NaCl Stress on Antioxidant Enzymes

Compared with CK, CAT, and POD activities of Tartary buckwheat were significantly increased under 5H + N treatment (Figure 6A,B); POD increased nearly 2-fold (189.76%) (Figure 6B), while SOD activity was not significantly increased (Figure 6C), and APX activity was significantly reduced by 47.32% (Figure 6D). Under NaCl stress, CAT, POD, and APX were significantly reduced, and APX decreased the most by 0.09 times what it was under the CK treatment (Figure 6A–D). Compared with NaCl, CAT, POD, SOD, and APX were significantly increased under 5H + N treatment (Figure 6A–D). POD increased the most (11.27-fold) (Figure 6B). POD and APX were significantly increased under 10H + N treatment (Figure 6B,D), while CAT and SOD were not significantly decreased or increased (Figure 6A,C).

Compared with the CK, the GR activity and GSSG content of Tartary buckwheat decreased significantly under NaCl stress, while the AsA content did not increase significantly (Figure 6E–G). GR and AsA were significantly increased under 5H + N treatment, by 70.86% and 54.96%, respectively, while GSSG did not differ significantly from what it was under the CK treatment (Figure 6E–G). AsA and GSSG were significantly decreased under 10H + N treatment, and GR was obviously decreased (Figure 6E–G). Compared with the NaCl stress, GR and GSSG significantly increased under 5H + N treatment, and GR reached a maximum value (0.299) (Figure 6E,G), while the 10H + N treatment significantly increased GR and significantly decreased AsA (Figure 6E,F).

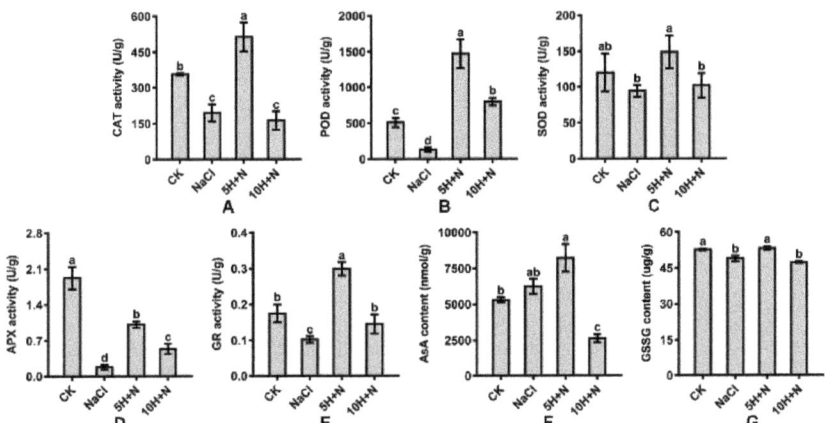

Figure 6. Effects of exogenous H_2O_2 on CAT (**A**), POD (**B**), SOD (**C**), APX (**D**), GR (**E**), AsA (**F**), and GSSG (**G**) of Tartary buckwheat under NaCl stress. CK: control, H_2O foliar spray + H_2O irrigation, NaCl: H_2O foliar spray + 150 mmol·L^{-1} NaCl irrigation, 5H + N: 5 mmol·L^{-1} H_2O_2 foliar spray + 150 mmol·L^{-1} NaCl irrigation, 10H + N: 10 mmol·L^{-1} H_2O_2 foliar spray + 150 mmol·L^{-1} NaCl irrigation. CAT: catalase activity, POD: peroxidase activity, SOD: superoxide dismutase activity, APX: ascorbate peroxidase activity, GR: glutathione reductase activity, AsA: ascorbic acid content, GSSG: glutathione oxidized content. All values are expressed as mean ± SD. According to Duncan's multiple comparisons, different letters represent significant differences among different treatments ($p < 0.05$).

2.6. Effect of Exogenous H_2O_2 and NaCl Stress on MDA and Phosphoenolpyruvate Carboxylase (PEPC)

The MDA content of Tartary buckwheat reached its maximum value (37.541) under NaCl stress, and its minimum value (17.194) under 5H + N treatment (Figure 7A). Compared with CK, the MDA content was significantly increased under NaCl stress, significantly decreased under 5H + N treatment, and not significantly increased under 10H + N treatment (Figure 7A). Compared with NaCl, the MDA content was significantly decreased by 54.20% and 28.18% under 5H + N and 10H + N treatments, respectively (Figure 7A).

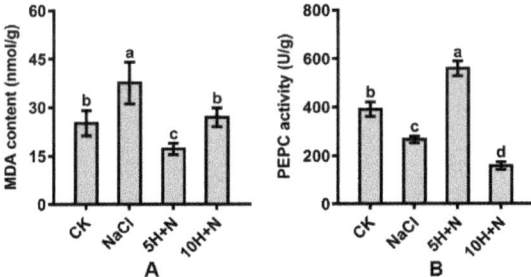

Figure 7. Effects of exogenous H_2O_2 on MDA content (**A**) and PEPC activity (**B**) of Tartary buckwheat under NaCl stress. CK: control, H_2O foliar spray + H_2O irrigation, NaCl: H_2O foliar spray + 150 mmol·L^{-1} NaCl irrigation, 5H + N: 5 mmol·L^{-1} H_2O_2 foliar spray + 150 mmol·L^{-1} NaCl irrigation, 10H + N: 10 mmol·L^{-1} H_2O_2 foliar spray + 150 mmol·L^{-1} NaCl irrigation. MDA: malondialdehyde content, PEPC: phosphoenolpyruvate carboxylase activity. All values are expressed as mean ± SD. According to Duncan's multiple comparisons, different letters represent significant differences among different treatments ($p < 0.05$).

The PEPC activity of Tartary buckwheat in the CK treatment was significantly higher than under NaCl stress and 10H + N treatment (Figure 7B). PEPC decreased most in the 10H + N treatment (59.72%). Under 5H + N treatment, PEPC increased significantly to 1.43 times what it was in the CK treatment (Figure 7B). Compared with NaCl, PEPC was significantly increased by 1.08 times under the 5H + N treatment, and significantly decreased by 40.85% under the 10H + N treatment (Figure 7B).

2.7. Effect of Exogenous H_2O_2 and NaCl Stress on Relative Expression of Stress Response Genes

As shown in Figure 8A, the relative expression levels of *FtNHX1*, *FtSOS1*, *FtNAC6*, *FtNAC9*, *FtWRKY46*, and *FtbZIP83* were highest in the 5H + N treatment. Relative expression levels of all these genes were greater than 10, with the expression of *FtNHX1* being the highest, followed by that of *FtbZIP83* and *FtNAC9*. The lowest expression of all genes was observed in *FtNAC6* (0.145) under NaCl stress (Figure 8A). Apart from *FtNHX1*, the relative expression levels of other genes were significantly higher in the 5H + N treatment than in the NaCl and 10H + N treatments (Figure 8A). Except for *FtNAC9*, the relative expression levels of other genes were significantly or not significantly lower under NaCl stress than under the 10H + N treatment (Figure 8A).

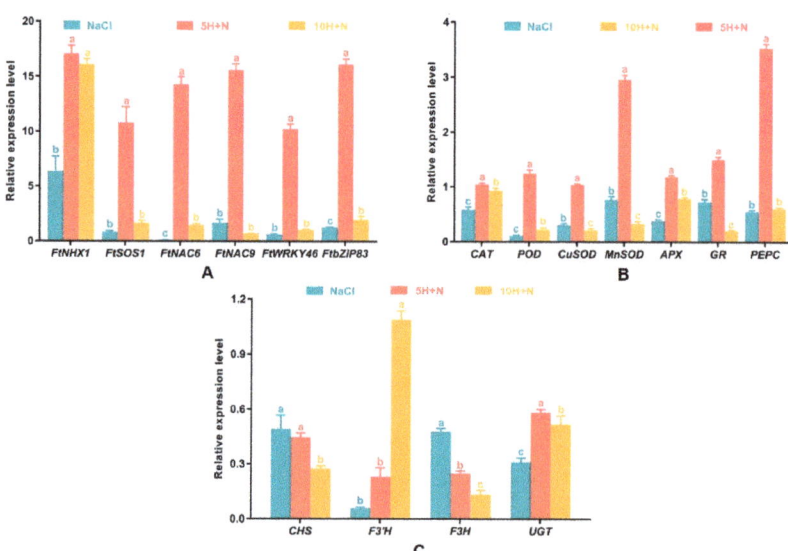

Figure 8. Effects of exogenous H_2O_2 on the relative expression levels of stress-related genes (*FtNHX1*, *FtSOS1*, *FtNAC6*, *FtNAC9*, *FtWRKY46*, and *FtbZIP8*) (**A**), enzyme-related genes (*CAT*, *POD*, *CuSOD*, *MnSOD*, *APX*, *GR*, and *PEPC*) (**B**), key genes (*CHS*, *F3′H*, *F3H*, and *UGT*) in the flavonoid synthesis pathway (**C**) of Tartary buckwheat under NaCl stress. NaCl: H_2O foliar spray + 150 mmol·L^{-1} NaCl irrigation, 5H + N: 5 mmol·L^{-1} H_2O_2 foliar spray + 150 mmol·L^{-1} NaCl irrigation, 10H + N: 10 mmol·L^{-1} H_2O_2 foliar spray + 150 mmol·L^{-1} NaCl irrigation. All values are expressed as mean ± SD. According to Duncan's multiple comparisons, different letters represent significant differences among different treatments ($p < 0.05$).

All genes reached their highest expression level under 5H + N treatment (Figure 8B). *PEPC* showed the highest relative expression (3.533), followed by *MnSOD* (2.962) and *GR* (1.503), and the relative expression levels of genes under 5H + N treatment were significantly higher than they were under NaCl stress and 10H + N treatment (Figure 8B). The relative expression levels of *CuSOD* and *MnSOD* under NaCl treatment were significantly higher (27.21% and 56.10%, respectively) than they were under 10H + N treatment (Figure 8B). For *CAT*, *POD*, and *APX*, the relative expression levels under 10H + N treatment were significantly higher (1.60-fold, 1.99-fold, and 2.07-fold, respectively) than they were under NaCl stress (Figure 8B). However, the relative expression level of *PEPC* did not differ significantly between NaCl stress and 5H + N treatment (Figure 8B).

2.8. Effect of Exogenous H_2O_2 and NaCl Stress on Relative Expression of Key Genes of the Flavonoid Synthesis Pathway

The relative expression levels of key genes in the flavonoid synthesis pathway were lower under each treatment than under the CK treatment (Figure 8C). The relative expression level of *F3′H* was lowest under NaCl stress (0.059), and highest under the 10H + N treatment (1.047); this was significantly higher than the relative expression levels in the NaCl and 5H + N treatments, by 18.47 times and 4.70 times, respectively. The relative expression levels of *F3H* and *CHS* showed a decreasing trend, and the relative expression levels were highest under NaCl stress (Figure 8C). The *UGT* relative expression level under 5H + N treatment was significantly higher than under the NaCl and 10H + N treatments (Figure 8C).

Cis-acting element analysis (Figure 9) showed that the four key genes in the flavonoid synthesis pathway (*CHS*, *F3′H*, *F3H*, and *UGT*) also contained abscisic-acid-responsive element (abscisic acid responsiveness), methyl-jasmonate-responsive element (MeJA responsiveness), low-temperature-responsive element (low-temperature responsiveness), and other hormones and stress elements, in addition to a large number of light responsiveness elements (light responsiveness). The *F3′H* promoter sequence contained two salt stress response elements (salt stresses) at 986 base pairs (bp) and 1908 bp, respectively (Figure 9). When Tartary buckwheat is subjected to salt stress, its relative expression level of *F3′H* may increase accordingly.

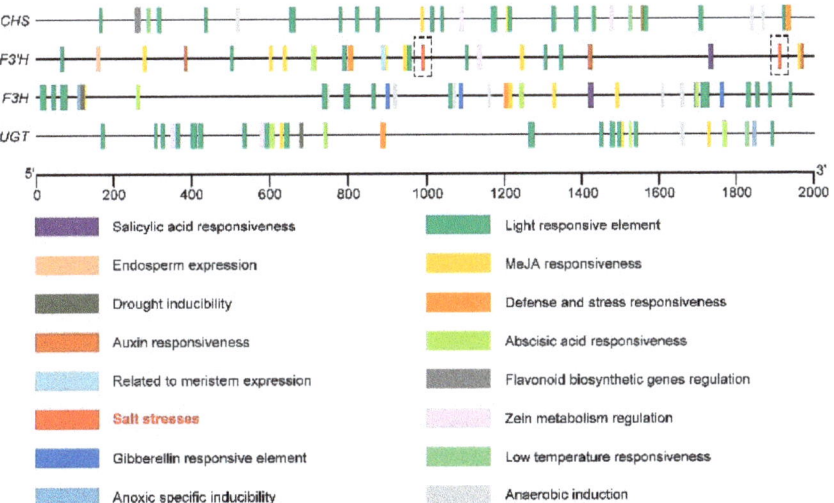

Figure 9. The cis-acting element of the promoter sequence (upstream 2000 bp) of the key genes *CHS*, *F3′H*, *F3H*, and *UGT* in the Tartary buckwheat flavonoid synthesis pathway.

2.9. Correlation Analysis, Hierarchical Cluster Analysis, and Principal Component Analysis

Figure 10A shows the results of correlation analysis of exogenous H_2O_2 on all growth, photosynthetic pigments, fluorescence parameters, different enzyme activities, and enzyme contents of Tartary buckwheat under NaCl stress. Most of the indicators in the figure are positively correlated, with the highest correlation coefficient between Fv and Fm (r = 0.99), followed by the correlations between Car and Chl b and between Fs and Fv′ (r = 0.97). (Figure 10A). Plant height (PH) was significantly positively correlated with most photosynthetic parameters, such as Fv and Fm ($p < 0.05$). Fm was significantly positively correlated with PH, RF/DW, SDW, and other growth parameters ($p < 0.05$), and PEPC was significantly positively correlated ($p < 0.05$) with leaf indices such as LL and LA (Figure 10A). Figure 10A shows that correlations of AsA and PH, LA, Chl a, and other indicators were not highly significant, and there was no significant negative correlation with Fm' ($p > 0.05$); SOD, GR, and fluorescence parameters such as Fo, Fv', Fm' were significantly lower, and POD showed a low correlation with most enzyme activity/content. In addition, Figure 10A shows that MDA was negatively correlated with all the indicators, and the indicators of enzyme activities and leaf growth, such as GR, CAT, POD, LL, LA, and others, were significantly higher.

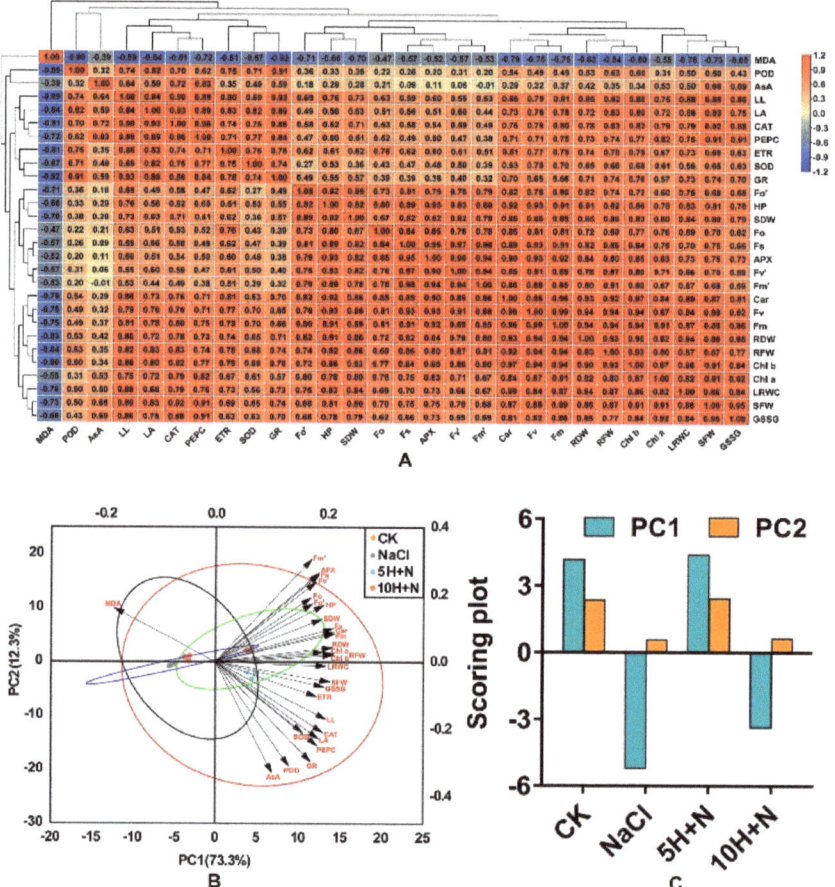

Figure 10. Correlation hierarchical cluster analysis of all indices. Positive number: positive correla–tion; negative number: negative correlation. Red frames indicate significant positive correlation ($p < 0.05$); blue frames indicate significant negative correlation ($p > 0.05$) (**A**). Principal component analysis (PCA) among all indices and treatments (**B**) and scoring plot of principal component anal–ysis among all treatments. (**C**) CK: control, H_2O foliar spray + H_2O irrigation, NaCl: H_2O foliar spray + 150 mmol·L^{-1} NaCl irrigation, 5H + N: 5 mmol·L^{-1} H_2O_2 foliar spray + 150 mmol·L^{-1} NaCl irrigation, 10H + N: 10 mmol·L^{-1} H_2O_2 foliar spray + 150 mmol·L^{-1} NaCl irrigation.

In this experiment, principal component analysis was carried out on 28 indices of morphology, physiology, and biochemistry. The two principal components extracted by PCA explained 73.3% and 12.3% of the variation (Figure 10B). The CK and

5H + N treatments showed a positive correlation in PC1 ($r > 0$), while NaCl and 10H + N were negatively correlated ($r < 0$). The 10H + N treatments were all on the negative axes of PC1 and PC2, and CK and 5H + N were close together but significantly separated from other treatments. Except for MDA, the other morphological, physiological, and biochemical indicators were positively correlated, which is consistent with the results of the correlation heat map (Figure 10A,B). Photosynthetic indicators such as photosynthetic pigments and fluorescence parameters were on the positive axes of PC1 and PC2, while most parameters, such as enzymes and leaf morphology, were on the positive axis of PC1 and the negative axis of PC2 (Figure 10B). Among the overall scores of the principal components PC1 and PC2, the scores of the CK treatment were closest to those of 5H + N, and the scores of NaCl were closest to those of 10H + N. For both PC1 and PC2, the scores of 5H + N were higher than those of CK.

3. Discussion

The impact of soil salinization is becoming increasingly serious on a global scale. About half of the world's cropland is predicted to be affected by salinization by 2050 [30]. Generally, salinization is due to elevated Na^+ and K^+ concentrations in the soil, which result in hyperosmotic conditions that prevent plants from absorbing water, causing adverse effects such as stunted growth [31], osmotic water loss [32], and nutritional imbalance [33]. Salt stress has predominantly inhibitory effects on plant morphogenesis, growth and development, and physiological metabolism. Mitigations and adaptations to the impact of this stress are more obvious in the shoot than in the root system, for example, the reduction in plant height [34], biomass decline [35], and leaf shrinkage [36]. H_2O_2 is a relatively stable, freely diffusing, and long-lived reactive oxygen molecule [37], which has a concentration-dependent effect. Exogenous H_2O_2 has different regulatory effects in different varieties of plants and different environments. In the present study, the growth of Tartary buckwheat under 5H + N treatment was significantly better than that under the CK, NaCl stress, and 10H + N treatments. Compared with NaCl stress, the PH, SFW/SDW, and RFW/RDW of Tartary buckwheat were significantly increased under 5H + N treatment ($p < 0.05$), particularly the RFW, which increased nearly 1.5-fold (Figures 1 and 2A–E). The above results of this study are consistent with those of Ellouzi et al. [38]. The results of this study indicate that when Tartary buckwheat was subjected to NaCl stress, it is difficult for the roots to absorb water through the conduit to transport it to the stem, leaves, and other tissues of the plant. Appropriate exogenous H_2O_2 concentration can alleviate the phenomenon of blocked water absorption under this stress, which is conducive to the growth of the root system and the accumulation of biomass. We also found that the PH, SDW, RFW, and RDW of Tartary buckwheat under 10H + N treatment were slightly higher than they were under NaCl stress, indicating that the treatment had a certain alleviation effect on NaCl stress, but the effect was not obvious. Our results show that in Tartary buckwheat under 5H + N treatment, both LA and LL were significantly increased compared to under CK and NaCl stress treatments, while LRWC was only significantly higher under NaCl stress (Figure 3A–C), which is consistent with Park et al. [39]. In response to salt stress and to protect plants, leaf tissue can reduce the transpiration of leaf water and improve the water retention of leaves by regulating the cuticle and leaf thickness [40]. Therefore, the results of this study show that when the leaf cells were subjected to NaCl stress, the water moved from cells with high water potential to those with low water potential, so that the extracellular water potential was higher than the intracellular water potential. To improve the regulation and adaptability of the plants, the cells reduced transpiration by closing stomata, so that the cells could not absorb further water from the outside.

Chlorophyll is not only an important photosynthetic pigment, but also enables preliminary judgment of leaf function under adverse conditions. Chlorophyll plays an important role in the absorption, transmission, and transformation of light energy, and chlorophyll content is the basis for measuring photosynthetic capacity [41]. Persistent salt stress induces the denaturation of important membrane proteins involved in photosynthesis, enhances the degradation of chlorophyll molecules by chlorophyll-degrading enzymes, and reduces the number of chloroplasts. This is not conducive to the formation and stability of chloroplast ultrastructure, thereby reducing the photosynthetic rate and inhibiting the growth of plants [42,43]. We found that under CK and 5H + N treatments, the Chl a, Chl b, and Car of Tartary buckwheat were significantly higher than under NaCl stress and 10H + N treatments, and values under the CK treatment were not significantly higher than under 5H + N treatment. The change trend of total chlorophyll content (Chl a + b) was the same as that of Chl a and Chl b (Figure 4A–C). This finding is consistent with that of Liu et al. [44]. We found that the Fv/Fm, Fv'/Fm', and ETR of Tartary buckwheat under 5H + N treatment were significantly increased compared with what they were under NaCl stress, but not significantly different to what they were with the CK treatment (Figure 5A,B,F). This is consistent with the results of He et al. [45]. Sun et al. showed [46] that chlorophyll fluorescence parameters can directly reflect the effects of stress on the activity, function, and electron transfer of PSII in plants. Stress destroys the chlorophyll photosynthetic system and reduces and inhibits the transformation efficiency and potential activity of PSII, resulting in the weakening of plant photosynthetic ability. We also found that the changes in PSII and qP of Tartary buckwheat were opposite to the changes in photosynthetic pigments and other parameters, while NPQ was significantly higher under 5H + N treatment than under NaCl stress (Figure 5C–E). The effect of salt stress on the chlorophyll fluorescence of plants is a complex process. Differences in salt types, plant species, and environmental conditions can affect the main stress mechanism and degree of plants to a large extent, resulting in inconsistent research results. Φ_{PSII} represents the actual photochemical efficiency of PSII under light, and qP reflects the level of photosynthetic activity. Under 150 mmol·L^{-1} NaCl stress, Tartary buckwheat was strongly inhibited. Under 5 or 10 mmol·L^{-1} H_2O_2 treatment, electron transfer in the PSII reaction center remained blocked, and the formation of ATP and NADPH assimilation was inhibited [47], while NPQ reflects the ability of plants to dissipate excess light energy into heat [48]. Therefore, the inhibition of photosynthetic fluorescence parameters of Tartary buckwheat under NaCl stress could not be significantly alleviated under H_2O_2 treatment. This result is similar to that of Zhang et al. [49].

The salt stress environment affects the normal metabolism of plants, destroys the cell membrane system, generates a large number of ROS, and causes lipid peroxidation of cell membranes. Plants contain antioxidants, including non-enzymatic antioxidants, that can scavenge a large number of ROS under stress conditions, which can alleviate the inhibitory effect of salt stress on plant organisms [50,51]. We found that the activities of CAT, POD, SOD, APX, and GR in Tartary buckwheat were significantly or not significantly higher under 5H + N treatment than under CK or NaCl stress (Figure 6A–E). This indicates that 5 mmol·L^{-1} H_2O_2 can significantly enhance the activity of antioxidant enzymes and effectively alleviate the toxic effect of NaCl stress on Tartary buckwheat. This is consistent with the findings of Bhardwaj et al. and Diao et al. [37,52]. We also found that the APX of Tartary buckwheat under 5H + N treatment was significantly lower than under the control treatment (Figure 6D), and the AsA content was significantly lower than that under NaCl stress (Figure 6F). This showed that APX activity and AsA content vary greatly with different plant species and different growth stages. We found that the MDA content of Tartary buckwheat was significantly increased under NaCl stress, but decreased significantly under 5H + N treatment (Figure 7A), which indicated that H_2O_2 reduced the formation of ROS in Tartary buckwheat under NaCl stress, attenuated the effect of cell membrane lipid peroxidation, and enhanced the stress resistance of Tartary buckwheat plants. PEPC plays an important role in the balance of carbon and nitrogen metabolism. Studies have shown that PEPC not only plays a key role in the photosynthesis of C4 and CAM plants [53], but also participates in the physiological metabolism of various stress in plants [53], regulates the synthesis and distribution of organic matter in plants, and enhances the carbon skeleton of plants [54]. We found that the PEPC activity of Tartary buckwheat was significantly increased under the 5H + N treatment (Figure 7B), indicating that the PEPC enzyme was involved in the photosynthesis and metabolism of Tartary buckwheat under NaCl stress.

In this study, we found that the relative expression of stress-related genes and related enzyme genes in Tartary buckwheat under 5H + N treatment was significantly increased compared to that under other treatments, and the expression of these genes followed a similar pattern to the morphological indicators and physiological and biochemical changes of Tartary buckwheat. In particular, the expression levels of *FtNHX1*, *FtNAC9*, *FtbZIP83*, *PEPC*, and *MnSOD* were high under 5H + N treatment (Figure 8A,B). These results show that H_2O_2 can improve the tolerance of Tartary buckwheat to NaCl stress by regulating the expression of related genes. The expression of *FtNHX1* was significantly increased when treated with 10H + N, which indicates that high concentrations of H_2O_2 may also significantly induce the expression of *FtNHX1*. The proteins related to ion transport under salt stress conditions include plasma membrane antiporter SOS1, potassium–sodium co-transporter HKT, and tonoplast antiporter NHX, which can jointly induce Na^+ accumulation and K^+ efflux [55,56]. Studies have shown that the expression of Na^+-induced *SOS1* in Arabidopsis is mediated by H_2O_2 induced by NADPH oxidase [57], and SOS3, as a Ca^{2+} sensor, can interact with SOS2 protein kinase. SOS2 compartmentalizes Na^+/K^+ in the vacuole by activating vacuolar H^+-ATPase and NHX, and this compartmentalization plays a crucial role in regulating plant tolerance to salt stress [58]. In this study, the analysis of key genes in the flavonoid synthesis pathway of Tartary buckwheat found that NaCl failed to increase the expression of *CHS*, *F3H*, and *UGT*, while *F3'H* was significantly increased under 10H + N treatment (Figure 8C), indicating that H_2O_2 can up-regulate the expression of *F3'H* in Tartary buckwheat under NaCl stress. Through the analysis of *F3H*, *CHS*, *UGT*, and *F3'H* cis-acting elements, we found that *F3'H* contains two salt stress elements (Figure 9), which is consistent with the observed relative expression of *F3'H*.

In this study, we found that the chlorophyll content of Tartary buckwheat was significantly positively correlated with fluorescence parameters ($r > 0$, $p < 0.05$; Figure 10A). There was a significant positive correlation between Fm and morphological indices ($r > 0$, $p < 0.05$), a significant positive correlation between PEPC activity and leaf morphology ($r > 0$, $p < 0.05$), and a significant negative correlation between MDA content and enzyme activity and leaf index ($r < 0$, $p < 0.05$). PC1 and PC2 identified through principal component analysis explained 73.3% and 12.3% of the variation, respectively (Figure 10B). CK and 5H + N treatments were positively correlated in PC1, and were closest. Except for MDA, the other morphological, physiological, and biochemical indices were clustered in the positive direction PC1, that is, positively correlated, which is consistent with the results of correlation analysis.

This study primarily discussed the regulation of exogenous H_2O_2 on the growth, physiology and biochemistry, and related genes of Tartary buckwheat under NaCl stress. Topics for further study include the mechanism by which exogenous H_2O_2 regulates the decomposition and distribution of nutrients during the growth of Tartary buckwheat, which metabolic pathways and related enzyme activities and gene expression are mainly regulated, and how to induce and regulate the physiological and biochemical mechanisms of Tartary buckwheat between exogenous H_2O_2 and NaCl stress.

4. Materials and Methods

4.1. Experiment Materials

The tested Tartary buckwheat variety, "Chuanqiao-2", was provided by the Alpine Crop Research Station (27.96° N, 102.20° E) of Xichang Institute of Agricultural Sciences, Liangshan Prefecture, Sichuan Province, China.

4.2. Experiment Design and Treatments

Tartary buckwheat variety "Chuanqiao-2" seeds with uniform size and no pests and diseases were selected. Seeds were rinsed with distilled water, sterilized with 1% NaClO for 10 min, then rinsed with double-distilled water (DDW) several times. The sterilized and rinsed seeds were sown on the surface of 2 layers of qualitative filter paper in a sterile Petri dish (90 mm diameter), and placed in a constant-temperature incubator (25 ± 1 °C during the day, 20 ± 1 °C at night, relative humidity of 75%) for 7 days. At 7 d after sowing (DAS), the Tartary buckwheat seedlings with uniform growth were transplanted into plant pots (diameter 25.5 cm, height =17.5 cm; 3 seedlings per pot) containing mixed nutrient soil (soil:substrate = 1:1) that had been sterilized by high temperature. At 20 d after transplanting (DAT), each pot was irrigated with 2 liters (L) of Hoagland nutrient solution containing 150 mmol·L^{-1} NaCl [59,60]. Based on previous research results [1], 5 or 10 mmol·L^{-1} H$_2$O$_2$ was used for foliar spraying in this experiment. The transplanted Tartary buckwheat was cultured in a culture room under specific conditions (25 ± 1 °C during the day, 20 ± 1 °C at night, relative humidity of 75%). At 40 days after transplanting (DAT), growth indices and photosynthesis parameters were measured, and sampling was used to measure different physiological and biochemical indices. The schematic diagram of the experimental design is shown in Figure 11. Three replicates were set for each treatment in this experiment, and each treatment was as follows:

$$CK: H_2O + H_2O$$

$$NaCl: 150\ mmol·L^{-1}\ NaCl + H_2O$$

$$5H + N: 5\ mmol·L^{-1}\ H_2O_2 + 150\ mmol·L^{-1}\ NaCl$$

$$10H + N: 10\ mmol·L^{-1}\ H_2O_2 + 150\ mmol·L^{-1}\ NaCl$$

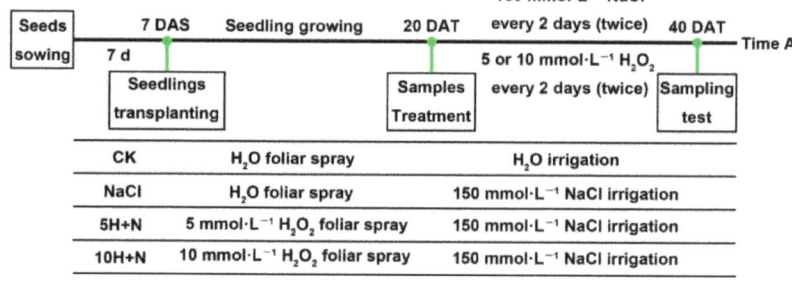

Figure 11. Schematic diagram of experimental design. 7 DAS: 7 days after sowing, 20 DAT: 20 days after transplanting. CK: control, H$_2$O foliar spray + H$_2$O irrigation, NaCl: H$_2$O foliar spray + 150 mmol·L^{-1} NaCl irrigation, 5H + N: 5 mmol·L^{-1} H$_2$O$_2$ foliar spray + 150 mmol·L^{-1} NaCl irrigation, 10H + N: 10 mmol·L^{-1} H$_2$O$_2$ foliar spray + 150 mmol·L^{-1} NaCl irrigation.

4.3. Measurement of Plant Height and Biomass

After the plants were uprooted, the surface soil was washed, and the plant height (PH) was measured directly using a ruler, and expressed in cm. The stem fresh weight (SFW) and root fresh weight (RFW) were measured using a thousand-point electronic balance, and then samples were placed in an oven at 105 °C for 30 min, and dried at 65 °C to a constant weight. Stem dry weight (SDW) and root dry weight (RDW) were measured with a thousand-point electronic balance and expressed in mg.

4.4. Determination of Leaf Phenotype and Relative Leaf Water Content

Leaf area (LA) and leaf length (LL) were determined with an LI-3000C portable leaf area analyzer (LI-COR, Lincoln, NE, USA), and are expressed in cm^2 and cm, respectively. Relative water content (LRWC) was determined according to Su et al. [61], using the calculation formula:

$$LRWC\ (\%) = [(FW - DW) \times 100] / [TW - DW] \qquad (1)$$

4.5. Determination of Photosynthetic Pigment Content

The contents of chlorophyll a (Chl a), chlorophyll b (Chl b), and carotenoids (Car) were determined according to Hossain et al. [62]. Briefly, 0.1 g of leaves was cut and placed into 95% (v/v) ethanol, soaked in dark conditions for 72 h, and then the absorbance at 663 nm, 645 nm, and 470 nm was measured using a microplate reader (Thermo scientific, Waltham, MA, USA). Chl a, Chl b, and Car were determined using the following formulae:

$$\text{Chl a (mg·g}^{-1}\text{ FW)} = [12.7 \times (OD663) - 2.69 \times (OD645)] \times V/(1000 \times W) \quad (2)$$

$$\text{Chl b (mg·g}^{-1}\text{ FW)} = [22.9 \times (OD645) - 4.68 \times (OD663)] \times V/(1000 \times W) \quad (3)$$

$$\text{Car (mg·g}^{-1}\text{ FW)} = [100 \times (OD470) - 3.27 \times \text{Chl a} - 104 \times \text{Chl b}]/227 \quad (4)$$

4.6. Measurement of Photosynthetic Fluorescence Parameters

The following parameters were measured in the third top-down leaf of the plant's fully expanded leaves according to the method of Lu et al. [63], using the LI-6400XT portable photosynthesis measurement system (LI-COR, USA): PSII maximum photochemical efficiency (Fv/Fm), minimal fluorescence under dark adaptation (Fo), effective quantum yield (Fv'/Fm'), minimal fluorescence under light (Fo'), stable fluorescence under dark adaptation (Fs), actual photochemical efficiency of PSII (ΦPSII), photochemical quenching coefficient (qP), non-photochemical quenching coefficient (NPQ), and effective electron transfer rate (ETR). The system parameters were set to a light intensity of 1400 µmol·m^{-2}·s^{-1}, and an atmospheric CO_2 concentration of 380 ± 5 mmol·mol^{-1}. The measurements were conducted between 10 a.m. and 2 p.m. on a sunny day. Parameters were calculated using the following formulae:

$$Fv/Fm = (Fm - Fo)/Fm \quad (5)$$

$$Fv'/Fm' = (Fm' - Fo')/Fm' \quad (6)$$

$$\Phi_{PSII} = (Fm' - Fs)/Fm' \quad (7)$$

$$qP = (Fm' - Fs)/(Fm' - Fo') \quad (8)$$

$$NPQ = Fm/Fm' - 1 \quad (9)$$

4.7. Estimation of Biochemical Indicators

Biochemical indicators were estimated using the BOXBIO kit (Beijing Boxbio Science & Technology Co., Ltd., Beijing, China) following the manufacturer instructions.

4.7.1. Antioxidant Enzyme Activity

Catalase activity (CAT), peroxidase activity (POD), superoxide dismutase activity (SOD), ascorbate peroxidase activity (APX), and glutathione reductase activity (GR) were estimated according to Kong et al. [64]. The corresponding extract was added to 0.1 g of sample, followed by 1 mL of extract, and the supernatant was collected. Identical tubes that were not illuminated served as blanks. Absorbance values for CAT, POD, SOD, APX, and GR were measured at 240 nm, 470 nm, 560 nm, 290 nm, and 340 nm, respectively, using a spectrophotometer (Unico Instrument Co., Ltd., Shanghai, China) and expressed in U·g^{-1}.

4.7.2. Antioxidant Enzyme Content

Ascorbic acid content (AsA) and oxidized glutathione content (GSSG) were measured according to Rama and Prasad [65]. Homogenization was conducted in an ice bath: the sample was centrifuged at 8000× g at 4 °C to collect the supernatant and then placed on ice for measurement. A microplate reader (Thermo scientific, Waltham, MA, USA) was used at 265 nm and 412 nm, respectively, after adding related reagents. The absorbance values for AsA and GSSG are expressed in nmol·g^{-1} and ug·g^{-1}, respectively.

4.7.3. Malondialdehyde Content

Malondialdehyde content was measured according to Hodges et al. [66] We took 0.1 g of sample and homogenized the supernatant for subsequent detection. The MDA content was calculated according to the corresponding absorbance and formula. The absorbance was measured at 450 nm, 532 nm, and 600 nm with a visible spectrophotometer (Unico Instrument Co., Ltd., Shanghai, China), and expressed in $nmol \cdot g^{-1}$.

4.7.4. Phosphoenolpyruvate Carboxylase (PEPC) Activity

PEPC activity was measured according to Aoyagi and Bassham [67]. The samples were processed according to the ratio of sampling mass and extract volume to 1:10, and the corresponding reagents were added for measurement after collecting the supernatant. Absorbance at 340 nm was measured with an ultraviolet spectrophotometer (Unico Instrument Co., Ltd., Shanghai, China) and expressed in $U \cdot g^{-1}$.

4.8. Total RNA Extraction and Reverse Transcription PCR (RT-PCR)

A 0.1 g sample of Tartary buckwheat leaves was fully ground in liquid nitrogen, and then total RNA was extracted using the E.Z.N.A. Plant RNA Kit (Omega Bio-tek, Inc., Norcross, GA, USA). The integrity of extracted RNA was checked by 1% agarose gel electrophoresis, and RNA purity and concentration were measured using a spectrophotometer (Beijing Kaiao Technology Development Co., Ltd., Beijing, China). RNA was reverse-transcribed into the first-strand cDNA using the Hiscript II Q RT Supermix for qPCR Kit (Vazyme Biotech Co., Ltd., Nanjing, China) in a 20 μL reaction system.

4.9. Quantitative Real-Time PCR (qRT-PCR) Analysis

Primer Premier 5.0 (Premier Corporation, Vancouver, BC, Canada) software was used to design specific primers for qPCR, with *FtH3* as the internal control gene, and the amplification primers are shown in Table 1. The ChamQ Universal SYBR qPCR Master Mix Kit (Vazyme Biotech Co., Ltd., Nanjing, China) was used with 1.0 μL of template cDNA, 10.0 μL of 2× SYBR Mix, and 0.4 μL each of Primer F and R, and the reaction volume was made up to 20 μL with ddH$_2$O. qPCR was performed using CFX96 Real-Time System (BIO-RAD, Hercules, CA, USA). The relative expression levels of target genes compared to the internal control gene were calculated using the $2^{-\Delta\Delta Ct}$ method [68].

Table 1. The primer sequences for qPCR.

Primer Name	Forward Primer (5'-3')	Reverse Primer (5'-3')	Functions
FtH3	AATTCGCAAGTACCAGAAG	CCAACAAGGTATGCCTCAGC	Actin
FtNHX1	CGTTGCTAGGACGCAATGTTCCA	ACAGTCCACGTCGGATGCCTTAT	Stress-related genes
FtSOS1	CCTTACACCGTACCCGCTC	CCGGAAGAAACACAGCCAACA	
FtNAC6	GATTCAATTCCCCGGCTCCA	AACGGGGACAACTCATTCCC	
FtNAC9	CTGAGGGTGTAATTCCGGGT	TCAACGGTAGGGGTAGAAGC	
FtWRKY46	TGTTCCGCCTTCTGATGGTT	CAGCACTGTGGGGTCATCAT	
FtbZIP83	ACCGAGTATTCCGCAAGCTC	AACTCTCCCCAAAACCCACC	
CAT	GAGTTTGGTTCCCTTGCTT	TTCATACACTTCACTGGCGT	Enzyme-related genes
POD	GTTCTGGTTGGGCTTGG	TTGTCCTCGTCTGTTGGTC	
CuSOD	ATGGTGCTCCTGACGATG	CCACTGCCCTTCCAATAAT	
MnSOD	GTCTACGGTCCTCCTTCTACAT	TAACAACAGCACACTTCTTTCT	
APX	TACTCCGAGGTTGTGTGCC	CAATCAAGGTGTTCCAGTCA	
GR	TTGAGTGGAGAGAAGGAAGG	CATAGTCGGCAAAGAAAGC	
PEPC	AAGTCTCCACATTCGGTCTC	ATCTCCAAGTGCCTGGTTAT	
CHS	GAGGAGATCAGAAAGGCACAAAGGG	GTCGGCTTGGTAGATACAGTTAGGC	Flavonoid-synthesis-related genes
F3'H	TCAAGGAGAATGGCGGAGTT	TGGGTGGTTCAGGAGGAGTG	
F3H	GCCTGTTGAGGGTGCCTTTGT	TGGCGATTGAAAGACGGCTGAAG	
UGT	CAGCTTCTTCACCACCGAATCCTC	TCTCGCCCGCTAACCCATCTTC	

4.10. Prediction of Cis-Acting Element

The PlantCare website (We accessed on 14 July 2021 to http://bioinformatics.psb.ugent.be/webtools/plantcare/html/) was used to predict the presence of cis-acting elements in promoter sequences (the upstream 2000 bp) of key genes of the flavonoid synthesis pathway.

4.11. Statistics and Analysis

Analysis of variance (ANOVA, $p < 0.05$), multiple comparisons (Duncan), and correlation analysis (Pearson) were performed with IBM SPSS Statistics 26.0 software (International Business Machines Corporation, New York, NY, USA), and the results are expressed as the mean ± SD. Column charts were drawn using GraphPad Prism7.0 software (GraphPad Software, LLC, San Diego, CA, USA). Principal component analysis plots were made using Origin 2019b software. Correlation heatmaps and cis-acting element visualizations were generated using TBtools v1.09876 [69].

5. Conclusions

In this study, compared with CK and NaCl treatments, spraying exogenous H_2O_2 could promote the growth of Tartary buckwheat under NaCl stress, increase the accumulation of chlorophyll content, enhance electron transfer and transformation during photosynthesis, effectively improve enzymatic reactions, reduce cell membrane lipid peroxidation, induce or activate the expression level of related genes, alleviate the toxic effect of NaCl stress on Tartary buckwheat, and promote normal physiological metabolism and biochemical reactions in Tartary buckwheat. A concentration of 5 mmol·L^{-1} H_2O_2 produced the optimal promoting effect on Tartary buckwheat under 150 mmol·L^{-1} NaCl stress. Appropriate concentrations of H_2O_2 can alleviate the inhibitory effect of salt stress, but the mechanisms and signaling pathways of H_2O_2-mediated salt tolerance still need to be further dissected in detail, as well as how H_2O_2 regulates related genes to activate defense systems to alleviate salt stress.

Author Contributions: Experimental design, X.Y., Y.P.; data analysis, X.Y., C.M., and W.W. (Weijiao Wu); validation, X.Y., J.R., and M.Z. charting, X.Y., W.W. (Wenfeng Weng); first draft, X.Y. and A.G.; edited and revised papers, J.R. and J.C. All authors have read and agreed to the published version of the manuscript.

Funding: This work was supported by the National Key R & D Project of China (2017YFE0117600, SQ2020YFF0402959), the National Science Foundation of China (32160669, 32161143005), and the Guizhou Science and Technology Support Program (Qiankehe Support [2020]1Y051).

Informed Consent Statement: Not applicable.

Data Availability Statement: Data are contained within the article.

Acknowledgments: We acknowledge the College of Agronomy, Guizhou University, Guiyang, China, for providing the experimental facilities and other necessary materials for this study. We are also thankful for the *Fagopyrum tataricum* breeders provided by the Alpine Crops Research Station of the Xichang Institute of Agricultural Sciences, Liangshan Prefecture, Sichuan Province, China.

Conflicts of Interest: The authors declare no conflict of interest.

References

1. Yao, X.; Zhou, M.; Ruan, J.; Peng, Y.; Yang, H.; Tang, Y.; Gao, A.; Cheng, J. Pretreatment with H_2O_2 Alleviates the Negative Impacts of NaCl Stress on Seed Germination of Tartary Buckwheat (*Fagopyrum tataricum*). *Plants* **2021**, *10*, 1784. [CrossRef] [PubMed]
2. Ren, K.; Tang, Y.; Fan, Y.; Li, W.; Lai, D.L.; Yan, M.L.; Zhang, K.X.; Zhou, M.L. Collection and taxonomic identification of rare germplasm resources of the genus Buckwheat in six provinces (regions) in western China. *J. Plant Genet. Resour.* **2021**, *22*, 963–970, (In Chinese with English abstract).
3. Tan, Y.R.; Tao, B.B.; Guan, Y.F.; Ming, J.; Zhao, G.H. Research status and prospect of flavonoids in tartary buckwheat. *Food Ind. Sci. Technol.* **2012**, *33*, 377–381, (In Chinese with English abstract).

4. Sun, Z.; Wang, X.; Liu, R.; Du, W.; Ma, M.; Han, Y.; Li, H.; Liu, L.; Hou, S. Comparative transcriptomic analysis reveals the regulatory mechanism of the gibberellic acid pathway of Tartary buckwheat (*Fagopyrum tataricum* (L.) Gaertn.) dwarf mutants. *BMC Plant Biol.* **2021**, *21*, 240. [CrossRef] [PubMed]
5. Fan, Y.; Ding, M.Q.; Zhang, K.X.; Yang, K.L.; Tang, Y.; Zhang, Z.W.; Fang, W.; Yan, J.; Zhou, M.L. Overview of Buckwheat Germplasm Resources. *J. Plant Genet. Resour.* **2019**, *20*, 813–828, (In Chinese with English abstract).
6. Holasova, M.; Fiedlerova, V.; Smrcinova, H.; Orsak, M.; Lachman, J.; Vavreinova, S. Buckwheat—the source of antioxidant activity in functional foods. *Food Res. Int.* **2022**, *35*, 207–211. [CrossRef]
7. Chua, L.S. A review on plant-based rutin extraction methods and its pharmacological activities. *J. Ethnopharmacol.* **2013**, *150*, 805–817. [CrossRef]
8. Ma, Z.; Liu, M.; Sun, W.; Huang, L.; Wu, Q.; Bu, T.; Li, C.; Chen, H. Genome-wide identification and expression analysis of the trihelix transcription factor family in tartary buckwheat (*Fagopyrum tataricum*). *BMC Plant Biol.* **2019**, *19*, 344. [CrossRef]
9. Martin, N.G.; Eaves, L.J.; Heath, A.C. Prospects for detecting genotype X environment interactions in twins with breast cancer. *Acta Genet. Med. Gemellol.* **1987**, *36*, 5–20. [CrossRef]
10. Chakrabarty, S.; Kravcov, N.; Schaffasz, A.; Snowdon, R.J.; Wittkop, B.; Windpassinger, S. Genetic Architecture of Novel Sources for Reproductive Cold Tolerance in Sorghum. *Front. Plant Sci.* **2021**, *12*, 772177. [CrossRef]
11. Parihar, P.; Singh, S.; Singh, R.; Singh, V.P.; Prasad, S.M. Effect of salinity stress on plants and its tolerance strategies: A review. *Environ. Sci. Pollut. Res. Int.* **2015**, *22*, 4056–4075. [CrossRef] [PubMed]
12. Munns, R. Genes and salt tolerance: Bringing them together. *New Phytol.* **2015**, *167*, 645–663. [CrossRef] [PubMed]
13. Karle, S.B.; Guru, A.; Dwivedi, P.; Kumar, K. Insights into the Role of Gasotransmitters Mediating Salt Stress Responses in Plants. *J. Plant Growth Regul.* **2021**, *40*, 2259–2275. [CrossRef]
14. Kumar, P.; Choudhary, M.; Halder, T.; Prakash, N.R.; Singh, V.; Vineeth, T.V.; Sheoran, S.; Ravikiran, K.T.; Longmei, N.; Rakshit, S.; et al. Salinity stress tolerance and omics approaches: Revisiting the progress and achievements in major cereal crops. *Heredity* **2022**, *128*, 497–518. [CrossRef] [PubMed]
15. Yin, M.Q.; Wang, D.; Wang, J.R.; Lan, M.; Zhao, J.; Dong, S.Q.; Song, X.E.; Alam, S.; Yuan, X.Y.; Wang, Y.G.; et al. Effects of exogenous nitric oxide on sorghum seed germination and starch transformation under salt stress. *Chin. Agric. Sci.* **2019**, *52*, 4119–4128, (In Chinese with English abstract).
16. Song, C.; Zhang, Y.; Chen, R.; Zhu, F.; Wei, P.; Pan, H.; Chen, C.; Dai, J. Label-Free Quantitative Proteomics Unravel the Impacts of Salt Stress on Dendrobium huoshanense. *Front. Plant Sci.* **2022**, *13*, 874579. [CrossRef]
17. Noreen, S.; Sultan, M.; Akhter, M.S.; Shah, K.H.; Ummara, U.; Manzoor, H.; Ulfat, M.; Alyemeni, M.N.; Ahmad, P. Foliar fertigation of ascorbic acid and zinc improves growth, antioxidant enzyme activity and harvest index in barley (*Hordeum vulgare* L.) grown under salt stress. *Plant Physiol. Biochem.* **2021**, *158*, 244–254. [CrossRef]
18. Li, Z.; Geng, W.; Tan, M.; Ling, Y.; Zhang, Y.; Zhang, L.; Peng, Y. Differential Responses to Salt Stress in Four White Clover Genotypes Associated with Root Growth, Endogenous Polyamines Metabolism, and Sodium/Potassium Accumulation and Transport. *Front. Plant Sci.* **2022**, *13*, 896436. [CrossRef]
19. Sheikh-Mohamadi, M.H.; Etemadi, N.; Aalifar, M.; Pessarakli, M. Salt stress triggers augmented levels of Na+, K+ and ROS alters salt-related gene expression in leaves and roots of tall wheatgrass (*Agropyron elongatum*). *Plant Physiol. Biochem.* **2022**, *183*, 9–22. [CrossRef]
20. Cembrowska-Lech, D. Tissue Printing and Dual Excitation Flow Cytometry for Oxidative Stress-New Tools for Reactive Oxygen Species Research in Seed Biology. *Int. J. Mol. Sci.* **2022**, *21*, 8656. [CrossRef]
21. Nazir, F.; Fariduddin, Q.; Khan, T.A. Hydrogen peroxide as a signalling molecule in plants and its crosstalk with other plant growth regulators under heavy metal stress. *Chemosphere* **2020**, *252*, 126486. [CrossRef] [PubMed]
22. Wu, J.; Shu, S.; Li, C.; Sun, J.; Guo, S. Spermidine-mediated hydrogen peroxide signaling enhances the antioxidant capacity of salt-stressed cucumber roots. *Plant Physiol. Biochem.* **2018**, *128*, 152–162. [CrossRef] [PubMed]
23. Wu, M.; Shan, W.; Zhao, G.P.; Lyu, L.D. H_2O_2 concentration-dependent kinetics of gene expression: Linking the intensity of oxidative stress and mycobacterial physiological adaptation. *Emerg. Microbes Infect.* **2022**, *11*, 573–584. [CrossRef] [PubMed]
24. Kumar, U.; Kaviraj, M.; Panneerselvam, P.; Priya, H.; Chakraborty, K.; Swain, P.; Chatterjee, S.N.; Sharma, S.G.; Nayak, P.K.; Nayak, A.K. Ascorbic acid formulation for survivability and diazotrophic efficacy of *Azotobacter chroococcum* Avi2 (MCC 3432) under hydrogen peroxide stress and its role in plant-growth promotion in rice (*Oryza sativa* L.). *Plant Physiol. Biochem.* **2019**, *139*, 419–427. [CrossRef]
25. Ren, Y.F.; He, J.Y.; Yang, J.; Wei, Y.J. Effects of exogenous H_2O_2 on seed germination and seedling physiological characteristics of Chinese cabbage under salt stress. *Acta Ecol. Sin.* **2019**, *39*, 7745–7756, (In Chinese with English abstract).
26. Lomovatskaya, L.A.; Kuzakova, O.V.; Goncharova, A.M.; Romanenko, A.S. Implication of cAMP in Regulation of Hydrogen Peroxide Level in Pea Seedling Roots under Biotic Stress. *Russ. J. Plant Physiol.* **2020**, *67*, 435–442. [CrossRef]
27. Li, Z.; Xu, J.; Gao, Y.; Wang, C.; Guo, G.; Luo, Y.; Huang, Y.; Hu, W.; Sheteiwy, M.S.; Guan, Y.; et al. The Synergistic Priming Effect of Exogenous Salicylic Acid and H_2O_2 on Chilling Tolerance Enhancement during Maize (*Zea mays* L.) Seed Germination. *Front. Plant Sci.* **2017**, *8*, 1153. [CrossRef]
28. Khan, M.I.; Khan, N.A.; Masood, A.; Per, T.S.; Asgher, M. Hydrogen Peroxide Alleviates Nickel-Inhibited Photosynthetic Responses through Increase in Use-Efficiency of Nitrogen and Sulfur, and Glutathione Production in Mustard. *Front. Plant Sci.* **2016**, *7*, 44. [CrossRef]

29. Yao, X.; Liu, T.T.; Ruan, J.J.; Peng, Y.; Tang, Y.; Yang, H.; Gao, A.J.; Cheng, J.P. Effects of seed soaking with H_2O_2 on root growth and antioxidant enzyme activity of tartary buckwheat seedlings. *Mol. Plant Breed.* **2021**, *9*, 1–17.
30. Ivushkin, K.; Bartholomeus, H.; Bregt, A.K.; Pulatov, A.; Kempen, B.; de Sousa, L. Global mapping of soil salinity change. *Remote Sens. Environ.* **2019**, *231*, 111260. [CrossRef]
31. Xu, Y.; Magwanga, R.O.; Cai, X.; Zhou, Z.; Wang, X.; Wang, Y.; Zhang, Z.; Jin, D.; Guo, X.; Wei, Y.; et al. Deep Transcriptome Analysis Reveals Reactive Oxygen Species (ROS) Network Evolution, Response to Abiotic Stress, and Regulation of Fiber Development in Cotton. *Int. J. Mol. Sci.* **2019**, *20*, 1863. [CrossRef] [PubMed]
32. Zhang, Y.; Fang, Q.; Zheng, J.; Li, Z.; Li, Y.; Feng, Y.; Han, Y.; Li, Y. GmLecRlk, a Lectin Receptor-like Protein Kinase, Contributes to Salt Stress Tolerance by Regulating Salt-Responsive Genes in Soybean. *Int. J. Mol. Sci.* **2022**, *23*, 1030. [CrossRef] [PubMed]
33. Wang, L.; Zhang, X.; Xu, S. Is salinity the main ecological factor that influences foliar nutrient resorption of desert plants in a hyper-arid environment? *BMC Plant Biol.* **2020**, *20*, 461. [CrossRef] [PubMed]
34. Abo-Elyousr, K.; Mousa, M.; Ibrahim, O.; Alshareef, N.O.; Eissa, M.A. Calcium-Rich Biochar Stimulates Salt Resistance in Pearl Millet (*Pennisetum glaucum* L.) Plants by Improving Soil Quality and Enhancing the Antioxidant Defense. *Plants* **2022**, *11*, 1301. [CrossRef] [PubMed]
35. Cocozza, C.; Bartolini, P.; Brunetti, C.; Miozzi, L.; Pignattelli, S.; Podda, A.; Scippa, G.S.; Trupiano, D.; Rotunno, S.; Brilli, F.; et al. Modulation of class III peroxidase pathways and phenylpropanoids in Arundo donax under salt and phosphorus stress. *Plant Physiol. Biochem.* **2022**, *183*, 151–159. [CrossRef] [PubMed]
36. Borrajo, C.I.; Sánchez-Moreiras, A.M.; Reigosa, M.J. Ecophysiological Responses of Tall Wheatgrass Germplasm to Drought and Salinity. *Plants* **2022**, *11*, 1548. [CrossRef]
37. Bhardwaj, R.D.; Singh, N.; Sharma, A.; Joshi, R.; Srivastava, P. Hydrogen peroxide regulates antioxidant responses and redox related proteins in drought stressed wheat seedlings. *Physiol. Mol. Biol. Plants* **2021**, *27*, 151–163. [CrossRef]
38. Ellouzi, H.; Oueslati, S.; Hessini, K.; Rabhi, M.; Abdelly, C. Seed-priming with H_2O_2 alleviates subsequent salt stress by preventing ROS production and amplifying antioxidant defense in cauliflower seeds and seedlings. *Sci. Hortic.* **2021**, *288*, 110360. [CrossRef]
39. Park, H.S.; Kazerooni, E.A.; Kang, S.M.; Al-Sadi, A.M.; Lee, I.J. Melatonin Enhances the Tolerance and Recovery Mechanisms in *Brassica juncea* (L.) Czern. Under Saline Conditions. *Front. Plant Sci.* **2021**, *12*, 593717. [CrossRef]
40. Verslues, P.E.; Kim, Y.S.; Zhu, J.K. Altered ABA, proline and hydrogen peroxide in an Arabidopsis glutamate: Glyoxylate aminotransferase mutant. *Plant Mol. Biol.* **2007**, *64*, 205–217. [CrossRef]
41. Chaves, M.M.; Flexas, J.; Pinheiro, C. Photosynthesis under drought and salt stress: Regulation mechanisms from whole plant to cell. *Ann. Bot.* **2009**, *103*, 551–560. [CrossRef] [PubMed]
42. Jawad Hassan, M.; Ali Raza, M.; Khan, I.; Ahmad Meraj, T.; Ahmed, M.; Abbas Shah, G.; Ansar, M.; Awan, S.A.; Khan, N.; Iqbal, N.; et al. Selenium and Salt Interactions in Black Gram (*Vigna mungo* L.): Ion Uptake, Antioxidant Defense System, and Photochemistry Efficiency. *Plants* **2020**, *9*, 467. [CrossRef] [PubMed]
43. Zulfiqar, F.; Ashraf, M. Nanoparticles potentially mediate salt stress tolerance in plants. *Plant Physiol. Biochem.* **2021**, *160*, 257–268. [CrossRef] [PubMed]
44. Liu, J.X.; Ou, X.B.; Wang, J.C. Effects of exogenous hydrogen peroxide on chlorophyll fluorescence parameters and photosynthetic carbon assimilation enzymes activities in naked oat seedlings under lanthanum stress. *Acta Ecol. Sin.* **2019**, *39*, 2833–2841. (In Chinese with English abstract).
45. He, X.; Richmond, M.; Williams, D.V.; Zheng, W.; Wu, F. Exogenous Glycinebetaine Reduces Cadmium Uptake and Mitigates Cadmium Toxicity in Two Tobacco Genotypes Differing in Cadmium Tolerance. *Int. J. Mol. Sci.* **2019**, *20*, 1612. [CrossRef]
46. Sun, L.; Zhou, Y.F.; Li, F.X.; Xiao, M.J.; Tao, Y.; Xu, W.J.; Huang, R.D. Effects of salt stress on photosynthesis and fluorescence characteristics of sorghum seedlings. *Chin. Agric. Sci.* **2012**, *45*, 3265–3272, (In Chinese with English abstract).
47. Rapacz, M. Chlorophyll a fluorescence transient during freezing and recovery in winter wheat. *Photosynthetica* **2007**, *45*, 409–418. [CrossRef]
48. Sixto, H.; Aranda, I.; Grau, J.M. Assessment of salt tolerance in Populus alba clones using chlorophyll fluorescence. *Photosynthetica* **2006**, *44*, 169–173. [CrossRef]
49. Zhang, X.; Zhang, L.; Sun, Y.; Zheng, S.; Wang, J.; Zhang, T. Hydrogen peroxide is involved in strigolactone induced low temperature stress tolerance in rape seedlings (*Brassica rapa* L.). *Plant Physiol. Biochem.* **2020**, *157*, 402–415. [CrossRef]
50. Guo, J.; Shan, C.; Zhang, Y.; Wang, X.; Tian, H.; Han, G.; Zhang, Y.; Wang, B. Mechanisms of Salt Tolerance and Molecular Breeding of Salt-Tolerant Ornamental Plants. *Front. Plant Sci.* **2022**, *13*, 854116. [CrossRef]
51. Ramakrishnan, M.; Papolu, P.K.; Satish, L.; Vinod, K.K.; Wei, Q.; Sharma, A.; Emamverdian, A.; Zou, L.H.; Zhou, M. Redox status of the plant cell determines epigenetic modifications under abiotic stress conditions and during developmental processes. *J. Adv. Res.* **2022**. [CrossRef] [PubMed]
52. Diao, M.; Ma, L.; Wang, J.; Cui, J.; Fu, A.; Liu, H. Selenium Promotes the Growth and Photosynthesis of Tomato Seedlings Under Salt Stress by Enhancing Chloroplast Antioxidant Defense System. *J. Plant Growth Regul.* **2014**, *33*, 671–682. [CrossRef]
53. Correia, P.; da Silva, A.B.; Vaz, M.; Carmo-Silva, E.; Marques da Silva, J. Efficient Regulation of CO_2 Assimilation Enables Greater Resilience to High Temperature and Drought in Maize. *Front. Plant Sci.* **2021**, *12*, 675546. [CrossRef] [PubMed]
54. Wang, N.; Zhong, X.; Cong, Y.; Wang, T.; Yang, S.; Li, Y.; Gai, J. Genome-wide Analysis of Phosphoenolpyruvate Carboxylase Gene Family and Their Response to Abiotic Stresses in Soybean. *Sci. Rep.* **2016**, *6*, 38448. [CrossRef]

55. Liu, Z.; Ma, C.; Hou, L.; Wu, X.; Wang, D.; Zhang, L.; Liu, P. Exogenous SA Affects Rice Seed Germination under Salt Stress by Regulating Na^+/K^+ Balance and Endogenous GAs and ABA Homeostasis. *Int. J. Mol. Sci.* **2022**, *23*, 3293. [CrossRef]
56. Nutan, K.K.; Kumar, G.; Singla-Pareek, S.L.; Pareek, A. A Salt Overly Sensitive Pathway Member from Brassica juncea BjSOS3 Can Functionally Complement Δ*Atsos3* in Arabidopsis. *Curr. Genomics* **2018**, *19*, 60–69. [CrossRef]
57. Chung, J.S.; Zhu, J.K.; Bressan, R.A.; Hasegawa, P.M.; Shi, H. Reactive oxygen species mediate Na^+-induced SOS1 mRNA stability in Arabidopsis. *Plant J.* **2008**, *53*, 554–565. [CrossRef]
58. Fan, W.; Deng, G.; Wang, H.; Zhang, H.; Zhang, P. Elevated compartmentalization of Na^+ into vacuoles improves salt and cold stress tolerance in sweet potato (*Ipomoea batatas*). *Physiol. Plant.* **2015**, *154*, 560–571. [CrossRef]
59. Nazir, F.; Hussain, A.; Fariduddin, Q. Hydrogen peroxide modulate photosynthesis and antioxidant systems in tomato (*Solanum lycopersicum* L.) plants under copper stress. *Chemosphere* **2019**, *230*, 544–558. [CrossRef]
60. Asgher, M.; Ahmed, S.; Sehar, Z.; Gautam, H.; Gandhi, S.G.; Khan, N.A. Hydrogen peroxide modulates activity and expression of antioxidant enzymes and protects photosynthetic activity from arsenic damage in rice (*Oryza sativa* L.). *J. Hazard. Mater.* **2021**, *401*, 123365. [CrossRef]
61. Su, X.; Fan, X.; Shao, R.; Guo, J.; Wang, Y.; Yang, J.; Yang, Q.; Guo, L. Physiological and iTRAQ-based proteomic analyses reveal that melatonin alleviates oxidative damage in maize leaves exposed to drought stress. *Plant Physiol. Biochem.* **2019**, *142*, 263–274. [CrossRef]
62. Hossain, M.S.; Li, J.; Sikdar, A.; Hasanuzzaman, M.; Uzizerimana, F.; Muhammad, I.; Yuan, Y.; Zhang, C.; Wang, C.; Feng, B. Exogenous Melatonin Modulates the Physiological and Biochemical Mechanisms of Drought Tolerance in Tartary Buckwheat (*Fagopyrum tataricum* (L.) Gaertn). *Molecules* **2020**, *25*, 2828. [CrossRef]
63. Lu, Y.; Wang, Q.F.; Li, J.; Xiong, J.; Zhou, L.N.; He, S.L.; Zhang, J.Q.; Chen, Z.A.; He, S.G.; Liu, H. Effects of exogenous sulfur on alleviating cadmium stress in tartary buckwheat. *Sci. Rep.* **2019**, *9*, 7397. [CrossRef]
64. Kong, L.; Huo, H.; Mao, P. Antioxidant response and related gene expression in aged oat seed. *Front. Plant Sci.* **2015**, *6*, 158. [CrossRef]
65. Devi, S.R.; Prasad, M.N. Copper toxicity in *Ceratophyllum demersum* L. (Coontail), a free floating macrophyte: Response of antioxidant enzymes and antioxidants. *Plant Sci.* **1998**, *138*, 157–165. [CrossRef]
66. Hodges, D.M.; DeLong, J.M.; Forney, C.F.; Prange, R.K. Improving the thiobarbituric acid-reactive-substances assay for estimating lipid peroxidation in plant tissues containing anthocyanin and other interfering compounds. *Planta* **1999**, *207*, 604–611. [CrossRef]
67. Aoyagi, K.; Bassham, J.A. Pyruvate orthophosphate dikinase in wheat leaves. *Plant Physiol.* **1983**, *73*, 853–854. [CrossRef]
68. Livak, K.J.; Schmittgen, T.D. Analysis of relative gene expression data using real-time quantitative PCR and the 2(-Delta Delta C(T)) Method. *Methods* **2001**, *25*, 402–408. [CrossRef]
69. Chen, C.; Chen, H.; Zhang, Y.; Thomas, H.R.; Frank, M.H.; He, Y.; Xia, R. TBtools: An Integrative Toolkit Developed for Interactive Analyses of Big Biological Data. *Mol. Plant* **2020**, *13*, 1194–1202. [CrossRef]

International Journal of *Molecular Sciences*

Article

Ectopic Expression of *AeNAC83*, a NAC Transcription Factor from *Abelmoschus esculentus*, Inhibits Growth and Confers Tolerance to Salt Stress in *Arabidopsis*

Xuan Zhao [†], Tingting Wu [†], Shixian Guo, Junling Hu and Yihua Zhan *

The Key Laboratory for Quality Improvement of Agricultural Products of Zhejiang Province, College of Advanced Agricultural Sciences, Zhejiang A & F University, Hangzhou 311300, China
* Correspondence: yhzhan@zafu.edu.cn
† These authors contributed equally to this work.

Abstract: NAC transcription factors play crucial roles in plant growth, development and stress responses. Previously, we preliminarily identified that the transcription factor *AeNAC83* gene was significantly up-regulated under salt stress in okra (*Abelmoschus esculentus*). Herein, we cloned the nuclear-localized AeNAC83 from okra and identified its possible role in salt stress response and plant growth. The down-regulation of *AeNAC83* caused by virus-induced gene silencing enhanced plant sensitivity to salt stress and increased the biomass accumulation of okra seedlings. Meanwhile, *AeNAC83*-overexpression *Arabidopsis* lines improved salt tolerance and exhibited many altered phenotypes, including small rosette, short primary roots, and promoted crown roots and root hairs. RNA-seq showed numerous genes at the transcriptional level that changed significantly in the *AeNAC83*-overexpression transgenic and the wild *Arabidopsis* with or without NaCl treatment, respectively. The expression of most phenylpropanoid and flavonoid biosynthesis-related genes was largely induced by salt stress. While genes encoding key proteins involved in photosynthesis were almost declined dramatically in *AeNAC83*-overexpression transgenic plants, and NaCl treatment further resulted in the down-regulation of these genes. Furthermore, DEGs encoding various plant hormone signal pathways were also identified. These results indicate that AeNAC83 is involved in resistance to salt stress and plant growth.

Keywords: okra; salt stress; growth; NAC transcription factor; flavonoid; photosynthesis

1. Introduction

Salinity is one of the most detrimental abiotic stresses which limits plant growth and reduces biomass and grain yield severely worldwide. Salt stress impairs the productivity of plants by affecting cell growth and metabolic process, causing irreversible damage to seed germination, seedling growth and crop yield [1]. To counter the adverse effects of environmental stress, plants have evolved complex mechanisms to cope with salt stress at both physiological and biochemical levels [2]. The improvement of morphological structure and physiological metabolism level is controlled by the expression of stress-response genes, and transcription factors (TFs) play a prominent part in regulating these genes. Numerous TFs, including NAC TFs, have been identified to be involved in plant growth and salt stress response.

The plant-specific NAC TF family is one of the largest TF families in plants, named after the initials of the *NAM* (no apical meristem) gene from *Petunia hybrida* [3], the *ATAF1/2* [4] and *CUC2* (cup-shaped cotyledon) [5] from *Arabidopsis thaliana*. In 1996, Souer et al. [3] cloned the first NAC transcription factor gene *NAM* from *Petunia*, which affects the formation and differentiation of *Petunia* apical meristem. Subsequently, NAC TFs were successively identified in *Arabidopsis* [6], rice [7], soybean [8], tomato [9] and other species. The N-terminal of all NAC TFs contains a conserved NAC domain composed of about

150 amino acid residues, and the C-terminal contains a highly variable transcriptional activation region (TAR) [6].

Many studies have shown that NAC TFs not only play an important role in plant growth and development, such as secondary wall formation [10–12], leaf senescence [13], and lateral root development [14,15], but also participate in response to abiotic stresses [16,17]. NAC TFs respond to salt stress mainly through maintaining intracellular Na^+ and K^+ concentrations and the relative homeostasis of the intracellular environment by activating the expression of stress response genes and protecting the stability of cell structure by enhancing the accumulation of osmoregulation substances such as soluble sugar, proline, and betaine. For example, compared with WT, transgenic rice overexpressing *ONAC022* with lower Na^+ accumulation and transpiration rate, and increased free proline and soluble sugar content, enhanced salt tolerance significantly [18], which were also observed in transgenic rice overexpressing *ONAC009* and *ONAC058* [19,20]. Wheat TaNAC29 improved plant salt tolerance by inducing the expression of stress-related genes, enhancing the antioxidant system, and scavenging reactive oxygen species [21]. Under salt stress, *NAC13*-overexpressing transgenic poplar showed enhanced salt tolerance and the *NAC13*-suppressing plants increased sensitivity to salt stress [22].

Okra (*Abelmoschus esculentus* L.), a medicinal and edible plant, has attracted increasing attention worldwide [23,24]. However, no molecular characterization of any NAC family member's impact on okra growth and salt stress response has been reported. In our previous study, an NAC TF gene *AeNAC83* from okra was up-regulated and exposed to salt stress [25]. In the present work, we cloned the coding sequence of AeNAC83 by using full-length isoforms obtained in okra and silenced the gene by the whole plant virus-induced gene silencing (VIGS). The *AeNAC83*-silenced okra seedlings exhibited a significantly higher biomass accumulation under normal conditions and were more sensitive to salt stress than that in Mock plants. Moreover, we further investigated its functions in transgenic *Arabidopsis* overexpressing *AeNAC83*. The overexpression of *AeNAC83* improved salt resistance and inhibited plant growth. According to the transcriptome analysis, the improved resistance and inhibited growth were related to the up-regulation of flavonoid biosynthesis-related genes and the down-regulation of photosynthesis-related genes in transgenic plants. The above work will provide important resources for molecular breeding of plant stress resistance.

2. Results

2.1. Isolation and Characterization of AeNAC83

We first cloned the coding sequence of AeNAC83 from okra, which encoded a protein of 255 amino acids. Then, multiple sequence alignment was performed with other NAC TFs from *Arabidopsis* and other species. AeNAC83 has a high homology with other NAC members, containing one NAC domain with five conserved regions (Figure 1A). In order to reveal the evolutionary relationship between AeNAC83 and other NAC TFs, a phylogenetic tree was constructed. All the selected NAC members were clustered into two distinct subgroups, and AeNAC83 belonged to the ATAF subgroup (Figure 1B).

To determine the subcellular location of AeNAC83, the 35Sp:: *AeNAC83*:GFP fusion protein was constructed and transiently expressed in tobacco epidermal cells. Fluorescence analysis showed that the control empty vector was expressed in both the nucleus and cytoplasm, while the protein was only localized in the nucleus (Figure 2A), indicating that AeNAC83 was localized in the nucleus.

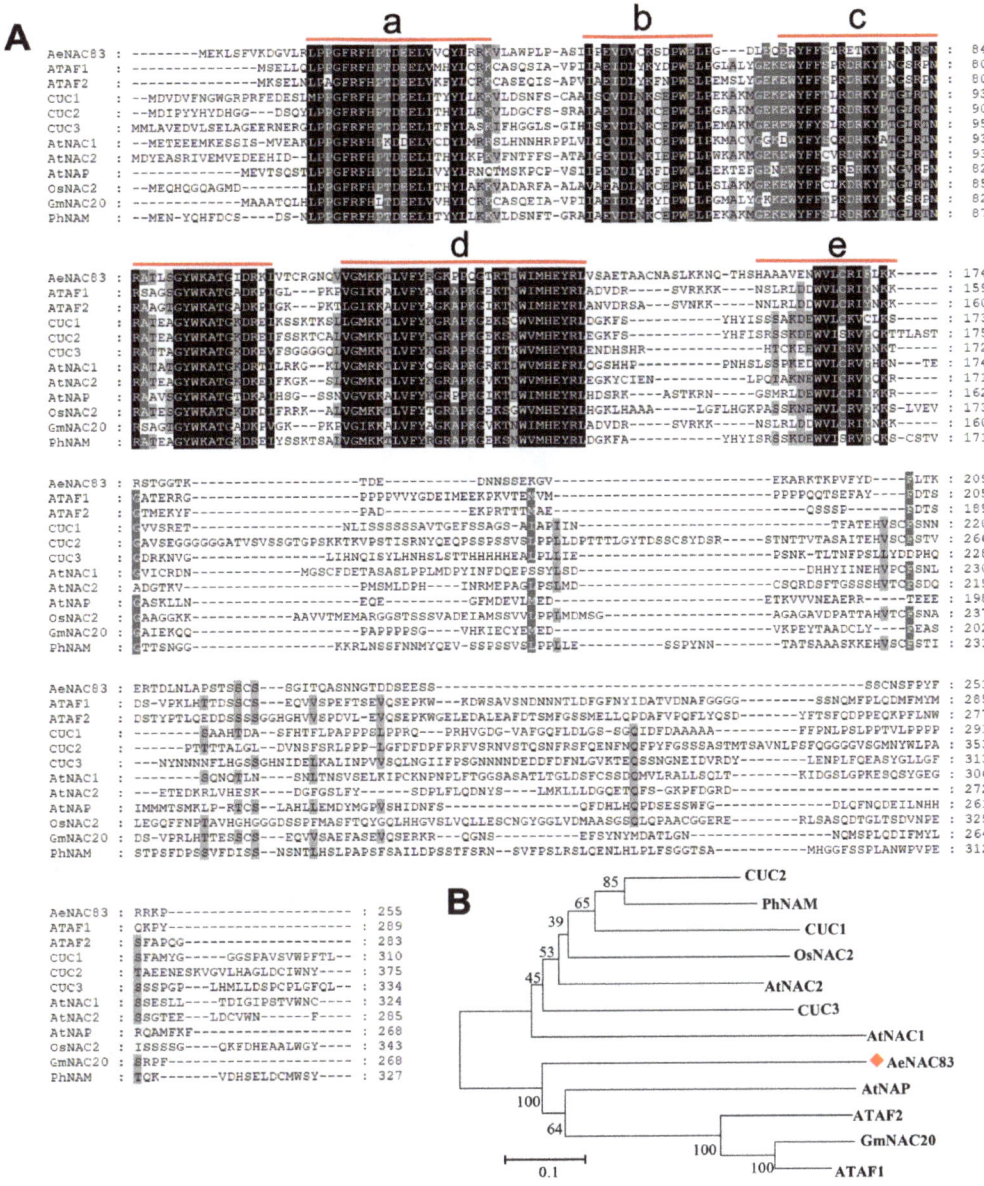

Figure 1. Characterization of AeNAC83: (**A**) Multiple sequence alignment of AeNAC83 and its homologous NAC proteins. The NAC domain with five conserved regions (a–e) are indicated by red lines; (**B**) Phylogenetic analysis of AeNAC83 with its homologous NAC proteins.

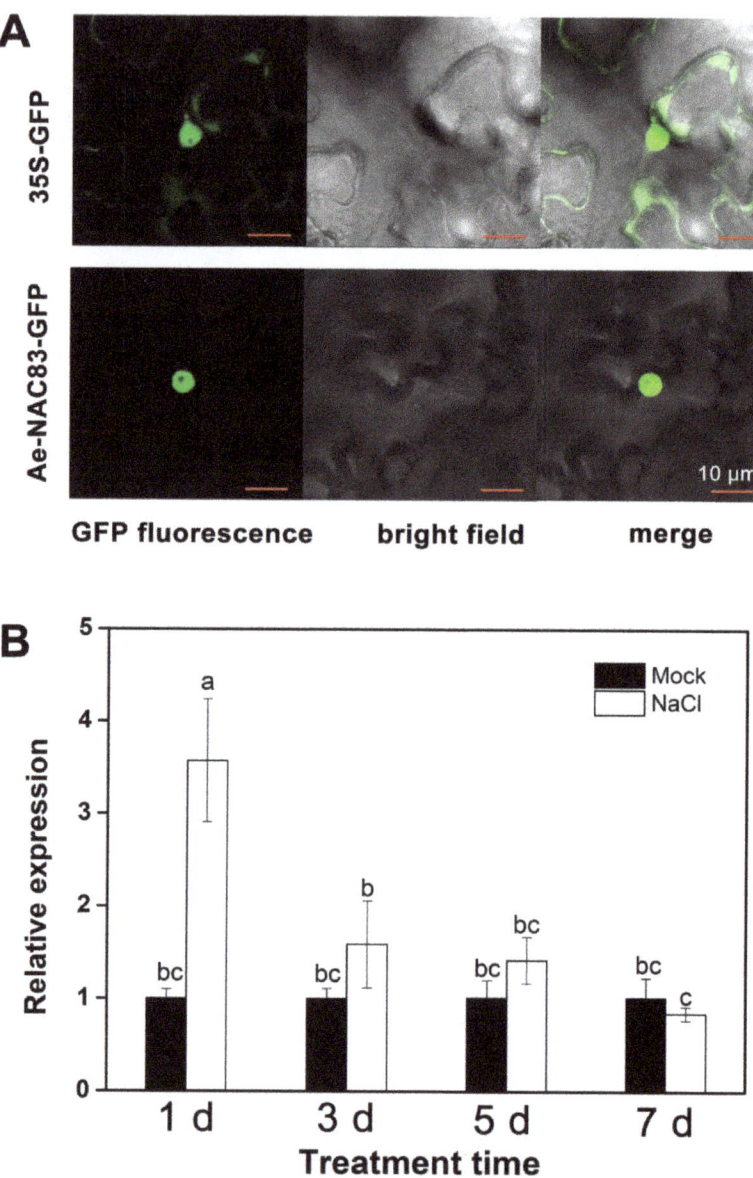

Figure 2. Subcellular localization and expression pattern of AeNAC83: (**A**) Subcellular localization of AeNAC83 in *N. benthamiana*. *N. benthamiana* cells were transformed with 35Sp:: *AeNAC83*:GFP or pCAMBIA1300-GFP. After incubating for 48 h, the transformed cells were observed under a confocal microscope. The photographs were taken under detecting GFP fluorescence, bright field, and in combination (merge), respectively. Empty vector (pCAMBIA1300-GFP) was used as a control; (**B**) Expression of *AeNAC83* in okra seedlings after 300 mM NaCl treatment by qRT-PCR. Total RNA for expression analysis was isolated from leaves of two-week-old seedlings after 300 mM NaCl treatment for 1, 3, 5, and 7 days. Data are presented as mean ± SD (n = 3). Different letters denote significant differences at $p < 0.05$, using ANOVA and Duncan's multiple tests.

Our previous analysis showed that salt stress induced the expression of *AeNAC83*. To further confirm whether AeNAC83 participates in the salt resistance, the two-week-old okra seedlings were irrigated with 300 mM NaCl and sampled at 1, 3, 5 and 7 days after treatment, and then the transcription level of the *AeNAC83* gene in the second true leaf of the seedlings was determined by qRT-PCR. We found that compared with the control, the expression of *AeNAC83* was up-regulated at three time points after salt treatment, and the expression was highest on the first day, then gradually decreased (Figure 2B). The result suggests that AeNAC83 may play a regulatory role in salt stress in okra.

2.2. Performance of AeNAC83-Silenced Plants under Salt Stress

The VIGS method was used to assess the *AeNAC83* gene function. We constructed the pTRV2-*NAC83* vector with a 336-bp fragment and then inoculated the *Agrobacterium* mixture containing the pTRV2-*NAC83* vector and pTRV1 into the cotyledons of okra. Negative control was performed by using empty vectors pTRV1 and pTRV2. The *AeNAC83* mRNA levels of newly grown young true leaves at 25-days post-inoculation were detected to assess the efficiency of VIGS. As shown in Figure 3A, the mRNA levels of AeNAC83 in silenced plants decreased dramatically compared with the control, accounting for about 60% of the control, indicating that the gene was partially silenced.

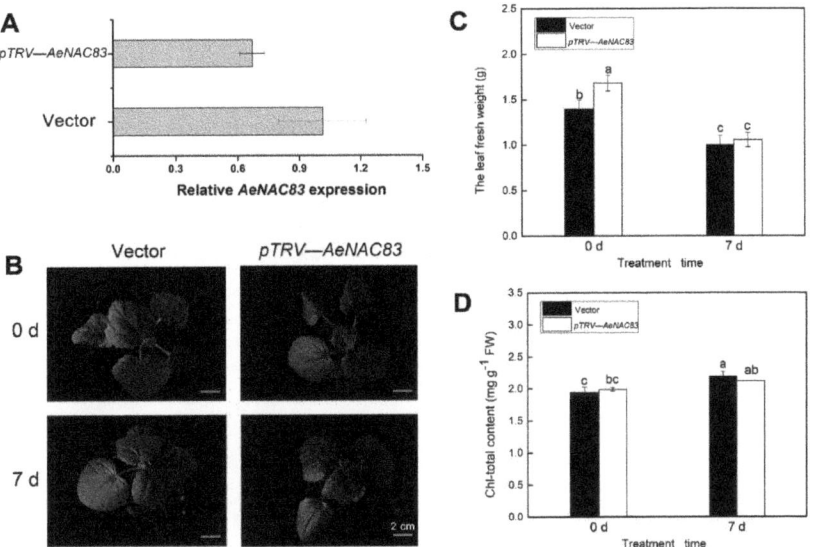

Figure 3. Phenotype analysis of *AeNAC83*-silenced okra plants produced by VIGS under salt stress. The *Agrobacterium tumefaciens* GV3101 cell culture harboring the pTRV2 or *AeNAC83*-pTRV2 together with pTRV1 were mixed with a ratio of 1:1 and syringe-infiltrated into okra cotyledons. At 25 d post-inoculation (dpi), the leaves were used for gene expression assay and 300 mM NaCl treatment for 7 days. (**A**) Expression analysis of *AeNAC83* in *AeNAC83*-silenced okra seedlings by qRT-PCR. (**B**) Images of plant phenotype. (**C**) The leaf fresh weight. (**D**) Total chlorophyll content. Data are presented as mean ± SD (n = 10). Different letters denote significant differences at $p < 0.05$, using ANOVA and Duncan's multiple tests.

To determine the roles of AeNAC83 in salt stress and plant growth, the *AeNAC83*-silenced okra seedlings produced by VIGS were irrigated with water or 300 mM NaCl solution for 7 days. Under normal conditions, the total leaf fresh weight of *AeNAC83*-silenced seedlings was significantly higher than Mock plants, indicating that AeNAC83 may be involved in plant growth. After the 7-day treatment, the seedling growth was inhibited obviously (Figure 3B), and leaf fresh weight of *AeNAC83*-silenced plants decreased significantly than that of Mock plant (Figure 3C). The total chlorophyll content in *AeNAC83*-silenced plants was not significantly different from that in the control before salt treatment, whereas after salt treatment the chlorophyll content increased in the Mock plants (Figure 3D). The results showed that *AeNAC83*-silenced plants were more sensitive to NaCl treatment, indicating that AeNAC83 may play a positive regulatory role in salt stress.

2.3. Overexpression of AeNAC83 in Arabidopsis Inhibits Plant Growth and Improves Salt Tolerance

To further investigate the role of AeNAC83 in salt stress and growth, over-expression of *AeNAC83* gene transgenic Arabidopsis plants were produced. The lines OX3 and OX7 with high expression of *AeNAC83* were selected for further phenotypic analysis from at least 10 homozygous transgenic lines (Figure 4A).

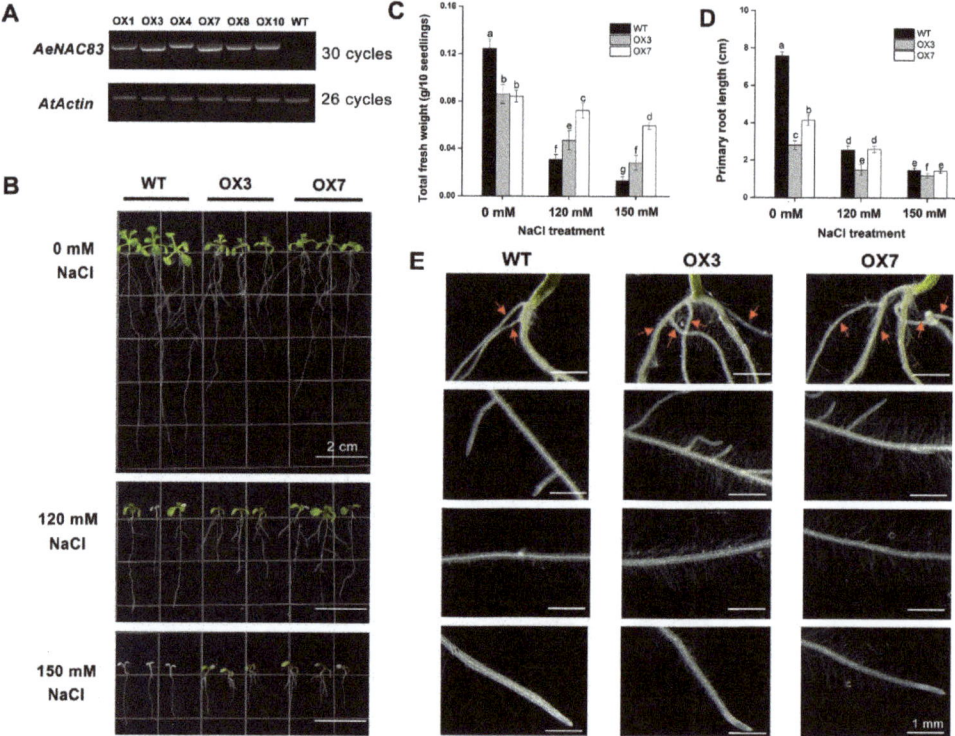

Figure 4. Roles of AeNAC83 in salt tolerance and in growth in *Arabidopsis* transgenic plants. Four-day-old seedlings of wild-type (WT) and two transgenic lines (OX3 and OX7) were transplanted on 1/2 MS medium supplemented with 0, 120, or 150 mM NaCl for 10 days. (**A**) Semi-quantitative analysis of *AeNAC83* gene expression in wild-type (WT) and *AeNAC83*-overexpression transgenic plants. The first bands show *AeNAC83* gene expression (30 cycles) and the bands below show *AtActin* gene expression (26 cycles) used as internal control. (**B**) Phenotypes of seedlings on 1/2 MS medium supplemented with 0, 120, or 150 mM NaCl, respectively. (**C,D**) Fresh weight and primary root length of seedlings at the end of the experiment in (**B**). Data are presented as mean ± SD (n = 10). Different letters denote significant differences at $p < 0.05$, using ANOVA and Duncan's multiple tests. (**E**) Root phenotype of seedlings on 1/2 MS medium without NaCl treatment. The red arrows indicated the crown root.

The *Arabidopsis thaliana* ecotype Columbia (wild-type, WT) and transgenic seedlings were grown in 1/2 MS medium for 4 days and then transplanted to medium with 0, 120 or 150 mM NaCl for 10 days to observe the phenotypic difference. Under normal conditions, significant differences existed in the growth between the WT and transgenic plants. Compared with WT, *AeNAC83*-overexpression transgenic plants exhibited many altered phenotypes (Figure 4B), including small rosette, short primary roots and promoted crown roots and root hairs (Figure 4E). The fresh weight (Figure 4C) and primary root length (Figure 4D) of the transgenic plants were significantly declined than those of wild-type plants. After NaCl treatment for 10 days, better performance was observed for transgenic lines than WT. Most WT leaves showed albinistic symptoms, and the number of crown roots of transgenic lines was greater than in control (Figure 4D). The fresh weight of the transgenic lines was greater than that of WT exposed to salt stress, and there was no obvious difference in primary root length between WT and OX7 transgenic lines. Together, these results suggest that over-expression of *AeNAC83* inhibited plant growth and enhanced tolerance to salt stress.

2.4. Identification and Functional Enrichment Analysis of Differential Expression Genes (DEGs) under Salt Stress between the AeNAC83-Overexpression Transgenic and the Wild Arabidopsis

To understand the molecular mechanism of AeNAC83 in plant growth and salt stress response, RNA was extracted from *AeNAC83*-overexpression transgenic (OX3) and the wild (WT) *Arabidopsis* treated with 120 mM NaCl (12 samples, three replicates for each treatment) and RNA-seq was carried out using Illumina sequencing platform. The Pearson correlation coefficient heat map showed that the samples had good repeatability (Figure S1). A total of 245.54 M reads was obtained, with a total of 73.35 Gb clean data. Clean data of each sample reached 5.74 Gb, and the percentage of Q30 was at least 93.91% (Table S1). We assembled and quantified reads compared with HISAT2 using StringTie and performed fragments per kilobase of transcript per million fragments mapped (FPKM) conversion to analyze gene expression level. The screening of threshold for DEGs was Padj < 0.05 and |log2FoldChange| > 1. The hierarchical clustering analysis of DEGs under different experimental conditions was shown in Figure 5A. The volcanoplots represent the overall distribution of DEGs (Figure 5B). Compared with WT-CK, there were 1917 DEGs in OX3-CK, including 1259 up- and 658 down-regulated genes, and there were 4285 DEGs in WT-N including 2098 up- and 2187 down-regulated genes. Compared with OX3-CK, 1360 DEGs were found in OX3-N, including 692 up and 668 down-regulated genes. Compared with WT-N, there were 1537 DEGs in OX3-N, including 775 up- and 762 down-regulated genes (Figure 5C, Table S2). Venn diagram showed that the number of DEGs between "WT-CK vs. OX3-CK", "WT-CK vs. WT-N", "OX3-CK vs. OX3-N", and "WT-N vs. OX3-N" were 1917, 4285, 1360 and 1537, respectively. A total of 105 common DEGs were identified in the four comparative groups (Figure 5D).

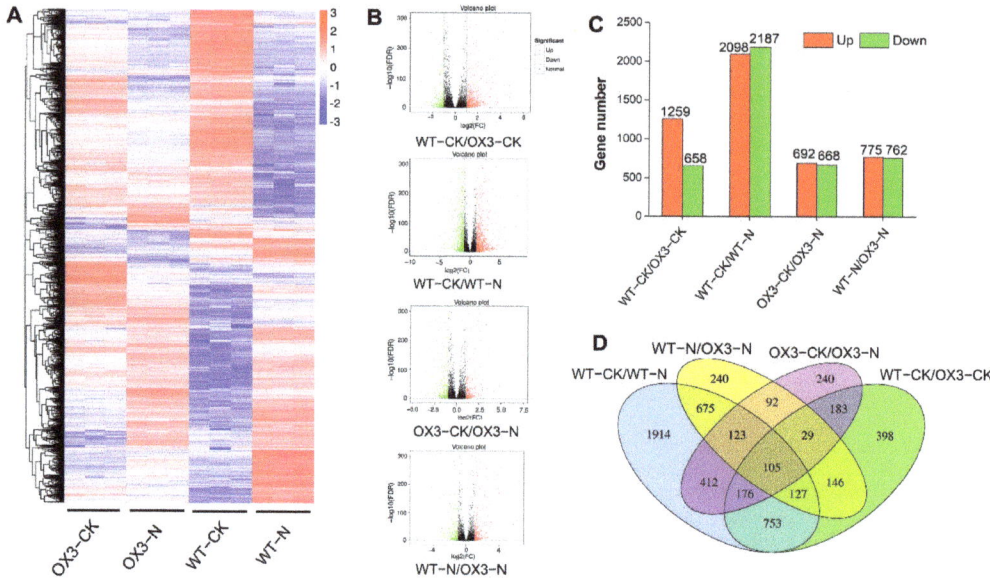

Figure 5. Transcriptional variations in *Arabidopsis* wild-type (WT) and transgenic plant (OX3) under NaCl treatment: (**A**) Expression profiles of the DEGs under NaCl treatment were shown by a heatmap; (**B**) Significance analysis of the DEGs in different comparisons by volcanoplots; (**C**) The number of up- and down-regulated genes in different comparisons; (**D**) Venn diagrams showed the proportions of the up- and down-regulated genes in four comparisons. WT-CK, OX3-CK: WT and OX3 grown under optimum conditions; WT-N, OX3-N: WT and OX3 were subjected to salt stress. Three replicates for each treatment.

2.5. Identification of DEGs Involved in the Phenylpropanoid and Flavonoid Biosynthesis Pathways

KEGG pathway enrichment analysis was performed to interpret the functions of these DEGs. The top three DEGs participated in "MAPK signaling pathway-plant", "plant-pathogen interaction" and "phenylpropanoid biosynthesis" in WT-CK vs. OX3-CK and WT-CK vs. WT-N comparison groups (Figure 6A,B). The top three DEGs participated in "phenylpropanoid biosynthesis", "MAPK signaling pathway", and "starch and sucrose metabolism" in OX3-CK vs. OX3-N comparison group (Figure 6C). The top three DEGs participated in "phenylpropanoid biosynthesis", "circadian rhythm-plant", and "starch and sucrose metabolism" in WT-N vs. OX3-N comparison group (Figure 6D).

The phenylpropanoid biosynthesis pathway was always found to contain significant enrichment of DEGs for all comparisons (Figure 6). Through integration, 13 gene families involved in phenylpropanoid biosynthesis were identified (Figure 7A), including 129 DEGs. The POD family was the most represented (49 DEGs) among these 13 gene families, whereas the C4H, F5H, and CSE family were the least represented, with only 1 DEG (Figure 7A). The phenylpropanoid biosynthesis pathway provides the precursors for the flavonoid biosynthesis pathway [26]. Hence, we analyzed the DEGs involved in flavonoid biosynthesis and 10 gene families were identified, among which the HCT family was the most represented (7 DEGs). The expression levels of most DEGs were up-regulated by NaCl treatment, except for 2 *CHI* genes and 1 *HCT* gene (Figure 7B). In addition, under normal conditions, nearly half of the flavonoid biosynthesis-related DEGs were significantly induced by overexpression of *AeNAC83*. *CHI*, *F3H*, *DFR*, and *ANS* are crucial structural genes involved in the anthocyanin synthesis pathway. Anthocyanins, a class of flavonoids distributed ubiquitously in the plant, play important roles in the growth and development of plants and stress response. The content of anthocyanins in OX3 was significantly higher than that of WT before salt treatment. Salt stress induced the accumulation of anthocyanins in WT, while there was no significant change in anthocyanin content in OX3 (Figure S2A).

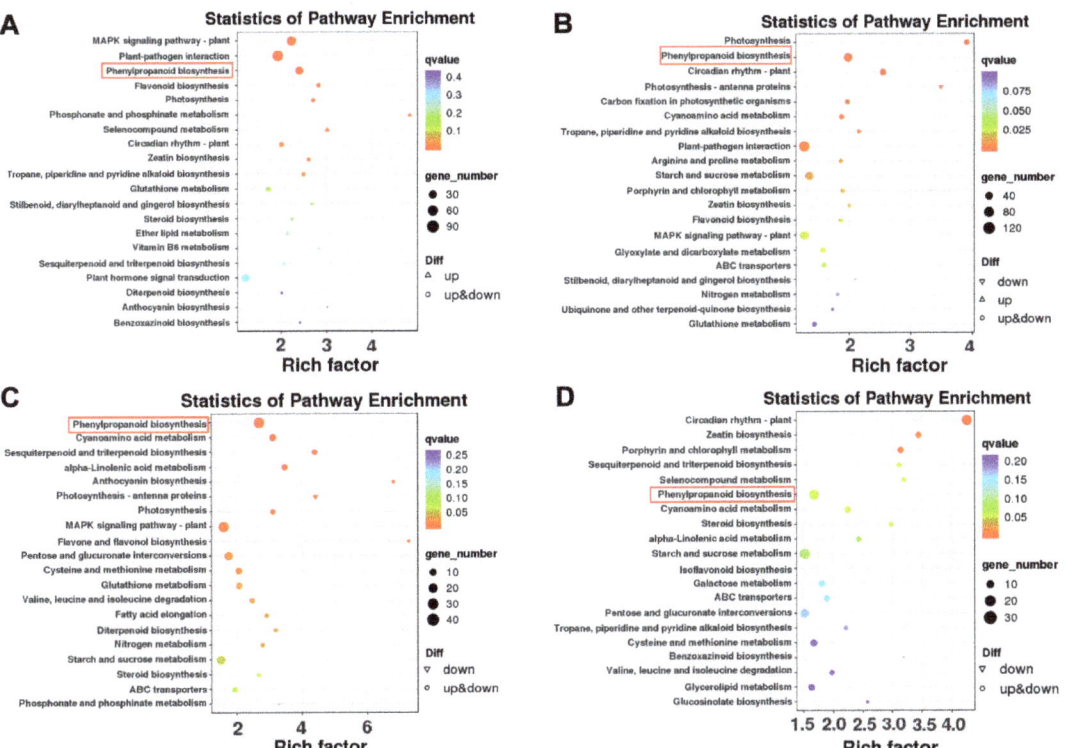

Figure 6. The top 20 enriched KEGG pathways enrichment analysis of DEGs in the four comparison groups. (**A–D**) KEGG pathway enrichment analysis of DEGs in the WT-CK vs. OX3-CK, WT-CK vs. WT-N, OX3-CK vs. OX3-N, and WT-N vs. OX3-N comparisons, respectively.

2.6. Identification of DEGs That Participate in Photosynthesis and Hormone Signal Transduction Pathways

Thylakoid membrane photosynthetic complexes consist of photosystem II (PSII), cytochrome b6f, photosystem I (PSI), light-harvesting antenna complexes, and ATP synthase [27]. Compared with WT-CK, genes encoding key proteins involved in these complexes were almost greatly decreased in *AeNAC83*-overexpression transgenic (OX3) *Arabidopsis*, and NaCl treatment further resulted in the down-regulation of these genes, except for 1 *PsbP* gene of PSII and 1 *gamma* gene of ATP synthase (Figure 8). Chlorophyll is one of the most important pigments in higher plants, which is an important substance for photosynthesis. The total chlorophyll content was similar between OX3 and WT before salt treatment. After salt treatment, the chlorophyll content decreased dramatically in both WT and OX3 plants (Figure S2B).

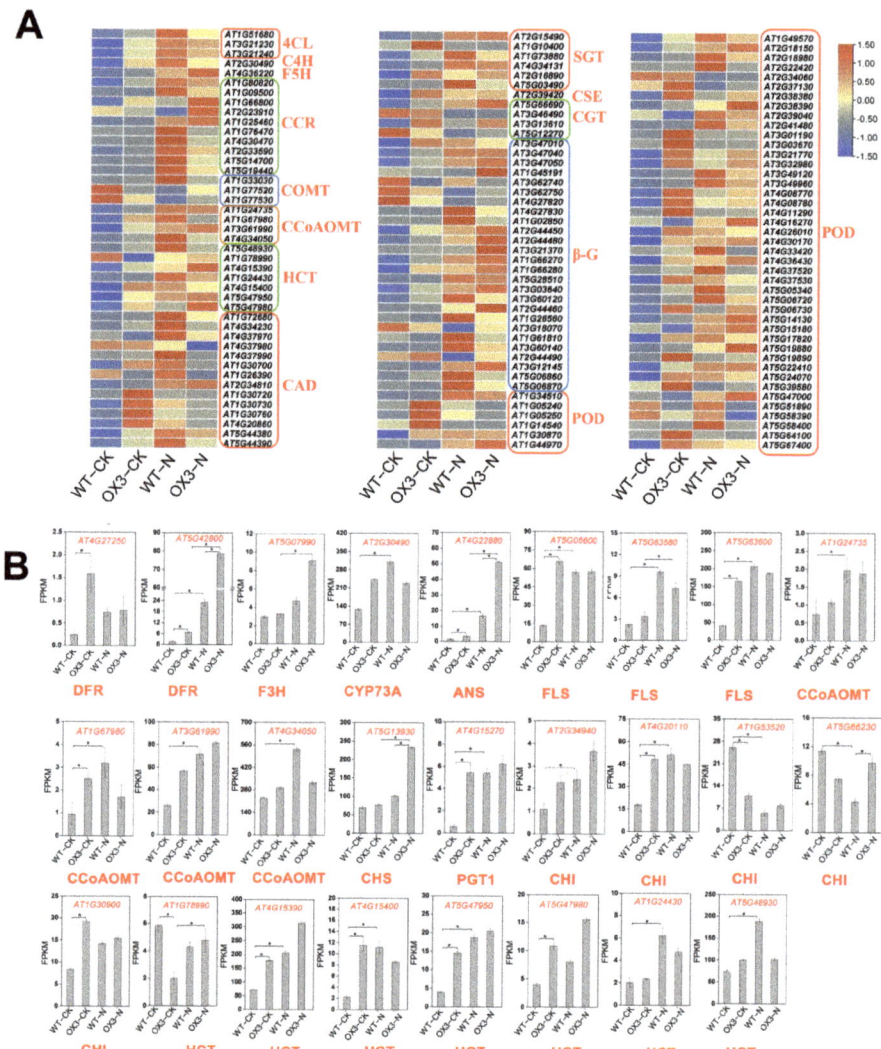

Figure 7. Transcript abundance changes of the phenylpropanoid and flavonoid biosynthesis pathway-related DEGs in *Arabidopsis* wild type (WT) and transgenic plant (OX3) under NaCl treatment. (**A**) Heat map of DEGs in phenylpropanoid biosynthesis pathway. The log2-transformed FPKM values of DEGs were used to generate the diagram. 4CL, 4-coumarate-CoA ligase; C4H, cinnamate-4-hydroxylase; F5H, ferulate-5-hydroxylase; CCR, cinnamoyl-CoA reductase; CCoAOMT, caffeoyl-CoA O-methyltransferase; COMT, caffeic acid 3-O-methyltransferase; HCT, hydroxyl cinnamoyl transferase; CAD, cinnamyl-alcohol dehydrogenase; SGT, scopoletin glucosyltransferase; CSE, caffeoylshikimate esterase; CGT, coniferyl-alcohol glucosyltransferase; β-G, beta-glucosidase; POD, peroxidase. (**B**) Gene expression of DEGs in flavonoid biosynthesis pathway. DFR: dihydroflavonol 4-reductase; F3H: flavanone 3-hydroxylase; CYP73A: trans-cinnamate 4-monooxygenase; ANS: anthocyanidin synthase; FLS: flavonol synthase; CHS: chalcone synthase; PGT1: phlorizin synthase; CHI: chalcone isomerase; HCT: shikimate O-hydroxycinnamoyltransferase. * indicate statistical significance based on two-tailed Student's t-test at p-values < 0.05.

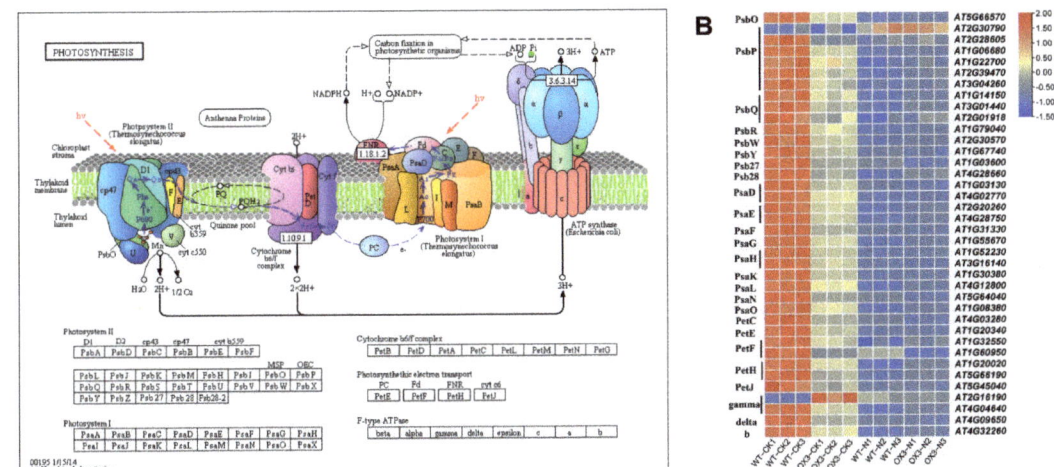

Figure 8. Transcript abundance changes of photosynthesis pathway-related DEGs in *Arabidopsis* wild-type (WT) and transgenic plant (OX3) under NaCl treatment. (**A**) Photosynthesis pathway (ko00195). Different letters indicated the different subunits of photosynthetic complexes. (**B**) Heat map of DEGs in photosynthesis pathway. The log2-transformed FPKM values of DEGs were used to generate the diagram.

In addition, the abundance of hormone-related DEGs changed significantly under salt stress. A large number of genes encoding AUX1, AUX/IAA, ARF, GH3 and SAUR for the auxin signaling pathway were identified, including 2 *AUX1* genes, 8 *AUX/IAA* genes, 1 *ARF* gene, 13 *GH3* genes and 21 *SAUR* genes; 1 *CRE1* gene, 2 *AHP* genes, 11 *B-ARR* genes and 2 *A-ARR* genes for the cytokinin signaling pathway were obtained; 3 *GID1* genes, 8 *DELLA* genes and 13 *TF* genes were identified in the GA signaling pathway; DEGs involved in ABA signaling pathway were identified, including 2 *PYR* genes, 8 *PP2C* genes, 4 *SnRK2* genes and 4 *AREB/ABF* genes; 1 *ETR*, 2 *CTR1* genes, 4 *SIMKK* genes, 1 *EBF1/2* gene and 4 *ERF1/2* genes for ET signaling pathway were analyzed; 9 *BAK1* genes, 4 *BRI1* genes, 2 *BSK* genes, and 1 *TCH4* gene related to the BR signaling pathway were identified; there were 5 *JAZ* and 9 *MYC2* DEGs in the JA signaling pathway; for the SA signaling pathway, 1 *NPR1*, 4 *TGA* and 8 *PR1* were identified (Figure 9). Twelve randomly selected genes of hormone signal transduction pathways, including 1 *AUX1*, 1 *ARF*, 2 *SnRK2*, 1 *AREB/ABF*, 2 *JAZ*, 1 *MYC2*, 1 *NPR1*, 1 *TGA* and 2 *PR1*, were verified by qRT-PCR (Figure S3). The results showed that the relative expression levels were basically consistent with RNA-seq data, which supported the accuracy and reliability of transcriptome analysis results.

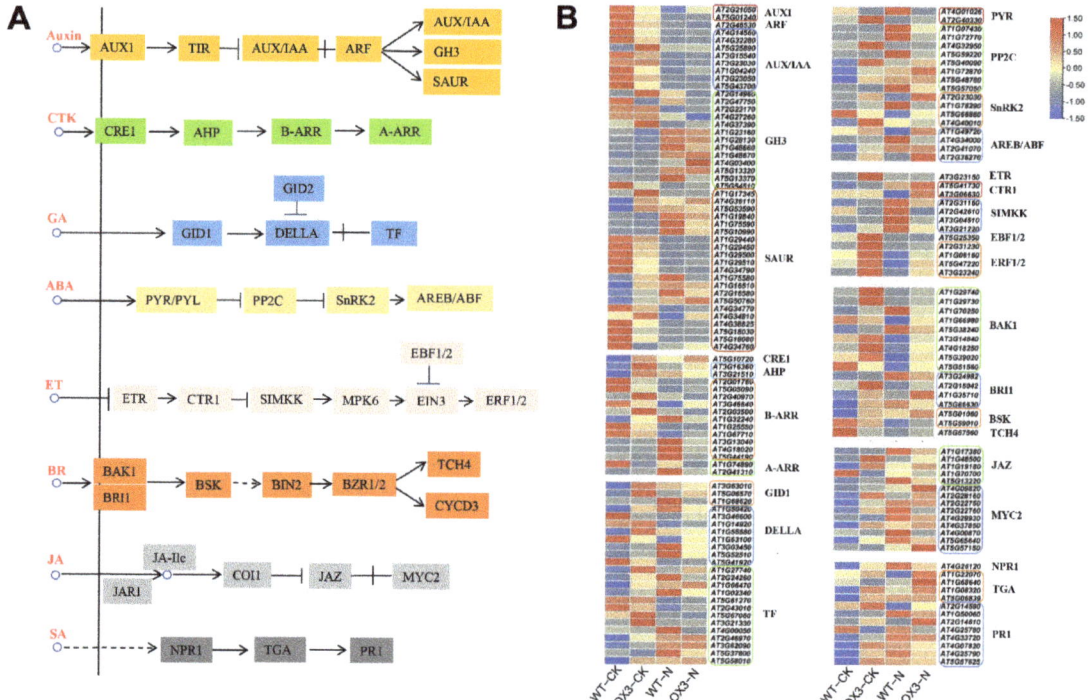

Figure 9. Transcript abundance changes of the various hormone signal transduction pathways-related DEGs in *Arabidopsis* wild-type (WT) and transgenic plant (OX3) under NaCl treatment. (**A**) Plant hormone signal transduction pathways. (**B**) Heat map of DEGs in various hormone signal transduction pathways. The log2-transformed FPKM values of DEGs were used to generate the diagram.

3. Discussion

Okra has important edible and medicinal value. However, due to the complexity and lack of the okra genome, it is difficult to assess the gene function of okra [28,29]. VIGS has been widely used as an effective functional genomics tool for gene function analysis [30]. NAC TFs play important roles in plant growth, development and stress responses. In this study, the role of AeNAC83 in the salt stress responses in okra was investigated. We found that the *AeNAC83*-silenced okra seedlings produced by VIGS were slightly more sensitive to salt stress (Figure 3). Further analysis found that the gene was not completely silenced, probably because VIGS sometimes silences only part of the target gene [31]. Hence, transgenic *Arabidopsis* plants over-expressing *AeNAC83* were generated to further determine the role of AeNAC83 in growth and salt stress. Over-expression of *AeNAC83* enhanced tolerance to salt stress and suppressed vegetative growth (Figure 4). Taken together, okra NAC TF AeNAC83 plays a pivotal role in mediating plant growth and defense response to salt stress.

Transcriptomic analyses showed that compared with WT-CK, there were 4285 DEGs in WT-N, while only 1360 DEGs were found in OX3-N compared with OX3-CK (Figure 5). The result showed that overexpression of *AeNAC83* resulted in the insensitivity of numerous genes to salt stress, which may be one of the reasons for enhanced resistance to salt stress. KEGG pathway enrichment analysis showed that these DEGs mainly participated in "MAPK signaling pathway-plant", "plant-pathogen interaction" and "phenylpropanoid biosynthesis", "starch and sucrose metabolism" and "circadian rhythm-plant" (Figure 6). In plants, the rapid activation of MAPK cascades has long been observed involved in growth and development, as well as in response to drought, salinity, wounding, heat, and cold [32]. Moreover, for all comparison groups, the phenylpropanoid biosynthesis pathway was always found to contain significant enrichment of DEGs.

The phenylpropanoid pathway that produces lignin, flavonoids, and other secondary metabolites [26], contributes to the defense and growth of plants. Fortifying cell walls by increasing their lignin content is one of the common plant defense mechanisms [33]. The 4CL, C4H and POD are core enzymes involved in the biosynthesis of flavonoids and lignin. From our analysis, the expressions of 4CL, C4H, CAD and POD-related DEGs were increased in OX3-CK and the NaCl-treated WT and OX3 samples (Figure 7A). This finding suggested that NaCl stress could cause an increase in lignin synthesis. Under normal conditions, the *AeNAC83*-overexpression plant may accumulate more lignin, which may be one of the reasons why it has stronger salt stress resistance than the wild type. In response to a variety of abiotic stress, flavonoids (including anthocyanins) play a major antioxidant role [34,35]. Flavonoids can reduce reactive oxygen species in plant tissues [36–38], which are usually produced by stress such as ultraviolet radiation, drought and salt stresses. We observed up-regulation of flavonoids genes in NaCl-treated seedlings (Figure 7B), indicating that the genes involved in stress metabolism are generally up-regulated. Some transcription factors have been identified to have regulatory functions in phenylpropanoid and flavonoid biosynthesis pathways [39]. Overexpression of *PvMYB4*, a suppress phenylpropanoid metabolism TF, caused a reduction in the lignin content and decreased recalcitrance in *Panicum virgatum* [40]. NAC is the second largest class of TFs in plants and has been shown as a key regulator of abiotic stresses [41]. In *Arabidopsis* and rice, NAC genes' overexpression enhanced drought and salt tolerance [20,42,43]. Additionally, overexpression of *AeNAC83* significantly induced the expression of most flavonoid biosynthesis-related DEGs under normal conditions. This result was further confirmed by the experiment of anthocyanin content determination (Figure S2A). These results showed that overexpression of *AeNAC83* improved the resistance of plants to salt stress, possibly by regulating the accumulation of lignin and flavonoids.

Chloroplast is the site of photosynthesis. High Na^+ levels can destroy the structure of chloroplast, damage the membrane system of plant and seriously degrade chlorophyll, resulting in the decline of photosynthesis. In the study, the expression levels of photosynthesis-related genes were significantly down-regulated in WT and transgenic *Arabidopsis* (OX-3) after NaCl treatment and the degree of decline of genes in WT was significantly higher than that of OX-3 (Figure 8). The results showed that *AeNAC83*-overexpression transgenic plants had stronger photosynthetic capacity under salt stress, which could ensure more organic accumulation and improve the salt tolerance of plants. The same phenomenon has been observed in transgenic *Arabidopsis* overexpressing wheat *TaNAC67*, the chlorophyll content and Fv/Fm of which were higher than those of the control [44]. In addition, under salt stress, *AeNAC83*-overexpression transgenic plants have larger roots, suggesting that *AeNAC83* may regulate roots to improve salt stress tolerance. Similarly, overexpression of soybean *GmNAC20* improved the tolerance to low temperature and salt and promoted lateral root formation [45]. Under high salt stress, *Arabidopsis thaliana AtNAC2* was highly expressed in roots, and lateral roots of transgenic plants overexpressing *AtNAC2* increased [15]. Compared with wild-type plants, *AeNAC83*-overexpression plants showed growth retardation, which may be due to the redistribution of energy between stress tolerance and normal growth and development, thereby improving the survival rate of plants under salt stress. This was consistent with transgenic rice overexpressing *ONAC022* [18].

Phytohormones, such as auxin, abscisic acid (ABA), ethylene, gibberellic acid (GA), and jasmonic acid (JA), also play central roles in salt stress response [46]. *Malus domestica MdNAC047* induced ethylene accumulation by increasing the expression of ethylene synthesis genes *MdACS1* and *MdACO1* and TF gene *MdERF3*, enhancing tolerance to salt stress by regulating ethylene response [47]. Soybean (*Glycine Max*) GmNAC109 promoted the formation of lateral roots of transgenic *Arabidopsis thaliana* and enhanced salt stress tolerance through positive regulation of auxin response gene *AtAIR3* and negative regulation of TF *AtARF2* [48]. To reveal the involvement of these hormones in the salt stress response and growth, numerous genes related to these hormones signaling pathways were identified (Figure 9). For example, in the ABA-signaling pathway, the core of salt- and drought-stress responses in plants [49], 1 *SnRK2* gene and 2 *AREB/ABF* genes were up-regulated and 1 *PYR* gene, 3 *PP2C* genes, 1 *SnRK2* gene and 1 *AREB/ABF* gene were down-regulated compared with WT-N. The expressions of JA- and SA-signaling pathways related to DEGs were increased in NaCl-treated WT and OX3 samples. These data indicated that AeNAC83 may enhance tolerance to salt stress by regulating various hormone signaling responses.

Interestingly, besides the increase in salt sensitivity in *AeNAC83*-silenced okra seedlings and salt tolerance in *AeNAC83*-overexpression transgenic *Arabidopsis*, the *AeNAC83*-silenced seedlings showed enhanced plant growth, while *AeNAC83*-overexpression lines exhibited the opposite phenotype. These data imply that AeNAC83 participates in the balance between defense responses and plant growth. In plants, the trade-off between defense and growth has attracted considerable attention, where enhanced resistance often impairs growth and development. For example, in rice, over-expression of *OsRCI-1* resulted in an increase in BPH resistance and a reduction in thousand-grain weight [50]. OsNAC2 negatively regulates root growth [51] and mediates abiotic stress tolerance [52].

4. Materials and Methods

4.1. Multiple Sequence Alignment and Phylogenetic Analysis

The amino acid sequences of NAC proteins were obtained from the TAIR (https://www.arabidopsis.org/) and NCBI. The accession number and references for other NACs: *Arabidopsis thaliana*, ATAF1 (AT1G01720), ATAF2 (AT5G08790) [4], CUC1 (AT3G15170), CUC2 (AT5G53950) [5], CUC3 (AT1G76420) [53], AtNAC1 (AF198054.1) [14], AtNAC2 (AT5G39610) [15], AtNAP (AJ222713) [54]; rice, OsNAC2 (Os04g0460600) [51,55]; soybean, GmNAC20 (EU440353.1) [45]; petunia, PhNAM (X92204) [3]. Multiple sequence alignment of AeNAC83 with other 11 NAC members was carried out using ClustalX 2.0 and modified with GeneDoc. The phylogenetic tree was generated using the MEGA7 software with the neighbor-joining (NJ) method, Poisson-corrected distances and 1000 replicates.

4.2. RNA Isolation and qRT-PCR

Mature seeds of okra (*Abelmoschus esculentus* L.) Cultivar "xian zhi" was used in this study. Plant growth conditions and treatment were performed according to our previous research [25]. Okra seedlings with only two true leaves under normal conditions were selected for salt treatment, irrigated with 300 mM NaCl solution, and the control was irrigated with water. At 1 d, 3 d, 5 d and 7 d after treatment, RNA was extracted from the second true leaf, and the expression of the *AeNAC83* gene was analyzed by qRT-PCR. The primers were listed in Table S3.

4.3. Subcellular Localization of GFP-AeNAC83 Fusion Protein

The coding sequence of AeNAC83 was amplified by RT-PCR with primers 5'-GAGCTC ATGGAGAAGCTTAGTTTTGT-3' and 5'-TCTAGAAGGTTTTCTTCTGAAGTAAG-3', and cloned into the SacI/XbaI sites of pCAMBIA1300-sGFP under the control of the CaMV 35S promoter, resulting in 35Sp:: AeNAC83:GFP construct. The constructed 35Sp:: AeNAC83:GFP vector was introduced into *Agrobacterium* strain GV3101, and transiently transformed in tobacco epidermal cells. The empty vector was used as a negative control. The distribution of fluorescence was imaged 48 h after the inoculation by a confocal laser-scanning microscope (Zeiss LSM 710).

4.4. Virus-Induced Gene Silencing (VIGS) of AeNAC83 and NaCl Treatment

A specific sequence of about 300bp from AeNAC83 was amplified by RT-PCR with primers 5'-GAATTCATGGAGAAGCTTAGTTTT-3' and 5'-GGTACCTTTCCTGTCGATTCC

GGT-3′, and cloned into the EcoRI/KpnI sites of pTRV2, resulting in TRV-NAC83 construct. The construct was transferred into *Agrobacterium tumefaciens* GV3101.

A single colony of the *Agrobacterium tumefaciens* containing TRV1, TRV-*NAC83* or empty TRV2 was inoculated into 3ml YEP liquid medium with 50 mg/L rifampicin and 50 mg/L kanamycin, and cultured at 28 °C for 12 h at 200 rpm. For secondary activation, 1 mL culture was added to 50 mL of YEP and grown until OD600 reached 1–1.5. Agrobacteria were resuspended in a monomethylamine (MMA) solution (20 μM acetylsyringone, 10 mM $MgCl_2$ and 10 mM MES, PH 5.6) to a final concentration of OD600 = 1 and placed without shaking at 28 °C for 3 h in dark. Then the cotyledons of okra seedlings that cotyledons were fully stretched and the true leaves had not yet grown were used to inoculate with the suspension of TRV1 and TRV-*NAC83* mixed at a ratio of 1:1 (v/v). pTRV2 and pTRV1 empty vectors were used as negative controls. The inoculated seedlings were maintained for 12 h in the dark and employed to a 16-h light/8 h dark cycle at 25 °C in a growth chamber. After 25 days, the *AeNAC83* mRNA level was measured and the phenotype of leaves was observed. Then the *AeNAC83*-silenced okra seedlings and the negative control were irrigated with 300 mM NaCl solution for 7 days. The leaf fresh weight and total chlorophyll content was measured [25].

4.5. Generation of AeNAC83 Transgenic Arabidopsis Plants

The *Agrobacterium* GV3101 containing the 35Sp:: *AeNAC83*:GFP construct was transformed into *Arabidopsis thaliana* ecotype Columbia (wild-type, WT) via the inflorescence infiltration method. Transformed lines were selected on 1/2 MS medium with 30 mg/L hygromycin and then confirmed by PCR. Two *AeNAC83* homozygous transgenic lines (OX-3 and OX-7) of the T4 generations with high *AeNAC83* expression were selected for further phenotypic analysis. The primers were listed in Table S3.

4.6. Performance of Transgenic Lines under Salt Stress Treatment

Seeds of WT and OX-3 and OX-7 were kept at 4 °C for 3 days and then plated on 1/2 MS medium under a 16-h light/8-h dark cycle at 23 °C in a growth chamber. Four-day-old seedlings were transplanted on 1/2 MS medium containing 0, 120 or 150 mM NaCl. After 10 days, the total fresh weight and primary root length were measured. Samples of whole plants of OX3 and WT treated with 0 or 120 mM NaCl were harvested and frozen in liquid nitrogen for Illumina sequencing assay. Seedlings were sown on 1/2 MS medium without NaCl treatment for two weeks, and the roots were imaged using a stereo microscope.

4.7. RNA Isolation and Illumina Sequencing

Total RNA was extracted from the leaves and roots of 14-day *AeNAC83*-overexpression transgenic (OX3) and the wild (WT) *Arabidopsis* treated with 120 mM NaCl. RNase-free DNase I (Takara, Dalian, China) was used to remove the DNA. cDNA libraries were constructed as described by our previous study [25]. The constructed libraries were sequenced on the Illumina platform. The resulting reads (clean reads) were mapped to the Arabidopsis reference genome using HISAT2. FPKM was used to estimate gene expression levels. Differential expression analysis of two groups was performed using the DESeq2 with an adjusted p-value < 0.05.

4.8. Gene Annotation and Enrichment Analysis

Based on a Wallenius non-central hyper-geometric distribution [56], GOseq R packages were used to analyze GO enrichment of DEGs. KOBAS software [57] was used to test whether DEGs were statistically enriched in KEGG pathways.

4.9. Measurement of Anthocyanin Content

Anthocyanins were extracted from seedlings of WT and OX3 grown in 1/2 MS medium supplemented with 0 or 120 mM NaCl. Leaf material from each treatment was weighed W (g), and then 1 mL of acidic methanol containing 1% HCl (v/v) was added and kept in the dark at 4 °C for 24 h. The leaching solution was centrifuged at 13,000 rpm for 10 min, and the absorbance was measured at the wavelength of 530 and 657 nm, respectively. The relative content of anthocyanins was calculated by the formula: $Q_{anthocyanins} = (A_{530} - 0.25 \times A_{657}) \cdot g^{-1}$ FW [58].

5. Conclusions

In this study, we demonstrate that the nucleus-located AeNAC83 participates in salt stress tolerance in okra. Additionally, AeNAC83 negatively regulates plant growth, indicating a possible node of trade-off between okra resistance and growth. Transcriptome analysis revealed that most of the genes involved in flavonoid biosynthesis and photosynthesis were up-regulated and down-regulated, respectively. Various plant hormone signaling pathway-related DEGs were also identified. Our study provides a comprehensive understanding of the molecular mechanism of AeNAC83 involved in plant growth and salt tolerance.

Supplementary Materials: The following are available online at https://www.mdpi.com/article/10.3390/ijms231710182/s1.

Author Contributions: Y.Z. conceived and designed the research, and drafted the manuscript; X.Z. and T.W. performed the experiments; S.G. and J.H. analyzed the data. All authors have read and agreed to the published version of the manuscript.

Funding: This research was funded by the Natural Science Foundation of Zhejiang Province (LQ22C130003) and Scientific Research Fund of Zhejiang A&F University (2020FR057).

Institutional Review Board Statement: Not applicable.

Informed Consent Statement: Not applicable.

Data Availability Statement: The transcriptional data was deposited in the NCBI Sequence Read Archive (BioProject: PRJNA857503). All data generated or analysed during this study are included in this published article and its Supplementary Materials files.

Conflicts of Interest: The authors declare no conflict of interest.

References

1. Hasegawa, P.M.; Bressan, R.A.; Zhu, J.K.; Bohnert, H.J. Plant Cellular And Molecular Responses To High Salinity. *Annu. Rev. Plant Physiol. Plant Mol. Biol.* **2000**, *51*, 463–499. [CrossRef] [PubMed]
2. Liang, W.; Ma, X.; Wan, P.; Liu, L. Plant salt-tolerance mechanism: A review. *Biochem. Biophys. Res. Commun.* **2018**, *495*, 286–291. [CrossRef] [PubMed]
3. Souer, E.; van Houwelingen, A.; Kloos, D.; Mol, J.; Koes, R. The no apical meristem gene of *Petunia* is required for pattern formation in embryos and flowers and is expressed at meristem and primordia boundaries. *Cell* **1996**, *85*, 159–170. [CrossRef]
4. Aida, M.; Ishida, T.; Fukaki, H.; Fujisawa, H.; Tasaka, M. Genes involved in organ separation in *Arabidopsis*: An analysis of the cup-shaped cotyledon mutant. *Plant Cell* **1997**, *9*, 841–857. [CrossRef] [PubMed]
5. Takada, S.; Hibara, K.; Ishida, T.; Tasaka, M. The CUP-SHAPED COTYLEDON1 gene of *Arabidopsis* regulates shoot apical meristem formation. *Development* **2001**, *128*, 1127–1135. [CrossRef] [PubMed]
6. Ooka, H.; Satoh, K.; Doi, K.; Nagata, T.; Otomo, Y.; Murakami, K.; Matsubara, K.; Osato, N.; Kawai, J.; Carninci, P.; et al. Comprehensive analysis of NAC family genes in *Oryza sativa* and *Arabidopsis thaliana*. *DNA Res.* **2003**, *10*, 239–247. [CrossRef] [PubMed]
7. Nuruzzaman, M.; Manimekalai, R.; Sharoni, A.M.; Satoh, K.; Kondoh, H.; Ooka, H.; Kikuchi, S. Genome-wide analysis of NAC transcription factor family in rice. *Gene* **2010**, *465*, 30–44. [CrossRef]
8. Le, D.T.; Nishiyama, R.; Watanabe, Y.; Mochida, K.; Yamaguchi-Shinozaki, K.; Shinozaki, K.; Tran, L.S. Genome-wide survey and expression analysis of the plant-specific NAC transcription factor family in soybean during development and dehydration stress. *DNA Res.* **2011**, *18*, 263–276. [CrossRef]
9. Jin, J.F.; Wang, Z.Q.; He, Q.Y.; Wang, J.Y.; Li, P.F.; Xu, J.M.; Zheng, S.J.; Fan, W.; Yang, J.L. Genome-wide identification and expression analysis of the NAC transcription factor family in tomato (*Solanum lycopersicum*) during aluminum stress. *BMC Genom.* **2020**, *21*, 288. [CrossRef]
10. Zhong, R.; Demura, T.; Ye, Z.H. SND1, a NAC domain transcription factor, is a key regulator of secondary wall synthesis in fibers of *Arabidopsis*. *Plant Cell* **2006**, *18*, 3158–3170. [CrossRef]

1. Zhong, R.; Richardson, E.A.; Ye, Z.H. Two NAC domain transcription factors, SND1 and NST1, function redundantly in regulation of secondary wall synthesis in fibers of *Arabidopsis*. *Planta* **2007**, *225*, 1603–1611. [CrossRef]
2. Mitsuda, N.; Ohme-Takagi, M. NAC transcription factors NST1 and NST3 regulate pod shattering in a partially redundant manner by promoting secondary wall formation after the establishment of tissue identity. *Plant J.* **2008**, *56*, 768–778. [CrossRef]
3. Guo, Y.; Gan, S. AtNAP, a NAC family transcription factor, has an important role in leaf senescence. *Plant J.* **2006**, *46*, 601–612. [CrossRef]
4. Xie, Q.; Frugis, G.; Colgan, D.; Chua, N.H. *Arabidopsis* NAC1 transduces auxin signal downstream of TIR1 to promote lateral root development. *Genes Dev.* **2000**, *14*, 3024–3036. [CrossRef]
5. He, X.J.; Mu, R.L.; Cao, W.H.; Zhang, Z.G.; Zhang, J.S.; Chen, S.Y. AtNAC2, a transcription factor downstream of ethylene and auxin signaling pathways, is involved in salt stress response and lateral root development. *Plant J.* **2005**, *44*, 903–916. [CrossRef]
6. Nakashima, K.; Takasaki, H.; Mizoi, J.; Shinozaki, K.; Yamaguchi-Shinozaki, K. NAC transcription factors in plant abiotic stress responses. *Biochim. Biophys. Acta* **2012**, *1819*, 97–103. [CrossRef]
7. Shao, H.; Wang, H.; Tang, X. NAC transcription factors in plant multiple abiotic stress responses: Progress and prospects. *Front. Plant Sci.* **2015**, *6*, 902. [CrossRef]
8. Hong, Y.; Zhang, H.; Huang, L.; Li, D.; Song, F. Overexpression of a Stress-Responsive NAC Transcription Factor Gene ONAC022 Improves Drought and Salt Tolerance in Rice. *Front. Plant Sci.* **2016**, *7*, 4. [CrossRef]
9. Chen, X.; Wang, Y.; Lv, B.; Li, J.; Luo, L.; Lu, S.; Zhang, X.; Ma, H.; Ming, F. The NAC family transcription factor OsNAP confers abiotic stress response through the ABA pathway. *Plant Cell Physiol.* **2014**, *55*, 604–619. [CrossRef]
10. Song, S.Y.; Chen, Y.; Chen, J.; Dai, X.Y.; Zhang, W.H. Physiological mechanisms underlying OsNAC5-dependent tolerance of rice plants to abiotic stress. *Planta* **2011**, *234*, 331–345. [CrossRef]
11. Xu, Z.; Gongbuzhaxi; Wang, C.; Xue, F.; Zhang, H.; Ji, W. Wheat NAC transcription factor TaNAC29 is involved in response to salt stress. *Plant Physiol. Biochem.* **2015**, *96*, 356–363. [CrossRef]
12. Zhang, X.; Cheng, Z.; Zhao, K.; Yao, W.; Sun, X.; Jiang, T.; Zhou, B. Functional characterization of poplar NAC13 gene in salt tolerance. *Plant Sci.* **2019**, *281*, 1–8. [CrossRef] [PubMed]
13. Chandra, S.; Saha, R.; Pal, P. Arsenic Uptake and Accumulation in Okra (*Abelmoschus esculentus*) as Affected by Different Arsenical Speciation. *Bull. Environ. Contam. Toxicol.* **2016**, *96*, 395–400. [CrossRef]
14. Islam, M.T. Phytochemical information and pharmacological activities of Okra (*Abelmoschus esculentus*): A literature-based review. *Phytother. Res.* **2019**, *33*, 72–80. [CrossRef]
15. Zhan, Y.; Wu, T.; Zhao, X.; Wang, Z.; Chen, Y. Comparative physiological and full-length transcriptome analyses reveal the molecular mechanism of melatonin-mediated salt tolerance in okra (*Abelmoschus esculentus* L.). *BMC Plant Biol.* **2021**, *21*, 180. [CrossRef]
16. Tohge, T.; Fernie, A.R. An Overview of Compounds Derived from the Shikimate and Phenylpropanoid Pathways and Their Medicinal Importance. *Mini Rev. Med. Chem.* **2017**, *17*, 1013–1027. [CrossRef]
17. Eberhard, S.; Finazzi, G.; Wollman, F.A. The dynamics of photosynthesis. *Annu. Rev. Genet.* **2008**, *42*, 463–515. [CrossRef]
18. Zhang, C.; Dong, W.; Gen, W.; Xu, B.; Shen, C.; Yu, C. De Novo Transcriptome Assembly and Characterization of the Synthesis Genes of Bioactive Constituents in *Abelmoschus esculentus* (L.) Moench. *Genes* **2018**, *9*, 130. [CrossRef]
19. Adelakun, O.E.; Oyelade, O.J.; Ade-Omowaye, B.I.; Adeyemi, I.A.; Van de Venter, M. Chemical composition and the antioxidative properties of Nigerian Okra Seed (*Abelmoschus esculentus* Moench) Flour. *Food Chem. Toxicol.* **2009**, *47*, 1123–1126. [CrossRef]
20. Purkayastha, A.; Dasgupta, I. Virus-induced gene silencing: A versatile tool for discovery of gene functions in plants. *Plant Physiol. Biochem.* **2009**, *47*, 967–976. [CrossRef]
21. Godge, M.R.; Purkayastha, A.; Dasgupta, I.; Kumar, P.P. Virus-induced gene silencing for functional analysis of selected genes. *Plant Cell Rep.* **2008**, *27*, 209–219. [CrossRef] [PubMed]
22. De Zelicourt, A.; Colcombet, J.; Hirt, H. The Role of MAPK Modules and ABA during Abiotic Stress Signaling. *Trends Plant Sci.* **2016**, *21*, 677–685. [CrossRef] [PubMed]
23. Nair, P.M.; Chung, I.M. A mechanistic study on the toxic effect of copper oxide nanoparticles in soybean (*Glycine max* L.) root development and lignification of root cells. *Biol. Trace Elem. Res.* **2014**, *162*, 342–352. [CrossRef]
24. Lai, J.P.; Lim, Y.H.; Su, J.; Shen, H.M.; Ong, C.N. Identification and characterization of major flavonoids and caffeoylquinic acids in three Compositae plants by LC/DAD-APCI/MS. *J. Chromatogr. B* **2007**, *848*, 215–225. [CrossRef]
25. Xie, Y.Y.; Yuan, D.; Yang, J.Y.; Wang, L.H.; Wu, C.F. Cytotoxic activity of flavonoids from the flowers of *Chrysanthemum morifolium* on human colon cancer Colon205 cells. *J. Asian Nat. Prod. Res.* **2009**, *11*, 771–778. [CrossRef]
26. Yang, Y.; Guo, J.; Cheng, J.; Jiang, Z.; Xu, M. Identification of UV-B radiation responsive microRNAs and their target genes in chrysanthemum (*Chrysanthemum morifolium* Ramat) using high-throughput sequencing. *Ind. Crop. Prod.* **2020**, *151*, 112484. [CrossRef]
27. Sobral-Souza, C.E.; Silva, A.R.P.; Leite, N.F.; Costa, J.G.M.; Menezes, I.R.A.; Cunha, F.A.B.; Rolim, L.A.; Coutinho, H.D.M. LC-MS analysis and cytoprotective effect against the mercurium and aluminium toxicity by bioactive products of *Psidium brownianum* Mart. ex DC. *J. Hazard. Mater.* **2019**, *370*, 54–62. [CrossRef]
28. Yu, C.; Zeng, H.; Wang, Q.; Chen, W.; Chen, W.; Yu, W.; Lou, H.; Wu, J. Multi-omics analysis reveals the molecular responses of *Torreya grandis* shoots to nanoplastic pollutant. *J. Hazard. Mater.* **2022**, *436*, 129181. [CrossRef]

39. Ambawat, S.; Sharma, P.; Yadav, N.R.; Yadav, R.C. MYB transcription factor genes as regulators for plant responses: An overview *Physiol. Mol. Biol. Plants* **2013**, *19*, 307–321. [CrossRef]
40. Shen, H.; Poovaiah, C.R.; Ziebell, A.; Tschaplinski, T.J.; Pattathil, S.; Gjersing, E.; Engle, N.L.; Katahira, R.; Pu, Y.; Sykes, R.; et al Enhanced characteristics of genetically modified switchgrass (*Panicum virgatum* L.) for high biofuel production. *Biotechnol. Biofuels.* **2013**, *6*, 71. [CrossRef]
41. Liu, C.; Wang, B.; Li, Z.; Peng, Z.; Zhang, J. TsNAC1 Is a Key Transcription Factor in Abiotic Stress Resistance and Growth. *Plant Physiol.* **2018**, *176*, 742–756. [CrossRef]
42. Tran, L.S.; Nakashima, K.; Sakuma, Y.; Simpson, S.D.; Fujita, Y.; Maruyama, K.; Fujita, M.; Seki, M.; Shinozaki, K.; Yamaguchi-Shinozaki, K. Isolation and functional analysis of *Arabidopsis* stress-inducible NAC transcription factors that bind to a drought-responsive cis-element in the early responsive to dehydration stress 1 promoter. *Plant Cell* **2004**, *16*, 2481–2498. [CrossRef]
43. Jeong, J.S.; Kim, Y.S.; Baek, K.H.; Jung, H.; Ha, S.H.; Do Choi, Y.; Kim, M.; Reuzeau, C.; Kim, J.K. Root-specific expression of OsNAC10 improves drought tolerance and grain yield in rice under field drought conditions. *Plant Physiol.* **2010**, *153*, 185–197. [CrossRef]
44. Mao, X.; Chen, S.; Li, A.; Zhai, C.; Jing, R. Novel NAC transcription factor TaNAC67 confers enhanced multi-abiotic stress tolerances in *Arabidopsis*. *PLoS ONE* **2014**, *9*, e84359. [CrossRef]
45. Hao, Y.J.; Wei, W.; Song, Q.X.; Chen, H.W.; Zhang, Y.Q.; Wang, F.; Zou, H.F.; Lei, G.; Tian, A.G.; Zhang, W.K.; et al. Soybean NAC transcription factors promote abiotic stress tolerance and lateral root formation in transgenic plants. *Plant J.* **2011**, *68*, 302–313. [CrossRef]
46. Waadt, R.; Seller, C.A.; Hsu, P.K.; Takahashi, Y.; Munemasa, S.; Schroeder, J.I. Plant hormone regulation of abiotic stress responses. *Nat. Rev. Mol. Cell Biol.* **2022**, *2022*, 1–15. [CrossRef]
47. An, J.P.; Yao, J.F.; Xu, R.R.; You, C.X.; Wang, X.F.; Hao, Y.J. An apple NAC transcription factor enhances salt stress tolerance by modulating the ethylene response. *Physiol. Plant* **2018**, *164*, 279–289. [CrossRef] [PubMed]
48. Yang, X.; Kim, M.Y.; Ha, J.; Lee, S.H. Overexpression of the Soybean NAC Gene GmNAC109 Increases Lateral Root Formation and Abiotic Stress Tolerance in Transgenic *Arabidopsis* Plants. *Front. Plant Sci.* **2019**, *10*, 1036. [CrossRef]
49. Zhu, J.K. Salt and drought stress signal transduction in plants. *Annu. Rev. Plant Biol.* **2002**, *53*, 247–273. [CrossRef]
50. Liao, Z.; Wang, L.; Li, C.; Cao, M.; Wang, J.; Yao, Z.; Zhou, S.; Zhou, G.; Zhang, D.; Lou, Y. The lipoxygenase gene OsRCI-1 is involved in the biosynthesis of herbivore-induced JAs and regulates plant defense and growth in rice. *Plant Cell Environ.* **2022**, *45*, 2827–2840. [CrossRef]
51. Mao, C.; He, J.; Liu, L.; Deng, Q.; Yao, X.; Liu, C.; Qiao, Y.; Li, P.; Ming, F. OsNAC2 integrates auxin and cytokinin pathways to modulate rice root development. *Plant Biotechnol. J.* **2020**, *18*, 429–442. [CrossRef] [PubMed]
52. Shen, J.; Lv, B.; Luo, L.; He, J.; Mao, C.; Xi, D.; Ming, F. The NAC-type transcription factor OsNAC2 regulates ABA-dependent genes and abiotic stress tolerance in rice. *Sci. Rep.* **2017**, *7*, 40641. [CrossRef] [PubMed]
53. Vroemen, C.W.; Mordhorst, A.P.; Albrecht, C.; Kwaaitaal, M.A.; de Vries, S.C. The CUP-SHAPED COTYLEDON3 gene is required for boundary and shoot meristem formation in *Arabidopsis*. *Plant Cell* **2003**, *15*, 1563–1577. [CrossRef]
54. Sablowski, R.W.; Meyerowitz, E.M. A homolog of NO APICAL MERISTEM is an immediate target of the floral homeotic genes APETALA3/PISTILLATA. *Cell* **1998**, *92*, 93–103. [CrossRef]
55. Mao, C.; Ding, W.; Wu, Y.; Yu, J.; He, X.; Shou, H.; Wu, P. Overexpression of a NAC-domain protein promotes shoot branching in rice. *New Phytol.* **2007**, *176*, 288–298. [CrossRef]
56. Young, M.D.; Wakefield, M.J.; Smyth, G.K.; Oshlack, A. Gene ontology analysis for RNA-seq: Accounting for selection bias. *Genome. Biol.* **2010**, *11*, R14. [CrossRef]
57. Mao, X.; Cai, T.; Olyarchuk, J.G.; Wei, L. Automated genome annotation and pathway identification using the KEGG Orthology (KO) as a controlled vocabulary. *Bioinformatics* **2005**, *21*, 3787–3793. [CrossRef]
58. Xie, Y.; Tan, H.; Ma, Z.; Huang, J. DELLA proteins promote anthocyanin biosynthesis via sequestering MYBL2 and JAZ suppressors of the MYB/bHLH/WD40 complex in *Arabidopsis thaliana*. *Mol. Plant* **2016**, *9*, 711–721. [CrossRef]

Article

LEAF TIP RUMPLED 1 Regulates Leaf Morphology and Salt Tolerance in Rice

Jiajia Wang [1,2,†], Yiting Liu [1,3,†], Songping Hu [3], Jing Xu [1], Jinqiang Nian [1], Xiaoping Cao [1], Minmin Chen [1], Jiangsu Cen [1], Xiong Liu [1], Zhihai Zhang [1], Dan Liu [3], Li Zhu [1], Jiang Hu [1], Deyong Ren [1], Zhenyu Gao [1], Lan Shen [1], Guojun Dong [1], Qiang Zhang [1], Qing Li [1], Sibin Yu [2], Qian Qian [1,*] and Guangheng Zhang [1,4,*]

1. State Key Laboratory of Rice Biology, China National Rice Research Institute, Hangzhou 310006, China
2. National Key Laboratory of Crop Genetic Improvement and National Center of Plant Gene Research, College of Plant Science and Technology, Huazhong Agricultural University, Wuhan 430070, China
3. Research Center of Plant Functional Genes and Tissue Culture Technology, College of Bioscience and Bioengineering, Jiangxi Agricultural University, Nanchang 330045, China
4. National Nanfan Research Institute (Sanya), Chinese Academy of Agricultural Sciences, Sanya 572024, China
* Correspondence: qianqian@caas.cn (Q.Q.); zhangguangheng@caas.cn (G.Z.); Tel.: +86-571-6337-1418 (Q.Q.); +86-571-6337-0211 (G.Z.)
† These authors contributed equally to this work.

Abstract: Leaf morphology is one of the important traits related to ideal plant architecture and is an important factor determining rice stress resistance, which directly affects yield. Wax layers form a barrier to protect plants from different environmental stresses. However, the regulatory effect of wax synthesis genes on leaf morphology and salt tolerance is not well-understood. In this study, we identified a rice mutant, *leaf tip rumpled 1* (*ltr1*), in a mutant library of the classic *japonica* variety Nipponbare. Phenotypic investigation of NPB and *ltr1* suggested that *ltr1* showed rumpled leaf with uneven distribution of bulliform cells and sclerenchyma cells, and disordered vascular bundles. A decrease in seed-setting rate in *ltr1* led to decreased per-plant grain yield. Moreover, *ltr1* was sensitive to salt stress, and *LTR1* was strongly induced by salt stress. Map-based cloning of *LTR1* showed that there was a 2-bp deletion in the eighth exon of *LOC_Os02g40784* in *ltr1*, resulting in a frameshift mutation and early termination of transcription. Subsequently, the candidate gene was confirmed using complementation, overexpression, and knockout analysis of *LOC_Os02g40784*. Functional analysis of *LTR1* showed that it was a wax synthesis gene and constitutively expressed in entire tissues with higher relative expression level in leaves and panicles. Moreover, overexpression of *LTR1* enhanced yield in rice and *LTR1* positively regulates salt stress by affecting water and ion homeostasis. These results lay a theoretical foundation for exploring the molecular mechanism of leaf morphogenesis and stress response, providing a new potential strategy for stress-tolerance breeding.

Keywords: *Oryza sativa* L.; leaf shape; salt stress; bulliform cells; aquaporin

1. Introduction

Leaves are the main photosynthetic organ of plants. Leaf morphology affects the effective photosynthetic area, which affects accumulation of photosynthetic products and subsequent crop yield. In rice, numerous genes associated with leaf morphogenesis have been mined and cloned, such as *SHALLOT-LIKE 1* (*SLL1*) [1], *HOMEODOMAIN CONTAINING PROTEIN4* (*OsHB4*) [2], *SEMI-ROLLED LEAF1*(*SRL1*) [3], *Rice outermost cell-specific gene 5* (*Roc5*) [4], *AGO1 homologs 1b* (*OsAGO1b*) [5], *Rice outermost cell-specific 8* (*Roc8*) [6], *PHOTO-SENSITIVE LEAF ROLLING 1* (*PSL1*) [7]. These genes regulate leaf morphogenesis through complex interactions among plant hormone signaling pathways, transcription factors, and microRNAs [8,9]. In addition, leaf morphology is also affected by genes associated with ribosomes synthesis, DNA repair, cell cycle process, cuticle development, ion homeostasis, and microtubule arrangement [9]. However, these genes alone are not sufficient to accurately outline the genetic regulatory network of rice leaf morphogenesis in detail. One of the main challenges in modern agriculture is to increase crop yields under different environmental conditions by cultivating ideal plant architecture [10]. Leaf morphology is an important component of plant architecture and improving it contributes to collaborative improvement of stress resistance and yield. In recent years, great progress has been made in the regulation mechanism of leaf morphology and stress resistance. In addition to the key regulatory roles in plant architecture and yield, many genes regulating leaf morphology also affect characteristics such as drought tolerance, nutrient utilization, and disease resistance. For example, *Ideal Plant Architecture 1* (*IPA1*) not only increases rice yield but also improves rice blast resistance, which counters the traditional view that a single gene cannot simultaneously increase yield and disease resistance [11–14]. *Dwarf 1* (*D1*) is involved in complex network affecting plant height, leaf size, and abiotic stress response [15–17]. *PSL1* regulates rice leaf cell wall development and drought tolerance [7]. Higher leaf temperature, respiration rate, lower transpiration rate, and stomatal conductance in *high temperature susceptibility* (*hts*) resulted in high temperature sensitivity of *hts* [18]. Thus, leaf morphology is closely associated with stress resistance, nutrient utilization, disease resistance, and yield.

Soil salinization is an increasingly serious agricultural problem worldwide [19,20], limiting plant growth and crop productivity in saline–alkali areas [20]. Poor irrigation practices, the improper application of fertilizers, and industrial pollution increased soil salinity in cultivated soil, resulting in aggravated soil salinization [21,22]. Most plants had to develop suitable mechanisms to adjust their physiological and biochemical processes to adapt to high salinity environments during their long evolutionary history due to their sessile nature [20]. Significant progresses have been made for salt tolerance mechanism in plants. They developed suitable strategies to regulate ion and osmotic homeostasis and minimize stress damage [23,24], including exclusion of Na^+ from leaf tissues, compartmentalization of Na^+ (mainly into vacuoles), and reducing water loss while maximizing water absorption [21,25,26]. However, few favorable genetic loci associated with salt resistance have been identified in the breeding practices of rice. Therefore, breeding potentially yield-penalty-free rice varieties with high salt tolerance is of great significance and an effective way to expand the adaptability and planting area of rice and improve the yield potential of rice in saline–alkali areas.

Wax is the outermost barrier that plays an important role in plant–environment interactions, including plant adaptation to drought environments and various abiotic and biotic stresses. It promotes resistance to ultraviolet (UV) radiation [27] and pests and diseases [28] and protects internal plant tissues from temperature stress [29]. Moreover, the epicuticle wax layer provides the necessary barrier for reducing non-stomatal water loss during drought stress; thereby significantly improve drought tolerance in rice [30,31]. For example, the wax synthesis regulator DROUGHT HYPERSENSITIVE (DHS) interacts with rice outermost cell-specific gene 4 (Roc4), regulating expression of *BODYGUARD* (*BDG*) and thus affecting rice drought tolerance [32,33]. The rice ethylene response factor *WAX SYNTHESIS REGULATORY GENE 1* (*OsWR1*) positively regulates rice wax synthesis and affects drought tolerance by regulating cuticle development and leaf water retention [34]. In addition, wax has a critical effect on the differentiation of plant tissues and organs, such as the morphological development of leaves, fruits, and pollen, thereby affecting plant fertility. Loss function of wax synthesis genes led to morphological abnormalities of flowers and leaves, such as *knb1* (*knobhead*), *bcf1* (*bicentifolia*), and *wax1* in *Arabidopsis* [35]. The *wax2* plants showed disordered leaf structure and fused floral organs in *Arabidopsis* [36]. In rice, most research on wax synthesis genes has focused on pollen development, panicle fertility, and drought resistance; there are few reports on the regulatory role of wax synthesis genes in leaf morphology. Here, we identified a rice mutant *ltr1* with abnormal leaf morphology. This mutant was obtained by ethyl methanesulfonate (EMS) mutagenesis of Nipponbare and was used to isolate and analyze the function of the candidate gene *LTR1* in regulating leaf morphology. We demonstrated that loss function of *LTR1* led to abnormal development of bulliform cells, vascular bundles, and sclerenchyma cells, and to rumpled leaves, decreases in the seed setting rate and yield, and high sensitivity to salt stress. We also confirmed that *LTR1* mediated regulatory activities of aquaporin and ion transporters result in altered water retention and ion homeostasis under salt stress. Hence, function analysis of *LTR1* in leaf morphology and response to salt stress could provide theoretical foundation for molecular mechanism of leaf morphogenesis and salt response in rice and contribute to breeding efforts to develop salt-tolerant varieties with ideal leaf morphology.

2. Results

2.1. Identification of the ltr1 Mutant

The *ltr1* mutant was successfully obtained by EMS mutagenesis of the NPB. Phenotypic observation indicated that *ltr1* exhibited abnormal leaf morphology with uneven distribution of bulliform cell on adaxial surface and sclerenchyma cells on abaxial surface and disordered vascular bundles (Figure 1a–c). The contents of chlorophyll *a*, chlorophyll *b*, and carotenoids were significantly higher in *ltr1* than in NPB, with increases of 19.23%, 24.96%, and 17.83%, respectively (Figure 1d). The SPAD (soil and plant analyzer development) value of *ltr1* was significantly higher than that of NPB (Figure 1e). The quantum efficiency of photosystem II (Fv/Fm) and leaf water content of *ltr1* were significantly lower than those of NPB, decreased by 8.69% and 5.26%, respectively (Figure 1f,g). These results showed that growth and development of *ltr1* were seriously impaired. The abnormal leaf morphology of *ltr1* was associated with lower light energy conversion efficiency of the PS II (Photosystem II) reaction center and poor leaf water retention.

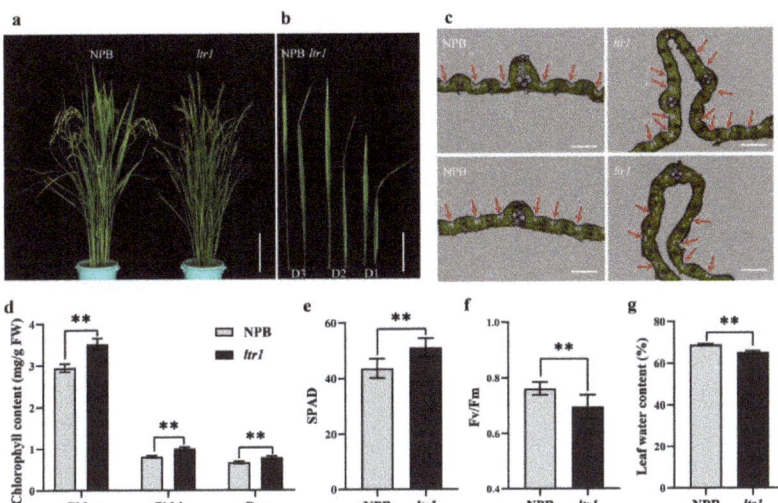

Figure 1. Phenotype analysis of NPB and *ltr1* plants. (**a**) Plant morphology (bar = 20.0 cm), (**b**) leaf morphology (bar = 4.0 cm), and (**c**) observation of frozen sections of NPB and *ltr1* (bar = 200 μm), red arrows in (**c**) represent bulliform cells. (**d**) Chlorophyll content, (**e**) SPAD, (**f**) Fv/Fm, and (**g**) leaf water content of NPB and *ltr1*. Data are given as means ± SD. Asterisks indicate significant difference based on the Student's *t*-test: ** in the figure represents significant difference at $p < 0.01$.

2.2. Effect of LTR1 on Photosynthetic Efficiency and Seed Setting Rate

According to our results, the panicle length (Figure 2a,b), seed setting rate (Figure 2f), secondary branch numbers (Figure 2e), and grain yield per plant (Figure 2h) were significantly lower in *ltr1* plants than in the wild type, by 14.50%, 95.54%, 16.67%, and 84.49%, respectively. The effective panicle number was significantly higher for *ltr1* than for NPB, with a 43.38% increase (Figure 2c). The primary branch numbers and 1000-grain weight showed no significant differences (Figure 2d,g). These results indicated that the decrease of yield per plant in *ltr1* was caused by the extremely low seed-setting rate and showed that *ltr1* had serious defects in leaf morphology and fertility.

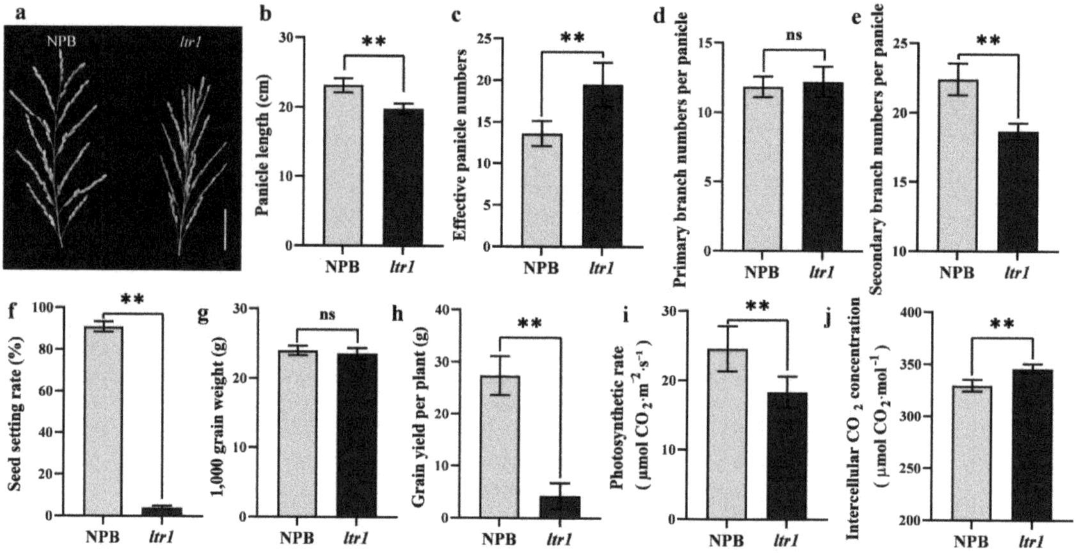

Figure 2. Comparisons of yield characters in NPB and *ltr1*. (**a**) Spike morphology, bar = 4 cm, (**b**) panicle length, (**c**) numbers of effective panicle, (**d**) number of primary branches, (**e**) number of secondary branches, (**f**) seed setting rate, (**g**) 1000-grain weight, (**h**) grain yield per plant, (**i**) photosynthetic efficiency, and (**j**) intercellular CO_2 concentration of NPB and *ltr1*. Data are given as means ± SD. Asterisks indicate significant difference based on the Student's *t*-test: ** in the figure represents significant difference at $p < 0.01$ and ns in the figure represents there is no significant different at $p < 0.05$.

Photosynthesis is the sum of a series of complex metabolic reactions [37]. Maintaining high chlorophyll content in leaves is not necessary to improve the effective photosynthetic rate. Light intensity under low-light conditions is a limiting factor for leaf photosynthesis, and high chlorophyll content is conducive to light absorption; the photosynthetic rate under saturated light intensity is mainly affected by the catalytic ability of the Rubisco enzyme, rather than the limitation of electron transfer rate in light reactions [38]. To explore whether the increased chlorophyll content and abnormal leaf morphology of *ltr1* affect photosynthetic efficiency, we measured the photosynthetic efficiency of NPB and *ltr1* in the field. Compared to NPB, the intercellular CO_2 concentration of *ltr1* was 4.91% higher, and the photosynthetic efficiency was 25.33% lower (Figure 2i,j). Although the photosynthetic pigment content of *ltr1* increased, the photosynthetic efficiency did not. The reasons for the decrease of photosynthetic efficiency in *ltr1* require further exploration.

2.3. Map-Based Cloning of LTR1

To explore the molecular mechanism of the phenotype in *ltr1*, an F_2 segregation population was developed by crossing *ltr1* and the *indica* cultivar TN1. The segregation of wild type and mutant phenotype displayed a ratio of 3:1 (Table S1), indicating that the mutant phenotype was controlled by a single recessive gene. Using 21 F_2 mutant individuals, the *LTR1* locus was first mapped to the region between RM6318 and RM1920 on the long arm of chromosome 2. The location was then narrowed down to a 13.5-kb genomic region between the markers N-12 and N-20 (Figure 3a). In this region, only one putative opening reading frame (ORF) was found based on data from the Rice Genome Annotation Project (http://rice.plantbiology.msu.edu accessed on 24 March 2021) database.

DNA sequence analysis of the ORF in *ltr1* and NPB revealed that a 2-bp deletion in exon 8 of *LOC_Os02g40784*, which resulted in a frameshift mutation and early termination of transcription (Figure 3b). *LOC_Os02g40784* includes ten exons and nine introns and encodes a polypeptide 619 amino acid in length. We therefore inferred that *LOC_Os02g40784* was the gene controlling the mutant phenotype of *ltr1*.

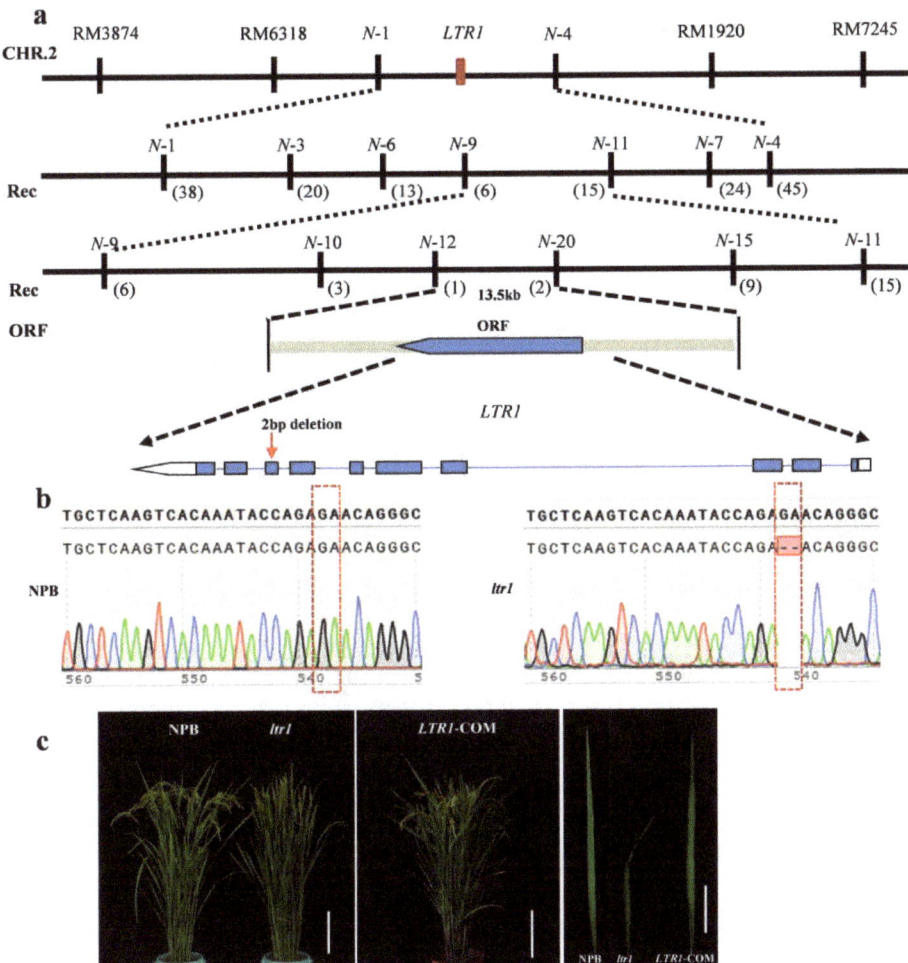

Figure 3. Map-based cloning of *LTR1*. (a) Fine mapping of *LTR1*; the red arrow represents the mutation site of *LTR1* in *ltr1*. (b) Sequence analysis of NPB and *ltr1*; the red box represents the mutation site in *ltr1*. (c) Complementary analysis of *LTR1* in *ltr1*; bar for plants and leaves was 20 cm and 5 cm, respectively.

To confirm that the phenotype of *ltr1* was attributable to the detected mutation in *LTR1*, we constructed a complementation vector with a NPB genomic fragment containing the entire coding region of *LTR1* and obtained complementary plants of *LOC_Os02g40784* under *ltr1* background. As expected, the complementary transgenic T_0 plants showed normal flat leaves: this indicated that the normal expression of *LOC_Os02g40784* in *ltr1* can complement the phenotype of the mutant (Figure 3c).

2.4. Overexpression and Targeted Deletion of LTR1

We next used CRISPR/Cas9 to generate mutant alleles of *LTR1* alleles in a NPB background. We obtained three independent transgenic lines that all carried homozygous mutants, including 3-bp, 4-bp, and 5-bp deletions in exon 3, respectively (Figure 4a,e). These lines had comparable phenotypes to those of *ltr1* with shrunken and distorted leaves, uneven distribution of bulliform cells on adaxial surface and sclerenchyma cells on abaxial surface, and disordered vascular bundles (Figure 4a–d). We also generated overexpression line of *LTR1* in the NPB background, which exhibited longer leaves and higher relative expression level (Figure 5a–c). These results showed that *LOC_Os02g40784* was *LTR1* and that the mutation in *LOC_Os02g40784* led to rumpled leaf phenotype in *ltr1*. Moreover, we found that compared with NPB, the grain yield per plant in overexpression of *LTR1* increased by 38.59% ($p < 0.05$), but the grain yields per plant in *ltr1* and *LTR1-KO* decreased by 82.13% and 76.31% ($p < 0.05$), respectively (Figure S6), suggesting that overexpression of *LTR1* enhanced yield in rice.

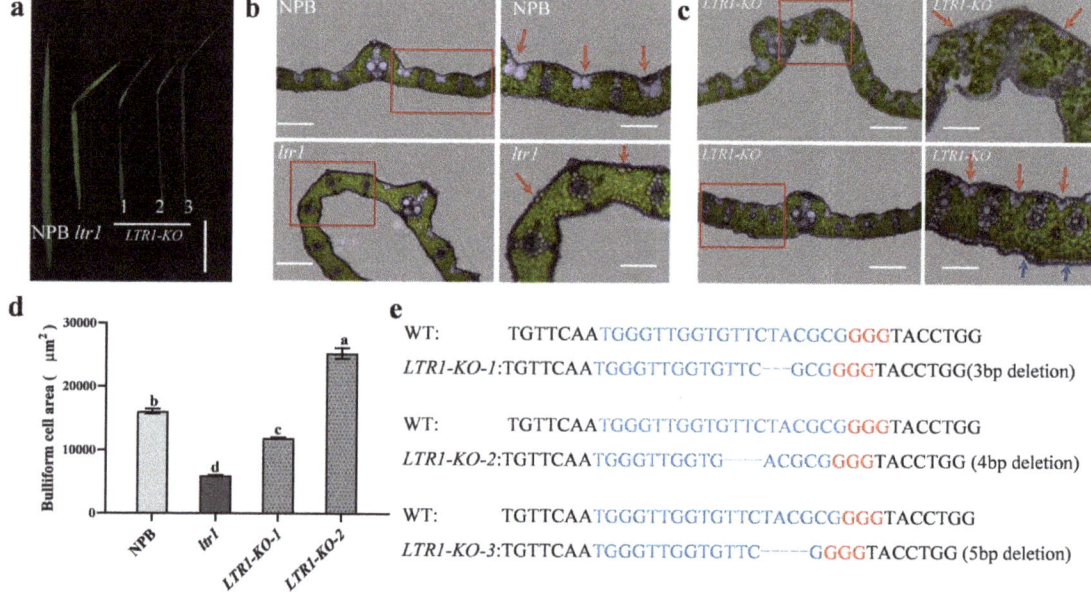

Figure 4. Phenotypic investigation of *LTR1* knockout lines. (**a**) Photos of leaves in NPB, *ltr1*, and *LTR1-KO* lines, bar = 8 cm. (**b**,**c**) Frozen section analysis of leaf in NPB, *ltr1*, and *LTR1-KO* lines; the red arrow represents bulliform cells, and the blue arrow represents the location of sclerenchyma cells. Right of (**b**) is the enlarged detail of red box in the left of (**b**), bar = 200 μm. Right of (**c**) is the enlarged detail of red box in the left of (**c**), bar = 100 μm. (**d**) The area of bulliform cells of *LTR1* knockout lines. (**e**) Sequence analysis of WT and *LTR1-KO*. Data are given as means ± SD. Significant differences were determined by Duncan's new multiple range test and indicated with different lowercase letters ($p < 0.05$).

To examine the expression pattern of *LTR1* in NPB, total RNA was extracted from roots, stem, leaf, sheath, and panicles. The qRT-PCR showed that *LTR1* was constitutively expressed in all of the tested tissues, with a dramatic increase in leaves and panicles (Figure 5d). The results were consistent with those of β-glucuronidase (GUS) staining (Figure 5e) and the decreased seed-setting rate of *ltr1* (Figure 2f), showing the important regulatory role of *LTR1* in leaf and panicle development.

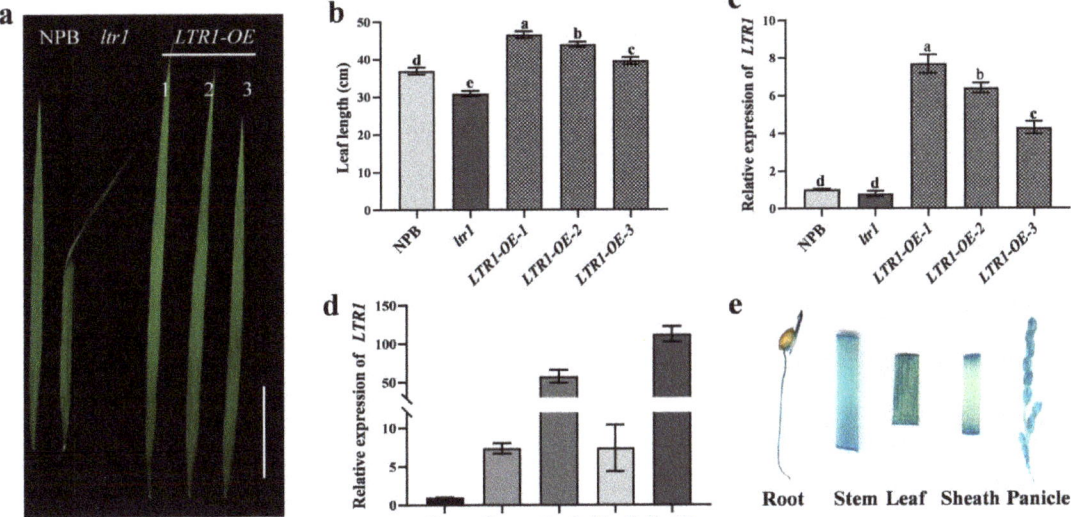

Figure 5. Overexpression and expression pattern analysis of *LTR1*. (**a**) Phenotypic investigation of overexpression lines of *LTR1*, bar = 6 cm. (**b**) Leaf length of overexpression lines. (**c**) The relative expression level of *LTR1* in overexpression lines. (**d**) The relative expression level of *LTR1* in different organs of NPB. (**e**) Promoter activities of *LTR1* in different organs of NPB as determined by promoter–GUS assays. Data are given as means ± SD. Significant differences were determined by Duncan's new multiple range test and indicated with different lowercase letters ($p < 0.05$).

2.5. Phylogenetic Analysis of LTR1

Protein domain predictions using NCBI CD Search (https://www.ncbi.nlm.nih.gov/Structure/cdd/wrpsb.cgi accessed on 10 August 2018) showed that LTR1 contained ERG3 (elicitor-responsive genes, ERG) and wax2_C domains. BLAST-P analysis of the NCBI database showed that LTR1 was highly conserved in higher plants including *Oryza brachyantha* (92.25%), *Brachypodium distachyon* (84.98%), *Aegilops tauschii* (84.87%), *Triticum aestivum* (83.84%), *Setaria italic* (82.90%), *Panicum hallii* (81.42%), *Sorghum bicolor* (82.23%), and *Zea mays* (78.33%) (Figure S1). To investigate the evolutionary relationships between LTR1 homologs, a phylogenic analysis was performed using the Text Neighbor-Joining Tree method [39]. The results showed that LTR1 is closely related to homologues in the grass family containing *Aegilops tauschii*, *Brachypodium distachyon*, and *Triticum aestivum* (Figure 6 and Figure S1). Overall, these analyses demonstrated that the LTR1 was highly conserved in plants.

2.6. LTR1 Participates in Water Transport and Ion Homeostasis

RNA-seq analysis showed that there were 6513 differentially expressed genes (DEGs) in NPB and *ltr1*, of which 3022 were up-regulated and 3480 were down-regulated (Figure S2a and Table S6). There were 118 DEGs related to leaf development, comprising 36 up-regulated and 82 down-regulated genes (Figure S3a and Table S7). A Kyoto Encyclopedia of Genes and Genomes (KEGG) pathway analysis showed that these DEGs were mainly enriched in plant hormone signal transduction pathways, which indicated that *LTR1* may regulate leaf development by participating in hormone signal transduction pathways (Figure S3b). For example, BR C-6 oxidase gene (*OsBR6ox*), *AUXIN RESPONSE FACTOR8* (*OsARF8*), *AUXIN RESPONSE FACTOR17* (*OsARF17*), *AUXIN RESPONSE FACTOR16* (*OsARF16*), *PHYTOSULFOKINE RECEPTOR 2* (*OsPSKR2*), and *PHYTOSULFOKINE RECEPTOR 3* (*OsPSKR3*) were up-regulated (Figure S3c) and *PENTATRICOPEPTIDE REPEAT PROTEIN* (*OsPPR6*), *RNA-dependent RNA polymerase 6* (*OsRDR6*), *RNA-directed RNA poly-*

merase 1 (*OsRDR1*), *INCREASED LEAF ANGLE1* (*ILA1*), *dwarf 11* (*d11*), and *GIBBERELLIN 20-OXIDASE GENE* (*OsGA20ox1*) were down-regulated in *ltr1* plants (Figure S3d).

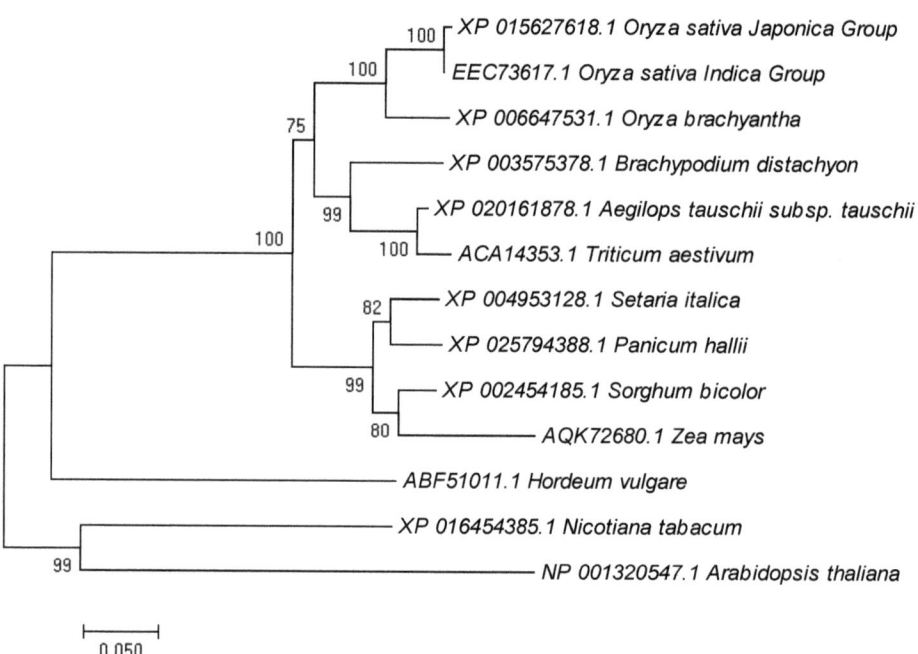

Figure 6. Phylogenic tree of LTR1 and its homologs. The tree was constructed using MEGA 7.0. Protein sequences are *Oryza sativa Japonica* Group (XP 015627618.1), *Oryza sativa Indica* Group (EEC 73617.1), *Oryza brachyantha* (XP 006647531.1), *Brachypodium distachyon* (XP 003575378.1), *Aegilops tauschii* (XP 020161878.1), *Triticum aestivum* (ACA 14353.1), *Setaria italic* (XP 004953128.1), *Panicum hallii* (XP 025794388.1), *Sorghum bicolor* (XP 002454185.1), *Zea mays* (AQK 72680.1), *Hordeum vulgare* (ABF 51011.1), *Nicotiana tabacum* (XP 016454385.1), *Arabidopsis thaliana* (NP 001320547.1). Scale represents percentage substitutions per site. Statistical support for the nodes is indicated.

A Gene Ontology (GO) term enrichment analysis was also conducted for DEGs between NPB and *ltr1*. The most highly enriched GO biological processes were in salt-stress response, stimulus response, and ABA response (Figure S2b). The most highly enriched GO molecular functions were ATP binding and protein binding, and the most enriched cell components were plasma membrane and nucleus (Figure S2c,d). These results suggested that *LTR1* was involved in the salt-stress response. It was previously reported that plant membrane transporters play key roles in resistance to biological and abiotic stresses; in particular, Na^+/K^+ transporters increase resistance to salt stress [40]. We further found that there were 259 up-regulated and 178 down-regulated DEGs related to the salt-stress response (Figure 7a and Table S8). In *ltr1*, most of the genes encoding aquaporin or related to Na^+/K^+ transporters were up-regulated, such as *PLASMA MEMBRANE INTRINSIC PROTEIN genes OsPIP1;1, OsPIP1;2, OsPIP1;3, OsPIP2;1, OsPIP2;2, OsPIP2;4, OsPIP2;4*; *TONOPLAST INTRINSIC PROTEIN* genes *OsTIP1;1, OsTIP1;2*; *HIGH-AFFINITY K^+ TRANSPORTERS* genes *OsHKT1;14, OsHKT2;3*, and *OsHKT1;5* (Figure 7b). These results suggested that *LTR1* may affect salt tolerance by regulating water transport and ion homeostasis in plants through aquaporin and Na^+/K^+ transporters. Given that many genes encoding aquaporins and ion transporter were differentially expressed in NPB and *ltr1*, we considered the possibility that *LTR1* may regulate salt tolerance by affecting water transport and ion homeostasis. Therefore, we measured the Na^+ content in solution and in tissues of NPB and *ltr1* under salt stress. After salt stress, Na$^+$ content in stems and leaves of *ltr1* were significantly higher than those of NPB, which increased by 28.24% and 45.75%, respectively ($p < 0.05$). There was no significant difference in Na$^+$ content in the roots of NPB and *ltr1* ($p < 0.05$) (Figure 7c). Furthermore, there was no significant difference in Na$^+$ content in the liquid media in which NPB and *ltr1* plants were grown after treatment in hydroponic solution for 1 d ($p < 0.05$). However, after treatment for 3 or 6 d, Na$^+$ content was lower in the solution in which *ltr1* plants were grown compared to NPB plants, decreased by 15.20% and 8.03%, respectively ($p < 0.01$) (Figure 7d). Under normal growth conditions (CK), the relative expression levels of *OsPIP1;1*, *OsPIP1;2*, *OsPIP2;1*, *OsPIP2;2*, *OsTIP1;1*, *OsTIP1;2*, *OsHKT1;1*, *OsHKT1;5*, and *OsHKT2;3* were significantly higher in *ltr1* than NPB leaves (Figure 7e), increased by 1.85, 2.36, 3.48, 3.46, 10.2, 2.57, 2.67, 2.52, 3.44 times, respectively ($p < 0.01$); which was consistent with the RNA-seq results. The genes encoding aquaporin and ion transporter in NPB and *ltr1* plants both were strongly induced by salt stress. However, the induction of these genes was stronger in *ltr1* than in leaves of NPB, leading to relative expression levels of *OsPIP1;1*, *OsPIP1;2*, *OsPIP2;1*, *OsPIP2;2*, *OsTIP1;1*, *OsTIP1;2*, *OsHKT1;5*, and *OsHKT2;3* that were significantly higher in *ltr1* than NPB leaves under salt stress ($p < 0.01$), especially the expression of *OsPIP2;1* and *OsHKT2;3* (Figure 7f). This was consistent with the finding that the Na$^+$ content in stems and leaves of *ltr1* were significantly higher than those of NPB.

2.7. LTR1 Regulates Salt Tolerance in Rice

To further explore the function of *LTR1* in the salt-stress response, we first screened a suitable salt concentration for treatment. NPB and *ltr1* were cultured in soil treated with 0 mM NaCl (CK treatment), 50 mM NaCl, 100 mM NaCl, or 150 mM NaCl at the five-leaf stage. Two weeks later, the survival rates of NPB treated with 150 mM NaCl was higher than that of *ltr1* plants (92.30% and 64.30%, respectively) ($p < 0.05$) (Figure S4). We then grew NPB plants in solution, treated them with 150 mM NaCl, and measured the relative expression level of *LTR1* at 0, 1, 3, 6, 12, and 24 h. The relative expression level of *LTR1* increased overtime; the relative expression level of *LTR1* increased by 6.95 times at 6 h and by 26.29 times at 24 h after treatment, indicating that *LTR1* was significantly induced by salt stress (Figure 8c). After 7 d of salt stress in hydroponic solution, the survival rate of NPB reached 93.05%, which was significantly higher than that of *ltr1* (43.52%) ($p < 0.05$) (Figure 8b). After 3 d of salt stress in hydroponic solution, H_2O_2 and MDA in levels of NPB and *ltr1* both accumulated, and the accumulation of MDA in the leaves of *ltr1* was significantly higher than that of NPB ($p < 0.05$) (Figure 8d–f). These results suggested that, compared with NPB, the membrane lipid peroxidation and plasma membrane damage in *ltr1* were more serious after salt stress, and that *ltr1* was more sensitive to salt stress (Figure 8a–f). Studies have shown that when plants are subjected to stress, the enzymatic protection system is initiated rapidly, and the activities of peroxidase (POD), ascorbate peroxidase (APX), and other enzymes increase significantly, which enhances the capacity for reactive oxygen species (ROS) scavenging and reduces damage [41–43]. In this study, after salt stress, the catalase (CAT) activity in NPB and *ltr1* decreased by 14.38% and 26.17%, respectively ($p < 0.05$). The decrease of CAT activity in *ltr1* was more significant (Figure 8g). Furthermore, the activities of POD and APX in NPB and *ltr1* both increased after stress, and the increases in POD and APX activities induced by stress in *ltr1* were weaker than that in NPB. After salt stress, the POD and APX activities of NPB increased by 32.31% and 81.62% compared with CK, while the POD and APX activities in *ltr1* increased by 16.97% and 18.01% ($p < 0.05$) (Figure 8h,i). These were consistent with the expression change of antioxidant system in leaves of NPB and *ltr1* (Figure 8j,k). Therefore, these results indicated that *ltr1* had an inferior ability to adapt to salt stress.

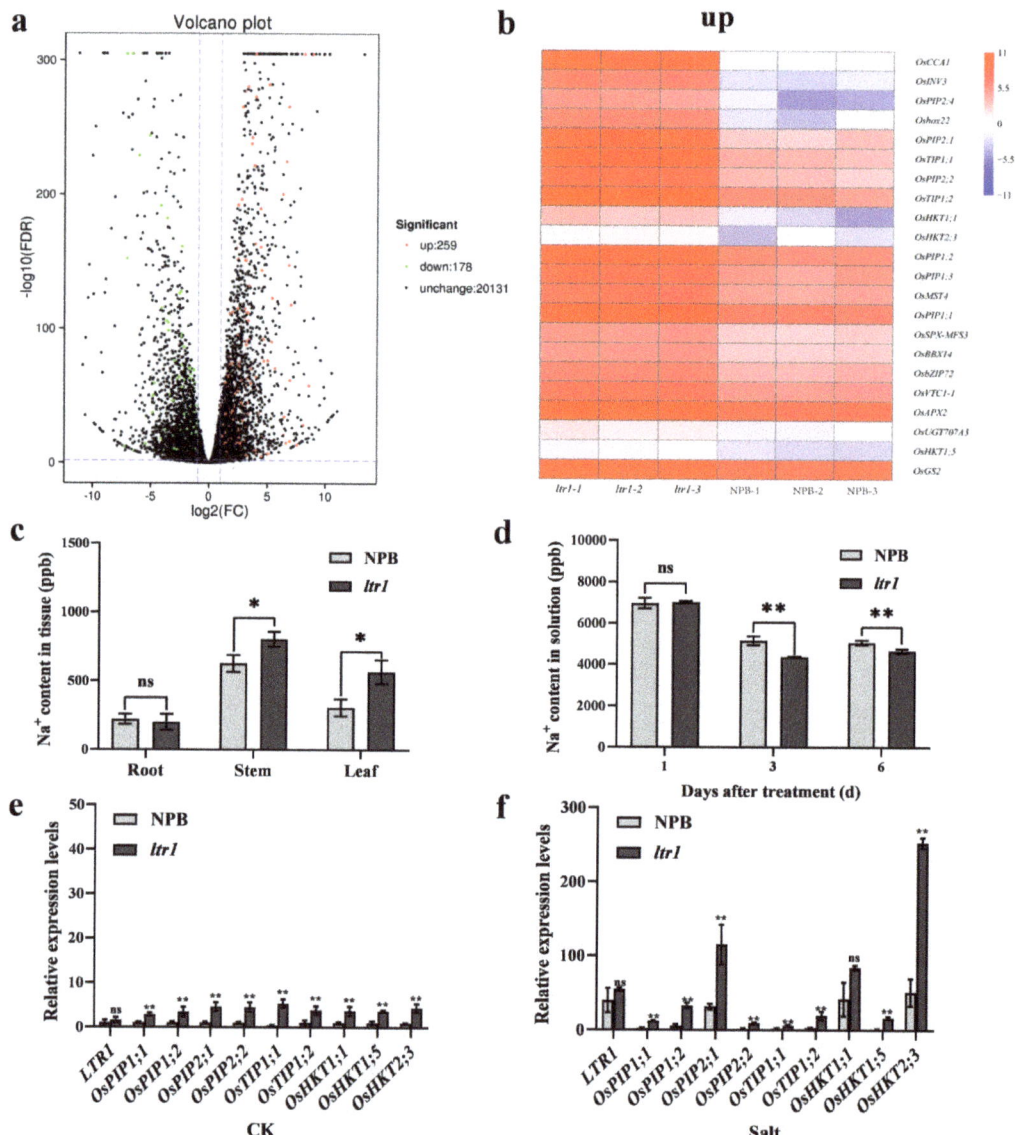

Figure 7. *LTR1* regulates salt-stress response by regulating genes encoding aquaporins and ion transporters. (**a**) Volcano plot of DEGs related to salt response between NPB and *ltr1*. (**b**) Heat map of significantly up-regulated DEGs encoding aquaporin and ion transporters between NPB and *ltr1*. (**c**) Na$^+$ content in different tissues of NPB and *ltr1*. (**d**) Na$^+$ content in solutions where NPB and *ltr1* were cultured after treatment for 1, 3, or 6 d. (**e**) Relative expression levels of *LTR1* and genes encoding aquaporin and ion transporters under normal condition (CK). (**f**) The relative expression levels of *LTR1* and genes encoding aquaporin and ion transporters under 150 mM NaCl (Salt). Data are given as means ± SD. Asterisks indicate significant difference based on the Student's *t*-test: * in the figure represents significant difference at $p < 0.05$; ** in the figure represents significant difference at $p < 0.01$ and ns in the figure represents there is no significant difference at $p < 0.05$.

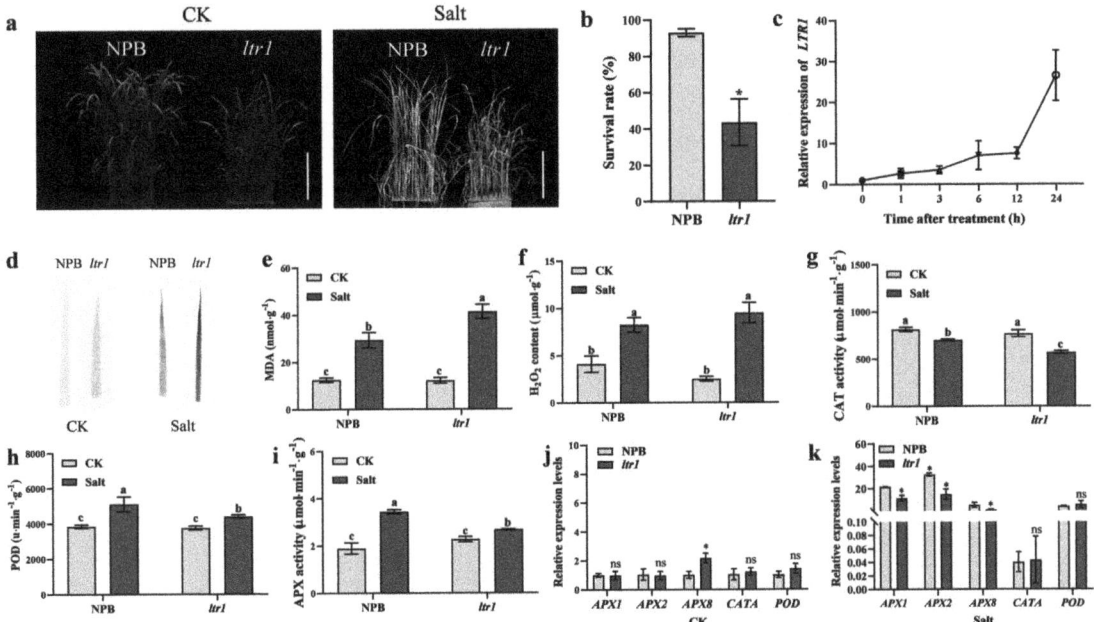

Figure 8. The response of *LTR1* to salt stress in NPB and *ltr1*. (**a**) Photos of NPB and *ltr1* under CK and Salt treatment, bar = 10.5 cm. (**b**) The survival rate of NPB and *ltr1* after treatment for 7 d. (**c**) The relative expression level of *LTR1* after treatment for 0, 1, 3, 6, 12, 24 h. (**d**) DAB staining in leaves of NPB and *ltr1* under CK and salt treatment. (**e,f**) MDA and H_2O_2 content in leaves of NPB and *ltr1* under CK and Salt treatment. (**g–i**) CAT, POD, and APX activity in leaves of NPB and *ltr1* under CK and salt treatment. (**j,k**) The relative expression level of genes related to antioxidant system in leaves of NPB and *ltr1* under CK and Salt treatment, $n = 4$. Data are given as means ± SD. Asterisks indicate significant difference based on the Student's *t*-test: * in the figure represents significant difference at $p < 0.05$ and ns in the figure represents there is no significant difference at $p < 0.05$. Different lowercase letters indicate significant differences based on the Duncan's new multiple range test ($p < 0.05$).

3. Discussion

3.1. LTR1 Encodes a Wax Synthesis Gene and Regulates Leaf Morphology

Cell structure is a key factor regulating leaf morphology. Many cloned genes regulated leaf morphology through affecting the normal development of vascular bundles, sclerenchyma cells, bulliform cells, epidermis, and cell walls [9]. However, few of these genes that affect leaf shape are involved in wax synthesis. In this study, we cloned a leaf shape gene, *LEAF TIP RUMPLED1* (*LTR1*), which is an allele of the wax synthesis gene *OsGL 1-4* [44]. *LTR1* regulated leaf morphology, and loss function of *LTR1* led to rumpled leaves with the abnormal development of bulliform cells, vascular bundles, and sclerenchyma cell. These indicated that *LTR1* affected leaf morphology by regulating the development of bulliform cells, vascular bundles, and sclerenchyma cell. BR signal and auxin metabolism pathway played important roles in leaf morphogenesis [8,45]. *OsBR6ox*, which participates in brassinosteroid (BR) biosynthesis and signal transduction pathway, regulated normal development of organs and then induced abnormal vascular tissue and twisted leaves in its loss-of-function mutant [46]. *OsARF16* [47] and *OsARF17* [48] participate in the auxin response, affecting auxin polar transport and vascular tissue development. The RNA-dependent RNA polymerase OsRDR6 participates in formation of trans-acting small interfering RNA (ta-siRNA) [49], and ta-siRNA inhibits ARF3/ARF4 expression and thus inhibits maintenance of abaxial polarity [50]. *OsAGO7*, a *ZIP/Ago7* homolog in *Arabidopsis thaliana*, is a critical member of the ta-siRNA-ARF3/ARF4-OsAGO7 complex and participates in regulation of leaf rolling [51]. In this study, *OsBR6ox*, *OsARF16*, *OsARF17* and *OsRDR6* were found to be differentially expressed in NPB and *ltr1*. We therefore speculated that *LTR1* may affect leaf morphology by participating in plant hormone signal transduction pathway, while the detailed regulatory network involved requires further study.

3.2. LTR1 Has Multiple Effects on Plant Growth and Development

There are 11 Glossy1 (*GL1*) homologous genes in rice, *OsGL1-1* through *OsGL1-11*, which vary expression levels between rice tissues and organs. Most are induced by abiotic stress and play key roles in wax synthesis and stress tolerance [44]. It was reported that *OsGL1-1*, *OsGL1-2*, *OsGL1-3*, and *OsGL1-6* affect the leaf water loss rate by controlling the wax content in the leaf epidermis, thereby controlling drought resistance in rice [31,44,52,53]. In the present study, we found that *LTR1*, an allele of *OsGL1-4*, was also involved in the regulation of salt tolerance with *LTR1* strongly induced by salt stress. The *ltr1* plants showed high sensitivity to salt stress compared to the wild-type, with more serious membrane lipid peroxidation and plasma membrane damage. Moreover, in rice, many humidity-sensitive genic male sterile lines (HGMS) were obtained by identifying wax synthesis genes involved in regulating pollen development and affecting panicle fertility. Previous studies have shown that most wax synthesis genes, such as *DROUGHT HYPERSENSITIVE* (*DPS1*) [32], *SUBTILISIN-LIKE SERINE PROTEASE 1* (*SUBSrP1*) [54], *HMS1-INTERACTING PROTEIN* (*HMS1I*) [55], *HUMIDITY-SENSITIVE GENIC MALE STERILITY 1* (*HMS1*) [56], and *OsGL1-5* [44] were involved in the regulation of panicle fertility. Loss functions of these genes resulted in abnormal pollen development and a decrease in the seed setting rate at low humidity but a normal seed setting rate at high humidity. Based on this mechanism, the corresponding mutants can be used as HGMSs. It has also been reported that *OsGL1-4* controls male sterility in rice by affecting pollen adhesion and water cooperation under ambient humidity [57]. We here found that loss function of *LTR1* resulted in a severe decrease in the seed setting rate and grain yield per plant, and significant changes in the number of branches and effective panicles in *ltr1*. What's more, overexpression of *LTR1* enhances yield in rice. These results indicated that *LTR1* had pleiotropic functions in rice growth and development.

3.3. LTR1 Regulated Salt Tolerance by Altering Plant Water Status and Ion Homeostasis

Plant aquaporins play very important roles in water transport of transmembrane and form a large protein family [58]. Great progress has been made in functional studies of plasma membrane intrinsic proteins (PIPs) and tonoplast intrinsic proteins (TIPs), which have shown that their main physiological function is to promote transmembrane transport of osmotic water [58]. The expression regulation of *PIPs* varies with differing experimental conditions [59]. *OsPIP1;1* showed low water channel activity in *Xenopus oocytes*, but the permeability of OsPIP1;1 improved significantly when it was co-expressed with *OsPIP2.1* [60]. In the present study, the relative expression level of *OsPIP2;1* was much higher than that of *OsPIP1;1*, *OsPIP1;2*, and *OsPIP2;2*). This indicated that the upregulation of *OsPIP2;1* resulted in enhanced leaf permeability and poor water retention in *ltr1*. Class I HKT transporters play an important role in removing sodium ions from the xylem [61,62]. Because the accumulation of K^+ in plant cells homeostasis the toxicity of Na^+ accumulation, stable acquisition and distribution of K^+ are required during salt-stress conditions [63]. The OsHKT transporter is involved in Na^+ transport in rice, and OsHKT1 specifically mediates Na^+ uptake by rice roots under conditions of K^+ deficiency [64]. *OsHKT1;5* controls the transport of K^+ and Na^+ from roots to shoots. Under salt stress, *OsHKT1;5* refluxes of excessive Na^+ from shoots to roots by unloading it from the xylem, thereby reducing Na^+ toxicity and enhancing salt tolerance [61]. However, we here found that high expression of *OsHKT1;5* under high salt conditions did not reduce the accumulation of Na^+ in *ltr1* leaves. Thus, the excessive accumulation of Na^+ in *ltr1* under salt stress may be regulated by other factors. Under salt stress, the relative expression of *HKT2;3* was significantly higher than the expression of other genes encoding ion transporters. Meanwhile, overexpression of the aquaporin gene *OsPIP2;1* led to enhanced water permeability and poor water retention in *ltr1*. More Na$^+$ was absorbed by *ltr1* than NPB roots and transported to aboveground parts; thus, the Na$^+$ content was significantly higher in stems and leaves of *ltr1* than NPB. Furthermore, there were more white crystals on the stems of NPB than that of *ltr1* (Figure S5). These results suggested that over-accumulation of Na$^+$ in *ltr1* could not be reversed in a timely fashion, resulting in high sensitivity of *ltr1* to salt stress. Therefore, we speculated that *LTR1* affected the water status and ion homeostasis of plants by regulating the expression of genes encoding aquaporins and ion transporters, which ultimately regulated salt tolerance in plants.

3.4. Prospects

Wax, cuticle, and polysaccharide form the cuticle of epidermis, which is a self-protective barrier against biotic and abiotic stresses in plants [29,65,66]. Wax affects canopy temperature and water transport in plants, which further affect plants adaptation to harmful environmental factors such as heat/drought/salt stress and pest/pathogen damage [29,32]. Here, we found that the wax synthesis gene *LTR1* regulates leaf morphology by affecting the normal development of bulliform cells, vascular bundles, and sclerenchyma cells. Moreover, overexpression of *LTR1* enhanced yield in rice and *LTR1* positively regulates salt stress by affecting water and ion homeostasis in plants. However, the regulatory and response mechanism by which *LTR1* affected leaf morphogenesis, water retention, and ion transport between the root and shoot requires further analysis. The differences in ion transport (ion flow rate, ion transport efficiency) and horizontal balance ability between NPB and *ltr1*, together with their regulatory mechanisms need to be further analyzed. How wax content affects cell structure, tissue moisture, and ion balance need further exploration. Identifying proteins that directly interact with LTR1 and analyzing the molecular mechanism of their interaction in regulating leaf shape and salt tolerance will further supplement the known genetic regulation network that governs leaf shape and salt tolerance, providing a theoretical foundation for breeding high-yield rice varieties with high salt tolerance. In addition, identification and application of favorable alleles of *LTR1*, which confers resistances without negative effects on yield, can potentially be used to breed high-yield and high-resistance rice varieties through the combination of multi-omics and bioinformatics. Therefore, according to the insights uncovered in this study, *LTR1* can be considered as a potentially highly valuable gene resource for the improvement of leaf morphology and stress resistance in rice breeding. Manipulating genes associated with leaf morphology and stress resistance individually or in combination makes it possible in the "precision breeding" to breed rice varieties with ideal plant architecture and high resistances without yield penalties. Thus, our results illustrate innovative approaches for developing potentially high stress resistant crop varieties with ideal plant architecture and carry significant implications for breeding application of high yield and stress-resistance-related genetic resources.

4. Materials and Methods

4.1. Plant Materials and Growth Conditions

In this study, the *ltr1* mutant was isolated from a population of the *Oryza sativa* ssp. *japonica* variety Nipponbare (NPB) mutagenized with a 1% ethyl methanesulfonate (EMS) solution using a forward genetic screen for altered leaf shape. Rice plants were grown under natural environmental conditions in an experimental field at the China National Rice Research Institute in Fuyang District (Zhejiang province, China) and Lingshui (Hainan province, China).

Seedlings used in salt treatments were cultured in soil and hydroponic solution (1.25 mM NH$_4$NO$_3$, 0.3 mM KH$_2$PO$_4$, 0.35 mM K$_2$SO$_4$, 1 mM CaCl$_2$, 1 mM MgSO$_4$, 0.5 mM NaSiO$_3$·9H$_2$O, 20 µM Fe-EDTA, 9 µM MnCl$_2$·4H$_2$O, 0.39 µM (NH$_4$)$_6$Mo$_7$O$_{24}$·4H$_2$O, 20 µM

H_3BO_3, 0.77 μM $ZnSO_4·7H_2O$, 0.32 μM $CuSO_4·5H_2O$) in an artificial incubator with a 12 h/12 h light/dark at 70–80% humidity and a 25–30 °C/28 °C day/night temperature (MLR-352H-PC, Panasonic, Osaka, Japan). For salt-stress treatments, plants were cultivated in hydroponic media containing 0 mM NaCl (CK), 50 mM NaCl, 100 mM NaCl, or 150 mM NaCl (Salt), respectively.

4.2. Phenotypicl Characterization and Histological Analysis

To investigate whether the *LTR1* mutation affected rice yield, agronomic traits such as panicle length, effective panicle number, numbers of branches, grain numbers per panicle, seed setting rate, grain yield per plant, and 1000-grain weight were measured for each of 5 or 6 biological replicates at the mature stage. The panicle length, number of branches, and grain numbers per panicle were obtained from measurements of the main panicle.

For frozen cross-section assays, the leaves were immersed in frozen embedding agent (Tissue-Tek® O.C.T. Compound, Sakura, Tokyo, Japan) for 2–3 h at −20 °C. Sections (15 μm) were cut with a freezing microtome (Leica CM1950, Wetzlar, Germany) and placed on microscope slides. Slices were observed and photographed using a microscope (Leica DM4 B). The areas of bulliform cells were calculated using Image J software.

4.3. Measurements of Chlorophyll Content and Photosynthetic Parameters

Chlorophyll *a*, Chlorophyll *b*, and carotenoid (Car) content were measured in three biological replicates using the methods described by Sartory and Grobbelaar [67].

SPAD values were determined for ten biological replicates using a SPAD-502 PLUS. Chlorophyll fluorescence was measured for ten biological replicates with a FluorPen FP100. The QY (Fv/Fm) was determined after a 20 min dark adaptation period.

The net photosynthesis rate, stomatal conductance, and transpiration rate of NPB and *ltr1* plants were evaluated for eight biological replicates with a Li-COR 6400 portable system. All measurements were conducted under the following conditions: photosynthetic photon flux density of 1200 $μmol·m^{-2}·s^{-1}$, ambient CO_2 (400 $μmol·mol^{-1}$), 6 cm^2 of leaf area, 500 $μmol·s^{-1}$ flow speed, and ambient temperature.

4.4. Map-Based Cloning and Complementation Assay

To fine-map the mutated gene, an F_2 population was constructed from a cross between *ltr1* and a wild-type *indica* variety, TN1, with flat leaves. Plants from this population that exhibited rumpled leaves were selected for gene mapping. The locus was first mapped to an interval between the two markers RM6318 and RM1920 (Table S2) on the long arm of chromosome 2, then was further narrowed down to a 13.5-kb DNA region. There was only one open reading frame (ORF) in this region. Genomic DNA fragments in this region were amplified using primers listed in Table S3 from NPB and *ltr1*.

An 8628-bp genomic DNA fragment containing the coding region of *LOC_Os02g40784*, plus 2060-bp upstream, 5450-bp of the coding region and 1118-bp downstream regions, was amplified from NPB (primers for this process are shown in Table S4) and then was cloned into the binary vector pCAMBIA1300 by homologous recombination. The resulting construct pCAMBIA1300-LTR1 was transformed into *ltr1* calli to obtain complementary transgenic plants.

4.5. Gene Editing and Overexpression

For generation of knockout plants using CRISPR/Cas9 technology, gene-specific guide sequences (primers are listed in Table S4) targeting *LTR1* were designed to create single guide RNAs (sgRNAs), after which the sgRNA–Cas9 sequences were cloned into pYLCRISPR/Cas9-MH [68].

Full-length cDNA of *LTR1* amplified (primers are listed in Table S4) from NPB was cloned into the Gateway entry vector pDONR ZEO (Invitrogen, Carlsbad, CA, USA), then recombined into the pUbi::attR-GFP-3×FLAG vector using the Gateway cloning system (Invitrogen). The resulting construct was transformed into NPB calli to obtain overexpression lines of *LTR1*.

4.6. Histological GUS Assay

The promoter of *LTR1* (2094-bp upstream of the start codon) was amplified from NPB genomic DNA (primers are listed in Table S4) and inserted into the *Eco*RI and *Nco*I sites of the binary vector pCAMBIA1305.1. This resulted in a fusion of the promoter and the GUS reporter gene (*pLTR1::GUS*). The recombinant vector was then introduced into NPB calli to obtain transgenic plants.

For GUS staining, different tissues of transgenic plants were incubated in X-Gluc buffer (0.1 mol·L^{-1} K$_2$HPO$_4$ (pH 7.0), 0.1 mol·L^{-1} KH$_2$PO$_4$ (pH 7.0), 5 mmol·L^{-1} K$_3$Fe (CN)$_6$, 5 mmol·L^{-1} K$_4$Fe (CN)$_6$·3H$_2$O, 0.1% Triton X-100, 20% methanol, and 1 mg·mL^{-1} X-Gluc) at 37 °C for 2 h [69]. Stained samples were cleared of chlorophyll by dehydration with ethanol, then scanned using a Microtek Scan Maker i800 plus.

4.7. RNA-seq and Data Analysis

Plants were harvested for total RNA extraction at the booting stage. Three biological replicates were used for RNA-seq analysis. The RNA-seq libraries were constructed and sequenced using an Illumine HiSeq. Each sample obtained approximately 20,000,000 clean reads, which were mapped to NPB reference genome based on the genome information by HISAT2 (http://ccb.jhu.edu/software/hisat2/index.shtml accessed on 10 July 2022). Differential expression analysis for NPB and *ltr1* was performed with DESeq2 using thresholds of FDR < 0.01 and |log$_2$ (fold change)| ≥ 2). A GO enrichment analysis was implemented with the GOseq R packages. The KEGG pathway analysis of DEGs was conducted using the KEGG database (http://www.genome.jp/kegg/ accessed on 24 March 2021).

4.8. Determination of Stress-Related Physiological Index

For 3,3′-diaminobenzidine (DAB) staining, 0.1 g DAB was fully dissolved in ddH$_2$O by adjusting pH to 5.8. The samples were immersed into a tube containing 1 mg/mL DAB solution overnight at 28 °C under dark conditions. Stained samples were cleared of chlorophyll by dehydration with 80% ethanol, then scanned using a Microtek Scan Maker i800 plus. The contents of hydrogen peroxide (H$_2$O$_2$) and malondialdehyde (MDA), activities of catalase (CAT), peroxidase (POD), and ascorbate peroxidase (APX) were measured using appropriate kits from Geruisi (http://www.geruisi-bio.com/ accessed on 18 May 2021) following the manufacturer's instructions with four biological replicates per sample.

4.9. Measurement of Na$^+$ Content

For each sample, a total of 0.05 g of dry tissues power were weighed and immersed in 4 mL concentrated nitric acid with 2 mL 30% H$_2$O$_2$ overnight, and then were decocted with temperature gradient (60 °C for 1 h, 120 °C for 1 h, 160 °C for 1 h, 190 °C until the solution was clarified) using a graphite digestion instrument (DigiBlock ED54, Beijing, China). The content of extracted Na$^+$ was measured by inductively coupled plasma-mass spectrometry (ICP-MS) (iCAP RQ, Thermo Fisher Scientific, 168 Third Avenue, Waltham, MA, USA) after acid catching, constant volume and filtration with three biological replicates per sample.

4.10. RNA Extraction and Quantitative Real-Time PCR

Total RNA was extracted using the Total RNA Miniprep kit (Axygen, Hangzhou, China) following the manufacturer's instructions. First-strand cDNA was synthesized using the ReverTra Ace qPCR-RT kit (Toyobo, Osaka, Japan) as instructed by the manufacturer, using 2 µg of total RNA for each reaction. qRT-PCR analyses were performed using SYBR Premix Ex Taq (Takara, Kusatsu, Japan) and gene-specific primers on a CFX96TM real-time system. Three or four biological replicates were performed for all experiments. The primers used are listed in Table S5 in the Supplementary Materials.

4.11. Quantification and Statistical Analysis

Quantification analyses on all the measurements were conducted in GraphPad Prism 8. Significant differences were determined with Student's *t*-test and Duncan's new multiple range tests.

Supplementary Materials: The following supporting information can be downloaded at: https://www.mdpi.com/article/10.3390/ijms23158818/s1.

Author Contributions: Project administration, Q.Q. and G.Z.; funding acquisition, J.X., Q.Q., G.Z.; supervision, G.Z.; writing—original draft preparation, J.W., Y.L., J.X.; investigation, J.W., Y.L., J.N., M.C., J.C.; data curation, J.W., Y.L., X.L., Z.Z., D.L.; methodology, L.Z., J.H., D.R., Z.G., L.S., G.D., Q.Z., Q.L., S.Y., X.C., S.H. All authors have read and agreed to the published version of the manuscript.

Funding: This research was funded by grants from the National Natural Science Foundation of China (31901483, 31861143006 and 32188102), Nanfan special project of CAAS (ZDXM06), Special Support Program for Distinguished Talents of CAAS (NKYCLJ-C-2021-015) and Zhejiang Province (2019R52031), and Hainan Yazhou Bay Seed Laboratory (B21HJ0220-02).

Institutional Review Board Statement: Not applicable.

Informed Consent Statement: Not applicable.

Data Availability Statement: Not applicable.

Conflicts of Interest: The authors declare no conflict of interest.

References

1. Zhang, G.H.; Xu, Q.; Zhu, X.D.; Qian, Q.; Xue, H.W. SHALLOT-LIKE1 is a KANADI transcription factor that modulates rice leaf rolling by regulating leaf abaxial cell development. *Plant Cell* **2009**, *217*, 19–35. [CrossRef] [PubMed]
2. Li, Y.Y.; Shen, A.; Xiong, W.; Sun, Q.L.; Luo, Q.; Song, T.; Li, Z.L.; Luan, W.J. Overexpression of *OsHox32* results in pleiotropic effects on plant type architecture and leaf development in rice. *Rice* **2016**, *94*, 6. [CrossRef] [PubMed]
3. Xiang, J.J.; Zhang, G.H.; Qian, Q.; Xue, H.W. *Semi-rolled leaf1* encodes a putative glycosylphosphatidylinositol-anchored protein and modulates rice leaf rolling by regulating the formation of bulliform cells. *Plant Physiol.* **2012**, *1591*, 1488–1500. [CrossRef] [PubMed]
4. Zou, L.P.; Sun, X.H.; Zhang, Z.G.; Liu, P.; Wu, J.X.; Tian, C.J.; Qiu, J.L.; Lu, T.G. Leaf rolling controlled by the homeodomain leucine zipper class IV gene *Roc5* in rice. *Plant Physiol.* **2011**, *1561*, 589–1602.
5. Li, Y.; Yang, Y.; Liu, Y.; Li, D.; Zhao, Y.; Li, Z.; Liu, Y.; Jiang, D.; Li, J.; Zhou, H.; et al. Overexpression of *OsAGO1b* induces adaxially rolled leaves by affecting leaf abaxial sclerenchymatous cell development in rice. *Rice* **2019**, *126*, 60. [CrossRef] [PubMed]
6. Sun, J.; Cui, X.; Teng, S.; Kunnong, Z.; Wang, Y.; Chen, Z.; Sun, X.; Wu, J.; Ai, P.; Quick, W.P.; et al. HD-ZIP IV gene *Roc8* regulates the size of bulliform cells and lignin content in rice. *Plant Biotechnol. J.* **2020**, *182*, 2559–2572. [CrossRef]
7. Zhang, G.; Hou, X.; Wang, L.; Xu, J.; Chen, J.; Fu, X.; Shen, N.; Nian, J.; Jiang, Z.; Hu, J.; et al. PHOTO-SENSITIVE LEAF ROLLING 1 encodes a polygalacturonase that modifies cell wall structure and drought tolerance in rice. *New Phytol.* **2021**, *2298*, 890–901. [CrossRef]
8. Xu, P.; Ali, A.; Han, B.; Wu, X. Current advances in molecular basis and mechanisms regulating leaf morphology in rice. *Front. Plant Sci.* **2018**, *91*, 528. [CrossRef] [PubMed]
9. Wang, J.J.; Xu, J.; Qian, Q.; Zhang, G.H. Development of rice leaves: How histocytes modulate leaf polarity establishment. *Rice Sci.* **2020**, *274*, 468–479.
10. Guo, W.; Chen, L.; Herrera-Estrella, L.; Cao, D.; Tran, L.P. Altering plant architecture to improve performance and resistance. *Trends Plant Sci.* **2020**, *251*, 1154–1170. [CrossRef] [PubMed]
11. Springer, N. Shaping a better rice plant. *Nat. Genet.* **2010**, *424*, 475–476. [CrossRef]
12. Lu, Z.; Yu, H.; Xiong, G.; Wang, J.; Jiao, Y.; Liu, G.; Jing, Y.; Meng, X.; Hu, X.; Qian, Q.; et al. Genome-wide binding analysis of the transcription activator ideal plant architecture1 reveals a complex network regulating rice plant architecture. *Plant Cell* **2013**, *253*, 3743–3759. [CrossRef]
13. Wang, L.; Ming, L.; Liao, K.; Xia, C.; Sun, S.; Chang, Y.; Wang, H.; Fu, D.; Xu, C.; Wang, Z.; et al. Bract suppression regulated by the miR156/529-SPLs-NL1-PLA1 module is required for the transition from vegetative to reproductive branching in rice. *Mol. Plant* **2021**, *141*, 1168–1184. [CrossRef]
14. Wang, J.; Zhou, L.; Shi, H.; Chern, M.; Yu, H.; Yi, H.; He, M.; Yin, J.; Zhu, X.; Li, Y.; et al. A single transcription factor promotes both yield and immunity in rice. *Science* **2018**, *3611*, 1026–1028. [CrossRef] [PubMed]
15. Fujisawa, Y.; Kato, T.; Ohki, S.; Ishikawa, A.; Kitano, H.; Sasaki, T.; Asahi, T.; Iwasaki, Y. Suppression of the heterotrimeric G protein causes abnormal morphology, including dwarfism, in rice. *Proc. Natl. Acad. Sci. USA* **1999**, *967*, 7575–7580. [CrossRef] [PubMed]

6. Jangam, A.P.; Pathak, R.R.; Raghuram, N. Microarray analysis of rice *d1* (*RGA1*) mutant reveals the potential role of g-protein alpha subunit in regulating multiple abiotic stresses such as drought, salinity, heat, and cold. *Front. Plant Sci.* **2016**, *71*, 11. [CrossRef]
7. Zhu, Y.; Li, T.; Xu, J.; Wang, J.; Wang, L.; Zou, W.; Zeng, D.; Zhu, L.; Chen, G.; Hu, J.; et al. Leaf width gene *LW5/D1* affects plant architecture and yield in rice by regulating nitrogen utilization efficiency. *Plant Physiol. Biochem.* **2020**, *1573*, 359–369. [CrossRef] [PubMed]
8. Li, G.; Zhang, C.; Zhang, G.; Fu, W.; Feng, B.; Chen, T.; Peng, S.; Tao, L.; Fu, G. Abscisic acid negatively modulates heat tolerance in rolled leaf rice by increasing leaf temperature and regulating energy homeostasis. *Rice* **2020**, *131*, 8. [CrossRef] [PubMed]
9. Wang, Z.; Chen, Z.; Cheng, J.; Lai, Y.; Wang, J.; Bao, Y.; Huang, J.; Zhang, H. QTL analysis of Na^+ and K^+ concentrations in roots and shoots under different levels of NaCl stress in rice (*Oryza sativa* L.). *PLoS ONE* **2012**, *7*, e51202. [CrossRef] [PubMed]
10. Yang, Y.; Guo, Y. Unraveling salt stress signaling in plants. *Integr. Plant Biol.* **2018**, *607*, 796–804. [CrossRef] [PubMed]
11. Deinlein, U.; Stephan, A.B.; Horie, T.; Luo, W.; Xu, G.; Schroeder, J.I. Plant salt-tolerance mechanisms. *Trends Plant Sci.* **2014**, *193*, 371–379. [CrossRef] [PubMed]
12. Tester, M.; Davenport, R. Na^+ tolerance and Na^+ transport in higher plants. *Ann. Bot.* **2003**, *915*, 503–527. [CrossRef] [PubMed]
13. Zhu, J.K. Salt and drought stress signal transduction in plants. *Annu. Rev. Plant Biol.* **2002**, *532*, 247–273. [CrossRef] [PubMed]
14. Zhu, J.K. Genetic analysis of plant salt tolerance using *Arabidopsis*. *Plant Physiol.* **2000**, *1249*, 941–948. [CrossRef]
15. Van Zelm, E.; Zhang, Y.; Testerink, C. Salt tolerance mechanisms of plants. *Annu. Rev. Plant Biol.* **2020**, *714*, 403–433. [CrossRef]
16. Blumwald, E. Sodium transport and salt tolerance in plants. *Curr. Opin. Cell Biol.* **2000**, *124*, 431–434. [CrossRef]
17. Long, L.M.; Patel, H.P.; Cory, W.C.; Stapleton, A.E. The maize epicuticular wax layer provides UV protection. *Funct. Plant Biol.* **2003**, *307*, 75–81. [CrossRef]
18. Uppalapati, S.R.; Ishiga, Y.; Doraiswamy, V.; Bedair, M.; Mittal, S.; Chen, J.; Nakashima, J.; Tang, Y.; Tadege, M.; Ratet, P.; et al. Loss of abaxial leaf epicuticular wax in *Medicago truncatula irg1/palm1* mutants results in reduced spore differentiation of anthracnose and nonhost rust pathogens. *Plant Cell* **2012**, *243*, 353–370. [CrossRef]
19. Kan, Y.; Mu, X.R.; Zhang, H.; Gao, J.; Shan, J.X.; Ye, W.W.; Lin, H.X. TT2 controls rice thermotolerance through SCT1-dependent alteration of wax biosynthesis. *Nat. Plants* **2022**, *85*, 53–67. [CrossRef]
20. Zhu, X.; Xiong, L. Putative megaenzyme DWA1 plays essential roles in drought resistance by regulating stress-induced wax deposition in rice. *Proc. Natl. Acad. Sci. USA* **2013**, *1101*, 17790–17795. [CrossRef]
21. Zhou, X.; Li, L.; Xiang, J.; Gao, G.; Xu, F.; Liu, A.; Zhang, X.; Peng, Y.; Chen, X.; Wan, X. OsGL1-3 is involved in cuticular wax biosynthesis and tolerance to water deficit in rice. *PLoS ONE* **2015**, *10*, e116676.
22. Wang, Z.; Tian, X.; Zhao, Q.; Liu, Z.; Li, X.; Ren, Y.; Tang, J.; Fang, J.; Xu, Q.; Bu, Q. The E3 ligase DROUGHT HYPERSENSITIVE negatively regulates cuticular wax biosynthesis by promoting the degradation of transcription factor ROC4 in rice. *Plant Cell* **2018**, *302*, 228–244. [CrossRef] [PubMed]
23. Wei, J.; Choi, H.; Jin, P.; Wu, Y.; Yoon, J.; Lee, Y.S.; Quan, T.; An, G. GL2-type homeobox gene *Roc4* in rice promotes flowering time preferentially under long days by repressing *Ghd7*. *Plant Sci.* **2016**, *2521*, 133–143. [CrossRef] [PubMed]
24. Wang, Y.; Wan, L.; Zhang, L.; Zhang, Z.; Zhang, H.; Quan, R.; Zhou, S.; Huang, R. An ethylene response factor OsWR1 responsive to drought stress transcriptionally activates wax synthesis related genes and increases wax production in rice. *Plant Mol. Biol.* **2012**, *782*, 275–288. [CrossRef]
25. Jenks, M.A.; Rashotte, A.M.; Tuttle, H.A.; Feldmann, K.A. Mutants in *Arabidopsis thaliana* altered in epicuticular wax and leaf morphology. *Plant Physiol.* **1996**, *1103*, 377–385. [CrossRef]
26. Chen, X.; Goodwin, S.M.; Boroff, V.L.; Liu, X.; Jenks, M.A. Cloning and characterization of the *WAX2* gene of *Arabidopsis* involved in cuticle membrane and wax production. *Plant Cell* **2003**, *151*, 1170–1185. [CrossRef]
27. Horton, P. Prospects for crop improvement through the genetic manipulation of photosynthesis: Morphological and biochemical aspects of light capture. *J. Exp. Bot.* **2000**, *51*, 475–485. [CrossRef]
28. Sage, R.F. A model describing the regulation of ribulose-15, -bisphosphate carboxylase, electron transport, and triose phosphate use in response to light intensity and CO_2 in C_3 plants. *Plant Physiol.* **1990**, *941*, 1728–1734. [CrossRef]
29. Qiu, Z.; Chen, D.; He, L.; Zhang, S.; Yang, Z.; Zhang, Y.; Wang, Z.; Ren, D.; Qian, Q.; Guo, L.; et al. The rice *white green leaf 2* gene causes defects in chloroplast development and affects the plastid ribosomal protein S9. *Rice* **2018**, *113*, 9. [CrossRef]
30. Schroeder, J.I.; Delhaize, E.; Frommer, W.B.; Guerinot, M.L.; Harrison, M.J.; Herrera-Estrella, L.; Horie, T.; Kochian, L.V.; Munns, R.; Nishizawa, N.K.; et al. Using membrane transporters to improve crops for sustainable food production. *Nature* **2013**, *497*, 60–66. [CrossRef]
31. Dumanović, J.; Nepovimova, E.; Natić, M.; Kuča, K.; Jaćević, V. The significance of reactive oxygen species and antioxidant defense system in plants: A concise overview. *Front. Plant Sci.* **2020**, *115*, 52969. [CrossRef] [PubMed]
32. Hasanuzzaman, M.; Bhuyan, M.; Zulfiqar, F.; Raza, A.; Mohsin, S.M.; Mahmud, J.A.; Fujita, M.; Fotopoulos, V. Reactive oxygen species and antioxidant defense in plants under abiotic stress: Revisiting the crucial role of a universal defense regulator. *Antioxidants* **2020**, *9*, 681. [CrossRef] [PubMed]
33. Gill, S.S.; Tuteja, N. Reactive oxygen species and antioxidant machinery in abiotic stress tolerance in crop plants. *Plant Physiol. Biochem.* **2010**, *489*, 909–930. [CrossRef] [PubMed]
34. Islam, M.A.; Du, H.; Ning, J.; Ye, H.; Xiong, L. Characterization of *Glossy1*-homologous genes in rice involved in leaf wax accumulation and drought resistance. *Plant Mol. Biol.* **2009**, *704*, 443–456. [CrossRef]

45. Xu, J.; Wang, J.J.; Xue, H.W.; Zhang, G.H. Leaf direction: Lamina joint development and environmental responses. *Plant Cell Environ.* **2021**, *442*, 2441–2454. [CrossRef]
46. Hong, Z.; Ueguchi-Tanaka, M.; Shimizu-Sato, S.; Inukai, Y.; Fujioka, S.; Shimada, Y.; Takatsuto, S.; Agetsuma, M.; Yoshida, S.; Watanabe, Y.; et al. Loss-of-function of a rice brassinosteroid biosynthetic enzyme, C-6 oxidase, prevents the organized arrangement and polar elongation of cells in the leaves and stem. *Plant J.* **2002**, *324*, 495–508. [CrossRef]
47. Shen, C.; Wang, S.; Zhang, S.; Xu, Y.; Qian, Q.; Qi, Y.; Jiang, D. OsARF16, a transcription factor, is required for auxin and phosphate starvation response in rice (Oryza sativa L.). *Plant Cell Environ.* **2013**, *366*, 607–620. [CrossRef]
48. Huang, G.; Hu, H.; van de Meene, A.; Zhang, J.; Dong, L.; Zheng, S.; Zhang, F.; Betts, N.S.; Liang, W.; Bennett, M.J.; et al. *AUXIN RESPONSE FACTORS 6* and *17* control the flag leaf angle in rice by regulating secondary cell wall biosynthesis of lamina joints. *Plant Cell* **2021**, *333*, 3120–3133. [CrossRef]
49. Toriba, T.; Suzaki, T.; Yamaguchi, T.; Ohmori, Y.; Tsukaya, H.; Hirano, H.Y. Distinct regulation of adaxial-abaxial polarity in anther patterning in rice. *Plant Cell* **2010**, *221*, 1452–1462. [CrossRef]
50. Hasson, A.; Blein, T.; Laufs, P. Leaving the meristem behind: The genetic and molecular control of leaf patterning and morphogenesis. *C. R. Biol.* **2010**, *3333*, 350–360. [CrossRef]
51. Shi, Z.; Wang, J.; Wan, X.; Shen, G.; Wang, X.; Zhang, J. Over-expression of rice *OsAGO7* gene induces upward curling of the leaf blade that enhanced erect-leaf habit. *Planta* **2007**, *2269*, 99–108. [CrossRef] [PubMed]
52. Qin, B.X.; Tang, D.; Huang, J.; Li, M.; Wu, X.R.; Lu, L.L.; Wang, K.J.; Yu, H.X.; Chen, J.M.; Gu, M.H.; et al. Rice *OsGL1-1* is involved in leaf cuticular wax and cuticle membrane. *Mol. Plant* **2011**, *49*, 985–995. [CrossRef] [PubMed]
53. Zhou, L.; Ni, E.; Yang, J.; Zhou, H.; Liang, H.; Li, J.; Jiang, D.; Wang, Z.; Liu, Z.; Zhuang, C. Rice *OsGL1-6* is involved in leaf cuticular wax accumulation and drought resistance. *PLoS ONE* **2013**, *8*, e65139. [CrossRef] [PubMed]
54. Ali, A.; Wu, T.; Zhang, H.; Xu, P.; Zafar, S.A.; Liao, Y.; Chen, X.; Zhou, H.; Liu, Y.; Wang, W.; et al. A putative *SUBTILISIN-LIKE SERINE PROTEASE 1 (SUBSrP1)* regulates anther cuticle biosynthesis and panicle development in rice. *J. Adv. Res.* **2022**. [CrossRef]
55. Wang, X.; Guan, Y.; Zhang, D.; Dong, X.; Tian, L.; Qu, L.Q. A β-Ketoacyl-CoA Synthase Is Involved in Rice Leaf Cuticular Wax Synthesis and Requires a CER2-LIKE Protein as a Cofactor. *Plant Physiol.* **2017**, *1739*, 944–955. [CrossRef]
56. Chen, H.; Zhang, Z.; Ni, E.; Lin, J.; Peng, G.; Huang, J.; Zhu, L.; Deng, L.; Yang, F.; Luo, Q.; et al. HMS1 interacts with HMS1I to regulate very-long-chain fatty acid biosynthesis and the humidity-sensitive genic male sterility in rice (*Oryza sativa*). *New Phytol.* **2020**, *2252*, 2077–2093. [CrossRef]
57. Yu, B.; Liu, L.; Wang, T. Deficiency of very long chain alkanes biosynthesis causes humidity-sensitive male sterility via affecting pollen adhesion and hydration in rice. *Plant Cell Environ.* **2019**, *423*, 3340–3354. [CrossRef]
58. Maurel, C.; Boursiac, Y.; Luu, D.T.; Santoni, V.; Shahzad, Z.; Verdoucq, L. Aquaporins in Plants. *Physiol. Rev.* **2015**, *951*, 1321–1358. [CrossRef]
59. Lian, H.L.; Yu, X.; Lane, D.; Sun, W.N.; Tang, Z.C.; Su, W.A. Upland rice and lowland rice exhibited different *PIP* expression under water deficit and ABA treatment. *Cell Res.* **2006**, *166*, 651–660. [CrossRef]
60. Liu, C.; Fukumoto, T.; Matsumoto, T.; Gena, P.; Frascaria, D.; Kaneko, T.; Katsuhara, M.; Zhong, S.; Sun, X.; Zhu, Y.; et al. Aquaporin OsPIP1;1 promotes rice salt resistance and seed germination. *Plant Physiol. Biochem.* **2013**, *631*, 151–158. [CrossRef]
61. Ren, Z.H.; Gao, J.P.; Li, L.G.; Cai, X.L.; Huang, W.; Chao, D.Y.; Zhu, M.Z.; Wang, Z.Y.; Luan, S.; Lin, H.X. A rice quantitative trait locus for salt tolerance encodes a sodium transporter. *Nat. Genet.* **2005**, *371*, 1141–1146. [CrossRef] [PubMed]
62. Sunarpi; Horie, T.; Motoda, J.; Kubo, M.; Yang, H.; Yoda, K.; Horie, R.; Chan, W.Y.; Leung, H.Y.; Hattori, K.; et al. Enhanced salt tolerance mediated by AtHKT1 transporter-induced Na unloading from xylem vessels to xylem parenchyma cells. *Plant J.* **2005**, *449*, 928–938. [CrossRef] [PubMed]
63. Schroeder, J.I.; Ward, J.M.; Gassmann, W. Perspectives on the physiology and structure of inward-rectifying K^+ channels in higher plants: Biophysical implications for K^+ uptake. *Annu. Rev. Biophys. Biomol. Struct.* **1994**, *234*, 441–471. [CrossRef] [PubMed]
64. Garciadeblás, B.; Senn, M.E.; Bañuelos, M.A.; Rodríguez-Navarro, A. Sodium transport and HKT transporters: The rice model. *Plant J.* **2003**, *347*, 788–801. [CrossRef] [PubMed]
65. Shepherd, T.; Wynne Griffiths, D. The effects of stress on plant cuticular waxes. *New Phytol.* **2006**, *1714*, 469–499. [CrossRef]
66. Lewandowska, M.; Keyl, A.; Feussner, I. Wax biosynthesis in response to danger: Its regulation upon abiotic and biotic stress. *New Phytol.* **2020**, *2276*, 698–713. [CrossRef] [PubMed]
67. Sartory, D.P.; Grobbelaar, J.U. Extraction of chlorophyll a from freshwater phytoplankton for spectrophotometric analysis. *Hydrobiologia* **1984**, *1141*, 177–187. [CrossRef]
68. Ma, X.; Zhang, Q.; Zhu, Q.; Liu, W.; Chen, Y.; Qiu, R.; Wang, B.; Yang, Z.; Li, H.; Lin, Y.; et al. A robust CRISPR/Cas9 system for convenient, high-efficiency multiplex genome editing in monocot and dicot plants. *Mol. Plant* **2015**, *81*, 274–1284. [CrossRef]
69. Jefferson, R.A.; Kavanagh, T.A.; Bevan, M.W. GUS fusions: Beta-glucuronidase as a sensitive and versatile gene fusion marker in higher plants. *Embo J.* **1987**, *63*, 3901–3907. [CrossRef]

Insights into the Genomic Regions and Candidate Genes of Senescence-Related Traits in Upland Cotton via GWAS

Qibao Liu [†], Zhen Feng [†], Chenjue Huang, Jia Wen, Libei Li *[ ] and Shuxun Yu *

College of Advanced Agriculture Sciences, Zhejiang A&F University, Hangzhou 311300, China; liuqibao566@163.com (Q.L.); fengzhen@zafu.edu.cn (Z.F.); huang_chenjue@163.com (C.H.); 17773326468@163.com (J.W.)
* Correspondence: libeili@zafu.edu.cn (L.L.); yushuxun@zafu.edu.cn (S.Y.)
† These authors contributed equally to this work.

Abstract: Senescence is the last stage of plant development and is controlled by both internal and external factors. Premature senescence significantly affects the yield and quality of cotton. However, the genetic architecture underlying cotton senescence remains unclear. In this study, genome-wide association studies (GWAS) were performed based on 3,015,002 high-quality SNP markers from the resequencing data of 355 upland cotton accessions to detect genomic regions for cotton senescence. A total of 977 candidate genes within 55 senescence-related genomic regions (SGRs), SGR1–SGR55, were predicted. Gene ontology (GO) analysis of candidate genes revealed that a set of biological processes was enriched, such as salt stress, ethylene processes, and leaf senescence. Furthermore, in the leaf senescence GO term, one candidate gene was focused on: *Gohir.A12G270900* (*GhMKK9*), located in SGR36, which encodes a protein of the MAP kinase kinase family. Quantitative real-time PCR (qRT-PCR) analysis showed that *GhMKK9* was up-regulated in old cotton leaves. Overexpression of *GhMKK9* in *Arabidopsis* accelerated natural leaf senescence. Virus-induced gene silencing (VIGS) of *GhMKK9* in cotton increased drought tolerance. These results suggest that *GhMKK9* is a positive regulator and might be involved in drought-induced senescence in cotton. The results provide new insights into the genetic basis of cotton senescence and will be useful for improving cotton breeding in the future.

Keywords: GWAS; upland cotton; senescence; genomic region; candidate gene; *GhMKK9*

1. Introduction

Cotton (*Gossypium* spp.) is an important industrial crop worldwide that offers renewable natural fibers, oil, and animal feed [1]. The genomes of the genus *Gossypium* are extraordinarily diverse, including approximately 45 diploid species (2n = 2x = 26) and seven tetraploid (2n = 4x = 52) species [2,3]. *Gossypium hirsutum* L. (also known as upland cotton), one of the seven tetraploid cotton species, is the most widely cultivated species worldwide because of its adaptability, high yield, and moderate fiber quality [4,5]. Although upland cotton makes a significant contribution to revenue in several countries [4], cotton yield is reduced due to senescence when it is induced prematurely under adverse environmental stresses [6].

Senescence is the last stage of plant development and is accompanied by a transition from nutrient assimilation to nutrient remobilization [7,8]. During plant senescence, many major macromolecules are degraded, including proteins, lipids, and nucleic acids, but the most visible symptom is leaf yellowing owing to the catabolism of chlorophyll [9,10]. The onset and progression of senescence are regulated by both internal and external factors. Internal factors include various phytohormones [7,11] that play diverse roles in leaf development. For example, ethylene, abscisic acid (ABA), and salicylic acids (SA) are acknowledged as senescence-promoting hormones [12–16]. Additionally, multiple external environmental factors, including abiotic and biotic stresses, can trigger changes of hormones, which form a complex regulatory network of senescence [8]. Interestingly, the mitogen-activated protein kinase (MAPK) cascades play an important role in conveying endogenous and exogenous signals [17].

Senescence is a complex, quantitative trait, and many studies have reported the genetic basis of leaf senescence in plants. Under various stress conditions, several quantitative trait loci (QTL) associated with senescence were discovered using linkage mapping in crop plants, such as rice [18,19], wheat [20–23], barley [24], maize [25], sorghum [26–29], and potato [30]. Although these studies are helpful for understanding the genetic architecture of senescence, it is difficult to identify the underlying genes owing to a lack of resolution. In the past decade, genome-wide association studies (GWAS) have become a powerful method for detecting quantitative trait loci and candidate genes at the genome-wide level [31–34]. In a recent study, 25 candidate genes for chlorophyll content (CC) and stay-green (SG) traits were identified using a diverse population of 368 rice accessions via GWAS [35]. *OsSG1* is considered a pleiotropic gene regulating CC, SG, and chlorophyll accumulation [35]. In another GWAS study, 64 candidate genes associated with maize senescence were identified using the maize diversity panel, of which 14 genes were involved in senescence-related processes, such as proteolysis and sink activity, and eight candidate genes were supported by a regulatory network [36]. Furthermore, our previous study revealed 50 genomic regions associated with cotton senescence via a multi-locus GWAS based on 185 upland cotton accessions and SLAF-seq data [37]. The candidate gene, *GhCDF1*, was identified as a negative regulator of cotton senescence. However, further studies are needed to understand the mechanisms underlying cotton senescence.

Here, a genome-wide association study was conducted to dissect the genetic basis of senescence in cotton. The association panel consisted of 355 upland cotton accessions planted in multiple environments, and chlorophyll content indices were measured as indicators of senescence. Using resequencing data, 55 senescence-related genomic regions (SGRs) were discovered based on GWAS, and 977 potential candidate genes associated with cotton senescence were identified. The function of candidate gene *GhMKK9* was then analyzed, and it was found that *GhMKK9* silencing improves the drought resistance of cotton, whereas *GhMKK9* overexpression accelerates senescence in *Arabidopsis*. These results provide a foundation for the breeding and the genetic improvement of cotton.

2. Results

2.1. Analysis of Phenotypic Variations

To evaluate the variability of senescence in the GWAS panel, the relative chlorophyll levels of 355 upland cottons were investigated with the SPAD-502 m during two periods, the flowering and boll-setting period (FBP) and the boll-opening period (BOP), in multiple environments, including Anyang (AY) and Huanggang (HG) in 2016 and 2017, designated as SPAD_FBP_AY16, SPAD_FBP_AY17, SPAD_FBP_HG16, SPAD_FBP_HG17, SPAD_BOP_AY16, SPAD_BOP_AY17, SPAD_BOP_HG16, and SPAD_BOP_HG17. To assess the rate of leaf senescence, the diurnal variation of SPAD was calculated, including D_SPAD_AY1, D_SPAD_AY17, D_SPAD_HG16, and D_SPAD_HG17. Additionally, the absolute chlorophyll concentrations and diurnal variation were determined at AY in 2017 (see the Methods section).

The investigated traits followed approximately normal distributions (Figures 1 and S1–S3) and exhibited wide variation among different years and locations (Supplementary Table S1). In the FBP period, the average SPAD values in AY and HG in 2016 were 49.12 and 46.27, respectively, compared to 55.10 and 48.87 in 2017. In the BOP period, the average SPAD in AY in 2016 was higher than that in 2017, at 52.01 and 48.77, respectively, whereas the average SPAD in HG in 2016 was 42.52, lower than that in 2017 (50.19). The standard deviation of SPAD values in the FBP period was distributed from 2.25 to 3.66, compared with the range of 3.54–12.84 in the BOP period. In addition, the average variations of the index D_SPAD ranged from −0.19 to 0.19. Furthermore, the ANOVA result indicated that genotype, environment, and the genotype-by-environment interaction had significant effects on SPAD ($p < 0.01$), while heritability of SPAD in the FBP period was higher than that in the BOP period (0.65 and 0.41, respectively) (Supplementary Table S2). These results indicate that cotton senescence is significantly influenced by environmental factors, particularly in the BOP period.

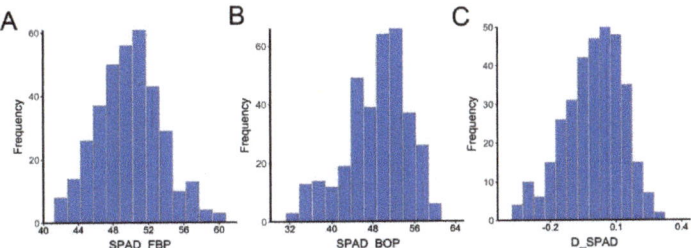

Figure 1. Frequency distributions of the mean values of SPAD. (**A**) The mean value of SPAD in the FBP period. (**B**) The mean value of SPAD in the BOP period. (**C**) The mean value of diurnal variation of SPAD.

Pearson's product–moment correlation coefficients and test statistics were used to evaluate traits. Although there were significant positive correlations ($p < 0.001$) among chlorophyll contents, the diurnal variations of chlorophyll content were more related to the BOP period ($|r| = 0.00–0.35$) than the FBP period ($|r| = 0.01–0.93$) (Supplementary Figure S4).

2.2. GWAS for Cotton Senescence and Identified Genomic Regions

A total of 3,015,002 high-quality single-nucleotide polymorphisms (SNPs) were identified after a strict filtering pipeline. GWAS was then performed for both single traits across different environments and the best linear unbiased prediction (BLUP) values across all environments using a linear mixed model by EMMAX [38] (Supplementary Figure S5–S7). Given the significant thresholds ($p < 10^{-6}$ or $p < 10^{-5}$ in at least two environments), 380 significant signals were identified (Supplementary Table S3).

Because the majority of GWAS signals are usually located in noncoding or intergenic regions, functional variations are rarely identified by association tests from SNPs [39]. Therefore, significant signals were integrated, and 55 senescence-related genomic regions (SGRs) obtained, namely, SGR1–SGR55. (Table 1). The total span of SGRs was approximately 18.09 megabases (Mb), of which 27 were over 1 kb in length. In the A subgenome, 37 SGRs were distributed across all 13 chromosomes (A01–A13) with a total length of 9.49 Mb, while 18 SGRs were distributed across only nine chromosomes of the D subgenome, with a total length of 8.60 Mb. Interestingly, there was an extremely long genomic region on the D12 chromosome, SGR52, which spanned 4.33 Mb and accounted for half of the total length of SGRs in the D subgenome. In addition, forty-three SGRs (78.18%) were detected at least twice, indicating that the results were stable and reliable.

Table 1. Summary of senescence-related genomic regions.

SGR	Chr	Start (bp)	End (bp)	Trait
SGR1	A01	115,568,753	115,568,865	Ratio_ab_FBP
SGR2	A02	6,556,304	6,653,102	SPAD_BOP_AY17, D_SPAD_blup, D_SPAD_AY17
SGR3	A02	155,73,340	15,573,343	D_total_ab, D_chla
SGR4	A02	82,886,748	82,925,332	SPAD_BOP_AY16, D_SPAD_AY16
SGR5	A03	5,877,926	6,672,551	D_SPAD_blup, SPAD_BOP_AY17
SGR6	A03	82,626,806	84,562,267	SPAD_BOP_AY16, D_SPAD_AY16

Table 1. Cont.

SGR	Chr	Start (bp)	End (bp)	Trait
SGR7	A03	113,682,897	113,683,008	SPAD_BOP_AY16, D_SPAD_AY16
SGR8	A04	69,311,017	69,315,686	Ratio_ab_BOP
SGR9	A05	9,695,614	9,695,614	Ratio_ab_FBP
SGR10	A05	31,628,900	31,628,910	SPAD_BOP_AY17, D_SPAD_AY17
SGR11	A05	61,940,854	61,940,864	D_chlb, D_ratio_ab
SGR12	A06	1,128,396	2,179,285	Chlb_BOP, D_chlb, Total_ab_BOP, Ratio_ab_BOP, D_ratio_ab
SGR13	A06	49,902,382	49,941,259	D_ratio_ab, D_chla, D_total_ab, Chla_BOP, Total_ab_BOP
SGR14	A06	61,589,693	61,589,748	Ratio_ab_FBP
SGR15	A06	63,390,768	63,390,795	D_chla, Chla_BOP, Total_ab_BOP
SGR16	A06	113,048,017	115,046,117	D_SPAD_HG17, Ratio_ab_BOP
SGR17	A07	12,602,257	12,602,280	SPAD_BOP_AY17, D_SPAD_AY17
SGR18	A07	61,313,631	61,313,636	Chla_BOP, Total_ab_BOP
SGR19	A08	18,726,393	18,727,072	D_SPAD_AY16, SPAD_BOP_AY16
SGR20	A08	48,942,300	50,338,088	SPAD_BOP_AY17, D_SPAD_AY17
SGR21	A08	85,414,259	85,415,664	Ratio_ab_BOP, D_SPAD_HG17
SGR22	A09	40,190,612	40,190,642	D_SPAD_AY17, D_SPAD_blup
SGR23	A09	71,543,578	71,543,578	Ratio_ab_FBP
SGR24	A09	83,855,283	84,822,913	D_SPAD_AY16, SPAD_BOP_AY16
SGR25	A10	5,521,694	5,526,321	D_SPAD_AY16, SPAD_BOP_AY16
SGR26	A10	27,875,320	28,364,565	BOP_blup, D_SPAD_blup
SGR27	A10	92,752,436	92,766,905	SPAD_FBP_HG16, Ratio_ab_FBP
SGR28	A11	5,387,130	5,588,967	D_SPAD_blup, D_SPAD_AY17, SPAD_BOP_AY17
SGR29	A11	106,469,948	106,477,914	D_SPAD_AY16, SPAD_BOP_AY16
SGR30	A12	4,116,205	4,126,358	Chlb_FBP, Total_ab_FBP
SGR31	A12	10,742,829	10,742,829	FBP_blup
SGR32	A12	12,591,095	12,594,279	D_SPAD_blup, D_SPAD_AY17, SPAD_BOP_AY17
SGR33	A12	57,300,054	57,307,338	D_SPAD_AY16, SPAD_BOP_AY16
SGR34	A12	62,551,488	62,556,405	D_ratio_ab, D_ratio_ab
SGR35	A12	70,740,474	70,740,474	D_SPAD_blup, D_ratio_ab, D_SPAD_AY17
SGR36	A12	108,514,660	108,934,586	SPAD_BOP_AY17, D_SPAD_AY17, D_SPAD_blup
SGR37	A13	77,580,822	77,580,838	BOP_blup, SPAD_BOP_AY16
SGR38	D01	51,414,839	51,415,298	BOP_blup, D_SPAD_blup
SGR39	D01	57,675,353	57,675,364	Ratio_ab_FBP, Chla_FBP
SGR40	D01	60,671,107	60,671,202	D_SPAD_blup, D_SPAD_AY17
SGR41	D02	23,837,463	23,837,463	Ratio_ab_FBP
SGR42	D02	32,314,480	32,314,480	D_SPAD_AY17, SPAD_BOP_AY17
SGR43	D02	67,476,213	67,476,218	D_ratio_ab
SGR44	D06	55,810,689	56,177,650	Ratio_ab_BOP, D_ratio_ab, D_SPAD_HG17
SGR45	D07	510,384	510,546	Ratio_ab_BOP
SGR46	D07	10,915,964	11,432,705	SPAD_FBP_HG17, SPAD_FBP_HG17
SGR47	D07	26,125,297	26,125,366	Ratio_ab_FBP, D_ratio_ab, Chlb_FBP
SGR48	D08	43,939,104	43,939,104	Chlb_FBP
SGR49	D10	65,317,192	67,265,400	SPAD_BOP_AY17, BOP_blup, D_SPAD_AY17, FBP_blup, SPAD_FBP_HG16
SGR50	D11	9,657,009	10,060,413	FBP_blup, SPAD_FBP_HG16
SGR51	D12	23,540,471	23,540,583	D_SPAD_AY16
SGR52	D12	37,359,379	41,685,124	D_total_ab, Chla_BOP, BOP_blup, SPAD_BOP_HG16, SPAD_FBP_HG16
SGR53	D12	55,716,433	56,755,391	Ratio_ab_BOP, D_ratio_ab
SGR54	D13	26,198,781	26,198,804	D_SPAD_AY17, SPAD_BOP_AY17
SGR55	D13	64,827,390	64,827,424	Ratio_ab_FBP

SGR: senescence-related genomic region; Chr: chromosome.

2.3. Prediction of Candidate Genes

In this study, all the genes located in the 55 SGRs were identified as candidate senescence-related genes. Subsequently, 977 candidate genes were identified

(Supplementary Table S4). Of these, 853 candidate genes were annotated as orthologs in *Arabidopsis*. Notably, 156 genes were recorded in the leaf senescence database LSD 3.0, such as *EIN3* (*Gohir.A03G034800/Gohir.A03G034800*), *WRKY6* (*Gohir.D07G088100*), and *PPH* (*Gohir.D12G102900*) (Supplementary Table S5). This result suggests that our approach to dissecting the genetic basis of cotton senescence was effective. Furthermore, enrichment analysis of gene ontology (GO) biological processes (BPs) showed that the significant enrichments ($p < 0.05$) of these genes were associated with plant senescence-related processes, such as response to salt stress, ethylene processes, and leaf senescence (Figure 2). For example, *Gohir.D12G208700* (*GhRCD1*) is a homolog of *AT1G32230* in *Arabidopsis*, encoding a protein belonging to the (ADP-ribosyl) transferase domain-containing subfamily of the WWE protein–protein interaction domain protein family, and RCD1 was reported to be involved in superoxide-induced cell death [40,41]. *Gohir.A12G270200* (*GhJAZ3*) encodes jasmonate zim-domain protein 3, which negatively regulates *AtMYC2*, a key transcriptional activator of JA responses [42]. Most strikingly, we focused on the candidate gene *Gohir.A12G270900* (*GhMKK9*), which is a homolog of *AT1G73500* (*AtMKK9*), a member of the MAP kinase kinase family that was reported to play a positive role in leaf senescence of *Arabidopsis* [43].

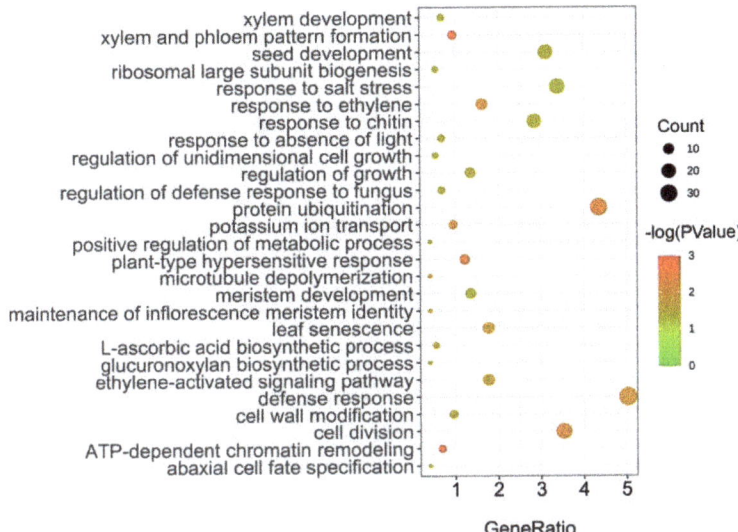

Figure 2. GO enrichment analysis of candidate genes associated with cotton senescence.

GhMKK9 is located in SGR36, which spans approximately 420 kb and is associated with three phenotypic values, D_SPAD_AY17, SPAD_BOP_AY17, and D_SPAD_blup (Figure 3A). In the genomic region, we discovered a non-synonymous SNP (A12_108859102) within the CDS region of *GhMKK9*, which causes a change in the base from C to T, as well as a change in amino acid from alanine (GCC) to valine (GTC) (Figure 3B). This SNP and another synonymous SNP (A12_108860059), also located in the CDS region, form two haplotypes, TG (Hap1) and GA (Hap2). In the associated panel, 158 cotton accessions carried Hap1, and 197 accessions carried Hap2. Although the SPAD values (FBP_blup) of Hap1 and Hap2 were not significantly different in the FBP period, the BOP_blup and D_SPAD_blup values of Hap1 were significantly higher than those of Hap2 (Figure 3C), indicating that Hap1 is a favorable haplotype for delaying cotton senescence.

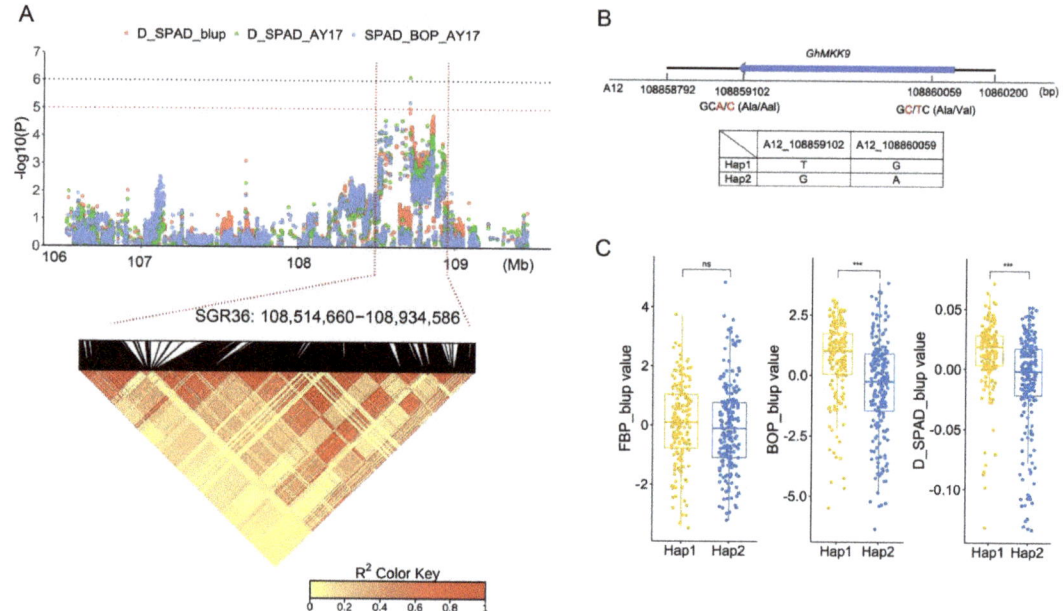

Figure 3. GWAS identification of candidate gene in the SGR36. (**A**) Manhattan plot (upper) and LD heat map (lower) of SGR36. (**B**) Gene structure and haplotypes of the candidate gene *GhMKK9*. (**C**) Phenotypes of different haplotypes. There are 158 accessions for Hap1 and 197 accessions for Hap2. Asterisks indicate significance levels (*** $p < 0.001$); ns, not significant.

2.4. GhMKK9, A Positive Regulator of Cotton Senescence

Quantitative real-time PCR (qRT-PCR) analysis showed that the expression level of *GhMKK9* in old cotton leaves was significantly higher than that in young cotton leaves (Figure 4A). Furthermore, we silenced the expression of *GhMKK9* in cotton using virus-induced gene silencing (VIGS) (Figure 4C). After one week of drought treatment, the CK group showed an obvious leaf wilting phenotype, whereas the VIGS-silenced plants (pTRV2-GhMKK9) only showed a barely visible wilting phenotype (Figure 4B). The SPAD value of cotton leaves in the CK group after drought treatment was also significantly lower than that of the VIGS-silenced plants (Figure 4D). Moreover, to further examine the function of *GhMKK9*, we overexpressed *GhMKK9* under the control of the 35S promoter (35S::GhMKK9) in *Arabidopsis* and obtained two transgenic lines (OE7 and OE14), which were confirmed by qRT-PCR (Figure 4F). After six weeks of culture under normal conditions, the overexpressing *Arabidopsis* lines OE7 and OE14 exhibited more severe senescence phenotypes than wild-type *Arabidopsis*, such as rosette leaf wilting and a higher degree of yellowing (Figure 4E). In addition, we determined the transcript levels of two senescence-marked genes, *AtSAG12* (up-regulated during senescence) [44,45] and *AtCAT2* (down-regulated during senescence) [46,47]. The transcript level of *AtSAG12* in the transgenic plants was significantly higher than that in the WT plants (Figure 4G), whereas the transcript level of *AtCAT2* in the transgenic plants was significantly lower than that in the WT plants (Figure 4H). Taken together, these results suggest that *GhMKK9* is a positive regulator of leaf senescence and may also be involved in drought-stress-induced senescence.

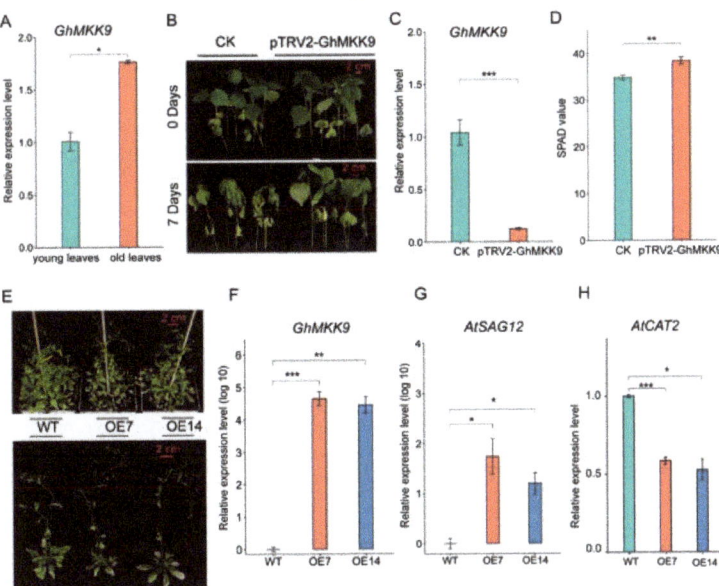

Figure 4. Functional analysis of the candidate gene *GhMKK9*. (**A**) Expression of *GhMKK9* in young and old cotton leaves by qRT-PCR. (**B**) Phenotypes of empty control (CK) and VIGS cotton plants (pTRV2-GhMKK9) under drought stress. After four weeks, the CK and VIGS cotton plants were treated with water shortage for 7 days. (**C**) Expression levels of *GhMKK9* in the CK and VIGS cotton plants. (**D**) SPAD value of the CK and VIGS plants under drought stress. (**E**) Phenotypes of six-week-old WT and transgenic *Arabidopsis* plants (OE7 and OE14). (**F**) Expression levels of *GhMKK9* in the WT and transgenic *Arabidopsis* plants. (**G,H**) Expression levels of senescence-marked genes *AtSAG12* and *AtCAT2* in the WT and transgenic *Arabidopsis* plants. Asterisks indicate significance levels (*** $p < 0.001$, ** $p < 0.01$, and * $p < 0.05$).

3. Discussion

The senescence process of plant leaves is a very complex biological regulation process which first depends on age and is also affected by external environmental signal stimuli [11]. Therefore, internal genetic and external environmental factors together determine the onset and rate of senescence. The senescence process in plants involves the remobilization and reutilization of nutrients from senescing parts as sinks [7,48], which is particularly important for crop plant products. Cotton fiber is one of the most important industrial textile fibers worldwide. Senescence has an important impact on the quality and yield of cotton fiber [6]. Compared with other crops, such as rice, wheat, and corn, cotton has the habit of indeterminate growth, which blurs the lines between growth, maturation, and senescence. Nevertheless, the flowering and boll period (FBP) is considered to be an important developmental stage of cotton because the plant undergoes a transition from vegetative to reproductive growth in which the level of plant endogenous hormones reaches a peak, photosynthesis is enhanced, and the activity of the "sink" is also enhanced. Then, in the boll-opening period (BOP), cotton senescence, such as chlorosis, is visible. Therefore, these two periods were chosen to study the regulation of senescence in cotton. Although senescence has received increasing attention in cotton breeding, research on the genetic basis of cotton senescence remains limited. In this study, chlorophyll content indices were selected as indicators to evaluate the senescence performance of the upland cotton population. Due to the combined action of genetic and environmental factors, the chlorophyll content varied widely across different planting locations and years. The SPAD value in the BOP period had a larger range of variation than that in the FBP period.

Moreover, the SPAD value in the FBP period showed higher heritability than that in the BOP period (0.65 and 0.41, respectively), which is similar to the results of the previous study [37]. These results indicate that environmental factors have a more significant impact on later cotton development.

A GWAS was performed based on 3,015,002 high-quality SNP markers from the resequencing data of 355 accessions to detect the genetic structure of cotton senescence. A total of 380 significant signals were identified. Given that functional variations are usually rare in GWAS [49], significant SNPs were integrated into genomic regions (GWAS loci). In the previous study, 50 genomic regions associated with cotton senescence were revealed based on SLAF-seq data of 185 accessions, which spanned a total of 51.50 Mb [37]. In the present study, 55 senescence-related genomic regions (SGRs) spanning approximately 18.09 Mb were identified. Compared with SLAF-seq-based GWAS, the resequencing data greatly increased the fine-mapping resolution. Six SGRs (SGR29, SGR39, SGR40, SGR43, SGR44, and SGR49) were located within ~1 Mb of the genomic regions reported in the previous study. (Figure 5). Interestingly, these SGRs were located in the D subgenome (except for SGR29) and were associated with the chlorophyll content in the BOP period and/or the diurnal variation of chlorophyll content (excepted for SGR39). These results suggest that the D subgenome plays an important role in the regulation of senescence in cotton. A range of abiotic and biotic stressors, such as drought, salt, and pathogen infection, can accelerate the onset and/or progression of plant senescence [7,8,50], and the D subgenome was reported to make an important contribution to stress tolerance in allotetraploid cotton [51]. This provides a possible explanation for the results.

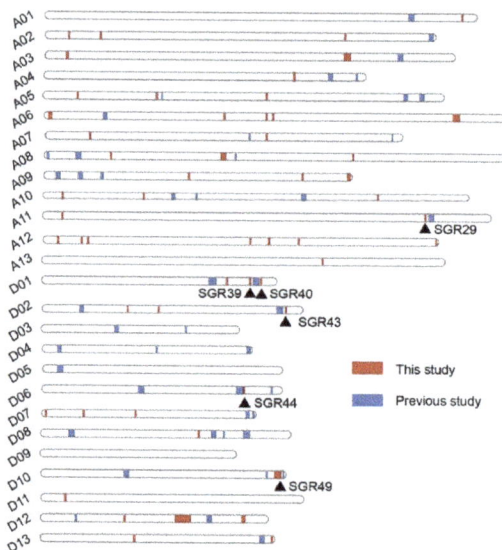

Figure 5. Distribution of senescence-related genomic regions at chromosomes from this and previous studies. Red vertical bars represent genomic regions from this study. Blue vertical bars represent genomic regions from previous study. Black triangles indicate SGRs located within ~1 Mb of the genomic regions reported by previous study.

Of the 55 SGRs, a total of 977 candidate genes were annotated. Among them, 156 genes were also recorded in the leaf senescence database LSD 3.0, and GO analysis revealed a set of biological processes, such as salt stress, ethylene processes, and leaf senescence. This suggests that the theory used in this study was effective. Interestingly, focus was given to a candidate gene, *Gohir.A12G270900*, which is homologous to *AT1G73500* and encodes an MKK9 protein in *Arabidopsis*. *AtMKK9* plays an important role in the regulation of *Arabidop-*

sis senescence [43]. There are many signaling pathways in plants that involve responses to external stimuli, and one of the most common is the MAPK signaling pathway. In eukaryotes, the MAPs cascade signaling pathway is a highly conserved signaling module [52,53]. Each MAPKs cascade signaling module is composed of three protein kinases that act in sequence: MPK, MKK, and MKKK. In *Arabidopsis*, there are 20 MPK genes, 10 MKK genes, and 69 MKKK genes [54]. In upland cotton, there may be 52 GhMKs, 23 GhMKKs, and 166 GhMKKKs genes [55]. The candidate gene *Gohir.A12G270900* (*GhMKK9*) is a member of the GhMKK family. *GhMKK9* is located in SGR36, which is associated with multiple senescence phenotypes, indicating high repeatability and reliability. Interestingly, a non-synonymous SNP (A12_108859102) and synonymous SNP (A12_108860059) were observed in the exon region of *GhMKK9*. The SNP A12_108859102 changed the amino acid from alanine (GCC) to valine (GTC), which may affect the function of the GhMKK9 protein. In addition, these two SNPs formed two haplotypes, Hap1 and Hap2 (Figure 3C). The BOP_blup and D_SPAD_blup values of the Hap1 are significantly higher than those of the Hap2. These results suggest that Hap1 is a favorable haplotype for delaying senescence and that the *GhMKK9* gene may play an important role in the regulation of cotton senescence.

The function of the *GhMKK9* gene was further verified. By qRT-PCR analysis, it was found that the expression level of *GhMKK9* was significantly higher in old cotton leaves than that in young cotton leaves, and overexpression of *GhMKK9* gene in *Arabidopsis thaliana* promoted the senescence process of *Arabidopsis* leaves, indicating that *GhMKK9* is a positive regulator of plant senescence, which is consistent with the results of a previous study [43,56]. In *Rosa hybrida*, *RhMKK9* silencing significantly delayed petal senescence in flowers [57]. The MKK9–MPK6 module was reported to play an important role in the regulation of the senescence process [43], in which MPK3/MPK6 could be activated by MKK9 to induce ethylene biosynthesis [56–58]. Furthermore, the endogenous *GhMKK9* gene in cotton was silenced using VIGS. *GhMKK9* gene-silenced plants were found to have enhanced drought tolerance compared with the control plants (CK), indicating that *GhMKK9* may be involved in drought-stress-induced senescence in cotton. MKK9 is widely involved in the transmission of environmental signals, but its effects on plant stress tolerance remain controversial. For example, Yoo et al. [59] and Shen et al. [60] showed that *AtMKK9* is a positive regulator of salt tolerance in *Arabidopsis*, which is contrary to the results reported by Alzwiy and Morris [61] and Xu et al. [56]. Similarly, although there is no obvious difference between WT and *mkk9* mutant *Arabidopsis* plants under drought stress [61], this study shows that the silencing of *GhMKK9* enhances drought tolerance in cotton. These discrepant results may be attributed to different experimental methods and functionally redundant genes [60].

4. Materials and Methods

4.1. Plant Materials

The association mapping panels consisted of 355 upland cotton accessions (Supplementary Table S6), and the germplasm resources were obtained from the Institute of Cotton Research of Chinese Academy of Agricultural Sciences (ICR-CAAS). These materials are geographically widespread across China, including in the Yellow River Region (YRR), the Yangtze River Region (YZRR), the Northwest Inland Region (NIR), and the Northern Specific Early-Maturity Region (NSER), and a few were from abroad (e.g., the United States) [62]. In 2016 and 2017, 355 upland cotton accessions were planted in Anyang (AY), Henan (36°08' N, 114°48' E), and Huanggang (HG), Hubei (31°14' N, 114°78' E), respectively. Three replicates were planted in each environment, except Anyang in 2017, where two replicates were used.

4.2. Phenotyping and Data Analysis

The relative chlorophyll level (SPAD) of association panels was measured with the chlorophyll meter SPAD-502 (Konica Minolta, Japan) in four environments in the flowering and boll-setting period (FBP) and boll-opening period (BOP). The third parietal leaf from the top was selected after topping to measure the chlorophyll level, and the average SPAD of at least three individuals for each accession was recorded. The absolute chlorophyll concentration of the materials planted in Anyang in 2017 was also measured. Three discs of 0.6 cm diameter were cut by punch from the third parietal leaf, and these leaf discs were mixed from at least three individuals for each accession. Chlorophyll concentration was estimated using the method described by Arnon [63]. Four chlorophyll concentration indices were obtained: chlorophyll a (Chla), chlorophyll b (Chlb), total chlorophyll (Total_ab), and chlorophyll a/b (Ratio_ab). In addition, the diurnal variation of chlorophyll content was calculated using the following formulae: D (%) = (chlorophyll content of BOP − chlorophyll content of FBP)/(chlorophyll content of BOP × days between FBP and BOP) × 100%, which included D_SPAD, D_chla, D_chlb, D_total_ab, and D_ratio_ab.

The best linear unbiased predictions (BLUPs) and broad-sense heritability (H^2) of SPAD values in the four environments were calculated using the R package sommer [64]. Broad-sense heritability was defined as $H^2 = \sigma_g^2/(\sigma_g^2 + \sigma_{gl}^2/l + \sigma_{gy}^2/y + \sigma_e^2/rly)$, where σ_g^2 is the genotypic variance; σ_{gl}^2 is the interactions of genotype with location; σ_{gy}^2 is the interactions of genotype with year; σ_e^2 is the error variance; and l, y, and r are the number of locations, years, and replications, respectively. Statistical and correlation analyses were performed using the R package Hmisc [65] and visualized using the package corrplot [66].

4.3. SNP Genotyping

The resequencing data of 355 upland cotton accessions were reported in a previous study [62]. The quality of paired-end reads from 355 accessions was evaluated using FastQC v.0.11.9 [67] and was controlled using Trimmomatic v.0.39 [68]. All high-quality clean reads were mapped to the *Gossypium hirsutum* v1.1 reference genome [69] with BWA mem v.0.7.17 [70]. The mapping results were sorted and converted to the BAM format using Picard tools (http://broadinstitute.github.io/picard). GATK v.4.1.8 [71] was used to detect variants following the best-practice workflows. High-quality SNPs were filtered with: "QD < 2.0 QUAL < 30.0 FS > 60.0 MQ < 40.0 MQRankSum < −12.5 ReadPosRankSum < −8.0", missing rate < 50%, and MAF > 0.05.

4.4. GWAS and Identification of Genomic Regions

A linear mixed model was used to perform GWAS on 355 upland cotton accessions, implemented in the EMMAX software [38]. Before conducting the GWAS, the SNPs were imputed using Beagle v.5.1 [72]. Both trial values of the single environment and BLUPs were used for the GWAS. Because a high correlation between SNPs always leads to information redundancy, PLINK was used to detect the number of genome-wide, effective SNPs. The parameters for pruning were as follows: within a 500 bp sliding window, $r^2 \geq 0.2$, and a step of 100 bp. After pruning, 925,819 SNPs were obtained, and the genome-wide significance cutoff for GWAS was selected as $p = 1 \times 10^{-6}$ (1/925819). Significant SNPs were then determined using the following criteria: (1) $p < 10^{-6}$ or (2) $p < 10^{-5}$ in at least two environmental trail values owing to the stability. To identify senescence-related genomic regions (SGRs), we selected independent, significant SNPs ($r^2 < 0.6$). If $r^2 > 0.1$, the SNP with $p < 10^{-3}$ and independent significant SNP were merged into the same genomic region. In addition, if the distance between two genomic regions was less than 900 kb, they were merged into one genomic region. The R packages CMplot [73], LDheatmap [74], and ggplot2 [75] were used to visualize the GWAS results.

4.5. Prediction of Candidate Genes

All genes located in SGRs were selected as putative candidate genes based on the *Gossypium hirsutum* v1.1 reference genome [69]. Homologs of these genes in *Arabidopsis thaliana* were determined using BLAST [76], and GO enrichment was performed on the database for annotation, visualization, and integrated discovery (DAVID) to identify enriched biological themes [77,78].

4.6. RNA Extraction and qRT-PCR

To determine the expression level of *GhMKK9*, the cotton accession "CRI 10" was planted in a greenhouse, and two-week-old (young) and eight-week-old (old) leaves were sampled from eight individuals with three biological replicates in each group. Total RNA was extracted using an RNA Purification Kit (Tiangen, Beijing, China), and the RNA was reverse transcribed using the PrimeScript RT Reagent Kit (TAKARA, Dalian, China) following the manufacturer's instructions. Quantitative real-time PCR (qRT-PCR) was performed on a Roche Applied Science LightCycler 480 using the NovoStart® SYBR qPCR SuperMix Plus (Novoprotein, Shanghai, China). The qRT-PCR was conducted as follows: pre-denaturation at 95 °C for 60 s; 40 cycles of 95 °C for 20 s and 60 °C for 60 s. Three technical replicates were performed for each sample, and the relative expression of genes was calculated using the $2^{-\Delta\Delta Ct}$ method [79]. The primers are listed in Supplementary Table S7.

4.7. VIGS

For the VIGS assays, one fragment of *GhMKK9* amplified from the cDNA of "CRI 10" was integrated into the pTRV2 vector (pTRV2-GhMKK9) using the nimble cloning method [80] and then the recombinant vector was introduced into *Agrobacterium tumefaciens* GV3101. *Agrobacterium* strains harboring the pTRV2-GhMKK9 and pTRV2 (negative control) vectors combined with strains harboring the pTRV1 vector were co-transferred into the cotyledons of 2-week-old cotton plants following previously described methods [81]. The injected plants were kept in darkness for 24 h and transferred to a greenhouse at 25 °C with 16 h light/8 h dark cycle. Four weeks after injection, plants injected with pTRV2 and pTRV2-GhMKK9 were subjected to drought treatment, and SPAD values were determined. The primers used for the construction of the VIGS vector and qRT-PCR are listed in Supplementary Table S7.

4.8. Genetic Transformation of Arabidopsis Thaliana

The ORF of *GhMKK9* was inserted into the binary expression vector pNC-Cam2304 to generate the 35S::GhMKK9 construct using the nimble cloning method [80]. The 35::GhMKK9 construct was introduced into *Agrobacterium tumefaciens* GV3101 and then transformed into *Arabidopsis* ecotype Columbia using the floral dip method [82]. The positive plants were screened out using 1/2 MS medium containing kanamycin (100 mg/L) and confirmed via qRT-PCR. The T_3 homozygous generation plants were used for phenotypic observation of senescence. To observe the performance of transgenic plants under normal conditions, seeds of WT and two independent 35S::GhMKK9 lines (OE7 and OE14) were germinated on 1/2 MS agar medium. After two weeks, the seedlings were transplanted into the soil. Phenotypic characteristics were observed, and the rosette leaves at position six from six-week-old plants were sampled for qRT-PCR. Primers used for the construction of 35::GhMKK9 and qRT-PCR are listed in Supplementary Table S7. The primer specificity of GhMKK9 were confirmed (Supplementary Figure S8).

Supplementary Materials: The following supporting information can be downloaded at: https://www.mdpi.com/article/10.3390/ijms23158584/s1.

Author Contributions: Conceptualization, Q.L., Z.F. and S.Y.; methodology, Q.L. and L.L.; software, Q.L.; validation, Q.L., C.H. and J.W.; formal analysis, Q.L.; investigation, Q.L. and L.L.; resources, S.Y.; data curation, Q.L.; writing—original draft preparation, Q.L. and Z.F.; writing—review and editing, L.L.; visualization, Q.L.; supervision, Z.F.; project administration, Z.F.; funding acquisition, S.Y. All authors have read and agreed to the published version of the manuscript.

Funding: This research was sponsored by the Program for Research and Development of Zhejiang A&F University (2021LFR005).

Institutional Review Board Statement: Not applicable.

Informed Consent Statement: Not applicable.

Data Availability Statement: The datasets presented in this study can be found in online repositories. The names of the repository/repositories and accession number(s) can be found in the Supplementary Materials.

Acknowledgments: We would like to thank Pu Yan (Institute of Tropical Bioscience and Biotechnology, Chinese Academy of Tropical Agricultural Sciences) for providing the vectors.

Conflicts of Interest: The authors declare no conflict of interest.

References

1. Shahrajabian, M.H.; Sun, W.; Cheng, Q. Considering White Gold, Cotton, for its Fiber, Seed Oil, Traditional and Modern Health Benefits. *J. Biol. Environ. Sci.* **2020**, *14*, 25–39.
2. Gallagher, J.P.; Grover, C.E.; Rex, K.; Moran, M.; Wendel, J.F. A New Species of Cotton from Wake Atoll, Gossypium Stephensii (Malvaceae). *Syst. Bot.* **2017**, *42*, 115–123. [CrossRef]
3. Grover, C.; Zhu, X.; Grupp, K.; Jareczek, J.; Gallagher, J.; Szadkowski, E.; Seijo, J.G.; Wendel, J. Molecular Confirmation of Species Status for the Allopolyploid Cotton Species, Gossypium Ekmanianum Wittmack. *Genet. Resour. Crop Evol.* **2015**, *62*, 103–114. [CrossRef]
4. Fang, D.D.; Jenkins, J.N.; Deng, D.D.; McCarty, J.C.; Li, P.; Wu, J. Quantitative Trait Loci Analysis of Fiber Quality Traits Using a Random-Mated Recombinant Inbred Population in Upland Cotton (*Gossypium hirsutum*, L.). *BMC Genom.* **2014**, *15*, 397. [CrossRef]
5. Hulse-Kemp, A.M.; Lemm, J.; Plieske, J.; Ashrafi, H.; Buyyarapu, R.; Fang, D.D.; Frelichowski, J.; Giband, M.; Hague, S.; Hinze, L.L.; et al. Development of a 63K SNP Array for Cotton and High-Density Mapping of Intraspecific and Interspecific Populations of *Gossypium* spp. *G3 Genes Genomes Genet.* **2015**, *5*, 1187–1209. [CrossRef]
6. Chen, Y.; Dong, H. Mechanisms and Regulation of Senescence and Maturity Performance in Cotton. *Field Crops Res.* **2016**, *189*, 1–9. [CrossRef]
7. Lim, P.O.; Kim, H.J.; Nam, H.G. Leaf Senescence. *Annu. Rev. Plant Biol.* **2007**, *58*, 115–136. [CrossRef]
8. Guo, Y.; Ren, G.; Zhang, K.; Li, Z.; Miao, Y.; Guo, H. Leaf Senescence: Progression, Regulation, and Application. *Mol. Hortic.* **2021**, *1*, 1–25. [CrossRef]
9. Diaz, C.; Saliba-Colombani, V.; Loudet, O.; Belluomo, P.; Moreau, L.; Daniel-Vedele, F.; Morot-Gaudry, J.F.; Masclaux-Daubresse, C. Leaf Yellowing and Anthocyanin Accumulation are Two Genetically Independent Strategies in Response to Nitrogen Limitation in Arabidopsis Thaliana. *Plant Cell Physiol.* **2006**, *47*, 74–83. [CrossRef]
10. Woo, H.R.; Kim, H.J.; Lim, P.O.; Nam, H.G. Leaf Senescence: Systems and Dynamics Aspects. *Annu. Rev. Plant Biol.* **2019**, *70*, 347–376. [CrossRef]
11. Woo, H.R.; Kim, H.J.; Nam, H.G.; Lim, P.O. Plant Leaf Senescence and Death—Regulation by Multiple Layers of Control and Implications for Aging in General. *J. Cell Sci.* **2013**, *126*, 4823–4833. [CrossRef]
12. Jing, H.-C.; Schippers, J.H.; Hille, J.; Dijkwel, P.P. Ethylene-Induced Leaf Senescence Depends on Age-Related Changes and OLD Genes in Arabidopsis. *J. Exp. Bot.* **2005**, *56*, 2915–2923. [CrossRef]
13. Zhang, K.; Xia, X.; Zhang, Y.; Gan, S.-S. An ABA-Regulated and Golgi-Localized Protein Phosphatase Controls Water Loss during Leaf Senescence in Arabidopsis. *Plant J.* **2012**, *69*, 667–678. [CrossRef]
14. Piao, W.; Kim, S.-H.; Lee, B.-D.; An, G.; Sakuraba, Y.; Paek, N.-C. Rice Transcription Factor OsMYB102 Delays Leaf Senescence by Down-Regulating Abscisic Acid Accumulation and Signaling. *J. Exp. Bot.* **2019**, *70*, 2699–2715. [CrossRef]
15. Zhang, Y.; Wang, Y.; Wei, H.; Li, N.; Tian, W.; Chong, K.; Wang, L. Circadian Evening Complex Represses Jasmonate-Induced Leaf Senescence in Arabidopsis. *Mol. Plant* **2018**, *11*, 326–337. [CrossRef]
16. Zhang, Y.; Ji, T.-T.; Li, T.-T.; Tian, Y.-Y.; Wang, L.-F.; Liu, W.-C. Jasmonic Acid Promotes Leaf Senescence through MYC2-Mediated Repression of CATALASE2 Expression in Arabidopsis. *Plant Sci.* **2020**, *299*, 110604. [CrossRef]
17. Zhang, M.; Su, J.; Zhang, Y.; Xu, J.; Zhang, S. Conveying Endogenous and Exogenous Signals: MAPK Cascades in Plant Growth and Defense. *Cell Signal. Gene Regul.* **2018**, *45*, 1–10. [CrossRef]
18. Abdelkhalik, A.F.; Shishido, R.; Nomura, K.; Ikehashi, H. QTL-Based Analysis of Leaf Senescence in an Indica/Japonica Hybrid in Rice (*Oryza sativa*, L.). *Theor. Appl. Genet.* **2005**, *110*, 1226–1235. [CrossRef]
19. Singh, U.M.; Sinha, P.; Dixit, S.; Abbai, R.; Venkateshwarlu, C.; Chitikineni, A.; Singh, V.K.; Varshney, R.K.; Kumar, A. Unraveling Candidate Genomic Regions Responsible for Delayed Leaf Senescence in Rice. *PLoS ONE* **2020**, *15*, e0240591. [CrossRef]
20. Vijayalakshmi, K.; Fritz, A.K.; Paulsen, G.M.; Bai, G.; Pandravada, S.; Gill, B.S. Modeling and Mapping QTL for Senescence-Related Traits in Winter Wheat under High Temperature. *Mol. Breed.* **2010**, *26*, 163–175. [CrossRef]
21. Bogard, M.; Jourdan, M.; Allard, V.; Martre, P.; Perretant, M.R.; Ravel, C.; Heumez, E.; Orford, S.; Snape, J.; Griffiths, S.; et al. Anthesis Date Mainly Explained Correlations between Post-Anthesis Leaf Senescence, Grain Yield, and Grain Protein Concentration in a Winter Wheat Population Segregating for Flowering Time QTLs. *J. Exp. Bot.* **2011**, *62*, 3621–3636. [CrossRef] [PubMed]
22. Pinto, R.S.; Lopes, M.S.; Collins, N.C.; Reynolds, M.P. Modelling and Genetic Dissection of Staygreen under Heat Stress. *Theor. Appl. Genet.* **2016**, *129*, 2055–2074. [CrossRef] [PubMed]
23. Chapman, E.A.; Orford, S.; Lage, J.; Griffiths, S. Capturing and Selecting Senescence Variation in Wheat. *Front. Plant Sci.* **2021**, *12*, 638738. [CrossRef] [PubMed]

4. Wehner, G.G.; Balko, C.C.; Enders, M.M.; Humbeck, K.K.; Ordon, F.F. Identification of Genomic Regions Involved in Tolerance to Drought Stress and Drought Stress Induced Leaf Senescence in Juvenile Barley. *BMC Plant Biol.* **2015**, *15*, 125. [CrossRef]
5. Zhang, J.; Fengler, K.A.; Van Hemert, J.L.; Gupta, R.; Mongar, N.; Sun, J.; Allen, W.B.; Wang, Y.; Weers, B.; Mo, H.; et al. Identification and Characterization of a Novel Stay-Green QTL That Increases Yield in Maize. *Plant Biotechnol. J.* **2019**, *17*, 2272–2285. [CrossRef]
6. Xu, W.; Subudhi, P.K.; Crasta, O.R.; Rosenow, D.T.; Mullet, J.E.; Nguyen, H.T. Molecular Mapping of QTLs Conferring Stay-Green in Grain Sorghum (*Sorghum Bicolor* L. Moench). *Genome* **2000**, *43*, 461–469. [CrossRef]
7. Sanchez, A.; Subudhi, P.; Rosenow, D.; Nguyen, H. Mapping QTLs Associated with Drought Resistance in Sorghum (*Sorghum bicolor* L. Moench). *Plant Mol. Biol.* **2002**, *48*, 713–726. [CrossRef]
8. Harris, K.; Subudhi, P.; Borrell, A.; Jordan, D.; Rosenow, D.; Nguyen, H.; Klein, P.; Klein, R.; Mullet, J. Sorghum Stay-Green QTL Individually Reduce Post-Flowering Drought-Induced Leaf Senescence. *J. Exp. Bot.* **2007**, *58*, 327–338. [CrossRef]
9. Kiranmayee, K.U.; Hash, C.T.; Sivasubramani, S.; Ramu, P.; Amindala, B.P.; Rathore, A.; Kishor, P.K.; Gupta, R.; Deshpande, S.P. Fine-Mapping of Sorghum Stay-Green QTL on Chromosome10 Revealed Genes Associated with Delayed Senescence. *Genes* **2020**, *11*, 1026. [CrossRef]
10. Hurtado, P.X.; Schnabel, S.K.; Zaban, A.; Veteläinen, M.; Virtanen, E.; Eilers, P.H.; Van Eeuwijk, F.A.; Visser, R.G.; Maliepaard, C. Dynamics of Senescence-Related QTLs in Potato. *Euphytica Neth. J. Plant Breed.* **2012**, *183*, 289–302.
11. Visscher, P.M.; Wray, N.R.; Zhang, Q.; Sklar, P.; McCarthy, M.I.; Brown, M.A.; Yang, J. 10 Years of GWAS Discovery: Biology, Function, and Translation. *Am. J. Hum. Genet.* **2017**, *101*, 5–22. [CrossRef]
12. Kermanshahi, F.; Ghazizadeh, H.; Hussein, N.A.; Amerizadeh, F.; Samadi, S.; Tayefi, M.; Khodabandeh, A.K.; Moohebati, M.; Ebrahimi, M.; Esmaily, H.; et al. Association of a Genetic Variant in the AKT Gene Locus and Cardiovascular Risk Factors. *Cell. Mol. Biol.* **2020**, *66*, 57–64. [CrossRef]
13. Shamari, A.-R.; Mehrabi, A.-A.; Maleki, A.; Rostami, A. Association Analysis of Tolerance to Dieback Phenomena and Trunk Form Using ISSR Markers in Quercus Brantii. *Cell. Mol. Biol.* **2018**, *64*, 116–124. [CrossRef]
14. Akan, G.; Kisenge, P.; Sanga, T.S.; Mbugi, E.; Adolf, I.; Turkcan, M.K.; Janabi, M.; Atalar, F. Common SNP-Based Haplotype Analysis of the 9p21. 3 Gene Locus as Predictor Coronary Artery Disease in Tanzanian Population. *Cell. Mol. Biol.* **2019**, *65*, 33–43. [CrossRef]
15. Zhao, Y.; Qiang, C.; Wang, X.; Chen, Y.; Deng, J.; Jiang, C.; Sun, X.; Chen, H.; Li, J.; Piao, W.; et al. New Alleles for Chlorophyll Content and Stay-Green Traits Revealed by a Genome Wide Association Study in Rice (*Oryza sativa*). *Sci. Rep.* **2019**, *9*, 1–11. [CrossRef]
16. Sekhon, R.S.; Saski, C.; Kumar, R.; Flinn, B.S.; Luo, F.; Beissinger, T.M.; Ackerman, A.J.; Breitzman, M.W.; Bridges, W.C.; de Leon, N.; et al. Integrated Genome-Scale Analysis Identifies Novel Genes and Networks Underlying Senescence in Maize. *Plant Cell* **2019**, *31*, 1968–1989. [CrossRef]
17. Liu, Q.; Li, L.; Feng, Z.; Yu, S. Uncovering Novel Genomic Regions and Candidate Genes for Senescence-Related Traits by Genome-Wide Association Studies in Upland Cotton (*Gossypium hirsutum* L.). *Front. Plant Sci.* **2021**, *12*, 809522. [CrossRef]
18. Kang, H.M.; Sul, J.H.; Service, S.K.; Zaitlen, N.A.; Kong, S.; Freimer, N.B.; Sabatti, C.; Eskin, E. Variance Component Model to Account for Sample Structure in Genome-Wide Association Studies. *Nat. Genet.* **2010**, *42*, 348–354. [CrossRef]
19. Maurano, M.T.; Humbert, R.; Rynes, E.; Thurman, R.E.; Haugen, E.; Wang, H.; Reynolds, A.P.; Sandstrom, R.; Qu, H.; Brody, J.; et al. Systematic Localization of Common Disease-Associated Variation in Regulatory DNA. *Science* **2012**, *337*, 1190–1195. [CrossRef]
20. Overmyer, K.; Tuominen, H.; Kettunen, R.; Betz, C.; Langebartels, C.; Sandermann, H., Jr.; Kangasjärvi, J. Ozone-Sensitive Arabidopsis Rcd1 Mutant Reveals Opposite Roles for Ethylene and Jasmonate Signaling Pathways in Regulating Superoxide-Dependent Cell Death. *Plant Cell* **2000**, *12*, 1849–1862. [CrossRef]
21. Overmyer, K.; Brosché, M.; Pellinen, R.; Kuittinen, T.; Tuominen, H.; Ahlfors, R.; Keinänen, M.; Saarma, M.; Scheel, D.; Kangasjärvi, J. Ozone-Induced Programmed Cell Death in the Arabidopsis Radical-Induced Cell Death1 Mutant. *Plant Physiol.* **2005**, *137*, 1092–1104. [CrossRef]
22. Chini, A.; Fonseca, S.; Chico, J.M.; Fernández-Calvo, P.; Solano, R. The ZIM Domain Mediates Homo-and Heteromeric Interactions between Arabidopsis JAZ Proteins. *Plant J.* **2009**, *59*, 77–87. [CrossRef]
23. Zhou, C.; Cai, Z.; Guo, Y.; Gan, S. An Arabidopsis Mitogen-Activated Protein Kinase Cascade, MKK9-MPK6, Plays a Role in Leaf Senescence. *Plant Physiol.* **2009**, *150*, 167–177. [CrossRef]
24. Noh, Y.-S.; Amasino, R.M. Identification of a Promoter Region Responsible for the Senescence-Specific Expression of SAG12. *Plant Mol. Biol.* **1999**, *41*, 181–194. [CrossRef]
25. Xiao, S.; Dai, L.; Liu, F.; Wang, Z.; Peng, W.; Xie, D. COS1: An Arabidopsis *coronatine insensitive1* Suppressor Essential for Regulation of Jasmonate-Mediated Plant Defense and Senescence. *Plant Cell* **2004**, *16*, 1132–1142. [CrossRef]
26. Zimmermann, P.; Heinlein, C.; Orendi, G.; Zentgraf, U. Senescence-Specific Regulation of Catalases in Arabidopsis THALIANA (L.) Heynh. *Plant Cell Environ.* **2006**, *29*, 1049–1060. [CrossRef]
27. Wang, C.; Li, T.; Liu, Q.; Li, L.; Feng, Z.; Yu, S. Characterization and Functional Analysis of GhNAC82, A NAM Domain Gene, Coordinates the Leaf Senescence in Upland Cotton (*Gossypium hirsutum* L.). *Plants* **2022**, *11*, 1491. [CrossRef]
28. Schippers, J.H.; Schmidt, R.; Wagstaff, C.; Jing, H.-C. Living to Die and Dying to Live: The Survival Strategy behind Leaf Senescence. *Plant Physiol.* **2015**, *169*, 914–930. [CrossRef]

49. Watanabe, K.; Taskesen, E.; Van Bochoven, A.; Posthuma, D. Functional Mapping and Annotation of Genetic Associations with FUMA. *Nat. Commun.* **2017**, *8*, 1–11. [CrossRef]
50. Guo, Y.; Gan, S. Leaf Senescence: Signals, Execution, and Regulation. *Curr. Top. Dev. Biol.* **2005**, *71*, 83–112. [CrossRef] [PubMed]
51. Zhang, T.; Hu, Y.; Jiang, W.; Fang, L.; Guan, X.; Chen, J.; Zhang, J.; Saski, C.A.; Scheffler, B.E.; Stelly, D.M.; et al. Sequencing of Allotetraploid Cotton (*Gossypium hirsutum* L. Acc. TM-1) Provides a Resource for Fiber Improvement. *Nat. Biotechnol.* **2015**, *33*, 531–537. [CrossRef] [PubMed]
52. Ichimura, K.; Shinozaki, K.; Tena, G.; Sheen, J.; Henry, Y.; Champion, A.; Kreis, M.; Zhang, S.; Hirt, H.; Wilson, C.; et al. Mitogen-Activated Protein Kinase Cascades in Plants: A New Nomenclature. *Trends Plant Sci.* **2002**, *7*, 301–308. [CrossRef]
53. Widmann, C.; Gibson, S.; Jarpe, M.B.; Johnson, G.L. Mitogen-Activated Protein Kinase: Conservation of a Three-Kinase Module from Yeast to Human. *Physiol. Rev.* **1999**, *79*, 143–180. [CrossRef] [PubMed]
54. Zhang, M.; Zhang, S. Mitogen-Activated Protein Kinase Cascades in Plant Signaling. *J. Integr. Plant Biol.* **2022**, *64*, 301–341. [CrossRef]
55. Yin, Z.; Zhu, W.; Zhang, X.; Chen, X.; Wang, W.; Lin, H.; Wang, J.; Ye, W. Molecular Characterization, Expression and Interaction of MAPK, MAPKK and MAPKKK Genes in Upland Cotton. *Genomics* **2021**, *113*, 1071–1086. [CrossRef]
56. Xu, J.; Li, Y.; Wang, Y.; Liu, H.; Lei, L.; Yang, H.; Liu, G.; Ren, D. Activation of MAPK Kinase 9 Induces Ethylene and Camalexin Biosynthesis and Enhances Sensitivity to Salt Stress in Arabidopsis. *J. Biol. Chem.* **2008**, *283*, 26996–27006. [CrossRef]
57. Chen, J.; Zhang, Q.; Wang, Q.; Feng, M.; Li, Y.; Meng, Y.; Zhang, Y.; Liu, G.; Ma, Z.; Wu, H.; et al. RhMKK9, a Rose MAP KINASE KINASE Gene, Is Involved in Rehydration-Triggered Ethylene Production in Rose Gynoecia. *BMC Plant Biol.* **2017**, *17*, 51. [CrossRef]
58. Meng, Y.; Ma, N.; Zhang, Q.; You, Q.; Li, N.; Ali Khan, M.; Liu, X.; Wu, L.; Su, Z.; Gao, J. Precise Spatio-Temporal Modulation of ACC Synthase by MPK 6 Cascade Mediates the Response of Rose Flowers to Rehydration. *Plant J.* **2014**, *79*, 941–950. [CrossRef]
59. Yoo, S.-D.; Cho, Y.-H.; Tena, G.; Xiong, Y.; Sheen, J. Dual Control of Nuclear EIN3 by Bifurcate MAPK Cascades in C_2H_4 Signalling. *Nature* **2008**, *451*, 789–795. [CrossRef]
60. Shen, L.; Zhuang, B.; Wu, Q.; Zhang, H.; Nie, J.; Jing, W.; Yang, L.; Zhang, W. Phosphatidic Acid Promotes the Activation and Plasma Membrane Localization of MKK7 and MKK9 in Response to Salt Stress. *Plant Sci.* **2019**, *287*, 110190. [CrossRef]
61. Alzwiy, I.A.; Morris, P.C. A Mutation in the Arabidopsis MAP Kinase Kinase 9 Gene Results in Enhanced Seedling Stress Tolerance. *Plant Sci.* **2007**, *173*, 302–308. [CrossRef]
62. Li, L.; Zhang, C.; Huang, J.; Liu, Q.; Wei, H.; Wang, H.; Liu, G.; Gu, L.; Yu, S. Genomic Analyses Reveal the Genetic Basis of Early Maturity and Identification of Loci and Candidate Genes in Upland Cotton (*Gossypium hirsutum* L.). *Plant Biotechnol. J.* **2021**, *19*, 109–123. [CrossRef]
63. Arnon, D.I. Copper Enzymes in Isolated Chloroplasts. Polyphenoloxidase in Beta Vulgaris. *Plant Physiol.* **1949**, *24*, 1. [CrossRef]
64. Zhang, B.; Tieman, D.M.; Jiao, C.; Xu, Y.; Chen, K.; Fei, Z.; Giovannoni, J.J.; Klee, H.J. Chilling-Induced Tomato Flavor Loss Is Associated with Altered Volatile Synthesis and Transient Changes in DNA Methylation. *Proc. Natl. Acad. Sci. USA* **2016**, *113*, 12580–12585. [CrossRef]
65. Harrell, F.E.; Dupont, C. Hmisc: Harrell Miscellaneous, Version 4.5-0. 2021. Available online: https://cran.r-project.org/package=Hmisc (accessed on 1 December 2021).
66. Wei, T.; Simko, V. R Package "Corrplot": Visualization of a Correlation Matrix, Version 0.84. 2021. Available online: https://githubcom/taiyun/corrplot (accessed on 3 January 2022).
67. Andrews, S. *FastQC: A Quality Control Tool for High Throughput Sequence Data*; Babraham Institute: Cambridge, UK, 2010; Available online: http://www.bioinformatics.babraham.ac.uk/projects/fastqc (accessed on 15 March 2021).
68. Bolger, A.M.; Lohse, M.; Usadel, B. Trimmomatic: A Flexible Trimmer for Illumina Sequence Data. *Bioinformatics* **2014**, *30*, 2114–2120. [CrossRef]
69. Chen, Z.J.; Sreedasyam, A.; Ando, A.; Song, Q.; De Santiago, L.M.; Hulse-Kemp, A.M.; Ding, M.; Ye, W.; Kirkbride, R.C.; Jenkins, J.; et al. Genomic Diversifications of Five Gossypium Allopolyploid Species and Their Impact on Cotton Improvement. *Nat. Genet.* **2020**, *52*, 525–533. [CrossRef]
70. Li, H. Aligning Sequence Reads, Clone Sequences and Assembly Contigs with BWA-MEM. *arXiv* **2013**, arXiv:13033997.
71. Van der Auwera, G.A.; O'Connor, B.D. *Genomics in the Cloud: Using Docker, GATK, and WDL in Terra*; O'Reilly Media: Sebastopol, CA, USA, 2020.
72. Browning, B.L.; Zhou, Y.; Browning, S.R. A One-Penny Imputed Genome from next-Generation Reference Panels. *Am. J. Hum. Genet.* **2018**, *103*, 338–348. [CrossRef]
73. Yin, L.; Zhang, H.; Tang, Z.; Xu, J.; Yin, D.; Zhang, Z.; Yuan, X.; Zhu, M.; Zhao, S.; Li, X.; et al. RMVP: A Memory-Efficient, Visualization-Enhanced, and Parallel-Accelerated Tool for Genome-Wide Association Study. *Genom. Proteom. Bioinform.* **2021**, *19*, 618–628. [CrossRef]
74. Shin, J.-H.; Blay, S.; McNeney, B.; Graham, J. LDheatmap: An r Function for Graphical Display of Pairwise Linkage Disequilibria between Single Nucleotide Polymorphisms. *J. Stat. Softw.* **2006**, *16*, 1–9. [CrossRef]
75. Wickham, H. *Ggplot2: Elegant Graphics for Data Analysis*; Springer: New York, NY, USA, 2016.
76. Camacho, C.; Coulouris, G.; Avagyan, V.; Ma, N.; Papadopoulos, J.; Bealer, K.; Madden, T.L. BLAST+: Architecture and Applications. *BMC Bioinform.* **2009**, *10*, 421. [CrossRef] [PubMed]

77. Huang, D.W.; Sherman, B.T.; Lempicki, R.A. Systematic and Integrative Analysis of Large Gene Lists Using DAVID Bioinformatics Resources. *Nat. Protoc.* **2009**, *4*, 44–57. [CrossRef] [PubMed]
78. Huang, D.W.; Sherman, B.T.; Lempicki, R.A. Bioinformatics Enrichment Tools: Paths toward the Comprehensive Functional Analysis of Large Gene Lists. *Nucleic Acids Res.* **2009**, *37*, 1–13. [CrossRef] [PubMed]
79. Livak, K.J.; Schmittgen, T.D. Analysis of Relative Gene Expression Data Using Real-Time Quantitative PCR and the $2^{-\Delta\Delta CT}$ Method. *Methods* **2001**, *25*, 402–408. [CrossRef]
80. Yan, P.; Zeng, Y.; Shen, W.; Tuo, D.; Li, X.; Zhou, P. Nimble Cloning: A Simple, Versatile, and Efficient System for Standardized Molecular Cloning. *Front. Bioeng. Biotechnol.* **2020**, *7*, 460. [CrossRef]
81. Gao, X.; Britt Jr, R.C.; Shan, L.; He, P. Agrobacterium-Mediated Virus-Induced Gene Silencing Assay in Cotton. *J. Vis. Exp. JoVE* **2011**, *54*, 2938. [CrossRef]
82. Bent, A. Arabidopsis Thaliana Floral Dip Transformation Method. *Agrobacterium Protoc.* **2006**, *343*, 87–104. [CrossRef]

Article

Five Rice Seed-Specific *NF-YC* Genes Redundantly Regulate Grain Quality and Seed Germination via Interfering Gibberellin Pathway

Huayu Xu [1,†], Shufan Li [1,†], Bello Babatunde Kazeem [1], Abolore Adijat Ajadi [1], Jinjin Luo [1], Man Yin [1], Xinyong Liu [1], Lijuan Chen [1], Jiezheng Ying [1], Xiaohong Tong [1], Yifeng Wang [1], Baixiao Niu [2], Chen Chen [2], Xiaoshan Zeng [3,*] and Jian Zhang [1,*]

[1] State Key Lab of Rice Biology, China National Rice Research Institute, Hangzhou 311400, China; xuhuayu202107@163.com (H.X.); shufanli1994@163.com (S.L.); tunlapa2k13@gmail.com (B.B.K.); threetriplea@yahoo.com (A.A.A.); luojinjin998220@163.com (J.L.); 13545663721@163.com (M.Y.); liuxinyong1234@gmail.com (X.L.); chenlijuan0723@163.com (L.C.); yingjiezheng@caas.cn (J.Y.); tongxiaohong@caas.cn (X.T.); wangyifeng@caas.cn (Y.W.)
[2] College of Agriculture, Yangzhou University, Yangzhou 225009, China; bxniu@yzu.edu.cn (B.N.); chenchen@yzu.edu.cn (C.C.)
[3] Hunan Rice Research Institute, Hunan Academy of Agricultural Sciences, Changsha 410125, China
* Correspondence: zengxiaoshan2@163.com (X.Z.); zhangjian@caas.cn (J.Z.); Tel./Fax: +86-731-86491768 (X.Z.); +86-571-63370277 (J.Z.)
† These authors contributed equally to this work.

Citation: Xu, H.; Li, S.; Kazeem, B.B.; Ajadi, A.A.; Luo, J.; Yin, M.; Liu, X.; Chen, L.; Ying, J.; Tong, X.; et al. Five Rice Seed-Specific *NF-YC* Genes Redundantly Regulate Grain Quality and Seed Germination via Interfering Gibberellin Pathway. *Int. J. Mol. Sci.* 2022, 23, 8382. https://doi.org/10.3390/ijms23158382

Academic Editor: Cristina Martínez-Villaluenga

Received: 26 May 2022
Accepted: 26 July 2022
Published: 29 July 2022

Publisher's Note: MDPI stays neutral with regard to jurisdictional claims in published maps and institutional affiliations.

Copyright: © 2022 by the authors. Licensee MDPI, Basel, Switzerland. This article is an open access article distributed under the terms and conditions of the Creative Commons Attribution (CC BY) license (https://creativecommons.org/licenses/by/4.0/).

Abstract: NF-YCs are important transcription factors with diverse functions in the plant kingdoms including seed development. *NF-YC8*, *9*, *10*, *11* and *12* are close homologs with similar seed-specific expression patterns. Despite the fact that some of the *NF-YCs* are functionally known; their biological roles have not been systematically explored yet, given the potential functional redundancy. In this study, we generated pentuple mutant *pnfyc* of *NF-YC8-12* and revealed their functions in the regulation of grain quality and seed germination. *pnfyc* grains displayed significantly more chalkiness with abnormal starch granule packaging. *pnfyc* seed germination and post-germination growth are much slower than the wild-type NIP, largely owing to the GA-deficiency as exogenous GA was able to fully recover the germination phenotype. The RNA-seq experiment identified a total of 469 differentially expressed genes, and several GA-, ABA- and grain quality control-related genes might be transcriptionally regulated by the five NF-YCs, as revealed by qRT-PCR analysis. The results demonstrated the redundant functions of *NF-YC8-12* in regulating GA pathways that underpin rice grain quality and seed germination, and shed a novel light on the functions of the seed-specific *NF-YCs*.

Keywords: rice (*Oryza sativa* L.); gibberellins; abscisic acid; NF-YCs

1. Introduction

Rice (*Oryza sativa* L.) is a major cereal crop in the world, as it is consumed as a staple food by more than half of the world's population [1]. Rice seed is a complex organ that is comprised of a maternal caryopsis coat, a diploid embryo and a triploid endosperm. The nutrients such as starch, protein and lipids are accumulated in the endosperm underpinning seed germination or grain yield and quality for human consumption. It has been known that phytohormones are extensively involved in the regulation of plant seed development [2–5]. The action of GAs and ABA on seed development is strictly correlated and antagonistic [2]. Through the investigation of the rice seed hormonal dynamics during the grain filling stage, Yang et al. (2001) revealed that GAs play key roles in embryogenesis, while the ABA content reached the peak at a much later stage, thus it seemed to be more relevant to the seed maturation [6]. So far, numerous pieces of literature about genes controlling rice seed development have been published, and these genes are involved in transcriptional regulation, the ubiquitin–proteasome pathway, plant hormone response, and so on [7–9]. Specifically, Yao et al. (2017) found that *NF-YC8* to *NF-YC12* are five important genes involved in starch synthesis, seed storage protein and the stress response [7]. The study shows that *NF-YC12* is a key transcription factor in regulating endosperm development [8] and storage material accumulation in rice seeds [10]. A novel transcription factor subunit *NF-YC13* was identified in indica rice, which can respond to salt stress signals by interacting with the B-subunit [11]. It is studied that NF-YC2 and NF-YC4 proteins can interact with three flowering-time genes to regulate the photoperiodic flowering response under long sunlight conditions [12].

Nuclear Factor Y (NF-Y) is a family of transcription factors that are found in vast quantities in higher eukaryotes. The NF-Y protein complex consists of three subunits: NF-YA (CBF-B/HAP2), NF-YB (CBF-A/HAP3), and NF-YC (CBF-C/HAP5) which usually forms a heterotrimer to regulate the transcription of the target genes [13,14]. For yeast and animals, each NF-Y subunit is encoded by a single gene. However, the situation in plants is more complicated with multiple members of each subunit, which dramatically expanded the diversity of *NF-Ys'* gene function in the plant kingdom [13,15]. In rice, each NF-Y subunit covers more than 10 gene members as reported, and many of them have been identified to participate in extensive developmental processes like nutrient accumulation in the endosperm, flowering regulation and ABA signal response [16–19]. *NF-YBs* are well-documented among those three subunits and have been implicated in plant height regulation, grain yield, carbon assimilation, photoperiodic flowering and other processes [20–24]. For example, *NF-YB2*, *NF-YB3* and *NF-YB4* are close homologs that are functionally redundant in regulating chloroplast biogenesis in rice [25]. It is noteworthy that several *NF-YBs* and *NF-YCs* were found to be specifically expressed in rice endosperm. Some seed-specific *NF-YBs* control rice seed development by affecting the nutrition accumulation and the loading of sucrose into developing seeds [26–28]. In addition, endosperm-specific *NF-YBs* and *NF-YCs* may also form heterotrimer complexes with other non-NF-Y transcription factors, hence regulating grain filling and quality via the ubiquitin–proteasome pathway [8,29]. For example, AtNF-YB9-YC12-bZIP67 can activate the expression of *SUS2* and promote seed development [30]; OsNF-YB1 interacts with OsNF-YC12, and OsNF-YC12 can bind to the promoter of *FLO6* and *OsGS1;3* to regulate grain weight and chalky endosperm [10].

Several previous works have revealed that OsNF-YC8 (LOC_Os01g01290), OsNF-YC9 (LOC_Os01g24460), OsNF-YC10 (LOC_Os01g39850), OsNF-YC11 (LOC_Os10g23910) and OsNF-YC12 (LOC_Os05g11580) are close homologs with similar seed-specific expression pattern, implying that they play a role in rice seed development [7,8,29,31]. So far, it is known that the NF-YC9 controls cell proliferation to influence grain width, and NF-YC11 regulates the accumulation of storage substances in rice seeds [10,32]. In 2019, our lab reported that NF-YC12 forms a heterotrimer complex with NF-YB1 and bHLH144 to regulate rice grain quality [8]. Nevertheless, knowledge about the function of the five genes is rather fragmented, given that the high similarity of the genes may give rise to functional redundancy. Here, we report the systematic functional analysis of NF-YC8, 9, 10, 11 and 12 using single gene or pentuple gene mutants. The five genes may work redundantly to regulate ABA and GA response, thus determining grain quality and seed germination. This work sheds new insight into the functional roles of the seed-specific OsNF-YCs.

2. Results and Discussion

2.1. Five Seed-Specific NY-YCs Works Redundantly to Regulate Grain Quality

To specify the biological functions of *OsNF-YC8*, *OsNF-YC9*, *OsNF-YC10*, *OsNF-YC11* and *OsNF-YC12*, we generated a single mutant of each gene in the background of Kitaake (*Oryza sativa* ssp. Japonica), respectively. Sanger sequencing further confirmed the mutations of the corresponding genes with insertion or deletions, which should have shifted the open reading frame and disrupted the resulting protein functions (Figure S1). Compared with Kitaake, *nfyc8* exhibited increased percentage of grains with chalkiness (PGWC), *nfyc12* had higher degree of chalkiness (DEC), while *nfyc9* and *nfyc10* exhibited increased PGWC and higher DEC (Figure S2A–C). We used the CRISPR/Cas9 technology to simultaneously knock out all the five *NF-YC* genes in Nipponbare (NIP, *Oryza sativa* ssp. japonica) to generate the pentuple *nf-yc* mutants (hereafter referred to as *pnfyc*) to assess the potential functional redundancy among the genes. Two representative lines *pnfyc-1* and *pnfyc-2* were selected for the followed genotyping and genetic analysis. As shown in Figure S3, Sanger sequencing detected various types of homozygous insertion or deletion mutations in each of the *NF-YC8-12*, suggesting all the five genes were successfully knocked-out. During the vegetative growth stage, no visible differences were observed in major agronomic traits such as plant height, flowering date, seed setting and spikelets per panicle in the *pnfyc* lines (Table S1). However, the milled grains of *pnfyc* lines showed obvious chalkiness. As revealed by the cross-sections of the *pnfyc* seeds, the starchy endosperm of *pnfyc* was floury-white when compared with NIP. Scanning electron microscopy (SEM) images of transverse sections indicated that the starch granule of NIP and *pnfyc* grains had different morphologies, shape and packaging densities. Unlike the regular shape of the starch granule of NIP, *pnfyc* had irregular, loosely packed starch granules, which might be responsible for the observed chalkiness (Figure 1A). Furthermore, we examined the contents of storage substances in the brown seeds, and found that the total starch and amylose contents of *pnfyc* were significantly lower than that of the NIP. Conversely, *pnfyc* had relatively higher crude protein contents than NIP (Figure 1B–D). The PGWC in *pnfyc* reached over 90%, while that of NIP was less than 10% (Figure 1E). Following the change in the starch contents, differential scanning calorimetry (DSC) analysis demonstrated that the gelatinization characteristics including the onset, peak as well as end gelatinization temperatures of *pnfyc* were also significantly altered (Figure 1F and Table S2). The results above indicated that *OsNF-YC8, 9, 10, 11* and *12* work redundantly to positive regulate rice grain quality, particularly the grain chalkiness.

2.2. pnfyc Are GA-Deficient with Retarded Seed Germination

The altered storage substance proportion in *pnfyc* lines provoked us to test the roles of the five *NF-YCs* in seed germination. We carried out germination assays on the *nfyc* single mutants, *pnfyc* and NIP lines on $\frac{1}{2}$ MS medium for 4 days. All the five *nfyc* single mutants showed slightly lower germination rates compared with Kitaake, which was followed by retarded post-germination growth (Figure S4A–C). However, the *pnfyc* seeds showed much retarded germination (56.7–60.0%) than the NIP (90.0–96.7%) at day 4, which further confirmed the functional redundancy among the five NF-YCs. Given the key roles of ABA and GA in seed germination regulation, we subsequently investigated the seed germination under exogenous ABA, GA and PAC (GA biosynthesis inhibitor) treatments (Figure 2A–C). The application of 2 μM exogenous ABA significantly restrained the seed germination of both *pnfyc* and NIP seeds. To evaluate the relative ABA sensitivity of the seeds, we calculated the relative germination rate of the seeds under mock and ABA treatments. The results showed that the WT relative germination rates of 2 μM ABA/mock and 5 μM ABA/mock were 72.2% and 65.5%, respectively. However, for the *pnfyc* seeds, 84.5% and 51.4% were obtained. Therefore, *pnfyc* is hypersensitive to exogenous ABA treatment in seed germination ($p < 0.05$) (Figure 2D). Similar hypersensitivity to ABA inhibition effects was also observed in the post-germination growth of *pnfyc* seedlings. In contrast to the ABA treatments, the application of 2 μM exogenous GA significantly recovered the seed germination of *pnfyc*, and a more intense recovering effect was observed when 5 μM exogenous GA was applied, suggesting the GA deficiency might be the major reason for the retarded seed germination in *pnfyc*. Exogenous PAC displayed similar inhibitory effects as ABA, as 10 μM exogenous PAC decreased the NIP germination rate and post-germination growth to the *pnfyc* level. In addition, we quantified the GA3 contents in the germinative seeds of NIP and *pnfyc-1*. The result showed that the GA3 content in *pnfyc-1* was reduced to only 30% of the NIP ($p < 0.01$), which is in agreement with the recovered germination of *pnfyc* by GA (Figure 2E).

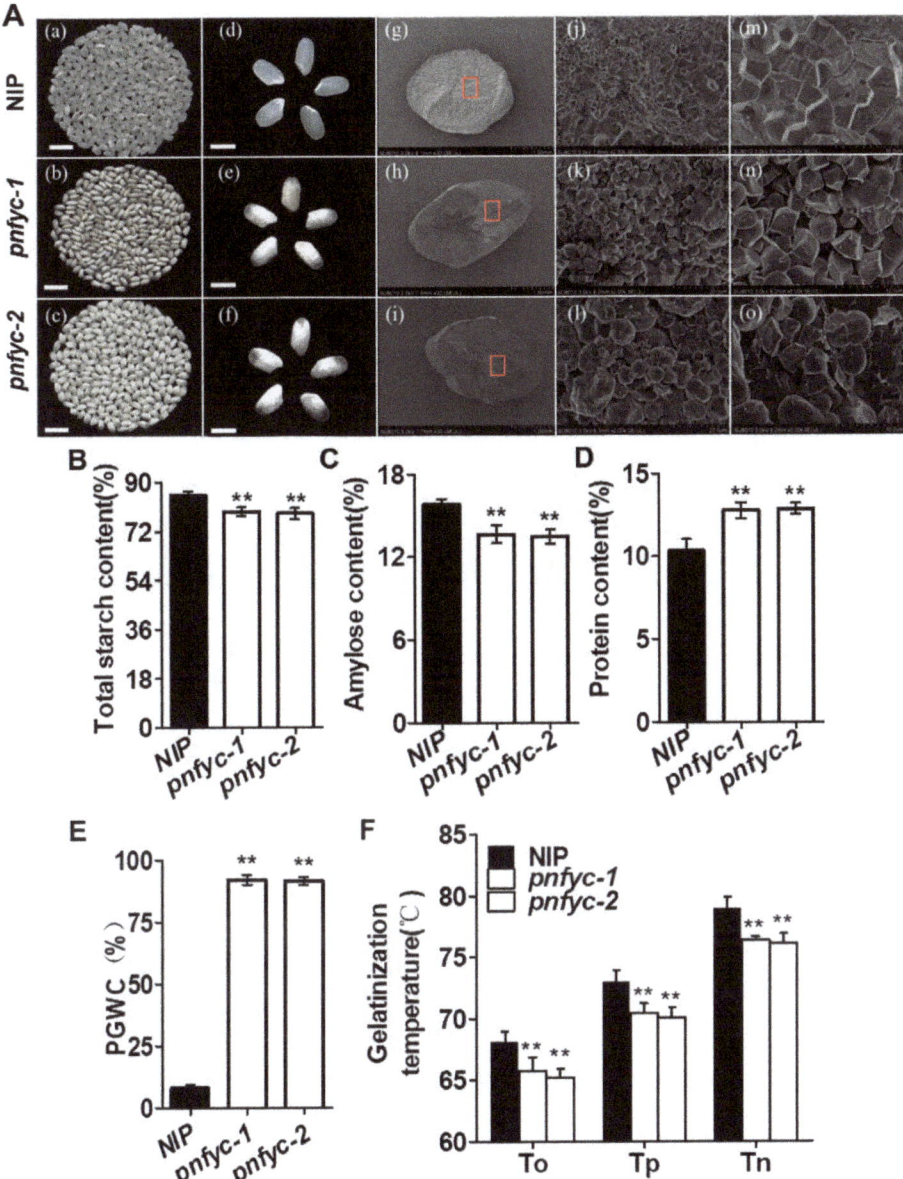

Figure 1. (**A**) (a–o) Grain chalkiness phenotypical characterization (a–f), bar = 2 mm. Scanning electron microscopy (SEM) analysis (g–o). The central areas shown are indicated as red squares. The magnification is 30 times in (g,h,i); 2000 times in (j,k,l), and 5000 times in (m,n,o). (**B–F**) Quality trait parameters of mature seeds from *pnfyc* lines and NIP. Data are shown as means ± SD of at least three biological replicates. (** $p < 0.01$ by two-tailed Student's *t*-test).

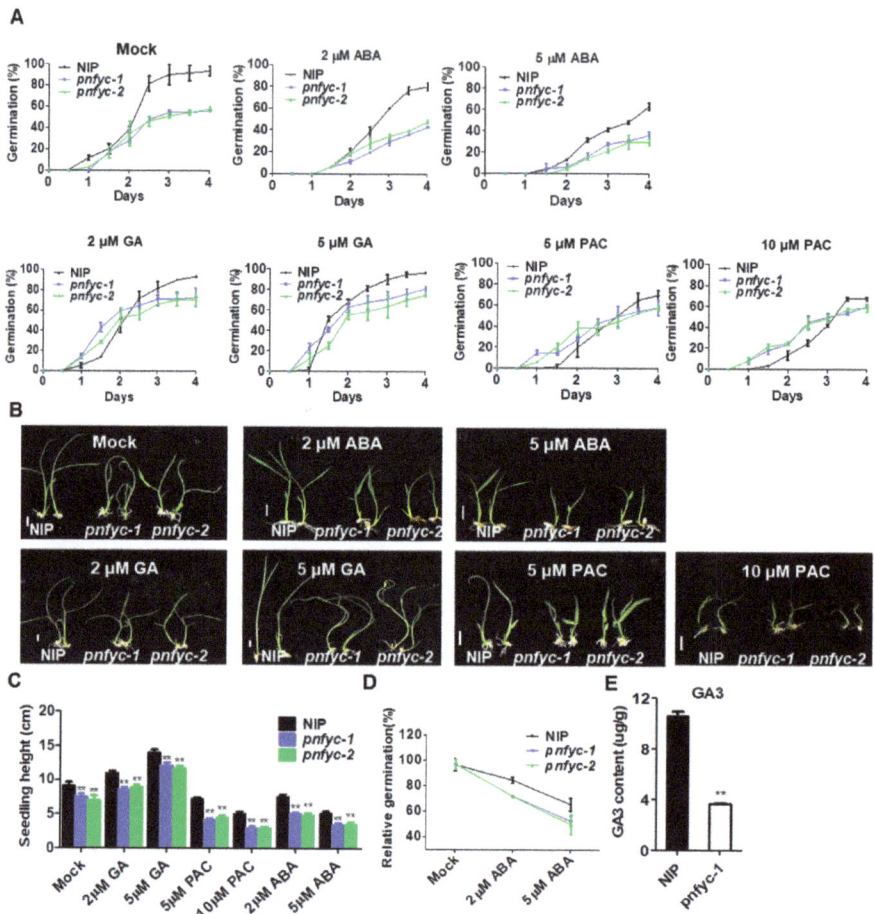

Figure 2. (**A**) The germination rate of NIP and *pnfyc* lines under different concentrations of exogenous hormones. (**B**) The plant morphology of NIP and *pnfyc* lines treated with different exogenous hormones at 7 days after germination. (**C**) The seedling height of NIP and *pnfyc* lines under different exogenous hormones at 7 days after germination. Bar = 1 cm. (**D**) The relative germination of the NIP and *pnfyc* seeds under ABA treatments were determined after 4 days and expressed as a percentage of those grown under 'mock' conditions. (**E**) Quantification of GA3 derivatives in NIP and *pnfyc* seeds germinated for 6 h was analyzed with liquid chromatography-tandem mass spectrometry. Data are shown as means ± SD of at least three biological replicates. (** $p < 0.01$ by two-tailed Student's *t*-test).

To test the involvement of the five *NF-YCs* in ABA and GA biosynthesis and signaling, we examined the transcription of the genes in the germinative seeds of various ABA or GA-related genetic lines by qRT-PCR. *SAPK8, 9* and *10* are core elements of ABA signaling, and over-expression of the genes conferred plants ABA hypersensitivity [33]. We found that *NF-YC10* was significantly up-regulated by *SAPK8* and *10*, while *NF-YC8* and *NF-YC12* were up-regulated by *SAPK10* and *SAPK9*, respectively. However, *NF-YC9* and *NF-YC11* were down-regulated in all the three *OxSAPK* lines (Figure 3A). In the GA-deficient mutant *sd1* [34], the transcription of *NF-YC8, 9, 10* and *11* were all severely repressed, indicating the five *NF-YCs* are highly responsive to endogenous GA level (Figure 3B). Taken together, we proposed that *NF-YC8, 9, 10, 11* and *12* may serve as key regulators mediating the balance of GA and ABA.

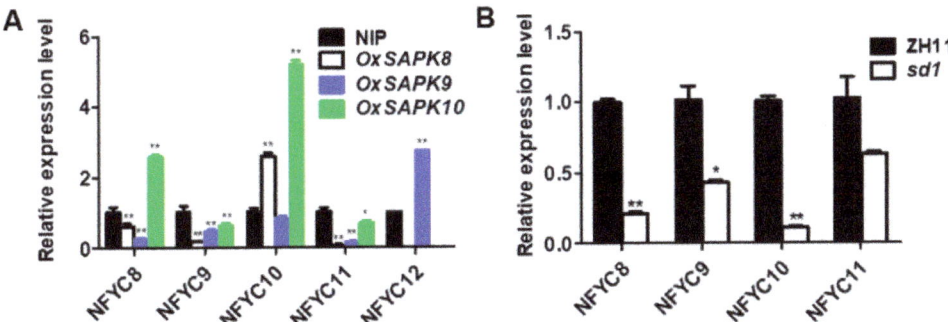

Figure 3. (**A**) The expression level of *NFYC8* to *NFYC12* in the seeds of wild-type-NIP, *OxSAPK8*, *OxSAPK9*, and *OxSAPK10* mutants that germinated for 6 h. (**B**) The expression level of *NFYC8* to *NFYC11* in the seeds of wild-type-ZH11 and *sd1* mutant that germinated for 6 h ($*p < 0.05$, $** p < 0.01$ by two-tailed Student's *t*-test).

Although *NF-YC10* and *NF-YC12* have been reported as key regulators of rice seed development [8,10,32], the roles of the five seed-specific NY-YCs are still not very clear so far, given the potential functional redundancy among them. By simultaneously knocking out the five genes, we revealed their functions in grain quality and seed germination, and the GA-deficiency in *pnfyc* might be the major reason for the observed phenotype. Aside from the well-known function as a seed germination promoter, GA has been recently found to regulate endosperm development as well. Cui et al. (2020) reported that application of exogenous GA_{4+7} significantly altered the content of other phytohormones such as auxin, zeatin and ABA, increased the activities of superoxide dismutase, catalases, and peroxidases and reduced the malondialdehyde content, which finally improved grain filling and yield in maize [35]. In Arabidopsis, *NF-YC3*, *NF-YC4* and *NF-YC9* have redundant roles in the regulation of GA-ABA-mediated seed germination [36]. NF-YCs bind RGL2, a repressor during GA signaling, and then in the form of NF-YC–RGL2 module targets *ABI5*, a key factor in ABA signaling [37,38]. NF-YC can bind to the CCAAT-box on the *ABI5* promoter to regulate ABI5 gene expression. The NF-YC–RGL2–ABI5 module integrates ABA and GA signaling to regulate seed germination [36].

2.3. NF-YCs Regulates the Transcription of ABA and GA Pathway Genes

RNA sequencing experiments on the 6 HAI (6 h after imbibition) germinating seeds of *pnfyc* and WT were carried out to identify the potential target genes of the five NF-YCs. As a result, there were a total of 469 differentially expressed genes (DEGs) as shown in (Supplementary Table S1). KEGG analysis revealed predominant enrichment of the DEGs on the pathways of 'phenylpropanoid biosynthesis', 'protein processing in endoplasmic reticulum' and 'plant hormone signal transduction' (Figure S5A). We further carried out a qRT-PCR analysis on eight randomly selected DEGs to verify the RNA-seq results, and the results showed that seven of those genes had a similar transcriptional level inclination to that detected by the RNA-seq, indicating the high reliability of our RNA-seq results (Figure S5B). Notably, a series of genes reported as critical regulators of the ABA signal pathway were found to be down-regulated in *pnfyc* (Figure 4B), including positive ABA signaling factors like *OsbZIP46* [39,40], *OsbZIP12* [41,42] and *OsNAC52* [43] as well as the *OsHSP24.1* encoding an ABA-responsive heat shock protein [44]. Furthermore, we conducted the qRT-PCR and results showed that the expression of most GA biosynthesis genes was down-regulated in *pnfyc* plants, while the expression of ABA biosynthesis and negative signal pathway-related genes was up-regulated (Figure 4A,B).

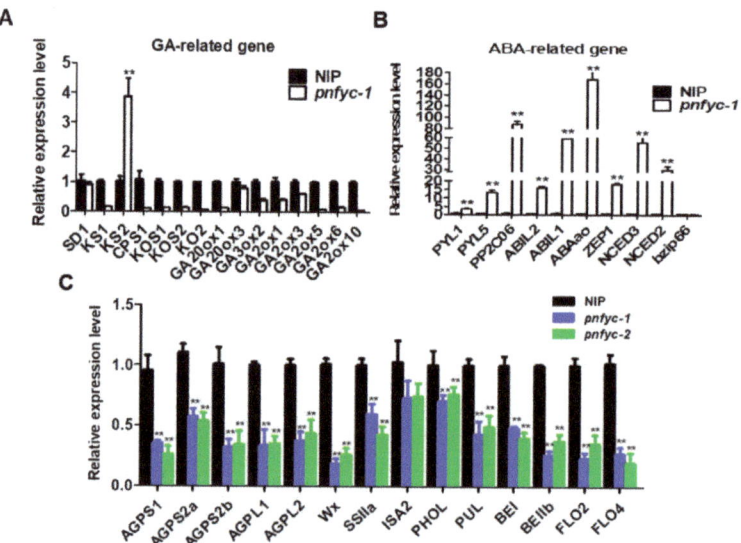

Figure 4. (**A**) The expression level of GA synthesis and metabolism-related genes in the seeds of wild-type-NIP and *pnfyc* mutants that germinated for 6 h. (**B**) The expression level of ABA biosynthesis and negative signal pathway-related genes in the seeds of wild-type-NIP and *pnfyc* mutants that germinated for 6 h. (**C**) The expression level of starch synthesis-related genes in wild-type-NIP and *pnfyc* mutant seeds 7 days after pollination. Data are shown as means ± SD of at least three biological replicates. (** $p < 0.01$ by two-tailed Student's *t*-test).

Given the severely affected starch qualities in *pnfyc* seeds, we also examined the transcriptional levels of several starch biosynthesis enzyme or regulator genes in the developing seeds of *pnfyc* and NIP. It was found that, except for *ISA2*, all of these ADP-glucose pyrophosphorylase, granule-bound starch synthase, starch synthase and starch branching enzyme were mostly down-regulated in 7 DAP endosperm of *pnfyc* lines (Figure 4C) [45–48].

2.4. Potential Interactive Proteins of the Five NF-YCs

We tested the protein–protein interaction between the 5 NF-YCs and 10 SAPKs which are ABA signaling components. A total of 50 NF-YC-SAPK combinations were tested by yeast-two-hybrid, and results showed that NF-YC10 binds to SAPK4, 6 and 10, while all the other combinations were negative. Hence, the suggestion is that NF-YC10 perceives the ABA signal from SAPK4, 6 and 10 (Figure 5A).

Our previous study has demonstrated that NF-YB1-YC12 dimer binds to bHLH144 to form a heterotrimer complex that regulates rice grain quality [8]. To identify other components that may interact with NF-YB1-YC12 dimer, we performed yeast-three hybrid (Y3H) experiments to screen a seed-derived prey library using NF-YB$_1$-YC$_{12}$-pBRIDGE as bait. We finally obtained three interactive proteins LOC_Os01g68950, LOC_Os07g46160 and LOC_Os11g38670 which are annotated as ubiquitin domain-containing protein, BTB/POZ domain-containing protein and dead-box ATP-dependent RNA helicase, respectively. Interestingly, the interactions are valid only on the SD/-Met/-Leu/-Trp/-Ade/-His/+X-α-Gal medium, in which the drop-out of methionine drove the expression of NF-YC12 under Met25 promoter. Meanwhile, the interactions were compromised on SD/-Leu/-Trp/-Ade/-His/+X-α-Gal medium, in which NF-YC12 was suppressed by the supplemented methionine in the medium. Hence, the binding of NF-YB1-YC12 is necessary for the formation of heterotrimer complexes with the three proteins (Figure 5B–D).

In conclusion, we report a rice pentuple gene mutant *pnfyc*, which knocked out five homologous genes *OsNF-YC8*, *OsNF-YC9*, *OsNF-YC10*, *OsNF-YC11* and *OsNF-YC12* simultaneously. The expression of starch synthesis genes decreased in *pnfyc*, resulting in the decrease of starch content and the increase of protein content, the change of grain quality and the significant increase of chalkiness trait. The results showed that *NF-YC8* to *NF-YC12* could regulate grain quality traits by regulating starch synthesis. In addition, *NF-YC8-12* also inhibited seed germination by affecting the expression of GA-related genes, and the phenotype is significantly restored by applying exogenous GA. Finally, the expression of ABA-related genes in *pnfyc* increased, and *pnfyc* seeds were hypersensitive to exogenous ABA. NFYC10 could interact with SAPK to regulate ABA expression.

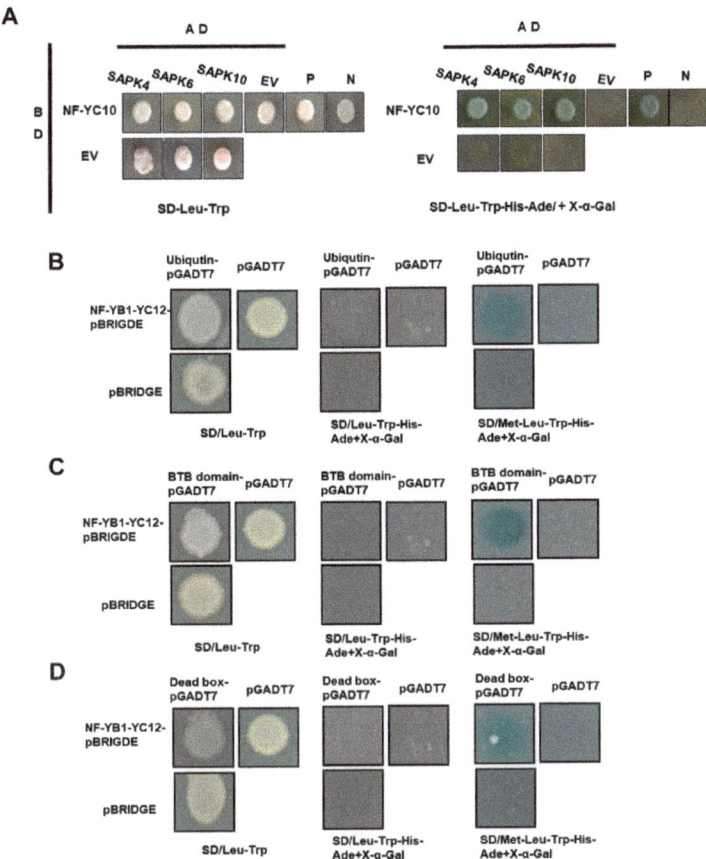

Figure 5. (**A**) Y2H assay of interaction between NF-YC10 and SAPKs. BD: pDEST32; AD: pDEST22; EV: empty vector, pDEST32 or pDEST22; P: positive control, pGBKT7-53/pGADT7-T; N: positive control, pGBKT7-Lam/pGADT7-T. (**B**) Y3H analysis of NF-YB1, NF-YC12, and Ubiquitin protein domain. (**C**) Y3H analysis of NF-YB1, NF-YC12, and BTB/POZ protein domain. (**D**) Y3H analysis of NF-YB1, NF-YC12, and Dead-box protein domain.

3. Materials and Methods

3.1. Plant Growth Conditions and Phenotype Measurement

To generate the CRISPR/Cas9-derived knock-out mutants, the specific target small-guide RNA (sgRNA) of each gene was designed and assembled in a pYLCRISPR/CAS9-MH vector system according to a previous report [49], and subsequently transformed into Nipponbare and Kitaake (*Oryza sativa*, ssp. japonica) backgrounds. All the plants were grown in the experimental greenhouse and field of the China National Rice Research

Institute (CNRRI). Agronomic traits were analyzed with 10 replicates. Panicle length, number of primary branch panicles, number of effective panicles per plant, seed setting rate (%), and plant height were measured manually. Rice seed grains were harvested and air-dried at room temperature for at least 2 weeks. The thousand-grain-weight, seed length, width and chalkiness were examined by a seed phenotyping system (Wan Sheng, Hangzhou, China). Grain thickness was determined at the same time for each grain using an electronic digital calliper.

3.2. Physicochemical Properties of Seed Grain

Total starch content of the dried brown seeds was measured using a starch assay kits Megazyme K-TSTA and KAMYL (Megazyme, Ireland, UK, http://www.megazyme.com/ accessed on 6 May 2020). The total amylose and protein contents in the grains were measured by following a previous report [50]. The content is expressed as the percentage of total sample weight on an oven-dry basis. To analyze the gelatinization temperature, DSC assay was conducted on a differential scanning calorimeter DSC1 STARe system (METTLER-TOLEDO, Zurich, Switzerland). Briefly, 5 mg rice powder was sealed and placed in an aluminum sample cup, mixed with 10 µL distilled water, and then the samples were analyzed by the differential scanning calorimeter (METTLER-TOLEDO, Zurich, Switzerland). The heating rate was 10 $^\circ$C min^{-1} over a temperature range of 40 $^\circ$C to 100 $^\circ$C [51].

3.3. Scanning Electron Microscopy (SEM) Assay

Prepare two types of milled rice, one was wild-type NIP and the other was the *pnfyc* mutant. The whole grains were cut transversely with a sharp blade and then sputtered with gold in order to increase electrical conductivity. Fractured rice grains were mounted on the copper stage and then viewed with a scanning electron microscope at 30, 2000 and 5000 times magnification. The analysis was performed based on three biological replicates at least. The experiment was conducted in institute of Agriculture and Biotechnology, ZheJiang University as described previously using a HITACHI S-3400N scanning electron microscope (HITACHI, Tokyo, Japan).

3.4. Seed Germination and Phenotypic Assay

Briefly, 100 dehusked seeds were surface-sterilized in 75% ethanol for 2 min, then in 50% bleach for 30 min, and then cleaned with sterilized ddH$_2$O 5–8 times for 3 min each time. The sterilized seeds were air dried and sown on a $\frac{1}{2}$ strength MS medium containing different concentrations of ABA (0, 2, 5 µM), GA (0, 2, 5 µM) and PAC (0, 2, 5, 10 µM). Germination rates were recorded every 12 h. The seedlings' height above ground was measured and the growth status of seedlings was photographed after 7 days. Germination is established with the appearance of the emergence of 2 mm embryos through the seed coat. The data are the mean of 3 biological triplicates.

3.5. RNA-Seq Analysis and RT-PCR Analysis

For RNA-seq, total RNA of germinative seeds at 6 h-after-imbibition was extracted using Trizol as instructed (Yeasen, Shanghai, China). The high-throughput sequencing was performed using the Illumina HiSeq™ 2500 platform and the KEGG pathway analysis of the DEGs was ultimately done by Personalgene Technology Co (Personal, Shanghai, China). DEGs were defined as genes with |log2Fold change| $\geq$ 1 and FDR < 0.01 using EBSeq [52]. The endosperm of 7 days after fertilization and seeds of 6 h after germination were collected as samples for the extraction of RNA to detect the expression level of genes in starch biosynthesis and hormone synthesis, and the RNA was extracted by Trizol according to the kit manufacturer's instructions (Yeasen, Shanghai, China).

For the RT-PCR analysis, the first-strand cDNA was synthesized using M-MLV reverse transcriptase according to the manufacturer's instructions (Takara, Dalian, China). The expression levels of different samples were determined using CFX96 touch real-time PCR detection system (Bio-Rad, Hercules, CA, USA). Expression was assessed by evaluating threshold cycle (CT) values. The relative expression level of tested genes was normalized to ubiquitin gene and calculated by the $2^{-\Delta\Delta CT}$ method. The experiment was performed in two biological replicates with three technical triplicates of each. Primer sequences are listed in Supplementary Table S3.

3.6. Yeast-Two-Hybrid Assay

The yeast two-hybrid assay was conducted based on the manufacturer's protocol (Invitrogen, Carlsbad, CA, USA). The coding sequence of *NF-YC8* to *NF-YC12* and *SAPK1* to *SAPK10* were amplified and cloned into bait vector pdest32 and prey vector pdest22, respectively. Two vectors, pDEST32-NFYCs and pDEST-SAPKs, were co-transfected into the Y2H Gold strain, and then yeast cells were grown on SD/-Trp/-Leu and SD/-Trp/-Leu/-His/-Ade medium for screening. pGBKT7-53 and pGADT7-T were used as positive controls, pGBKT7-Lam and pGADT7-T were used as negative controls. The primers used are listed in Supplementary Table S3.

3.7. Yeast-Three-Hybrid Assay

NF-YB1 CDS was cloned to fuse with GAL4 BD domain, and NF-YC12 was driven by a methionine-responsive promoter Met25 in pBRIDGE (Clontech, Dalian, China). NF-YB1-NF-YC12-pBRIDGE in strain Y2H Gold was mated with an AD domain-fused seed cDNA library in Y187 strain. The mated transformants were first selected on SD/-Leu/-Trp. Positive colonies were then transferred to SD/-Leu/-Trp/-His/-Ade/-Met/+X-a-Gal and SD/-Leu/-Trp/-His/-Ade/+X-a-Gal, respectively. The interaction was confirmed by the visualization of blue colonies on the medium.

Supplementary Materials: The following supporting information can be downloaded at: https://www.mdpi.com/article/10.3390/ijms23158382/s1.

Author Contributions: Conceptualization, J.Z.; Formal analysis, S.L.; Investigation, J.L.; Methodology, L.C., X.L., M.Y. and X.T.; Project administration, J.Z.; Resources, B.B.K., J.Y. and J.Z.; Software, C.C., B.N., A.A.A. and X.L.; Supervision, Y.W. and X.Z.; Writing—original draft, H.X.; Writing—review & editing, J.Z. All the authors read and approved the final manuscript. All authors have read and agreed to the published version of the manuscript.

Funding: This research was funded by Natural Science Foundation of Zhejiang province (Grant No. 343 LZ21C130001), National Natural Science Foundation of China (Grant No. 32071986 and U20A2030), CNRRI key research and development project (CNRRI-2020-01), and ASTIP program of CAAS.

Institutional Review Board Statement: Not applicable.

Informed Consent Statement: Not applicable.

Data Availability Statement: Data is contained within the article and within Supplementary Material.

Conflicts of Interest: The authors declare no conflict of interest.

References

1. Wang, N.L.; Long, T.; Yao, W.; Xiong, L.Z.; Zhang, Q.F.; Wu, C.Y. Mutant resources for the functional analysis of the rice genome. *Mol. Plant* **2013**, *6*, 596–604. [CrossRef] [PubMed]
2. Locascio, A.; Roig-Villanova, I.; Bernardi, J.; Varotto, S. Current perspectives on the hormonal control of seed development in Arabidopsis and maize: A focus on auxin. *Front Plant Sci.* **2014**, *5*, 412. [CrossRef] [PubMed]
3. Rijavec, T.; Dermastia, M. Cytokinins and their function in developing seeds. *Acta Chim. Slov.* **2010**, *57*, 617–629.
4. Abbas, M.; Alabadi, D.; Blazquez, M.A. Differential growth at the apical hook: All roads lead to auxin. *Front. Plant Sci.* **2013**, *4*, 441. [CrossRef] [PubMed]
5. Hedden, P.; Thomas, S.G. Plant Hormone Signaling. *Annu. Plant Rev.* **2006**, *24*, 229–255.
6. Yang, J.; Zhang, J.; Wang, Z.; Zhu, Q.; Wang, W. Hormonal changes in the grains of rice subjected to water stress during grain filling. *Plant Physiol.* **2001**, *127*, 315–323. [CrossRef]

7. Yang, W.; Lu, Z.; Xiong, Y.; Yao, J. Genome-wide identification and co-expression network analysis of the OsNF-Y gene family in rice. *Crop J.* **2017**, *5*, 21–31. [CrossRef]
8. Bello, B.K.; Hou, Y.X.; Zhao, J.; Jiao, G.A.; Wu, Y.W.; Li, Z.Y.; Wang, Y.F.; Tong, X.H.; Wang, W.; Yuan, W.Y.; et al. NF-YB1-YC12-bHLH144 complex directly activates Wx to regulate grain quality in rice (*Oryza sativa* L.). *Plant Biotechnol. J.* **2019**, *17*, 1222–1235. [CrossRef]
9. Yamakawa, H.; Hirose, T.; Kuroda, M.; Yamaguchi, T. Comprehensive expression profiling of rice grain filling-related genes under high temperature using DNA microarray. *Plant Physiol.* **2007**, *144*, 258–277. [CrossRef]
10. Xiong, Y.F.; Ren, Y.; Li, W.; Wu, F.S.; Yang, W.J.; Huang, X.L.; Yao, J.L. NF-YC12 is a key multi-functional regulator of accumulation of seed storage substances in rice. *J. Exp. Bot.* **2019**, *70*, 3765–3780. [CrossRef]
11. Manimaran, P.; Venkata Reddy, S.; Moin, M.; Raghurami Reddy, M.; Yugandhar, P.; Mohanraj, S.S.; Balachandran, S.M.; Kirti, P.B. Activation-tagging in indica rice identifies a novel transcription factor subunit, NF-YC13 associated with salt tolerance. *Sci. Rep.* **2017**, *7*, 9341. [CrossRef] [PubMed]
12. Kim, S.K.; Park, H.Y.; Jang, Y.H.; Lee, K.C.; Chung, Y.S.; Lee, J.H.; Kim, J.K. OsNF-YC2 and OsNF-YC4 proteins inhibit flowering under long-day conditions in rice. *Planta* **2016**, *243*, 563–576. [CrossRef] [PubMed]
13. Laloum, T.; De Mita, S.; Gamas, P.; Baudin, M.; Niebel, A. CCAAT-box binding transcription factors in plants: Y so many? *Trends Plant Sci.* **2013**, *18*, 157–166. [CrossRef] [PubMed]
14. Zhao, H.; Wu, D.; Kong, F.; Lin, K.; Zhang, H.; Li, G. The Arabidopsis thaliana nuclear factor Y transcription factors. *Front. Plant Sci.* **2017**, *7*, 2045. [CrossRef] [PubMed]
15. Swain, S.; Myers, Z.A.; Siriwardana, C.L.; Holt, B.F., 3rd. The multifaceted roles of NUCLEAR FACTOR-Y in Arabidopsis thaliana development and stress responses. *Biochim. Biophys. Acta Gene Regul. Mech.* **2017**, *1860*, 636–644. [CrossRef] [PubMed]
16. Petroni, K.; Kumimoto, R.W.; Gnesutta, N.; Calvenzani, V.; Fornari, M.; Tonelli, C.; Holt, B.F.; Mantovani, R. The promiscuous life of plant NUCLEAR FACTOR Y transcription factors. *Plant Cell* **2012**, *24*, 4777–4792. [CrossRef] [PubMed]
17. Thirumurugan, T.; Ito, Y.; Kubo, T.; Serizawa, A.; Kurata, N. Identification, characterization and interaction of HAP family genes in rice. *Mol. Genet. Genom.* **2008**, *279*, 279–289. [CrossRef]
18. Lee, D.K.; Kim, H.I.; Jang, G.; Chung, P.J.; Jeong, J.S.; Kim, Y.S.; Bang, S.W.; Jung, H.; Do Choi, Y.; Kim, J.K. The NF-YA transcription factor OsNF-YA$_7$ confers drought stress tolerance of rice in an abscisic acid independent manner. *Plant Sci.* **2015**, *241*, 199–210. [CrossRef]
19. Li, L.; Zheng, W.; Zhu, Y.; Ye, H.; Tang, B.; Arendsee, Z.W.; Jones, D.; Li, R.; Ortiz, D.; Zhao, X. QQS orphan gene regulates carbon and nitrogen partitioning across species via NF-YC interactions. *Proc. Natl. Acad. Sci. USA* **2015**, *112*, 14734–14739. [CrossRef] [PubMed]
20. Adachi, S.; Yoshikawa, K.; Yamanouchi, U.; Tanabata, T.; Sun, J.; Ookawa, T.; Yamamoto, T.; Sage, R.F.; Hirasawa, T.; Yonemaru, J. Fine mapping of carbon assimilation rate 8, a quantitative trait locus for flag leaf nitrogen content, stomatal conductance and photosynthesis in rice. *Front. Plant Sci.* **2017**, *8*, 60. [CrossRef]
21. Yan, W.H.; Wang, P.; Chen, H.X.; Zhou, H.J.; Li, Q.P.; Wang, C.R.; Ding, Z.H.; Zhang, Y.S.; Yu, S.B.; Xing, Y.Z. A major QTL, Ghd8, plays pleiotropic roles in regulating grain productivity, plant height, and heading date in rice. *Mol. Plant* **2011**, *4*, 319–330. [CrossRef] [PubMed]
22. Wei, X.; Xu, J.; Guo, H.; Jiang, L.; Chen, S.; Yu, C.; Zhou, Z.; Hu, P.; Zhai, H.; Wan, J. DTH8 suppresses flowering in rice, influencing plant height and yield potential simultaneously. *Plant Physiol.* **2010**, *153*, 1747–1758. [CrossRef] [PubMed]
23. Zhang, J.J.; Xue, H.W. OsLEC1/OsHAP3E participates in the determination of meristem identity in both vegetative and reproductive developments of rice F. *J. Integr. Plant Biol.* **2013**, *55*, 232–249. [CrossRef] [PubMed]
24. Ito, Y.; Thirumurugan, T.; Serizawa, A.; Hiratsu, K.; Ohme-Takagi, M.; Kurata, N. Aberrant vegetative and reproductive development by overexpression and lethality by silencing of OsHAP3E in rice. *Plant Sci.* **2011**, *181*, 105–110. [CrossRef] [PubMed]
25. Miyoshi, K.; Ito, Y.; Serizawa, A.; Kurata, N. OsHAP3 genes regulate chloroplast biogenesis in rice. *Plant J.* **2003**, *36*, 532–540. [CrossRef] [PubMed]
26. Sun, X.C.; Ling, S.; Lu, Z.H.; Ouyang, Y.D.; Liu, S.S.; Yao, J.L. OsNF-YB$_1$, a rice endosperm-specific gene, is essential for cell proliferation in endosperm development. *Gene* **2014**, *551*, 214–221. [CrossRef] [PubMed]
27. Bai, A.N.; Lu, X.D.; Li, D.Q.; Liu, J.X.; Liu, C.M. NF-YB1-regulated expression of sucrose transporters in aleurone facilitates sugar loading to rice endosperm. *Cell Res.* **2016**, *26*, 384. [CrossRef]
28. Nie, D.M.; Ouyang, Y.D.; Wang, X.; Zhou, W.; Hu, C.-G.; Yao, J. Genome-wide analysis of endosperm-specific genes in rice. *Gene* **2013**, *530*, 236–247. [CrossRef]
29. Xu, J.J.; Zhang, X.F.; Xue, H.W. Rice aleurone layer specific OsNF-YB1 regulates grain filling and endosperm development by interacting with an ERF transcription factor. *J. Exp. Bot.* **2016**, *67*, 6399–6411. [CrossRef]
30. Yamamoto, A.; Kagaya, Y.; Toyoshima, R.; Kagaya, M.; Takeda, S.; Hattori, T. Arabidopsis NF-YB subunits LEC1 and LEC1-LIKE activate transcription by interacting with seed-specific ABRE-binding factors. *Plant J.* **2009**, *58*, 843–856. [CrossRef] [PubMed]
31. E, Z.G.; Li, T.T.; Zhang, H.Y.; Liu, Z.H.; Deng, H.; Sharma, S.; Wei, X.F.; Wang, L.; Niu, B.X.; Chen, C. A group of nuclear factor Y transcription factors are sub-functionalized during endosperm development in monocots. *J. Exp. Bot.* **2018**, *69*, 2495–2510. [CrossRef] [PubMed]
32. Jia, S.Z.; Xiong, Y.F.; Xiao, P.P.; Wang, X.; Yao, J.L. OsNF-YC$_{10}$, a seed preferentially expressed gene regulates grain width by affecting cell proliferation in rice. *Plant Sci.* **2019**, *280*, 219–227. [CrossRef] [PubMed]

33. Kobayashi, Y.; Murata, M.; Minami, H.; Yamamoto, S.; Kagaya, Y.; Hobo, T.; Yamamoto, A.; Hattori, T. Abscisic acid-activated SNRK$_2$ protein kinases function in the gene-regulation pathway of ABA signal transduction by phosphorylating ABA response element-binding factors. *Plant J.* **2005**, *44*, 939–949. [CrossRef] [PubMed]
34. Ayele, B.T.; Magome, H.; Lee, S.; Shin, K.; Kamiya, Y.; Soh, M.S.; Yamaguchi, S. GA-sensitive dwarf1-1D (gsd1-1D) Defines a New Mutation that Controls Endogenous GA Levels in Arabidopsis. *J. Plant Growth Regul.* **2014**, *33*, 340–354. [CrossRef]
35. Cui, W.; Song, Q.; Zuo, B.; Han, Q.; Jia, Z. Effects of Gibberellin (GA$_{4+7}$) in Grain Filling, Hormonal Behavior, and Antioxidants in High-Density Maize (*Zea mays* L.). *Plants* **2020**, *9*, 978. [CrossRef]
36. Liu, X.; Hu, P.; Huang, M.; Tang, Y.; Li, Y.; Li, L.; Hou, X. The NF-YC-RGL2 module integrates GA and ABA signalling to regulate seed germination in Arabidopsis. *Nat. Commun.* **2016**, *7*, 12768. [CrossRef] [PubMed]
37. Finkelstein, R.R.; Lynch, T.J. The Arabidopsis abscisic acid response gene ABI5 encodes a basic leucine zipper transcription factor. *Plant Cell* **2000**, *12*, 599–609. [CrossRef]
38. Carles, C.; Bies-Etheve, N.; Aspart, L.; Léon-Kloosterziel, K.M.; Koornneef, M.; Echeverria, M.; Delseny, M. Regulation of Arabidopsis thaliana Em genes: Role of ABI5. *Plant J.* **2002**, *30*, 373–383. [CrossRef] [PubMed]
39. Yang, X.; Yang, Y.N.; Xue, L.J.; Zou, M.J.; Liu, J.Y.; Chen, F.; Xue, H.W. Rice ABI5-Like1 regulates abscisic acid and auxin responses by affecting the expression of ABRE-containing genes. *Plant Physiol.* **2011**, *156*, 1397–1409. [CrossRef]
40. Tang, N.; Zhang, H.; Li, X.; Xiao, J.; Xiong, L. Constitutive activation of transcription factor OsbZIP46 improves drought tolerance in rice. *Plant Physiol.* **2012**, *158*, 1755–1768. [CrossRef]
41. Joo, J.; Lee, Y.H.; Song, S.I. Overexpression of the rice basic leucine zipper transcription factor OsbZIP12 confers drought tolerance to rice and makes seedlings hypersensitive to ABA. *Plant Biotechnol. Rep.* **2014**, *8*, 431–441. [CrossRef]
42. Hossain, M.A.; Lee, Y.; Cho, J.-I.; Ahn, C.-H.; Lee, S.-K.; Jeon, J.-S.; Kang, H.; Lee, C.-H.; An, G.; Park, P.B. The bZIP transcription factor OsABF1 is an ABA responsive element binding factor that enhances abiotic stress signaling in rice. *Plant Mol. Biol.* **2010**, *72*, 557–566. [CrossRef] [PubMed]
43. Gao, F.; Xiong, A.; Peng, R.; Jin, X.; Xu, J.; Zhu, B.; Chen, J.; Yao, Q. OsNAC52, a rice NAC transcription factor, potentially responds to ABA and confers drought tolerance in transgenic plants. *Plant Cell Tissue Organ Cult. (PCTOC)* **2010**, *100*, 255–262. [CrossRef]
44. Zou, J.; Liu, A.; Chen, X.; Zhou, X.; Gao, G.; Wang, W.; Zhang, X. Expression analysis of nine rice heat shock protein genes under abiotic stresses and ABA treatment. *J. Plant Physiol.* **2009**, *166*, 851–861. [CrossRef]
45. Lee, S.-K.; Hwang, S.-K.; Han, M.; Eom, J.-S.; Kang, H.-G.; Han, Y.; Choi, S.-B.; Cho, M.-H.; Bhoo, S.H.; An, G. Identification of the ADP-glucose pyrophosphorylase isoforms essential for starch synthesis in the leaf and seed endosperm of rice (*Oryza sativa* L.). *Plant Mol. Biol.* **2007**, *65*, 531–546. [CrossRef]
46. Tang, X.J.; Peng, C.; Zhang, J.; Cai, Y.; You, X.-M.; Kong, F.; Yan, H.-G.; Wang, G.-X.; Wang, L.; Jin, J. ADP-glucose pyrophosphorylase large subunit 2 is essential for storage substance accumulation and subunit interactions in rice endosperm. *Plant Sci.* **2016**, *249*, 70–83. [CrossRef] [PubMed]
47. Wei, X.; Jiao, G.; Lin, H.; Sheng, Z.; Shao, G.; Xie, L.; Tang, S.; Xu, Q.; Hu, P. GRAIN INCOMPLETE FILLING 2 regulates grain filling and starch synthesis during rice caryopsis development. *J. Integr. Plant Biol.* **2017**, *59*, 134–153. [CrossRef] [PubMed]
48. Ryoo, N.; Yu, C.; Park, C.-S.; Baik, M.-Y.; Park, I.M.; Cho, M.-H.; Bhoo, S.H.; An, G.; Hahn, T.-R.; Jeon, J.-S. Knockout of a starch synthase gene OsSSIIIa/Flo5 causes white-core floury endosperm in rice (*Oryza sativa* L.). *Plant Cell Rep.* **2007**, *26*, 1083–1095. [CrossRef]
49. Ma, X.; Zhang, Q.; Zhu, Q.; Liu, W.; Chen, Y.; Qiu, R.; Wang, B.; Yang, Z.; Li, H.; Lin, Y. A robust CRISPR/Cas9 system for convenient, high-efficiency multiplex genome editing in monocot and dicot plants. *Mol. Plant* **2015**, *8*, 1274–1284. [CrossRef]
50. Kang, H.G.; Park, S.; Matsuoka, M.; An, G. White-core endosperm floury endosperm-4 in rice is generated by knockout mutations in the C-type pyruvate orthophosphate dikinase gene (OsPPDKB). *Plant J.* **2005**, *42*, 901–911. [CrossRef]
51. Nishi, A.; Nakamura, Y.; Tanaka, N.; Satoh, H. Biochemical and genetic analysis of the effects of amylose-extender mutation in rice endosperm. *Plant Physiol.* **2001**, *127*, 459–472. [CrossRef] [PubMed]
52. Leng, N.; Dawson, J.A.; Thomson, J.A.; Ruotti, V.; Rissman, A.I.; Smits, B.M.; Haag, J.D.; Gould, M.N.; Stewart, R.M.; Kendziorski, C. EBSeq: An empirical Bayes hierarchical model for inference in RNA-seq experiments. *Bioinformatics* **2013**, *29*, 1035–1043. [CrossRef] [PubMed]

Article

Deciphering the Genetic Basis of Root and Biomass Traits in Rapeseed (*Brassica napus* L.) through the Integration of GWAS and RNA-Seq under Nitrogen Stress

Nazir Ahmad [1,†], Bin Su [1,†], Sani Ibrahim [1,2], Lieqiong Kuang [1], Ze Tian [1], Xinfa Wang [1,3], Hanzhong Wang [1,3,*] and Xiaoling Dun [1,*]

1 Oil Crops Research Institute of the Chinese Academy of Agricultural Sciences/Key Laboratory of Biology and Genetic Improvement of Oil Crops, Ministry of Agriculture and Rural Affairs, Wuhan 430062, China; nazir_aup@yahoo.com (N.A.); su13297095753@163.com (B.S.); sibrahim.bot@buk.edu.ng (S.I.); kuanglieqiong@163.com (L.K.); tianze0825@163.com (Z.T.); wangxinfa@caas.cn (X.W.)
2 Department of Plant Biology, Faculty of Life Sciences, College of Physical and Pharmaceutical Sciences, Bayero University, P.M.B. 3011, Kano 700006, Nigeria
3 Hubei Hongshan Laboratory, Wuhan 430070, China
* Correspondence: wanghz@oilcrops.cn (H.W.); dunxiaoling@caas.cn (X.D.)
† These authors contributed equally to this work.

Citation: Ahmad, N.; Su, B.; Ibrahim, S.; Kuang, L.; Tian, Z.; Wang, X.; Wang, H.; Dun, X. Deciphering the Genetic Basis of Root and Biomass Traits in Rapeseed (*Brassica napus* L.) through the Integration of GWAS and RNA-Seq under Nitrogen Stress. *Int. J. Mol. Sci.* **2022**, *23*, 7958. https://doi.org/10.3390/ijms23147958

Academic Editors: Jian Zhang and Zhiyong Li

Received: 3 June 2022
Accepted: 16 July 2022
Published: 19 July 2022

Publisher's Note: MDPI stays neutral with regard to jurisdictional claims in published maps and institutional affiliations.

Copyright: © 2022 by the authors. Licensee MDPI, Basel, Switzerland. This article is an open access article distributed under the terms and conditions of the Creative Commons Attribution (CC BY) license (https://creativecommons.org/licenses/by/4.0/).

Abstract: An excellent root system is responsible for crops with high nitrogen-use efficiency (NUE). The current study evaluated the natural variations in 13 root- and biomass-related traits under a low nitrogen (LN) treatment in a rapeseed association panel. The studied traits exhibited significant phenotypic differences with heritabilities ranging from 0.53 to 0.66, and most of the traits showed significant correlations with each other. The genome-wide association study (GWAS) found 51 significant and 30 suggestive trait–SNP associations that integrated into 14 valid quantitative trait loci (QTL) clusters and explained 5.7–21.2% phenotypic variance. In addition, RNA sequencing was performed at two time points to examine the differential expression of genes (DEGs) between high and low NUE lines. In total, 245, 540, and 399 DEGs were identified as LN stress-specific, high nitrogen (HN) condition-specific, and HNLN common DEGs, respectively. An integrated analysis of GWAS, weighted gene co-expression network, and DEGs revealed 16 genes involved in rapeseed root development under LN stress. Previous studies have reported that the homologs of seven out of sixteen potential genes control root growth and NUE. These findings revealed the genetic basis underlying nitrogen stress and provided worthwhile SNPs/genes information for the genetic improvement of NUE in rapeseed.

Keywords: rapeseed; root and biomass traits; nitrogen stress; GWAS; RNA sequencing

1. Introduction

Nitrogen (N) is one of the most important macronutrients required for plant growth and development. It is the basic component of proteins, nucleic acids, chlorophyll, and several hormones [1,2]. Although global agricultural N consumption has increased sevenfold in the last half-century, most crops utilize only 30–40% of the supplied N [3,4]. A high N fertilizer application causes serious environmental problems, such as water eutrophication, acid rain, and soil acidification [5]. To ensure the sustainability of agriculture, it is imperative to breed crop varieties with a higher nitrogen use efficiency (NUE) [6].

Rapeseed (*Brassica napus* L.) is the third-largest oil crop in the world, following soybeans and palm. However, while absorbing a significant amount of N from the soil, rapeseed is generally regarded as a low NUE crop, with seed yield per unit N applied about half that of other cereals [7]. Therefore, addressing the genetic architecture of low nitrogen stress tolerance and boosting rapeseed's NUE is vital for the rapeseed's economic competitiveness [8,9]. Root system architecture (RSA) plays a critical role in N acquisition, both in terms of absorption capacity and soil exploration potential. A deeper understanding of how the RSA adapts to N availability seems to be a potential lever for optimizing N acquisition [10–12]. As a primary step, determining the key traits underpinning genotype variability and RSA adaptability to N availability is required. For example, in most elite cultivars, increasing the root-to-shoot ratio increases the uptake of N from the deep soil, or the longer roots provide optimum root nutrient storage in shoots by taking advantage of the soil's spatial characteristics [13–15]. Furthermore, the genetic improvement of root morphological traits was reported to affect crop yield [16]. Rice yields have been shown to increase under drought stress when *DRO1*, a QTL associated with both root depth and root development angle, is expressed at a higher level [17]. The *big root biomass* (*BRB*) gene was reported previously to affect shoot traits and seed yield in sesame [18]. In rapeseed, coarse root length promotes soil exploration and phosphorus uptake, which increases seed yield [19]. As a result, optimizing root-related parameters could be a potential strategy for promoting the development of cultivars with high NUE and yield.

Genome-wide association study (GWAS) is an effective method for identifying loci/genes associated with complex traits in a genetically diverse crop population, including root architectural traits in rice, wheat, soybean, maize, rapeseed, etc. [20–26]. Transcriptome analysis has proven to be valuable for identifying candidate genes, especially differentially expressed genes (DEGs), between samples with contrasting traits. The weighted gene co-expression network analysis (WGCNA) method was designed to explore the system-level functionality of the transcriptome and is widely used in plants to identify gene modules related to the uncovering of potential transcriptional regulation [27,28]. The combination of GWAS, transcriptome sequencing, and/or WGCNA has proven to be a quick and effective strategy for detecting major candidate genes regulating complex traits. In response to cadmium stress, three hub-genes (*OsHSP*, *OsHSFC2A*, and *OsDJA5*) were identified through RNA sequencing and WGCNA in rice [29]. Through the integration of GWAS, WGCNA, and differential expression analysis, four and eight important candidate genes related to root growth in *B. napus* were identified during the persistent and specific stages, respectively [30].

Due to the hidden feature of roots, the main challenge in studying root traits is developing robust phenotypic evaluation methods. Different artificial systems have been used to evaluate root traits, including sand, germination paper, and hydroponic-based cultures [31]. Hydroponic culture with digital imaging could quickly and precisely detect a variety of root traits in large populations and has been used to examine variations in root architecture among different crops, including rice, wheat, maize, soybean, and rapeseed [25,30,32–35]. In the present study, 13 RSA traits were investigated in hydroponic culture under control and low nitrogen (LN) treatments in an association panel of 327 *B. napus* cultivars, which was genotyped by the 50 K *Brassica* Infinium SNP array [25]. RNA-seq was performed in the high and low nitrogen efficient groups at two developmental stages to determine the expression levels of candidate genes. The goals of this study were to examine the phenotypic variations of 13 root and biomass traits under both control and LN treatments within a rapeseed association panel, to identify significant SNPs associated with root and biomass traits related to NUE, and to use a GWAS and RNA-seq approach to determine potential candidate genes associated with root response to LN stress.

2. Results

2.1. Phenotypic Analysis of Root and Shoot Biomass Traits under the LN Stress

A total of 13 root and shoot biomass traits across 327 accessions in two treatment conditions (control and LN stress) were evaluated under hydroponics, including five root morphological traits (MT) and eight biomass traits (BT) (Table 1).

For the investigated traits in the association panel, extensive variations were observed among genotypes in control and LN-treated conditions (Table 2). The coefficient of variation (CV) of the 13 traits ranged from 14.5–45.6% and 12.8–31.9%, respectively (Table 2). In both treatment conditions, moderate to high broad-sense heritability was observed for all the studied traits, with the value of 0.53 to 0.70. Overall, these results showed that all 13 traits were inherited in a stable manner under hydroponics in both control and LN stress conditions. LN treatment affected all investigated root and biomass traits compared with the control (Table 2). On average, LN treatment significantly increased length, number, and area of roots in the association population, resulting in an increase in RFW (6.4%) and RDW (14.2%). However, LN treatment significantly reduced the formation of aboveground biomass, with SFW and SDW decreased by 80.3% and 24.7%, respectively. Therefore, the root–shoot ratio increased significantly under LN treatment, RSRF and RSRD by 48.9% and 33.8%, respectively.

Table 1. Description of the 13 examined traits.

Classification	Trait Description	Abbreviations	Units
Root morphological traits (MT)	Primary root length	PRL	cm
	Total root length	TRL	cm
	Total root surface area	TSA	cm^2
	Total root volume	TRV	cm^3
	Total number of roots	TNR	number
Biomass-related traits (BM)	Shoot fresh weight	SFW	g
	Root fresh weight	RFW	g
	Shoot dry weight	SDW	g
	Root dry weight	RDW	g
	Total fresh weight	TFW	g
	Total dry weight	TDW	g
	Fresh root–shoot ratio	RSRF	RFW/SFW
	Dry root–shoot ratio	RSRD	RDW/SDW

Table 2. Descriptive statistics for investigated traits under control and low nitrogen treatment in association panel.

Trait	Control					LN					Control × N Stress	N Stress Impact (%)
	Min	Max	Mean	CV (%)	h^2	Min	Max	Mean	CV (%)	h^2		
PRL (cm)	12.2	34.3	24.3	14.5	0.62	16.3	36.5	25.9	12.8	0.58	**	6.4
TRL (cm)	434.3	1435.7	843.8	20.7	0.66	330.7	2104.9	912.1	21.6	0.58	*	7.5
TSA (cm^2)	18.4	99.9	56.5	23.6	0.62	24.6	158.0	65.8	23.1	0.55	*	14.1
TRV (cm^3)	0.062	0.604	0.311	29.9	0.57	0.154	0.719	0.319	25.4	0.53	ns	2.4
TRN	673.5	4804.5	1654.2	45.6	0.55	667	6213.9	2166.8	31.9	0.53	**	23.7
SFW (g)	1.291	4.889	3.243	20.1	0.72	0.858	4.062	1.799	18.7	0.66	**	−80.3
SDW (g)	0.234	1.377	0.741	26.5	-	0.266	0.992	0.594	20.8	-	**	−24.7
RFW (g)	0.261	0.788	0.503	19.2	0.60	0.227	1.479	0.538	22.2	0.60	**	6.4
RDW (g)	0.013	0.134	0.070	28.2	-	0.033	0.165	0.081	23.4	-	**	14.2
TFW (g)	1.617	5.599	3.746	19.3	0.70	1.077	5.541	2.345	18.3	0.65	**	−59.7
TDW (g)	0.277	1.476	0.809	25.9	-	0.302	1.157	0.676	19.6	-	**	−19.6
RSRF	0.102	0.248	0.159	15.9	0.70	0.199	0.470	0.311	16.1	0.55	**	48.9
RSRD	0.031	0.298	0.095	27.5	-	0.046	0.250	0.143	23.5	-	**	33.8

CV is coefficient of variation; h^2 represents heritability; ns, not significant *, $p < 0.05$; **, $p < 0.01$, significance based on the analysis of variance.

Consistent with the correlations under the control condition [30], SFW were positively and significantly correlated with RFW under LN stress ($r = 0.67$, $p < 0.01$). Between aboveground biomass and root morphological traits, SFW displayed the highest correlations with TRL ($r = 0.52$, $p < 0.01$), indicating that root morphology contributed significantly to the formation of aboveground biomass. Negative correlations were also observed between shoot biomass traits and root–shoot ratios, with the values of −0.29 between SFW and RSRF and −0.45 between SDW and RSRD, respectively (Figure 1A). Furthermore, root and biomass traits showed positive and significant correlations between CK and LN stress treatment, and the correlation of SFW ($r = 0.63$, $p < 0.01$) was the highest, while TRN ($r = 0.12$, $p < 0.05$) had the lowest correlation (Figure 1B). It further illustrated that it is essential to consider multiple inter-related traits to comprehensively assess nitrogen efficiency concerning RSA traits. In addition, the frequency distributions for 13 parameters under LN stress were nearly continuous and normal, demonstrating that the examined accessions were appropriate for subsequent association study (Figure 1A).

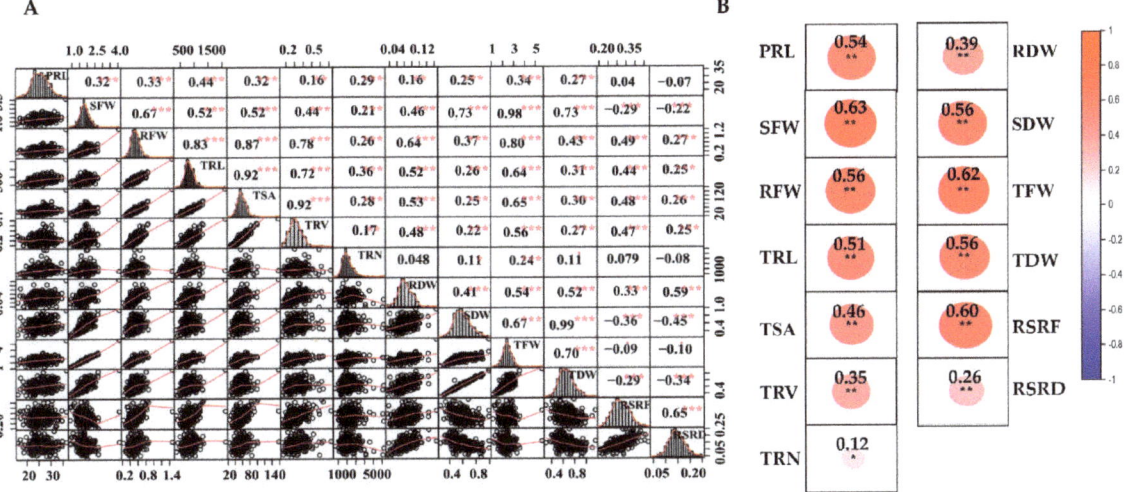

Figure 1. Correlation analysis of the investigated traits. (**A**) Correlations of studied traits under low nitrogen stress. Frequency distribution for each trait was displayed on the diagonal. The upper and lower parts represent the correlation coefficient and scatter plots between two diagonal traits, respectively. (**B**) Correlations of each investigated trait between control and low nitrogen stress. Red and blue indicate positive and negative correlations, respectively. ***, ** and * denote significance at the 0.1%, 1% and 5% levels of probability, respectively.

2.2. Marker–Trait Association Analysis for Root and Biomass Traits under LN Stress

The association panel was genotyped using the Brassica 50 K Illumina Infinium SNP array containing 45,708 SNPs. As a result of SNP filtering, 20,131 SNP markers were used to further identify trait–SNP associations [25]. This study only performed genome-wide association analyses (GWAS) with BLUE values from three trials under LN stress (Figure 2A–F), since the results of association analysis under normal condition were shown in the previous study [30]. We grouped SNPs with close proximity (within 1 MB) and an LD $r^2 > 0.2$ together, since they were found to be a part of the same QTL [36].

To avoid missing SNPs due to the complex nature of RSA traits and the strict criteria of MLM, we defined suggestive trait–SNP associations ($3.50 < -\log_{10} p \leq -\log_{10} 1/20,131$). This resulted in 51 significant trait–SNP associations ($-s s \log_{10} > 4.30$, $-\log_{10} 1/20,131$), with 24 significant SNP markers and 30 suggestive trait–SNP associations integrated into 14 valid QTL clusters (Table 3 and Table S1), most of which included at least two investigated root and biomass-related traits. Genetic variation explained by these QTL clusters varied from 5.7 to 21.2%. These QTLs were detected for 11 RSA traits, except PRL and RSRF (Table 3). The highest number of loci were identified on A09 and C03, containing 28 and 18 loci (Table S1).

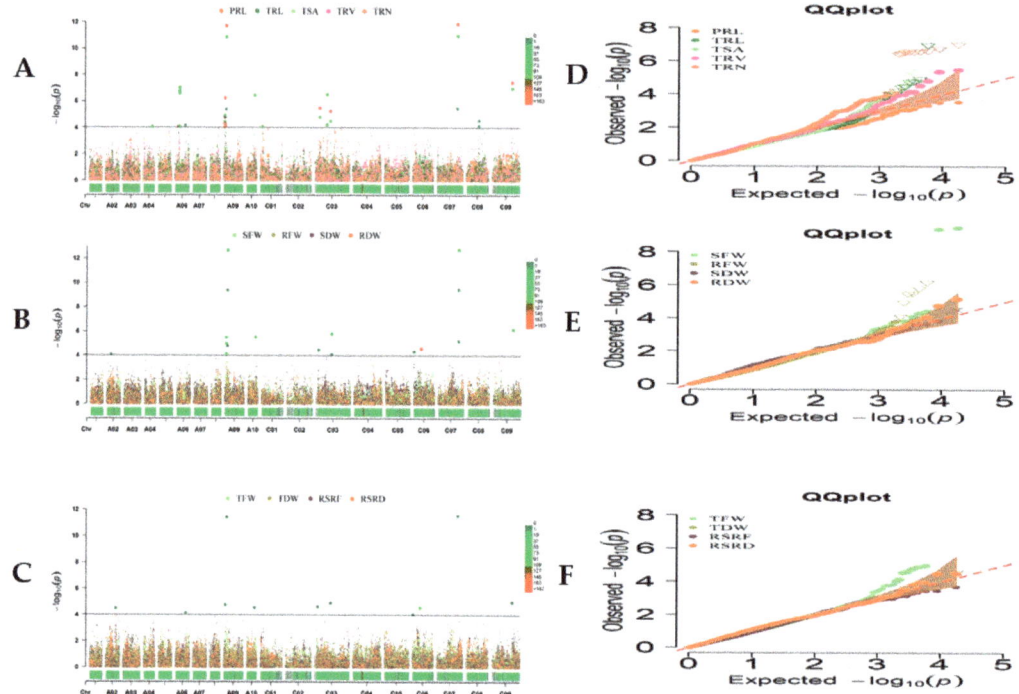

Figure 2. (**A–C**) Manhattan plots of the phenotype–genotype association analysis for 13 root and shoot biomass traits of *B. napus* by MLM with BLUE values. The x-axis displays the chromosome label, and the y-axis displays $-\log_{10}$ (p-value). The solid gray line shows significant associations between SNPs and phenotype value with threshold level of p-value ($-\log_{10} 1/20{,}131 = 4.30 \times 10^{-5}$). The color dots above the threshold values indicate the significant SNPs for root and shoot biomass traits. (**D–F**) QQ plots represent MLM analysis of the above 13 traits.

Table 3. List of important QTL clusters for investigated traits under LN stress in association panel.

QTL Cluster	No. of SNPs	Peak SNP	Chr.	Position	PVE (%)	$-\log_{10}P$	Haplotype Block (Mb)	Traits
RT-A02-1	4	Bn-A02-p18994312	A02	17,997,697	7.0	4.5	17.95–18.25	RSRD
RT-A06-1	8	seq-new-rs31601	A06	7,556,204	10.5	6.8	7.25–7.72	TRN
RT-A09-1	19	Bn-A09-p1552993	A09	2,375,212	9.9	6.2	2.18–2.51	TRL, TSA, RFW, RDW, TRV, TFW, SFW
RT-A09-2	9	seq-new-rs41996	A09	4,405,703	21.2	12.7	4.35–4.41	TRL, TFW, RFW, TSA, SFW, TRV, RDW
RT-A10-1	4	Bn-A10-p11396195	A10	12,709,829	8.9	5.5	12.67–12.72	RFW, TFW, TRL, SFW
RT-A10-2	1	Bn-A10-p13659996	A10	13,700,523	10.0	6.5	13.56–13.70	TRN
RT-C03-1	2	seq-new-rs49231	C03	4,066,558	7.0	4.6	4.00–4.18	TFW, SFW
RT-C03-2	3	seq-new-rs28219	C03	8,417,394	8.4	5.5	8.32–8.41	TSA, TRL, TRV
RT-C03-3	2	seq-new-rs39672	C03	21,177,143	10.0	6.5	20.85–21.30	TRN
RT-C03-4	11	seq-new-rs41373	C03	27,421,341	9.0	5.8	27.42–27.46	TRV, TSA, TFW, TRL, RFW, SFW
RT-C06-1	2	seq-new-rs23016	C06	12,887,745	7.1	4.6	12.80–13.08	TDW, SDW
RT-C07-1	7	seq-new-rs46512	C07	35,123,112	20.4	12.8	35.12–35.17	RFW, TSA, TFW, TRL, SFW, TRV, RDW
RT-C08-1	4	seq-new-rs29850	C08	21,323,655	5.7	4.5	21.23–21.32	TRV, TSA, TFW
RT-C09-1	5	seq-new-rs34959	C09	34,541,392	11.9	7.4	34.38–34.54	TSA, TRL, RFW, TFW, SFW

Since root and biomass traits have exhibited considerable and strong correlations, several pleiotropic genetic loci were identified, including QTL clusters RT-A09-1, RT-A09-2,

RT-A10-1, RT-C03-4, RT-C07-1, and RT-C09-1, which affected both root development and aboveground biomass formation. In particular, the SNP seq-new-rs41996 in the QTL cluster RT-A09-2 was associated with both BT and RMT (RFW, TFW, SFW, RDW, TRL, TSA, and TRV), with the highest phenotypic contribution (R^2) of 21.2% for RFW. Similarly, the SNP seq-new-rs46512 loci were also detected as pleiotropic on RT-C07-1 for RFW, TSA, TFW, TRL, SFW, TRV, and RDW, with the highest R^2 of 20.4% for RFW (Table S1). After validation, these identified loci simultaneously influencing root and shoot biomass traits could be potential loci for marker-assisted breeding.

2.3. Differentially Expressed Genes (DEGs) between High and Low Nitrogen Efficient Group

According to further phenotypic investigation, root tissues of 10 lines with extremely high SFW and 10 lines with extremely low SFW were selected as a high nitrogen-efficient group (HN group) and low nitrogen-efficient group (LN group), respectively, at two developmental points, T1 (7 days after transplantation) and T2 (14 days after transplantation) under both control and LN stress for RNA sequencing analysis (Figure 3A). Consequently, 24 libraries, including three biological replicates of the HN and LN groups under the low nitrogen stress, and HNCK and LNCK groups under the control condition, at T1 and T2 time points were generated. The total, mapped, and unique mapped reads to the reference *B. napus* genome are shown in Table S2. After filtering and trimming, the Illumina RNA-seq analysis yielded 1,153,760,000 clean reads. The average guanine–cytosine (GC) content was 46.97%, and all of the Phred quality scores (Q30) were above 94.35%. According to the principal component analysis (PCA) and correlation analysis based on gene expression levels, the correlation between individuals within the same groups was greater than the correlation between individuals within different groups (Figure S1A,B), indicating that the three biological repeats used in the experimental design were sufficiently accurate.

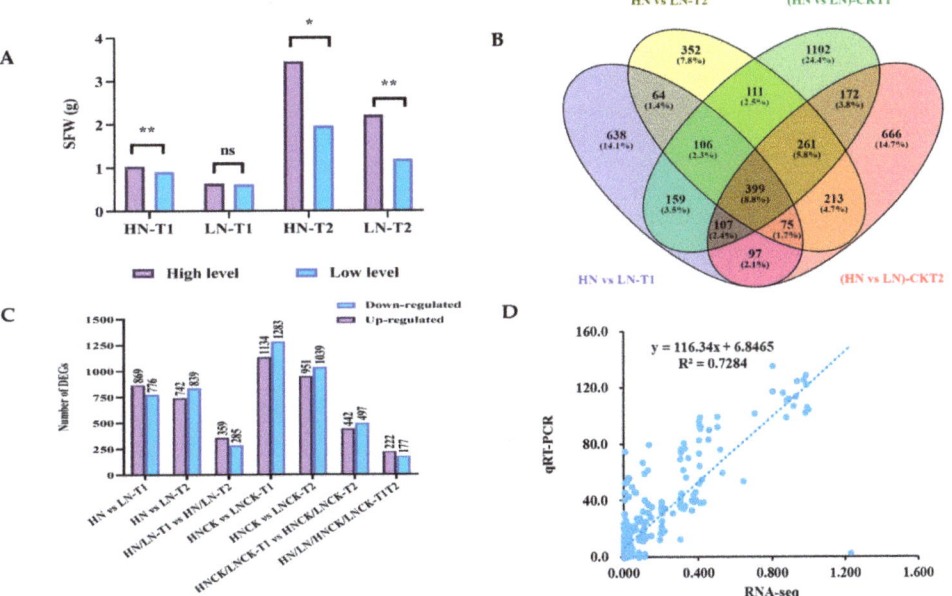

Figure 3. Differential gene expression analysis. (**A**) Phenotypic performance of two groups (HN and LN) at T1 and T2. (**B**) Venn diagram of the DEGs in the selected groups. (**C**) Up- and downregulated DEGs in different groups. (**D**) Correlation between qRT-PCR and RNA-seq data. ** and * denote significance at the 1% and 5% levels of probability, respectively. ns, not significant.

Using a pairwise approach, we first identified the DEGs of the HN and LN groups at T1 and T2 under control and LN stress, respectively, including HN/LN-T1 vs. HN/LN-T2 and HNCK/LNCK-T1 vs. HNCK/LNCK-T2. Then, the common DEGs under both control and LN stress conditions were identified between these two categories (HN/LN-T1/T2 vs. HNCK/LNCK-T1/T2) (Figure 3B). The DEGs between these groups were determined using a false discovery rate (FDR) ≤ 0.05, and an absolute value of |log2 (fold change)| was used as the threshold. In HN/LN-T1 vs. HN/LN-T2, 644 DEGs were identified, including 359 upregulated and 285 downregulated DEGs (Figure 3C). Similarly, in group HNCK/LNCK-T1 vs. HNCK/LNCK-T2, 939 DEGs were identified (442 upregulated and 497 downregulated). Furthermore, 399 DEGs (222 upregulated and 177 downregulated) were regarded as common DEGs for HN/LN/HNCK/LNCK-T1/T2 (Figure 3B,C). This meant that there were 245 DEGs specific to the HN/LN group under the LN stress condition, and 540 DEGs were specific under the control condition (Table S3).

Details of DEGs, their full names and FPKM values in each group, and corresponding description information are presented in Table S3. A heatmap was constructed using normalized FPKM values ranging from -1 to 1 to classify high/low N-specific DEGs based on expression profile similarity and diversity (Figure S2). The heatmap clearly exhibited the upregulated and downregulated clusters for the gene expression patterns of DEGs. Furthermore, qRT-PCR for 12 DEGs in all of the samples was strikingly similar to the RNA-Seq data, showing that the RNA-Seq data were accurate (Figure 3D).

2.4. Functional Classification of DEGs Involved in High and Low Nitrogen Efficiency

To further determine the functional significance of the DEGs in each group, gene ontology (GO) classifications were performed. In total, 245 specific DEGs under the LN stress condition, 540 specific DEGs under the control condition, and 399 common under both conditions were significantly assigned to 174, 128, and 159 GO terms, respectively (Table S4). Interestingly, all three groups of DEGs were enriched in different pathways. For LN-stress-specific DEGs, significant GO terms in molecular function were chitinase activity, chitin binding, medium-chain-(S)-2-hydroxy-acid oxidase activity, ATPase binding, and nutrient reservoir activity; in the cellular component category were mitochondrial small ribosomal subunit, signal recognition particle, and vacuolar proton-transporting V-type ATPase activity; in the biological function category were chitin catabolic process, cell wall macromolecule catabolic process and polysaccharide catabolic process, oxidative photosynthetic carbon pathway, maintenance of root meristem identity, and cellular response to reactive oxygen species (Figure 4A).

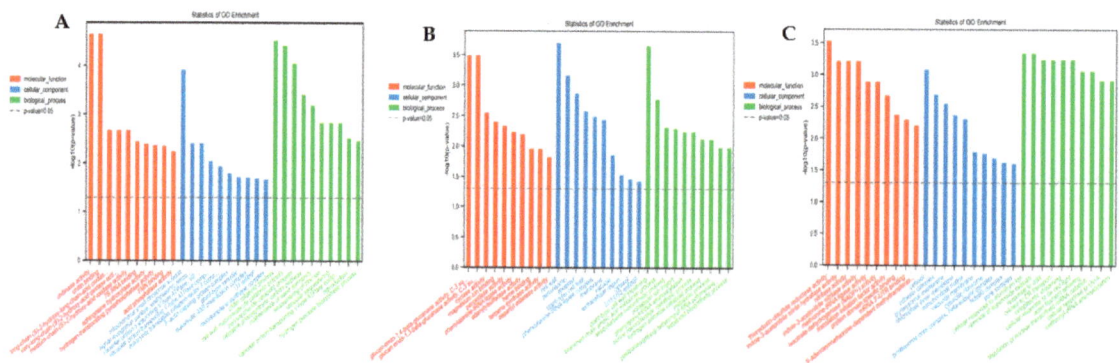

Figure 4. Gene ontology (GO) analysis of differentially expressed genes. (**A–C**) GO terms correspond to HN/LN group, HNCK/LNCK group, and HNLN/HNCK.LNCK group, respectively. Y-axis is $-\log_{10}$ (*p*-value). The relevant *p*-value decreases as the bar chart height increases. Red, blue, and green colors correspond to molecular function, cellular component, and biological process.

According to the GO classification of the specific DEGs under the control condition, significant terms of molecular function were glucan endo-1-4-beta-glucanase activity, glucan endo-1-3-beta-glucanase activity, cyclase activity, and carbohydrate binding; the over-represented terms in the cellular function category were cell wall, apoplast, followed by peroxisome; in the biological function category, plant-type cell wall loosening, branched-chain amino acid metabolic process, sesquiterpene biosynthetic process, cell wall modification, negative regulation of cell division, reactive oxygen species metabolic process, maintenance of root meristem identity, and glucose metabolic process were the most over-represented terms (Figure 4B).

For the common DEGs under both LN stress and control conditions, the significant GO terms of molecular function were thioredoxin-disulfide reductase activity, indole-3-acetonitrile nitrile hydratase activity, followed by nitrilase activity; the over-represented terms in the cellular function category were cytosol, cohesion complex, and glyoxysomal membrane, while, in the biological function category, the significant GO terms were cellular response to aluminum ion, removal of superoxide radicals followed by response to cold, NADP metabolic process, glutamate metabolic process, root hair elongation, regulation of cell shape, photosynthesis, nitrogen compound metabolic process, and cellular response to gravity (Figure 4C).

By analyzing the common enrichment pathway of these three DEG groups of differential genes, it was found that 14 significant pathways composed of 35 DEGs were identified as common pathways for HN/LN efficiency. Most of these pathways were related to hydrogen-translocating pyrophosphatase activity (GO:0009678), acid phosphatase activity (GO: 0003993), translation initiation factor activity (GO: 0003743), magnesium ion binding (GO: 0000287), 3-isopropylmalate dehydrogenase activity (GO: 0003862), oxidoreductase activity (GO: 0016702), chloroplast envelope (GO: 0009941), maintenance of root meristem identity (GO: 0010078), cellular response to reactive oxygen species (GO: 0034614), fatty acid alpha-oxidation (GO: 0001561), cellular response to aluminum ion (GO: 0071275), (R)-2-hydroxy-alpha-linolenic acid biosynthetic process (GO: 1902609), and translational initiation (GO: 0006413) (Table S4). Thus, the regulation of these genes might play an important role in the N-efficient utilization in *B. napus*.

2.5. Gene Co-Expression Network Construction and Analysis (WGCNA)

In order to investigate the gene regulatory network during LN stress, WGCNA was used to determine co-expression gene modules from 83,232 identified expressed genes with $p > 0.05$. The dendrogram revealed a total of 17 modules based on gene correlations (Figure 5A), and the relationships between modules and samples are depicted in Figure 5B. In total, 48,385 genes were identified to be involved in these 17 modules, ranging from 46 in the "MEgrey" module to 18,376 in the "MEturquoise" module (Figure 5C). The MEpink, MEsalmon, MEtan, MEmagenta, MEred, and MEcyan modules were highly correlated with HN-T1, HN-T2, HNCK-T1, LN-T1, LN-T2, and LNCK-T1, respectively. MEgreen modules were found to be highly correlated with HNCK-T2 and LNCK-T2. In addition, the MEsalmon module showed a consistent correlation with all samples of the high nitrogen-efficient group (HN-T1, HN-T2, HNCK-T1, and HNCK-T2). Likewise, the MEbrown module revealed a high correlation with all samples of the low nitrogen-efficient group (LN-T1, LN-T2, LNCK-T1, and LNCK-T2) (Figure 5B). The heatmaps revealed that the genes contained within a single module were significantly expressed in samples that were strongly correlated with the module (Figure S3).

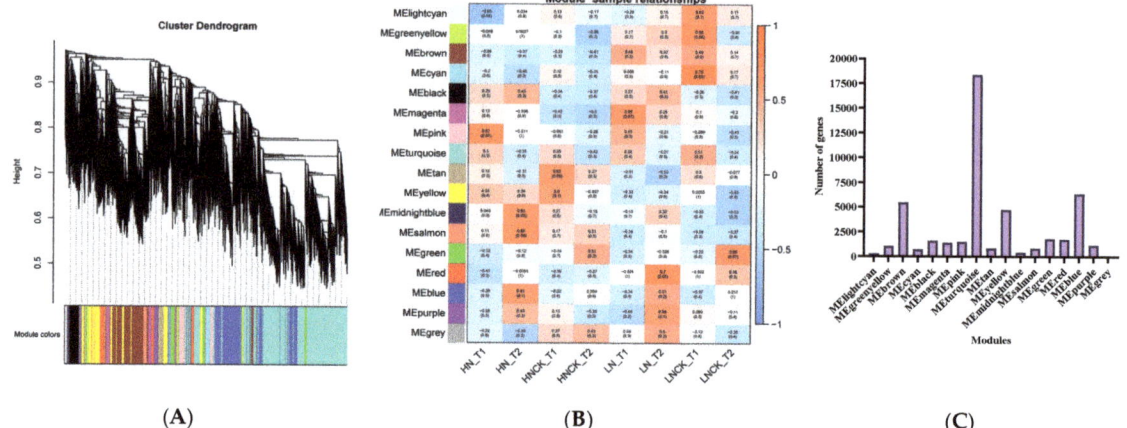

Figure 5. WGCNA of gene expression matrix. (**A**) Gene-based co-expression network analysis dendrogram. (**B**) Module–sample association; each row represents a module labeled with the same color as in (**A**), and each column represents a sample. (**C**) Overview of identified genes corresponds to each module.

The GO and KEGG analysis suggested that the significantly enriched GO terms of genes in the MEsalmon module were related to mRNA processing, meristem development, NADPH-hemoprotein reductase activity, cytokinin biosynthetic process and response to oxidative stress, NADH pyrophosphatase activity, and root hair elongation (Table S5). Meanwhile, pyruvate metabolism, glycolysis, and oxidative phosphorylation pathways were significantly enriched in the "MEsalmon" module by KEGG pathway enrichment analysis (Table S6). Furthermore, significant GO terms in the "MEbrown" module were mRNA binding, gibberellin-mediated signaling pathway, cell communication, NAD metabolic process, regulation of carbon utilization, NADP biosynthetic process, and regulation of auxin-mediated signaling pathway (Table S5). The most enriched KEGG pathway in the "MEbrown" module was RNA transport, nicotinate, and nicotinamide metabolism, lysine biosynthesis, and N-Glycan biosynthesis aminoacyl-tRNA biosynthesis (Table S6). These important pathways played a crucial role in the nitrogen metabolism and assimilation process.

2.6. Candidate Genes' Prediction and Prioritization by Integrating GWAS, DEGs, and WGCNA

Genes within 300 kb upstream and downstream of significant lead SNPs associated with each trait were revealed using the decay of the LD approach [37,38]. As a result, GWAS results revealed a total of 1378 genes around each peak SNP from 14 QTL clusters within the 300 kb region (up and down) (Table S7). The substantial and consistent correlation of WGCNA genes with each module allowed us to explore four potential genes from GWAS and WGCNA overlapped genes (Figure 6A,B). Among the four candidate genes, two genes with high and consistent correlation to the MEsalmon and two with MEbrown modules were highly expressed at all stages of the high nitrogen-efficient group and low nitrogen-efficient group, respectively (Table 4). Furthermore, we identified 12 genes simultaneously detected as common candidate genes by integrating GWAS and DEGs (Table 4). Within these 16 genes, we identified some potential candidate genes related to nitrogen use efficiency, nitrogen utilization, assimilation, and root growth and development. These findings assessed the efficiency of an approach for screening candidate genes that integrated GWAS, WGCNA, and differential expression analysis.

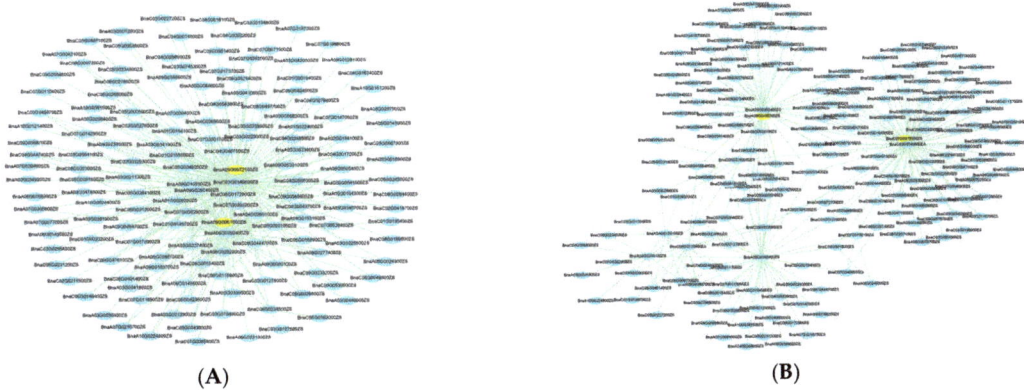

Figure 6. Network of genes in salmon and brown modules. (**A**,**B**) Correlation of networks in salmon and brown modules, respectively. Yellow color in the network indicates the candidate genes overlapped by GWAS and WGCNA.

Table 4. List of candidate genes identified through the integration of GWAS, WGCNA, differential expression analysis, and their FPKM values.

Darmor_ID	HN-T1	HN-T2	HNCK-T1	HNCK-T2	LN-T1	LN-T2	LNCK-T1	LNCK-T2	QTN/QTL Cluster	Distance from Lead SNP (Kb)	Description
Hub Genes in the MEsalmon Module (GWAS + WGCNA)											
BnaA09g04260D	7.22	27.94	9.28	37.08	7.39	25.75	6.49	34.92	RT-A09-1	273.55	Sugar transport protein 13
BnaA09g05270D	4.92	1.84	5.39	1.88	7.99	2.68	7.09	3.29	RT-A09-1	−205.92	Probable carboxylesterase 18
Hub genes in the MEbrown module (GWAS+WGCNA)											
BnaA09g08440D	1.93	4.34	1.84	3.41	2.12	4.73	1.70	3.23	RT-A09-2	249.99	BTB/POZ domain-containing protein
BnaC07g30970D	13.56	13.81	15.98	12.58	15.33	13.72	18.75	14.97	RT-C07-1	−118.79	Exocyst complex component
Significant DEGs from GWAS + RNA-sequencing data											
BnaA09g04280D	0.72	0.65	0.55	0.68	1.69	1.43	1.10	2.25	RT-A09-1	264.71	TRAF-like family protein
BnaA09g08450D	9.19	13.66	11.39	15.52	12.12	12.26	9.48	12.54	RT-A09-2	241.43	MVB pathway protein
BnaA10g16560D	0.00	0.08	0.08	3.80	7.23	0.23	0.00	0.00	RT-A10-1	147.27	/
BnaA10g17620D	5.56	6.81	6.50	6.40	4.37	3.03	4.04	4.40	RT-A10-1	−268.85	hydrolase II (PTH2) family protein
BnaA10g19550D	3.06	3.46	2.79	4.28	5.05	3.37	2.45	3.75	RT-A10-2	−150.98	unknown protein
BnaA10g19700D	46.34	27.83	38.00	18.89	16.09	12.86	18.94	12.44	RT-A10-2	−202.00	xyloglucan hydrolase 5 (XTH5)
BnaC03g34740D	1.22	1.04	1.49	0.92	1.15	0.96	1.05	1.08	RT-C03-3	105.06	GNS1/SUR4 membrane protein
BnaC03g42900D	16.87	15.62	28.99	14.44	26.24	17.19	27.69	21.44	RT-C03-4	−217.19	lipid-transfer protein
BnaC06g11180D	1.10	1.63	1.28	1.72	0.06	0.11	1.42	0.10	RT-C06-1	−262.51	/
BnaC06g11230D	0.81	0.50	1.57	1.02	0.83	1.10	1.22	1.01	RT-C06-1	−280.51	nicotinamidase 1 (NIC1)
BnaC07g30250D	2.51	1.76	2.50	1.18	1.07	0.60	0.89	0.43	RT-C07-1	295.55	/
BnaC07g30400D	5.52	3.82	10.18	8.55	3.83	5.54	9.20	10.00	RT-C07-1	223.14	SLAC1 homologue 3 (SLAH3)

3. Discussion

Nitrogen stress is a major limiting factor for crop production worldwide, and plant RSA is of great significance in nutrient stress tolerance [39]. In recent years, the use of RSA is predicted to result in a "second green revolution" in agriculture [13,40]. Different phenotyping methodologies for early crops' RSA screening were used, expecting that genotypes with diverse root architecture at the seedling stage would respond similarly at the adult stage when water and/or nutrients became limited for grain yield [12]. In this study, we studied root and shoot behaviors at the seedling stage under N-limited conditions in a modified hydroponics growth system, which was deemed a valid way to examine root system changes compared to field conditions [41]. The significant variations were observed for different root and biomass traits among the genotypes of the association panel due to their diverse genetic background and wide geographical distributions. In agreement with the previous studies [26,42], seedlings grown under LN conditions showed reduced SFW and TFW but increased RSRF, SDW, and RSRD than seedlings grown under control (CK) conditions. This finding shows that N-deficient plants transport more carbon in order to promote root development and, hence, mine the substrate for more nitrogen [12]. Greater TRL, TRV, TSA, and TRN indicate an increased ability to acquire more N from the nutrient solution. Under control and low-N conditions, the broad-sense heritability of the examined traits was moderate to high (0.55–0.72) and (0.53–0.66), respectively (Table 2), indicating that these root and biomass parameters are more genetically governed and, thus, more responsive to genetic improvement. Similar findings were reported by other researchers [26,43].

Correlations between root and biomass traits demonstrate the balance of root and shoot organs and resource partitioning between above- and below-ground plant tissues [44]. In the present study, a high correlation between shoot and root weights observed might be due to the supply of nutrients from roots to the shoot parts, as reported in rapeseed [25,26,30,43,45]. Furthermore, the strong correlation between root and biomass traits aids in successful soil exploration by intercepting nutrients and communicating stress signals [46]. Understanding the mechanisms underlying root-related traits in crops may be an effective strategy for developing high-quality root cultivars via marker-assisted selection [41]. Based on GWAS results, 81 trait–SNP associations (51 significant and 30 suggestive) were identified that were integrated into 14 valid QTL clusters (Table 3, Table S1). This revealed the complex genetic control of root and biomass traits at an early stage of crop growth. In addition, some pleiotropic QTLs were found (Table S1), revealing that different traits from specific QTLs may be addressed separately to increase RSA in rapeseed [47].

Transcriptome analysis is a robust approach that helps the identification of differentially expressed genes with their expression level and regulatory mechanisms [48]. The present study identified 245 and 540 DEGs specific to HN/LN and HNCK/LNCK, respectively, while 399 DEGs were considered common DEGs for HN/LN/HNCK/LNCK-T1T2 (Table S3, Figure 3B). DEGs associated with significant GO terms might have a critical role in root growth and nitrogen efficiency/stress tolerance. For example, BnaC02G0384400ZS (AtGLO4) associated with GO-term medium-chain-(S)-2-hydroxy-acid oxidase activity has been reported to play a crucial role in carbon and nitrogen metabolism [49]. BnaC06G0457600ZS (AtGLYK), corresponding to the oxidative photosynthetic carbon pathway, has been recently reported to regulate nitrogen assimilation [50]. Similarly, BnaC02G0071500ZS (AtNLA), associated with GO term glucan endo-1-4-beta-glucanase activity, has been reported to play a key role in nitrogen remobilization and promote root elongation under nitrogen stress [51,52]. BnaC02G0065900ZS (AtGMII), associated with carbohydrate binding, has been reported to play a key role in root growth and development [53]. These findings indicate that DEGs encoding several metabolic, regulatory, signaling, and structural proteins involved in arginine and proline metabolism, galactose metabolism, and tryptophan metabolism may be primarily responsible for the differences in nitrogen use efficiency between these groups [54].

The integration of GWAS, WGCNA, and differential expression analysis has already been used to identify candidate genes in many crops [55]. We identified 12 and 4 candidate genes related to root growth and nitrogen stress by integrating GWAS, WGCNA, and DEGs. For example, BnaA09g04260D, a major hub gene in the MEsalmon module and located in the region of the QTL cluster qcA09-1, encodes a major facilitator superfamily protein, whose homologous AtSTP13 had a potential role in root growth responses and nitrogen uptake under nitrogen-starved conditions [56–58]. Another hub gene in the MEsalmon module and the qcA09-1 region, BnaA09g05270D, has been reported to have a crucial role in regulating root gravitropism and elongation against various environmental stresses [59,60]. An overlapped candidate gene between the MEbrown module and qcA09-2, BnaA09g08440D, a member of the NPY gene family (AtNPY1), had a crucial role in root gravitropism in *A.thaliana* [61]. Four potential candidate genes out of twelve were identified through the integration of GWAS, and DEGs have been reported to function as central regulators in root development and nitrogen stress. BnaA10g16560D (AtGSR2), encoding glutamine synthetase, has been reported to function in nitrogen assimilation, and thus improve nitrogen use efficiency [62,63]. BnaA10g19550D (AtLAZY1/AtANR1), an important candidate gene identified in RT-A10-2, has a crucial role in root gravitropism and nitrate regulation of root development [64,65]. BnaA10g19700D (AtXTH5) encoding endoxyloglucan transferase has been reported to regulate root cap during nitrogen stress [66]. Another important gene, BnaC07g30400D (AtSLAH3), plays a potential role in nitrogen uptake and assimilation during nitrogen-deprived conditions [67–69].

The aforementioned results revealed that these candidate genes played an important role in root growth and nitrogen utilization efficiency. A better understanding of nitrogen stress tolerance and root growth was acquired by identifying potential associated SNPs and promising candidate genes of nitrogen stress tolerance, which will serve a crucial role in the elite's rapeseed breeding programs. However, further research based on these putative candidate genes will comprehensively elucidate the significance of these genes in rapeseed NUE, root development, and growth.

4. Materials and Methods

4.1. Plant Materials and Growth Conditions

Based on the Rapeseed Research Network in China, 327 *B. napus* lines were used in this study, including 191 semi-winter (population 1; P1), 34 winter (P2), and 102 spring accessions (P3). A total of 327 rapeseed germplasm accessions were studied, with 222 from China's Yangtze River, 52 from other places/unknown origins, 23 from northwestern China, 16 from Europe, and 14 from Australia. All the accessions were strictly self-crossed.

The root-related traits of 327 *B. napus* accessions were evaluated using the previously described hydroponic setup [41]. After two days in the dark on the medical gauze of the germination device, uniform and robust rapeseed seeds were exposed to light (180 μmol photons $m^{-2}s^{-1}$) and grew for four days in a greenhouse (60–80 percent relative humidity) under 16/8 h day/night cycles at 24 °C [70]. The Hoagland's solution (the concentration of N was 15 Mm; control treatment) was composed of: 5 mM Ca $(NO_3)_2 \cdot 4 H_2O$, 5 mM KNO_3, 2 mM $MgSO_4 \cdot 7H_2O$, 1mM KH_2PO_4, 0.05 mM EDTA-Fe, 46 μM H_3BO_3, 9.14 μM $MnCl_2 \cdot 4H_2O$, 0.77 μM $ZnSO_4 \cdot 7H_2O$, 0.37 μM $NaMoO_4 \cdot 2H_2O$, and 0.32 μM $CuSO_4.5H_2O$. Six days after planting, seedlings were shifted to a growth device containing a quarter of Hoagland's solution (two treatments, control and LN), as described by [43]. Each basin had 30 seedlings of five different lines (five seedlings for each line). Once a week, the nutrient solution was replaced. Each week, the 1/4 solution was changed to a 1/2 solution, and then 100% solution until harvesting. The N content was decreased to 0.5 mM for LN treatment by lowering KNO_3 and replacing $Ca(NO_3)_2$ with $CaCl_2$. A completely randomized design was applied to three independent hydroponic culture trials conducted at Oil Crops Research Institute, Chinese Academy of Agricultural Sciences-Wuhan, China.

4.2. Phenotypic Investigation

Three plants from each genotype were collected during harvest, and each plant was divided into root and shoot sections. Five root morphology traits (RMT) viz. total root length (TRL), total root surface area (TSA), total root volume (TRV), and total root number (TRN) were captured through images using a scanner (EPSON V700, Japan) and further analyzed by WinRHIZO software (Pro 2012b, Canada), while primary root length (PRL) was measured manually using a ruler. Eight biomass-related traits (BT), including root fresh weight (RFW) and shoot fresh weight (SFW), were measured manually by using a weighing balance. Root dry weight (RDW) and shoot dry weight (SDW) were measured after oven drying at 80 °C until a consistent weight was reached. Total dry weight (TDW) and total fresh weight (TFW) were estimated as SDW + RDW and SFW + RFW, respectively. The ratio of root-to-shoot fresh weight (RSRF) and ratio of root-to-shoot dry weight (RSRD) were measured as the ratio between RFW and SFW and ratio between RDW and SDW, respectively.

4.3. Statistical Analysis

Statistical analysis was conducted using BLUE values for 13 traits studied under nitrogen stress across three trials. Statistically significant differences between treatments were estimated using a paired samples t-test, with $\alpha = 0.05$ as a significant threshold. Basic statistics and broad-sense heritability were calculated using QTL Ici mapping 4.2 [36,71]. The "PerformanceAnalytics" package in R software was used to calculate Pearson correlation at a significance level of ($p < 0.05$). The response of each trait to LN was represented by the increase or decrease of LN relative to CK, calculated as (LN-CK)/LN × 100%.

4.4. Association Analysis

Using the new *B. napus* 50 K Illumina Infinium SNP array, 327 *B. napus* lines were genotyped. After filtering, there were 20,131 SNP markers for further investigation [25]. The trait–SNP association was investigated utilizing Best Linear Unbiased Estimates (BLUE) values of three LN trials via Tassel 5.0 software using a mixed linear model (MLM) with (Q + K) matrix [72]. To find marker–trait associations, an arbitrary cutoff value of 1/20,131 SNPs ($-\log_{10}(p) = 4.30$) was used. The Manhattan and Quantile–Quantile (Q-Q) plots were generated using the qqman and ggplot2 tools, respectively [73,74]. The four-gamete criterion was used to identify marker haplotypes at each linked locus using Haploview software [75].

4.5. Candidate Gene Prediction

The complete gene list in the QTL cluster region was scanned using the *B. napus* "Darmor" reference genome information [76]. Potential candidate genes for nitrogen efficiency/LN nitrogen tolerance were identified using gene ontology (GO terms) from the TAIR website and gene functions recovered from prior studies [76].

4.6. Transcriptome Sequencing and Analysis

Based on the differences in SFW, 20 accessions (S5, S39, S46, S49, S64, S78, S118, S129, S140, S170, S189, S226, S251, S275, S283, S289, S291, S303, S313, and S324) with extremely high and 20 accessions (S18, S32, S90, S104, S106, S124, S145, S149, S161, S176, S193, S197, S205, S237, S252, S256, S265, S272, and S326) with extremely low SFW were selected from the association panel. The phenotypic values divided these accessions mainly into high nitrogen efficiency (HN) and low nitrogen efficiency (LN). These accessions were grown hydroponically under control and LN treatment conditions, and the same protocol of Hoagland's solution was applied for the transcriptome experiment as described above. Samples were collected at two time points, 7 and 14 days after transplanting (T1 and T2).

Total RNA was extracted from root tissue of high-N and low-N efficiency accessions. Then, equal amounts of total RNA from 20 high-N and 20 low-N efficiency accessions were separately pooled. Three biological replicates, each obtained from three independent plants, were collected for RNA sequencing (RNA-seq) for each sample.

Twenty-four RNA-seq libraries (one tissue × two groups × two treatments × two-time points × three biological replicates per sample) were prepared for total RNA extraction with IRIzol reagent (Invitrogen, USA). An Illumina HiseqTM 2500 platform was used by Oebiotech Company in Shanghai, China, to construct sequencing libraries and conduct Illumina sequencing. Raw readings with 150 paired-end base pairs (bp) were filtered and aligned [77].

The clean reads were mapped using HISAT v2.0.4 and the *B. napus* ZS11 reference genome (https://www.genoscope.cns.fr/brassicanapus/data/, accessed on 22 April 2022) [76]. The WGCNA was conducted using the WGCNA package in R [78]. The "DESeq" R package was utilized to identify DEGs using ≤ 0.05 for the false discovery rate (FDR) and $|\log2 \text{ ratio}| \geq 1$ as criteria.

4.7. Validation of DEGs by Quantitative Real-Time Polymerase Chain Reaction (qRT-PCR)

Twelve differentially expressed candidate genes were assessed by qRT-PCR to measure the reliability of the RNA-seq data, as previously described [79]. The primer sequences are presented in Table S8. The SYBR qPCR Master Mix (Vazyme) was used with the CFX96 for qRT-PCR analysis (BIO-RAD). Each sample was subjected to three technical replications. The $2^{-\Delta\Delta CT}$ method was utilized to determine the relative expression of target genes using *B. napus* ACTIN2 as an internal control. [80].

Supplementary Materials: The following supporting information can be downloaded at: https://www.mdpi.com/article/10.3390/ijms23147958/s1.

Author Contributions: X.D. planned and supervised the research; N.A., B.S., S.I., L.K., Z.T. and X.W. performed root traits investigation and analyzed the data; N.A. wrote the manuscript; X.D. and H.W. contributed to modify the manuscript. All authors have read and agreed to the published version of the manuscript.

Funding: The study was supported by the Major Project of Hubei Hongshan Laboratory (2021HSZD004), Agricultural Science and Technology Innovation Project (CAAS-ZDRW202109), Central Public-interest Scientific Institution Basal Research Fund (No.1610172020004), Agricultural Science and Technology Innovation Project (CAAS-ASTIP-2013-OCRI), and China Agriculture Research System of MOF and MARA (CARS-12).

Institutional Review Board Statement: Not applicable.

Informed Consent Statement: Not applicable.

Data Availability Statement: The datasets generated or analyzed during the present study are available from the corresponding authors on reasonable request.

Conflicts of Interest: The authors declare no conflict of interest.

References

1. The, S.V.; Snyder, R.; Tegeder, M. Targeting Nitrogen Metabolism and Transport Processes to Improve Plant Nitrogen Use Efficiency. *Front. Plant Sci.* **2021**, *11*, 628366. [CrossRef] [PubMed]
2. Müller, M.; Munné-Bosch, S. Hormonal impact on photosynthesis and photoprotection in plants. *Plant Physiol.* **2021**, *185*, 1500–1522. [CrossRef] [PubMed]
3. Hirel, B.; Le Gouis, J.; Ney, B.; Gallais, A. The challenge of improving nitrogen use efficiency in crop plants: Towards a more central role for genetic variability and quantitative genetics within integrated approaches. *J. Exp. Bot.* **2007**, *58*, 2369–2387. [CrossRef] [PubMed]
4. Sharma, L.K.; Bali, S.K. A review of methods to improve nitrogen use efficiency in agriculture. *Sustainability* **2017**, *10*, 51. [CrossRef]
5. Liu, C.W.; Sung, Y.; Chen, B.C.; Lai, H.Y. Effects of nitrogen fertilizers on the growth and nitrate content of lettuce (*Lactuca sativa* L.). *Int. J. Environ. Res. Public Health* **2014**, *11*, 4427–4440. [CrossRef]
6. Ali, J.; Jewel, Z.A.; Mahender, A.; Anandan, A.; Hernandez, J.; Li, Z. Molecular Genetics and Breeding for Nutrient Use Efficiency in Rice. *Int. J. Mol. Sci.* **2018**, *19*, 1762. [CrossRef]
7. Lammerts van Bueren, E.T.; Struik, P.C. Diverse concepts of breeding for nitrogen use efficiency. A review. *Agron. Sustain. Dev.* **2017**, *37*, 50. [CrossRef]

8. Raboanatahiry, N.; Li, H.; Yu, L.; Li, M. Rapeseed (*Brassica napus* L.): Processing, utilization, and genetic improvement. *Agronomy* **2021**, *11*, 1776. [CrossRef]
9. Snowdon, R.J.; Wittkop, B.; Chen, T.W.; Stahl, A. Crop adaptation to climate change as a consequence of long-term breeding. *Theor. Appl. Genet.* **2021**, *134*, 1613–1623. [CrossRef]
10. Vazquez-Carrasquer, V.; Laperche, A.; Bissuel-Bélaygue, C.; Chelle, M.; Richard-Molard, C. Nitrogen Uptake Efficiency, Mediated by Fine Root Growth, Early Determines Temporal and Genotypic Variations in Nitrogen Use Efficiency of Winter Oilseed Rape. *Front. Plant Sci.* **2021**, *12*, 641459. [CrossRef]
11. Williams, S.T.; Vail, S.; Arcand, M.M. Nitrogen Use Efficiency in Parent vs. Hybrid Canola under Varying Nitrogen Availabilities. *Plants* **2021**, *10*, 2364. [CrossRef] [PubMed]
12. Lecarpentier, C. Genotypic diversity and plasticity of root system architecture to nitrogen availability in oilseed rape. *PLoS ONE* **2021**, *16*, e0250966. [CrossRef] [PubMed]
13. Calleja-Cabrera, J.; Boter, M.; Oñate-Sánchez, L.; Pernas, M. Root Growth Adaptation to Climate Change in Crops. *Front. Plant Sci.* **2020**, *11*, 544. [CrossRef] [PubMed]
14. Guo, Y.; Xu, F.K.Y. QTL mapping for seedling traits in wheat grown under varying concentrations of N, P and K nutrients. *Theor. Appl. Genet.* **2012**, *124*, 851–865. [CrossRef]
15. Ober, E.S.; Alahmad, S.; Cockram, J.; Forestan, C.; Hickey, L.T.; Kant, J.; Maccaferri, M.; Marr, E.; Milner, M.; Pinto, F.; et al. Wheat root systems as a breeding target for climate resilience. *Theor. Appl. Genet.* **2021**, *134*, 1645–1662. [CrossRef]
16. Siddiqui, M.N.; León, J.; Naz, A.A.; Ballvora, A. Genetics and genomics of root system variation in adaptation to drought stress in cereal crops. *J. Exp. Bot.* **2021**, *72*, 1007–1019. [CrossRef]
17. Kitomi, Y.; Hanzawa, E.; Kuya, N.; Inoue, H.; Hara, N.; Kawai, S.; Kanno, N.; Endo, M.; Sugimoto, K.; Yamazaki, T. Root angle modifications by the DRO1 homolog improve rice yields in saline paddy fields. *Proc. Natl. Acad. Sci. USA* **2020**, *117*, 21242–21250. [CrossRef]
18. Dossa, K.; Zhou, R.; Li, D.; Liu, A.; Qin, L.; Mmadi, M.A.; Su, R.; Zhang, Y.; Wang, J.; Gao, Y.; et al. A novel motif in the $5'$-UTR of an orphan gene 'Big Root Biomass' modulates root biomass in sesame. *Plant Biotechnol. J.* **2021**, *19*, 1065–1079. [CrossRef]
19. Duan, X.; Jin, K.; Ding, G.; Wang, C.; Cai, H.; Wang, S.; White, P.J.; Xu, F.; Shi, L. The impact of different morphological and biochemical root traits on phosphorus acquisition and seed yield of *Brassica napus*. *Field Crops Res.* **2020**, *258*, 107960. [CrossRef]
20. Xu, F.; Chen, S.; Yang, X.; Zhou, S.; Wang, J. Genome-Wide Association Study on Root Traits Under Different Growing Environments in Wheat (*Triticum aestivum* L.). *Front. Genet.* **2021**, *12*, 646712. [CrossRef]
21. Xu, X.; Ye, J.; Yang, Y.; Zhang, M.; Xu, Q.; Feng, Y.; Yuan, X.; Yu, H. Genome-Wide Association Study of Rice Rooting Ability at the Seedling Stage. *Rice* **2020**, *13*, 59. [CrossRef] [PubMed]
22. Mandozai, A.; Moussa, A.A.; Zhang, Q.; Qu, J.; Du, Y. Genome-Wide Association Study of Root and Shoot Related Traits in Spring Soybean (*Glycine max* L.) at Seedling Stages Using SLAF-Seq. *Front. Plant Sci.* **2021**, *12*, 568995. [CrossRef] [PubMed]
23. Moussa, A.A.; Mandozai, A.; Jin, Y.; Qu, J.; Zhang, Q.; Zhao, H.; Anwari, G. Genome-wide association screening and verification of potential genes associated with root architectural traits in maize (*Zea mays* L.) at multiple seedling stages. *BMC Genom.* **2021**, *22*, 558. [CrossRef] [PubMed]
24. Kiran, A.; Wakeel, A.; Snowdon, R.; Friedt, W. Genetic dissection of root architectural traits by QTL and genome-wide association mapping in rapeseed (*Brassica napus* L.). *Plant Breed.* **2019**, *138*, 184–192. [CrossRef]
25. Ibrahim, S.; Li, K.; Ahmad, N.; Kuang, L.; Sadau, S.B.; Tian, Z.; Huang, L.; Wang, X.; Dun, X.; Wang, H. Genetic Dissection of Mature Root Characteristics by Genome-Wide Association Studies in Rapeseed (*Brassica napus* L.). *Plants* **2021**, *10*, 2569. [CrossRef]
26. Wang, J.; Dun, X.; Shi, J.; Wang, X.; Liu, G.; Wang, H. Genetic dissection of root morphological traits related to nitrogen use efficiency in *brassica napus* L. Under two contrasting nitrogen conditions. *Front. Plant Sci.* **2017**, *8*, 1709. [CrossRef]
27. De Kesel, J.; Bonneure, E.; Mangelinckx, S. The Use of PTI-Marker Genes to Identify Novel Compounds that Establish Induced Resistance in Rice. *Int. J. Mol. Sci.* **2020**, *21*, 317. [CrossRef]
28. Yuan, Y.; Zhang, B.; Tang, X.; Zhang, J.; Lin, J. Comparative Transcriptome Analysis of Different Dendrobium Species Reveals Active Ingredients-Related Genes and Pathways. *Int. J. Mol. Sci.* **2020**, *21*, 861. [CrossRef]
29. Wang, Q.; Zeng, X.; Song, Q.; Sun, Y.; Feng, Y.; Lai, Y. Identification of key genes and modules in response to Cadmium stress in different rice varieties and stem nodes by weighted gene co-expression network analysis. *Sci. Rep.* **2020**, *10*, 9525. [CrossRef]
30. Li, K.; Wang, J.; Kuang, L.; Tian, Z.; Wang, X.; Dun, X.; Tu, J.; Wang, H. Genome-wide association study and transcriptome analysis reveal key genes affecting root growth dynamics in rapeseed. *Biotechnol. Biofuels* **2021**, *14*, 178. [CrossRef]
31. Tracy, S.R.; Nagel, K.A.; Postma, J.A.; Fassbender, H.; Wasson, A.; Watt, M. Crop Improvement from Phenotyping Roots: Highlights Reveal Expanding Opportunities. *Trends Plant Sci.* **2020**, *25*, 105–118. [CrossRef] [PubMed]
32. Soriano, J.M.; Alvaro, F. Discovering consensus genomic regions in wheat for root-related traits by QTL meta-analysis. *Sci. Rep.* **2019**, *9*, 10537. [CrossRef] [PubMed]
33. Song, L.; Chen, W.; Yao, Q.; Guo, B.; Valliyod, B.; Wang, Z.; Nguyen, H.T. Genome-wide transcriptional profiling for elucidating the effects of brassinosteroids on Glycine max during early vegetative development. *Sci. Rep.* **2019**, *9*, 16085. [CrossRef]
34. Sandhu, N.; Subedi, S.R.; Yadaw, R.B.; Chaudhary, B.; Prasai, H. Root Traits Enhancing Rice Grain Yield under Alternate Wetting and Drying Condition. *Front. Plant Sci.* **2017**, *8*, 1–13. [CrossRef] [PubMed]

35. Gao, H.; Xie, W.; Yang, C.; Xu, J.; Li, J.; Wang, H.; Chen, X.; Huang, C. transporter, is required for Arabidopsis root growth under manganese deficiency. *New Phytol.* **2018**, *217*, 179–193. [CrossRef] [PubMed]
36. Liu, S.; Fan, C.; Li, J.; Cai, G.; Yang, Q.; Wu, J.; Yi, X.; Zhang, C.; Zhou, Y. A genome-wide association study reveals novel elite allelic variations in seed oil content of *Brassica napus*. *Theor. Appl. Genet.* **2016**, *129*, 1203–1215. [CrossRef]
37. Xu, L.; Hu, K.; Zhang, Z.; Guan, C.; Chen, S.; Hua, W.; Li, J.; Wen, J.; Yi, B.; Shen, J.; et al. Genome-wide association study reveals the genetic architecture of flowering time in rapeseed (*Brassica napus* L.). *DNA Res.* **2015**, *23*, 43–52. [CrossRef]
38. Qian, L.; Qian, W.; Snowdon, R.J. Sub-genomic selection patterns as a signature of breeding in the allopolyploid *Brassica napus* genome. *BMC Genom.* **2014**, *15*, 1170. [CrossRef]
39. Anas, M.; Liao, F.; Verma, K.K.; Sarwar, M.A.; Mahmood, A.; Chen, Z.L.; Li, Q.; Zeng, X.P.; Liu, Y.; Li, Y.R. Fate of nitrogen in agriculture and environment: Agronomic, eco-physiological and molecular approaches to improve nitrogen use efficiency. *Biol. Res.* **2020**, *53*, 47. [CrossRef]
40. Wu, B.; Ren, W.; Zhao, L.; Li, Q.; Sun, J.; Chen, F.; Pan, Q. Genome-Wide Association Study of Root System Architecture in Maize. *Genes* **2022**, *13*, 181. [CrossRef]
41. Wang, J.; Kuang, L.; Wang, X.; Liu, G.; Dun, X.; Wang, H. Temporal genetic patterns of root growth in *Brassica napus* L. revealed by a low-cost, high-efficiency hydroponic system. *Theor. Appl. Genet.* **2019**, *132*, 2309–2323. [CrossRef] [PubMed]
42. Kupcsik, L.; Chiodi, C.; Moturu, T.R.; De Gernier, H.; Haelterman, L.; Louvieaux, J.; Tillard, P.; Sturrock, C.J.; Bennett, M.; Nacry, P.; et al. Oilseed Rape Cultivars Show Diversity of Root Morphologies with the Potential for Better Capture of Nitrogen. *Nitrogen* **2021**, *2*, 491–505. [CrossRef]
43. Dun, X.; Tao, Z.; Wang, J.; Wang, X.; Liu, G.; Wang, H. Comparative transcriptome analysis of primary roots of *brassica napus* seedlings with extremely different primary root lengths using RNA sequencing. *Front. Plant Sci.* **2016**, *7*, 1238. [CrossRef] [PubMed]
44. Robinson, D.; Peterkin, J.H. Clothing the Emperor: Dynamic Root-Shoot Allocation Trajectories in Relation to Whole-Plant Growth Rate and in Response to Temperature. *Plants* **2019**, *8*, 212. [CrossRef]
45. Kuang, L.; Ahmad, N.; Su, B.; Huang, L.; Li, K.; Wang, H.; Wang, X.; Dun, X. Discovery of Genomic Regions and Candidate Genes Controlling Root Development Using a Recombinant Inbred Line Population in Rapeseed (*Brassica napus* L.). *Int. J. Mol. Sci.* **2022**, *23*, 4781. [CrossRef]
46. Falik, O.; Mordoch, Y.; Ben-Natan, D.; Vanunu, M.; Goldstein, O.; Novoplansky, A. Plant responsiveness to root-root communication of stress cues. *Ann. Bot.* **2012**, *110*, 271–280. [CrossRef]
47. Canales, J.; Verdejo, J.; Carrasco-Puga, G.; Castillo, F.M.; Arenas-M, A.; Calderini, D.F. Transcriptome analysis of seed weight plasticity in *Brassica napus*. *Int. J. Mol. Sci.* **2021**, *22*, 4449. [CrossRef]
48. Liu, H.; Li, X.; Zhang, Q.; Yuan, P.; Liu, L.; King, G.J.; Ding, G.; Wang, S.; Cai, H.; Wang, C.; et al. Integrating a genome-wide association study with transcriptomic data to predict candidate genes and favourable haplotypes influencing *Brassica napus* seed phytate. *DNA Res.* **2021**, *28*, dsab011. [CrossRef]
49. Zhang, M.; Wang, Y.; Chen, X.; Xu, F.; Ding, M.; Ye, W.; Kawai, Y.; Toda, Y.; Hayashi, Y.; Suzuki, T.; et al. Plasma membrane H^+-ATPase overexpression increases rice yield via simultaneous enhancement of nutrient uptake and photosynthesis. *Nat. Commun.* **2021**, *12*, 735. [CrossRef]
50. Samuilov, S.; Brilhaus, D.; Rademacher, N.; Flachbart, S.; Arab, L.; Alfarraj, S.; Kuhnert, F.; Kopriva, S.; Weber, A.P.M.; Mettler-Altmann, T.; et al. The photorespiratory BOU gene mutation alters sulfur assimilation and its crosstalk with carbon and nitrogen metabolism in *Arabidopsis thaliana*. *Front. Plant Sci.* **2018**, *9*, 1709. [CrossRef]
51. Liao, Q.; Tang, T.; Zhou, T.; Song, H.; Hua, Y. Integrated Transcriptional and Proteomic Profiling Reveals Potential Amino Acid Transporters Targeted by Nitrogen Limitation Adaptation. *Int. J. Mol. Sci.* **2020**, *2*, 2171. [CrossRef] [PubMed]
52. Xu, P.; Cai, X.T.; Wang, Y.; Xing, L.; Chen, Q.; Xiang, C. Bin HDG11 upregulates cell-wall-loosening protein genes to promote root elongation in Arabidopsis. *J. Exp. Bot.* **2014**, *65*, 4285–4295. [CrossRef] [PubMed]
53. Veit, C.; König, J.; Altmann, F.; Strasser, R. Processing of the terminal alpha-1,2-linked mannose residues from oligomannosidic n-glycans is critical for proper root growth. *Front. Plant Sci.* **2018**, *9*, 1807. [CrossRef] [PubMed]
54. Dellero, Y. Manipulating Amino Acid Metabolism to Improve Crop Nitrogen Use Efficiency for a Sustainable Agriculture. *Front. Plant Sci.* **2020**, *11*, 602548. [CrossRef]
55. Li, H.; Cheng, X.; Zhang, L.; Hu, J.; Zhang, F.; Chen, B.; Xu, K.; Gao, G.; Li, H.; Li, L.; et al. An integration of genome-wide association study and gene co-expression network analysis identifies candidate genes of stem lodging-related traits in *Brassica napus*. *Front. Plant Sci.* **2018**, *9*, 796. [CrossRef]
56. Lu, B.; Wen, S.; Zhu, P.; Cao, H.; Zhou, Y.; Bie, Z.; Cheng, J. Overexpression of melon tonoplast sugar transporter cmtst1 improved root growth under high sugar content. *Int. J. Mol. Sci.* **2020**, *21*, 3524. [CrossRef]
57. Schofield, R.A.; Bi, Y.M.; Kant, S.; Rothstein, S.J. Over-expression of STP13, a hexose transporter, improves plant growth and nitrogen use in *Arabidopsis thaliana* seedlings. *Plant Cell Environ.* **2009**, *32*, 271–285. [CrossRef]
58. Wang, D.; Xu, T.; Yin, Z.; Wu, W.; Geng, H.; Li, L.; Yang, M.; Cai, H.; Lian, X. Overexpression of OsMYB305 in Rice Enhances the Nitrogen Uptake Under Low-Nitrogen Condition. *Front. Plant Sci.* **2020**, *11*, 369. [CrossRef]
59. Vandenbussche, F.; Fierro, A.C.; Wiedemann, G.; Reski, R.; Van Der Straeten, D. Evolutionary conservation of plant gibberellin signalling pathway components. *BMC Plant Biol.* **2007**, *7*, 65. [CrossRef]

20. Rui, C.; Peng, F.; Fan, Y.; Zhang, Y.; Zhang, Z.; Xu, N.; Zhang, H.; Wang, J.; Li, S.; Yang, T.; et al. Genome-wide expression analysis of carboxylesterase (CXE) gene family implies GBCXE49 functional responding to alkaline stress in cotton. *BMC Plant Biol.* **2022**, *22*, 194. [CrossRef]
21. Li, Y.; Dai, X.; Cheng, Y.; Zhao, Y. NPY genes play an essential role in root gravitropic responses in Arabidopsis. *Mol. Plant* **2011**, *4*, 171–179. [CrossRef] [PubMed]
22. Bi, Y.M.; Wang, R.L.; Zhu, T.; Rothstein, S.J. Global transcription profiling reveals differential responses to chronic nitrogen stress and putative nitrogen regulatory components in Arabidopsis. *BMC Genom.* **2007**, *8*, 281. [CrossRef] [PubMed]
23. Moison, M.; Marmagne, A.; Dinant, S.; Soulay, F.; Azzopardi, M.; Lothier, J.; Citerne, S.; Morin, H.; Legay, N.; Chardon, F.; et al. Three cytosolic glutamine synthetase isoforms localized in different-order veins act together for N remobilization and seed filling in Arabidopsis. *J. Exp. Bot.* **2018**, *69*, 4379–4393. [CrossRef]
24. Gan, Y.; Bernreiter, A.; Filleur, S.; Abram, B.; Forde, B.G. Overexpressing the ANR1 MADS-box gene in transgenic plants provides new insights into its role in the nitrate regulation of root development. *Plant Cell Physiol.* **2012**, *53*, 1003–1016. [CrossRef] [PubMed]
25. Furutani, M.; Hirano, Y.; Nishimura, T.; Nakamura, M.; Taniguchi, M.; Suzuki, K.; Oshida, R.; Kondo, C.; Sun, S.; Kato, K.; et al. Polar recruitment of RLD by LAZY1-like protein during gravity signaling in root branch angle control. *Nat. Commun.* **2020**, *11*, 76. [CrossRef]
26. Kumar, N.; Iyer-Pascuzzi, A.S. Shedding the last layer: Mechanisms of root cap cell release. *Plants* **2020**, *9*, 308. [CrossRef] [PubMed]
27. Massaro, M.; De Paoli, E.; Tomasi, N.; Morgante, M.; Pinton, R.; Zanin, L. Transgenerational Response to Nitrogen Deprivation in Arabidopsis thaliana. *Int. J. Mol. Sci.* **2019**, *20*, 5587. [CrossRef]
28. Sun, D.; Fang, X.; Xiao, C.; Ma, Z.; Huang, X.; Su, J.; Li, J.; Wang, J.; Wang, S.; Luan, S.; et al. Kinase SnRK1.1 regulates nitrate channel SLAH3 engaged in nitrate-dependent alleviation of ammonium toxicity. *Plant Physiol.* **2021**, *186*, 731–749. [CrossRef]
29. Lhamo, D.; Luan, S. Potential Networks of Channels and Transporters in Arabidopsis Roots at a Single Cell Resolution. *Front. Plant Sci.* **2021**, *12*, 689545. [CrossRef]
30. Hoagland, D.R. Optimum nutrient solutions for plants. *Science* **1920**, *52*, 562–564. [CrossRef]
31. Meng, L.; Li, H.; Zhang, L.; Wang, J. QTL IciMapping: Integrated software for genetic linkage map construction and quantitative trait locus mapping in biparental populations. *Crop J.* **2015**, *3*, 269–283. [CrossRef]
32. Bradbury, P.J.; Zhang, Z.; Kroon, D.E.; Casstevens, T.M.; Ramdoss, Y.; Buckler, E.S. TASSEL: Software for association mapping of complex traits in diverse samples. *Bioinformatics* **2007**, *23*, 2633–2635. [CrossRef]
33. Gómez-Rubio, V. ggplot2—Elegant Graphics for Data Analysis (2nd Edition). *J. Stat. Softw.* **2017**, *77*, 3–5. [CrossRef]
34. Turner, S.D. qqman: An R package for visualizing GWAS results using Q-Q and manhattan plots. *J. Open Source Softw.* **2018**, *3*, 2–3. [CrossRef]
35. Wei, L.; Jian, H.; Lu, K.; Filardo, F.; Yin, N.; Liu, L.; Qu, C.; Li, W.; Du, H.; Li, J. Genome-wide association analysis and differential expression analysis of resistance to Sclerotinia stem rot in *Brassica napus*. *Plant Biotechnol. J.* **2016**, *14*, 1368–1380. [CrossRef]
36. Chalhoub, B.; Denoeud, F.; Liu, S.; Parkin, I.A.P.; Tang, H.; Wang, X.; Chiquet, J.; Belcram, H.; Tong, C.; Samans, B.; et al. Early allopolyploid evolution in the post-neolithic *Brassica napus* oilseed genome. *Science* **2014**, *345*, 950–953. [CrossRef]
37. Liu, S.; Huang, H.; Yi, X.; Zhang, Y.; Yang, Q.; Zhang, C.; Fan, C.; Zhou, Y. Dissection of genetic architecture for glucosinolate accumulations in leaves and seeds of *Brassica napus* by genome-wide association study. *Plant Biotechnol. J.* **2020**, *18*, 1472–1484. [CrossRef]
38. Langfelder, P.; Horvath, S. WGCNA: An R package for weighted correlation network analysis. *BMC Bioinform.* **2008**, *9*, 559. [CrossRef]
39. Lu, K.; Guo, W.; Lu, J.; Yu, H.; Qu, C.; Tang, Z. Genome-Wide Survey and Expression Profile Analysis of the Mitogen-Activated Protein Kinase (MAPK) Gene Family in *Brassica rapa*. *PLoS ONE* **2015**, *10*, e0132051. [CrossRef]
40. Livak, K.J.; Schmittgen, T.D. Analysis of Relative Gene Expression Data Using Real-Time Quantitative PCR and the $2^{-\Delta\Delta CT}$ Method. *Methods* **2001**, *25*, 402–408. [CrossRef]

Review

Studies on Lotus Genomics and the Contribution to Its Breeding

Huanhuan Qi, Feng Yu, Jiao Deng[†] and Pingfang Yang *

State Key Laboratory of Biocatalysis and Enzyme Engineering, School of Life Science, Hubei University, Wuhan 430062, China; qihuanhuan0911@163.com (H.Q.); yufeng@hubu.edu.cn (F.Y.); ddj613@163.com (J.D.)
* Correspondence: yangpf@hubu.edu.cn; Tel.: +86-27-88661837
† Current Address: Research Center of Buckwheat Industry Technology, School of Life Sciences, Guizhou Normal University, Guiyang 550001, China.

Abstract: Lotus (*Nelumbo nucifera*), under the Nelumbonaceae family, is one of the relict plants possessing important scientific research and economic values. Because of this, much attention has been paid to this species on both its biology and breeding among the scientific community. In the last decade, the genome of lotus has been sequenced, and several high-quality genome assemblies are available, which have significantly facilitated functional genomics studies in lotus. Meanwhile, re-sequencing of the natural and genetic populations along with different levels of omics studies have not only helped to classify the germplasm resources but also to identify the domestication of selected regions and genes controlling different horticultural traits. This review summarizes the latest progress of all these studies on lotus and discusses their potential application in lotus breeding.

Keywords: lotus; genome; variant; germplasm; breeding; omics

Citation: Qi, H.; Yu, F.; Deng, J.; Yang, P. Studies on Lotus Genomics and the Contribution to Its Breeding. *Int. J. Mol. Sci.* **2022**, *23*, 7270. https://doi.org/10.3390/ijms23137270

Academic Editors: Zhiyong Li and Jian Zhang

Received: 26 May 2022
Accepted: 27 June 2022
Published: 30 June 2022

Publisher's Note: MDPI stays neutral with regard to jurisdictional claims in published maps and institutional affiliations.

Copyright: © 2022 by the authors. Licensee MDPI, Basel, Switzerland. This article is an open access article distributed under the terms and conditions of the Creative Commons Attribution (CC BY) license (https://creativecommons.org/licenses/by/4.0/).

1. Introduction

Food shortage has become a rising challenge with the increase of the world's population and decrease of natural resources. It is incredibly significant to breed crops with high yields, of good quality, and high-stress resistance to ascertain food security since crops provide a staple food supply for the world. To achieve this, it is necessary to obtain a deeper understanding of the crops' genetic background, especially their genome information. Since the first flower plant, *Arabidopsis thaliana* genome was sequenced in 2000 [1], more and more plant whole genomes have been sequenced and deposited in databases, which are available to the public [2], including *Nelumbo* genome database (http://nelumbo.biocloud.net/nelumbo/home) [3]. Third-generation sequencing, which can produce long sequence read, has shown its advantages over next-generation sequencing (NGS) in generating high continuity reference genome assemblies [4].

Genomics is the cornerstone of breeding, and studies based on whole-genome sequencing and genome-wide association study have greatly driven forward genomics-assisted breeding in many crops [5,6]. Cloning and functional analysis of genes associated with important agronomic traits in rice (*Oryza Sativa*), soybean (*Glycine max*), and tomato (*Solanum lycopersicum*) have also demonstrated that high-quality genomes are prerequisite to clarify variations in each species [7–11]. However, population genetic analysis relied on a single reference genome that lost variant information, especially in the highly polymorphic region. Pan-genome contains the totality of genome sequence information of the target species and covers more comprehensive variant information. Pan-genomes have been constructed in many plants, such as rice, maize, brassica, and soybean, and applied to identify causal genes [12–15]. Pan-genome or graph pan-genome is obtaining new references along with the upgrading of sequencing. The information on genome maps, domestication, improvement-related genes, and regulation pathway promotes the understanding of plant evolution and accelerate breeding [15].

Lotus is one of the relict plants retaining the original morphology of its ancestors, as well as *Ginkgo biloba*, *Liriodendron*, and *Metasequoia glyptostroboides*. It belongs to the

Nelumbo genus of the Nelumbonaceae family, which includes two species, namely Asian lotus (*Nelumbo nucifera* Gaertn.) and American lotus (*Nelumbo lutea* Pear.). The two species are named for their different geographical distributions. Asian lotus is mainly distributed in Asia and the north of Oceania, while American lotus is distributed in North America and South America. The plant morphology differs between them. Asian lotus is a tall plant, with oval leaves and seeds, and red or white flower colors, whereas American lotus is a short plant, nearly round and with dark green leaves, spherical seeds, and yellow flowers [16]. There is no strict reproductive isolation between them, and the life cycles are similar at about five months. Asian lotus is commonly called lotus and has more than 3000 years of cultivation history as a horticultural crop [17]. Lotus seeds and rhizomes have rich nutritional value and unique health-care function. Lotus seeds contain starch, proteins, amino acids, polysaccharides, polyphenols, alkaloids, and mineral elements. Lotus rhizome has a high vitamin C content. During the long period of domestication and artificial selection, about 4500 lotus cultivars have been obtained up till the present [18]. These cultivars have been planted to produce edible vegetables, snacks, beverages, restorative materials, and ornamental flowers, which impact human life and economic development. The lotus industry is also important for rural revitalization in the Yangtze River, Pearl River, and Huang Huai river basins. The cultivated lotus is generally divided into rhizome lotus, seed lotus, and flower lotus based on their different usage. The notable feature of the rhizome lotus is the enlarged rhizome but with few flowers. It can be divided into power and crisp type according to the taste of the rhizome. Different varieties were bred to meet the taste of the different regions of people or for further usage. The main breeding goal of rhizome lotus is to improve the yield and quality of the rhizome. Seed lotus is mainly for lotus seed production, with high yield, good quality, and disease resistance being the breeding goals. Flower lotus is preferred for ornamental use, and it has distinct flower colors and shapes. During long cultivation, ornamental lotus with different flower morphologies were obtained, including few-petaled, double-petaled, petaloidy, and thousand-petalled flowers. Red, pink, yellow, and white are the main flower colors. Currently, the breeding objective is mainly aimed at flower shape and color, yield or quality of lotus seed and rhizome, and wide adaptability.

As a basal eudicot species, lotus plays an essential role in studying plant evolution and phylogeny. It is adapted to the aquatic environment, while its relatives are shrubs or trees living on land. Water lily lies at the phylogenetic position of the base angiosperm and has similar living conditions and flowers. However, its genomes are vastly different [19]. Lotus has unique features such as water-repellent self-cleaning function, multi-seed production, and flower thermogenesis, which may relate to flower protogyny or provide a warm environment for pollination [16]. Because of its importance in plant phylogeny and wide application, lotus has gained increasing attention from the scientific community. Since the release of the first version of two lotus reference genomes [20,21], genome-based investigations have been conducted continuously. Subsequently, the high-resolution genetic map and BioNano optical map were applied to improve the accuracy and assembly of the lotus genome [22]. A hybrid assembly was completed using PacBio sequencing data and previously published short reads [23]. High-quality genome assembly of "Taikonglian NO. 3" and American lotus genome were also recently generated [24,25]. High-throughput re-sequencing of different lotus cultivars has been utilized to identify numerous molecular markers, promoting marker-assisted selection. Moreover, "omics" approaches such as transcriptomics, proteomics, and metabolomics were applied in elucidating molecular regulatory networks of yield, quality, and response to stress in lotus. Here, we briefly review the latest progress of studies on the lotus genome, and how genome information could be used in lotus breeding. Meanwhile, the existing challenge and potential prospects are also discussed.

2. Sequencing, Assembly and Annotation of Lotus Genome

Lotus occupies a crucial phylogenetic position in flowering plants. The high-quality reference genome of lotus plays a vital role in studying the origin of eudicot and lotus molecular breeding. In the last decade, some lotus varieties were sequenced by different platforms, which resulted in a different version of the genome assembly and annotation (Table 1). Based on NGS, a wild lotus, "China antique (CA)", was successfully sequenced and assembled [20]. The total sequenced genome length of "CA" is 804 Mb, of which 543.4 Mb (67.6%) were anchored to nine megascaffolds. The contig N50 was 38.8 Kb and the scaffold N50 was 3.4 Mb. The heterozygosity of "CA" genome is 0.03%, and the repetitive sequence is about 57%. A total of 26,685 protein-coding genes were predicted, with the average length of a gene being 6561 bp. Simultaneously, another wild strain of lotus, "Chinese Taizi" was assembled through NGS technology. The final assembled genome size is 792 Mb with the contig N50 39.3 Kb and scaffold N50 986.5 Kb [21]. The length of transposable elements is 392 Mb (49.48%), and 36,385 protein-coding genes were annotated. One WGD event -λ in lotus instead of the paleo-hexaploid arrangement (γ WGD) event that occurred in core eudicots was predicted [20,21]. These two genomes were further anchored to eight pseudo-chromosomes by constructing a higher resolution genetic map and physical maps [22].

Table 1. Comparison of assembled lotus genomes.

Items	Year	Sequencing Technology	Final Assembly (Mb)	Contig N50	Number of Genes	Repeat Sequences	Ref.
China Antique v1.0	2013	Illumina, 454	804	38.8 Kb	26685	57%	[20]
Taizi	2013	Illumina Hiseq2000	792	39.3 Kb	40348	49.48% (TEs)	[21]
China Antique v2.0	2020	Pacbio Sequel, Illumina	821.2	484.3 Kb	32124	58.50%	[23]
Taikonglian NO.3	2022	Nanopore	807	5.1 Mb	28274	63.11%	[24]
American lotus	2022	Pacbio RSII, Hi-C	843	1.34 Mb	31382	81.00%	[25]

With the advent of a new sequencing platform, the genome of "CA" was re-assembled using 11.9 Gb long-read data from PacBio Sequel, and 94.2 Gb previously sequenced short-read data [23]. The new assembly of "CA" is 807.6 Mb with the contig N50 being 484.3 Kb, which has significantly increased the quality of the genome. The ratio of repetitive sequence (58.5%) was similar to the first version. Moreover, a cultivated lotus, "Taikonglian NO. 3", was also assembled using the Oxford Nanopore sequencing platform (57.9 Gb raw data) with the contig N50 being 5.1 Mb, and eight chromosomes were anchored based on high-throughput chromatin conformation capture (Hi-C) data [24]. Another lotus species, American lotus, was recently assembled using PacBio RSII (74.6 Gb raw data) and Hi-C (50.32 Gb raw data), and the total length is 843 Mb while contig N50 is 1.34 Mb [25]. These data demonstrate that long-read sequencing technology has greatly improved the quality of the lotus genome. The successful assembly of the genome in Asian lotus, including wild and cultivar varieties, and American lotus will assist the investigation of functional genomics as well as molecular breeding in lotus.

3. Study on the Potential Adaptive Evolution and Domestication of Lotus

The availability of lotus reference genome information has facilitated the resequencing of different lotus germplasms. Several studies were conducted on how the lotus genome was subjected to adaptive evolution and artificial selection. Although it is known that there are only two species of lotus, namely Asian lotus and American lotus, except for the difference in flower color, their plant architecture and morphology are very similar. Based on molecular phylogeny analysis, significant genetic differentiation between American and Asian lotus was verified [25–28]. De-novo deep sequencing of the American lotus showed that its genome size is 843 Mb, and an approximate 81% repeat sequence was identified (Table 1), which is larger than the genome of Asian lotus. It is interesting to investigate the dramatic difference in repeat sequence between them because most protein-coding genes show a one-to-one synteny pattern. A total of 29,533 structure variations (SVs) were detected between two lotus species, with the SV-associated genes overexpressed in 'regulation of mitotic cell cycle', and 'protein transporter activity' [25]. Meanwhile, this study also showed that the selection on an *MYB* gene might contribute to the color difference between Asian and American lotus [25]. It is still an open question about when the two species diverged during the evolution and how they could keep high similarity in the independent geographical evolution. The wild lotus is distributed widely worldwide and maintains higher genomic diversity than cultivated lotus. Tropical and temperate lotus are the two ecotypes of Asian lotus. The comparison of the genome of these two ecotypes showed that a total of 453 genes were subjected to selection, including *cyp714a* genes that may relate to rhizome morphogenesis and a 10-Mb region in chromosome 1 that might play key roles in environmental adaption; including a homolog gene of *at5g2394* in *Arabidopsis* encoding an acyltransferase protein [24]. By comparing their expressional patterns, the genes encoding granule-bound starch synthases, storage organ development, *COSTAN-like* gene family, vernalization, as well as cold response genes may relate to ecotypic differentiation [26].

It is very important to know the genetic backgrounds of parental lines in breeding. The origin, classification, and evolution of cultivated lotus were investigated through population re-sequence analysis. A total of 18 lotus accessions, including categories of American, seed, rhizome, flower, wild, and Thai lotus, were re-sequenced, based on which phylogenetic tree was constructed. The results indicated that the rhizome lotus had a closer relationship with wild lotus. In contrast, seed and flower lotus were admixed [26], which could be supported by re-sequencing of an enlarged population containing 296 accessions of different germplasm (58 wild, 163 rhizome, 39 flower, 32 seed lotus varieties) [28]. Further re-sequencing of 69 lotus accessions showed that flower lotus might mix with rhizome or seed lotus [27]. All studies showed a low genetic variation in rhizome lotus, while higher genetic diversity in seed lotus. The origin of different subgroups is controversial, which is possible because the same accession of lotus has other names which were then divided into different subgroups by various people. Based on this genomic diversity, the potential domestication signals of cultivar subgroups could be speculated because the selected genomic regions had lower nucleotide diversity. When subgroups of seed, rhizome, and flower lotus were compared with wild lotus subgroup, a total of 1214, 95, and 37 artificially selected regions containing 2176, 77, and 24 genes were identified in seed, rhizome, and flower lotus, respectively [27]. Several of these selected genes were involved in key developmental processes associated with different organs. For example, a *SUPER-MAN like* gene affecting seed weight and size and a *legumin A-like* gene involved in storage protein synthesis were identified in the subgroup of seed lotus, while an *expansin-A 13-like* gene was identified in the subgroup of rhizome lotus [27]. These specifically selected genes controlling agronomic traits in different subgroups are also possible targets for lotus breeding. Meanwhile, different types of molecular markers have been developed, which may further facilitate the clarification of the relationship between different subgroups and maker-assisted breeding of new lotus varieties [29–38].

4. Identification of Genes with Potential Application in Lotus Breeding

As the largest aquatic vegetable in China, lotus is mainly bred through traditional cross-breeding and physical and chemical mutation as supplementation, based on which thousands of varieties have been obtained [39]. However, the selection of high-quality varieties was mainly based on the breeders' experience, because the mechanisms underlying each economic trait remained unclear. With the development of genomics and molecular genetics of the lotus, genome-based breeding is gradually becoming an effective method for lotus. Causal genes regulating essential traits, such as flower color and shape, rhizome yield, and seed quality, have been widely studied.

Flower color, shape, and flowering time are important traits that determine the ornamental value of lotus. There are three different colors in lotus, red and white in Asian lotus and yellow in American lotus. The red color in Asian lotus is determined by the contents of anthocyanin [40,41], which is controlled by key enzyme encoding genes, and their regulating transcription factors (TFs) such as *MYB*, *basic-Helix-Loop-Helix* (*bHLH*), *WD40* in its biosynthetic pathway. Among all the enzyme encoding genes in this pathway, *NnANS* and *NnUFGT* seem to be the decisive two genes [42,43]. Several TFs including 5 *MYB*, 2 *bHLH*, and one *WD-repeat* genes, may be involved in the regulation of anthocyanin biosynthesis in lotus based on a transcriptome analysis [43]. Among them, a *bHLH* gene *NnTT8* was verified to regulate anthocyanin biosynthesis [44], whereas the yellow color of American lotus is determined by carotenoid, and no anthocyanin was detected [25,45]. Further analysis indicated that the difference in the coding region between *NnMYB5* (Genbank accession, KU198697) and *NlMYB5* (Genbank accession, KU198698) is the main reason for the different colors in the two species. Flower morphology is another factor that determines the ornamental value of lotus. Flower development is controlled by intricate gene-regulatory networks, and many vital genes that control flowering time have been identified in flowering plant species. However, the molecular regulation mechanism has not been well characterized in lotus. Comparative transcriptomic analysis of different bud development stages in temperate and tropical lotus identified 147 lotus floweringtime associated genes that participate in photoperiod, gibberellic acid and vernalization pathways [46]. The *MADS-box* TFs are widely involved in plant growth and development. A total of 44 *MADS-box* genes were identified in lotus, and based on the selected candidates, *NnMADS14* (*SEPALLATA3* (*SEP3*) homolog gene) was identified to be related to floral organogenesis in lotus [47]. Lotus possesses distinct types of flower morphology, and the floral organ petaloid phenomenon is universal. Comparative transcriptomic analysis identified many hormonal signal transduction pathway genes and *MADS-box* genes; *AGAMOUS*(*AG*) was predicted as the candidate which was gene related to carpel petaloidy [48,49]. Genome-wide DNA methylation analysis showed that different flower organs exhibited different methylation levels, while *plant U-box* (*PUB33*) homolog gene might play crucial roles in the stamen petaloid [50]. Furthermore, *NnFTIP1* was proven to interact with *NnFT1* and regulate the flowering time in lotus [51].

The rhizome is the main edible part of lotus. It is important to explore the mechanisms underlying rhizome formation and expansion in rhizome lotus breeding. Comparative transcriptomic and proteomic analyses focusing on rhizome development have been conducted to dig out the key genes and pathways critical for the crucial physiological process [52–54]. Furthermore, re-sequencing of the natural and genetic F_2 populations has also identified several genetic regions and candidate genes that might be involved in lotus rhizome enlargement [55]. A systematic analysis was conducted on one candidate gene *CONSTANS-LIKE 5* (*COL5*). Functional analysis in the potato system indicated that *NnCOL5* might be positively associated with rhizome enlargement by regulating the expression of *CO-FT* genes and the GA signaling pathway [56]. In addition, one SNP was identified in another candidate gene *NnADAP* of *AP2* subfamily, which is closely associated with rhizome enlargement phenotype and the soluble sugar content [57]. There is a big difference between temperate and tropical lotus, especially the rhizome's morphology. Many genes were highly differentiated between them, such as *APL* homologs and granule-bound starch synthases genes [26]. Temperate lotus is distributed at 20° north latitude and shows a significant annual growth cycle, whereas tropical lotus is distributed south of 17° north latitude and exhibits perennial growth. Asian wild lotus can be further divided into temperate, subtropical, and tropical types and is distributed in northeast China, the Yangtze River and Pearl River Basin, Thailand and India [27]. Different lotus groups are subject to different selection pressure, such as light, temperature, UV, and soil types. The genes underlying selection were discussed by integrating population genetics and omics data. Several genes related to photosynthesis and DNA repair were selected, such as NAD + ADP ribosyltranferase, 8-oxoguanine-DNA glycosylase 1 and DNA polymerase epsilon subunit B2. The *vacuolar iron transporter* (*VIT*) family gene, *nodulin-like 21* gene encoding vacuolar iron transporter, may be related to metal ion metabolism [24]. The homolog gene of *Arabidopsis VIN3* in lotus was predicted to be related to flowering time and dormancy, with higher expression in temperate lotus than in tropical lotus [26].

Lotus seeds are rich in nutrients and functional compounds such as alkaloids, flavonoids, and polyphenols [58,59]. They are consumed "as both food and medicine" [60]. It is essential to increase the yield and nutrition of lotus seed. The main factors determining lotus seed yield are the seed size and the number of lotus seeds per seedpod. Transcriptome analysis on the cotyledon of "CA" and "Jianxuan-17 (JX-17)" seeds at different developmental stages identified 8437 differentially expressed genes (DEGs). Many DEGs are involved in the brassinosteroid biosynthesis pathway, and further analysis predicted two *AGPase* genes as candidate genes affecting lotus seed yield [61]. It seems that phytohormones are involved in lotus seed development. A combination of metabolomic and proteomic methods revealed that 15 DAP (Day After Pollination) was a switch time point from the physiological active to the nutrition accumulation stage [62]. Starch is the primary nutritional component in mature lotus seed [63]. Its contents and the proportion of amylose and amylopectin could largely determine the nutritional value and taste of lotus cotyledon, respectively. ADP-glucose pyrophosphorylase (AGPase) plays an important role in regulating starch biosynthesis. The evolution of *AGPase* genes experienced a purification selection, and *NnAGPL2a* and *NnAGPS1a* were the candidate genes related to starch content [64]. Starch branching enzyme (SBE) genes are key regulatory genes during starch synthesis, and *NnSBEI* and *NnSBEIII* were identified as related to the chain length of amylopectin in lotus [65]. In addition, comparative metabolomics between wild germplasm "CA" and domesticated cultivar "JX-17" indicated that the seed yield and the content of metabolites showed trade-offs [66]. For nutritional and medicinal values of lotus seed, the metabolomics-assisted strategy might be applied in lotus breeding in the future [67]. Seed dormancy is one of the domestication traits. The classical stay-green *G* gene controlling seed dormancy was cloned in domestication and as improvement genes in soybean, rice, and tomato. G gene interacts with *NCEDS* and *SPY* and in turn, regulates abscisic acid (ABA) synthesis [68]. *NnDREB1* and *NnPER1* were identified from lotus and may be involved in the ABA signal transduction pathway and then modulate longevity and dormancy [69].

Except for the above breeding objectives, there are other diversified breeding objectives, such as resistance to submerging and high antioxidant content. Lotus has evolved novel features to adapt to aquatic lifestyle. Many putative copper-dependent proteins, especially *COG2132* gene family, expand in lotus and form a separate phylogenetic clade having functions distinct from *Arabidopsis* [20]. Research has shown that although lotus grows in water, it is actually "afraid" of water. A time-course submergence experiment and RNA-seq analysis showed lotus has a low tolerance to complete submergence stress, and took two major strategies to cope with submergence stress in different stages. In the early stage (3~6 h) it initiates a low oxygen "escape" strategy (LOES), with the rapid accumulation of ethylene, rapid elongation of petioles, and significantly increases the density of aerenchyma and *ERF-VII* genes while lotus innate immunity genes become elevated; In the later stage (24~120 h), it starts a "breath holding" mode to limit its anaerobic respiration to the lowest level [70]. Flooding is serious abiotic stress affecting plant growth and can be classified into waterlogging and submergence. During the rainy season, the lotus is vulnerable to submergence. It is necessary to cultivate lotus varieties that are resistant to flooding to promote economic value.

WRKY TFs play key roles in modulating plant biotic and abiotic stress response and secondary metabolic regulations. A total of 65 *WRKY* genes were identified in lotus, and they were regulated by salicylic acid (SA) and jasmonic acid (JA), of which *NnWRKY40a* and *NnWRKY40b* were significantly induced by JA and promoted benzylisoquinoline alkaloid (BIA) biosynthesis [71]. Lotus predominantly accumulates BIA, and the leaf and embryo have different alkaloid components that may be caused by two cluster *CYP80* genes synthetic bis-BIAs, and aporphine-type BIAs, respectively. Five TFs (3 *MYBs*, one ethylene-response factor, and one *bHLH*) were identified as the regulator involved in the BIA biosynthetic pathway in lotus [72].

5. Conclusions and Perspective

The new varieties of lotus with high yield, wide adaptability, and stress resistance play a vital role in improving the economic value of this important horticulture crop. The variations identification, functional gene cloning, and metabolites alterations among diverse germplasm resources were investigated in the past decades, driven by the progressively improved genome information which could facilitate breeding practices in lotus (Figure 1). However, a high-quality reference genome is the limiting factor that will affect the molecular breeding process. Improvement of the lotus reference genome will be a requisite in the future, directly affecting the accuracy of molecular markers and the efficiency of cloning functional genes. Gapless reference genomes and pan-genomes have become the new reference, based on which plentiful information of genomes such as open chromatin and more variant information can be explored. With the explosive growth of large-omics data, deep learning can be used to mine biological information and decipher gene regulation networks. Moreover, a sound genetic transformation system has not yet been well established in lotus, which still restricts the validation of gene function and genome-based gene editing, further hindering breeding strategies. Few studies on epigenomics, such as histone marks, accessible chromatin regions, and genomic interactions, have been conducted and are needed in future. Based on these investigations, collection of wild lotus germplasm and classification of both wild and cultivated germplasm, analysis of domestication, and identification of molecular markers and genes closely linked to important agronomic traits, should be systematically conducted in the coming years. Combining these and developing multiple breeding targets will speed up the breeding efficiency in lotus.

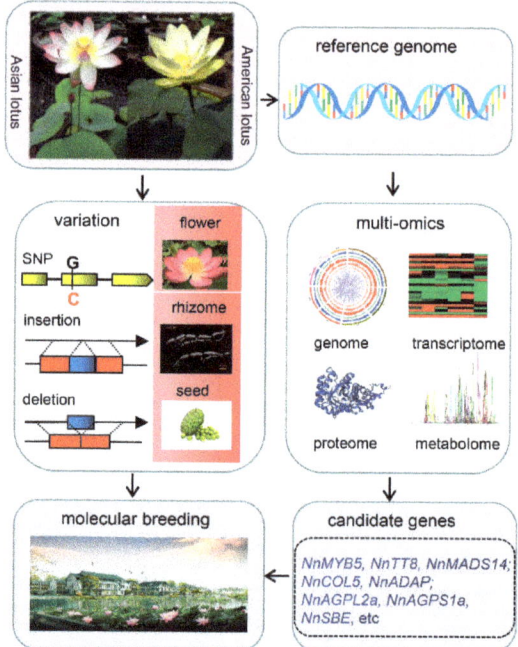

Figure 1. Flowchart of the molecular breeding process of lotus.

Author Contributions: H.Q.: Original draft preparation and writing. F.Y.: review and editing; J.D.: discussion and editing; P.Y.: editing, review, conceptualization, supervision. All authors have read and agreed to the published version of the manuscript.

Funding: This research was supported by the National Natural Science Foundation of China (NSFC no. 32102422).

Institutional Review Board Statement: Not applicable.

Informed Consent Statement: Not applicable.

Data Availability Statement: Not applicable.

Acknowledgments: We thank all the colleagues who have been involved in the studies on lotus.

Conflicts of Interest: The authors declare no conflict of interest.

References

1. Arabidopsis Genome Initiative. Analysis of the genome sequence of the flowering plant *Arabidopsis thaliana*. *Nature* **2000**, *408*, 796–815. [CrossRef] [PubMed]
2. Chen, F.; Dong, W.; Zhang, J.; Guo, X.; Chen, J.; Wang, Z.; Lin, Z.; Tang, H.; Zhang, L. The sequenced angiosperm genomes and genome databases. *Front. Plant Sci.* **2018**, *9*, 418. [CrossRef] [PubMed]
3. Li, H.; Yang, X.; Zhang, Y.; Gao, Z.; Liang, Y.; Chen, J.; Shi, T. Nelumbo genome database, an integrative resource for gene expression and variants of *Nelumbo nucifera*. *Sci. Data* **2021**, *8*, 38. [CrossRef] [PubMed]
4. van Dijk, E.L.; Jaszczyszyn, Y.; Naquin, D.; Thermes, C. The Third Revolution in Sequencing Technology. *Trends Genet.* **2018**, *34*, 666–681. [CrossRef]
5. Liang, Y.; Liu, H.J.; Yan, J.; Tian, F. Natural Variation in Crops, Realized Understanding, Continuing Promise. *Annu. Rev. Plant Biol.* **2021**, *72*, 357–385. [CrossRef]
6. Huang, X.; Wei, X.; Sang, T.; Zhao, Q.; Feng, Q.; Zhao, Y.; Li, C.; Zhu, C.; Lu, T.; Zhang, Z.; et al. Genome-wide association studies of 14 agronomic traits in rice landraces. *Nat. Genet.* **2010**, *42*, 961–967. [CrossRef]
7. Huang, X.; Kurata, N.; Wei, X.; Wang, Z.X.; Wang, A.; Zhao, Q.; Zhao, Y.; Liu, K.; Lu, H.; Li, W.; et al. A map of rice genome variation reveals the origin of cultivated rice. *Nature* **2012**, *490*, 497–501. [CrossRef]
8. Liu, S.; Zhang, M.; Feng, F.; Tian, Z. Toward a "Green Revolution" for Soybean. *Mol. Plant* **2020**, *13*, 688–697. [CrossRef]
9. Tomato Genome Consortium. The tomato genome sequence provides insights into fleshy fruit evolution. *Nature* **2012**, *485*, 635–641. [CrossRef]
10. Bolger, A.; Scossa, F.; Bolger, M.E.; Lanz, C.; Maumus, F.; Tohge, T.; Quesneville, H.; Alseekh, S.; Sørensen, I.; Lichtenstein, G.; et al. The genome of the stress-tolerant wild tomato species Solanum pennellii. *Nat. Genet.* **2014**, *46*, 1034–1038. [CrossRef]
11. Lin, T.; Zhu, G.; Zhang, J.; Xu, X.; Yu, Q.; Zheng, Z.; Zhang, Z.; Lun, Y.; Li, S.; Wang, X.; et al. Genomic analyses provide insights into the history of tomato breeding. *Nat. Genet.* **2014**, *46*, 1220–1226. [CrossRef]
12. Tay Fernandez, C.G.; Nestor, B.J.; Danilevicz, M.F.; Gill, M.; Petereit, J.; Bayer, P.E.; Finnegan, P.M.; Batley, J.; Edwards, D. Pangenomes as a resource to accelerate breeding of under-utilised crop species. *Int. J. Mol. Sci.* **2022**, *23*, 2671. [CrossRef] [PubMed]
13. Gao, L.; Gonda, I.; Sun, H.; Ma, Q.; Bao, K.; Tieman, D.M.; Burzynski-Chang, E.A.; Fish, T.L.; Stromberg, K.A.; Sacks, G.L.; et al. The tomato pan-genome uncovers new genes and a rare allele regulating fruit flavor. *Nat. Genet.* **2019**, *51*, 1044–1051. [CrossRef] [PubMed]
14. Liu, Y.; Du, H.; Li, P.; Shen, Y.; Peng, H.; Liu, S.; Zhou, G.A.; Zhang, H.; Liu, Z.; Shi, M. Pan-Genome of Wild and Cultivated Soybeans. *Cell* **2020**, *182*, 162–176. [CrossRef] [PubMed]
15. Huang, X.; Huang, S.; Han, B.; Li, J. The integrated genomics of crop domestication and breeding. *Cell*, **2022**; *Advance online publication*. [CrossRef]
16. Lin, Z.; Zhang, C.; Cao, D.; Damaris, R.N.; Yang, P. The latest studies on lotus (*Nelumbo nucifera*)-an emerging horticultural model plant. *Int. J. Mol. Sci.* **2019**, *20*, 3680. [CrossRef]
17. Shen-Miller, J. Sacred lotus, the long-living fruits of China Antique. *Seed Sci. Res.* **2002**, *12*, 131–143. [CrossRef]
18. Liu, L.; Li, Y.; Min, J.; Xiang, Y.; Tian, D. Analysis of the cultivar names and characteristics of global lotus (*Nelumbo*). *Hans J. Agric. Sci.* **2019**, *9*, 163–181.
19. Zhang, L.; Chen, F.; Zhang, X.; Li, Z.; Zhao, Y.; Lohaus, R.; Chang, X.; Dong, W.; Ho, S.; Liu, X.; et al. The water lily genome and the early evolution of flowering plants. *Nature* **2020**, *577*, 79–84. [CrossRef]
20. Ming, R.; VanBuren, R.; Liu, Y.; Yang, M.; Han, Y.; Li, L.T.; Zhang, Q.; Kim, M.J.; Schatz, M.C.; Campbell, M.; et al. Genome of the long-living sacred lotus (*Nelumbo nucifera* Gaertn.). *Genome Biol.* **2013**, *14*, R41. [CrossRef]
21. Wang, Y.; Fan, G.; Liu, Y.; Sun, F.; Shi, C.; Liu, X.; Peng, J.; Chen, W.; Huang, X.; Cheng, S.; et al. The sacred lotus genome provides insights into the evolution of flowering plants. *Plant J.* **2013**, *76*, 557–567. [CrossRef]

22. Gui, S.; Peng, J.; Wang, X.; Wu, Z.; Cao, R.; Salse, J.; Zhang, H.; Zhu, Z.; Xia, Q.; Quan, Z.; et al. Improving *Nelumbo nucifera* genome assemblies using high-resolution genetic maps and BioNano genome mapping reveals ancient chromosome rearrangements. *Plant J.* **2018**, *94*, 721–734. [CrossRef] [PubMed]
23. Shi, T.; Rahmani, R.S.; Gugger, P.F.; Wang, M.; Li, H.; Zhang, Y.; Li, Z.; Wang, Q.; Van de Peer, Y.; Marchal, K.; et al. Distinct expression and methylation patterns for genes with different fates following a single whole-genome duplication in flowering plants. *Mol. Biol. Evol.* **2020**, *37*, 2394–2413. [CrossRef] [PubMed]
24. Zheng, X.; Wang, T.; Cheng, T.; Zhao, L.; Zheng, X.; Zhu, F.; Dong, C.; Xu, J.; Xie, K.; Hu, Z.; et al. Genomic variation reveals demographic history and biological adaptation of the ancient relictual, lotus (*Nelumbo* Adans). *Hortic. Res.* **2022**, *9*, uhac029. [CrossRef] [PubMed]
25. Zheng, P.; Sun, H.; Liu, J.; Lin, J.; Zhang, X.; Qin, Y.; Zhang, W.; Xu, X.; Deng, X.; Yang, D.; et al. Comparative analyses of American and Asian lotus genomes reveal insights into petal color, carpel thermogenesis and domestication. *Plant J.* **2022**, *110*, 1498–1515. [CrossRef] [PubMed]
26. Huang, L.; Yang, M.; Li, L.; Li, H.; Yang, D.; Shi, T.; Yang, P. Whole genome re-sequencing reveals evolutionary patterns of sacred lotus (*Nelumbo nucifera*). *J. Integr. Plant Biol.* **2018**, *60*, 2–15. [CrossRef]
27. Li, Y.; Zhu, F.L.; Zheng, X.W.; Hu, M.L.; Dong, C.; Diao, Y.; Wang, Y.W.; Xie, K.Q.; Hu, Z.L. Comparative population genomics reveals genetic divergence and selection in lotus, *Nelumbo nucifera*. *BMC Genom.* **2020**, *21*, 146. [CrossRef]
28. Liu, Z.; Zhu, H.; Zhou, J.; Jiang, S.; Wang, Y.; Kuang, J.; Ji, Q.; Peng, J.; Wang, J.; Gao, L.; et al. Resequencing of 296 cultivated and wild lotus accessions unravels its evolution and breeding history. *Plant J.* **2020**, *104*, 1673–1684. [CrossRef]
29. Liu, Z.; Zhu, H.; Liu, Y.; Kuang, J.; Zhou, K.; Liang, F.; Liu, Z.; Wang, D.; Ke, W. Construction of a high-density, high-quality genetic map of cultivated lotus (*Nelumbo nucifera*) using next-generation sequencing. *BMC Genom.* **2016**, *17*, 466. [CrossRef]
30. Chen, Y.; Zhou, R.; Lin, X.; Wu, K.; Qian, X.; Huang, S. ISSR analysis of genetic diversity in sacred lotus cultivars. *Aquat. Bot.* **2008**, *89*, 311–316. [CrossRef]
31. Guo, H.B.; Li, S.M.; Peng, J.; Ke, W.D. Genetic diversity of *nelumbo* accessions revealed by RAPD. *Genet. Resour. Crop Evol.* **2007**, *54*, 741–748. [CrossRef]
32. Han, Y.C.; Teng, C.Z.; Sheng, Z.; Zhou, M.Q.; Hu, Z.L.; Song, Y.C. Genetic variation and clonal diversity in populations of *Nelumbo nucifera* (nelumbonaceae) in central china detected by ISSR markers. *Aquat. Bot.* **2007**, *86*, 69–75. [CrossRef]
33. Hu, J.; Pan, L.; Liu, H.; Wang, S.; Wu, Z.; Ke, W.; Ding, Y. Comparative analysis of genetic diversity in sacred lotus (*Nelumbo nucifera* Gaertn.) using AFLP and SSR markers. *Mol. Biol. Rep.* **2012**, *39*, 3637–3647. [CrossRef] [PubMed]
34. Li, Z.; Liu, X.; Gituru, R.W.; Juntawong, N.; Zhou, M.; Chen, L. Genetic diversity and classification of *nelumbo* germplasm of different origins by RAPD and ISSR analysis. *Sci. Hortic.* **2010**, *125*, 724–732. [CrossRef]
35. Pan, L.; Quan, Z.; Li, S.; Liu, H.; Huang, X.; Ke, W.; Ding, Y. Isolation and characterization of microsatellite markers in the sacred lotus (*Nelumbo nucifera* gaertn.). *Mol. Ecol. Notes* **2007**, *7*, 1054–1056. [CrossRef]
36. Yang, M.; Xu, L.; Liu, Y.; Yang, P. RNA-Seq Uncovers SNPs and alternative aplicing events in Asian lotus (*Nelumbo nucifera*). *PLoS ONE* **2015**, *10*, e0125702.
37. Hu, J.; Gui, S.; Zhu, Z.; Wang, X.; Ke, W.; Ding, Y. Genome-wide identification of SSR and SNP markers based on whole-genome re-sequencing of a Thailand wild sacred lotus (*Nelumbo nucifera*). *PLoS ONE* **2015**, *10*, e0143765. [CrossRef]
38. Zhang, Q.; Zhang, X.; Liu, J.; Mao, C.; Chen, S.; Zhang, Y.; Leng, L. Identification of copy number variation and population analysis of the sacred lotus (*Nelumbo nucifera*). *Biosci. Biotechnol. Biochem.* **2020**, *84*, 2037–2044. [CrossRef]
39. Zhang, X.; Wang, Q. A preliminary investigation on the morphological and biological characteristics of lotus varieties. *Acta Hortic. Sin.* **1966**, *5*, 89–100.
40. Yang, R.Z.; Wei, X.L.; Gao, F.F.; Wang, L.S.; Zhang, H.J.; Xu, Y.J.; Li, C.H.; Ge, Y.X.; Zhang, J.J.; Zhang, J. Simultaneous analysis of anthocyanins and flavonols in petals of lotus (*Nelumbo*) cultivars by high-performance liquid chromatography-photodiode array detection/electrospray ionization mass spectrometry. *J. Chromatogr. A* **2009**, *1216*, 106–112. [CrossRef]
41. Deng, J.; Chen, S.; Yin, X.; Wang, K.; Liu, Y.; Li, S.; Yang, P. Systematic qualitative and quantitative assessment of anthocyanins, flavones and flavonols in the petals of 108 lotus (*Nelumbo nucifera*) cultivars. *Food Chem.* **2013**, *139*, 307–312. [CrossRef]
42. Deng, J.; Fu, Z.; Chen, S.; Damaris, R.N.; Wang, K.; Li, T.; Yang, P. Proteomic and epigenetic analyses of lotus (*Nelumbo nucifera*) petals between red and white cultivars. *Plant Cell Physiol.* **2015**, *56*, 1546–1555. [CrossRef] [PubMed]
43. Deng, J.; Li, J.; Su, M.; Chen, L.; Yang, P. Characterization of key genes involved in anthocyanins biosynthesis in *Nelumbo nucifera* through RNA-Seq. *Aquat. Bot.* **2021**, *21*, 103428. [CrossRef]
44. Deng, J.; Li, J.; Su, M.; Lin, Z.; Chen, L.; Yang, P. A bHLH gene *NnTT8* of *Nelumbo nucifera* regulates anthocyanin biosynthesis. *Plant Physiol. Biochem.* **2021**, *158*, 518–523. [CrossRef] [PubMed]
45. Sun, S.; Gugger, P.; Wang, Q.; Chen, J. Identification of a R2R3-MYB gene regulating anthocyanin biosynthesis and relationships between its variation and flower color difference in lotus (*Nelumbo* Adans.). *PeerJ* **2016**, *4*, e2369. [CrossRef]
46. Yang, M.; Zhu, L.; Xu, L.; Liu, Y. Comparative transcriptomic analysis of the regulation of flowering in temperate and tropical lotus (*Nelumbo nucifera*) by RNA-Seq. *Ann. Appl. Biol.* **2014**, *165*, 73–95. [CrossRef]
47. Lin, Z.; Cao, D.; Damaris, R.N.; Yang, P. Genome-wide identification of *MADS-box* gene family in sacred lotus (*Nelumbo nucifera*) identifies a *SEPALLATA* homolog gene involved in floral development. *BMC Plant Biol.* **2020**, *20*, 497. [CrossRef]
48. Lin, Z.; Damaris, R.N.; Shi, T.; Li, J.; Yang, P. Transcriptomic analysis identifies the key genes involved in stamen petaloid in lotus (*Nelumbo nucifera*). *BMC Genom.* **2018**, *19*, 554. [CrossRef]

49. Lin, Z.; Cao, D.; Damaris, R.N.; Yang, P. Comparative transcriptomic analysis provides insight into carpel petaloidy in lotus (*Nelumbo nucifera*). *PeerJ* **2021**, *9*, e12322. [CrossRef]
50. Lin, Z.; Liu, M.; Damaris, R.N.; Nyong'a, T.M.; Cao, D.; Ou, K.; Yang, P. Genome-wide DNA methylation profiling in the lotus (*Nelumbo nucifera*) flower showing its contribution to the stamen petaloid. *Plants* **2019**, *8*, 135. [CrossRef]
51. Zhang, L.; Zhang, F.; Liu, F.; Shen, J.; Wang, J.; Jiang, M.; Zhang, D.; Yang, P.; Chen, Y.; Song, S. The lotus *NnFTIP1* and *NnFT1* regulate flowering time in *Arabidopsis*. *Plant Sci.* **2021**, *302*, 110677. [CrossRef]
52. Cheng, L.; Li, S.; Yin, J.; Li, L.; Chen, X. Genome-wide analysis of differentially expressed genes relevant to rhizome formation in lotus root (*Nelumbo nucifera* Gaertn). *PLoS ONE* **2013**, *8*, e67116. [CrossRef]
53. Yang, M.; Zhu, L.; Pan, C.; Xu, L.; Liu, Y.; Ke, W.; Yang, P. Transcriptomic analysis of the regulation of rhizome formation in temperate and tropical lotus (*Nelumbo nucifera*). *Sci. Rep.* **2015**, *5*, 13059. [CrossRef]
54. Cao, D.; Damaris, R.N.; Zhang, Y.; Liu, M.; Li, M.; Yang, P. Proteomic analysis showing the signaling pathways involved in the rhizome enlargement process in *Nelumbo nucifera*. *BMC Genom.* **2019**, *20*, 766. [CrossRef] [PubMed]
55. Huang, L.; Li, M.; Cao, D.; Yang, P. Genetic dissection of rhizome yield-related traits in *Nelumbo nucifera* through genetic linkage map construction and QTL mapping. *Plant Physiol. Biochem.* **2021**, *160*, 155–165. [CrossRef] [PubMed]
56. Cao, D.; Lin, Z.; Huang, L.; Damaris, R.N.; Li, M.; Yang, P. A *CONSTANS-LIKE* gene of *Nelumbo nucifera* could promote potato tuberization. *Planta* **2021**, *253*, 65. [CrossRef] [PubMed]
57. Cao, D.; Lin, Z.; Huang, L.; Damaris, R.N.; Yang, P. Genome-wide analysis of AP2/ERF superfamily in lotus (*Nelumbo nucifera*) and the association between *NnADAP* and rhizome morphology. *BMC Genom.* **2021**, *22*, 171. [CrossRef] [PubMed]
58. Limwachiranon, J.; Huang, H.; Shi, Z.; Li, L.; Luo, Z. Lotus flavonoids and phenolic acids, health promotion and safe consumption dosages. *Compr. Rev. Food Sci. Food Saf.* **2018**, *17*, 458–471. [CrossRef]
59. Yu, Y.; Wei, X.; Liu, Y.; Dong, G.; Hao, C.; Zhang, J.; Jiang, J.; Cheng, J.; Liu, A.; Chen, S. Identification and quantification of oligomeric proanthocyanidins, alkaloids, and flavonoids in lotus seeds, A potentially rich source of bioactive compounds. *Food Chem.* **2022**, *379*, 132124. [CrossRef]
60. Zhang, Y.; Lu, X.; Zeng, S.; Huang, X.; Guo, Z.; Zheng, Y.; Tian, Y.; Zheng, B. Nutritional composition, physiological functions and processing of lotus (*Nelumbo nucifera* gaertn.) seeds, a review. *Phytochem. Rev.* **2015**, *14*, 321–334. [CrossRef]
61. Li, J.; Tao, S.; Huang, L.; He, D.; Maraga, N.T.; Yang, P. Systematic transcriptomic analysis provides insights into lotus (*Nelumbo nucifera*) seed development. *Plant Growth Regul.* **2018**, *86*, 339–350. [CrossRef]
62. Wang, L.; Fu, J.; Li, M.; Fragner, L.; Weckwerth, W.; Yang, P. Metabolomic and proteomic profiles reveal the dynamics of primary metabolism during seed development of lotus (*Nelumbo nucifera*). *Front. Plant Sci.* **2016**, *7*, 750. [CrossRef]
63. Shen-Miller, J.; Mudgett, M.B.; Schopf, J.W.; Clarke, S.; Berger, R. Exceptional seed longevity and robust growth, ancient sacred lotus from china. *Am. J. Bot.* **1995**, *82*, 1367–1380. [CrossRef]
64. Sun, H.; Li, J.; Song, H.; Yang, D.; Deng, X.; Liu, J.; Wang, Y.; Ma, J.; Xiong, Y.; Liu, Y.; et al. Comprehensive analysis of *AGPase* genes uncovers their potential roles in starch biosynthesis in lotus seed. *BMC Plant Biol.* **2020**, *20*, 457. [CrossRef] [PubMed]
65. Zhu, F.; Sun, H.; Diao, Y.; Zheng, X.; Xie, K.; Hu, Z. Genetic diversity, functional properties and expression analysis of *NnSBE* genes involved in starch synthesis of lotus (*Nelumbo nucifera* Gaertn.). *PeerJ* **2019**, *7*, e7750. [CrossRef] [PubMed]
66. Qi, H.; Yu, F.; Damaris, R.N.; Yang, P. Metabolomics analyses of cotyledon and plumule showing the potential domestic selection in lotus breeding. *Molecules* **2021**, *26*, 913. [CrossRef]
67. Zhu, M.; Liu, T.; Guo, M. Current advances in the metabolomics study on lotus Seeds. *Front. Plant Sci.* **2016**, *7*, 891. [CrossRef]
68. Wang, M.; Li, W.; Fang, C.; Xu, F.; Liu, Y.; Wang, Z.; Yang, R.; Zhang, M.; Liu, S.; Lu, S.; et al. Parallel selection on a dormancy gene during domestication of crops from multiple families. *Nat. Genet.* **2018**, *50*, 1435–1441. [CrossRef]
69. Liu, M.; Lin, Z.; Yang, P.; He, D. Research progress on the longevity mechanism of *Nelumbo nucifera* seeds. *Plant Sci. J.* **2019**, *37*, 396–403.
70. Deng, X.; Yang, D.; Sun, H.; Liu, J.; Song, H.; Xiong, Y.; Wang, Y.; Ma, J.; Zhang, M.; Li, J. Time-course analysis and transcriptomic identification of key response strategies to complete submergence in *Nelumbo nucifera*. *Hortic. Res.* **2022**, *9*, uhac001. [CrossRef]
71. Li, J.; Xiong, Y.; Li, Y.; Ye, S.; Yin, Q.; Gao, S.; Yang, D.; Yang, M.; Palva, E.T.; Deng, X. Comprehensive Analysis and Functional Studies of WRKY Transcription Factors in *Nelumbo nucifera*. *Int. J. Mol. Sci.* **2019**, *20*, 5006. [CrossRef]
72. Deng, X.; Zhao, L.; Fang, T.; Xiong, Y.; Ogutu, C.; Yang, D.; Vimolmangkang, S.; Liu, Y.; Han, Y. Investigation of benzylisoquinoline alkaloid biosynthetic pathway and its transcriptional regulation in lotus. *Hortic. Res.* **2018**, *5*, 29. [CrossRef] [PubMed]

MDPI
St. Alban-Anlage 66
4052 Basel
Switzerland
www.mdpi.com

International Journal of Molecular Sciences Editorial Office
E-mail: ijms@mdpi.com
www.mdpi.com/journal/ijms

Disclaimer/Publisher's Note: The statements, opinions and data contained in all publications are solely those of the individual author(s) and contributor(s) and not of MDPI and/or the editor(s). MDPI and/or the editor(s) disclaim responsibility for any injury to people or property resulting from any ideas, methods, instructions or products referred to in the content.